MINING FOR ANCIENT COPPER

ESSAYS IN MEMORY OF BENO ROTHENBERG

MONOGRAPH SERIES OF THE

SONIA AND MARCO NADLER INSTITUTE OF ARCHAEOLOGY

TEL AVIV UNIVERSITY

NUMBER 37

MINING FOR ANCIENT COPPER

Essays in Memory of Beno Rothenberg

Editor: Erez Ben-Yosef

EISENBRAUNS
University Park, Pennsylvania

EMERY AND CLAIRE YASS
PUBLICATIONS IN ARCHAEOLOGY OF
THE INSTITUTE OF ARCHAEOLOGY
Tel Aviv University

Monograph Series under the auspices of the
Friends of the Institute of Archaeology of Tel Aviv University

Library of Congress Cataloging-in-Publication Data

Names: Ben-Yosef, Erez, editor. | Rothenberg, Benno, honouree.
Title: Mining for ancient copper : essays in memory of Beno Rothenberg / editor, Erez Ben-Yosef.
Other titles: Monograph series (Makhon le-arkhe'ologyah 'a. sh. Sonyah u-Marko Nadler)
Description: University Park, Pennsylvania : Eisenbrauns, 2018. | Series: Monograph series of the Sonia and Marco Nadler Institute of Archaeology | Includes bibliographical references and index.
Summary: "A collection of new studies dedicated to Professor Beno Rothenberg, focused on copper in antiquity in the Near East, the eastern Mediterranean, and the British Isles"—Provided by publisher.
Identifiers: LCCN 2018031301 | ISBN 9781575069647 (cloth : alk. paper)
Subjects: LCSH: Copper mines and mining, Prehistoric. | Metal-work, Prehistoric. | Copper mines and mining, Prehistoric—Middle East. | Metal-work, Prehistoric—Middle East.
Classification: LCC GN799.M4 M55 2018 | DDC 671.09/01—dc23
LC record available at https://lccn.loc.gov/2018031301

Beno Rothenberg
1914–2012

CONTENTS

SECTION III: FAYNAN, THE NEGEV AND BEYOND

SECTION IV: BEYOND THE SOUTHERN LEVANT: CYPRUS, OMAN, GREECE AND BRITAIN

SECTION V: METALWORKING

CONTRIBUTORS

Amzallag, Nissim
Ben-Gurion University of the Negev, Israel

Arav, Reuma
Technion – Israel Institute of Technology, Israel

Avner, Uzi
Dead Sea-Arava Science Center, Israel

Bason, Uri
GeoSense Ltd, Israel

Ben-Dov, Rachel
Hebrew Union College, Israel

Ben-Yosef, Erez
Tel Aviv University, Israel

Benoist, Anne
Centre National de la Recherche Scientifique, France

Bernardi, Elena
University of Bologna, Italy

Beyth, Michael
Geological Survey of Israel, Israel

Bruins, Hendrik J.
Ben-Gurion University of the Negev, Israel

Chiavari, Cristina
University of Bologna, Italy

Czarnowicz, Marcin
Jagiellonian University, Krakow, Poland

Dar, Shimon
Bar Ilan University, Israel

Davey, Christopher John
University of Melbourne, Australia

Degli Esposti, Michele
University of Pisa, Italy

Drenka, Alexandra
Independent scholar, Israel

Filin, Sagi
Technion – Israel Institute of Technology, Israel

Frumkin, Amos
Hebrew University, Israel

Garagnani, Gian Luca
University of Ferrara, Italy

Ginat, Hanan
Dead Sea and Arava Science Center and Ben-Gurion University of the Negev, Israel

Gottlieb, Yulia
Tel Aviv University, Israel

Goy, Julie
University of Paris 1 Panthéon-Sorbonne, France

Haiman, Moti
Israel Antiquities Authority and Bar Ilan University, Israel

Jones, Ian W.N.
University of California, San Diego, USA

Kassianidou, Vasiliki
University of Cyprus, Cyprus

Kolska-Horwiz, Liora
Hebrew University, Israel

Langford, Boaz
Hebrew University, Israel

Le Carlier de Veslud, Cécile
University of Rennes 1, France

Lernau, Omri
University of Haifa, Israel

Levene, Dan
University of Southampton, UK

Levy, Thomas E.
University of California, San Diego, USA

Löffler, Ingolf
Deutsches Bergbau-Museum, Germany

Malamidou, Dimitra
Ministry of Culture, Kavala, Greece

Marshall, Peter
English Heritage, UK

Martini, Carla
University of Bologna, Italy

Meeshly, Dagan
Dead Sea-Arava Science Center, Israel

Melfos, Vasilios
Aristotle University of Thessaloniki, Greece

Meshel, Ze'ev
Tel Aviv University, Israel

Muhly, James D.
University of Pennsylvania, USA

Najjar, Mohammad
University of California, San Diego, USA

Pavlides, Spyros
Aristotle University of Thessaloniki, Greece

Peters, Ilana
Tel Aviv University, Israel

Ponting, Matthew
University of Liverpool, UK

Pusch, Edgar B.
University College London – Qatar, Qatar

Rademakers, Frederik W.
University College London, UK

Rehren, Thilo
The Cyprus Institute, Cyprus

Sapir-Hen, Lidar
Tel Aviv University, Israel

Scott-Cummings, Linda
PaleoResearch Institute, USA

Segal, Irina
Geological Survey of Israel, Israel

Segev, Amit
Geological Survey of Israel, Israel

Shalev, Sariel
University of Haifa, Israel

Shamir, Omer
GeoSense Ltd, Israel

Shaw, Tim
Imperial College of Science and Technology, UK

Shem-Tov, Rachamim
Dead Sea-Arava Science Center, Israel

Shilstein, Sana
Weizmann Institute of Science, Israel

Shugar, Aaron N.
Buffalo State College, USA

Sweeney, Deborah
Tel Aviv University, Israel

Szudy, Jamie M.
University of Vienna, Austria

Tauxe, Lisa
Scripps Institution of Oceanography, USA

Timberlake, Simon
University of Cambridge, UK

van der Plicht, Johannes
University of Groningen and Leiden University, The Netherlands

Vavelidis, Michail
Aristotle University of Thessaloniki, Greece

Vaxevanopoulos, Markos
Aristotle University of Thessaloniki, Greece

Williams, Robert Alan
University of Liverpool, UK

Yahalom-Mack, Naama
Weizmann Institute of Science and Hebrew University, Israel

Zucconi, Laura M.
Stockton University, USA

PREFACE

There is no better way to honor the memory of Beno Rothenberg (1914–2012) than by publishing a volume dedicated to new studies on copper in antiquity. Rothenberg's pioneering work in the Timna Valley, which was the center of his academic career, focused on ancient copper mining and smelting technologies, and paved the way to other studies of this metal and its role in ancient societies around the world.[1] Rothenberg's work is considered by many to be a cornerstone in the development of archaeometallurgy as an integrative research discipline; as such, the study of ancient metal and metal production technologies is based on a synthesis of various avenues of investigation from the natural and social sciences and the humanities, the main objective of which is the study of the people and societies behind the artifacts and technologies.[2] This book celebrates just such an approach with a collection of studies that includes, in addition to contributions on technologies, results of research on various aspects of the production and use of copper in ancient societies: from the geological settings of copper mines to the diet of metalworkers and the characteristics of metal trade systems. The studies range from Oman to the British Isles, with a special emphasis on the southern Levant and the Arabah Valley. They testify not only to the current prosperity of research in the geographical region whose systematic study was pioneered by Rothenberg, but also to the growth and vitality of the research discipline that Rothenberg fundamentally helped to advance (cf., Thornton 2012, Roberts and Thornton 2014).

The book's publication follows the international conference on "Copper in Antiquity" held at the Timna Park (southern Arabah, the Eilot Regional Council) in 2013. The conference, also in memory of Rothenberg, was organized by Tel Aviv University and the Timna Park with the help and support of other organizations,[3] and steered by E. Ben-Yosef with the help of Y. Goren, H. Ginat and A. Holzer. Some of the contributions are based on presentations given at the conference, while others were written especially for the book.

The 37 chapters of the book, contributed by 66 scholars, present a wide array of topics. They are organized in five sections—the first four are divided by geography, while the final section includes studies related specifically to metalworking. The geographic sections are organized according to their proximity to Timna, which, as mentioned above, was at the core of Rothenberg's academic work. The book commences with Timna itself (Section I), goes on to nearby Nahal ʿAmram (Section II), a smaller copper ore district located ca. 5 km to the south of Timna, also within the general region of the southern Arabah Valley. The next section (III) deals with the Faynan copper ore district in the northern Arabah Valley, together with contributions on the Negev and southern Canaan. The last geographic section (IV) contains contributions related to various locations, from Oman to the British Isles, through Cyprus and Greece. This wide geographic spectrum helps to contextualize the intense research in the southern

1 On Rothenberg's work and its contribution to archaeometallurgical research, see Pigott (1996), Ben-Yosef (2012).

2 Rothenberg's research into the metallurgical aspects of the ancient copper ore district of Timna was part of his broader quest for a better understanding of the archaeology and history of the Negev and the Arabah, which started even before his independent projects, while he participated in the expeditions of N. Glueck and Y. Aharoni. This background helps to explain Rothenberg's integrative approach to archaeometallurgical research, and his keen interest in questions related to the society behind the technology— or as he himself put it, archaeometallurgy helps us understand "not only how men made metal, but also how metal made men" (via Bachmann 1990).

3 These organizations include the Dead Sea and Arava Science Center, the Institute of Archaeo-Metallurgical Studies (IAMS) at University College London, the Jewish National Fund, the Eilot Regional Council, the Israel Government Tourist Corporation and the Economic Corporation for the Tourism Development in Hevel Eilot.

Levant presented in the previous sections, and in general emphasizes common denominators in the study of copper across diverse cultures and space.

Section I, "Timna Valley," consists of nine chapters.

- Chapter 1 presents the geological settings of the copper ore, which is a necessary background to the archaeometallurgical research of the region (mining and smelting technologies, distribution of sites, etc.). It also provides a basic background to the ore bodies of Faynan, the Jordanian counterpart of Timna, and Umm Bogma in southern Sinai, both heavily exploited in antiquity.
- Chapter 2 presents an overview of the rather tumultuous debates over the date of the earliest evidence of smelting in the valley, and over the chronology of the main phase of copper exploitation there. The latter is related to the question of "King Solomon's Mines," a subject that is again part of the scholarly discourse as a result of recent discoveries.[4]
- Chapter 3 introduces the Central Timna Valley Project, which commenced in 2012 and has focused since that time on investigating the Late Bronze and Iron Age s (13th– 9th centuries BCE) mining and smelting sites of the region, with emphasis on technological developments and social processes of the people responsible for the copper industry (the early phase of the Edomite Kingdom).
- Chapter 4 presents new data on the diet of the Late Bronze and Iron Age metalworkers, based on remains of mammalian and fish bones from the main smelting sites in Timna. These data are presented together with a summary of previously published materials from Timna and Faynan in order to assess the social status, ethnicity, and other aspects of the people directly engaged in the smelting activities in these periods.
- Chapter 5 is a detailed report of Rothenberg's last excavations at Timna (2001–2002), in the complex shafts and gallery system of Mine T (dated to the Chalcolithic period). The report integrates all the data from the earlier excavations at the mine (1974-1976) into a comprehensive presentation of the research and its results.
- Chapter 6 provides a fresh look at the Egyptian inscription that was found in 1972 on the cliffs above the "Hathor Temple," and in particular at Ramessesempere, the head of the Egyptian expedition to the mines in the days of Ramesses III.
- Chapter 7 presents preliminary results of an archaeomagnetic study of pottery sherds from the Yotvata Fortress. Located just above the nearest permanent water source to Timna, the fortress has been associated with the Egyptian phase of copper production and interpreted as part of the efforts to maintain water supply at this time. The results suggest Late Bronze Age activities, thus corroborating the excavator's dating and supporting his interpretation of the site.
- Chapter 8 presents a fresh interpretation of the later (Iron Age) phase of the "Hathor Temple," with an intriguing suggestion that the place served for the worship of YHWH, the deity of the Israelites, whose source may have been in the south and in connection to ancient metallurgy.
- Chapter 9 presents an analysis of the genealogy of Esau (Edom), in light of our current understanding of the region and the rather fluid role of tribes and clans in forming political alliances. Such alliance is probably behind the Iron Age copper exploitation in Timna and the northern Arabah.

---

4 Forty years after the excavations of the "Hathor Temple" in the center of the Timna Valley and the consequent revision in the dates of all major smelting and mining sites, new radiocarbon dates indicated that one of the smelting camps was most active during the 10th – 9th centuries BCE, and not earlier than the second half of the 12th century BCE (Ben-Yosef, Shaar, Tauxe and Ron 2012). This in turn triggered new research in the valley, which has demonstrated that the peak in production was indeed in the early Iron Age and after the Egyptians left the region (Ben-Yosef, this volume); while possible connections to Jerusalem are still debated, the new chronological framework and evidence of long-distance copper trade necessitate reconsideration of the region's role in this formative period in the history of the southern Levant.

Section II, "Nahal ᶜAmram," consists of six chapters, all of which present results of a recent interdisciplinary project led by Uzi Avner and focused on the history of mining and smelting in this region.

- Chapter 10 is an overview of the Nahal ᶜAmram Project, and provides the dating skeleton of the different sites based on artifact typologies and a large suite of new radiocarbon dates. The main periods of activity were found to be Late Bronze–Iron Ages, Nabataean–Byzantine and Early Islamic.
- Chapter 11 presents volume and mass estimates of mining dumps and slag piles. In addition to enhancing our understanding of the scale of mining and smelting in the Nahal ᶜAmram area and nearby sites, the chapter contributes to the development of research methods by the introduction of high-resolution terrestrial laser scans as an efficient technique to tackle such problems.
- Chapter 12 presents a study of paleo-floods based on their record within Nahal ᶜAmram's mines. The galleries preserve unique evidence of mega-floods, whose contexts provide important information on their frequencies in the past.[5]
- Chapter 13 presents the diet of the miners based on faunal remains found within the galleries, and concludes that their food was surprisingly rich.
- Chapter 14 presents a detailed report of the comprehensive underground survey of the mines in Nahal ᶜAmram, including new mapping of the entire system.
- Chapter 15 presents a preliminary chemical analysis of slag in an attempt to assess developments in smelting technologies through time.

Section III, "Faynan, the Negev and Beyond," consists of eight chapters.

- Chapter 16 presents an overview of the intensive, large-scale anthropological archaeology project in Faynan, Jordan, directed by Thomas Levy and Mohammad Najjar. Commenced in 1997, the project has shed new light on the history of copper ore exploitation in the northern Arabah from the Neolithic to the Late Islamic period, with a substantial contribution to the early Iron Age archaeology of the region. The latter is the focus of the chapter, which discusses the most recent finds and their interpretation by the excavation team (and addresses some criticism).
- Chapter 17 also focuses on Faynan. It presents a new study on the technological developments and organizational structure in the Bronze Age, based primarily on the finds of the comprehensive archaeometallurgical project of the Deutsches Bergbau-Museum (under the direction of Andreas Haputmann, 1983–1993).
- Chapter 18 presents an overview of the Early Bronze Age IV settlement wave in the Negev, with emphasis on its connection to the copper trade between the northern Arabah (Faynan) and Egypt.
- Chapter 19 presents a new study on copper ore fragments found in the Chalcolithic (Ghassulian) site of Abu Matar in the Beer-sheba Valley. While the majority of the ore fragments correspond with the mining site of Faynan, one type offers closer association with ore formations in Anatolia and the Caucasus.
- Chapter 20 presents a lead isotope study of a 12th century BCE bronze chisel from Horvat Haluqim in the Negev. The results suggest that the copper originated in Faynan and that an active metal trade network existed in the south in that period.
- Chapter 21 presents a brief summary of the discovery of Kuntillet ᶜAjrud in northeastern Sinai, a unique, possibly cultic site, near the road between the Mediterranean Sea and Elath (Darb al-Ghaza).

---

5 The ancient mining landscape of the southern Arabah holds important evidence of the paleo-environment and young geomorphological processes; see, for example, the previous studies of Hauptmann and Horowitz (1980) and Shlomi *et al.* (2015).

Rothenberg's visit to the site in 1967 was an important milestone in its research. The site probably served as an important road station; however, it is not clear if it was related to the copper trade.[6]

- Chapter 22 presents a new study on the provenance of copper in Canaan during the second half of the second millennium BCE. Based on chemical and lead isotope analyses of final copper-based objects, it is demonstrated that Timna played an important role in the copper trade after the Egyptians left the region, and in particular during the 11th century BCE.
- Chapter 23 is a comprehensive overview of our current knowledge of Islamic copper production in the Arabah Valley. While in the Early Islamic period the copper mines of the southern Arabah (Timna and Nahal ʿAmram) were most active, in the Late Islamic period mining activities are documented only in Faynan.

Section IV, "Beyond the southern Levant: Cyprus, Oman, Greece and Britain," contains six chapters, all related to primary copper production (i.e., ore mining and smelting), as evidenced in the archaeological record and historical documentation.

- Chapter 24 presents evidence from a Late Bronze Age (13th century BCE) miners' settlement in the Apliki ore district of Cyprus, which was exposed by modern exploitation of the region.
- Chapter 25 is also focused on Cyprus, albeit in a much later period. It presents the case of King Herod's exploitation of the Cypriot mines (1st century BCE), as relayed by Flavius Josephus, and suggests that this was the major source for Herod's wealth, which enabled his grandiose construction enterprises in Judea and many cities around the Mediterranean.
- Chapter 26 presents preliminary results of an ongoing archaeometallurgical research on Iron Age copper production in the northern al-Hajjar Mountains in Oman. This research is part of a multi-faceted project, which includes surveys, excavations and complementary laboratory work.
- Chapter 27 presents a new study of one of the mines in Mount Pangaeon in northeastern Greece, and at a nearby smelting site (Valtouda). The study includes documentation of mining technologies from the Roman period to the days of the Ottoman Empire, and an assessment of the complex history of exploitation of the multi-metallic (gold-silver-copper) ore body.
- Chapter 28 presents preliminary results of a study on the Bronze Age Great Orme copper mine in north Wales. The study attempts to establish a robust geochemical signature for the ore body, in order to enhance the quality of provenance studies and in turn to reassess the scale of the mining activities (and its geographical impact) and the importance of the site in Bronze Age Britain.
- Chapter 29 is also concerned with Britain in the Bronze Age. It presents new copper mining sites and discusses the main phase of Bronze Age copper exploitation in Britain based on reexamination of radiocarbon dates. The study demonstrates that widespread small-scale mining activities took place in western Britain between ca. 2000–1500 BCE (the termination possibly related to the beginning of copper importation from Europe).

Section V, "Metalworking," consists of eight chapters, all related to secondary copper production (e.g., alloying, recycling, mending and casting) and final copper objects.

- Chapter 30 discusses the transition from copper to iron in the southern Levant. Based on the archaeological evidence, it seems that Judah adopted iron-working earlier than its northern counterpart,

6 Although Kuntillet ʿAjrud is located near one of the main roads between Gaza and Timna, its accepted dating to the 8th century BCE precludes the possibility that it was connected to the copper trade, as the Arabah copper industry ceased at the end of the 9th century BCE (e.g., Ben-Yosef, Shaar, Tauxe and Ron 2012). However, if the beginning of the occupation at Kuntillet ʿAjrud were to be dated earlier, such a connection should be considered (cf., Schniedewind 2017, contra the opinion of the excavator).

the Kingdom of Israel. Various possible reasons for this discrepancy are discussed, among them the stronger affiliation of Judah to metalworking and trends in metallurgical developments based on its connection to the flourishing early Iron Age copper industry of the Arabah.

- Chapter 31 presents a detailed report on the early Iron Age metallurgical workshops at Tel Dan, with evidence for bronze-working and recycling.
- Chapter 32 presents the results of a study on metal recycling procedures based on archaeological finds and Jewish literary sources of late antiquity. The study demonstrates that contrary to the notion that recycling resulted in poor control over composition and quality, the management of scrap metal was actually a developed and sophisticated industry that provided reliable products.
- Chapter 33 presents a study of Early Bronze Age copper refining. Based on evaluation of the technology, the author suggests identifying crucibles depicted in Egyptian Old Kingdom tombs and the Old Babylonian site of Tell edh-Dhiba'i as refining vessels.
- Chapter 34 presents new evidence of Late Bronze Age bronze production in Qantir–Pi-Ramesse. The evidence indicates a rather high technological variability, as alloying was achieved by different methods, including mixing fresh metals (copper and tin ingots?), recycling and cassiterite cementation.
- Chapter 35 presents copper harpoons of Pre-Dynastic Egypt and discusses their significance in the Naqada culture. An overview of the finds and relevant artistic depictions suggest that they were not only weapons, but also an important symbol that played several roles in ancient Egyptian society.
- Chapter 36 presents new data on bronze working at Sumhuram (Oman). The alloying process was evaluated based on the analysis of metal items dated from the 3rd century BCE to the 4th century CE and other evidence from the site. No diachronic trends or correlation between shape/function and chemical composition were found, suggesting low standardization.
- Chapter 37 discusses the use of copper for the production of weapons at times when iron was the dominant metal of choice. The case study of socketed copper alloy arrowheads of the 7th century BCE demonstrates that copper still had an advantage when certain forms were desired, and even more so given the ability to mass produce by casting.

The wide-ranging contents of this volume demonstrates the importance of copper in the shaping of human history. Since the dawn of metallurgy more than 7,000 years ago, copper has been used to produce a wide assortment of objects with different functions in ancient societies, from ornaments, cult and art to agricultural and domestic tools, weapons and coinage.[7] This in part is what makes copper, and the evidence related to the efforts invested in its production, so well suited for deciphering social meaning and extracting knowledge about the past.

This book, in memory of Beno Rothenberg, also commemorates his friend and colleague, Professor Tim Shaw (1934-2017), who contributed, together with Alexandra Drenka, a comprehensive chapter on Mine T in Timna (Chapter 5). Shaw was Professor of Mining Engineering at Imperial College London, who became fascinated with the archaeology of ancient mines (*mining archaeology/Montanarchäologie*)[8] through his work with Rothenberg in Timna. He was engaged in archaeological research and in teaching archeaologists for many years, including at the archeaometallurgy summer school of the Institute of Archaeo-metallurgical Studies (IAMS) at University College London. Shaw's research achievments will undoubtedly continue to be part of the investigation of the Timna mines and other ancient mining districts around the world.

7 On this, in relation to the modern exploitation of copper, see the recent publication of Golding and Golding (2017).
8 This research field, sometimes included under the broader discipline of "archaeometallurgy," has been attracting a growing interest in recent years (cf., Stöllner 2014).

ACKNOWLEDGMENTS

The book is published with the support of the Institute of Archaeology of Tel Aviv University, the Marie Curie Actions (FP7-PEOPLE-2012-CIG grant #334274 to E.B.-Y.), and the Israel Science Foundation (grant #1880/17 to E.B.-Y.). Special thanks are due to Myrna Pollak, head of the Institute of Archaeology's Publications Department, to Noa Evron, graphics editor and Nitsan Shalom and Brett Cohen, assistant editors, for their work on the text, layout and production of the book. Many individuals contributed towards the book's creation, from its seeds at the Timna International Conference to the final print. Among them are the local "Timnaeans," Hagit Gal (Park Manager), Dubi Goldman, Hanan Ginat, Assaf Holzer and Guy Markman, who helped with the conference organization and later with various aspects of the research and publications related to the southern Arabah. The editor is grateful to Yuval Goren, who was pivotal to the initiation of the renewed archaeological research in Timna, as well as to the success of the Timna Conference. Thanks are also due to the Israel Antiquities Authority and its representatives in the south, Tali Erickson-Gini and Yoram Haimi, for their help in promoting research in one of the most remote regions of Israel. Lastly, thanks are due to Aaron Greener for his assistance in the editorial process, and to the Central Timna Valley Project's staff, whose devotion and enthusiasm for the exploration of Timna also propelled the work on this book.

Erez Ben-Yosef, Editor
2018

REFERENCES

Bachmann, H.G. 1990. Introduction. In: Rothenberg, B., ed. *The Ancient Metallurgy of Copper*, Vol. 2. London: xvii-xxi.

Ben-Yosef, E. 2012. Beno Rothenberg: Obituary. *Israel Exploration Journal* 62(1): 244–246.

Ben-Yosef, E., Shaar, R., Tauxe, L. and Ron, H. 2012. A New Chronological Framework for Iron Age Copper Production in Timna (Israel). *Bulletin of the American Schools of Oriental Research* 367: 31–71.

Golding, G. and Golding, S.D. 2017. *Metals, Energy and Sustainability: The Story of Doctor Copper and King Coal*. Cham, Switzerland.

Hauptmann, A. and Horowitz, A. 1980. Zur geomorphologie und palaomorphologie des Modell-gebietes. In: Conrad, H.G. and Rothenberg, B., eds. *Antikes Kupfer im Timna-Tal*. Bochum: 57–67.

Pigott, V. 1996. Near Eastern Archaeometallurgy: Modern Research and Future Directions. In: Cooper, J.S. and Schwartz, G.M., eds. *The Study of the Ancient Near East in the 21st Century*. Winona Lake: 139–176.

Roberts, B.W. and Thornton, C.P., eds. 2014. *Archaeometallurgy in Global Perspective*. New York.

Schniedewind, W.M. 2017. An Early Iron Age Phase to Kuntillet ʻAjrud? In: Greenspahn, F. and Rendsburg, G.A., eds. *Le-maʻan Ziony: Essays in Honor of Ziony Zevit*. Eugene, Oregon: 134–146.

Shlomi, Y., Ginat, H., Meron, T. and Holtzer, A. 2015. Pace of Degradation in Streams within Sandstone in a Region with Copper Mines (Timna Valley, southern Israel). *Negev, Dead Sea and Arava Studies* 7(2): 38–50 (in Hebrew).

Stöllner, T.R. 2014. Methods in Mining Archaeology (Montanarchäologie). In: Roberts, B.W. and Thornton, C.P., eds. *Archaeometallurgy in Global Perspective: Methods and Syntheses*. New York: 133–159.

Thornton, C.P. 2012. Archaeometallurgy in the 21st Century. *Reviews in Anthropology* 41 (3): 173–187.

SECTION I

TIMNA VALLEY

CHAPTER 1

STRATIGRAPHY AND STRUCTURE OF THE TIMNA VALLEY AND ADJACENT ANCIENT MINING AREAS

Michael Beyth, Amit Segev and Hanan Ginat

The most ancient core in the Timna Valley is composed of shallow, intrusive rocks that originated in the Neoproterozoic Age; they comprise Mount Timna and are the northern tip of the Arabian Nubian Shield. Uplift and peneplanation during the Early Cambrian marked the final stage of this Neoproterozoic sequence. Early Cambrian sedimentation began with fluviatile subarkoses of the Amudei Shelomo Formation, which is overlain by dolostones, sandstones and siltstones of the Timna Formation, representing an Early to Middle Cambrian marine ingression followed by Middle Cambrian subarkoses and siltstones of the Shehoret Formation deposited in coastal environments. Lower Cretaceous quartz-arenites of the Amir, Avrona and Samar Formations overlie the Cambrian sequence. These Cambrian and Lower Cretaceous formations are cut in the northern Timna Valley by a basaltic plug and funnel-shaped vents with volcanoclastic fill of the Early Cretaceous Age. The Lower Cretaceous clastic units are overlain by a thick series of platform carbonates (Judea Group), which build the upper part of the escarpment surrounding the Timna Valley.

The major structural feature controlling the region is the Dead Sea Transform, which is an active plate boundary separating the Arabian Plate in the east from the Sinai Peninsula, a sub-plate of the African Plate in the west. Tectonic activity of this transform includes sinistral faulting and rifting. Faynan in Jordan, the ancient copper mines on the eastern margins of the Dead Sea Transform, was separated from the Timna Valley, situated on the western margins, by this sinistral movement of ca. 105 km. The ancient Egyptian turquoise mines discovered around the temple of Sarabit el-Khadim and the recent manganese mines of Umm Bogma in southwestern Sinai are located on the eastern margins of the Suez Rift. This rift is an active plate boundary between the Sinai sub-plate and the African plate. The stratigraphy of the Cambrian sediments in the Umm Bugma area is identical to that described in the Timna Valley and Elat area.

The copper-manganese-uranium mineralization and enrichment took place in four cycles controlled by the stratigraphic and tectonic evolution in the course of ca. 600 million years. The copper in ancient times was mined from the Lower Cretaceous—top Amir and base Avrona Formations in the Timna Valley—and from the equivalent of the Cambrian Timna and Shehoret Formations—the Burj and Umm Ishrin Formations in Faynan. In Umm Bugma traces of copper mineralization were mapped in the Timna Formation, the turquoise was mined from the Late Cambrian Netafim Formation and the manganese from the Carboniferous Umm Bogma Formation.

INTRODUCTION

Since the 1980s a great deal of new geological-mineralogical research has been carried out in the Timna Valley (Fig. 1.1). The 1:50,000 scale geological map of the Be'er Orah Sheet, including the Timna Valley, was published by the Geological Survey of Israel (Beyth, Segev and Bartov 1999; Fig. 1.2), as well as a summary of the geology with emphasis on the copper and manganese mineralization by Segev, Beyth and Bar-Matthews (1992). A 1:50,000 scale geological map of Al Qurayqira, including the Faynan area and Khirbat en-Nahas with explanatory notes, was published by the Jordanian Natural Resources Authority (Rabbaᶜ 1994). A detailed mapping of the Umm Bogma area in southwest Sinai was carried out by the Geological Survey of Israel (Beyth 1978 and 1986). Beno Rothenberg did the pioneering archeological studies of two of these three ancient mining areas (Timna Valley and Sinai) and always cooperated with the geologists (e.g., Rothenberg and Bachmann 1988). This paper in the volume dedicated to Beno Rothenberg presents an up-dated geological background of the copper, manganese and uranium mineralization mainly of the Timna Valley, and a brief description of the Mount ᶜAmram, Faynan and Umm Bugma areas.

PHYSIOGRAPHY

The Timna Valley, 9 × 10 sq km in size, is a semi-crater shaped valley, truncated in the east by the Dead Sea Transform and surrounded by 500-m-high cliffs of Cambrian to Cretaceous sedimentary rocks. The core of the Timna Valley is the high structural block of Mount Timna (5 × 3 sq km in size, 453 m asl, which is the highest topographic feature in the Timna Valley). The present climate in the Timna Valley is extremely arid, with short rain events with an average annual rainfall of 25 mm/yr (Survey of Israel 1985: 80). The regional water divide runs along the summits of the western cliffs of the Timna Valley. In the east short, steep channels drain the cliffs, whereas in the west wide, moderate gradient streams drain the elevated plateau. The Timna Valley is drained by ephemeral streams that flow eastwards toward the Yotvata and Avrona playas, which are the internal base level for the surrounding region within the Dead Sea Transform (Fig. 1.1). They are (from south to north): Nahal (river in Hebrew) Raham and its tributaries Nahal Gadna and Nahal Nimra; Nahal Nehushtan, which drains Mount Timna from the south; Nahal Timna, which drains Mount Timna from the north; Nahal Mangan and Nahal Sasgon north of Mount Mikhrot (Fig. 1.1).

PREVIOUS WORK

Many geological/mineralogical studies were carried out in the Timna Valley from the 1950s on, initiated and supported by the State of Israel and the Timna Copper Mines, partly in the framework of an academic program. A summary of the previous studies up to 1991 was presented by Segev, Beyth and Bar-Matthews (1992 and references therein); also in Arad and Bartov (1994). The Timna Valley was mapped as part of the southern Negev by Shaw (1947), Bentor and Vroman (1955), Garfunkel (1970) and specific mapping was done by Mart and Horowitz (1981), Segev and Beyth (1986), as well as others. A pioneering study of the igneous core of Mount Timna was done by Zlatkin and Wurzburger (1957), and of Cambrian Timna and Shehoret Formations by Bartura (1966), Weissbrod (1969, 1981) and Segev (1986). Some of the pioneering studies of the copper and manganese ores were done by Bentor (1952), Wurzburger (1967a, b), Bar-Matthews (1986) and Segev (1986). A correlation between the Timna Valley and the Faynan area was suggested by Quennel (1958), Freund (1965) and Segev (1984), and that of the Timna Valley and the Umm Bogma mineralization was presented by Beyth (1978 and 1987).

STRATIGRAPHY (FIG. 1.3)

Three major rock sequences can be found in the Timna Valley: (1) the Neoproterozoic, high level, shallow plutons and dykes; (2) the Cambrian to Eocene sedimentary rocks and (3) the young clastic

Fig. 1.1: (a) Location map, Timna Valley (images from Google Earth); (b) Regional location map, Faynan, Timna Valley and Umm Bugma.

rocks deposited during the evolution of the Dead Sea Transform since the Miocene. In the Timna Valley the second sequence consists of three groups: Yam Suf, Kurnub and Judea, exposed in belts surrounding Mount Timna and at the Timna Cliffs. The great lateral extension and slow facies and thickness changes of these sedimentary units on a regional scale suggest deposition over an extensive stable platform area with a generally flat morphology at the northern tip of the Arabian Nubian Shield and the neighboring areas.

The Early Cambrian (Beyth and Heimann 1999) erosion surface, a peneplain on a regional scale, separates the first two sequences and the Oligocene erosion surface separates between the second and the third sequences (Avni *et al.* 2012).

NEOPROTEROZOIC ROCKS

The Timna Igneous Complex (Fig. 1.4) at the northern tip of the Arabian Nubian Shield is composed of felsic to mafic high level plutons (Vapnik 1998) and dykes (Beyth and Peltz 1992) of the late orogenic extensional, exhumation phase of the East African Orogen (EAO). The Timna alkali granite formed by

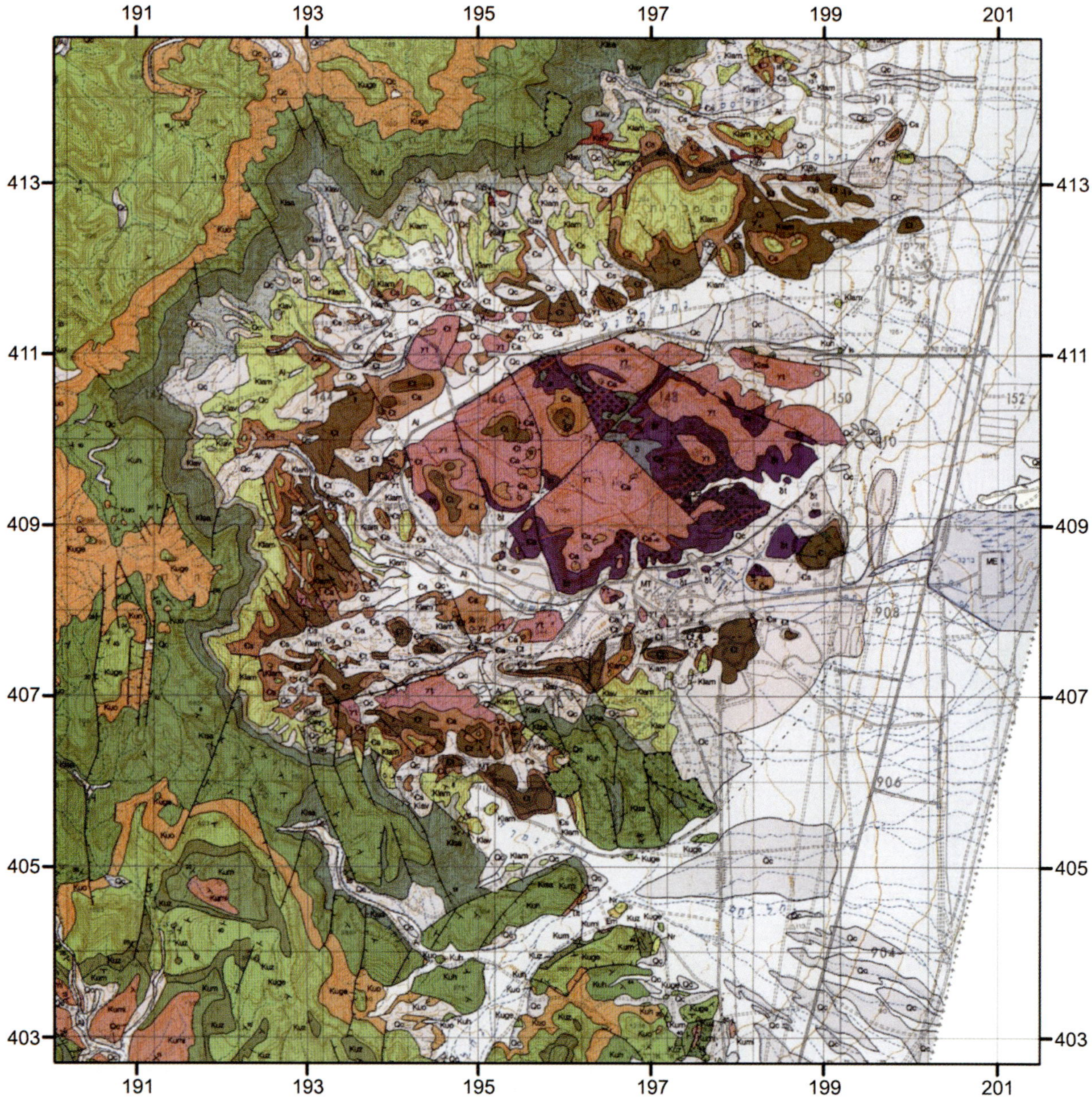

Fig. 1.2: The geological map, Timna Valley (Beyth, Segev and Bartov 1999); see detailed stratigraphy and rock units in Fig. 1.3.

fractional crystallization and the olivine norite as cumulate both from a mantle derived monzodioritic magma in a quasi-stratified magmatic cell. This magmatic cell was formed ca. 610 Ma ago in the 625 Ma old calc-alkaline, porphyritic granite (Shpitzer *et al.* 1992; Beyth *et al.* 1994a). Two distinct phases of rhyolite, andesite and composite dyke (Fig. 1.5: 1) intrusions are evident in the Timna Igneous Complex; the north–south ones intrude the Late Neoproterozoic, 625-610 Ma porphyritic granite and the monzodiorite, olivine nurite to alkali granites, predating the quartzmonzodiorite of 600 Ma (Beyth and Reischmann 1997) and the ENE swarms, which are ca. 593 Ma old (Katzir *et al.* 2007) and post-date this quartzmonzodiorite. The youngest dolerite dyke strikes N40W and is of 532 Ma old (Beyth

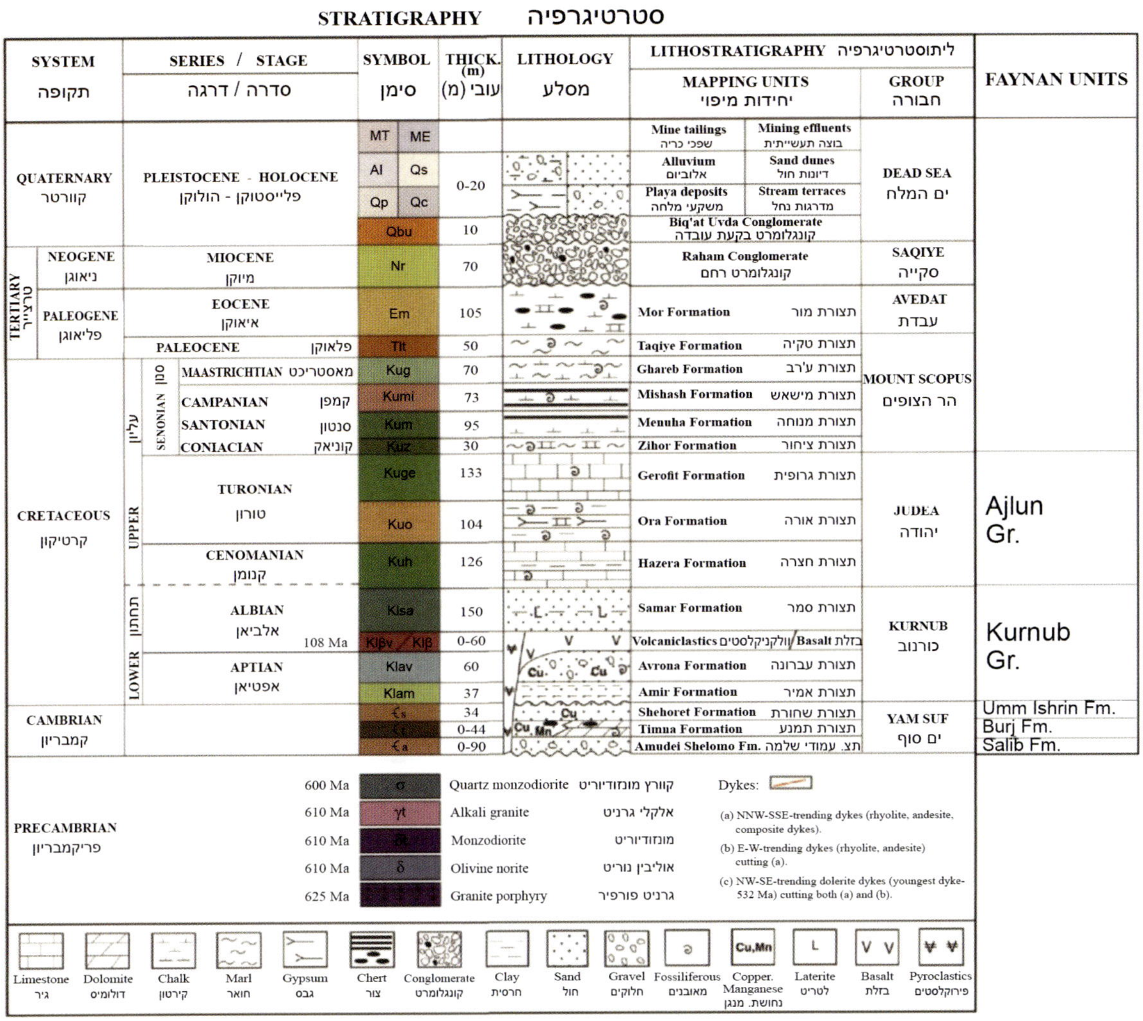

Fig. 1.3: Stratigraphy and geological legend, Timna Valley (Beyth, Segev and Bartov 1999) and the equivalent Faynan units.

and Heimann 1999). No igneous activity was documented between 590 Ma and 532 Ma. The Mount Timna dykes record transformation from arc-like geochemical signatures (silica oversaturated, low Ti content and strong Nb depletion) to one which is overwhelmingly intraplate in character (Beyth *et al.* 1994b; marginally saturated in silica, high Ti content and no Nb depletion). This doleritic dyke directly underlies the Early Cambrian sandstones which do not exhibit any sign of contact metamorphism. Thus the peneplain separating the Timna Igneous Complex from the overlying sedimentary Cambrian rocks is of Early Cambrian age. Similar doleritic dykes were found 12 km south of the Timna Valley intruding on Mount ʿAmram Late Neoproterozoic volcanic rocks and in the southeast Sinai Peninsula.

THE CAMBRIAN YAM SUF GROUP

The Cambrian Yam Suf Group in the Timna Valley is comprised of three mainly sandy formations: Amudei Shelomo, Timna and Shehoret, which were described by Weissbrod (1969, 2005) and Segev (1986). The Amudei Shelomo Formation of Early Cambrian Age is composed of sub-arkosic

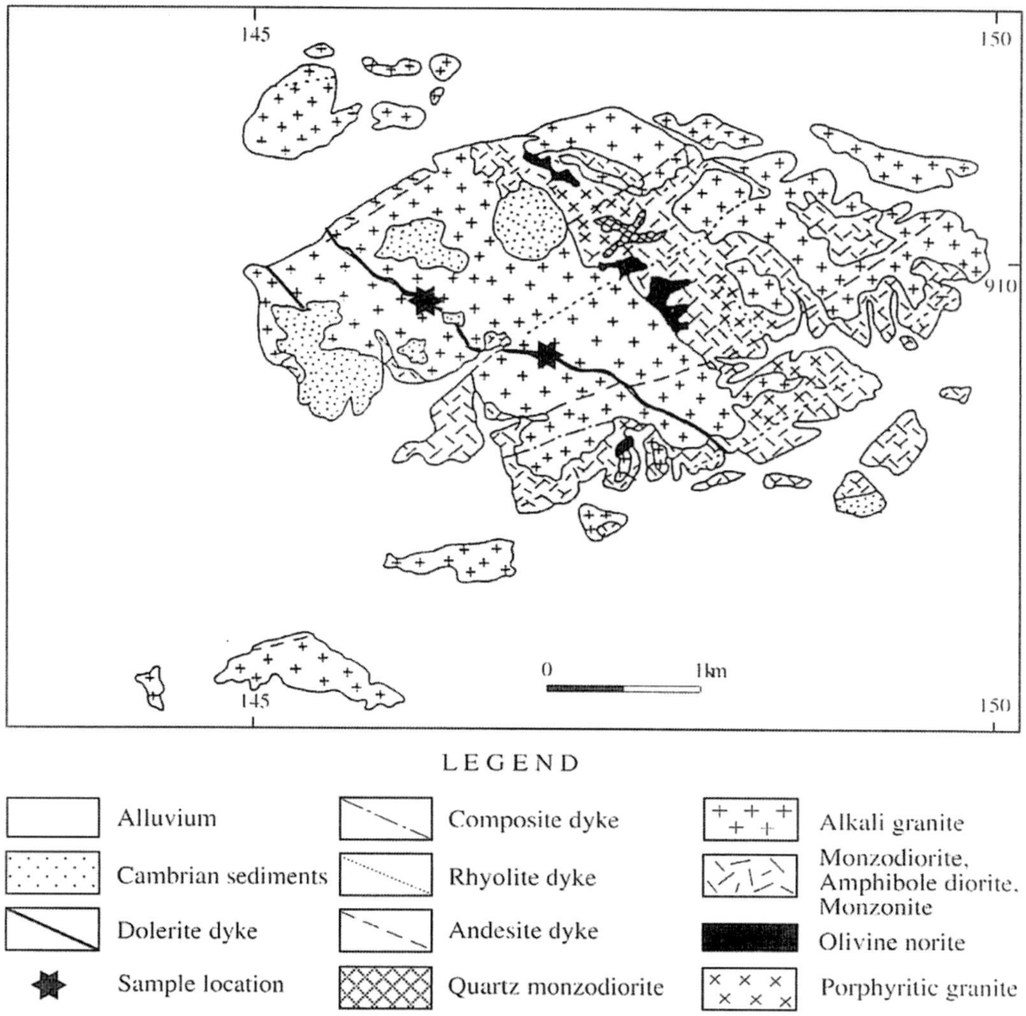

Fig. 1.4: The Timna Igneous Complex, Mount Timna (Beyth *et al.* 1994a).

sandstones and conglomerates that fill a moderate relief (Fig. 1.5: 2–3), which formed along a drainage pattern oriented north and northwest (Karcz and Key 1966). The base of this formation consists of a dark-brown silty-clay bed a few meters thick. The Timna Formation is exposed in two basins, the northern Timna basin and the southern Elat basin which are separated by the east–west trending ʿAmram-Neshef Ridge (Segev 1984). This formation is divided into two members: the lower Hakhlil Member is built of silt and sandstone with rhyolitic breccia at the base, and the upper Sasgon Member of three lithofacies: Dolomitic, averaging 0.06% Cu, Sand with copper mineable ore of 4–8 meters thick and copper content of 1.1–1.5% and Shaly Lithofacies with manganese nodules (Segev 1986). Segev and Sass (1989a) suggested that the lateral transitions between the dolomitic and the sandy lithofacies are due to epigenetic dissolution (karstification) of the dolomite by circulating thermal brines along Dead Sea Transform-related faults and fractures. In the Timna Valley the sandy lithofacies of the Sasgon Member is enriched in copper mineralization mainly south of Mount Timna and in manganese mineralization mainly around Mount Mikhort, forming two types of manganese nodules (Bar-Matthews 1986). The dolomite in the Elat basin is dolomitic thrombolite. The sand is sub-arkosic

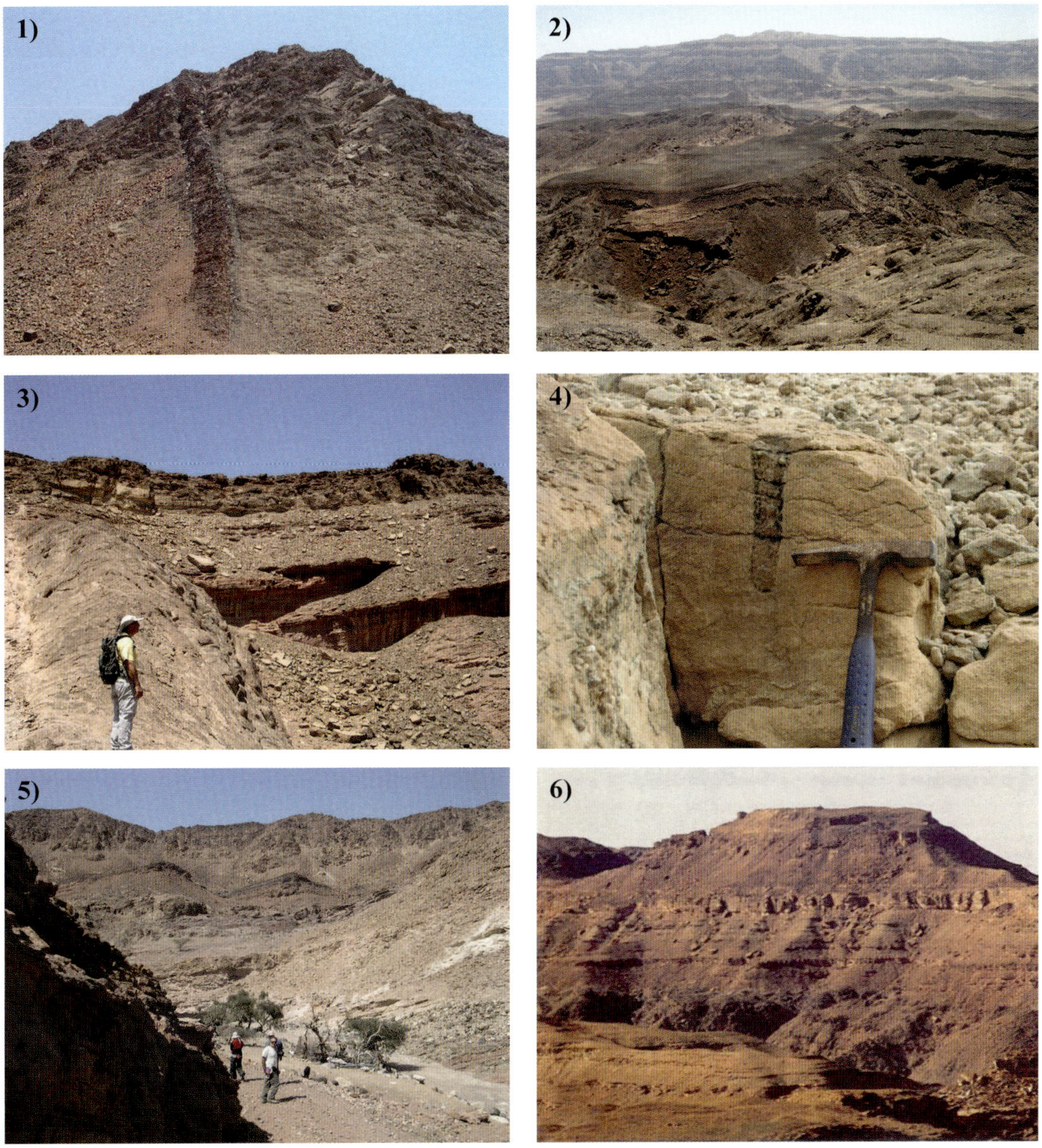

Fig. 1.5: Photographs of important elements in the geology of the Timna region: (1) Composite Neoproterozoic dyke (ca. 580 Ma) with gray andesite on both margins and pink rhyolite at the center (~ 7 m thick) intruding monzodiorite, Mount Timna. (2) The Amudei Shelomo and Timna Formations onlap the early Cambrian paleo-relief, Mount Timna. The Timna Cliffs in the background with the sandstones of the Kurnub Group in the lower part of the cliff and the carbonates of the Judea Group in the upper part. (3) Paleo-alluvial channel of Early Cambrian age, in alkali granite, filled with reddish sandstones and conglomerates of the Amudei Shelomo Formation. Well bedded silts and dolomite of the Timna Formation overlaying it, forming the top plateau of Mount Timna. (4) A burrow groove and copper mineralization along the fracture, top Amir Formation, Timna Valley. (5) The geological columnar section at the ancient copper mining area of Faynan with Salib Formation (equivalent to Amudei Shelomo) forming small rounded cliffs at the bottom, tilted, well bedded Burj Formation (equivalent to Timna) and Umm Ishrin Formation (equivalent to Shehoret and Netafim) forming the cliff on top. (6) Neoproterozoic migmatite and Early Cambrian peneplain at the base, the Cambrian Amudei Shelomo, well bedded Timna, Shehoret and Netafim Formations with the Carboniferous Umm Bugma Formation on top with a manganese mine, Umm Bugma.

with sandy columnar stromatolites (Soudry and Weissbrod 1995). The age of the Timna Formation is Early-Middle Cambrian according to trilobites (Parnes 1971; Elicki and Geyer 2013) and brachiopods (Cooper 1976) forming a transgressive-regressive lagoonal cycle. The Shehoret Formation consists of silt and sandstone deposited in a coastal environment. It was divided into three members (Weissbrod 2005) of which only the lower sub-arkosic one, 30–40 m thick, is preserved in the Timna Valley with copper mineralization ("Copper B" Bentor 1952; Weissbrod 1981). The top Netafim Formation is also missing in the Timna Valley.

Commonly the erosion surface at the top of the magmatic basement is flat-a peneplain-and the basal Early Cambrian beds are parallel to it. However where residual relief remained, like at Mount Timna, they onlap the paleo-relief. A paraconformity, represented by a stratigraphic gap of ca. 400 million years separates the top Yam Suf Group—Shehoret Formation—from the Aptian Amir Formation. This gap resulted from erosion that removed approximately 2,500 m of mainly lower Paleozoic and minor upper Paleozoic and lower Mesozoic rocks as a result of three uplift events (Weissbrod 2005).

THE EARLY CRETACEOUS KURNUB GROUP

The Kurnub Group (previously called the Hatira Formation) of the Lower Cretaceous period is comprised of the Amir, Avrona and Samar Formations. The Amir Formation is built of white friable, fine to medium grain quartz kaolinitic sandstone with recumbent cross-bedding, trace fossils, burrows (Fig. 1.5: 4) and tidal channels at the top of the unconformity with the Avrona Formation (Weissbrod and Sneh 1997) covered by a typical yellow-brown patina The age of the Amir Formation is between Late Barremian and Early Aptian and it is about 40 m thick. The Avrona Formation is up to 60 m thick and consists of medium to coarse-grained sandstones with quartz pebbles of 5–10 cm in diameter. It is characterized by a morphology of gray rounded cliffs, arches and caves and large-scale cross bedding. Ferruginous plant debris scattered in a bed 10–20 m thick at the lower part of the Avrona Formation (Ayalon, Beyth, and Vulkan 1983) and nodules of sulfidic copper mineralization with secondary green mineralization coating are common at the top of the Amir Formation. Copper mineralization was also found along fractures of the same formations at Mount ʿAmram close to Amudei ʿAmram. The Samar Formation of Middle Albian to Late Albian Age is 150 m thick and built of quartz sandstone of fine to medium sized grains with lateritic beds. The Kurnub Group is alluvial in origin and forms the soft geomorphological unit below the carbonate cliffs of the Judea Group. During Late Barremian–Albian times the southern margins of the Tethys Sea transgressed over the Arabian–Nubian continent in pulses evidenced by shallow marine sediments in between terrestrial clastics. Since the Early Cenomanian the sea covered the Timna–Elat region and gradually progressed southwards into Sinai. An Early Cretaceous basaltic plug (ca. 100 Ma) associated with phreatovolcanic rocks (previously described as the "Crazy Wall") was mapped east of Mount Mikhrot (Beyth and Segev 1983). Additional studies of those volcanoclastic rocks (Weissbrod *et al.* 1989; Segev *et al.* 1995; Weissbrod and Segev 2003) suggest a fossil explosion pipe associated with an east–west trending shallow gabbroic intrusion ca. 10 km long and ca. 1.5 km wide, and designate this whole volcanic complex as the Barak Formation.

LATEST ALBIAN TO CONIACIAN JUDEA GROUP

The Judea Group of Latest Albian–Coniacian Age comprises four marine formations: Hazera, Ora, Gerofit and Zihor, which were described by Bartov *et al.* (1972). The Hazera Formation, which is 120 m thick, is divided into four members. The lower Hevyon Member forms a light brown dolomite cliff 30 m thick with an (up to) 10 m thick yellow glauconitic–dolomitic sandstone bed at the base, unconformably overlying the Samar Formation. The En Yorqe'am Member, 30 m thick, is built of green marl and clayey limestone rich in bivalves, gastropods and echinoderms. The gray limestone with chert nodules with *Nerinea* gastropods of the Zafit Member forms a 30 m high cliff. The double cliff of the lower Hevyon Member and the upper Zafit Member—with the marly En Yorqe'am Member in between—is an important marker for geological mapping at the top of the Timna Cliffs. The upper Yotvata Member is built of white to gray well-bedded limestone at the lower parts, paper limestone in the middle, and white limestone containing the ammonite *Neolobites Vibrayeanus* indicative of the Late Cenomanian. A small block of the Yotvata Member is exposed in the Nahal Timna graben. A thin red-brown limonitic-dolomite bed at the top of the Yotvata Member marks the top Hazera Formation. During the Late Cenomanian–Early Turonian elongated shallow basins striking northeast–southwest developed in the Negev (Freund 1962). The sediments of the Ora Formation (previously called Ora Shales) were deposited in the biggest basin in the Timna–Elat area, which had a depth of 50 to 100 m. The basin was situated between the coast in the south and the elevated carbonate platform belt (Ramon–Judea) in the north. Relative sea-level fluctuations temporarily turned the basin into a lagoon with brackish water during influx of runoff water with plants, or hypersaline conditions depositing dolomite and gypsum. The Ora Formation is ca. 104 m thick, covers the Timna Cliffs and is divided into three members. The lower member is built mainly of green to gray shales with gypsum veins and thin beds of yellow nodular limestone with Early Turonian ammonites. The middle member is known as the fossiliferous dolostone of the "Vroman Bank" in the Makhtesh Ramon area. The upper member is built of red and green beds of shale, a 5 m thick massive gypsum bed and thin beds of limestone and dolostone. The soft morphology and the red and green shales with the gypsum bed are good mapping markers. The Gerofit Formation of the Late Turonian Age is ca. 130 m thick and built of gray, finely crystallized limestone precipitated in a shallow sea. The lower part is built of limestone interbedded with green clay. Disharmonic structure and block rotation attest to horizontal shearing along the Dead Sea Transform rift margins. The middle part of the formation is built of a white chalky limestone rich in ostracoda and foraminifera. The upper part is composed of well-bedded limestone and dolostone with thin beds of chert and chert nodules and common crustaceae burrows. The upper part forms a distinct brown cliff above the soft white carbonate. The rudist *Hippurites* of Turonian Age occurs at the top of the Gerofit Formation indicating upward shallowing during the deposition of the formation. The Gerofit Formation forms the plains to the north and west of the Timna Cliffs. The Zihor Formation of the Late Turonian-Coniacian Age was mapped by Bartov (1967) as the Eteq Formation south of the Timna Valley and is ca. 30 m thick. It is built of limestone, marl and dolostone rich in fossils such as stromatoporoids and oyster beds.

THE YOUNG CLASTIC (DEAD SEA GROUP)

The young clastic rocks (previously called the Upper Clastic Unit) of Miocene-Pliocene-Pleistocene Age were deposited as the result of the Dead Sea Transform rifting. At Nahal Raham, south of the Timna Valley, the Miocene Raham Conglomerate was mapped (Garfunkel *et al.* 1974). It is a poorly sorted, coarse conglomerate in which the boulders are rounded-angular and are composed mainly of the Eocene Avdat Group carbonate rocks with nummulites without magmatic and metamorphic boulders. Avni, Segev and Ginat (2012) suggested that the Raham Conglomerate is of the Eocene Age. Three fluvial terraces of the Pliestocene to Holocene Age were mapped in the Timna Valley (Ginat *et al.* 2009): the high, Old River Terraces (Q1), 30–60 m above the present channels of Nahal Timna; the Middle River Terraces (Q2), 10–30 m above the present channels; and the younger Lower River Terraces (Q3), 5–10 m high which are part of the recent drainage system.

The alluvial sediments in this extremely arid zone were studied as a marker for paleo-climatic changes by Finkelstein-Grossman (1982) and Grossman and Gerson (1987). Dating of six levels of river terraces was carried out at the Nahal Shehoret alluvial fan 20 km south of the Timna Valley by Porat *et al.* (2010) using the OSL method whereby the oldest age was found to be >231,000 and the youngest 14,000 years. In the southern Arava, east of the Timna Valley margins, the young clastic rocks are over 1000 m thick (Frieslander 2000).

STRUCTURE

The structures of the Timna Valley are of three types: (1) the ancient Neoproterozoic magmatic ones; (2) those related to the Themed E–W dextral fault system (Bartov 1974); and (3) the major young ones which are related to the Dead Sea Transform and its margins. The ca. E70N striking Neoproterozoic fractures in Mount Timna control the propagation of the rhyolite, andesite and composite dyke swarms and the propagation of the dolerite dyke (Baer, Beyth and Reches 1994). The dextral Themed Fault which is part of the Sinai-Negev Shear Zone (Bartov 1967) strikes ENE and is cut by the Dead Sea Transform east of Beʾer Orah, ca. 4 km south of the Timna Valley. Garfunkel (1970) suggested that the Themed Fault is a reactivated pre-rift marginal Neoproterozoic fault. The young-recent faults were first mapped by Quennel (1958) and the general pattern of the active faults along the Dead Sea Transform that cross the Arava alluvial fans, the Avrona Playa and the Elat Sabkha were presented by Garfunkel (1981). Frieslander (2000) defined two major subsurface faults at the eastern margin of the Timna Valley—the Timna and Samar Faults striking NNE. Amit *et al.* (1999) concluded that there were at least six earthquake events of magnitude 6 during the last 14,000 years at the Avrona Playa. During the last 1,000 years a relief inversion occurred in the Avrona Playa as a result of an uplift of the surface by 1 m due to a 6.1 magnitude earthquake. This event opened the previously closed drainage system and the Avrona Playa which is presently drained by Nahal Avrona to the Elat Sabkha. Faults of two systems were defined in the alluvial fans south of the Timna Valley, the oldest one, striking northwest, was active during the Pliocene to Late Pleistocene, and the youngest, striking northeast, was active during the Holocene (year>11,000). Most of the ancient copper shafts and the karstification of the dolomitic lithofacies in the Timna Valley are along the northwest striking faults. The internal structure of the Timna Valley was divided by Segev, Beyth and Bar-Matthews (1992) into eleven tectonic blocks from southeast to northeast: Nahal Raham; Mount Hakhlil; Nahal Nimra; Nahal Gadna; "Step Faults"; Mount Timna which is the central block; Arava; Nahal Timna Graben; Nahal Timna; NW Timna Valley; Mount Mikhrot. Some of the faults striking east northeast delimiting the Mount Timna, Mount Hakhlil and "Step Faults" blocks are reverse faults. The recent uplift of Mount Timna is normal to the main compressional stress direction of the Dead Sea Transform (Eyal and Reches 1983). Geomorphic evidence for young tectonic activity, uplift of Mount Timna and subsidence of the Yotvata Playa—such as the abandoned talus remnants on the northeast margin and the change of stream direction in the east margins of Mount Timna—was suggested by Ginat *et al.* (2009).

Four partly overlapping tectonic stages were suggested by Ginat *et al.* (2009) for the formation of the Timna Valley since the Late Eocene, the final regression of the Tethys Ocean:

1. Late Eocene uplift of Mount Timna adjacent to the embryonic Dead Sea Transform;
2. Intensive weathering creating two erosion surfaces, one at the summit of Mount Timna, approximately 200 m of Early Pliocene Age, and the second surrounding Mount Timna at approximately 110 m above the present day drainage system of the Late Pliocene;
3. Continuous subsidence of the Yotvata Playa during the Pleistocene forming the river terraces (Q1–Q3) associated with a 1–3 km retreat of the marginal cliffs which enlarged the area of the Timna Valley

from 40 sq km to 80 sq km. During this stage the direction of the water flow in the ephemeral streams east of the present-day water divide at the top of the Timna Cliffs, changed from westward to eastward;

4. A continuous eastward tectonic tilting during the Late Pleistocene through the Holocene due to the ongoing uplift of Mount Timna and/or subsidence of the Yotvata Playa.

MINERALIZATION

Copper mineralization in the Timna Valley was found from the Neoproterozoic Timna Igneous Complex through the Cambrian sediments up to the Early Cretaceous sandstone. In the Cambrian Timna Formation it is associated with manganese and uranium (Ayalon *et al.* 1985b). In ancient times copper was exploited mainly from the lower part of the Avrona and the top Amir Formations in the northwestern areas of the Timna Valley (Rothenberg and Bachmann 1988), whereas modern mining activities produced copper from the Cambrian host rocks south of Mount Timna. Copper was exploited in ancient times outside the Timna Valley along the margins of the Dead Sea Transform at Faynan (Hauptmann 2007), Wadi Khusheiba (Kind 1965) and Mount ʿAmram (Avner *et al.*, this volume). Other mineralization occurrences were reported from along the western margins of the Dead Sea Transform, such as the gold mineralization from Nahal Shehoret and Nahal Roded (Bogoch *et al.* 1993) and manganese barium and lead mineralization from Wadi Umm Zariq in southeast Sinai (Ayalon *et al.* 1985a). A summary of the various occurrences of copper mineralization along the southern Dead Sea Transform are presented in Table 1.1 (Wurzburger 1967a and 1970b; Segev, Beyth and Bar-Matthews 1992 and references therein). Chalcocite, chalcopyrite, bornite and digenite are the primary copper sulfides and djurleite is a copper-silver sulfide mineral. Malachite, pseudomalachite (contain PO_4), paratacamite, chrysocolla, cuprite, dioptase and bisbeeite are hydroxides, oxides, chlorite and silicate minerals of copper which are

Table 1.1: Summary of Copper Occurrences from Both Sides of the DST Excluding the Cenozoic Mineralization Timing (After Segev, Beyth and Bar-Matthews 1992, and References Therein)

Stratigraphy			*Main Copper Minerals*	*Mineralization*	
				Type	*Timing*
Lower Cretaceous	Avrona Fm.		Cha, cov, mal, para	Epigenetic, Hypogene	Lower Cretaceous?
	Amir Fm.		Secondary copper minerals	Epigenetic, Hypogene	Lower Cretaceous?
Middle (?) Cambrian	Shehoret Fm.	White Mbr	Cha, bor, chp, cup, mal	Synsedimentary? and Epigenetic	Cambrian?
		Multicolor Mbr	Chr, bis, mal, para		
	Timna Fm.	Sandy+Shaly lithofacies	Chr, bis, mal, Para, pse	Epigenetic, multi-stages hydrothermal	Tertiary, Lower Cretac.?
		Dolomitic lithofacies	Dju, para, mal	Synsedimentary	Cambrian
Precambrian		Alkaline volcanics	Cha, chp, cov, cup, ten, mal, chr	1) Primary volcanogenic/ magmatic	Precambrian
		Granite, rhyolitic dikes, gabbro	Cha, chp, chr, dig, cov, mal, para	2) Epigenetic hydrothermal alteration	Tertiary

Cha- chalcocite; chp- chalcopyrite; dju- djurleite; ten- tenorite; cov- covellite; bor- bornite; cup- cuprite; dig- digenite; chr- chrysocolla; bis- bisbeite; mal- malachite; pse-pseudomalachite; para-paratacamite.

secondary alteration products. The copper and manganese mineralization was formed in four major cycles during ca. 600 Ma (Beyth 1987; Segev and Sass 1989b; Shlomovitch *et al.* 1999; Asael 2010; Segev and Beyth 2011):

Stage I: hydrothermal copper and manganese enrichment in the final magmatic stage of the Timna Igneous Complex at ca. 600 Ma. The copper sulfidic minerals, chalcopyrite, dijenite, chalcocite and covelite which were partly altered to paratacamite, malachite and chrysocolla were reported by Wurzburger (1967a, b).

Stage II: sedimentation of erosional products of the volcanic Timna Igneous Complex rocks and precipitation of enriched copper and manganese dolomite in Cambrian lagoons at ca. 520 Ma. Main copper sulfide minerals are djurleite and covellite with minor chalcocite, digenite and anilite containing silver as a significant minor element, deposited in an early diagenetic stage (Shlomovitch *et al.* 1999). Also diagenetic type A manganese nodules composed of pyrolusite and hollandite and uranium–enriched, stratiform manganese (Bar-Matthews 1987) were deposited.

Stage III: as a result of the phreatovolcanic event of the Lower Cretaceous Barak Formation, those thermal brines dissolved the Cu and Mn enriched dolomite. Due to oxidation the manganese, mainly coronadite nodules, epigenetic type B uranium-depleted (Bar-Matthews 1987), precipitated in the Sasgon Member. K-Ar dating of three manganese nodules from the shaly lithofacies suggests Lower Cretaceous Age (Segev, Beyth and Bar-Matthews 1992). The copper which was transported up-section to the lower part of the Avrona top Amir Formations and precipitated as sulfides which later partly oxidized to malachite and atacamite at ca. 100 Ma. Keidar (1984) identified three types of copper mineralization in the Avrona Formation: In the ferruginous plant debris associated with lead and zinc and to a lesser extent molybdenum (the mineral wulfenite $PbMoO_4$ was identified) and uranium (Ayalon *et al.* 1983); concentrations and nodules; impregnation and cementation of sandstones.

Stage IV: epigenetic dissolution (karstification) of the enriched dolomitic lithofacies by circulating thermal brines along Dead Sea Transform related faults and fractures (Weber *et al.* 2009) mainly the northwest striking ones. As a result of the insoluble residue, the sandy lithofacies were enriched by copper and phosphate-uranium mineralization from ca. 14 Ma to the present. The present Timna Valley ground water constitutes thermal (surface T ~ 30°C) Ca-chloritic brines with extreme high Ca/Mg ratio, low $^{87}Sr/^{86}Sr$ isotopic ratio and enrichment in copper, iron and manganese (Beyth, Starinsky and Lazar 1983). These brines were most probably involved in the epigenetic alteration of the Timna Igneous Complex (Marco *et al.* 1993; Beyth *et al.* 1997) as well as in the karstification and mineral enrichment. Indeed, the main copper mineralization is concentrated within the sandy and shaly lithofacies as epigenetic secondary minerals in the following paragenetic order: chrysocola, plancheite, pseudomalachite and dioptase (Shlomovitch *et al.* 1999).

MOUNT ᶜAMRAM

Mount ᶜAmram is an east–west striking block, 12 km south of the Timna Valley, consisting of alkali granite, rhyolites and volcanoclastic rocks intruded by hypabyssal monzonite and syenite intrusions and by a few east–west striking andesite and composite dykes (Mushkin *et al.* 2003; Beyth *et al.* 2011) of ca. 600 Ma. (Be'eri-Shlevin 2008). Ancient copper mining activity was reported from the southwestern margins around the ᶜAmram Pillars (Rothenberg and Bachmann 1988; Willies 1991) in the Amir Formation which was preserved in a north–south striking graben. Copper mineralization was also mapped in small rhyolite outcrops at the upper part of Nahal Amir in the western margins of Mount ᶜAmram.

FAYNAN–SOUTH

Reconstruction of the 105 km left lateral movement of the Dead Sea Transform serves as the basis for the paleogeographic reconstruction of the Timna–Faynan marine Cambrian sedimentary basin (Segev 1984: Fig. 7, and references therein). Two major sinistral faults controlled this movement, the major Dead Sea Transform (85 km) and the Faynan or Al Quwayra fault zone at the eastern margins (24 km. Beyth *et al.* 2014). These suggest that the east–west striking Dana Fault in Jordan is the eastward continuation of the Themed Fault (Bartov 1974). In the Faynan area the Neoproterozoic and Cambrian rocks with copper mineralization or slugs can be found at Jabal Hamra Faddan, Jabal Hanna, Khubrat al Nuhas, Wadi Dana and Khirbat Faynan. The stratigraphy and lithology of the Cambrian to Eocene sedimentary rocks is almost identical: Salib Arkosic Sandstone is the equivalent of Amudei Shelomo; Burj Dolomite-Shale to Timna Formation; Umm Ishrin Sandstone to Shehoret and Netafim Formations: Kurnub to Kurnub Group (Fig. 1.3). Faynan is known for its intense ancient copper mining activity from the Burj and Umm Ishrin Formations. Like in the Timna Valley, the association of copper, manganese and uranium mineralization was reported (Rabbaᶜ 1994, and references therein).

UMM BUGMA–SOUTHWESTERN SINAI

The ancient Egyptian temple of Sarabit el-Khadim in southwestern Sinai was the center of turquoise (gemstone mineral of hydroxyl copper aluminum phosphate composition) mining in the Umm Bugma region on the eastern margins of the Gulf of Suez rift (Garfunkel and Bartov 1977). This mining continues in modern times by Bedouin. Additionally, manganese and kaolinite are mined by the Egyptian government. The stratigraphy of the Cambrian sedimentary rocks, the Amudei Shelomo through Netafim Formations overlying the Neoproterozoic metamorphic igneous complex, is an almost identical section to the one defined in the Timna Valley and Elat area. Carboniferous dolomite and marls rich in brachiopods and coral fossils of the Umm Bugma Formation and the Ataqa Formation (previously Abu Thora) of sandstones intruded by partly kaolonized basaltic sills are overlying the top of the Cambrian Formations (Fig. 1.5: 6). The Budra Formation from the Triassic Age and the sandstones of the Kurnub Group overlie it. Manganese-iron oxide mineralization is present at the base of the Umm Bugma Formation in the shape of lenses. The turquoise veins occur at the upper part of the Netafim Formation and traces of copper mineralization at the Timna Formation along the northeast striking faults (Beyth 1978, 1981; Weissbrod 2005). As in the Timna Valley and Faynan, the final stage of enrichment at Umm Bugma was Late Cenozoic epigenetic-hydrothermal (Beyth 1987; Segev 1984).

SUMMARY

The Timna Valley and Mount ᶜAmram on the western margins (Fig. 1.6), and the Faynan area on the eastern margins of the Dead Sea Transform, as well as Umm Bugma on the eastern margins of the Gulf of Suez Rift, contain ancient mines along plate boundaries with almost identical stratigraphy of the Cambrian sedimentary rocks. The enrichment process from the primary magmatic source rock took place over the past ca. 600 Ma. The final and most important stage was epigenetic-hydrothermal enrichment along the plate boundaries. Mineralization along the Dead Sea Transform plate boundary is known from additional locations such as the gold from Nahal Shehoret and Nahal Roded, and the manganese-barium-lead from Wadi Um-Zariq.

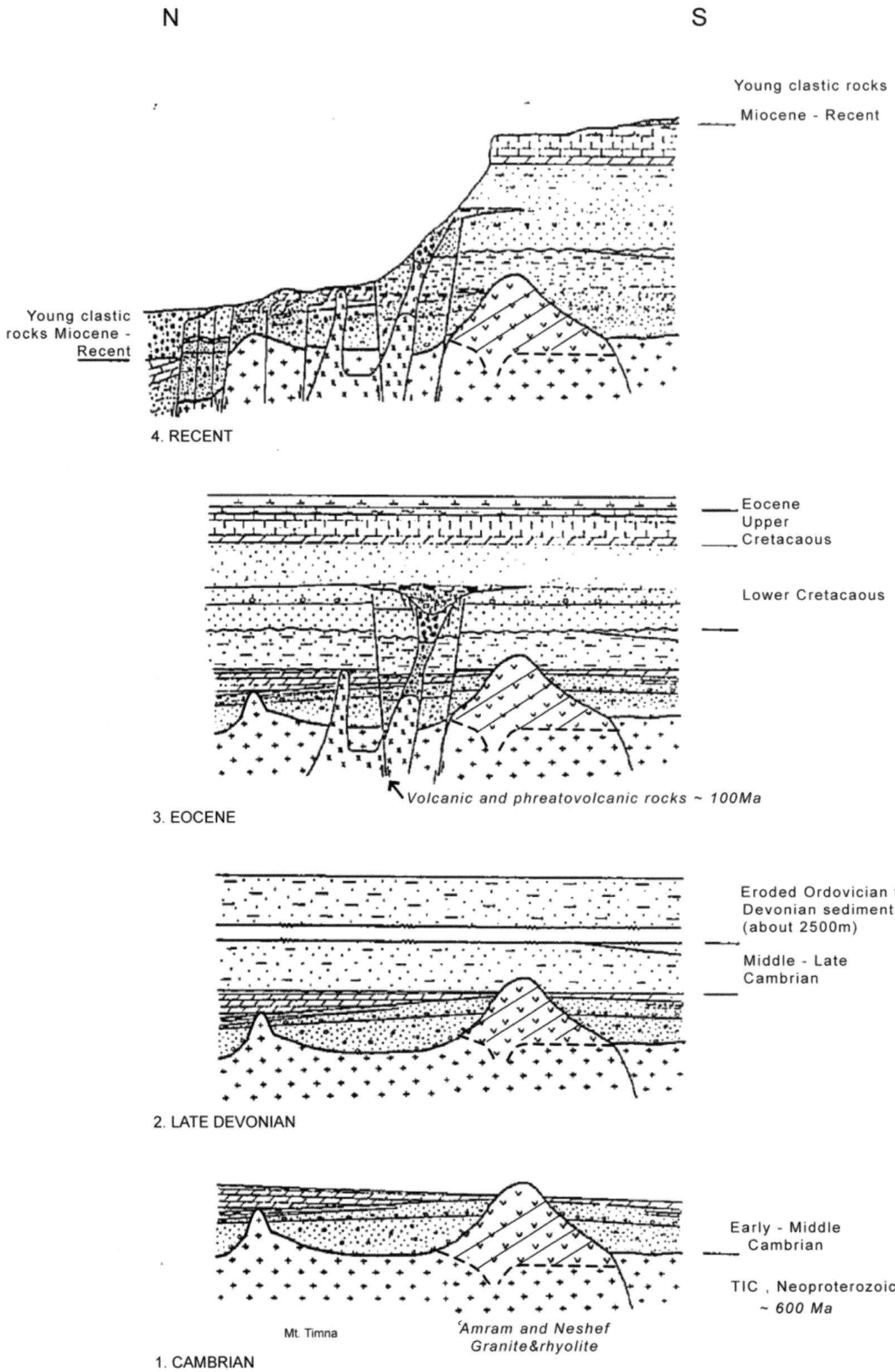

Fig. 1.6: Major geological events, Timna Valley and Mount ʿAmram (after Segev, Beyth and Bar-Matthews 1992).

ACKNOWLEDGMENTS

We thank Zeev Lewy, Bibi Katz and Erez Ben-Yosef for editing the manuscript, Batsheva Cohen and Nili Almog for supporting the drafting, and Shelomo Ashkenazi for supporting the field work.

REFERENCES

Amit, R., Zilberman, E., Porat, N. and Enzel, Y. 1999. Relief Inversion in the Avrona Playa as Evidence of Large-Magnitude Historical Earthquakes, Southern Arava Valley, Dead Sea Rift. *Quaternary Research* 52: 76–91.

Arad, V. and Bartov, Y. 1994. *Arava Region Israel and Jordan: Bibliography of Geological Research (Geological Survey of Israel Report/GSI/94)*. Jerusalem.

Asael, D. 2010. *Copper Stable Isotopes Fractionation in Low Temperature Geological Systems* (Ph.D. dissertation, The Hebrew University). Jerusalem.

Avni, Y., Segev, A. and Ginat, H. 2012. Oligocene Regional Denudation of the Northern Afar Dome: Pre- and Syn-Breakup Stages of the Afro-Arabian Plate. *Geological Society of America Bulletin* 124 (11/12): 1871–1897.

Ayalon, A., Beyth, M. and Vulkan, U. 1983. *Geochemistry and Radioactive Mineralization Occurrences in the Timna Valley* (Geological Survey of Israel Report ME-10/83). Jerusalem.

Ayalon, A., Beyth, M. and Vulkan, U. 1985a. Geochemistry of Radioactive Mineralization Occurrences in the Timna Valley: Transactions of the Institution of Mining and Metallurgy, Section B. *Applied Earth Science* 94: B197–B201.

Ayalon, A., Beyth, M., Zilberfarb, A. and Bar-Matthews, M. 1985b. Manganese, Barium and Lead Mineralization in Wadi Um Zariq, Southeastern Sinai. *Israel Journal of Earth Sciences* 34: 31–37.

Baer, G., Beyth, M. and Reches, Z. 1994. Dike Emplaced into Fractured Fasement, Timna Igneous Complex. *Israel Journal of Geophysical Research* 99: 24039–24050.

Bar-Matthews, M. 1986. *Mineralization of Uranium and other Metals in the Timna Formation, Timna Valley* (Ph.D. dissertation, The Hebrew University). Jerusalem (Hebrew with English abstract).

Bar-Matthews, M. 1987. The Genesis of Uranium in Manganese Phosphorite Assemblages, Timna Basin, Israel. *Geological Magazine* 127: 1–12.

Bartov, Y. 1967. *The Geology of Be'er Ora Region* (M.Sc. thesis, Hebrew University of Jerusalem). Jerusalem (Hebrew).

Bartov, Y. 1974. *A Structural and Paleogeographical Study of the Central Sinai Faults and Domes* (Ph.D. dissertation, The Hebrew University). Jerusalem (Hebrew with English abstract).

Bartov, J., Eyal, Y., Garfunkel, Z. and Steinitz, G. 1972. Late Cretaceous and Tertiary Stratigraphy and Paleogeography of Southern Israel. *Israel Journal of Earth Sciences* 21: 69–97.

Bartura, Y. 1966. Type Sections of the Paleozoic Formations in the Timna Area. *Geological Survey of Israel (Stratigraphic Section N3*). Jerusalem.

Be'eri-Shlevin, Y. 2008. *The Origin and Evolution of Neoproterozoic Magmatism in the Northern Arabian-Nubian Shield (Sinai Peninsula, Egypt and Southern Israel): Evidence from Stable and Radiogenic Isotope Record* (Ph.D. dissertation, Ben-Gurion University). Beer-Sheva.

Bentor, Y.K. 1952. *Report on the Copper and Manganese Deposits in Wadi Menayeh* (Geological Survey of Israel Report 34). Jerusalem.

Bentor, Y.K. and Vroman, A. 1955. *The Geological Map of the Negev, 1:100,000, Sheet 24, Elat* (Geological Survey of Israel). Jerusalem.

Beyth, M. 1978. *The Geological Map of the Umm Bogma Area (1:25,000), Explanatory Notes* (Geological Survey of Israel Report MP-575/78). Jerusalem (Hebrew).

Beyth, M. 1981. Paleozoic Vertical Movements in Um Bugma Area, Southwestern Sinai. *Bulletin of the American Association of Petroleum Geologists* 65: 160–165.

Beyth, M. 1986. The Geological Map of Umm Bugma. *GSI Current Research* 8: 32–40.

Beyth, M. 1987. Mineralization Related to Rift Systems: Examples from the Gulf of Suez and the Dead Sea Rift. *Tectonophysics* 141: 191–197.

Beyth, M., Eyal, Y. and Garfunkel, Z. 2011. *The Geology of the Elat Sheet, Explanatory Notes* (Geological Survey of Israel Report GSI/22/2011, updated March 2013). Jerusalem.

Beyth, M., Ginat, H., Hijazi, H., Al Khraisha, S. 2014. Accurate Displacement Along the Arava Segment of Dead Sea Transform: Preliminary Results. *Israel Geological Society Annual Meeting, Ein Boqeq*, 2014 (Abstract).

Beyth, M. and Heimann, A. 1999. The Youngest Igneous Event in the Crystalline Basement of the Arabian-Nubian Shield, Timna Igneous complex. *Israel Journal of Earth Sciences* 48: 113–120.

Beyth, M. and Peltz, S. 1992. *Petrology and Major-Element Geochemistry of the Dykes at Har Timna, Southern Israel* (Geological Survey of Israel Report, GSI/13/92). Jerusalem.

Beyth, M., Longstaff, F.J., Ayalon, A. and Matthew, A. 1997. Epigenetic Alteration of the Precambrian Igneous Complex of Mount Timna, Southern Israel: Oxygen-Isotope Studies. *Israel Journal of Earth Science* 46: 1–11.

Beyth, M. and Reischmann, T. 1997. The Age of the Quartzmonzodiorite, the Youngest Plutonic Intrusion in the Timna Igneous Complex. *Israel Journal of Earth Sciences* 45: 223–226.

Beyth, M. and Segev, A. 1983. Lower Cretaceous Basaltic Plug in the Timna Valley. *Israel Journal of Earth Sciences* 32: 165–166.

Beyth, M., Segev, A. and Bartov, Y. 1999. *The Geological Map of Be'er Ora, 1:50,000, Sheet 25-IV* (Geological Survey of Israel). Jerusalem.

Beyth, M., Starinsky, A. and Lazar, B. 1983. Low Temperature Geothermal Waters, Timna, Southern Israel. *GSI Current Research* 4: 17–20.

Beyth, M., Stern, R.J., Altherr, R. and Kroener, H. 1994a. The Late Precambrian Timna Igneous Complex, Southern Israel: Evidence for Comagmatic Type Sanukitoid Monzodiorite and Alkali Granite Magma. *Lithos* 31: 103–124.

Beyth, M., Stern, R., Altherr, R., Peltz, S. and Heimann, A. 1994b. Petrochemistry of Dolerite Dykes from Mt. Timna, Southern Israel: Implications for Plate Tectonic Setting. *GSI Current Research* 9: 24–26.

Bogoch, R., Shirav, M., Beyth, M. and Halicz, L. 1993. Geochemistry of Ephemeral Stream Sediments in the Precambrian Mountainous Arid Terrain of Southern Israel. *Journal of Geochemical Exploration* 46: 349–364.

Cooper, G.A. 1976. Lower Cambrian Brachiopods from the Rift Valley (Israel and Jordan). *Journal of Paleontology* 50: 269–289.

Elicki, O. and Geyer, G. 2013. The Cambrian Trilobites of Jordan—Taxonomy, Systematic and Stratigraphic Significance. *Acta Geologica Polonica* 63: 1–56.

Eyal, Y. and Reches, Z. 1983. Tectonic Analysis of the Dead Sea Rift Region since the Late Cretaceous based on Mesostructures. *Tectonics* 2: 167–185.

Finkelstein-Grossman, S. 1982. *Alluvial Sediments in Extremely Arid Zones—as a Marker for Environmental Changes* (M.Sc. thesis, The Hebrew University). Jerusalem (Hebrew with English abstract).

Freund, R. 1962. *Problems in the Stratigraphy of the Cenomanian-Turonian in Israel* (Ph.D. dissertation, The Hebrew University). Jerusalem (Hebrew with English abstract).

Freund, R. 1965. A Model of Structural Development of Israel and Adjacent Areas since Upper Cretaceous Times. *Geological Magazine* 102(3): 189–205.

Frieslander, U. 2000. *The Structure of the Dead Sea Transform Emphasizing the Arava Using New Geophysical Data* (Ph.D. dissertation, The Hebrew University). Jerusalem (Hebrew with English abstract).

Garfunkel, Z. 1970. *The Tectonics of the Western Margins of the Southern Arava* (Ph.D. dissertation, The Hebrew University). Jerusalem (Hebrew with English abstract).

Garfunkel, Z. 1981. Internal Structure of the Dead Sea Leaky Transform (Rift) in Relation to Plate Kinematics. *Tectonophysics* 80: 81–108.

Garfunkel, Z. and Bartov, Y. 1977. The Tectonics of the Suez Rift. *Geological Survey of Israel Bulletin* 71: 44.

Garfunkel, Z., Bartov, J., Eyal, Y. and Steinitz, G. 1974. Raham Conglomerate—New Evidence for Neogene Tectonism in the Southern Part of the Dead Sea Rift. *Geological Magazine* 111: 55–64.

Ginat, H., Beyth, M. and Crouvi, O. 2009. Geomorphic Evidence for Young Tectonic Activity around Mt. Timna. *Israel Journal of Earth Sciences* 57: 213–230.

Grossman, S. and Gerson, R. 1987. Fluviatile Deposits and Morphology of Alluvial Surfaces as Indicators of Quaternary Environmental Changes in the Southern Negev, Israel. In: Frostick, L.K. and Reid, I., eds. *Desert Sediments: Ancient and Modern* (Israel Geological Society Special Publication 35). London: 17–29.

Hauptmann, A. 2007. *The Archaeometallurgy of Copper—Evidence from Faynan, Jordan*. Berlin.

Karcz, I. and Key, C.A. 1966. Note on the Pre-Paleozoic Morphology of the Basement in the Timna Area (Southern Israel). *Israel Journal of Earth Sciences* 15: 47–56.

Katzir, Y., *et al*. 2007. Interrelations Between Coeval Mafic and A-type Silicic Magmas from Composite Dykes in a Bimodal Suite of Southern Israel, Northernmost Arabian-Nubian Shield: Geochemical and Isotope Constraints. *Lithos* 97: 336–364.

Kind, H.D. 1965. Antike Kupfergewinnung zwischen Rotem und Totem Meer. *Zeitschrift des Deutschen Palästina-Vereins* 81: 56–73.

Marco, S., Ron, H., Matthews, A., Beyth, M. and Navon, O. 1993. Chemical Remnant Magnetism Related to the Dead Sea Rift: Evidence from Precambrian Igneous Rocks of Mount Timna, Southern Israel. *Journal of Geophysical Research* 98: 16001–16012.

Mart, Y. and Horowitz, A. 1981. The Tectonics of the Timna Region in Southern Israel, and the Evolution of the Dead Sea Rift. *Tectonophysics* 79: 165–199.

Mushkin, A., Navon, O., Halicz, L., Hartmann, G. and Stein, M. 2003. The Petrogenesis of A-type Magmas from the Amram Massif, Southern Israel. *Journal of Petrology* 44 (5): 815–832.

Parnes, A. 1971. Late Lower Cambrian Trilobites from the Timna Area and Har Amram (Southern Negev, Israel). *Israel Journal of Earth Sciences* 20: 179–205.

Porat, N., *et al*. 2010. Abandonment Ages of Alluvial Landforms in the Hyperarid Negev Determined by Luminescence Dating. *Journal of Arid Environments* 74: 861–869.

Quennel, A.M. 1958. The Structural and Geomorphic Evolution of the Dead Sea Rift. *Quarterly Journal of the Geological Society* 114: 1–24.

Rabbaᶜ, I. 1994. The Geology of the Al Qurayqira (Jabal Hamra Faddan): Map Sheet No. 3051 II. *Natural Resources Authority, Ministry of Energy and Mineral Resources (The Hashemite Kingdom of Jordan) Bulletin* 28: 60.

Rothenberg, B. and Bachmann, H.G. 1988. *The Egyptian Mining Temple at Timna*. London.

Segev, A. 1984. Lithostratigraphy and Paleogeography of the Marine Cambrian Sequence in Southern Israel and Southwestern Jordan. *Israel Journal of Earth Sciences* 33: 26–33.

Segev, A. 1986. *Lithofacies Relations and Mineralization Occurrences in the Timna Formation, Timna Valley* (Ph.D. dissertation, The Hebrew University). Jerusalem (Hebrew with English abstract).

Segev, A. and Beyth, M. 1986. *Geological Map of Timna Valley, 1:20,000, Sheet 24* (Geological Survey of Israel). Jerusalem.

Segev, A., Beyth, M. and Bar-Matthews, M. 1992. The Geology of the Timna Valley with Emphasis on Copper and Manganese Mineralization—Updating and Correlation with the Eastern Margins of the Dead Sea Rift. *Geological Survey of Israel Report* GSI/14/92. Jerusalem.

Segev, A., Rybakov, M., Weissbrod, T. and Beyth, M. 1995. Lower Cretaceous Gabbroic Intrusion at Shallow Depth in the Timna Valley, Southern Israel: Magnetometric Evidence. *Israel Journal of Earth Sciences* 44: 137–148.

Segev, A. and Sass, E. 1989a. Lithofacies and Thickness Control by Epigenetic Dissolution—The Dolomitic Timna Formation, Cambrian, Southern Israel. *Sediment Geology* 63: 109–126.

Segev, A. and Sass, E. 1989b. Copper Enriched Syngenetic Dolostones as a Source for Epigenetic Copper Mineralization in Sandstones and Shales (Timna Israel). *Geological Association Canada, Special Publication* 36: 647–665.

Shaw, S.H. 1947. *Southern Palestine, Geological Map on a Scale of 1:250,000 with Explanatory Notes* (Government of Palestine, Government Printer). Jerusalem.

Shlomovitch, N., Bar-Matthews, M., Segev, A. and Matthews, A. 1999. Sedimentary and Epigenetic Copper Mineral Assemblages in the Cambrian Timna Formation, Southern Israel. *Israel Journal of Earth Sciences* 48: 195–208.

Shpitzer, M., Beyth, M. and Matthews, A. 1992. Igneous Differentiation in the Late Precambrian Plutonic Rocks of Mt. Timna. *Israel Journal of Earth Sciences* 40: 17–27.

Soudry, D. and Weissbrod, T. 1995. Morphogenesis and Facies Relationships of Thombolites and Siliciclastic Stromatolites in a Cambrian Tidal Sequence (Elat Area Southern Israel). *Palaeogeography, Palaeoclimatology, Palaeoecology* 114: 339–355.

Survey of Israel. 1985. *Atlas of Israel (3rd ed.)*. Tel Aviv.

Vapnik, Y. 1998. Fluid Inclusions in Metamorphic Rocks of the Elat Area. *Israel Journal of Earth Sciences* 42: 71–84.

Weber, M., *et al.* 2009. Anatomy of the Dead Sea Transform from Lithospheric to Microscopic Scall. *Review of Geophysics* 47: 1–44.

Weissbrod, T. 1969. The Paleozoic of Israel and Adjacent Countries (Part II). *Geological Survey of Israel Bulletin* 48: 1–32.

Weissbrod, T. 1981. *The Paleozoic of Israel and Adjacent Countries (Lithostratigraphic Study)* (Ph.D. dissertation, The Hebrew University). Jerusalem.

Weissbrod, T. 2005. The Paleozoic in Israel and Environs. In: Hall, J.K., Krasheninnikov, V.A., Hirsch, F., Benjamini, C. and Flexer, A., eds. *Geological Framework of the Levant*, Vol. II: *The Levantine Basin and Israel*. Jerusalem: 283–316.

Weissbrod, T. and Segev, A. 2003. Lower Cretaceous Magmatic Activity in the Timna Valley: Geological Setting and $^{40}Ar/^{39}Ar$ Dating. *Israel Journal of Earth Sciences* 52: 77–95.

Weissbrod, T., Segev, A., Beyth, M. and Theron, J.N. 1989. An Early Cretaceous Volcanoclastic Event: Reappraisal of the "Crazy Wall" at Timna. *Israel Geological Society Annual Meeting, Ramot 1989*: 166–167 (Abstract).

Weissbrod, T. and Sneh, A. 1997. Stratigraphy and Sedimentology of the Lower Cretaceous Amir Formation in Israel and Sinai: A Marine Clastic Deposit with Tidal Channels. *Israel Journal of Earth Sciences* 46: 79–93.

Willies, L. 1991. Ancient Copper Mining at Wadi Amram, Israel, an Archaeological Survey. *Bulletin of the Peak District Mines Historical Society* 11: 109–138.

Wurzburger, U. 1967a. *The Occurrence of Sulphides in the Timna Massif and Secondary Copper Minerals in the Timna Copper Deposits, Negev (Southern Israel)* (unpublished Report).

Wurzburger, U. 1967b. Copper Sikicates in the Timna Ore Deposit. *Israel Journal of Chemistry* 8: 443–457.

Zlatkin, A. and Wurzburger, U. 1957. Eruptive Rocks of Timna (Negev). *Geological Survey of Israel Bulletin* 14: 41.

CHAPTER 2

BENO ROTHENBERG AND THE CHRONOLOGY OF COPPER SMELTING AT TIMNA

James D. Muhly

In 1984 I published my interpretation of the chronology of copper smelting at Timna and the hypothesis that the copper mines and the copper smelting installations at that site could be identified with the fabled copper mines of King Solomon. My article,"Timna and King Solomon"(Muhly 1984), was a review published at a time when there was a great deal of interest in King Solomon's Mines and in the popular books that Rothenberg had published on that subject, especially, Were These King Solomon's Mines? Excavations in the Timna Valley. *It is fair to say that my review convinced many scholars and infuriated many others. It was also written years before the great surge of scholarly interest in dating the reigns of Solomon and David and the re-interpretation of the historical reality of an "Age of Solomon" in the 10th century BCE.*

The occasion of the 2013 conference, held in memory and appreciation of the pioneering contribution that Beno Rothenberg made to the field of archaeometallurgy, held on-site at Timna Park, was an excellent opportunity to take another look at outstanding questions regarding Solomon and Timna. The proper treatment of the issues involved, and the extensive recent literature, demands a scholarly monograph; I hope I am able to call attention to some of the central issues and to highlight how our understanding of all aspects of the controversy have changed due to the scholarship of the past thirty years. I feel that Beno would be delighted by the on-going interest in almost every facet of his scholarly career.

Beno Rothenberg and I go back some 40 years, fighting all the way. We had strong differences of opinion over the years, but because of a sense of mutual respect we always remained at least on speaking terms. The last time I saw him was at a conference in Cyprus, in 2002. During a visit to Skouriotissa, on that occasion, he suggested that we needed to find something new to fight about. In this paper I will try to show that Timna still provides much material for extended discussion.

I have a special interest in the relationship between copper production in Cyprus, on the one hand, and that in the southern Negev and the Nahal Arabah on the other. This relationship, it seems to me, is now clearer than ever before. In the area of Faynan the first main period of copper smelting is Late Chalcolithic—Early Bronze and Middle Bronze—including a remarkable Early Bronze A metalworking site at Khirbat Hamra Ifdan (Levy *et al.* 2002). There is very little going on in Cyprus during this period, essentially a period that I would call "incipient copper production," best seen at the site of Makri *Alonia* (Knapp 2008: 74–77; 2013: 299–302). In the Late Bronze Age we see a serious decline in Nahal Arabah copper production, with limited activity at Timna, as Rothenberg's New Kingdom Egyptian industry, seen especially at Site 30, is now re-dated to the Early Iron Age (Ben-Yosef *et al.* 2012). There is, as

I will argue, still evidence for Late Bronze Age smelting activity, especially at Site 2, but the general interpretation of copper production at Timna has now undergone a profound transformation.

The Late Bronze Age, on the other hand, is the major period of copper production in Cyprus. This is the great age of the distinctive copper "oxhide" ingots, whose production we now realize began in the 16th century BCE and whose distribution, in the 13th–12th centuries BCE extended from southern France to southern Mesopotamia (Muhly 2009). By ca. 1000 BCE this Cypriot industry seems to have gone into serious decline, as part of a development that must be connected with the collapse of Late Bronze Age empires in the late 13th century BCE (Singer 2011; Strobel 2011) and the emergence, beginning in the late 12th century BCE, of a new technology of working with iron—the development of a technology involving the production and export of iron knives that had been carburized, quenched and tempered. As Susan Sherratt argued back in 1994, this represented a brilliantly conceived export industry (Sherratt 1994; for details on these developments see Muhly 2006).

This is the situation in Greece, Cyprus and Israel, but not in the Negev and the Nahal Arabah. Here we have, beginning in the 11th century BCE, the development of a major copper production industry. This can be seen at Site 30 at Timna and at Khirbat en-Nahas in Jordan. A new copper smelting technology emerges, making use of shaft furnaces and large tuyères that resulted in the production of slag cakes weighing up to 25 kg. This does not seem to be a development connected with the United Monarchy or with the divided kingdoms of Israel and Judah. It was a southern, Negebite development associated with the emerging kingdom of Edom. Level I at Timna's Site 30 is an Edomite smelting site and its technology, best represented at Khirbat en-Nahas, must have been borrowed from lands farther to the east. I cannot believe that this advanced copper smelting technology was developed by pastoral nomads or by Shasu nomads. Smelting copper ores represents a very complex technology. Robert Maddin, in his review of Rothenberg's (1990) *The Ancient Metallurgy of Copper*, after pointing out all the metallurgical problems involved with modern smelting experiments, concluded that "I am still amazed that copper was produced in Antiquity" (Maddin 1992: 87). To my knowledge no sophisticated copper smelting technology has ever been developed by pastoral nomads. Regarding copper production itself, my interpretation of the relationship between Cyprus and the Nahal Arabah is in basic agreement with that proposed by Israel Finkelstein and Eli Piasetzky (2008).

What of the Midianites and their beautiful painted pictorial pottery, Rothenberg's skilled metalworkers at Timna? It does seem reasonable to derive this pottery from the Hejaz of northwest Arabia, so it is best called Qurayya Ware. Rothenberg designated these "Midianites," known from Timna only from their pottery, as "superb metallurgists with their own pyrotechnological tradition," back in 1998 (Rothenberg 1998: 203), but he never bothered to explain where these traditions might have come from. For me this pottery, which is an Early Iron Age fabric, has no Late Bronze Age context and nothing whatsoever to do with Mycenaean pictorial pottery. It must, in fact, be connected to the other important development in the early Iron Age—the spice and incense trade between Arabia and the Levant and the famous account in 1 Kings 10:1–13 of the visit of the Queen of Sheba to the court of Solomon, king of a United Israel. This romantic account (see Pritchard 1974) has long been confined to the pages of historical fiction, but I would argue that it has a historical basis and is a memory of the international trade across the Red Sea that began in the 11th century BCE. We are dealing here with both maritime and overland trade and it is very important to keep them separate.

The Queen of Sheba is never given a name because what was negotiated must have been a diplomatic trade agreement. Queens of Arabia are mentioned in Neo-Assyrian sources from the 8th and early 7th centuries BCE, but they never appear again. This means that the Solomonic story must be at least pre-Deuteronomic. As Kenneth Kitchen has pointed out (2003: 117), no one would have invented a story about a queen of Arabia after the early 7th century BCE. Overland trade must have involved the use of camels, and we now know that the camel was domesticated by at least the 11th century BCE (Grigson 2012). Sheba is, of course the ancient kingdom of Saba (Greek Sava), in southwest Arabia, a land long famous as the source of frankincense and myrrh. Recent scholarship puts the beginnings of this kingdom and of writing in the Sabaean alphabet in the 12th–11th century BCE (De Maigret 2005; Robin 2005). In other words, the elements of this story now fit together in ways that could not have been known to Beno Rothenberg or to Nelson Glueck.

What we have in the book of Kings is a literary account of the heroic era that saw the founding of the Kingdom of Israel. If you are impressed by the details recording the wealth of Solomon (as, for example, Kitchen 1989; Millard 1989), I would simply suggest that you read Homer. Epic literature revels in minute descriptions of wealth and authority. As for the kingdom itself, Kenneth Kitchen argues (2003: 157) that the empire of David and Solomon is nothing more than one of the many "mini-empires" of the early Iron Age, having excellent parallels in contemporary Neo-Hittite and Aramean kingdoms (for an historical background see also Fales 2011).

Does this mean that we are back in the days of Nelson Glueck, with King Solomon's Mines and Tell el-Kheleifeh as the "Pittsburgh of Palestine"? Of course not. What it does mean is that we are dealing with epic traditions that go back to some sort of historical reality (Mazar 2014). We no longer believe that Agamemnon's Mycenae can be directly compared with the archaeological Mycenae of the 13th century BCE, but both the literary tradition and the archaeological evidence deal with the same historical site. The biblical account in I Kings is somewhat more diffuse in that Ezion-Geber is almost certainly the island of Jezirat Fara'un ("Pharaoh's Island") and Tell el-Kheleifeh is ancient (and modern) Elath/Eloth (Bartlett 1990: 1–16; Flinder 1977; 1989). Anson Rainey, on the other hand, believed that we know nothing about biblical Elath. In his opinion "Elath was founded in 1949. No trace of the Biblical site was discovered by 19th-century surveyors" (Rainey 1985: 17).

Ezion-Geber was a Red Sea harbor and Elath the contemporary settlement. It was from Ezion-Geber that the fleet of Solomon and his ally, Hiram, King of Tyre, sailed to Ophir, returning with gold, exotic woods and precious stones (1 Kings 9:26–28; 10:11). No mention is made of copper and Solomon is never associated with any sort of trade or production of copper in any biblical text. The biblical text, however, does correctly point out that Ezion-Geber was in the land of Edom (1 Kings 9:26). What has dropped out of this historical tradition is the Edomite control of copper production, because the Edomites left no written records. Ophir is a historic land in southwest Arabia, famous as a source of gold, and this trade represents the bulk maritime trade in gold, copper and timber, the goods best carried by sea. The famous ostracon from Tell Qasile of uncertain date, records 30 shekels of gold of Ophir delivered to Beth-Horon (Mazar 1993: 1206, Ostracon No. 2). The Queen of Sheba came to Jerusalem with a large entourage that included a camel caravan loaded with gold, spices and precious stones. This was the overland trade, by donkey and camel, which became famous through the trade in frankincense and myrrh and represents a trade that survived long after the days of Solomon and Hiram. Surely this spice trade with Arabia is reflected in the distribution of the so called "Midianite" pottery, found at all the Iron I–II sites in the Negev and the Nahal Arabah (Tebes 2007). It was the combined copper and spice trade that brought wealth to the kingdoms of Israel and Edom.

The exact relationship between Israel and Edom remains to be worked out in detail, but it must be remembered that the long-standing idea that a kingdom of Edom developed only in the 7th century BCE, was based upon nothing more than the inscription on a seal from Umm el-Biyara, excavated by Crystal Bennett (Bennett 1966). The inscription on the seal reads "Qos-Gabr, King of Edom," a ruler mentioned by Ashurbanipal, King of Assyria (668–612 BCE), in the account of his first campaign. We have always been expected to believe that nothing happened in Edom before the 7th century BCE, even though the Book of Genesis (36:31–39) preserves an Edomite king list and both Saul (1 Samuel 14:47) and David

(2 Samuel 8:13–14) fought against Edom. Edom seems to have been recognized as a historical entity already in the time of Merneptah (1213–1203 BCE) (see also Bartlett 1992).

So what can now be said about the overall history of copper production at Timna? In a real sense it could be said that everything Beno Rothenberg ever wrote about Timna has now been turned upside down. It is very important to remember, however, that without his dedicated years of research and study at Timna, his archaeometallurgical and scientific investigations and his establishment of IAMS in London, we would not be having this conference. Nor, I would suggest, would Andreas Hauptmann and his colleagues, and Tom Levy together with his colleagues, ever have been inspired to carry out their own investigations on the eastern side of the Jordan River. One could argue today that Rothenberg's excavation techniques left much to be desired, but such criticism really misses the tremendous importance of the whole Timna Project. It was a pioneering project, carried out at a time when very little fieldwork had been done at prehistoric mining and smelting sites. The only real predecessor has to be the research of the mining engineers K. Zschocke and E. Preuschen in the Mitterberg area of the Austrian Alps, south of Salzburg, published in 1932 by the Anthropological Society of Vienna in a volume very hard to find in any British or American research library (Zschocke and Preuschen 1932).

The heated arguments of the past have now become historical curiosities. Site F2, considered by Rothenberg to be sixth millennium BCE, can in fact be no earlier than the 14th century BCE. The calibrated radiocarbon date from the site is 1386–1215 BC BM 1368 (Ben-Yosef *et al.* 2010: 730, Table 1). Site 39 has long been considered a Chalcolithic smelting site and dated by Rothenberg and Merkel to the fifth millennium BCE (Rothenberg and Merkel 1998: 1–3) on the basis of a new radiocarbon date (OxA 7632: 4355–4327 BC) of a charcoal sample from a very dubious context (still supported by John Merkel, in his contribution to this conference, but with the emphasis now on Site 39a). This is the occasion that prompted Beno to send me a message on the arch-metals list (June 13, 1998), suggesting that I needed a new hat and volunteering to take up a collection to buy me one. I declined his offer. Radiocarbon measurements on charcoal from Smelting Site 39b (BM 1116) gave readings between 400 BCE and 450 CE (Hauptmann and Wagner 2007: 69).

The main evidence for the new interpretation of Timna as an early Iron Age copper production site comes from the renewed work at the site carried out by Erez Ben-Yosef and his colleagues (Ben-Yosef *et al.* 2012). This argument is based, almost entirely, upon a re-examination of the evidence from Site 30. We should not, however, make the mistake of seeing Timna as an exclusively early Iron Age operation. Consider the evidence from Site 2, a Late Bronze Age–Iron I smelting camp excavated in 1964 and 1966 (Rothenberg 1983: 6–8; 1988: 7; 1990: 14). We now have a re-evaluation of 10 radiocarbon dates from this site (Ben-Yosef *et al.* 2010: 729–730, Table 1), documenting smelting activity in the 15th, 14th, 13th and 12th centuries BCE. This activity must be related to the construction of the Egyptian Mining Temple at Site 200 and to the voyage "to the land of Atika, to the great copper mines which are there" according to Papyrus Harris I, 78, from the time of Ramesses III (1184–1153 BCE). Ramesses goes on to say that "Their mines were found and [they] yielded copper which was loaded by tens of thousands into their ships" (Grigson 2012: 95). It seems to me that this text must, to a significant degree, document Egyptian exploitation of the Timna copper mines in the early 12th century BCE (see also Levene 1998).

The parallel with the Mitterberg is most instructive, although the ore being mined in the two regions was totally different: basically malachite at Timna and copper sulphide in the Mitterberg. Both mining operations involved the creation of mysterious circular depressions known as *pingen* (already Muhly 1984: 281). We now realize that the vast number of these circular depressions represent ancient mine shafts, dug into the ground in search of veins of copper ore and filled by drifting sand over the course of time. I attach great significance to these *pingen* because they represent a relentless search for new veins of copper mineralization. In the Mitterberg they are spread right across the side of the mountain, often in parallel rows. Rothenberg associated these *pingen* with operations at Site 30 (Rothenberg 1983: 4; 1988: 9), which means that Site 30 has to be seen as a serious copper production site. Why this search for new copper mineralization, if Site 30 was a New Kingdom Egyptian operation, given the explosive expansion of the Cypriot copper industry during the Late Bronze Age, and evidence for the use of the oxhide ingots, from both Syria and Israel? (For more about Levantine ingots, see Galili, Gale and Rosen 2011; 2013.)

This is one of the main reasons why an early Iron Age date for Site 30 is so attractive. Such a date certainly seems to be supported by the recent re-calibration of some 15 radiocarbon dates (Ben-Yosef *et al.* 2012: 55, Table 4; Grigson 2012: 84). With one outlier (sample HAM-216: 1689-1531 cal. BCE, from a charcoal from Layer I), these dates support operations in the 12th, 11th and 10th centuries BCE. Layers III–II at Site 30 seem to be Iron I, with Layer I being Iron IIA. I admit that the radiocarbon dates do not allow for a clear separation between these two periods, but that is not surprising. In spite of significant recent advances in all aspects of the procedure, radiocarbon dating remains a very imperfect dating technique and I still have my doubts about Bayesian modeling.

The archaeometallurgical evidence seems to support this new chronology. Layers III–II made use of small, clay-lined bowl furnaces and small tuyères that produced a slag that had to be crushed in order to remove the entrapped copper prills. In Layer I we have the introduction of what Rothenberg called "a new highly developed extractive metallurgical technique" (Rothenberg 1983: 11). This technology used a shaft furnace made of slag-tempered clay, large tuyères (15–20 cm in diameter), and produced a molten slag tapped into slag pits (for small and large tuyères, see Rothenberg 1990: 53, Fig. 76). These pits contained slag "cakes" that were 40–60 cm in diameter and weighed ca. 15–25 kg (Rothenberg 1990: 49, 55, Fig. 80).

This technology is identical to that excavated by Tom Levy and his colleagues at Khirbat en-Nahas, where a major series of calibrated radiocarbon dates places metallurgical operations securely within the parameters of ca. 1000–800 BC (Levy, Najjar and Higham 2010: 841–842, Fig. 3). For me the conclusion is obvious: Layer I at Timna represents the arrival of Edomite copper smelters, following upon a short break at the end of Layer II, probably the result of an earthquake. In 2007 T. Levy, M. Najjar and T. Higham published a very important article on "Iron Age Complex Societies, Radiocarbon Dates and Edom: Working with the Data and Debates." There they propose the following:

> We suggest that there was a cluster of at least three 10th–9th century BC fortresses involved in the movement of Iron Age copper produced in the Faynan district. (Levy, Najjar and Higham 2007: 19)

These fortresses were: Khirbat en-Nahas, Tell el-Kheleifeh and Hatzevah. To this list I would add a fourth site: Layer I at Timna which, according to Rothenberg (1983: 9), was fortified with a strong defensive wall with two large towers guarding its gate (see also Rothenberg 1988: 72, n. 48).

This reconstruction obviously involved a re-dating of Tell el-Kheleifeh and a rejection of the work of Gary Pratico at that site (Pratico 1993). I am sure that Levy's, Najjar's and Higham (2010) dating is correct, and I also accept their idea that, whereas Hatzevah dealt with Western trade to the Mediterranean, Tell el-Kheleifeh was in charge of the Red Sea trade with Arabia (2007: 20). Was Glueck somehow right for all the wrong reasons? However Solomon and Sheba might eventually fit into all of this once we can settle what we mean by the "Age of Solomon" (for a recent discussion of this problem, see Garfinkel 2011; see also Finkelstein and Piasetzky 2003), my feeling is that Glueck had a correct sense of what he saw happening in Palestine in the early Iron Age, on the basis of the archaeological evidence but, with his total commitment to the biblical text, he could only understand such developments within the context of a great Kingdom of David and Solomon. Now we live in a different world where, as Ronald

Hendel puts it in his contribution to the Dever Festschrift (2006: 225): "Even if Solomon is plausibly a historical figure, this figure is lost in the sands of time." The imposing physical and geological remains from Timna will hopefully endure for decades to come as a testimony and tribute to the memory of the late Beno Rothenberg.

REFERENCES

Bartlett, J.R. 1990. "Ezion-Geber, which is near Elath on the Shore of the Red Sea" (I Kings IX 26). In: Van der Woude, A.S., ed. *In Quest of the Past. Studies on Israelite Religion, Literature and Prophetism* (Oudtestamentische Studiën XXVI). Leiden: 1–16.

Bartlett, J.R. 1992. Edom: Edom in History. In: Friedman, D.N., ed. *The Anchor Bible Dictionary*, Vol. 2. New York: 287–295.

Bennett, C. 1966. Fouilles d'Umm el-Biyara. Rapport préliminaire. *Revue biblique* 73: 372–403.

Ben-Yosef, E., Levy, T.E., Higham, T., Najjar, M. and Tauxe, L., eds. 2010. The Beginnings of Iron Age Copper Production in the Southern Levant: New Evidence from Khirbat al-Jariya, Faynan, Jordan. *Antiquity* 84: 724–746.

Ben-Yosef, E., Tauxe, L., Shaar, R. and Ron, H. 2012. A New Chronological Framework for Iron Age Copper Production at Timna (Israel). *Bulletin of the American Schools of Oriental Research* 367: 31–71.

De Maigret, A. 2005. The Dawn of History in Yemen's Interior. In: Gunter, A.C., ed. *Caravan Kingdoms: Yemen and the Ancient Incense Trade*. Washington: 2–6.

Fales, F.M. 2011. Transition: The Assyrians on the Euphrates between the 13th and the 12th century BC. In: Strobel, K., ed. *Empires after the Empire: Anatolia, Syria and Assyria after Suppiluliuma II (ca. 1200–800/700 B.C.)* (Eothen. Collana di Studi Sulle Civiltà dell'Oriente Antico 17). Florence: 9–59.

Finkelstein, I. and Piasetzky, E. 2003. Recent Radiocarbon Results and King Solomon, *Antiquity* 77/298: 771–779.

Finkelstein, I. and Piasetzky, E. 2008. Radiocarbon and the History of Copper Production at Khirbet en-Nahas. *Tel Aviv* 35: 85–95.

Flinder, A. 1977. The Island of Jezirat Fara'un. *International Journal of Nautical Archaeology* 6: 127–139.

Flinder, A. 1989. Is This Solomon's Seaport? *Biblical Archaeology Review* 15/4: 31–43.

Galili, E., Gale, N. and Rosen, B. 2011. Bronze Age Metal Cargoes off the Israeli Coast. *Skyllis* 11(2): 64–73.

Galili, E., Gale, N. and Rosen, B. 2013. A Late Bronze Age Shipwreck with a Metal Cargo from Hishuley Carmel, Israel. *International Journal of Nautical Archaeology* 42:1: 2–23.

Garfinkel, Y. 2011. The Davidic Kingdom in Light of the Finds at Khirbet Qeiyafa. In: *City of David Studies of Ancient Jerusalem*, Vol. 6. Jerusalem: 13*–35*.

Grigson, C. 2012. Camels, Copper and Donkeys in the Early Iron Age of the Southern Levant: Timna Revisited. *Levant* 44: 82–100.

Hauptmann, A. and Wagner, I. 2007. Prehistoric Copper Production at Timna: Thermoluminescence (TL) Dating and Evidence from the East. In: La Niece, S., Hock, D. and Craddock, P., eds. *Metals and Mines. Studies in Archaeometallurgy*. London: 67–75.

Hendel, R. 2006. The Archaeology of Memory: King Solomon, Chronology and Biblical Representation. In: Gitin, S., Wright, J.E. and Dessel, J.P., eds. *Archaeological and Historical Essays on Ancient Israel in Honor of William G. Dever*. Winona Lake: 219–230.

Kitchen, K.A. 1989. Where Did Solomon's Gold Go? *British Archaeological Reports* 15/3: 30.

Kitchen, K.A. 2003. *On the Reliability of the Old Testament*. Grand Rapids, MI.

Knapp, A.B. 2008. *Prehistoric and Protohistoric Cyprus: Identity, Insularity and Connectivity*. Oxford.

Knapp, A.B. 2013. *The Archaeology of Cyprus: From Earliest Prehistory through the Bronze Age*. Cambridge.

Levene, D. 1998. Expedition to Atika. In: Esmael, F.A., ed. *Proceedings of the First International Conference on Ancient Egyptian Mining and Metallurgy and Conservation of Metallic Artifacts*. Cairo: 365–378.

Levy, T.E., Adams, R.B., Prange, M., Schmitt-Strecker, S. and Najjar, M. 2002. Early Bronze Age Metallurgy: A Newly Discovered Copper Manufactory in Southern Jordan. *Antiquity* 76(292): 425–437.

Levy, T.E., Najjar, M. and Higham, T. 2007. Iron Age Complex Societies, Radiocarbon Dates and Edom: Working with the Data and Debates. *Antiguo Oriente* 5: 13–34.

Levy, T., Najjar, M. and Higham, T. 2010. Ancient Texts and Archaeology Revisited—Radiocarbon and Biblical Dating in the Southern Levant. *Antiquity* 84(325): 834–847.

Maddin, R. 1992. Review of B. Rothenberg, The Ancient Metallurgy of Copper (1990). *Archeomaterials* 6: 85–90.

Mazar, A. 1993. Tell Qasile. In: Stern, E., ed. *The New Encyclopedia of Archaeological Excavations in the Holy Land*, Vol. 4. Jerusalem: 1204–1212.

Mazar, A. 2014. Archaeology and the Bible: Reflections on historical memory in the Deuteronomistic history. In: Maier, C.M., ed. *Congress Volume Munich 2013*. Leiden/Boston: 347–369.

Millard, A.R. 1989. Does the Bible Exaggerate King Solomon's Wealth? *Biblical Archaeology Review* 15(3): 20–34.

Muhly, J.D. 1984. Timna and King Solomon. *Bibliotheca Orientalis* 41: 276–292.

Muhly, J.D. 2006. Texts and Technology. The Beginnings of Iron Metallurgy in the Eastern Mediterranean. In: Tassios, T.P. and Palyvou, C., eds. *Proceedings of the 2nd International Conference on Ancient Greek Technology*. Athens: 19–31.

Muhly, J.D. 2009. Oxhide Ingots in the Aegean and in Egypt. In: Lo Schiavo, F., Muhly, J.D., Maddin, R. and Giumlia-Mair, A., eds. *Oxhide Ingots in the Central Mediterranean* (Biblioteca di Antichità Cipriote 8). Rome: 17–39.

Pratico, G.D. 1993. *Nelson Glueck's 1938–1940 Excavations at Tell el-Kheleifeh: A Reappraisal*. Atlanta.

Pritchard, J.B., ed. 1974. *Solomon and Sheba*. London.

Rainey, A.F. 1985. Review of G.W. Bowersock. "Roman Arabia" (1983). *Biblical Archaeology Review* 11(4): 15–17.

Robin, C. 2005. Saba and the Sabaeans. In: Gunter, A.C., ed. *Caravan Kingdoms: Yemen and the Incense Trade*. Washington, D.C.: 8–19.

Rothenberg, B. 1983. Explorations in the Arabah Valley (Israel). *Bulletin of the Metals Museum (Sendai, Japan)* 8: 1–16.

Rothenberg, B. 1988. *The Egyptian Mining Temple at Timna* (Researches in the Arabah 1959–1984 1). London.

Rothenberg, B., ed. 1990. *The Ancient Metallurgy of Copper: Archaeology – Experiment – Theory* (Researches in the Arabah 1959–1984 2). London.

Rothenberg, B. 1998. Who were the "Midianite" Copper Miners of the Arabah? In: Rehren, T., Hauptmann, A., Muhly, J.D., eds. *Metallurgica Antiqua, in Honour of Hans-Gert Bachmann and Robert Maddin*. Deutsches Bergbau-Museum, Bochum: 197–212.

Rothenberg, B. and Merkel, J. 1998. Chalcolithic, 5th Millennium BC Copper Smelting at Timna: New Radiocarbon Dating Evidence for Timna Site 39. *Institute for Archaeo-Metallurgical Studies Newsletter* 20: 1–3.

Sherratt, S. 1994. Commerce, Iron and Ideology: Metallurgical Innovations in 12th–11th Century Cyprus. In: Karageorghis, V., ed. *Cyprus in the 11th Century B.C.* Nicosia: 59–106.

Singer, I. 2011. *Calm before the Storm: Selected Writings of Itamar Singer on the Late Bronze Age in Anatolia and the Levant* (SBL Writings from the Ancient World Supplements 1). Atlanta.

Strobel, K. 2011. The Crucial 12th Century BC: The "Fall of Empires" Revisited. In: Strobel, K., ed. *Empires after the Empire: Anatolia, Syria and Assyria after Suppiluliuma II (ca. 1200–800/700 B.C.)* (Eothen. Collana di studi sulle civiltà dell'Oriente antico 17). Florence: 167–254.

Tebes, J.M. 2007. Pottery Makes and Premodern Exchange in the Fringes of Egypt: An Approximation to the Distribution of Iron Age Midianite Pottery. *Buried History* 43: 11–26.

Zschocke, K. and Preuschen, E. 1932. *Das urzeitliche Bergbaugebiet von Mühlbach-Bischofshofen* (Materialien zur Urgeschichte Österreichs 6). Vienna.

CHAPTER 3

THE CENTRAL TIMNA VALLEY PROJECT: RESEARCH DESIGN AND PRELIMINARY RESULTS

Erez Ben-Yosef

The Tel Aviv University Central Timna Valley (CTV) Project is a multi-year, multidisciplinary research endeavor that attempts to elucidate various aspects of ancient copper exploitation in the southern Arabah Valley. The first phase of the project (2012–2016) included new surveys and excavations of several of the main copper mining and smelting sites in Timna. The field work was designed to address key issues in Late Bronze and Iron Age archaeology of the region, including evolution of copper production technologies and the introduction of iron, historical questions concerning the nature of 13th—9th century BCE desert societies and the impact of intense copper production on social processes, regional and global interactions and the economy of the southern Levant at this juncture. Here we present an overview of the project and preliminary results of its first phase.

INTRODUCTION

The Timna Valley (Fig. 3.1) is one of the best preserved ancient copper ore districts in the world, with dozens of sites that represent over six millennia of copper mining and smelting (Rothenberg 1990b). The area was first reported as an ancient copper production district by John Petherick (1861), a mining engineer and the British consul of Khartoum. It was later visited by early scholars such as Fritz Frank (1934), Nelson Glueck (1935) and others. Beno Rothenberg was the first systematic researcher in the valley; he began investigating the area in 1959 and acknowledged the importance of Timna for the study of ancient copper metallurgy by organizing an interdisciplinary team of experts to help with various aspects of research (Pigott 1996; Ben-Yosef 2012a). The work of Rothenberg and his team ("the Aravah Expedition") lasted several decades (1959–1984; with some later field work in 2001/2 [Shaw and Drenka, this volume] and 2005 [J. Merkel, personal communication 2013]), and was fundamental to the shaping of archaeometallurgy as a well-recognized research discipline. Rothenberg's pioneering work on copper mining and smelting technologies was the basis for further research in the copper mines of the Arabah Valley (e.g., Hauptmann 2007; Ben-Yosef 2010; Levy, Najjar and Ben-Yosef 2014) and is the starting point of the current project, which builds on the published and unpublished materials of the Aravah Expedition.

The first phase of Tel Aviv University's Central Timna Valley Project (CTV, 2012–2016)[1] focused on the main mining and smelting sites in the valley, and included surveys, excavations and complementary laboratory work. The project was triggered by the results of the 2009 excavations at the large smelting camp of Site 30 (Ben-Yosef *et al.* 2012) that called for a substantial revision of the accepted chronological framework for the main phase of copper production in Timna. Until the renewed excavations, Site 30 was

1 See: http://archaeology.tau.ac.il/ben-yosef/CTV/

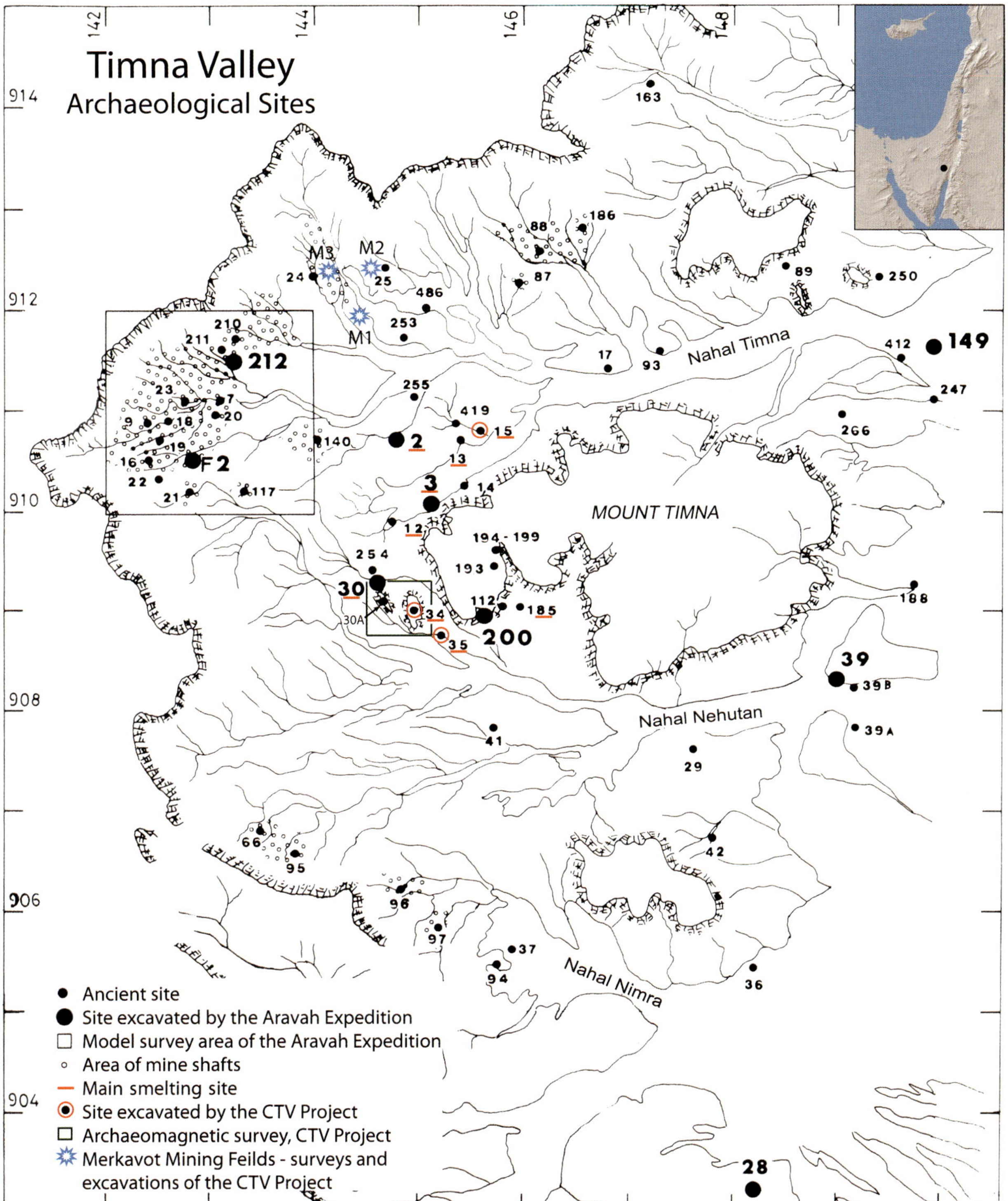

Fig. 3.1: Map of the Timna Valley and the archaeological sites recorded by the Aravah Expedition (based on Rothenberg 1990); the main smelting sites are indicated, as well as sites excavated by the Aravah Expedition and the CTV Project (the map's grid is based on the Israel Cassini Soldner [ICS], and the distance between the grid marks is 1 km).

dated predominantly to the Late Bronze Age (ca. 1300–1150 BCE, Rothenberg 1980)[2]—similar to all other major smelting and mining sites in the valley (cf., Rothenberg and Glass 1992). This dating is based on the results of the excavations at the Hathor Shrine (Site 200, Rothenberg 1988), which yielded numerous Egyptian finds including cartouches of pharaohs from Seti I to Ramesses V. However, radiocarbon results from the 2009 excavations demonstrate that Site 30 is exclusively Iron Age (late 12th–9th century BCE), and that the Egyptian finds from the Hathor Shrine cannot serve as a chronological anchor to all major sites in the valley; this revision has significant implications for our understanding of the copper industry and its background, as the suggested link between large scale production and the Egyptian New Kingdom of the 19th and 20th Dynasties (e.g., Rothenberg 1972) should now be reconsidered, and the capabilities (and complexity) of local sociopolitical entities should be reevaluated.

Above all, the recent revision in the chronological framework of the major sites at Timna demonstrates that these sites can no longer be treated as a homogeneous group; it became evident that the main period of copper exploitation in the valley covers ca. 500 years, roughly between 1300–900 BCE, with changing sociopolitical and historical settings. Thus, higher time resolution and careful examination of each site are necessary in order to achieve meaningful insights on the history of the region at this time, and to address questions concerning technological evolution, social processes and more.

In light of these new developments in the research of Timna, one of the CTV Project's main goals has been to establish absolute chronology for the main mining and smelting sites in the valley. In addition to multiple radiocarbon dates on short-lived samples, we examined the potential of other analytic methods to provide age constraints on mining and smelting contexts, including optically stimulated luminescence (OSL) dating and archaeomagnetism (below). Based on the foundation of absolute chronology, the first phase of the CTV project aimed at achieving the following goals (cf., Ben-Yosef *et al.* 2016):

1. Investigating technological evolution and innovation in the transitional period between the Bronze and Iron Ages, focusing on mining and smelting technologies and the possible link to the development of iron production technologies.
2. Investigating social, cultural and ethnic aspects of production systems, including questions regarding the role of Egypt, the formation of Edom, identity of the Shasu people, and more.
3. Contextualizing Timna in wider regional and historical perspectives, including its relation to Faynan, the Negev Highland and the Hijaz, its connections with potential markets such as Philistia, Moab and ancient Israel, and its possible role in historical events such as the military campaign of Shoshenk I to the southern Levant.

The CTV Project is based on new surveys and excavations of several of the main copper mining and smelting sites in Timna, and includes various experts that study different aspects of the finds. In addition, as part of the new project we work on old materials of the Aravah Expedition that were never published. These include both archival documents and excavated artifacts that provide more insights on the archaeology of Timna and contribute to the goals of the renewed research.[3]

The layout and preliminary results of the field work activities of the CTV Project are described in the next two sections (divided into smelting and mining contexts, Table 3.1). This is followed by a discussion of the significance of the results and their implications on the research of Timna and beyond.

2 Site 30 Layer I was the only context in Timna dated by Rothenberg to the 10th century BCE (attributed to the Shoshenk I campaign to the southern Levant).

3 Access to unpublished materials of the Aravah Expedition was kindly granted by J. Gavish (Rothenberg's personal collections) and the Israel Antiquities Authority (with the help of U. Dahari, M. Sebbane, A. Rochman-Halperin and O. Sion).

Table 3.1: Field Seasons of the First Phase of the CTV Project (2012–2016)

Season	*Mining Sites*	*Smelting Sites*
2012	-	Survey of slag mounds (Sites 34, 30A; Supervisor: I. Peters)
2013	Excavations at Merkavot 1, 2 (Supervisor: C. Smitheram) Geophysical survey of Merkavot 1 (Supervisor: U. Basson, no results)	Excavations at Site 34 Area 19 (Supervisors: W. Ondricek and I. Peters)
2014	Excavations at Merkavot 2 (Supervisor: C. Smitheram) Survey of Merkavot 3 (Supervisor: C. Smitheram) Geophysical survey of Merkavot 2, 3 (Supervisor: N. Wechsler)	Excavations at Site 34 Areas 19 and G (Supervisors: W. Ondricek and I. Peters) Survey of Site 34 (hilltop only; Supervisors: O. Yagel and V. Workman)
2015	-	Excavations at Site 34 Areas 13 and 21 (supervisors: W. Ondricek and V. Workman) Excavations at Site 15 Areas A, B, C and M (Supervisors: M. Cavanaugh and O. Yagel) Survey of Site 15 (Supervisor: C. Smitheram) Survey of Site 3 (Supervisor: O. Yagel)
2016	Geophysical survey of Merkavot 3 (Supervisor: N. Wechsler)	Excavations at Site 34 Areas G, 13 and 21 (supervisors: V. Workman and I. Peters) Excavations at Site 35 Areas A, B and C (supervisors: W. Ondricek and M. Cavanaugh) Survey of Site 35 (Supervisors: O. Yagel and C. Smitheram)

SURVEYS AND EXCAVATIONS OF COPPER SMELTING SITES

One of the most prominent archaeological remains in the Timna Valley is slag deposits. In fact, Timna was recognized as an ancient mining district based on these deposits (Petherick 1861), and it took more than a century of research for the mines themselves to be discovered and documented (Conrad and Rothenberg 1980). The Aravah Expedition documented nine sites with significant accumulation of slag at Timna (2, 3, 12, 13, 15, 30, 34, 35, 185)[4] and treated them as smelting camps that mostly represent simultaneous activities (Rothenberg 1967, 1972; Conrad and Rothenberg 1980; Rothenberg 1990b; 1999b, and see above). These sites were at the core of the Aravah Expedition's research; they were all surveyed and three of them (Sites 2, 30 and 3) were excavated in several seasons of field work.[5] The excavated materials were the basis for technological reconstructions (Bachmann 1980; Rothenberg 1990b), dating, and studies related to the history of the region and its inhabitants at the peak of copper production in the southern Arabah Valley (Rothenberg 1999b, and references therein). The main smelting sites of Timna are also the focus of the first phase of the CTV Project, which aims, among other objectives, to contextualize Rothenberg's extensive work in a refined, chronological framework. Together with the new data from the 2009 excavations at Site 30 (Ben-Yosef *et al.* 2012) and the recent excavations at Site 2 (Erickson-Gini 2014), the CTV Project has already deepened our understanding of Timna in the Late Bronze and Iron Ages, including substantial revisions of previous research (e.g., Ben-Yosef *et al.* 2016; Ben-Yosef 2016).

4 Glueck (1935) reported seven smelting camps in Timna (Sites 3, 12, 13, 15, 30, 34 and 35 of the Aravah Expedition [cf., Rothenberg 1988: 2, although the reference to Site 2 is an error]); these are probably the same seven camps reported by Frank (1934).

5 Small probes, like the one reported by Rothenberg (1990a: 12) at Site 185, were probably carried out by the Aravah Expedition at other sites, possibly without any publication. One of the aims of the current project is to track down information and materials from such probes.

While the mining fields of Timna are located around the perimeter of the valley and below the cliffs, the main smelting sites are found closer to the center at a distance of 1–3 km (Fig. 3.1). This separation facilitated both ore channeling from a wide area of mining and wood supply, as vegetation is more abundant in the wadi plains in the center of the valley. The CTV Project started with a survey of slag mounds at Site 34 and its vicinity and continued with excavations at Sites 34, 15 and 35. In addition, all of the main smelting camps (except Site 2) were surveyed and mapped, and materials from the Aravah Expedition's excavations at Site 3 were analyzed and processed for publication.

THE 2012 SURVEY OF SLAG MOUNDS AND ARCHAEOMAGNETIC DATING

In partial preparation for excavations we conducted a survey of slag mounds at Site 34, one of the largest smelting sites in Timna (below), and its vicinity (Figs. 3.1, 3.2). The slag mounds were documented and slag samples were collected as part of an archaeomagnetic study that aimed at providing age constraints to the individual slag deposits (Peters 2015, Peters, Tauxel and Ben-Yosef 2017). Building on previous archaeomagnetic studies at nearby Site 30 (Shaar *et al.* 2011) and other sites in the region (Ben-Yosef *et al.* 2008), the current research yielded important insights about the chronology of smelting at the center of the Timna Valley. The results demonstrate that the slag mounds on the hilltop of Site 34 represent (exclusively) early Iron Age activities, in contrast to the slag mounds at the hill bottom (Early Islamic), but similar to the slag deposits of nearby Sites 30 and 30A (the latter investigated here for the first time). Furthermore, the results demonstrate the potential of archaeomagnetism to provide age constraints on slag deposits, with time resolution of up to a few decades (e.g., Site 34 Slag Mound #3, 980±10 BCE).

As part of the current project we also applied archaeomagnetism on pottery from the Fortress at Yotvata that has been associated with the main period of copper production in Timna (Meshel 1993). Preliminary results suggest that the fortress is related to the Late Bronze Age activities in Timna (Peters, Ben-Yosef and Tauxe, this volume); however, additional research should be carried out in order to establish a firm date for the site, particularly in light of the recent revision in the chronology of the main sites in Timna.

SITE 34 (SLAVES' HILL)

Although Site 34 (Figs. 3.1, 3.2) is one of the largest smelting camps in Timna, Rothenberg did not excavate it and "left it intentionally undisturbed for the next generation of researchers" (Rothenberg, personal communication 2009). The site has been at the center of the CTV Project, with field work conducted there every season since the first one (Table 3.1; Ben-Yosef 2016). The site is located on a sandstone plateau (of ca. 3 ha.) surrounded by sheer cliffs averaging 20 m in height, and has only one access on the northwest, where an access path was cut into the bedrock. This location indicates a need for defense, which is also evident by a wall that protects the northern part of the site where the slopes are more accessible. This wall was first reported by Glueck (1935),[6] and later surveyed by Rothenberg (1962, 1967), who also described a gatehouse protecting the entrance to the site. Inside the wall, Site 34 contains numerous slag mounds (~1000 tons) and architectural features that are related to different aspects of the metallurgical activities (Fig. 3.3).

The CTV Project excavations concentrated in four areas (Fig. 3.3): a slag mound and associated installations (Area 19, Fig. 3.4), the gatehouse complex (Area G, Fig. 3.5; Ben-Yosef, Langgut and Sapir-

6 Glueck (1935) considered the site's location and the wall as a means to "forcibly retain the drafted laborers"; in turn, this description gave the site its popular name—"Slaves' Hill." Although slaves probably did not operate the furnaces on the hill (Sapir-Hen and Ben-Yosef 2014), the romantic epithet is very much in use, and even found its way into popular literature (such as the young adult fiction of Havron 1958–1959).

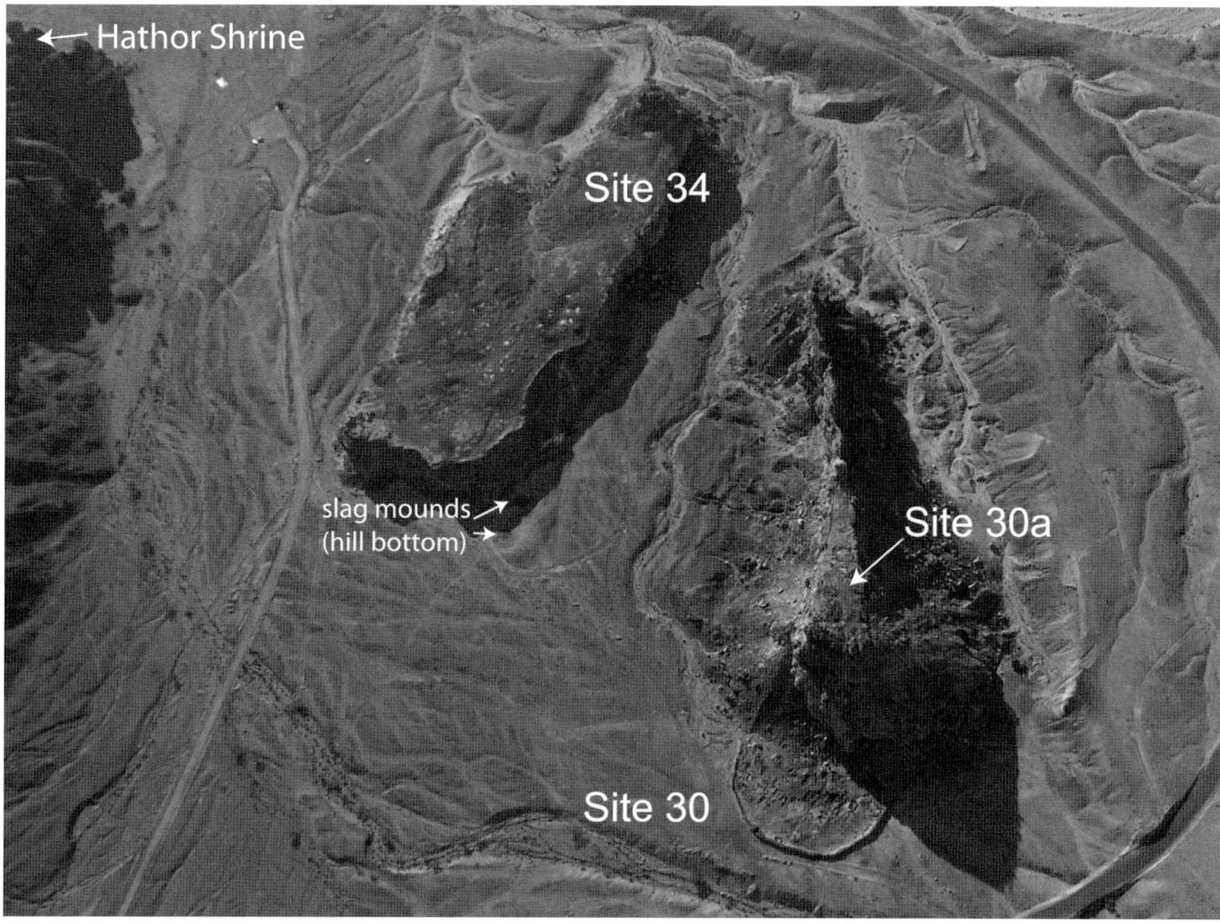

Fig. 3.2: The two largest smelting camps in Timna, Sites 30 and 34; note the proximity of the Hathor Shrine. Photo courtesy of U. Avner.

Hen 2017), metallurgical installations of various qualities (Area 13), and a rock shelf with metallurgical remains (Area 21). As mentioned above, Rothenberg dated the site to the Late Bronze, like the other main smelting camps in Timna (not surprisingly, given its proximity to the Hathor Shrine, cf., Fig. 3.2). Following the recent revision in the chronology of nearby Site 30 (Ben-Yosef *et al.* 2012) and in accordance with the goals of the current project, we put an emphasis on establishing absolute chronology for the site in the first two excavation seasons (2013 and 2014). Fourteen short-lived samples from well-defined contexts were sent for radiocarbon dating, and the results fixed the chronology of the site between the late 11th and late 10th centuries BCE, in agreement with the archaeomagnetic age constraints received from slag samples (above). Thus, Site 34 represents a relatively narrow time window within the five centuries of substantial copper production at Timna, which corresponds to Site 30 Layer II (or only the later part of it; Ben-Yosef *et al.* 2012) and Khirbat en-Nahas (KEN) Stratum IV (and possibly also Stratum V; Levy *et al.* 2014); both also demonstrate a need for defense at this exact same time (the fortress at KEN and the wall at Site 30). The abandonment of Site 34 corresponds to disruptions in the stratigraphy at Site 30 and KEN, and to abandonment of other sites in the Arabah Valley including, e.g., Khirbat al-Jariya (Ben-Yosef, Najjar and Levy 2014). The disruptions and abandonment of sites were attributed to the campaign of Shoshenk I to the region in the second half of the 10th century BCE (Ben-Yosef *et al.* 2010; Levy, Munger and Najjar 2014; Levy, Najjar and Ben-Yosef 2014).

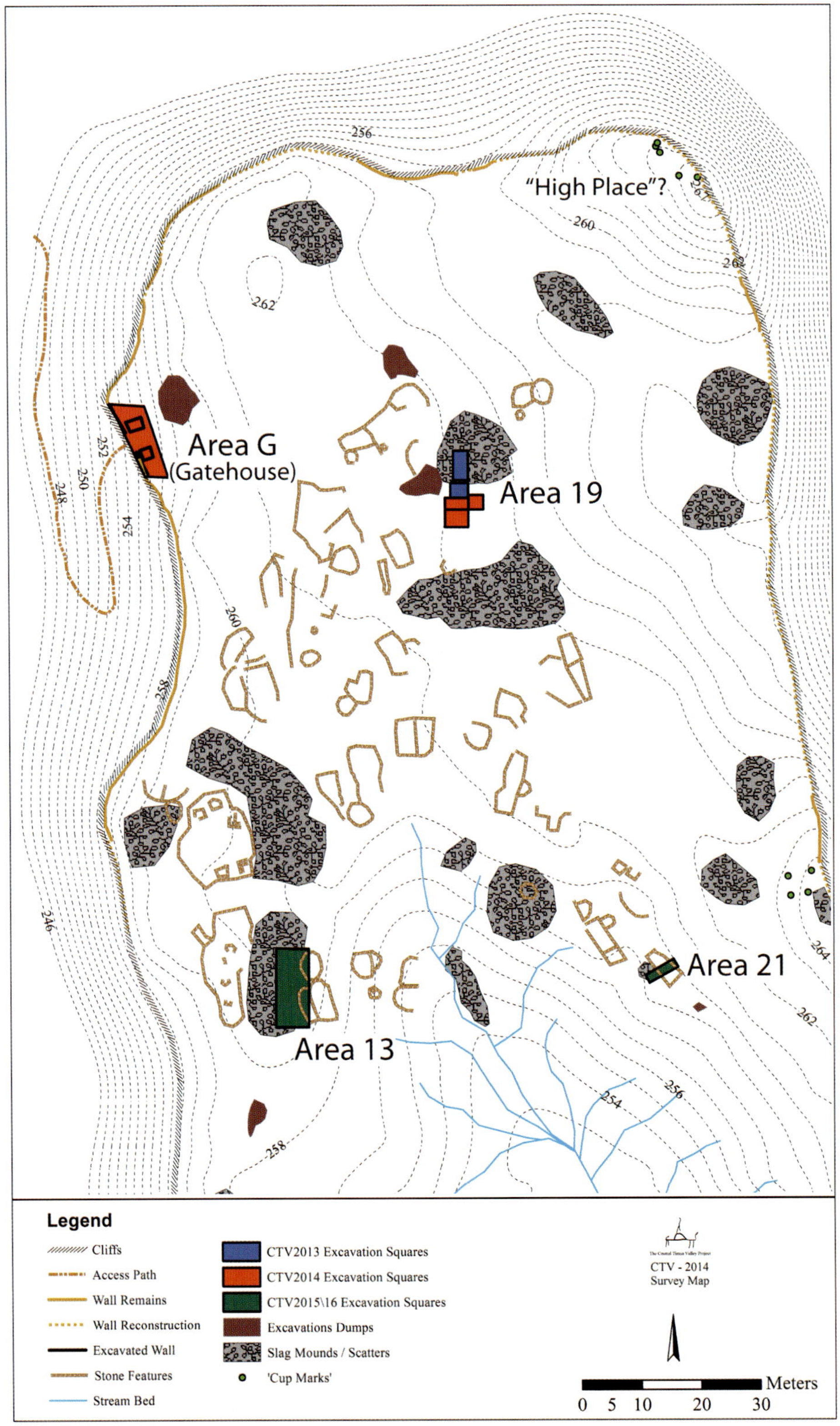

Fig. 3.3: Detailed map of Site 34 and the excavation areas of the CTV Project (the map is based on the 2014 survey supervised by O. Yagel and V. Workman).

Fig. 3.4: Site 34, Area 19A; sampling the western section of the excavation pit for seeds (^{14}C dating), charcoal (species identification, fuel sources), slag (chemistry/mineralogy, smelting technology), textile pieces (fabric materials, weaving technology) and more. Nine radiocarbon samples from this section fixed the date of Site 34 around the 10th century BCE.

Fig. 3.5: Site 34, Area G; the gatehouse complex at the end of the 2014 season; the two rooms abut solid extensions of the site's wall (probably platforms or bases of towers); the rooms are flanked by extensive patches of dung and other organic materials, representing donkey (or mule) or other livestock pens.

The excavated materials and the artifacts recorded during the intense surface survey of 2014 are the basis of various studies. These include:

1. Pottery: the Site 34 assemblage consists of hundreds of diagnostic sherds of all types defined previously for the main smelting sites in Timna ("Negebite" pottery, "normal" wheel-made pottery, and Quarrayyah Painted Ware [QPW]), excluding Egyptian pottery.[7] In addition to providing insights on the society of Timna during the 10th century BCE, the exceptionally rich assemblage coupled with tight absolute chronology enables refining the typological scheme for the Arabah Valley in the Iron Age (Kleiman, Kleiman and Ben-Yosef 2017).
2. Animal bones: faunal remains provide insights on the diet of the metal workers and their social status. Analysis of the goat/sheep bone assemblage (the most abundant find) by context indicated that the craftspersons working at the furnaces were treated to the best parts of meat, probably reflecting their higher social status (Sapir-Hen and Ben-Yosef 2014). The presence of fish (in all contexts) indicates significant food procuring efforts by the inhabitants of the site. Surprisingly, the fish assemblage does not correspond to the nearby Red Sea (25 km to the south), but rather to the Mediterranean Sea and fresh water rivers (Sapir-Hen, Lernau and Ben-Yosef, this volume). This implies trade connections (direct or indirect) with coastal societies and probably other peoples to the north. The bones also provide insights on one of the key components of the copper production system—the draught animals. At Site 34, only donkey (or mule) bones were found. Together with the tight dating and analyses of other assemblages from smelting sites in the Arabah Valley, it seems that camels were introduced to the region only in the late 10th century BCE, as part of the reorganization of the industry in the aftermath of Shoshenk I's campaign to the region (Sapir-Hen and Ben-Yosef 2013). The animal bone assemblage of Site 34 is yet to be fully studied, and more insights are expected based on careful, context-based analysis.
3. Seeds: remains of thousands of uncharred seeds were unearthed in almost all of the excavated contexts. The seeds not only provide excellent material for radiocarbon dating, but also add to our knowledge of the diet of the site inhabitants, including the non-humans. They include all of the biblical Seven Species as well as doum palm, pistachio, lentils and more. None of these grows in Timna, and except dates and the doum palm that may have grown in the oases of the southern Arabah Valley, their origin has to have been much further to the north in the Mediterranean regions. In addition, in certain contexts seeds were found mixed with animal dung, indicating the diet of these animals; dung was found in large quantities on both sides of the gatehouse's rooms, piled up against the inner side of the wall (Area G, Fig. 3.5). It seems that this was the location where donkeys were held together with other livestock animals (goat/sheep), and that their dung was deliberately piled, probably to be used in the operation of the furnaces (the initial heating stage). Thousands of grape seeds were found in this dung, indicating that the donkeys were fed with grape pomace, a high calorie food that suits their pivotal role in the industry (Ben-Yosef, Langgut and Sapir-Hen 2017).
4. Ground stones: more than 1000 ground stones of various types were documented on the site's surface during the 2014 survey and a few dozen were unearthed in the excavations (of all areas). This large quantity is typical at smelting sites (e.g., Levy, Bettilyon and Burton 2016) and reflects the important role of ground stones in the *chaîne opératoire* of copper production. The association

7 In light of the new chronological framework for the main smelting sites in Timna, the identification of certain pottery types and wares as "Egyptian" by the Aravah Expedition (e.g., Rothenberg 1980: Plate 211) should be reexamined, including the petrographic identification of some wares as "Nilotic." If the identification proves to be valid, it implies that Egypt had also been involved in copper production during the Iron Age.

of the new assemblage from Site 34 primarily with copper production[8] allowed for systematic investigation of the function of each type within the process of copper production (as domestic functions were generally excluded). The results of this investigation, with new insights on copper production processes, are presented in Greener and Ben-Yosef (2016).

5. Charcoal: the large scale copper production evident in the main smelting sites at Timna required large amounts of fuel, which, for successful smelting, had to be charcoal-based (cf. Merkel 1990). As wood is scarce in the arid environment of the Arabah Valley, the source of charcoal has intrigued many visitors to the region (e.g., Glueck 1935: 26); however, even when charcoal was excavated in abundance in the smelting sites of Timna, it has never been systematically studied prior to the current research. As part of the CTV Project excavations at Site 34, thousands of charcoal fragments were collected; hundreds of them were analyzed together with charcoal samples from Site 30, resulting in a database of plant species used as fuel sources during ca. 200 years (11th—9th centuries BCE).[9] Based on the assumption that the climate of Timna and the natural flora were not significantly different than they are today, changes in the assemblage through time were interpreted as reflecting human impact on the environment (Cavanagh 2016). The most dominant species in the assemblage was acacia and white broom (*Retama raetam*), two main sources of wood that grow locally—but rejuvenate slowly. The later part of the assemblage is characterized by increasing frequencies of less woody species, and even imported wood (*Phoenix*, *Juniperus*, *Pistacia*), indicating increasing stress on the natural environment that probably contributed to the cessation of production in Timna during the 9th century BCE. Moreover, fuel sources in Timna were probably a limiting factor for the industry, and one of the main reasons for the differences in scale of production between Timna and Faynan (the latter is larger in order of magnitude [Ben-Yosef 2012b]).
6. Furnace related materials: the excavations at Site 34 unearthed various artifacts that are the basis for reconstructing smelting technologies. These include copper ore and flux (iron ore), furnace fragments (including an intact furnace bottom uncovered in Area 19B), tuyère fragments, slag, crucibles (for secondary refining/melting), metal chunks and fragments of copper ingots. Even though similar artifacts were excavated by the Aravah Expedition, the high time resolution of the CTV Project's excavations allows for tracking fine technological changes and trends (within the time span represented at Site 34 and within a broader diachronic context by comparison to other [well-dated] sites). Moreover, some of the artifacts are rare, and have very little representation in the metallurgical assemblages of the Arabah Valley even after decades of excavations and surveys (see overview in Ben-Yosef and Levy 2014); future studies of these artifacts, and in particular the ingots, crucibles and some of the furnace-related features, are expected to shed more light on copper production technologies as well as other related aspects (e.g., establishing provenance signature and tracking trade networks). Slag is also a key for technological reconstructions, as its texture, chemistry and mineralogy provide invaluable information on parameters such as furnace temperature and redox conditions, ingredients of the smelting mixture, efficiency and standardization of the smelting cycles, and more (cf., e.g., Bachmann 1980, 1982). As part of the CTV Project, dozens of slag samples from Site 34 Area 19 and from Site 3 (below) were analyzed for their chemistry (ICP-MS/OES)[10] and

8 The architectural remains and other characteristics of the site rule out its use for dwelling; thus, probably only minor domestic activities (food preparation, etc.) were carried out at the site itself, and most of these were conducted in the dwelling area, which had to be in tents located at the base of the hill.

9 The analysis was done by M. Cavanaugh under the supervision of D. Langgut at Langgut's laboratory in the Institute of Archaeology of Tel Aviv University.

10 Inductively Coupled Plasma (ICP) analyses were conducted by O. Yagel in Y. Erel's laboratory at the Institute of Earth Sciences of the Hebrew University.

compared to previously published data from the Arabah Valley (Yagel 2016). The slag from Site 34 shows a striking similarity in efficiency and standardization to slag studied by Ben-Yosef (2010) from other 10th century BCE sites in the Arabah Valley, which were part of the identification of a distinct "socio-technological" system (System #2) that operated in the entire Arabah Valley at this time.

7. Textile, cordage, basketry and leather: the extraordinary preservation of organic materials in Timna is unparalleled even at other desert sites, including Faynan (with few exceptions such as Kuntilet ʿAjrud and caves in the Judean Desert). Thus, the organic remains at Timna provide a window into various aspects of past societies that are usually transparent in archaeology. The rich assemblage of textile and related finds from Site 34 is in full agreement with other evidence from the site that indicate a complex and hierarchical society that engaged in a sophisticated system of production and trade. The finds include clothes of varying qualities, with some presenting high quality weaving and dyeing (in red and blue; see Workman 2016; Sukenik *et al.* 2017; O. Shamir, personal communication)[11] Textile and leather fragments, as well as ropes and cords, probably represent disintegrated bags, saddles, tents, leather bags (which served as bellows and water containers), leather bellow pipes, and more; however, additional work should be done in order to decisively identify their original use.
8. Miscellaneous small finds: metal objects (awls, needles, rings, etc.), beads (dozens, of many material types), and other small artifacts were excavated and reveal different aspects of the daily life at Site 34, as well as trade connections. The few metal objects that were analyzed were found to be made of bronze and were most probably not local products.

The CTV Project's efforts to retrieve as much material as possible from each excavated context (at the expense of excavation scale) resulted in large quantities of artifacts and ecofacts even from relatively small volumes of excavations.[12] Most of these fall into the categories above and reflect key aspects of the people of Site 34 and the copper production industry; reviewing other materials, such as pollen, tiny seeds of wild flora, mollusks, etc., are beyond the scope of the present paper.

In sum, the rich archaeology of Site 34 provides a strong basis for studying the society at Timna in the 10th century BCE, as well as various aspects of its engagement in copper production, including organization of production and production technologies. It should be noted, however, that the site represents only one part of a larger system that included the mines (and miners), and the dwellings, which had to be based on tents and located in the wadi plains surrounding the hill.

SITE 3

Site 3 is a relatively small (7,700 sq m) smelting camp located north of Mount Timna (Figs. 3.1, 3.6). It was excavated by the Aravah Expedition in 1979 and 1984 (and probed in 1959) and reinvestigated by the CTV Project, including a survey and analyses of materials from previous excavations (Yagel 2016; Yagel, Ben-Yosef and Craddock 2016). Contra to Site 34, Site 3 has no defense elements and it includes

11 The textiles are made of goat hair, wool or linen (the first two sometimes mixed); it is worth noting that the red and blue dyes and tassels (several fragments of which were found at Site 34) echo the depiction of the *Shasu* nomads in Egyptian drawings (e.g., the Shasu prisoner on Ramesses III's reliefs at Medinet Habu, and see an overview in Ward 1972), suggesting a possible connection between the term and the Iron Age nomadic population of the copper ore districts of the Arabah Valley (cf. Levy 2009).

12 The efforts to retrieve maximum materials were based on extensive sieving with down to a 1 mm mesh, floatation (done in the artificial Timna Lake), systematic sorting ("picking") at camp, and in cases application of other methods to retrieve microscopic finds (e.g., a pollen study led by Dr. Dafna Langgut in Area G).

rectangular buildings (Fig. 3.7), suggesting the possibility that it represents Late Bronze Age smelting activities that correspond to the occupation of the Hathor Shrine (Strata II–III). This triggered our interest in the site, as after the chronological revision of Sites 30 and 34 there was no secure smelting context in Timna from the "Egyptian phase," except Site 2 (Erickson-Gini 2014). As part of the CTV Project we sent three short-lived samples from Rothenberg's excavations for radiocarbon dating; the results cover quite a long time span—from the end of the 13th to the mid-10th centuries BCE; however, based on the overlap with the occupation of the Hathor Shrine and other similarities to the Late Bronze activities in Timna (e.g., high frequency of Qurrayyah Painted Ware) we suggest dating the site to the later part of Egyptian activities in the valley (mid 12th c. BCE), as indicated by the finds in the Hathor Shrine (Fig. 3.8).

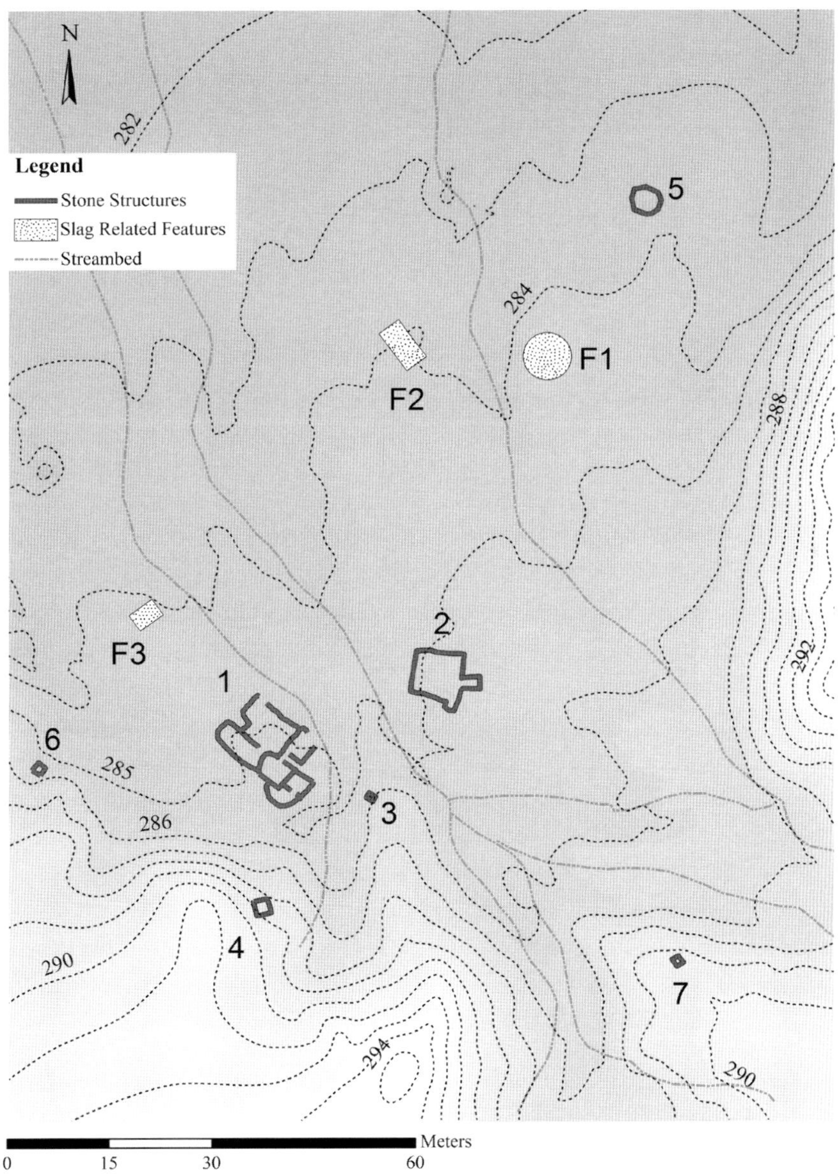

Fig. 3.6: Map of Site 3 (CTV Project 2015 Survey, mapping by O. Yagel; cf. Yagel *et al.* 2016).

Fig. 3.7: Aerial photograph of the southern part of Structure 1 (Site 3), including all of the areas excavated in the 1984 season of the Aravah Expedition.

Our new investigation of Site 3 strengthened Rothenberg's (1967: 30) suggestion that the site had an administrative role in the copper production system at the end of the Late Bronze Age. In addition, we focused on reconstructing smelting technologies based on chemical analysis of slag samples (n=28) and its comparison to samples from Site 34 (above) and previously published data (Yagel 2016). The results enabled linking Late Bronze and Iron Age technologies, and supported the assertion that the industry was based on a local labor force even under Egyptian control. After the Egyptians left Timna (and the southern Levant as a whole), the same technological traditions were used (and gradually improved upon); engagement in large scale copper production triggered and facilitated social processes, and served as a catalyst for the consolidation of local tribes and the formation of ancient Edom in the early Iron Age (if not earlier). Moreover, the continuity between the Late Bronze and Iron Ages at Timna suggests that the Egyptians had an important role in the formation of the early Edomite kingdom, as local institutions of control and exertion of power had probably already been established in the Late Bronze Age, whether under direct guidance and involvement of the Egyptians, or as a response to their demands.

SITE 15

Site 15 is located north of Mount Timna on one of the small tributaries of Nahal Timna (Fig. 3.1). Based on the site's similarity to Site 3—including no defense elements and some rectangular buildings (Figs. 3.9, 3.10)—we first hypothesized that it represents Late Bronze, Egyptian-controlled smelting. However, following the site's survey and excavations by the CTV Project (conducted in 2015), three short-lived samples were sent for radiocarbon dating. Two of the samples yielded ages in the range of 1131-934 (1

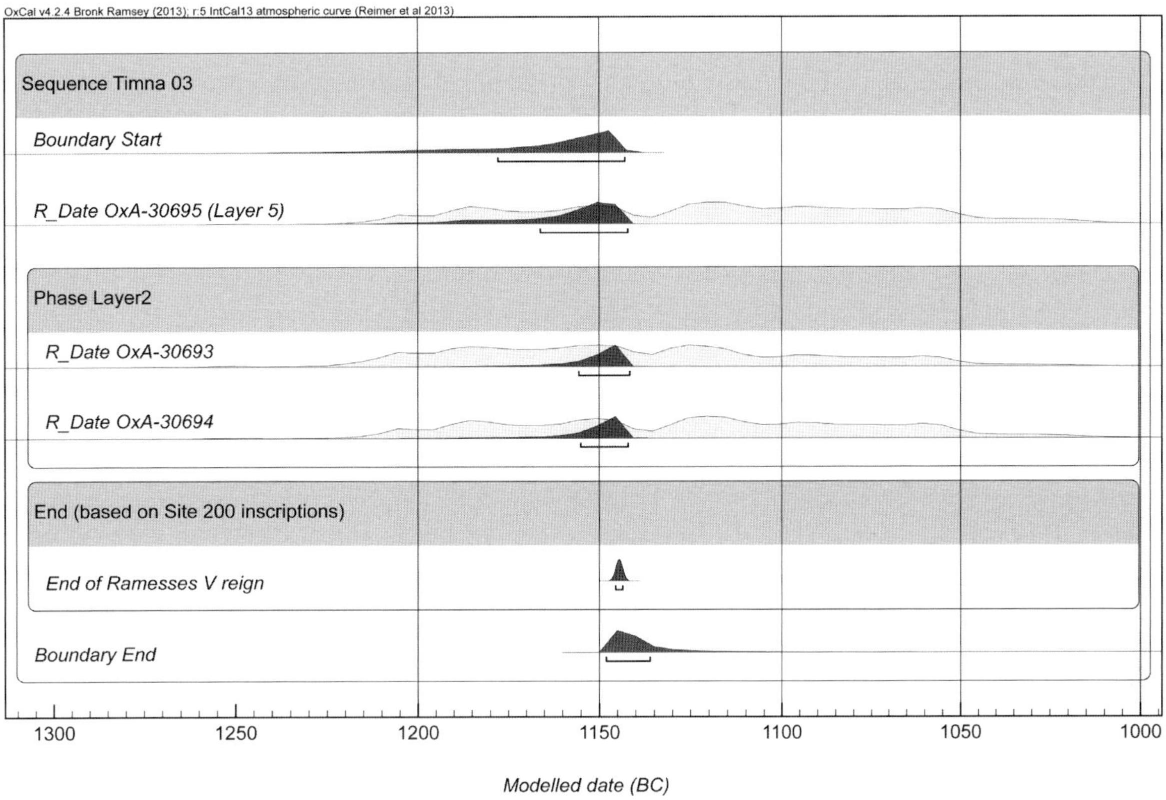

Fig. 3.8: Radiocarbon dates from Site 3 and Bayesian modelling based on the stratigraphy and the supposition that the site was abandoned at the same time as Site 200 Stratum III (the Hathor Shrine) (calibrated and modelled by OxCal v.4.2 and calibration curve IntCal13, © Bronk Ramsey 2013).

Fig. 3.9: Aerial photograph of Site 15 prior to the 2015 excavations.

sigma) Cal. BCE (the third failed). Thus, Site 15 reflects the new copper production system established in Timna immediately *after* the Egyptians left the region. This activity corresponds to the earliest contexts at Site 30 (Ben-Yosef *et al.* 2012) and to KEN Stratum VI (and possibly also to Stratum V; Levy *et al.* 2014).

The site sprawls over ca. 1 hectare and includes several thin scatters of slag and a few relatively well preserved architectural complexes (cf. Rothenberg 1962). The CTV Project's excavations concentrated on four small areas (Fig. 3.10):

1. Area A: a small probe in an open area (courtyard?) with a dense surface scatter of slag. The excavations here confirmed that the slag cover is thin, consisting of only the surface layer. A circular stone feature was unearthed half a meter below the surface (Fig. 3.11), indicating accumulation of sediments of varying qualities outside the main structures.
2. Area B: excavations in a rectangular room of one of the main architectural complexes (Fig. 3.12). The excavations reveal several stratigraphic phases and installations that probably relate to domestic activities (possibly a storage room and/or kitchen) within the room.
3. Area C: excavations in a room at the site's center (Fig. 3.13). The excavations revealed domestic and/or smelting activities and a phase that pre-dates the establishment of the structure's walls. Within the room, the natural alluvium had been dug into by the occupants creating a large (>1 m in diameter)

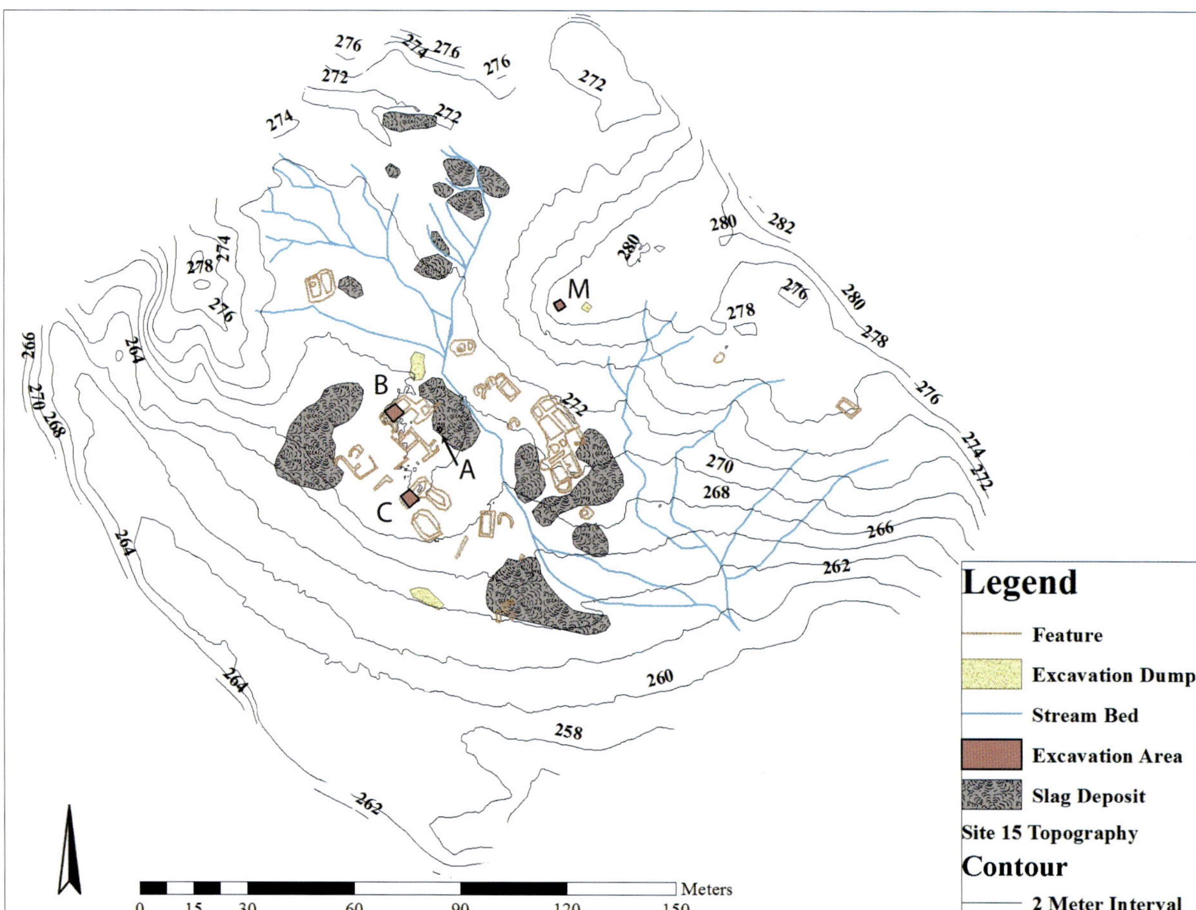

Fig. 3.10: Map of Site 15 and the 2015 excavation areas of the CTV Project (mapping by C. Smitheram).

Fig. 3.11: Circular stone feature (Locus 113) at the bottom of Area A, Site 15; note the lump of textile stuck between the stones above the scale (B. 9079).

Fig. 3.12: Site 15, Area B at the end of the 2015 excavations.

concavity. Evidence of high exposure to fire and smelting-related finds (such as technological ceramics, slag, charcoal and ore) suggest that this feature is directly related to copper production.

4. Area M: excavations of a small tumulus on the hill overlooking the main site. Although the excavations revealed an elliptical structure that is probably a tomb, we did not find any human bones or distinct burial-related artifacts. It is worth noting that similar structures were surveyed in Timna by E. Cohen-Sason (personal communication 2013) and by the Aravah Expedition, including in Site 3 (above). In the latter, a small probe was conducted by Rothenberg in 1959—refuting Glueck's claim that these were early furnaces and supporting their function as graves.

The other main structure on the hill overlooking the main site (Fig. 3.10) was interpreted by Yekutieli and Cohen-Sason (2010) as an observation post—part of a "surveillance system" used by the Egyptians in order to maintain their control over the operation of the mines and smelting sites. Although the functional interpretation of this structure, as well as other similar features in the Timna Valley (including Sites 2, 3, 12, 30, 34 and similar features within the mining areas) might be correct, their association with Egyptian "imperialism" has to be revised in light of the new, and much more complex chronological framework (see discussion below).

As at other sites in Timna, the preservation of organic materials at Site 15 is extraordinary. In addition to exceptionally large pieces of textile (Fig. 3.14), the excavations revealed hundreds of seeds of various kinds, including almonds (Fig. 3.15), which are the first to be reported from Timna so far.

Fig. 3.13: Site 15, Area C, during the 2015 excavations (the black line indicates boundaries of excavation square).

In addition, a well preserved, intact bronze needle (Fig. 3.16A) and a small iron tool (Fig. 3.16B) were found, the latter being a rare find for this period, when iron was only at its naissance as material for utilitarian tools (cf. Gottlieb, this volume).

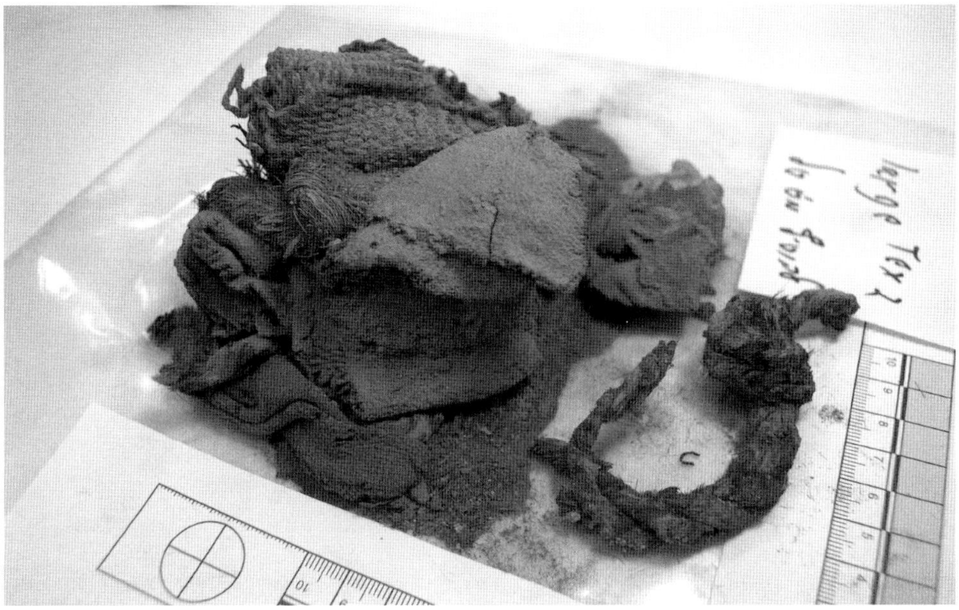

Fig. 3.14: Textile and ropes as found at Site 15 (B. 9068).

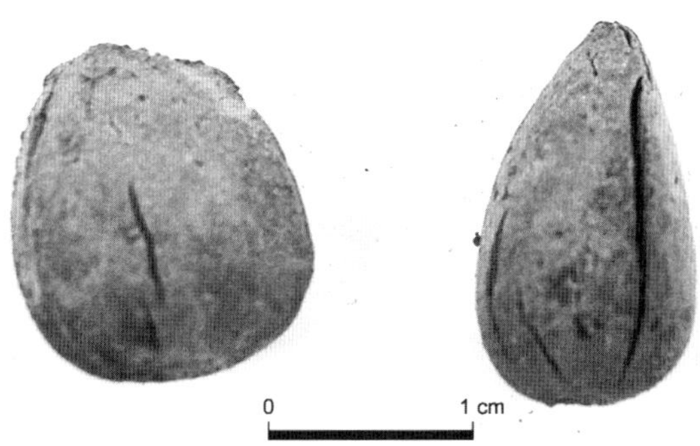

Fig. 3.15: Almonds found at Site 15, Area C.

Fig. 3.16: Metal objects from Site 15: (a) bronze needle, (b) iron (awl?).

SITE 35

Site 35 is located west of Mount Timna in a relatively flat area within the Nahal Nehushtan basin (Fig. 3.1). The site covers an area of ca. 125 × 100 m, and is dominated by architectural complexes (Figs. 3.17, 3.18). Between the architectural remains there are several scatters of slag, similar in characteristics to Sites 3 and 15 (above). The differences between Site 35 and the nearby smelting camps of Sites 34 and 30 (located ca. 200/700 m to the northwest respectively) led Rothenberg (1962) to suggest that the former was the location of the workers' dwellings and workshops. However, these differences most probably reflect a chronological discrepancy (see above), and although we agree that the workers at Sites 30 and 34 did not dwell at the sites themselves, it seems (based on various types of evidence) that their dwellings were at transient tent camps, and not based on architectural complexes such as the ones at Site 35. Thus, the working hypothesis of the CTV Project has been that Site 35 is contemporaneous with the nearby Hathor Shrine (located less than 200 m to the northeast) and represents Late Bronze Age activities; however, without radiocarbon dates, this dating remains tentative.

The first excavations at the site were conducted by the CTV Project in 2016 and included three areas (Figs. 3.18, 3.19):

1. Area A, a large building complex and a courtyard. This complex contains several rectangular rooms, stone pavements, stone benches and to its south a courtyard that was probably used as a pen (Fig. 3.20). The excavations exposed two distinct layers; the lower consists of ashy sediments and predates the architectural remains. The rooms include several small installations, and beneath one of the floors a pit containing a rather intact and unburned partial skull of a caprine was uncovered.

Fig. 3.17: Overview of Site 35 during excavations (2016).

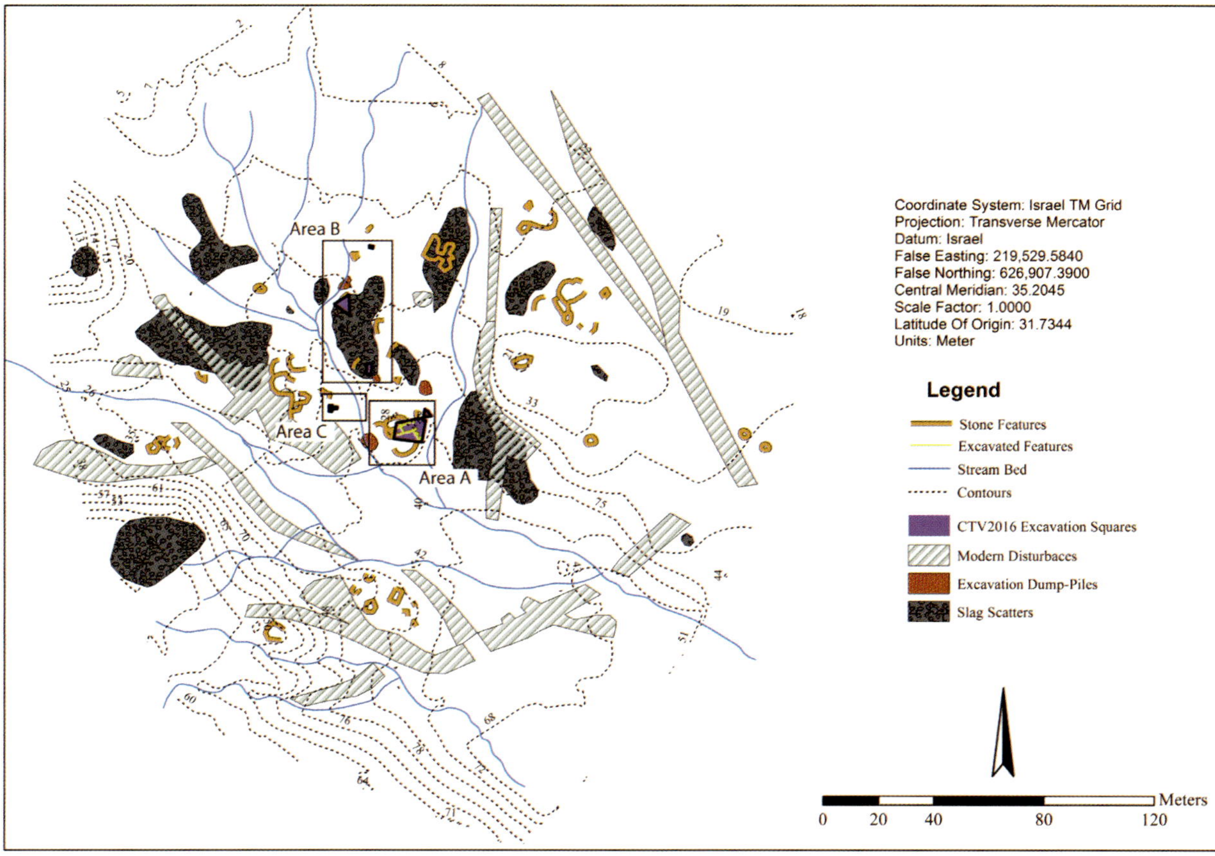

Fig. 3.18: Map of Site 35 and the 2016 excavation areas (mapping by O. Yagel and C. Smitheram).

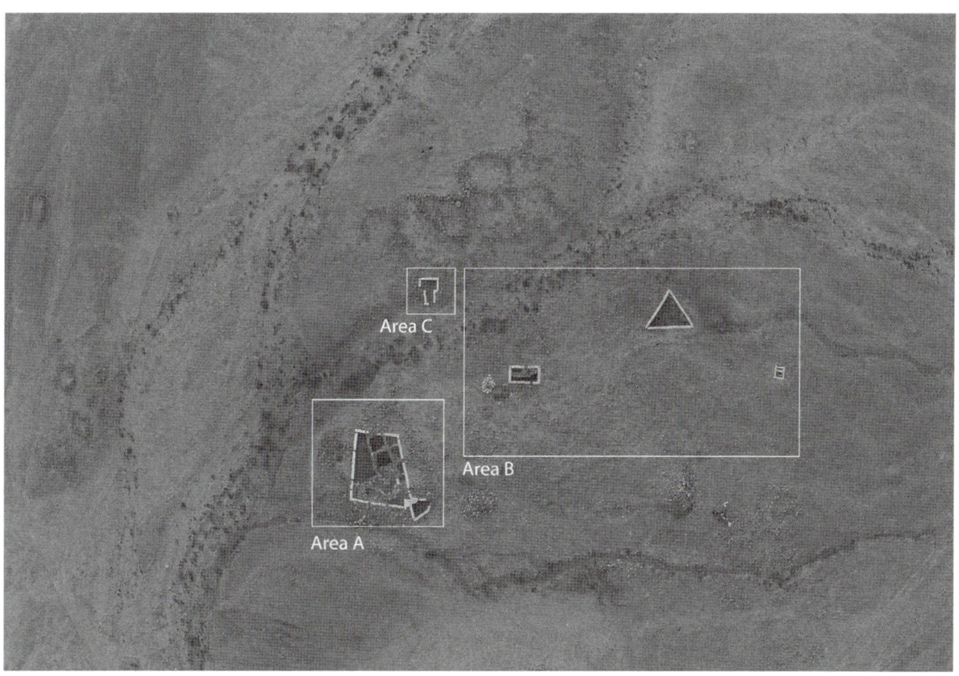

Fig. 3.19: Aerial photograph of Site 35 and the 2016 excavation areas (at the end of the season).

Fig. 3.20: Site 35, Area A at the end of the 2016 season. The two main stratigraphic phases are clearly visible: an earlier, organic-rich ashy layer pre-dates the architectural remains. The area at the center of the photograph was probably a pen attached to the building complex from the south (it is partially excavated).

2. Area B, flat area to the north of the building complex of Area A. Prior to excavations, several stone features and slag scatters were visible on the surface. The excavations demonstrated that the slag appears as a thin layer with no significant accumulation, and that the entire area contains the ashy sediments that correspond to the lower layer in Area A. Furthermore, the excavations exposed several rounded stone-built installations dug into the ground, probably used as storage pits. These are quite common at other smelting sites in Timna (Fig. 3.21; cf. Site 3 [Yagel, Ben-Yosef and Craddock 2016: Fig. 10]). Also uncovered were architectural remains—specifically walls—that were not originally visible on the surface.
3. Area C, a small dense scatter of slag to the west of Area A and near a large architectural complex (Fig. 3.22). The excavations here revealed beneath the slag an accumulation of thin layers of ashy sediments sitting on top of scorched clayey material (Fig. 3.23). This material seems to represent bottoms of smelting furnaces and/or other metallurgical installations that were constructed here continuously (furnaces were deliberately broken at the end of the smelting process in order to extract the raw metal; thus typically only the bottoms are found).

The pottery assemblage was rather small, but included a significant quantity of QPW sherds (Fig. 3.24) together with the other types typical of the main smelting sites at Timna (namely "Negebite" pottery and wheel-made "normal" pottery; see above). No typical Egyptian New Kingdom pottery was identified; however, further typological research on the assemblage is needed in order for this observation to be conclusive, as well as for providing more ceramic-related insights.

In addition to pottery, other artifacts were unearthed, including tuyère fragments, crucibles, textile, seeds, charcoal, beads of several types, metal artifacts and a broken copper ingot (Fig. 3.25). The latter is a rare find in Timna (cf. Rothenberg 1990a: 63–66), and represents the final product of the copper production industry.

Fig. 3.21: Site 35, rounded stone features (storage pits?) dug into the ground in Area B: (a) rounded stone feature in Square 1; (b) rounded stone feature in Square 2; (c) rounded stone feature in Square 4; the natural sandstone formation of "Solomon's Pillars" is in the background.

Fig. 3.22: Excavations at Site 35, Area C; in the background, Site 34 ("Slaves' Hill").

Fig. 3.23: Site 35, Area C; the dark patches are burned clay, probably furnace bottoms and/or other metallurgical installations. The white patches are gypsum casts used in preparation for archaeomagnetic sampling of the burned clay.

Fig. 3.24: A sherd of Quarrayyah Painted Ware (QPW) excavated at Site 35 (Area A, Square 1, Locus 1247, Basket 13301).

The 2016 excavations were accompanied by an intense pedestrian survey of the site, including artifact documentation and collection (mostly ground stones and pottery) and detailed mapping of the architectural features. The assemblage of the surveyed pottery is rather slim, most probably the result of previous surveys at the site.[13]

SURVEYS AND EXCAVATIONS IN COPPER MINING SITES

The ancient mines of Timna exploited almost exclusively the copper-bearing sandstones of the Amir-Avrona stratigraphic boundary, which is accessible around the perimeter of the valley below the cliffs (Fig. 3.1, Beyth,

13 In addition to the Aravah Expedition, Site 35 was surveyed by a team from Ben-Gurion University (over a period of more than one season; Y. Yekutieli, personal communication), and possibly also by other archaeologists working in the Timna Park. Unfortunately, none of the previous surveys was published.

Fig. 3.25: Two sides of a broken copper ingot (65.59 g) excavated at Site 35 (Area A, Square 1, Locus 1232, Basket 13137); as the final product of the smelting process, ingots are extremely rare in the archaeological record of Timna —in all of Rothenberg's excavations, only a handful of copper artifacts were identified as ingots (Rothenberg 1990a: 63–66).

Segev and Ginat, this volume).[14] This location, far from the main smelting sites and the center of the valley, is probably the reason that the investigation of the mines started relatively late (Conrad and Rothenberg 1980). Moreover, although some mines were recorded in the early surveys of the Aravah Expedition in the early 1960s,

14 There is no significant evidence for ancient exploitation of the Timna Formation, the other major copper ore body in Timna (contrary to the situation in Faynan, where the equivalent of this formation [Burj] was the dominant source of copper ore; see Hauptmann 2007). This is probably related to its limited exposure on the surface (especially of the ore-bearing member). However, there is some evidence for ancient mining in the Timna Formation for manganese oxides that were used as flux in advanced copper smelting technologies, such as the one that was introduced in the second half of the 10th century BCE (Bachmann and Rothenberg 1980; Rothenberg 1980; Ben-Yosef 2010: Chapter 8).

it took more than a decade to identify the main mining fields, which are represented by thousands of "plate-like" depressions that cover extensive areas around the valley's perimeter (Figs. 3.1, 3.26). These depressions were documented in the early surveys of the Aravah Expedition (e.g., Rothenberg 1967), but it was only in the mid-1970s that it was first demonstrated that these were all blocked mining shafts (Cohen 1976). These mining fields were systematically investigated by the Aravah Expedition, including a detailed survey in the "Model Area" (Fig. 3.1) and excavations of some of the blocked shafts (e.g., Hauptmann and Horowitz 1980).

The Aravah Expedition defined three main types of mines in Timna (Rothenberg 2005):

- **Type A – open-pit mines (*placer*)** (Fig. 3.27): relatively simple technology based on extracting copper ore nodules by digging pits into wadi terraces (gravel). These nodules (and ore fragments) are the result of natural erosion of the ore-bearing rocks and re-deposition in the valleys located *topographically* and *stratigraphically* below. Based on the simplicity of the technology and the assumption that such mines would have been exhausted prior to the application of more demanding technologies, Rothenberg associated this type with Chalcolithic (and earlier) mining.
- **Type B – basic shafts and galleries** (Fig. 3.28): shafts and galleries of relatively large diameter cut into the sandstone with stone tools. This type pre-dates the use of metal chisels and was dated to the Chalcolithic/Early Bronze Age by the Aravah Expedition (or the "Sinai–Arabah Copper Age, Early and Middle Phases," cf. Rothenberg and Glass 1992).
- **Type C – complex shafts and galleries** (Fig. 3.28): A more advanced form of shaft and gallery mining technology identified by the use of metal tools to carve into the copper ore bearing sandstone.[15] Most of the blocked shafts (the "plate-like" depressions, Fig. 3.26) in the valley are associated with this type of mining technology, and based on the "Egyptian paradigm" of the Aravah Expedition were dated to the Late Bronze Age. In light of the recent chronological revision of the main smelting sites in the valley (above), these mines should now be dated to the Late Bronze and Iron Ages (13th–9th century BCE), with a probable peak in mining activities during the 10th century BCE.

The CTV Project conducted surveys and excavations in mines near the site of the chariots petroglyph ("Merkavot" in Hebrew; see Yekutieli 2016). This area, located outside of the "Model Area" of the Aravah Expedition (Fig. 3.1), was not systematically studied prior to the new project (Conrad and Rothenberg 1980; Rothenberg 2005; and see also Shaw and Drenka, this volume). The investigated area contains both blocked shafts and open-pit mines; while the former were noted already by the Aravah Expedition (e.g., Conrad and Rothenberg 1980: 21) the latter were first identified by the current project. The "typical" surface disruption of shallow open pit mines might seem, to the untrained eye, to be a result of natural processes, and thus the ancient mines can easily be overlooked (Fig. 3.27). Only based on previous identification of such mines at Timna (Areas A, C and G of the Aravah Expedition, Rothenberg 2005) and recently in Faynan (Ben-Yosef, Levy and Najjar 2009) that we were able to confidently identify "Merkavot 1" as an ancient mining field even prior to excavations.

The new research at the Merkavot mining fields focused on the following questions:

Absolute Chronology

As ancient mines typically lack dateable material culture of any sort, the problem of dating ancient mining activities prevails in world-wide mining archaeology (e.g., Stollner 2014: 151). Many ancient

15 The exact nature of the metal tools used to dig the mines is not clear and is currently based on the marks these tools left on the walls of shafts and galleries. The copper "picks" found by the Aravah Expedition (e.g., Hauptmann and Horowitz 1980: 84) might be spear buts (for parallels from the Roman period, see e.g., Stiebel and Magness 2007: Plate 29; Stiebel 2013: Plate 13.1), thus their connection to mining activity is questionable.

Fig. 3.26: Aerial photograph of "plates"—blocked mining (and prospection) shafts—in the area of "Merkavot 3" (note the wadi on the right, which cut through blocked shafts); thousands of such features were recorded on the perimeter of the Timna Valley.

Fig. 3.27: An overview of "Merkavot 1," an open-pit mining field in the alluvial terraces of a small tributary of Nahal Timna. The wavy surface is the result of intense mining activity that was aimed at culling the copper nodules from the gravel.

Fig. 3.28: A mining gallery dug with stone tools (above the hand) crossed by a later gallery dug with metal tools (below the hand); in the photograph, the late Professor Hagai Ron during a field trip, 2012.

mines, including those of Timna, were dated by indirect observations (such as the date of nearby habitation or production sites) and/or by technological development schemes, usually assuming unilinear advancement. In order to address this issue and following the success in dating placer mines in Faynan (Ben-Yosef 2010), we applied optically stimulated luminescence dating (OSL) to establish at least a firm *terminus ante quem* for the mining activity.

Technology and Organization of Production

The CTV Project's survey and excavations aimed at adding additional insights on previous observations of the Aravah Expedition concerning mining technologies and the social aspects of mining activities, including organization of production and economic considerations (cf. Stollner 2003; Stollner 2014).

Post-abandonment Processes

The CTV Project aimed at further examination of the sediments in the blocked shafts in order to understand the filling process (natural vs. anthropogenic, etc.), and at better understanding erosional processes that affected the current exposure of mines and underground galleries (cf. Hauptmann and Horowitz 1980; Shlomi *et al.* 2015).

The following provides a brief description of the surveyed and excavated sites, as well as some preliminary results (Smitheram 2016).

MERKAVOT 1 – OPEN PIT (PLACER) MINES

The mining field of Merkavot 1 is located in a small valley in one of the upper tributaries of Nahal Timna. Two excavation squares were dug into the alluvial terrace, one of which (Square B) spanned all the way to the sandstone of the Shehoret Formation below the terrace (Fig. 3.29). The excavations demonstrated that the entire terrace was intensely mined and that the original pits were naturally filled by thin layers of aeolian and colluvial sediments (resulting in a much flatter landscape). The excavations yielded very few copper ore fragments, indicating that the gravel in its current state is a mining waste; for systematic exploitation of the terrace and avoiding missing any potential area, mined materials from the active pit were thrown into an older pit nearby. Thus, in cases where the fine quartz grains had enough exposure to light while being moved during the mining process, the material from the gravel could potentially provide OSL ages that represent the date of the mining activities. In addition to the gravel, the entire section of fine sediments was sampled for OSL dating (Fig. 3.29). While the natural fill gave excellent ages that correspond to their stratigraphy, the ages from the gravel were more problematic.

The OSL dates from Merkavot 1 (Square A) confirmed Rothenberg's early dating and provide the first analytic evidence of mining activity in Timna during the fifth millennium BCE. However, they also demonstrate that the area was mined during more than one period, including the late second millennium BCE—contra to Rothenberg's assertion that pit mining was not practiced when more advanced mining technologies were available.

MERKAVOT 2 AND 3—SHAFT AND GALLERY MINES

Merkavot 2 is a mining field of blocked shafts located on the ridge directly to the west of the petroglyphs site of the same name (Figs. 3.1, 3.30). The blocked shafts and many associated small stone-built features (including standing stones [*massebot*]) were mapped and documented during the 2013 and 2014 field seasons of the CTV Project. One of these blocked shafts was excavated (Figs. 3.31–3.33), and was found to be a mine of Type B (above). The OSL ages demonstrate that this mine was active during the first half of the fourth millennium BCE *at the latest*, thus placing it in the Early Bronze Age or Chalcolithic period—in accordance with the chronology suggested by Rothenberg for this type of mine. The OSL ages also demonstrate that the upper part of the fill (above the layer of mud, Fig. 3.33) is a natural accumulation of

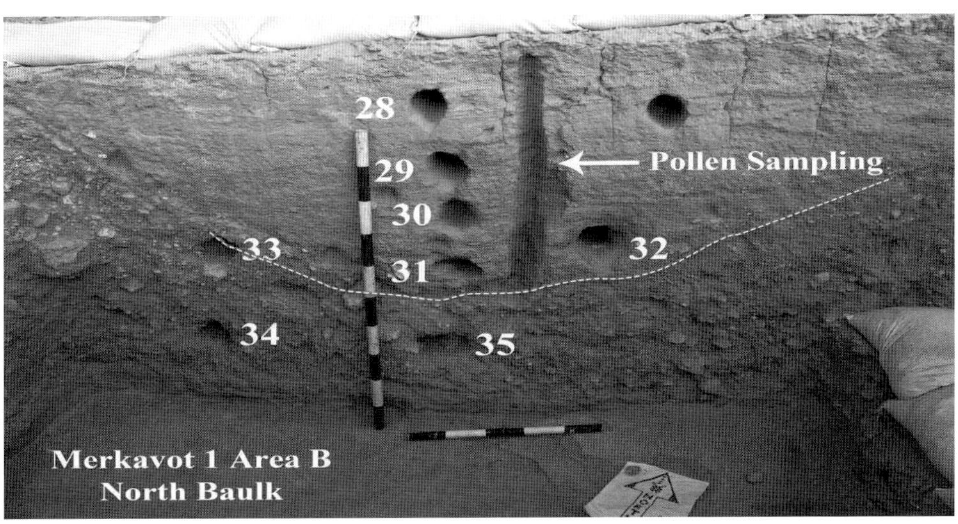

Fig. 3.29: Excavations at Merkavot 1, Area B; the wadi terrace sits stratigraphically and topographically *below* the ore-bearing sandstones of the Amir/Avrona boundary and on top of the red sandstone of the Shehoret Formation (visible in the photograph). The dashed line delineates the surface at the time the mine was abandoned; the thin layers on top of the dashed line are naturally accumulated aeolian (and colluvial) sediments. Also indicated are sample locations for optically stimulated luminescence (OSL) dating and pollen analysis.

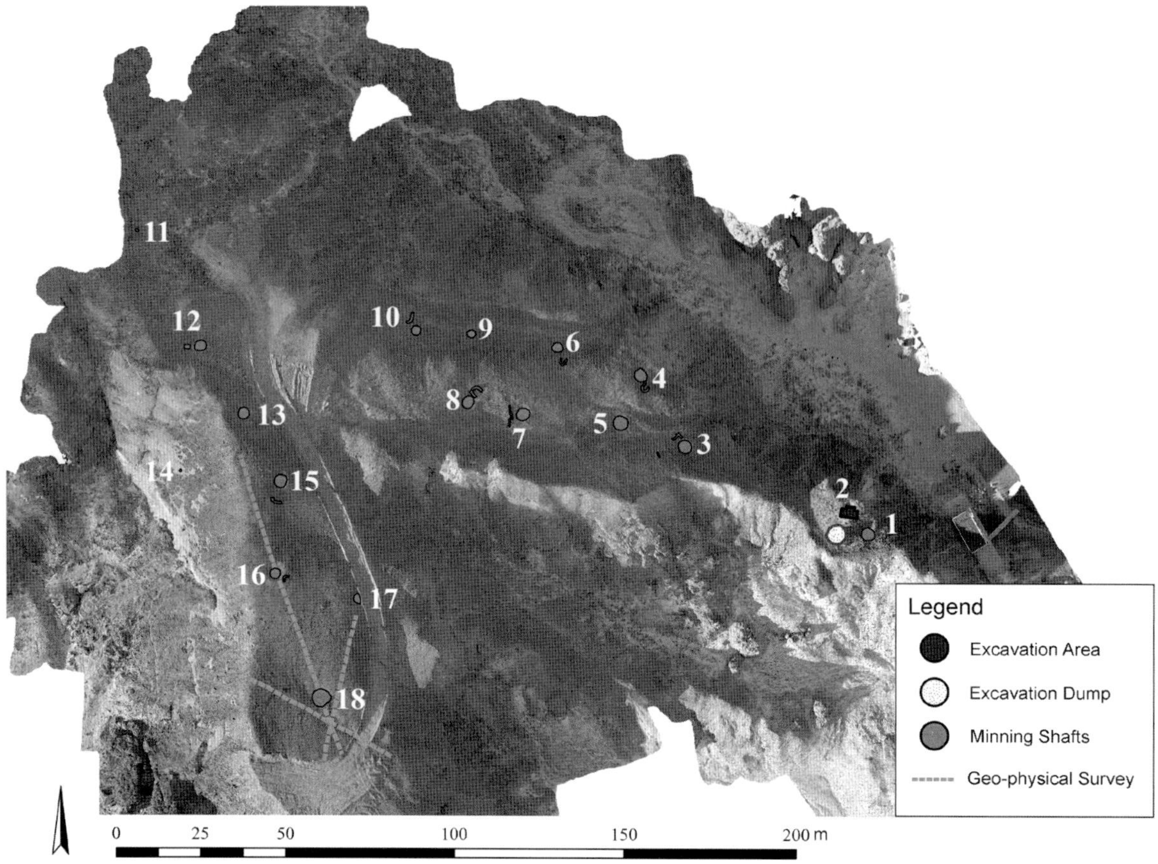

Fig. 3.30: Map of "Merkavot 2" mining field; note the many small stone-built features mapped by the CTV Project's 2015 survey near the blocked shafts (field supervisor: C. Smitheram).

Fig. 3.31: Merkavot 2, Shaft 2, before excavations.

Fig. 3.32: Excavating Shaft 2 in Merkavot 2 mining field (photograph by Sagi Bornstein).

fine aeolian and colluvial sediments. However, it seems that most of the fill of this shaft, as well as the other blocked shafts in this mining field and probably everywhere in the Timna Valley, was anthropogenic. We suggest a rather prosaic reason for this observation, namely that the "blocking" was a consequence of a systematic prospection survey of all the areas in the valley that bore potential for underground copper veins. As in "Merkavot 1," efforts were invested here as well to avoid missing any potential copper; thus not only waste from an active mine was dumped into a nearby shaft, but also the design of shaft digging was such that waste would not randomly accumulate as dump piles on the surface and hinder prospection beneath the dump locations. Moreover, it seems that the shafts were systematically dug from the lower to higher (topographically) areas, with progress towards the cliffs, where the ore bearing layer is deeper and thus deeper shafts are required.[16] The anthropogenic filling interpretation is also supported by a few shafts that are found empty (including Shaft 11 in Merkavot 2, Fig. 3.30), although their setting is similar to other nearby (and blocked) shafts. These empty shafts (some more than 40 m deep), among other observations, are the basis for our reconstruction of systematic prospection from lower to higher, as they are always located in the higher areas, closest to the cliffs (and representing the final shaft(s) in a designated mining field).

The Merkavot 2 mining field also provides insights on geomorphological processes, including rates of erosion and formation of wadi channels. Previous research has demonstrated an exceptionally high rate of erosion in the Timna Valley during the last 6000 years (4 m per millennium, and up to 30 m; Shlomi *et al.* 2015). This high rate was attributed to the mining activity itself, as the disrupted surface and the dense web of shafts and underground galleries accelerated erosional processes and provided conditions for wadi channels to cut deep into the sandstones (*ibid.*; Hauptmann and Horowitz 1980). In Merkavot 2, it is evident that at the time of the mining activity the copper bearing sandstone could have been reached only by shafts from above, and that the landscape was much flatter, with none of the wadis cutting through the sandstones.[17] The OSL dates from Merkavot 2 and the average level of the wadi channels today indicate

16 The sedimentary sequence in Timna slightly dips towards the valley's perimeter, thus the layers are deeper the farther they are from Mount Timna (in addition to higher topography).

17 An underground gallery is visible near the visitor's center at the parking lot of the Merkavot Site. This gallery, located beneath the shafts of Merkavot 2, was entirely underground when it was originally dug; the gallery was exposed by rapid erosional processes that created deep wadi channels that cut through the sandstone.

that during the last 6000 years the latter deepened about 15 m. Moreover, the chariot petroglyphs provide an additional time constraint on the erosional processes, as they are engraved into sandstones exposed only after the wadi channel has deepened. Based on their location and their accepted date (Late Bronze Age; Yekutieli 2016), it seems that the rate of erosion between 6000 and 3300 years before the present was slower than the average rate for the last 3000 years. A similar geomorphological process is clearly evident in the Merkavot 3 mining field, where a series of blocked shafts are cut by a deep wadi channel (Fig. 3.26), although here we do not have absolute dates to constrain rates.

As part of the CTV Project, a team from the department of geophysics at Tel Aviv University (supervised by N. Wechsler) carried out a geophysical survey (seismic diffraction) in Merkavot 2 and 3 in order to detect underground galleries (Fig. 3.34). The predominant aim of the survey was to demonstrate feasibility; in turn, the method may provide a potent tool for mapping underground mining systems, as a basis for insights on technology and scale (in particular evaluating the ratio of prospection vs. mining shafts). While at Merkavot 2 no anomalies were detected (possibly a consequence of the sloping

Fig. 3.33: Merkavot 2, Shaft 2, at the end of the 2014 excavations; a thick layer of mud at the bottom probably indicates the surface at the time of abandonment.

Fig. 3.34: Geophysical survey for detecting underground galleries in the area of Merkavot 2 (2014, cf. Fig. 30); the survey was conducted by the Department of Geophysics, Tel Aviv University (M. Reshef and N. Wechsler).

landscape rather than a real absence of galleries), at Merkavot 3 a gallery was detected at a depth of 7 m. However, additional research should be carried out, including field verification, in order to establish this method as a reliable tool for mapping the underground mines of Timna.

SUMMARY AND CONCLUSIONS

The first phase of the CTV Project focused on the main smelting sites in the Timna Valley. Following the recent revision in the chronology of these sites (Ben-Yosef *et al.* 2012), the results of the current project further demonstrate that the peak in copper production in the valley occurred during the 10th century BCE (Ben-Yosef 2016), similar to the situation at Faynan, the northern counterpart of Timna (Levy, Najjar and Ben-Yosef 2014). The multiple research avenues taken by the CTV Project provide new insights on all aspects of the *chaîne opératoire* of copper production during the ca. five centuries of substantial activities in the valley at the turn of the first millennium BCE (late Late Bronze Age and early Iron Age, ~1300–800 BCE). These include new observations on technological developments and innovations, from mining of ore and flux to smelting, refining and casting of copper ingots, as well as various related activities necessary for sustaining production on the scale evident at the mining and smelting sites. Given the logistically-challenging location of Timna deep in the desert and far from an agricultural hinterland, the latter had to involve major investment, in particular in tasks such as the constant provision of water and food (for the people, the industry [e.g., manufacturing of technological ceramic] and the livestock and draught animals). The nearest water source is more than 15 km away at the Yotvata (Ghadian) oasis (Peters, Ben-Yosef and Tauxe, this volume), and food had to be transported from a much greater distance.

In light of the archaeological evidence discussed above, it is not surprising that the main smelting and mining sites in Timna were attributed until recently to an empire (namely, the Egyptian New Kingdom). This was not only the view of Rothenberg and the Aravah Expedition (above); many other scholars accepted this research paradigm and even recognized mechanisms of "imperial control" in the archaeological evidence (Yekutieli and Cohen-Sason 2010). As the new chronological framework did not change the nature of evidence, the local society that operated the mines during the Iron Age has to be viewed as complex and centralized, despite its nomadic background (Ben-Yosef 2016). Moreover, the diachronic investigations of the CTV Project (within the transitional period between the Late Bronze and Iron Ages) resulted in strong evidence for continuity—in technology and most probably in the society that operated the mines. This society consolidated into an independent political entity during the early Iron Age (probably early Edom; see Ben-Yosef *et al.* 2010; Levy, Najjar and Ben-Yosef 2014) and was capable of organizing production on an even greater scale than under Egyptian control. However, it seems that the earlier "Egyptian phase" was fundamental to the formation of the new political entity, supporting the establishment of political institutions and facilitating early state formation processes.

Further assessment of the new data against contemporaneous evidence from Faynan and beyond, as well as broader contextualization within the culture history of the southern Levant and copper production technologies world-wide are beyond the scope of the current paper, which aims primarily at outlining the research design of the first phase of the CTV Project. The next phase of the CTV Project aims to investigate Timna and its vicinity in other periods (cf. Rothenberg 1999a), continuing to follow in Rothenberg's footsteps in exploring the ancient history of human exploitation of one of the most important natural resources in antiquity. As a tribute to Rothenberg's pioneering work, the CTV Project has adopted as its symbol a (rather abstract) petroglyph of a ridden camel from Slaves' Hill, echoing the symbol of the Aravah Expedition (suckling camel, Fig. 3.35).

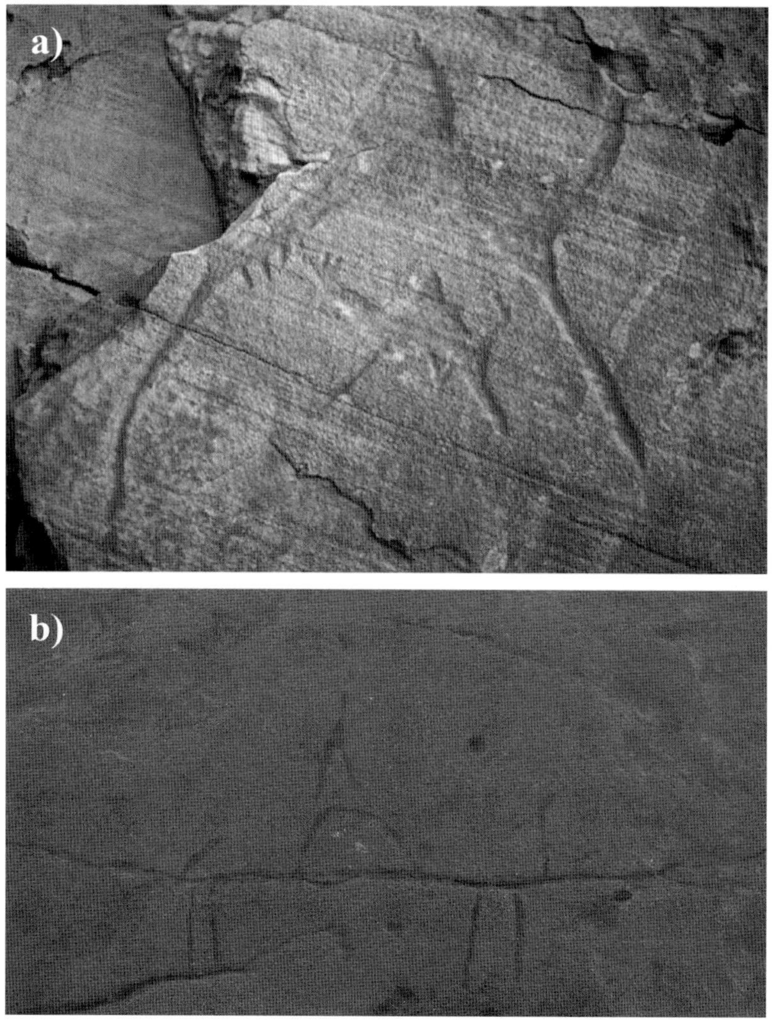

Fig. 3.35: Ancient petroglyphs of camels: (a) a suckling camel near Kibbutz Samar, the symbol of the Aravah Expedition (photo courtesy of U. Avner); (b) a ridden camel (?) engraved in the "High Place," Site 34. The petroglyph was adopted as the symbol of the CTV Project.

ACKNOWLEDGMENTS

Thanks are due to the people who make the CTV Project happen: students and volunteers, the project's staff (Willie Ondricek, Ilana Peters, Vanessa Workman, Mark Cavanagh, Omri Yagel, Craig Smitheram, Yitzhak Vassal, Duaa Abu Salah, Vanessa Linares and Aaron Greener), and the people of Timna Park and its manager Hagit Gal. In addition, materials published here benefited from the advice of various experts whom I wish to thank, including Lidar Sapir-Hen, Dafna Langgut, Omri Lernau, Yuval Goren, Yuval Sapir, Ehud Weiss, Shirly Ben-Dor Evian, Mario A.S. Martin, Michael Beyth, Assaf Kleiman, Sabine Metzer Kleiman, Orit Shamir, Naama Sukenik, Lisa Tauxe, Ron Shaar, Amotz Agnon, Assaf Holzer and Yigal Erel. Lastly I would like to thank Yoram Heimi and Tali Erickson-Gini of the Israel Antiquities Authority, and Oded Lipschits for their help with facilitating the field seasons. The excavations described here were carried out under permits numbers G-3/2013, G-4/2014, G-7/2015 and G-5/2016. The CTV Project is supported by the Marie Curie FP7-PEOPLE-2012-CIG grant #334274 and the Israel Science Foundation grant #1880/17.

REFERENCES

Bachmann, H.G. 1980. Early Copper Smelting Techniques in Sinai and in the Negev as Deduced from Slag Investigations. In: Craddock, P.T., ed. *Scientific Studies in Early Mining and Extractive Metallurgy.* London: 103–134.

Bachmann, H.G. 1982. *The Identification of Slags from Archaeological Sites*. London.

Bachmann, H.G. and Rothenberg, B. 1980. Die Verhüttungsverfahren von Site 30. In: Conrad, H.G. and Rothenberg, B., eds. *Antikes Kupfer im Timna-Tal*. Bochum: 215–236.

Ben-Yosef, E. 2010. *Technology and Social Process: Oscillations in Iron Age Copper Production and Power in Southern Jordan* (Ph.D. dissertation, University of California). San Diego.

Ben-Yosef, E. 2012a. Beno Rothenberg: Obituary. *Israel Exploration Journal* 62(1): 244–246.

Ben-Yosef, E. 2012b. Environmental Constraints on Ancient Copper Production in the Arava Valley: Implications of the Newly Discovered Site of Khirbat Manaiyah in Southern Jordan. *Tel Aviv* 39(2): 186–202.

Ben-Yosef, E. 2016. Back to Solomon's Era: Results of the First Excavations at Slaves' Hill (Site 34, Timna, Israel). *Bulletin of the American Schools of Oriental Research* 376: 169–198.

Ben-Yosef, E., Langgut, D. and Sapir-Hen, L. 2017. Beyond Smelting: New Insights on Iron Age (10th c. BCE) Metalworkers Community from Excavations at a Gatehouse and Associated Livestock Pens in Timna, Israel. *Journal of Archaeological Science Reports* 11: 411–426.

Ben-Yosef, E. and Levy, T.E. 2014. The Material Culture of Iron Age Copper Production in Faynan. In: Levy, T.E., Najjar, M. and Ben-Yosef, E., eds. *New Insights into the Iron Age Archaeology of Edom, Southern Jordan*, Vol. 2. Los Angeles: 887–959.

Ben-Yosef, E., Levy, T.E., Higham, T., Najjar, M. and Tauxe, L. 2010. The Beginning of Iron Age Copper Production in the Southern Levant: New Evidence from Khirbat al-Jariya, Faynan, Jordan. *Antiquity* 84 (325): 724–746.

Ben-Yosef, E., Levy, T.E. and Najjar, M. 2009. New Iron Age Copper Mine Fields Discovered in Southern Jordan. *Near Eastern Archaeology* 72(2): 98–101.

Ben-Yosef, E., Najjar, M. and Levy, T.E. 2014. New Iron Age Excavations at Copper Production Sites, Mines and Fortresses in Faynan, Jordan. In: Levy, T.E., Najjar, M. and Ben-Yosef, E., eds. *New Insights into the Iron Age Archaeology of Edom, Southern Jordan*, Vol. 2. Los Angeles: 767–885.

Ben-Yosef, E., Porat, N., Langgut, D. and Shaar, R. 2016. New Archaeological and Geo-Archaeological Studies in Timna Valley. In: Elisha, B., ed. *Israel Geological Society Annual Meeting Elat: Field Trips Guide.* Jerusalem: 22–44 (Hebrew).

Ben-Yosef, E., Shaar, R., Tauxe, L. and Ron, H. 2012. A New Chronological Framework for Iron Age Copper Production in Timna (Israel). *Bulletin of the American Schools of Oriental Research* 367: 31–71.

Ben-Yosef, E., *et al.* 2008. A New Approach for Geomagnetic Archaeointensity Research: Insights on Ancient Metallurgy in the Southern Levant. *Journal of Archaeological Science* 35: 2863–2879.

Cavanagh, M. 2016. *Sustainability of an Industry on the Fringe: A Dendroarchaeological Investigation into Fuel Sources at the Iron Age Copper Smelting Sites of the Timna Valley* (M.A. thesis, Tel Aviv University). Tel Aviv.

Cohen, E. 1976. The Mystery of "Plates" at Timna Valley. *Israel Land and Nature* 19: 15–17. (Hebrew)

Conrad, H.G. and Rothenberg, B., eds. 1980. *Antikes Kupfer im Timna-Tal*. Bochum.

Erickson-Gini, T. 2014. Timna Site 2 Revisited. In: Tebes, J.M., ed. *Unearthing the Wilderness: Studies on the History and Archaeology of the Negev and Edom in the Iron Age*. Leuven-Paris-Walpole: 47–84.

Frank, F. 1934. Aus der Araba I: Reiseberichte. *Zeitschrift des Deutschen Palästina-Vereins* 57: 191–280.

Glueck, N. 1935. Explorations in Eastern Palestine, II. *The Annual of the American Schools of Oriental Research* 15: 1–288.

Greener, A. and Ben-Yosef, E. 2016. The Ground-Stone Assemblage of a Metal Workers Community: An Unexplored Dimension of Iron Age Copper Production at Timna. *Journal of Lithic Studies* 3: 191–220.

Hauptmann, A. 2007. *The Archaeometallurgy of Copper: Evidence from Faynan, Jordan*. Berlin.

Hauptmann, A. and Horowitz, A. 1980. Zur Geomorphologie und Paläomorphologie des Modell-gebietes. In: Conrad, H.G. and Rothenberg, B., eds. *Antikes Kupfer im Timna-Tal*. Bochum: 57–67.

Havron, R. 1958–1959. *A Journey to the Negev: The Adventures of Amik and his Camel Kalregel*. Tel Aviv.

Kleiman, S., Kleiman, A. and Ben-Yosef, E. 2017. Metalworkers' material culture in the early Iron Age Levant: The ceramic assemblage of Site 34 ("Slaves' Hill") in the Timna Valley, Israel. *Tel Aviv 44* (2): 232–264.

Levy, T.E. 2009. Pastoral Nomads and Iron Age Metal Production in Ancient Edom. In: Szuchman, J., ed. *Nomads, Tribes, and the State in the Ancient Near East*. Chicago: 147–176.

Levy, T.E., Bettilyon, M. and Burton, M.M. 2016. The Iron Age Copper Industrial Complex: The Role of Ground Stone Tools at Khirbat en-Nahas, Jordan. *Journal of Lithic Studies* 3.

Levy, T.E., Munger, S. and Najjar, M. 2014. A Newly Discovered Scarab of Sheshonq I: Recent Iron Age Explorations in Southern Jordan. *Antiquity* 341 (Project Gallery Online).

Levy, T.E., Najjar, M. and Ben-Yosef, E. 2014. *New Insights into the Iron Age Archaeology of Edom, Southern Jordan*. Los Angeles.

Levy, T.E., *et al*. 2014. Excavations at Khirbat en-Nahas 2002–2009: Unearthing an Iron Age Copper Production Center in the Lowlands of Edom (Southern Jordan). In: Levy, T.E., Najjar, M. and Ben-Yosef, E., eds. *New Insights into the Iron Age Archaeology of Edom, Southern Jordan*, Vol. 1. Los Angeles: 89–245.

Merkel, J.F. 1990. Experimental Reconstruction of Bronze Age Copper Smelting Based on Archaeological Evidence from Timna. In: Rothenberg, B., ed. *Researches in the Arabah 1959–1984, Vol. 2: The Ancient Metallurgy of Copper.* London: 78–122.

Meshel, Z. 1993. Yotvata. In: Stern, E., ed. *The New Encyclopedia of Archaeological Excavations in the Holy Land*, Vol. 4. Jerusalem: 1517–1520.

Peters, I. 2015. *Determining Temporal Contexts of Technological Records: The Application of Archaeomagnetic Dating to Copper Slag Deposits in Timna* (M.A. thesis, Tel Aviv University). Tel Aviv.

Peters, I., Tauxe, L. and Ben-Yosef, E. 2017. Archaeomagnetic Dating of Pyrotechnological Contexts: A Case Study for Copper Smelting Sites in the Central Timna Valley, Israel. *Archaeometry. doi. 10.1111/arcm.12322*

Petherick, J. 1861. *Egypt, the Soudan and Central Afrika*. Edinburgh and London.

Pigott, V. 1996. Near Eastern Archaeometallurgy: Modern Research and Future Directions. In: Cooper, J.S. and Schwartz, G.M., eds. *The Study of the Ancient Near East in the 21st Century*. Winona Lake.

Rothenberg, B. 1962. Ancient Copper Industries in the Western Arabah. *Palestine Exploration Quarterly* 94: 5–71.

Rothenberg, B. 1967. *Negev: Archaeology in the Negev and the Arabah (in Hebrew)*. Tel Aviv.

Rothenberg, B. 1972. *Were these King Solomon's Mines? Excavations in the Timna Valley*. New York.

Rothenberg, B. 1980. Die Archäologie des Verhüttungslagers Site 30. In: Conrad, H.G. and Rothenberg, B., eds. *Antikes Kupfer im Timna-Tal*. Bochum: 187–213.

Rothenberg, B., ed. 1988. *Researches in the Araba 1959–1984, Vol. 1: The Egyptian Mining Temple at Timna*. London.

Rothenberg, B. 1990a. Copper Smelting Furnaces, Tuyeres, Slags, Ingot-moulds and Ingots in the Arabah: The Archeological Data. In: Rothenberg, B., ed. *The Ancient Metallurgy of Copper*. London: 1–77.

Rothenberg, B., ed. 1990b. *The Ancient Metallurgy of Copper: Archaeology - Experiment - Theory*. London.

Rothenberg, B. 1999a. Archaeo-metallurgical Researches in the Southern Arabah 1959–1990, Part 1: Late Pottery Neolithic to Early Bronze IV. *Palestine Exploration Quarterly* 131: 68–89.

Rothenberg, B. 1999b. Archaeo-metallurgical Researches in the Southern Arabah 1959–1990, Part 2: Egyptian New Kingdom (Ramesside) to Early Islam. *Palestine Exploration Quarterly* 131: 149–175.

Rothenberg, B. 2005. Explorations and Excavations in the Mines of the Timna Valley (Israel): Paleomorphology as Key to Major Problems in Mining Research. *Glasnik Srpskog arheološkog društva* 22(21): 133–148.

Rothenberg, B. and Glass, J. 1992. Beginnings and Development of Early Metallurgy and the Settlement and Chronology of the Western Arabah, from the Chalcolithic Period to Early Bronze Age IV. *Levant* 24: 141–157.

Sapir-Hen, L. and Ben-Yosef, E. 2013. The Introduction of Domestic Camels to the Southern Levant: Evidence from the Aravah Valley. *Tel Aviv* 40(2): 277–285.

Sapir-Hen, L. and Ben-Yosef, E. 2014. The Socioeconomic Status of Iron Age Metalworkers: Animal Economy in the 'Slaves' Hill', Timna, Israel. *Antiquity* 88(341): 775–790.

Shaar, R., Ben-Yosef, E., Ron, H., Tauxe, L., Agnon, A. and Kessel, R. 2011. Geomagnetic Field Intensity: How High Can It Get? How Fast Can It Change? Constraints from Iron Age Copper-slag from the Southern Levant. *Earth and Planetary Science Letters* 301: 297–306.

Shlomi, Y., Ginat, H., Meron, T. and Holtzer, A. 2015. Pace of Degradation in Streams within Sandstone in a Region with Copper Mines (Timna Valley, Southern Israel). *Negev, Dead Sea and Arava Studies* 7(2): 38–50 (Hebrew).

Smitheram, C. 2016. *OSL Dating of Timna's Copper Mines: Insights on Technological Evolution and Social Organization* (M.A. thesis, Tel Aviv University). Tel Aviv.

Stiebel, G.D. 2013. The Military Equipment from the Hasmonean and Herodian Palaces at Jericho and Cyprus. In: Bar-Nathan, R. and Gartner, J., eds. *Hasmonean and Herodian Palaces at Jericho—Final Reports of the 1973–1987 Excavations, Vol. V.* Jerusalem: 290–298.

Stiebel, G.D. and Magness, J. 2007. The Military Equipment from Masada. In: Aviram, J., Foerster, G., Netzer, E. and Stiebel, G.D., eds. *Masada VIII: The Yigael Yadin Excavations 1963–1965 Final Reports*. Jerusalem: 1–94.

Stollner, T. 2003. Mining and Economy: A Discussion of Spatial Organization and Structures. In: Stollner, T., Korlin, G., Steffens, G. and Cierny, J., eds. *Man and Mining (Mensch und Bergbau): Studies in Honour of Gerd Weisgerber on Occasion of his 65th Birthday*. Bochum: 415–446.

Stollner, T.R. 2014. Methods in Mining Archaeology (Montanarchäologie). In: Roberts, B.W. and Thornton, C.P., eds. *Archaeometallurgy in Global Perspective: Methods and Syntheses*. New York: 133–159.

Sukenik, N., *et al*. 2017. Early Evidence (Late 2nd Millennium BCE) of Plant-Based Dyeing of Textiles from Timna, Israel. *PLOS ONE* 12(6).

Ward, W.A. 1972. The Shasu "Bedouin": Notes on a Recent Publication. *Journal of the Economic and Social History of the Orient* 15: 35–60.

Workman, V. 2016. *The Fabric of Copper Production: The Textile and Cordage Artifacts from Iron Age Timna* (M.A. thesis, Tel Aviv University). Tel Aviv.

Yagel, O. 2016. *Late Bronze Age Copper Smelting Technology at Timna: Site 3 as a Case Study* (M.A. thesis, Tel Aviv University). Tel Aviv.

Yagel, O., Ben-Yosef, E. and Craddock, P. 2016. Late Bronze Age Copper Production in Timna: New Evidence from Site 3. *Levant* 48(1). 33–51.

Yekutieli, Y. 2016. The Chariots Engraving of Timna (Israel) Revisited. *Bulletin of the American Schools of Oriental Research* 375: 171–184.

Yekutieli, Y. and Cohen-Sason, E. 2010. Surveillance at Ancient Imperial Labor Camps in the Desert: a Southern Levantine Perspective. In: Hormann, R. and Mackenthun, G., eds. *Human Bondage in Cultural Contact Zone: Transdisciplinary Perspectives on Slavery and Its Discourses*. Münster: 33–62.

CHAPTER 4

THE DIET OF ANCIENT METAL WORKERS: THE LATE BRONZE AND EARLY IRON AGES IN THE ARABAH VALLEY (TIMNA AND FAYNAN)

Lidar Sapir-Hen, Omri Lernau and Erez Ben-Yosef

The nature of social organization in primary metal production sites is seldom studied from the zooarchaeological perspective. In this paper we focus on the diet of ancient metal workers in the Late Bronze/early Iron Ages of the southern Levant, a period of peak activity in the major copper production sites of the Arabah Valley. Based on the study of mammal and fish remains from new excavations, the re-examination of fish remains from previous research, and in light of other studies of faunal remains from Timna and Faynan, we present new insights regarding the social organization and cultural affiliation of the ancient societies. We demonstrate that the meat diet of metal workers is based primarily on livestock, supplemented by Mediterranean fish. We suggest that the people engaged in direct smelting enjoyed the highest status within the social system of copper producers, which included miners and various craftspersons of auxiliary activities. Finally, our research provides new insights regarding the ethnic identity of the metal workers, suggesting limited Egyptian presence in the Late Bronze Age Timna Valley and no significant Egyptian presence there in the early Iron Age.

INTRODUCTION

The turn of the first millennium BCE was a period of peak activity in the two major copper production districts of the southern Levant—Timna (southern Israel) and Faynan (ca. 100 km to the north, in southern Jordan) (Ben-Yosef *et al.* 2012; Levy, Najjar and Ben-Yosef 2014). It was a time of growing social complexity and political change, when new local polities appeared on the scene (Edom, ancient Israel), replacing the former Egyptian empire and its hegemony. Thus, fundamental questions in the studies of Timna and Faynan include the social organization of the metal workers and their cultural affiliation (Ben-Yosef 2010; Levy, Najjar and Ben-Yosef 2014). Animal economy is an important component of subsistence economy and is strongly related to ethnicity and culture. It is thus of prime relevance to the questions at hand. In this paper we review the current state of research of animal economy in these regions during the late Late Bronze and early Iron Ages (ca. 1300–800 BCE).

Understanding food provision systems is essential in order to comprehend the nature of the social organization of copper production (e.g., Muniz and Levy 2014). These are expressed through food production, distribution, consumption and disposal patterns, which are in turn reflected in the species frequency, mortality profiles and skeletal elements frequencies of the consumed fauna. Food habits are also correlated to the cultural affiliation of the population (Mintz and Du Bois 2002; Pearson 2003). This aspect is expressed through the identity of consumed animals (i.e., species) as well as in preparation and consumption practices (e.g., Sapir-Hen *et al.* 2014).

Based on the study of mammal and fish remains from the new excavations at Sites 30 (2009 season) and 34 (2013 season) (see also Sapir-Hen and Ben-Yosef 2013, 2014), the re-examination of fish remains from previous excavations at Sites 2, 30, and 200 (previously studied by Lernau 1988a), and in light of other studies of faunal remains from Timna and Faynan (Lernau 1988b; Grigson 2012; Muniz and Levy 2014; Bar-Oz and Erickson-Gini in press), we explore here the diet of ancient metal workers and its implications for understanding their social organization and cultural affiliation.

PREVIOUS STUDIES OF FAUNAL REMAINS FROM LATE BRONZE/IRON AGE FAYNAN AND TIMNA

The faunal remains from three sites in Timna and one site in Faynan have to some degree been previously analyzed and published. These sites include at Timna (Fig. 4.1): Site 200 (Lernau 1988b, 1988a), a predominantly Late Bronze Age temple dedicated to the Egyptian goddess Hathor (with complex stratigraphy, including an Iron Age layer [Stratum II], see Rothenberg 1988, Avner 2014); Site 2 (Bar-Oz and Erickson-Gini in press), a Late Bronze smelting camp (see, e.g., Rothenberg 1972, 1999; for a stratigraphic complication, see Avner and Magness 1998; radiocarbon dates and other finds confirm the Late Bronze dating of the published faunal assemblage [Erickson-Gini 2014]); and Site 30 (Grigson 2012), a smelting camp recently dated to the early Iron Age (Ben-Yosef *et al.* 2012). The mammalian remains from the Late Bronze Temple Site 200 were studied thoroughly and published as part of the final report of the site (Lernau 1988b), in addition to the fish remains from Sites 200, 2 and 30 (Lernau 1988a). The entire faunal assemblage was treated as a single group assigned to the Late Bronze Age, although Rothenberg acknowledged Iron Age occupation at Sites 200 and 30. The analysis of mammalian remains from the temple revealed only caprine bones, with a relatively equal representation of sheep (*Ovis aries*) and goats (*Capra hircus*). As sheep require a more humid environment than goats, the high frequencies of sheep should be noted. Caprines at the temple were young and represented by all body parts, suggesting slaughter for sacrifice near the temple (Lernau 1988b: 252). Site 2 and Site 30 analyses of mammalian remains are limited and incomplete. The smelting camp of Site 2 was excavated in 1964 by Rothenberg (e.g., Rothenberg 1999), and recently by Erickson-Gini (2014) who has conducted small scale excavations in the northern part of the site since 2005 (begun as a salvage excavation related to the construction of a visitor's center). The earlier excavations were never published as a final report and no mammalian faunal study is available (except for the note on one camel scapula, undated; Hakker-Orion 1984). The fauna from the recent excavations directed by Erickson-Gini is associated with contexts dated to the Late Bronze. This assemblage includes sheep, goat and pig (*Sus scrofa domesticus*) remains (Bar-Oz and Erickson-Gini in press). Here, the presence of pigs (including all body parts) should be noted, as they are considered unsuitable to the local dry climate. At Site 30, from the earlier excavations of 1974 and 1976, only the fish, camel (*Camelus dromedarius*) and donkey (*Equus asinus*) remains have been published, together with a qualitative species list, which also includes caprines and cattle (*Bos taurus*) (for the Iron Age date of the site, see Ben-Yosef *et al.* 2012; Grigson 2012; for reconsideration of the camel dating and its implications see Sapir-Hen and Ben-Yosef 2013).

In Faynan, the major site of Khirbat en-Nahas (KEN) was excavated by T.E. Levy and M. Najjar (Levy, Najjar and Ben-Yosef 2014). The faunal assemblages, dated to the 12th–9th centuries BCE, is significantly larger compared to the above-mentioned sites (NISP ca. 3800). The mammalian assemblage includes mainly caprines and cattle, with few remains of donkeys and camels, dogs (*Canis sp.*) and gazelles (*Gazella sp.*) (Muniz and Levy 2014). Muniz and Levy (*ibid.*) suggest that the site was provisioned with food supply and its population had equal access to food items (based on materials from

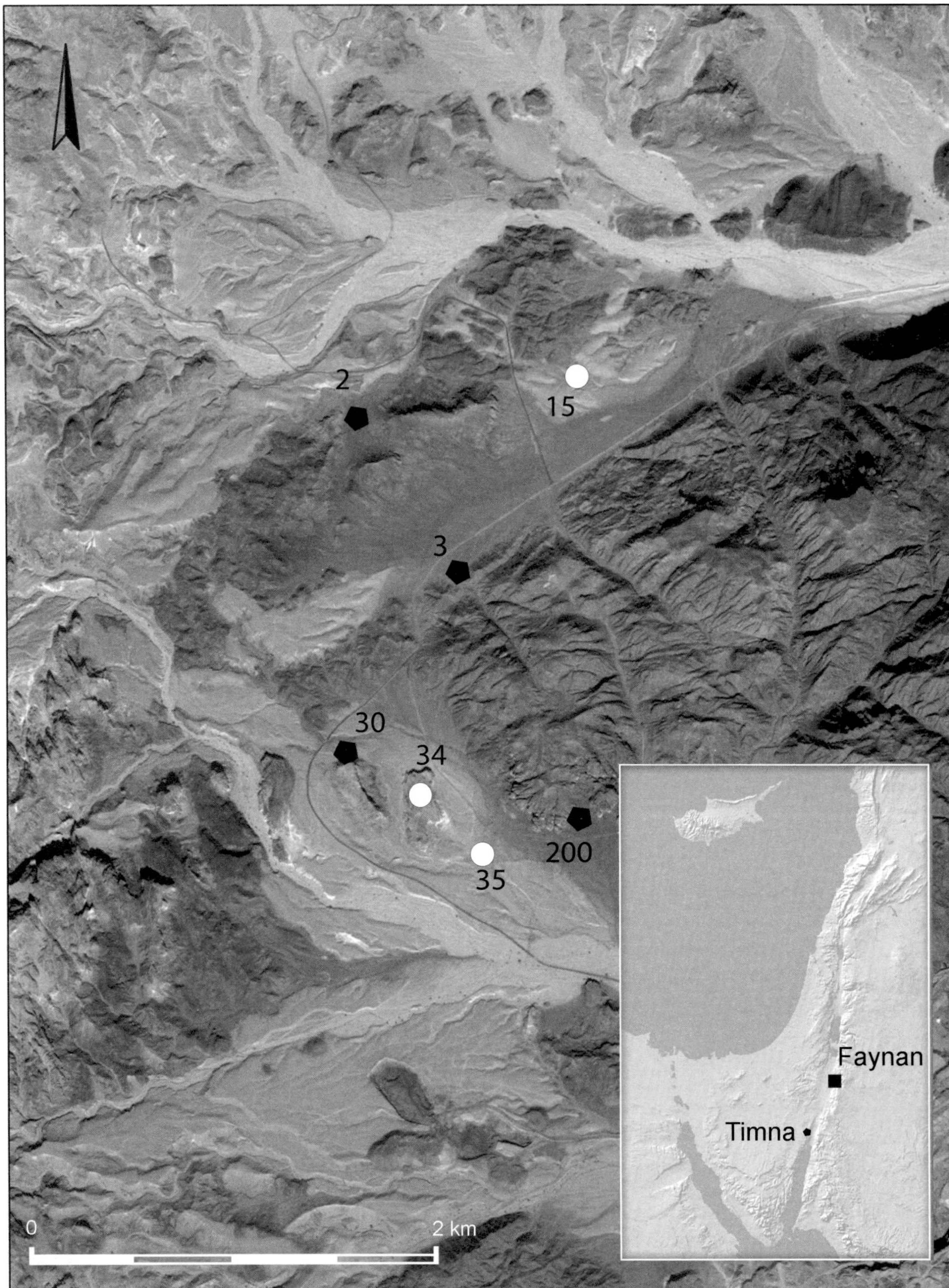

Fig. 4.1: Main copper production and related sites in Timna (polygons: sites excavated by the Aravah Expedition; circles: sites excavated by the Central Timna Valley Project). Inset indicates the location of Timna and Faynan, the two main copper ore deposits of the Arabah Valley.

various areas). A preliminary study of fish remains provided a species list that includes porgy (*Sparidae sp.*), parrot fish (*Scaridae sp.*), drum (*Sciaenidae sp.*), wrasse (*Labridae sp.*), and Nile perch. The authors suggest that these fish were brought dried as provisions for the workers, and that they originated from both the Red and Mediterranean Seas; however, as the authors themselves stress, further analysis is needed including identification of species to reach a decisive conclusions.[1] This is the only assemblage thus far that includes, to some extent, evidence of hunting wild game (gazelles, ~1% of the assemblage). The presence of gazelle implies there is some level of opportunistic hunting and the authors raise the possibility that this food is restricted to elites at KEN.

NEW DATA FROM SITES 30 AND 34

The faunal assemblages from the new excavations at Site 30 (Ben-Yosef *et al.* 2012) and Site 34 (Ben-Yosef, this volume) (see location in Fig. 4.1) were examined as part of the Central Timna Valley Project (CTV, directed by author E. B.-Y.). Both sites are smelting camps, the largest of their type in the Timna Valley, and both assemblages were dated to the early Iron Age, 11th–10th centuries BCE. The 2013 excavations at Site 34 offer an excellent case study for the research of culture and identity as reflected in the faunal remains, as they were specifically designed for best retrieval of animal bones accompanied by tight control over excavated contexts (see below).

MAMMALS

Methods

Retrieval of Finds

At Site 34, every excavated locus was dry sieved, and a sample was wet sieved. From the excavations, sediment was moved as is into buckets, and dry sieved through a 6 mm mesh which was placed over a wheelbarrow (Fig. 4.2a). Material that was left on the mesh was hand-picked in the field (Fig. 4.2b). Sediment that went through the mesh into the wheelbarrow was transferred into buckets lined with a 1 mm mesh for further wet sieving. Wet sieving took place in the nearby Timna Lake (Fig. 4.2c), an artificial body of brackish water resulting from modern mining activities. Subsequent micro-picking took place in the camp by the archaeology students (Fig. 4.2d).

In the 2009 Season at Site 30, all material was collected by dry sieving through a 6 mm mesh (without wet sieving).

Analysis

Mammal Remains from Site 30 (2009 Season) and Site 34

During the 2013 Season, two trenches were exposed at Site 34: Square 19A which includes a slag mound (Slag Mound 19) and Square 19B, which is located to the south of Square 19A and includes an elliptic stone-built installation (see Fig. 3.3, this volume). Most of the faunal remains were retrieved from Square 19A,

1 Note that if the identification of wrasse and Nile perch (*Lates niloticus*) is correct, the KEN assemblage represents connections with the Red Sea and the Nile, which are not represented in the Timna assemblage (below). This might reflect the fact that the KEN assemblage mainly represents the 9th century BCE, a period not represented in the Timna assemblage (except Site 30 Layer I). If so, the change in fish procurement reflects the major re-organization of production in the late 10th century BCE (Levy *et al.* 2014) that was triggered (and probably maintained) by Egyptian involvement in the region (cf. Sapir-Hen and Ben-Yosef 2013).

Fig. 4.2: Retrieval of finds at the 2013 excavations at Timna Site 34. (a) dry sieving: mesh placed over a wheelbarrow; (b) collecting material from the mesh; (c) wet sieving in Timna Lake: sediment that went through the dry sieving mesh to the wheelbarrow is wet sieved with a 1 mm mesh; (d) micro-picking: looking for the small fragments on the tray of sieved sediment.

which is characterized by a vertical sequence of layers representing ~100 years of activity. The alternating layers include those that were rich in pyrotechnological debris—slag, tuyère and furnace fragments—and hence were defined by us as "industrial," and those that were rich in organic material and hence were defined as "domestic" (Fig. 4.3; for loci distribution diagram [Harris matrix], see Sapir-Hen and Ben-Yosef 2014). Both contexts, however, represent complementary aspects of copper production activities carried out at the site. The domestic context most probably indicates activities such as preparation of ore, charcoal, food and other auxiliary tasks and the industrial context reflects smelting activities. The distinction between these two types of contexts was practiced in order to examine whether different activities are reflected in the faunal remains as well and whether they could testify to some form of social organization and specialization. Animal remains from these two contexts were divided into meat-rich and meat-poor parts, based on the values provided by Binford (1978) of sheep's Meat Utility Indices.

The assemblage was identified to skeletal element and species using the comparative collections stored at the Zooarchaeology Laboratory, Institute of Archaeology and at the Steinhardt Museum of

Fig. 4.3: Section of Timna Site 34 Slag Mound 19, showing alternating layers of "industrial" and "domestic" contexts.

Natural History, of Tel Aviv University. It was examined for aging of individuals (Payne 1973; Zeder 2006), correlation of survivorship to bone structural density (Symmons 2005), and for cut marks, burning signs and evidence of carnivore (scavenger) activity at the site (Lyman 1994).

Results

Sapir-Hen and Ben-Yosef published mammalian remains from Sites 30 (the 2009 season) and 34 (2013, 2014), and results from these papers are summarized here. We focus on the remains of Site 34, with a short description of the Site 30 assemblage.

At Site 34, the mammalian assemblage is dominated by domestic livestock—mainly caprines (sheep and goat)—followed by several donkey remains. These are prevalent in both the "industrial" and "domestic" contexts (above). Mouse bones (*Mus sp.*) were found only in the domestic layers (Table 4.1a). At Site 30, the small assemblage is dominated by caprines, followed by cattle (Table 4.1b). While the state of preservation of bones is poor, the fragmentation at Site 34 is not correlated to the bone structural density (Spearman's r=0.19, p=0.42) (for Spearman Rank Correlation, see Zar 2005). The presence of butchery marks and the low frequency of burnt bones testify to the fact that the assemblage reflects the leftovers of the meals of the metal workers, and the lack of gnawing marks suggests the bones were not altered by occasional carnivores.

Faunal remains from Slag Mound 19 (Area 19A) Site 34 were examined based on the two identified contexts mentioned above—industrial and domestic. Both unfused and fused elements of caprines were recorded (Table 4.2), and erupting as well as highly worn teeth were present (Table 4.3), indicating the presence of young animals as well as adults among the slaughtered fauna.

Table 4.1a: Species Frequencies (NISP) at Timna Site 34 (the 2013 Excavations), Divided by Context

Species	*Common Name*	*Industrial*		*Domestic*		*Undefined*	
		NISP	*MNI*	*NISP*	*MNI*	*NISP*	*MNI*
Ovis/Capra	Caprines	31	1	57	3	76	3
Ovis aries	Sheep			2	1	1	1
Capra hircus	Goat	3	1	1	1	8	1
Equus asinus	Donkey	2	1	3	1	6	1
Rodent				2	1	6	1
Mugilidae	Mullet	2		3		3	
Sparidae	Porgy			2			
Clariidae	Catfish	1					
Cypraea annulus	Cowrie			1			
Struthio camelus	Ostrich egg shell			1			
Total		*38*		*72*		*100*	

Table 4.1b: Species Frequencies (NISP) at Timna Site 30 (the 2009 Excavations)

Species	*Common Name*	*NISP*	*MNI*
Bos taurus	Cattle	4	1
Ovis/Capra	Caprines	13	1
Capra hircus	Goat	1	1
Ovis aries	Sheep	2	2
Rodent		1	1

Table 4.2: Fused and Unfused Elements of Caprines from Site 34

		Domestic		*Industrial*		*Not Defined*	
Fusion Age	*Element*	*Unfused*	*Fused*	*Unfused*	*Fused*	*Unfused*	*Fused*
0–6m	Prox radius				1		
6–12m	Dist humerus			1	1		2
6–12m	Pelvis		1		1		
6–12m	Scapula		1		1		
12–18m	1st phalanx		1		4		
12–18m	2nd phalanx		2		1		1
18–30m	Dist tibia				1		
18–30m	Dist metapod	1	1		1		
30–48m	Calcaneus		1				
30–48m	Prox femur			1			2
30–48m	Dist femur						
30–48m	Dist radius		1				
30–48m	Prox tibia		1				
30–48m	Prox ulna						
48+	Prox humerus						

Body part selection of caprines differs significantly between the two activity contexts (industrial vs. domestic, meat-poor vs. meat-rich parts; $\chi^2=5.56$, dF=1, p=0.01). An examination of the skeletal elements frequency (Fig. 4.4; Table 4.4) shows that the industrial contexts are dominated by meat-rich body parts, and a near absence of meat-poor body parts, while the domestic contexts include both groups of body parts—the meat-rich as well as the butchery waste (meat-poor).

Table 4.3: Tooth Wear Stages of Caprines from Site 34

Context	*Tooth*	*Wear Stage*
Undefined	M1	G
Undefined	M2	C
Domestic	M3	F
Domestic	dP4	D

Table 4.4: Frequencies (NISP) of Meat-rich and Meat-poor Parts in the "Domestic" and "Industrial" Contexts at Timna Site 34

Quality	*Body Part*	*Domestic*	*Industrial*
Meat-rich	Upper limbs	5	7
	Axial skeleton	5	6
Meat-poor	Extremities	14	7
	Lower limbs	3	2
	Head	11	1

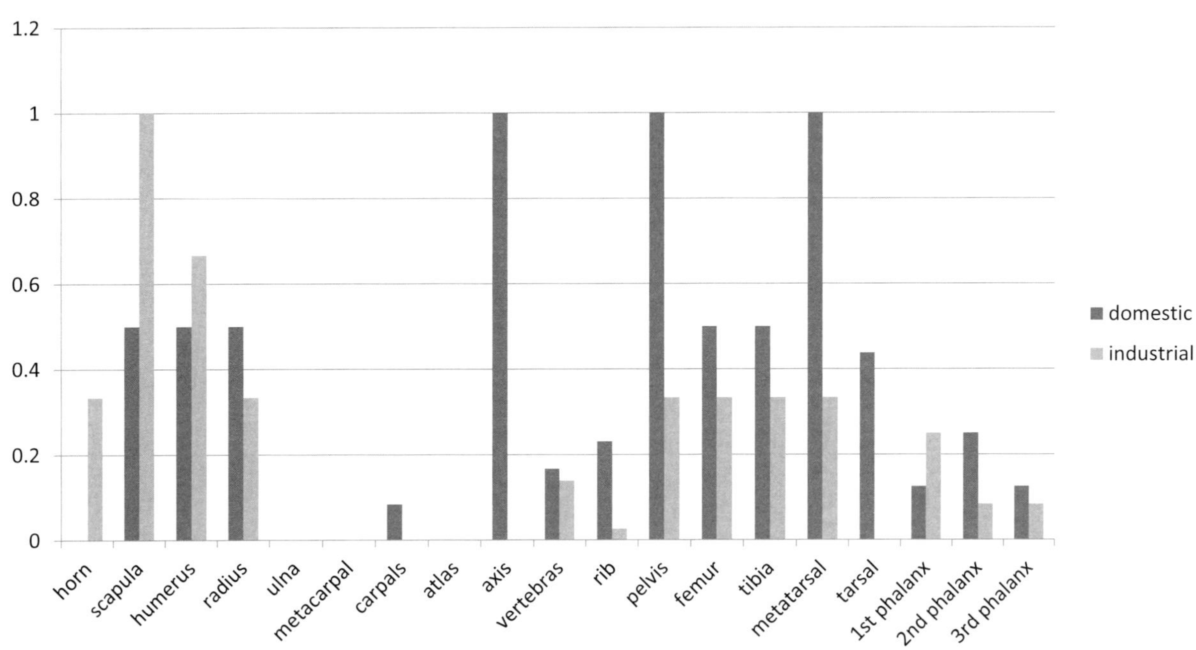

Fig. 4.4: Survivorship of skeletal elements of caprines in "domestic" and "industrial" contexts at Site 34. Skeletal part frequencies (R*i*) were calculated by dividing the observed MNE values of skeletal part *i* (N*i*) in the assemblage (per species), to the expected MNI value (the frequency of skeletal part *i* in one skeleton [E*i*], multiplied by MNI). R*i* = N*i*/(MNI)E*i*; adapted from Andrews (1990); method reviewed in Lyman (2008: 238).

FISH

Methods

All fish remains from Timna Sites 30, 2, 200 and 34 were examined (or re-examined) as part of the current project. Although most of the fish remains excavated by the Aravah Expedition (Sites 30, 2 and 200) were already published by H. Lernau (1988a), some of the original identifications needed revision. In addition, several bones found by the Aravah Expedition are published here for the first time, together with the Site 34 assemblage.

Bones were identified by comparing them to one of the author's (O. Lernau) reference collection of skeletons of modern fish. Names of skeletal elements follow Wheeler and Jones (1989). Information about the fish follows Golani and Darom (1997) and Golani, Öztürk and Basusta (2006). Measurements of selected well-preserved bones were obtained following Morales and Rosenlund (1979). A digital Mitutoyo caliper was used with a precision of 0.1 mm. These measurements allowed reconstructions of sizes of the original fish, using published data (Desse and Desse-Berset 1996) or via comparisons with the reference collection. While sizes have been stated in the following account in terms of exact numbers, the methodology allows for only crude estimates, with at least a 10% margin of error.

Results

The reported 137 fish remains from Timna include: 118 bones excavated by Rothenberg at the smelting camps of Sites 2 and 30, and the Temple of Hathor at Site 200; and 19 bones from the new excavations at Site 34. H. Lernau (1988a) examined fish remains from Sites 2, 200 and 30, assuming they all originated in Late Bronze strata. Based on the new dating of Site 30, it is attributed here to the early Iron Age.

Taxonomic identification was established for 119 bones (Table 4.5, Fig. 4.5a–f). The large majority belonged to the marine family Sparidae (porgies: 88 bones). These bones were further identified either to

Table 4.5: Fish Frequencies (NISP) in Late Bronze and Iron Age sites in Timna*

Site	*Family*	*NISP*
2	Shark	1
	Mullets	1
	Drums	15
	Groupers	1
	Porgies	39
200	Porgies	29
	Drums	1
30	Catfish	2
	Drums	1
	Porgies	18
319	Drums	1
34	Catfish	1
	Mullets	8
	Porgies	2
Total		*119*

* Sites 30, 2 and 200 were re-examined as part of the current study and include additional unpublished data after Lernau (1988a).

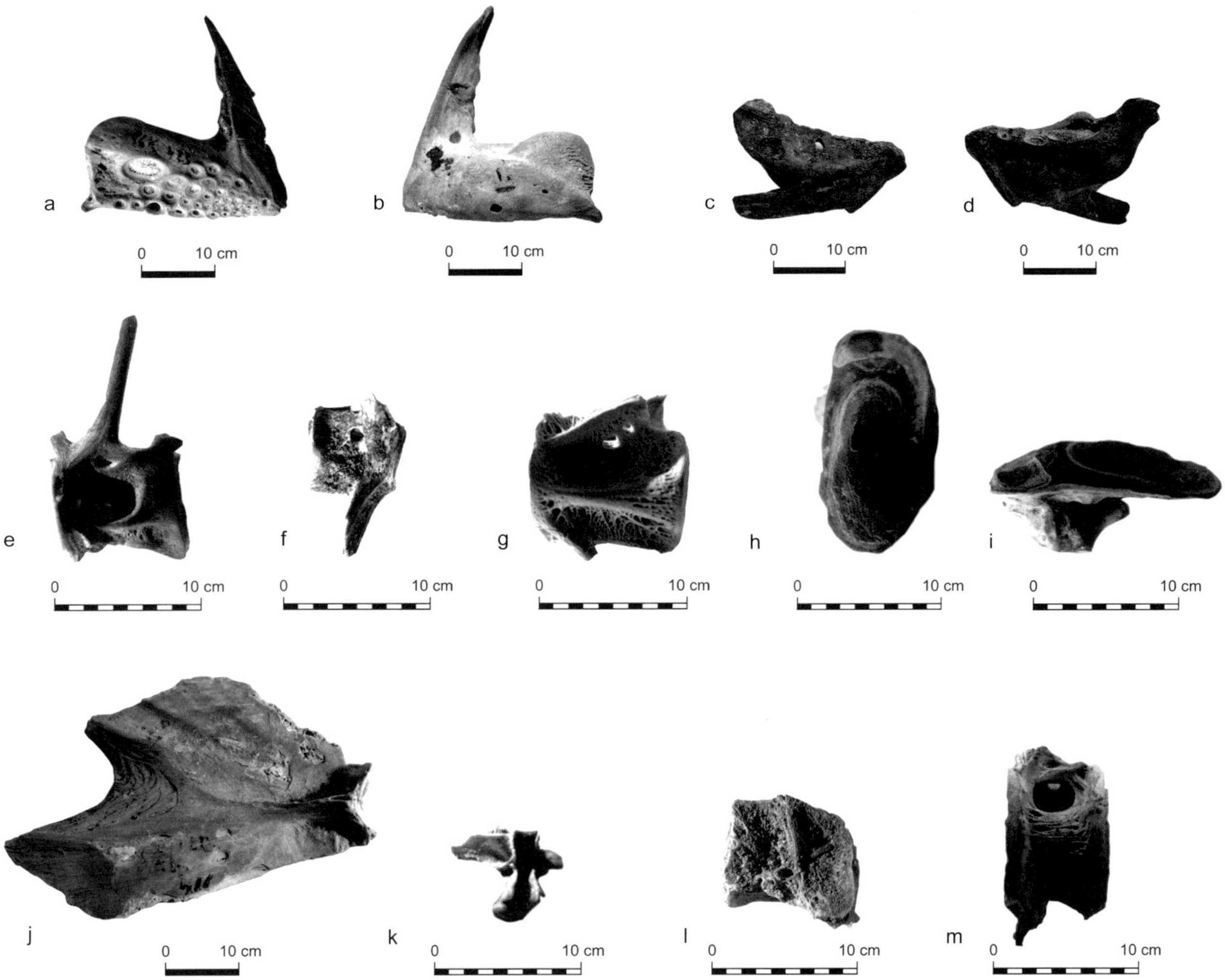

Fig. 4.5: Examples of fish remains from Timna:
(a) From Site 2: A well preserved left premaxilla (anterior part of the upper jaw of the fish) of a gilthead seabream, medial view. Note rows of shallow empty sockets of molariform teeth, with one typical large posterior kidney-shaped tooth.
(b) Same as (a) lateral view. Note the large posterior bulk where the new kidney-shaped molariform tooth is developing.
(c) From Site 30: A poorly preserved right dentary (anterior part of the lower jaw of the fish) of a gilthead seabream, lateral view.
(d) Same as (c), medial view. Note empty sockets of molariform teeth, again with a large posterior kidney-shaped tooth.
(e) From Site 2: Caudal vertebra of a porgy, left lateral view.
(f) From Site 34: A poorly preserved abdominal vertebra of a porgy, right lateral view.
(g) From Site 2: A caudal vertebra of meager, left lateral view.
(h) From Site 2: A well preserved left otolith (sagitta) of a meager, inner surface. Note typical tadpole-like engraving of the convex surface with a large "head" and a "tail" pointing to the left.
(i) Same as (h), lateral view.
(j) From Site 2: A poorly preserved, broken left opercular bone of a meager, medial view.
(k) From Site 34: A poorly preserved small abdominal vertebra of a mullet, superior view.
(l) From Site 34: A very poorly preserved and broken abdominal vertebra of a mullet, right lateral view.
(m) From Site 30. A caudal vertebra of a Nile catfish, left lateral view.

the species *Sparus auratus* (Linnaeus, 1758; gilthead sea bream), which inhabits the Mediterranean and may attain a maximum size of about 70 cm, or to *Rhabdosargus haffara* (Forsskal, 1775; haffara sea bream) which inhabits the Red Sea and is a smaller fish with a maximum size of about 35 cm. Measurements of 60 well preserved bones allowed size estimations of the original fish between 18 and 50 cm with an average size of 32 cm (Table 4.6). A definitive distinction between these species was not realized due to the morphological similarity of their bones. Nevertheless, according to their estimated sizes, at least half of the measured bones had to belong to the larger kind, *S. auratus*. This fish was a favored catch in ancient times, as it is also today, and was in fact the most common identified fish in many excavated sites in Israel in all periods, e.g., in Ashkelon (Lernau 2011) or in the City of David in Jerusalem (Lernau, Reich and Shukron 2008).

Bones of the Sciaenidae (drums) family were second in abundance. There were 17, which were further identified to the species *Argyrosomus regius* (Asso, 1801; Meagre) (Fig. 4.3.6g–j). They were mainly found at Site 2; one bone was found at Smelting Camp Site 30 (Table 4.5). Drums occur in the Mediterranean but not in the Red Sea.

There were nine bones of the Mugilidae (mullet) family, eight at Site 34 and one at Site 2 (Fig. 4.5k-l). Mullets have representative species both in the Mediterranean and in the Red Sea. Accurate taxonomic identification of the bones in order to differentiate the species was not realized.

One bone belonged to the Serranidae (grouper) family, and one calcified center of a vertebra belonged to a shark. This centrum had an irregular perforation through its middle, and most likely was employed as a bead in a necklace. Groupers and sharks inhabit both the Mediterranean and the Red Sea.

Three bones belonged to a freshwater fish, the Nile catfish *Clarias gariepinus* (Burchell, 1822) (Fig. 4.5m).

DISCUSSION: IMPLICATIONS FOR THE STUDY OF EARLY SOCIETIES ENGAGED IN METAL PRODUCTION

DIET AND FOOD PROVISIONING

Studies of mammalian remains from the smelting camps at Timna and Faynan show that the animal subsistence economy in the Late Bronze/Iron Ages was based on livestock animals, which is a common diet in the Bronze and Iron Ages in the southern Levant (Sasson 2010; Sapir-Hen, Gadot and Finkelstein 2014). Livestock animals from the various sites include those that are less suitable to the local climate–pig and cattle. These finds reinforce the perception that the decision of what to raise, in the Late Bronze–

Table 4.6: Measurements of Fish from Timna

Family		*NISP*	*Estimated Sizes of Fish, cm Average (range)*
Marine fish			
Sparidae	*Sparus auratus* or *Rhabdosargus haffara*	88	32 (18-50)
Sciaenidae	*Argyrosomus regius*	17	36 (24-46)
Mugilidae		9	22 (13-31)
Serranidae		1	65
Shark		1	
Freshwater fish			
Clariidae	*Clarias gariepinus*	3	32, 78
No identification		18	

Iron Age, is not dependent on the local climate but on human decision (Sapir-Hen, Gadot and Finkelstein 2014). The near-absence of wild game is also intriguing. At Timna sites, it might be attributed to the small sample that leads to the under-representation of rare species and thus resulting in a smaller diversity, but the situation is also similar in Faynan. The scarcity of occasional hunting could be related to the work force in the large-scale industrialized production, which is provisioned with food supply; a provisioned population is expected to have lower diversity of species available to the consumer, due to the indirect contact with herd or herders (Zeder 1991; see also Muniz and Levy 2014). This further supports the assumption that the copper smelting camps represent a designated area for industrial activities conducted by parts of the society, while most of the population, which probably did practice pastoralism, dwelled in tents in the vicinity of these smelting camps.

The presence of fish remains in the copper smelting sites indicates that fish were brought especially for the workers. Potential sources of fish for Timna were the Red Sea, about 30 km to the south, the Mediterranean and its coastal rivers, about 250–300 km to the north, the Jordan River system, about 300 km to the north, and the eastern branch of the Nile, which at the time was about 400 km to the east of Timna. Supplies brought to the copper mines from any of these had to be transported across long distances of desert, presumably by caravans of donkeys (or mules—not distinguishable in faunal remains) or camels.

Both the Red Sea and the Mediterranean carry a large number of different species of fish. Some of them inhabit both seas, some only one. This is also true for entire families of fish—some families have representatives in both seas, some only in one. It follows that an exact source can be allocated to a bone of a fish that has been identified taxonomically to the level of species, if this species occurs in only one of the seas; alternatively, if the bone could be identified only to the level of the family, its source can be decided provided that the family is confined to one sea. The five marine families identified in the Timna assemblage exist in the Mediterranean. Drums occur *only* in the Mediterranean. Mullets, Groupers and sharks occur in both seas, as do porgies. But the bones of porgies could be identified morphologically to two similar species, to *S. auratus*, which is restricted to the Mediterranean, or to *R. haffara*, which is restricted to the Red Sea. When applying the criterion of size, half the bones of porgies had to belong to *S. auratus*, as explained above. It stands to reason that all the porgies were obtained from the same source, which means that the shipment from the Mediterranean included smaller specimens of *S. auratus* along with the larger ones, as well as drums. If we accept that the trade with the northern coast probably also included Mediterranean species of mullet, shark and grouper, this would mean that Timna's supply of fish relied only on the Mediterranean. This conclusion is reinforced by the striking differences between the Timna fish assemblage and the one recently found in the Roman/Early Islamic mines of Nahal ʿAmram (Chapter 13, this volume). The latter represents an "expected" assemblage of Red Sea fish.

The origin of the bones of the freshwater catfish *C. gariepinus* presents a similar problem. This fish belongs to the Nilotic fauna and has spread to freshwater bodies along the Levant as far as northern Syria. It, therefore, may have been brought to Timna from the Nile, from the Jordan River system, or from coastal rivers along the Mediterranean. In compliance with the suggested scenario for supply of fish from the Mediterranean, we suggest that catfish was brought to Timna from coastal rivers, together with marine fish.

We chose Site 34 to study food distribution at a specific site. The approach employed in season 2013, of comprehensive retrieval of animal remains under tight control over context, allowed us to reconstruct social diversity on an intra-site level. The site was termed "Slaves' Hill" by N. Glueck (1935), who assumed that its inhabitants had been slaves engaged in copper production. The unique location of the site on a hilltop surrounded by cliffs and a wall was interpreted by Glueck as evidence of an effort to prevent the "slaves" from escaping into the open desert. Studying animal economy at the site provides a unique opportunity to examine this assumption.

The primary difference between the industrial and domestic contexts is evident not in the species diversity but in the choice of body parts consumed (see above), and two groups of people at the site can be discerned through the analysis of meat procurement methods: (1) the "industrial" contexts—workers engaged in direct smelting, were provided with the better cuts of meat, indicating that it represented highly skilled crafts persons who were treated as such by receiving selected portions of the higher quality meat. (2) The "domestic" contexts—workers engaged in auxiliary activities, probably including the preparation of food, as is evident by the presence of butchery waste as well. The resolution of faunal analysis did not allow for similar observations at the contemporaneous site of Khirbat en-Nahas in Faynan, although the architectural finds point to social stratification of the population (Muniz and Levy 2014). While the authors suggest that the site was provisioned with food, they found no evidence of differences between activity areas in terms of food distribution. As their otherwise insightful study does not provide full data on skeletal element frequencies (such as the frequencies of vertebras, ribs and skulls are not accounted for) and the analysis is based on site-level observations with emphasis on areas, patterns might be missed. As suggested by our study, potential differences between local contexts (rather than between areas) might relate to different levels of specialization among workers at the same site.

Our finds are in contrast to the common perception of furnace operators as a low class, cheap labor force (or slaves), as inferred by the common name of Site 34 – "Slaves' Hill." The contrast is also implied by the presence of fish remains and abundant seeds of cereals and fruits (including Mediterranean species such as grape and pistachio, see Ben-Yosef 2016) at the site, albeit in both contexts. These remains indicate a substantial effort to vary the diet of the workers by transporting food from a great distance, most probably using different food procurement systems via trade. The notion that the people engaged in smelting activities were not poor slaves but highly skilled crafts persons is also supported by ethnographic evidence of iron smelting, suggesting that primary smelting tasks require long apprenticeship and could be done only by specialists (e.g., de Barros 2000) (for level of specialization required for copper smelting in Late Bronze Age Timna, see Merkel 1990).

Magic and cult are also associated with metallurgical craftsmanship which is correlated to social status of metal workers (e.g., Budd and Taylor 1995). This spiritual aspect of the work is usually considered archaeologically invisible, but it might be reflected in the faunal assemblage of Site 34 as presented here. It is also worth noting that at Site 34 Rothenberg identified a "high place"—a rock platform with abundant cup marks at the northern edge of the site identified as a cult-related area (Rothenberg 1967: 27).

Our results imply a different purpose for the unique location of the site—a hilltop surrounded by cliffs and a wall—than that suggested by Glueck (1935). As copper was the raw material for preparing the most advanced weapons and agricultural tools of the time (for the slow advancement of iron in the early Iron Age see Gottlieb, this volume, and references therein), we suggest the wall was built to protect the valued metal from the rest of the population and from external enemies; it is also likely that the location played a role in segregating the most skilled workers, in order to further emphasize their importance as a distinct social stratum, and to protect valued technological knowledge which was probably a commodity in itself.

The copper ore of the Arabah Valley was exploited over eight millennia, but related faunal studies are rare. Except for the studies on Late Bronze/Iron Age assemblages summarized here, Muniz (2007) investigated an Early Bronze faunal assemblage from a secondary metal production camp in Faynan, and Kolska-Horwitz *et al.* (this volume) investigated an assemblage from within a mine in the mining complex of Nahal ʿAmram, 15 km south of Timna. The latter is dated to the Early Islamic period and originated from within a mine gallery, contra all of the Late Bronze/Iron Age assemblages discussed above, which originate in smelting contexts. The analysis of fish remains from Nahal ʿAmram reveals a different taxonomic pattern than the one in the Iron Age/Late Bronze smelting camps. They include four kinds of fish that occur in the Red Sea but not in the Mediterranean, and two families that are found in both seas. Thus, the assemblage implies that the workers in the copper mines at that time consumed fish caught in the Red Sea, and this might have been their only source of fish. This difference between the Late Bronze/Iron Age and Early Islamic assemblages should be attributed to different food procurement systems, and might reflect distinct trade connections, cultural and political ties, and/or different status of the working force. The study of mammal remains from the Nahal ʿAmram mine (*ibid.*) includes caprines and camel, both consumed within the mine. While the caprines are represented by the better body parts (although only NISPs are provided, hence it cannot be compared to our work), they also demonstrate evidence for human gnawing and bone fracturing, indicating intensive exploitation of the food (*ibid*).

ETHNICITY IN LATE BRONZE/EARLY IRON AGE COPPER PRODUCTION SITES IN THE ARABAH VALLEY

Observations from the surveyed data contribute to an ongoing debate regarding socio-political processes in the mining districts of the southern Levant at the turn of the first millennium BCE. While the temple of Site 200 (Rothenberg 1988) yielded clear evidence of Late Bronze Age Egyptian presence in the valley, undoubtedly related to copper production, the new radiocarbon dates from Sites 30 and 34 have suggested that most of the smelting sites in the region were operated by local societies during the early Iron Age (Ben-Yosef *et al.* 2012; Ben-Yosef 2016), implying that the role of Egyptians in operating the mines was considerably less important than previously assumed. However, new evidence suggests the possible involvement of Egyptians also during the Iron Age (22nd Dynasty, late tenth–ninth centuries BCE) (Fantalkin and Finkelstein 2006; Ben-Yosef *et al.* 2010; Munger and Levy 2014), and the presence of Egyptians at the sites is still an open question.

New insights regarding the ethnic identity of the metal workers can be gleaned from the data (Sapir-Hen and Ben-Yosef 2014). These are based on the notion that food-ways and cuisine are related to cultural affiliation. As caution should be practiced when connecting presence or absence of a species with ethnic identity (e.g., Sapir-Hen *et al.* 2013: and references therein), other factors influencing the species range should be accounted for, such as climate and site function (Hesse and Wapnish 1997). Examining the finds from a single period originating in a specific region with specific climatic conditions, and of a similar nature of activity (i.e., production of metal), supports the correlation between food choice and culture. This is highly applicable regarding the appearance of pigs which are not considered suitable for nomadic life or for the local desert climate (e.g., Zeder 1996; Hesse and Wapnish 1998; Grigson 2007). Pigs (domestic) are found only at the Late Bronze Timna Site 2 (Bar-Oz and Erickson-Gini in press, NISP=39, 79% caprines and 21% pigs), and not at other sites. During the Late Bronze, pork consumption is an Egyptian culinary habit, not practiced by the local population but only in Egyptian strongholds in Canaan (Sapir-Hen *et al.* 2013). Thus we suggest that presence of pigs at Late Bronze Age Site 2 in Timna indicate the presence of Egyptians, also corroborating the interpretation of that site as an Egyptian-controlled smelting camp (e.g., Rothenberg 1990; 1999). Since pigs were not included in Egyptian religious rituals and are considered to be the food of the less wealthy in Egypt (Hecker 1982), the absence of pigs in the contemporary temple at Site 200 (Lernau 1988b) does not imply the absence of

Egyptians there. Later, in early Iron Age Canaan, pork was still not consumed by the local population of Canaan (but only in the Philistine urban centers) (Sapir-Hen *et al.* 2013; Maeir, Hitchcock and Horwitz 2013). This corresponds to the absence of pigs from the other smelting sites (Sites 30, Site 34 and Khirbat en-Nahas), implying the absence of Egyptians in those locations in the early Iron Age.

CONCLUSIONS

In this paper we summarized the current state of research into the faunal remains in the Late Bronze/Iron Age copper production sites of the Arabah Valley. We demonstrated the potential of such research to illuminate various aspects of the ancient societies engaged in highly specialized technology. We also demonstrated the importance of careful contextual analysis for discerning fine patterns in food consumption and distribution. The Arabah Valley is an excellent proxy for the study of ancient metallurgy and social organization from synchronic and diachronic perspectives. Still, to date, aside from two exceptions(Horwitz, Avner and Lernau this volume: Roman and Early Islamic; Muniz 2007: Early Bronze) Roman and Early Islamic; Muniz 2007: Early Bronze) mining and smelting sites dated to other periods have not been studied from the faunal perspective. Our results call for further investigation of diet of highly specialized communities, and especially metal workers, through the lens of faunal remains. This study is part of an ongoing zooarchaeological project concerning archaeometallurgical sites across cultures and periods.

ACKNOWLEDGMENTS

We thank the staff of the Central Timna Valley Project and the management of the Timna Park for help and support of the field work. This research is supported by Marie Curie FP7-PEOPLE-2012-CIG grant #334274 and the Israel Science Foundation grant #1880/17.

REFERENCES

Andrews, P. 1990. *Owls, Caves and Fossils*. Chicago.

Avner, U. 2014. Egyptian Timna Reconsidered. In: Tebes, J.M., ed. *Unearthing the Wilderness: Studies on the History and Archaeology of the Negev and Edom in the Iron Age*. Leuven-Paris-Walpole: 103–162.

Bar-Oz, G. and Erickson-Gini, T. In press. Late Bronze faunal remains from Timna, Site 2. In: Bar-Oz, G. and Horwitz, L.K., eds. *Discovering Noah's Ark: Zooarchaeology of the Holyland*. Jerusalem.

Ben-Yosef, E. 2010. *Technology and Social Process: Oscillations in Iron Age Copper Production and Power in Southern Jordan* (Ph.D. dissertation, University of California). San Diego.

Ben-Yosef, E. 2016. Back to Solomon's Era: Results of the First Excavations at Slaves' Hill (Site 34, Timna, Israel). *Bulletin of the American Schools of Oriental Research* 376: 169–198.

Ben-Yosef, E., Levy, T.E., Higham, T., Najjar, M. and Tauxe, L. 2010. The Beginning of Iron Age Copper Production in the Southern Levant: New Evidence from Khirbat al-Jariya, Faynan, Jordan. *Antiquity* 84(325): 724–746.

Ben-Yosef, E., Shaar, R., Tauxe, L. and Ron, H. 2012. A New Chronological Framework for Iron Age Copper Production in Timna (Israel). *Bulletin of the American Schools of Oriental Research* 366: 1–41.

Binford, L.R. 1978. *Nunamiut Ethnoarchaeology*. New York.

Budd, P. and Taylor, T. 1995. The Faerie Smith Meets the Bronze Industry: Magic Versus Science in the Interpretation of Prehistoric Metal-Making. *World Archaeology* 27(1): 133–143.

de Barros, P. 2000. Iron Metallurgy: Sociocultural Context. In: Bisson, M.S., Childs, T.S., de Barros, P. and Holl, A.F.C., eds. *Ancient African Metallurgy: The Socio-Cultural Context*. Walnut Creek, CA: 147–198.

Desse, J. and Desse-Berset, N. 1996. Ostéométrie et archéozoologie de la daurade royale (Sparus aurata, Linné 1758) (Fiches d'ostéologie animale pour l'archéologie 9). Juan-les-Pins.

Erickson-Gini, T. 2014. Timna Site 2 Revisited. In: Tebes, J.M., ed. *Unearthing the Wilderness: Studies on the History and Archaeology of the Negev and Edom in the Iron Age*. Leuven-Paris-Walpole: 47–84.

Fantalkin, A. and Finkelstein, I. 2006. The Sheshonq I Campaign and the 8th Century BCE Earthquake: More on the Archaeology and History of the South in the Iron I–IIA. *Tel Aviv* 33: 18–42.

Glueck, N. 1935. Explorations in Eastern Palestine, II. *Annual of the American Schools of Oriental Research* 15. New Haven: 1–288.

Golani, D. and Darom, D. 1997. *Handbook of the Fishes of Israel*. Jerusalem.

Golani, D., Öztürk, B. and Basusta, N. 2006. *Fishes of the Eastern Mediterranean*. Istanbul.

Grigson, C. 2007. Culture, Ecology and Pigs from the 5th to 3rd Millennium BC around the Fertile Crescent. In: Albarella, U., Dobney, K., Ervynck, A. and Rowley, P., eds. *Pigs and Humans. 10000 Years of Interaction*. Oxford: 83–108.

Grigson, C. 2012. Camels, Copper and Donkeys in the Early Iron Age of the Southern Levant: Timna Revisited. *Levant* 44(1): 82–100.

Hakker-Orion, D. 1984. The Role of the Camel in Israel's Early History. In: Clutton-Brock, J. and Grigson, C., eds. *Early Herders and their Flocks* (Animals and Archaeology 3). Oxford: 207–212.

Hecker, H.M. 1982. A Zooarchaeological Inquiry into Pork Consumption in Egypt from Prehistoric to New Kingdom Times. *Journal of the American Research Center in Egypt* 19: 59–71.

Hesse, B. and Wapnish, P. 1997. Can Pig Remains Be Used for Ethnic Diagnosis in the Ancient Near East? In: Silberman, N.A. and Small, D., eds. *The Archaeology of Israel—Contructing the Past, Interpreting the Present*. Sheffield: 238–270.

Hesse, B. and Wapnish, P. 1998. Pig Use and Abuse in the Ancient Levant: Ethnoreligious Boundary-building with Swine. In: Nelson, S.M., ed. *Ancestors for the Pigs: Pigs in Prehistory*. Philadelphia.

Lernau, H. 1988a. Fish Remains. In: Rothenberg, B., ed. *Researches in the Arabah 1959–1984*, Vol. 1: *The Egyptian Mining Temple at Timna*. London: 241–246.

Lernau, H. 1988b. Mammalian Remains. In: Rothenberg, B., ed. *Researches in the Araba 1959–1984*, Vol. 1: *The Egyptian Mining Temple at Timna*. London: 246–253.

Lernau, O. 2011. Fish Remains. In: Stager, L.E., Master, D.M. and Schloen, J.D., eds. *Ashkelon 3: The Seventh Century B.C.* Winona Lake: 645–658.

Lernau, O., Reich, R. and Shukron, E. 2008. Bullae and Bones of Fish in the City of David, Jerusalem: A Preliminary Report. In: Béarez, P., Grouard, S. and Clavel, B., eds. *Archéologie du poisson. 30 ans d'archéo-ichthyologie au CNRS*. Antibes: 391–398.

Levy, T.E., Najjar, M. and Ben-Yosef, E. 2014. *New Insights into the Iron Age Archaeology of Edom, Southern Jordan—Surveys, Excavations and Research from the Edom Lowlands Regional Archaeology Project (ELRAP)*. Los Angeles.

Lyman, R.L. 1994. *Vertebrate Taphonomy*. Cambridge.

Lyman, R.L. 2008. *Quantitative Paleozoology*. Cambridge.

Maeir, A.M., Hitchcock, L.A. and Horwitz, L.K. 2013. On the Constitution and Transformation of Philistine Identity. *Oxford Journal of Archaeology* 32(1): 1–38.

Merkel, J.F. 1990. Experimental Reconstruction of Bronze Age Copper Smelting Based on Archaeological Evidence from Timna. In: Rothenberg, B., ed. *Researches in the Arabah 1959–1984*, Vol. 2: *The Ancient Metallurgy of Copper*. London: 78–122.

Mintz, S.W. and Du Bois, M. 2002. The Anthropology of Food and Eating. *Annual Review of Anthropology* 31: 99–119.

Morales, A. and Rosenlund, K. 1979. *Fish Bone Measurements*. Copenhagen.

Munger, S. and Levy, T.E. 2014. The Iron Age Egyptian Amulet Assemblage from the Edom Lowlands Regional Archaeology Project. In: Levy, T.E., Najjar, M. and Ben-Yosef, E., eds. *New Insights into the Iron Age Archaeology of Edom, Southern Jordan—Surveys, Excavations and Research from the Edom Lowlands Regional Archaeology Project (ELRAP)*, Vol. 2. Los Angeles: 741–765.

Muniz, A. 2007. *Feeding the Periphery: Modeling Early Bronze Age Economies and the Cultural Landscape of the Faynan District, Southern Jordan* (Ph.D. dissertation, University of California). San Diego.

Muniz, A. and Levy, T.E. 2014. Feeding the Iron Age Metalworkers at Khirbat en-Nahas—Zooarchaeological Perspectives. In: Levy, T.E., Najjar, M. and Ben-Yosef, E., eds. *New Insights into the Iron Age Archaeology of Edom, Southern Jordan—Surveys, Excavations and Research from the Edom Lowlands Regional Archaeology Project (ELRAP)*. Los Angeles: 627–664.

Payne, S. 1973. Kill-off Patterns in Sheep and Goats: The Mandibles from Asvan Kale. *Anatolian Studies* 23: 281–303.

Pearson, M.P. 2003. Food, Identity and Culture: An Introduction and Overview. In: Pearson, M.P., ed. *Food, Culture and Identity in the Neolithic and Early Bronze*. Oxford: 1–30.

Rothenberg, B. 1967. *Negev: Archaeology in the Negev and the Arabah*. Tel Aviv (Hebrew).

Rothenberg, B. 1972. *Were These King Solomon's Mines? Excavations in the Timna Valley*. New York.

Rothenberg, B., ed. 1988. *Researches in the Araba 1959–1984, Vol. 1: The Egyptian Mining Temple at Timna*. London.

Rothenberg, B., ed. 1990. *Researches in the Araba 1959–1984, Vol. 2: The Ancient Metallurgy of Copper*. London.

Rothenberg, B. 1999. Archaeo-metallurgical Researches in the Southern Arabah 1959–1990, Part 2: Egyptian New Kingdom (Ramesside) to Early Islam. *Palestine Exploration Quarterly* 131: 149–175.

Sapir-Hen, L., Bar-Oz, G., Gadot, Y. and Finkelstein, I. 2013. Pig Husbandry in Iron Age Israel and Judah: New Insights Regarding the Origin of the "Taboo." *Zeitschrift des Deutschen Palästina-Vereins* 129: 1–20.

Sapir-Hen, L., Bar-Oz, G., Sharon, I., Gilboa, A. and Dayan, T. 2014. Food, Economy and Culture at Tel Dor, Israel: A Diachronic Study of Faunal Remains from 15 Centuries of Occupation. *Bulletin of the American Schools of Oriental Research* 371: 83–101.

Sapir-Hen, L. and Ben-Yosef, E. 2013. The Introduction of Domestic Camels to the Southern Levant: Evidence from the Aravah Valley. *Tel Aviv* 40(2): 278–285.

Sapir-Hen, L. and Ben-Yosef, E. 2014. The Socioeconomic Status of Iron Age Metal Workers: Animal Economy in the "Slaves' Hill", Timna, Israel. *Antiquity* 88: 775–790.

Sapir-Hen, L., Gadot, Y. and Finkelstein, I. 2014. Environmental and Historical Impacts on Long Term Animal Economy: The Southern Levant in the Late Bronze and Iron Ages. *Journal of Economic and Social History of the Orient* 57(5): 703–744.

Sasson, A. 2010. *Animal Husbandry in Ancient Israel: A Zooarchaeological Perspective on Livestock Exploitation, Herd Management and Economic Strategies*. London.

Symmons, R. 2005. New Density Data for Unfused and Fused Sheep Bones, and a Preliminary Discussion on the Modeling of Taphonomic Bias in Archaeofaunal Age Profiles. *Journal of Archaeological Science* 32: 1691–1698.

Wheeler, A. and Jones, A.K. 1989. *Fishes,* Vol. 94. Cambridge.

Zar, J.H. 2005. Spearman Rank Correlation. In: Armitage, P. and Colton, T., eds. *Encyclopedia for Biostatistics*, Vol. 7. (DOI: 10.1002/0470011815.b2a15150)

Zeder, M.A. 1991. *Feeding Cities: Specialized Animal Economy in the Ancient Near East*. Washington, DC.

Zeder, M.A. 1996. The Role of Pigs in Near Eastern Subsistence: A View from the Southern Levant. In: Seger, J.D., ed. *Retrieving the Past: Essays on Archaeological Research and Methodology in Honor of Gus W. Van Beek*. Winona Lake: 297–312.

Zeder, M.A. 2006. Reconciling Rates of Long Bone Fusion and Tooth Eruption and Wear in Sheep (Ovis) and Goat (Capra). In: Ruscillo, D., ed. *Recent Advances in Ageing and Sexing Animal Bones*. Oxford: 297–312.

CHAPTER 5

THE SINAI-ARABAH COPPER AGE EARLY PHASE (CHALCOLITHIC) MINE T EXCAVATIONS

Tim Shaw and Alexandra Drenka

This paper presents a detailed report of the excavations of Mine T in the Timna Valley. The mine is dated to the Early Phase of the Sinai-Arabah Copper Age (Chalcolithic). The results are discussed in light of current knowledge of the history and technological developments of copper mining in the Timna Valley, including a comprehensive survey of previous work by the Aravah Expedition in the region.

INTRODUCTION

The southern Arabah is a unique museum of the history of mining. Here, preserved in one area, is the evidence for almost all periods of mining, from some of the earliest known copper mines on earth, dating to sometime in the sixth–fifth millennium BCE, right up to modern day copper mines. The interesting and unique feature of this preservation is that each subsequent period of work in the area has left at least some of the mines of its predecessors undisturbed or only slightly disrupted.

The earliest mines investigated thus far in the Timna Valley date to the "Sinai-Arabah Copper Age—Early Phase I" (Late Pottery Neolithic period). These mines were "pit mines" in relatively unconsolidated material, recovering eluvial copper ore nodules. These mines were in Areas A, F and G (Conrad and Rothenberg 1980) among others.

The next period of mining could be dated to a late stage of the Sinai-Arabah Copper Age—Early Phase II (Chalcolithic, fifth–fourth millennium BCE). There are numerous well-preserved mines of this period, particularly in Areas S and T. In some cases, there was a later overlay of workings by Egyptian New Kingdom mining expeditions of the 14th–12th centuries BCE, and sometimes by the Romans and even later expeditions. These are the first true underground mines, with quite extensive networks of underground activities, connected to the surface by numerous short shafts. One set of such workings,[1] Mine T (Fig. 5.1), has been partially cleared and investigated and are the focus of this publication. It is possible that this mining phase continued intermittently through to the end of the Chalcolithic period. The discovery of mine workings and smelting sites of the Sinai-Arabah Copper Age—Middle and Late Phase provided evidence for two further prehistoric periods of mining activities in this region.

In 1974–1976, the first mine clearances of the Timna mines by the Aravah Expedition were directed by H.G. Conrad, director of the German Mining Museum Bochum (Conrad and Rothenberg 1980), including mine workings in Area T (Figs. 5.2, 5.3, 5.4). In 2001, additional clearances were undertaken in Mine T, directed by Beno Rothenberg and supervised by Alexandra Drenka (2003), and further

1 For technical terms related to mining and mining archaeology, see glossary at the end of this paper.

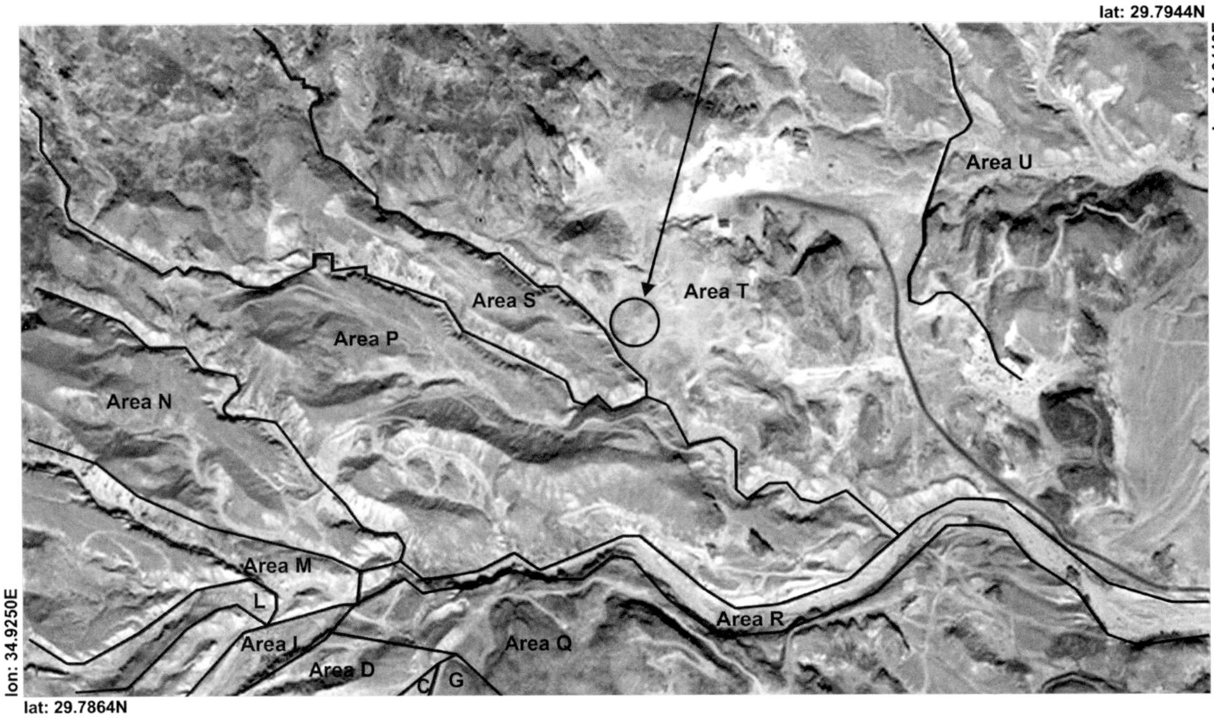

Fig. 5.1: Google photograph with areas defined by the Bochum excavations in the 1970s (Google Earth & CTS); the arrow indicates location of the T mines reported in this publication.

investigated in 2002 by C. Tim Shaw and Alexandra Drenka (Fig. 5.5). The following report covers both these clearances and investigations of 1974–1976 and 2001–2002.

SINAI–ARABAH COPPER AGE MINING—EARLY PHASE (CHALCOLITHIC)

This period of mining ties in with Site 39, the smelting site of the Sinai-Arabah Copper Age—Early Phase (Chalcolithic) at Timna. The dating of this smelting site is based on finds of pottery and flint (Rothenberg 1980: 181–185; Rothenberg and Glass 1992: 146), and was further confirmed by ^{14}C dating to the middle of the fifth millennium BCE (Rothenberg and Merkel 1998: 3; see Ben-Yosef *et al.* 2008 for a thorough discussion on this dating). On technological grounds as well, Mine T would appear to be a later period of mining than the "pit mining" in Areas E, F and G, since the Chalcolithic miners did mine underground for their ore and in the process created extensive networks of galleries.

THE ARCHAEOLOGICAL EXCAVATIONS IN AREA T (FIG. 5.5)

Area T is a plateau, which in places is covered by a thin layer of Neogene undifferentiated conglomerate. This layer is generally no more than 0.50 m thick and is located on top of the soft white sandstones of the Lower Cretaceous Avrona Formation.

In 1976, prior to the excavations, numerous "plates" could be seen on the surface of the plateau, conspicuous by their lighter color and the lack of rocks on their surfaces. Around many of these "plates" were dumps of rock fragments from underground mining operations. The plates in the area were initially numbered sequentially from T_1 to T_{45}, followed by a new sequence starting at T_{1000}. Codes for other designations, such as shaft "T_N," tunnel "j," gallery "e" and opening "a6" are used in the relevant figures.

Fig. 5.2: The cleared underground workings in Areas S and T (Conrad and Rothenberg 1980; Shaw 2003).

Research into the underground workings in Area T began during the Aravah Expedition's campaign of March–July 1976, when "plates" T1 and T31 were excavated and turned out to be sand-filled ancient mine shafts. The excavations were continued in the campaign of September–November of that year, with the collaboration of the German Mining Museum Bochum. During this second season, excavations continued at Shaft T31, and new Shafts T40, T42 and T43 were cleared. T41 was also excavated, but proved to be a short, uncompleted shaft of Egyptian New Kingdom style.

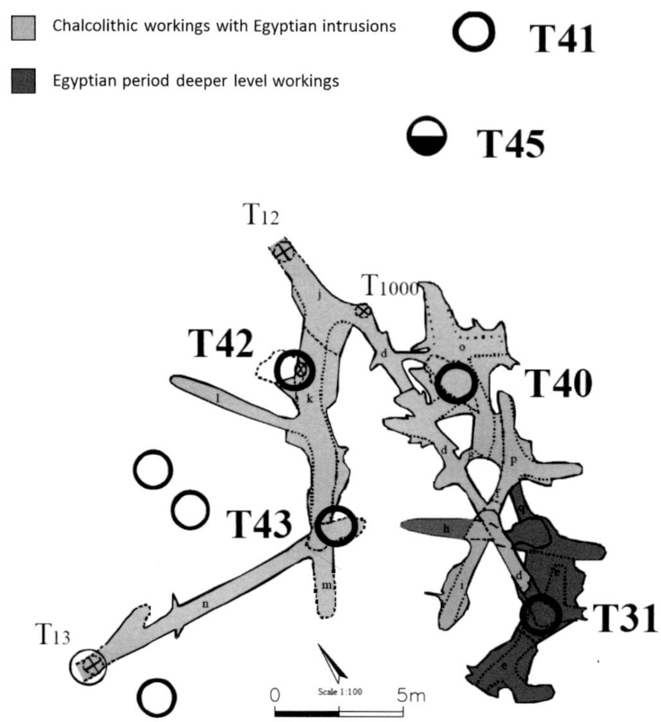

Fig. 5.3: TN mine workings as excavated in the 1970s (Conrad and Rothenberg 1980).

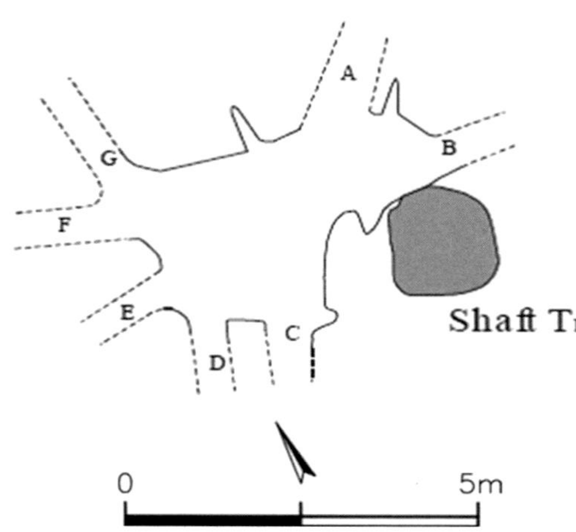

Fig. 5.4: Shaft T1 as excavated in the 1970s (Conrad and Rothenberg 1980).

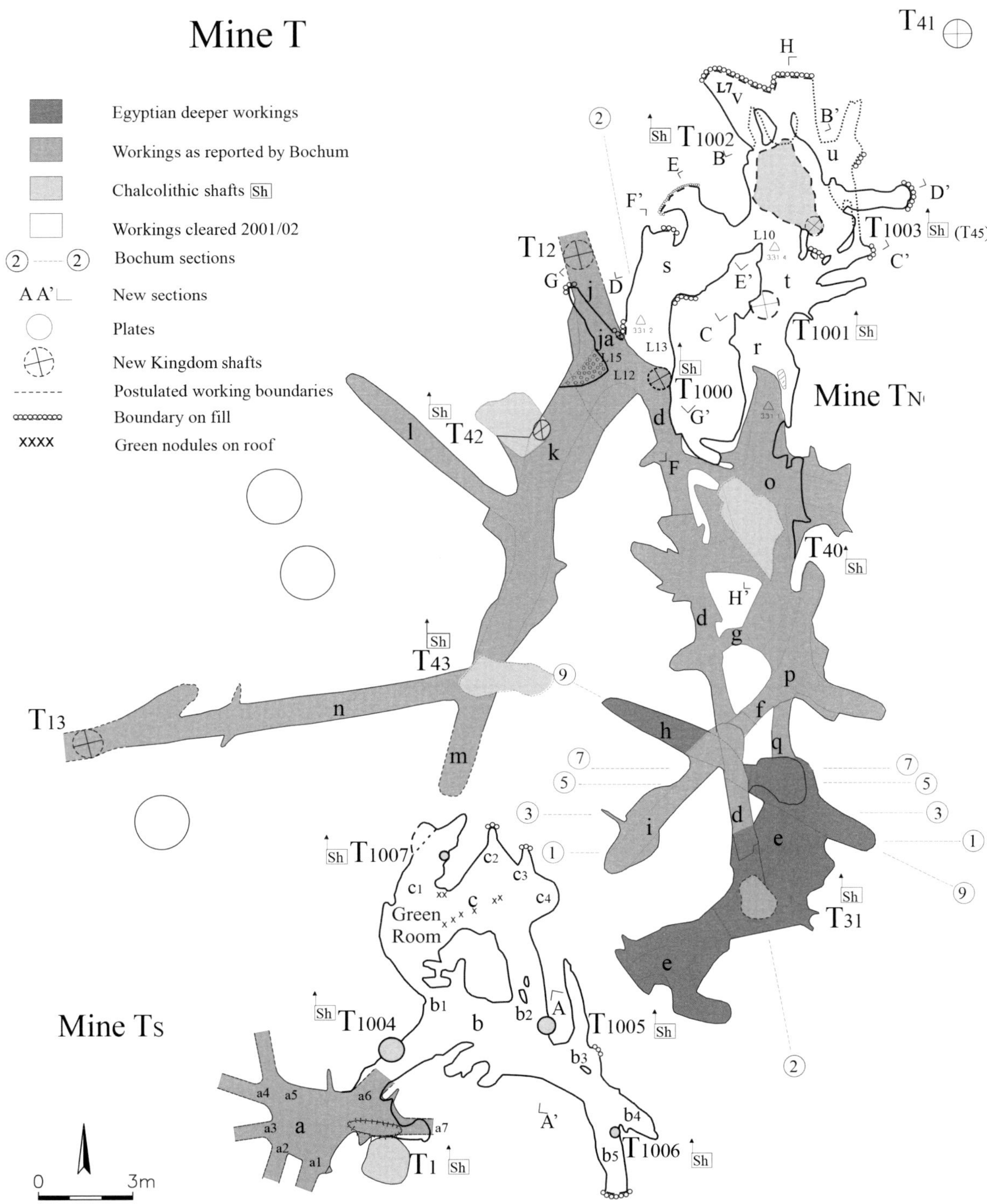

Fig. 5.5: The full extent of the clearing of Mines TN and TS; mapping by Shaw (partially based on unpublished reports of the Aravah Expedition).

In addition, the system of more or less horizontal underground tunnels or galleries, connected to T_{31}, and joining with T_{40}, T_{42}, and T_{43}, was partly cleared. This set of shafts and underground workings is here called T_N. A partial clearance, described below, was also carried out at T_I and in the underground workings radiating from that shaft (the T_S workings) (Fig. 5.4).

During the 1976 excavations, the entire area of the lower level workings E of Shaft T_{31} was unblocked to the footwall and it was demonstrated that there were no connections to the south of it (the extent of the workings cleared at that time is shown in Fig. 5.3 in dark gray). Continuing the clearing from Shaft T_{31} to the north, the workings that connected with Shafts T_{40}, T_{42} and T_{43} were opened and cleared of debris. Shaft T_{13}, as shown in Plan 3, was never reached or cleared. It is only an assumed shaft and it is not as yet known whether it actually exists. The same is true for Shaft T_{12}. There is a significant discrepancy in that area between the reports of the 1976 and the 2001–2002 excavations (cf. "j," "j_a," L15 in Fig. 5.5). Furthermore, tunnel "j_a," in which Shaft T_{12} is shown in the plan of 1976, could not be entered and so whether or not a shaft does exist remained uncertain.

Tunnel "m," which runs south from Shaft T_{43}, was also not fully cleared. In the excavation plan of 1976 (see Fig. 5.3) it is drawn in dotted lines, indicating that its length and dimensions were assumed and not measured; thus, it is possible that it continues in the direction shown to connect with the T_S workings. In the intervening years since the 1976 excavations, Shaft T_{43} has filled in again and is no longer open. It is still possible to reach the bottom of Shaft T_{43}, but the tunnel going south from there is now completely filled. Part of Tunnel "n," going west, is still open, but it is also not possible to investigate it further without re-clearing.

Shaft T_{1000} (see Fig. 5.5) was recognized in 1976 from fill in the hanging wall, but was not cleared. It has now been cleared and is listed as Shaft T_{1000}. In mine T_N, the gallery system cleared in 1976 was all on the same elevation, with the exception of workings "e" surrounding Shaft T_{31}. Around this shaft, there are two superimposed levels of workings. (The system of galleries shown in dark gray belongs to the lower elevation galleries. The higher elevation workings/galleries are shown in lighter gray). Apart from the connection at Shaft T_{31}, the lower and upper gallery systems are connected at two other points. An inclined gallery, "q," connects gallery "e" of the lower level to gallery "p" on the upper level. Another connection was achieved by winzing into the footwall at the intersection of galleries "d," "f" and "i" of the upper level workings, down to the level of gallery "h" and the top part of gallery "i," into the hanging wall into which a niche was cut.

It would appear that the upper level excavations were developed first, and according to the tool marks from hammerstones, these belonged to the Sinai-Arabah Copper Age—Early Phase (Chalcolithic) period. At a later date, the lower level was mined as an extension, and developed as the result of a successful exploration winze into the footwall at the intersection of galleries "d" and "e." According to the tool marks from metal chisels, this took place during the Egyptian New Kingdom. The inclined connection, "q," from "p" to "e," was probably put through at this time to improve the ventilation.

At the time of the end of the 1976 excavations, the stope around Shaft T_{40} was assumed to have been fully cleared, since the plan of these excavations (Fig. 5.3) shows it as closed off. However, the recent clearances show that this was not the case (Fig. 5.5).

THE 2001–2002 EXCAVATIONS AND INVESTIGATIONS

Mine T was further excavated and investigated in 2001 and 2002 (Figs. 5.5, 5.8). These new excavations revealed new parts of this large underground mine system and materially contributed to a better understanding of underground features and the mining technology used by the Chalcolithic miners.

Six weeks of digging, and the removal of tons of sand-deposit from a well-developed network of underground galleries, brought to light new mining tunnels and chambers, cleared eight new shafts and exposed numerous tool marks of different types and periods, preserved on the walls of the workings. It must be noted, however, that some of the shafts cleared in 1976 were no longer open in 2001.

Since the job of removing the fill was in many ways very similar to the work of the original miners, the team experienced similar difficult underground work conditions and, it is believed, had to come up with solutions for technical mining problems much in the same way as the original miners. From the methods devised to meet the conditions underground, the team learned about the conditions and problems encountered and the solutions devised by the ancient miners.

The aim of our excavations was to re-investigate, in detail, Mine T, and to try and establish the existence of extensions to the underground workings cleared in 1976. The primary goal was to investigate galleries leading northward from shaft T_1, in order to locate a connection between T_S and T_N. Our excavations in 2001 made it clear that T_S and T_N were almost certainly connected underground, though the actual connecting tunnel had not yet been located.

Shafts T_{31}, T_{40}, T_{42}, T_{43}, (T_{45}), T_{1000}, T_{1001}, T_{1002} and T_{1003} (see Fig. 5.5) were all interconnected by a single system of underground workings, the T_N workings. In T_S, the workings going north–northeast (marked "c_1"), have not yet been cleared. It may well be that they continue and connect to tunnel "m," coming southwards from T_{43}, which has also not yet been cleared. Shaft T_{43} would need to be re-cleared and tunnel "m," going southwards from it, would have to be checked to establish whether or not it continues. This seems at present to be the most likely connection, though Fig. 5.5 shows that there are other possibilities (for instance, "c4" to tunnel "I"). It is, however, clear that working "e," going west from Shaft T_{31}, does not connect to T_S, as this working (shown in extra dark gray) was completely cleared both in 1976 and again in 2001 and has no further connections.

The 2001 investigations of both the T_S and the T_N workings established that it is reasonable to assume that all the shafts on the T plateau are interconnected underground, and that they belong to the same underground mining complex of which the currently known areas (T_S and T_N) are merely parts. Both underground mining complexes (now shown as separate) have numerous tunnels ending in fill, which have not yet been cleared. The Mine T workings are clearly much more extensive.

However, it proved to be quite difficult to separate the workings conducted by the Sinai-Arabah Copper Age—Early Phase miners from the later, intrusive workings of the Egyptian New Kingdom. As in other mining systems in other parts of the world dated to similar times (flint mines for example), it is clear that the miners sank a shaft, mined around it, and then started another mine from another shaft nearby. Sometimes the workings interconnected and sometimes they didn't. It often seems as though such interconnections might have been accidental. So it may well be that during the Sinai-Arabah Copper Age—Early Phase, the workings in Area T were also mined as several adjoining but not necessarily interconnected workings. The later, intruding, Egyptian New Kingdom miners may have connected up the various old workings as part of their extensive shaft-and-gallery mining system. This may also be the story told by the tool marks preserved in the different parts of Mine T.

The workings surrounding Shaft T_{31}, for example, seem at present only to be connected to those of Shaft T_{40} by workings all of which could be Egyptian New Kingdom, and the Sinai-Arabah Copper Age—Early Phase (Chalcolithic) workings, surrounding these two shafts, may originally have been separate workings.

It also seems probable that the two tunnels recorded (see Fig. 5.2) as going into the side of the wadi north of Area T—designated T_{55} and T_{56} (and also several other openings to tunnels in this area, not yet recorded on the plan)—indeed led into this mining network. Judging from their position, and especially judging from the direction of the underground galleries starting from the already excavated underground workings in Area T, the other shafts in the area, such as T_{51}, T_{52} and T_{53}, may also belong to the same mine. However, since these latter three shafts are later, circular workings, with hammer-and-gad tool marks, they seem to belong to a later Egyptian New Kingdom extension of the T mining area.

From the excavations in 1974–1976 and 2001–2002, it can be inferred that the many other as yet unnumbered "plates" in Area T, some of which are still identifiable in the field and some of which are no longer identifiable, represent many other shafts, probably connected underground to the same mining complex.

EXCAVATION OF THE WORKINGS AROUND SHAFT T_{42}, MINE T_N

During the first excavations of the mining galleries of Mine T in 1976, approximately 3.30 m to the northeast of Shaft T_{42}, a circular feature in the rock-ceiling, filled with sand, was recorded as a shaft. This shaft has now been numbered as Shaft T_{1000}. The outline of the shaft was clearly marked with small round pebbles, which were in the fill layer. The fill was very compact and hard to dig, especially from below. Falling sand-fill, mixed with small to medium sized stones, made the work extremely unpleasant and difficult.

This work put us in the same conditions as the ancient miners had been in; it reinforced our opinion that they had seldom, if ever, excavated vertical shafts from the bottom upwards. For this reason we decided to try to open the shaft from the surface above. The problem was, of course, how to locate the shaft on the surface, in order to open it from above. Though it was relatively easy to determine the direction and distance from Shaft T_{42}, the collar of Shaft T_{1000} could not be found on the surface. A few prospecting holes were dug, but we could not locate the shaft collar. We therefore decided to use our sense of hearing and our sense of touch, assuming that sound would pass through the sand-fill more easily than through the rock. By listening for the sound of blows on the fill from below and by trying to feel and find the point where the vibration was strongest, a point was chosen to sink another prospecting hole. This time, the collar was successfully located and, using a pick, the fill was cleared from the shaft. Soon the darkness of the workings of the mine could be seen. Shaft T_{1000} was irregular, oval in shape, 0.82 × 0.85 m. It was probably used for transport of ore as well as for ventilation and light.

EXCAVATIONS NORTH OF SHAFT T_{40}, MINE T_N

An opening in the north wall of Shaft T_{40} was completely filled with sand. This opening was irregular in shape and roughly cut into the soft white sandstone formation. It was 1.70 m wide and at its highest point 1.72 m. We removed the fill, composed of light yellow sand and small pebbles. At a distance of about one meter from the entrance, the chamber widened slightly on its right side, forming a medium sized niche with numerous small cavities. These cavities were the result of simply following a vein of copper nodules and extracting the ore. At this point, the ancient miners decided to leave part of the rock intact and mined all around it, leaving a supporting pillar. This is a feature also known at other places in Mine T. Supporting pillars seemed to have played a very important role in the early stages of mining. They provided stability to the underground galleries and safety for the mines and miners. The surface of this pillar was covered with numerous deep and long chisel marks. In this case we did not date the original pillar by the tool marks, since these most probably reflect "pillar robbing" by the later, Egyptian New Kingdom miners.

We continued to remove the fill deposit, but about five meters from Shaft T_{40} difficult conditions were encountered. The archaeologists began to experience some of the difficulties associated with these methods of underground working. There was very limited space for the person clearing the sand and the difficulty in removing it in buckets slowed the work down significantly. This was soon intensified by lack of air and light. With each newly filled bucket conditions became more aggravated.

Obviously, the conditions we were experiencing must have been remarkably similar to those experienced by the ancient miners. After all, the means of removal of the fill were not very different from the original method of removal of the mined rock. It was interesting to speculate how they had overcome these very severe conditions. Evidently, they had continued digging despite the dust and the lack of air and light. Though they were probably much less spoiled than modern man, and much better trained, the evidence that they had encountered the same difficulties was actually visible above our heads, on the hanging wall of the tunnel, in an irregular circular formation, filled with sand and small pebbles. It was a feature that had already been noticed in Shaft T_{42} during the first survey of the mines in 1976, and was obviously a shaft. This shaft is marked on our plan as Shaft T_{1001} (see Fig. 5.5).

The distance between the entrance to the gallery and Shaft T_{1001} was 5 m. At this point the height of the gallery was 1.50 m. Looking from below, the shaft was an irregular oval with an inner diameter of 0.65–0.80 m. It was completely filled with blown-in sand and small stones. This fill was very compact and we could not possibly clean the shaft from below. Once again, this made it clear, that the ancient miners would not have mined shafts from the bottom up.

It was not practical, with the instruments available to us at the time, to try and determine underground the distance of the shaft from the entrance through the gallery, and its orientation. On the surface above, only an approximate position of the shaft could be obtained. Once again, listening while knocking from below, and with slowly moving hands on the surface, we tried to feel the point of strongest vibration produced by knocking; the results were amazing. Once the expected point had been determined, digging was begun and soon the outline of the shaft collar became visible. It was again obvious that the ancient miners had used a natural fault in the rock for cutting the shaft, and that the shaft was cut on one side of the fault, with the natural fault surface forming the other side of the shaft so that the rock there was almost left intact. This was probably the easiest way to mine a shaft. As a result of leaving one side and following the sloping face of the fault, this type of shaft was rarely precisely vertical. Shaft T_{1001} was probably used primarily as a ventilation shaft, but could also have been used for the transport of copper ore out of the mine, shortening the tramming distance underground.

While clearing the tunnel beyond Shaft T_{1001}, we noticed that the fill had two distinct layers. The upper layer consisted of alternating light and dark yellowish sand, while the lower layer consisted of light gray sand, mixed with a large quantity of small fragments of greenish ore nodules and small to medium sized pebbles. After we carefully cleaned this section of the tunnel, approximately 0.80 m north of the newly opened shaft T_{1001}, a large niche appeared on the right side. This was a place of very intensive activities, which had left behind numerous small to medium cavities in the rock and numerous tool marks of different types. A large part of the hanging wall was covered with round, shallow hammer marks, and just below these were deep and long pock-marked chisel marks. These tool marks (Fig. 5.13b) were clear evidence of the overlapping of two completely different mining techniques

Further removal of the gallery's sand fill (t) revealed the existence of yet another shaft, Shaft T_{1002}. The distance from the previous shaft (Shaft T_{1001}) was 3.30 m and the width of the gallery here was approximately 2.00 m (Fig. 5.5). Following the line of the bedrock, carefully excavating the relatively soft sand deposit, we re-opened this shaft from underground, since here the surface above was very close. Shaft T_{1002}, its dimensions 3.4 0× 2.40 m, and 2.00–2.30 m to the surface, was one of the large, irregular and roughly hammered out vertical shafts, probably of the Sinai-Arabah Copper Age—Early

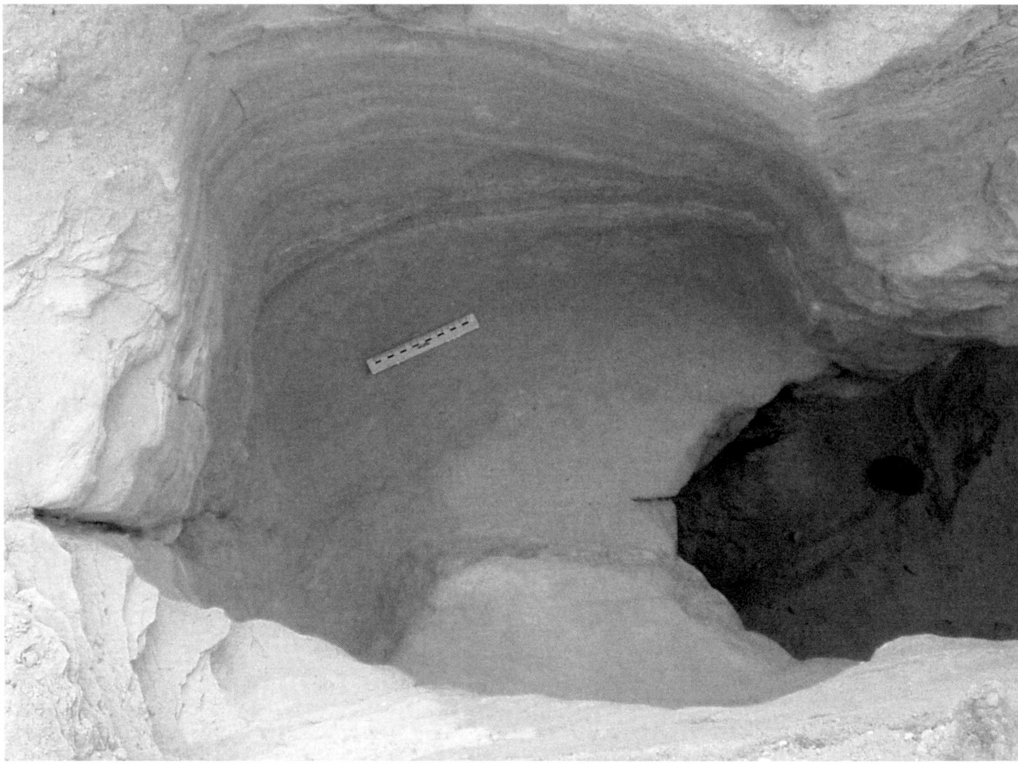

Fig. 6: Shaft T1003/T45 later intrusion on earlier workings.

Phase (Chalcolithic). After it was partly cleared, we noticed a feature on the eastern section of the shaft; after complete clearing, we identified it as another attached shaft, marked in Fig. 5.5 as Shaft T1003. Shaft T1003 was very carefully cut. Its diameter was approximately 0.80 m and its height 0.85–1.00 m. There were chisel-marks all over its walls (Fig. 5.6).

After completely clearing these shafts and exposing their outlines, it could be established that Shaft T1003 (numbered T45 in Conrad and Rothenberg 1980) was of a later date than the Sinai-Arabah Copper Age—Early Phase (Chalcolithic) Shaft T1002, which had been completely filled and no trace of which was visible on the surface. From the shapes and other characteristics of the shafts, we concluded that Shaft T1002 was opened and used by Chalcolithic miners as an entrance to underground mining galleries and for the transport of copper ore. Shaft T1003 was an Egyptian New Kingdom prospecting shaft, which led to the earlier underground workings, which the Egyptians then expanded, following the veins of copper ore. It should be noted that the bedrock between the workings and the surface around Shaft T1002 was very thin, less than half a meter, and it is probable that the present size of the shaft collar is larger than it was when originally mined, since it was probably enlarged westward due to rock falls in the past.

Just below Shaft T1003, there was a large irregular cavity, where shallow hammer marks and deep chisel marks were well preserved. The fill deposit in this cavity showed two layers: an upper layer of yellowish-brown sand and a lower layer of fine grayish sand, mixed with many small to medium sized nodules of copper ore and pebbles. After completely clearing the cavity, numerous small hollows appeared, showing a very complex, interconnected network of small, narrow workings, separated by small supporting pillars. Here, in about one square meter, was a miniature picture of the complex, sophisticated underground workings of the shaft-and-gallery mining technology typical of the Area T mines.

Fig. 5.7: Underground galleries from T_{1002}.

Further removal of the sand deposit from the northern section of Shaft T_{1002} revealed contours of what we at first believed to be two levels of underground galleries (See Fig. 5.7). Westwards and downwards from the shaft, the workings, L4, were filled with alternating layers of soft yellowish-brown sand and hard light gray sand with many small pebbles. This opening was 0.56 m wide and 0.90 m high. Half a meter from the "entrance," the cavity slightly narrowed to 0.40 m, then widened again to 0.60 m, and ended at a distance of 1.80 m from the shaft. Possible hammer marks and a vein of copper ore were visible on the walls of this elongated niche. The workings on the opposite eastern side, L5, with an opening of 0.50 × 0.95 m, were filled with two distinct layers of sand. The upper layer was of yellowish-brown sand with a few small pebbles. The lower layer was a light greenish-gray fill with lots of pebbles and larger stones and a significant quantity of small lumps of copper ore. These workings also ended at a distance of 1.20 m from the shaft, showing a possible connection with the upper gallery. However, after clearing the "lower galleries," it became obvious that this was not a case of mining in two gallery levels, but evidence of simply following the copper mineralization, which created a complex network of galleries.

The large opening over the galleries, described above, was the continuation of the workings and the entrance to the north gallery V. It was an irregular oval, 1.00 m wide and 1.40 m high. The sand fill was yellowish-brown, containing small to medium sized stones and a few small lumps of copper ore. To the west, at a distance of 0.90 m, the gallery suddenly widened to the east. We carefully cleared this section and in doing so, followed the tunnel as it went around a supporting pillar. A few deep elongated chisel marks were located in the pillar's west wall and in the floor.

Progressing further north, the fill gradually changed to alternating layers of yellowish-brown sand and layers of light yellow and grayish sand with many small pebbles and tiny bits of copper ore. Charcoal samples were also collected from this fill for radiocarbon dating. The distance from the beginning of the tunnel at Shaft T_{1002} to the edge of the excavation was approximately 4.00 m; the height at this point was 1.37 m.

At this point we cleared a relatively small cavity, L10, in the western wall of the gallery, just across from shaft T_{1002} (Fig. 5.5). The importance of completely removing the fill deposit was now becoming obvious. Our first impression was that this cavity was not very promising. 0.50 × 0.50 m in diameter, it seemed too narrow for us to hope that it could be extending anywhere. The fill was composed of hard compact yellowish-brown sand, which was not easy to cut and remove. We believed we were clearing yet another niche that had formed by extracting copper ore. However, approximately 0.70 m from the opening, the cavity widened on both sides and the color and composition of the fill gradually changed. It became light yellowish-gray sand with a few small to medium pebbles and was easier to dig out. The first worked stone found underground during our excavation came from this fill. There was also some charcoal collected for radiocarbon dating.

At a distance of 1.20 m from the opening of this working, a rock wall was reached but workings continued in two opposite directions, north and south. As work progressed it became more and more difficult, but it also became obvious that we had found a new mining tunnel. The fill composition had changed and consisted of whitish-gray sand, with lots of small and a few medium sized pebbles. There was also another stone tool. At this point the gallery widened into an irregular gallery with a few small cavities in its walls. As we continued to clear the gallery, it narrowed again to the shape of a normal gallery. The fill deposit consisted of whitish-gray and light-yellow sand. This fill was more compact and much harder to cut than the previous fill. Since we were approximately 5 m from the very narrow entrance of the gallery, working conditions became extremely difficult, with no natural light, heavy underground air and a lot of dust.

These almost impossible conditions led to an attempt to clear the opposite side of the fill. Here, there was a large fill, approximately 1.60 × 2 m, consisting of very compact and hard, light and dark yellowish-brown sand layers, with lots of small pebbles and greenish copper ore nodules. Four large rectangular boulders, each about 0.50 × 0.70 m, were removed from this fill, clearing the entrance to the tunnel going south, connecting Shaft T_{42} and Shaft T_{1002}. Just above the entrance, a few shallow hammer-marks were visible on the bedrock. These were the only tool marks documented in this mining gallery.

After we removed more of the sand fill, a large part of the eastern wall of this section of the mining gallery, running north from Shaft T_{42}, was cleared, revealing a large crack or fault in the rock in the hanging wall; this may be the reason the 1976 investigators marked this as a possible shaft. Beautiful greenish copper mineralization appeared at a few places on the hanging wall where the rock had broken away. When we removed more of the fill the entrance to a new tunnel was revealed (Fig. 5.5, location ja).

This tunnel had interesting characteristics that had not been met before in this part of Mine T. Though only partly cleared, there was a very precise circular entrance (diameter 0.40 m) and the tunnel was straight, circular (in cross-section), long and narrow. We assumed that this was an Egyptian New Kingdom working, since its resemblance to the Egyptian "borehole" techniques (Willies 1991: 114) cannot be by chance. We were left with the following questions: Was this tunnel intended for ventilation? What tools were used for mining such a long, narrow opening? Its precise circular shape suggests that a long stick might have been used for cutting the opening. But at an estimated 5 m in depth, this is hard to accept. If not done by a long stick, how was it done?

On the last day of our excavations, the fill in the tunnel was cleared to a distance of 3.85 m west of the pillar wall. In this part of the mining system no tool-marks were seen. The reason for this may be the very soft surfaces of the bedrock, which were falling apart and were probably also removed with the sand fill.

ADDITIONAL 2001 EXCAVATIONS AT SHAFT T_{31}, MINE T_N

The southern part of T_N was found completely filled with wind-blown sand. Though it had already been cleared in the 1976 excavations, there was a good reason to re-investigate it, assuming that the connection to T_S would be found here. A set of stairs was therefore cut into the fill, down to bedrock at a depth of almost 1.5 m. These new steps were cut down to the workings, which had been recorded in 1976 as lower workings. In 1976 this area was also cleared to bedrock. In the fill, which was of clean yellowish sand, we found a modern iron bar, which the previous excavators had used as a marker in their surveying. There was, therefore, no doubt that the fill was recent. However, from this fill one piece of flint was recovered.

The workings to the west of T_{31} were approximately 3.5 × 3.0 m. While removing the fill, numerous deep, long chisel-marks were exposed. After completely clearing this section of T_{31}, it became clear that there really was no connection to the workings of T_S from this part of Mine T_N.

THE 1976 EXCAVATIONS OF THE MINE WORKINGS OF T_S

In the mine clearances of 1976, Shaft T_1, another Sinai-Arabah Copper Age—Early Phase shaft was investigated. Initially, a "plate" on the surface was cleared, including a partial clearance of some of the mine workings at its bottom.

There was 0.5 m of conglomerate overlying the solid sandstone into which the shaft was sunk. The shaft itself was only 3 m deep. On either side of the shaft collar there were two other small pits in the sandstone. One was round and the other more square (Conrad and Rothenberg 1980: Fig. 182).

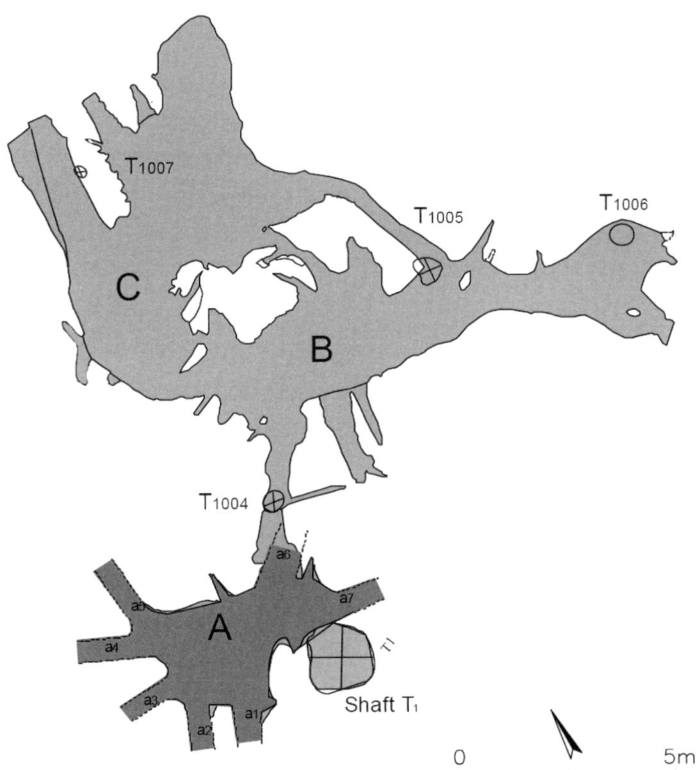

Fig. 5.8: Mine T_S 2002 clearance.

Fig. 5.9: Shaft T_1 in 2002.

The function of these small pits has not yet been established. T_1 was an inclined shaft of irregular shape (Fig. 5.9) and has rough steps from the surface down to the underground workings.

In 1976, the underground mine workings were only investigated in the immediate vicinity west of the shaft bottom and are shown in Fig. 5.8 in gray as a small stope at the bottom, west of Shaft T_1, with tunnels radiating from it in at least seven directions (though none of them was further investigated at the time). Seven hammerstones of the "hand-held" pebble type were recovered from this stope. Near this shaft, the modern mining company had made a road across the wadi and the bulldozer broke through into some underground workings. The tool marks in these workings, which our team inspected from the road, were of the hammer type only, i.e., the workings probably represent an extension of Chalcolithic Mine T_S.

2001 INVESTIGATIONS OF MINE T_S

During the 2001 excavation season, in order to establish whether the mine workings around the two shafts, T_1 and T_{31}, were connected underground, opening "a6" (see Fig. 5.8) in the northeast wall of the workings of T_S was cleared. This was an irregular opening to a gallery, almost oval in shape, 0.88 m wide and 0.72 m high (at the highest point). On both sides of this opening, long chisel-marks were visible. We, therefore, consider this to be a later enlargement by the Egyptian New Kingdom miners of an earlier and smaller Sinai-Arabah Copper Age—Early Phase gallery. The fill was of yellowish-brown sand, mixed with a large quantity of small to medium pebbles and some stones.

At a distance of 1.80 m from opening "a6," a rock-break with the apparent outlines of a shaft was observed on the hanging, which, when cleared, would provide the light and air so indispensable for our further progress. It was probably the same need that caused the Sinai-Arabah Copper Age—Early Phase

(Chalcolithic) miners to select this natural crack in the rock as the most convenient location for a new shaft. Previous work in Mine T_N taught us how to recognize a shaft, determine its outline and locate it from above. Shaft T_{1004} was found and opened (Fig. 5.10).

Its shape was an irregular oval and its measurements, taken from underground, were 0.50 × 0.40 m. Careful cleaning of the shaft-walls brought to light numerous round and shallow hammer marks, indicating the type of tools used for sinking the shaft. We assumed that this shaft was used for ventilation, supplying light and for the transport of copper ore in the Chalcolithic period.

Northeast of Shaft T_{1004}, the mining gallery expanded into relatively large Chamber "B," separated by an elongated supporting pillar from a second large Stope "C." Chamber "B" runs roughly east–west and formed a very rough irregular stope, approximately 4.50 m in length, 2.30 m wide and 0.95 m high (at the highest point). A number of shallow cavities in its walls were the result of the extraction of copper ore. It was evident that this area was very rich in ore as it was possible to follow some veins of copper mineralization. At a number of places, nodules of green copper ore still stuck out of the white sandstone formation. The large supporting pillar between the two rooms showed a few deep chisel marks—evidence of secondary, later mining activities.

The most impressive part of Mine T was found in Chamber "C," named by us "the Green Room." This chamber, a very large stope, 3.45 m in length, 2.50 m in width and 1.30 m high (at the highest point), had a large number of what looked like large copper ore nodules protruding from its ceiling (Fig. 5.10).

However, a study of these "nodules" (some examples of which were recovered from the fill) by Irina Segal of the Israel Geological Survey in Jerusalem, showed that they were not real ore, but green stained sand nodules. Most likely, these nodules were hidden by a layer of sandstone on the hanging at the time when the ancient mining took place, and eroded with time, probably as a result of humidity. When this layer of sandstone gradually fell, it mixed with fill already present in Chamber "C," which also contained some of the green stained nodules.

We found the "Green Room" partly filled with sand and broken off pieces of rock. A possible shaft-like feature ("Shaft T_{1007}" on Fig. 5.6) was observed on a rock fracture in the hanging. This would have been the only direct connection from the surface to the "Green Room," but it was too small (diameter approximately 0.20 m) to allow much blown-in sand or pieces of rock to fill the chamber. The fill found in the "Green Room" obviously originated from the ceiling of the gallery, especially since "Shaft T_{1007}" could well have been a small natural opening.

The floor of the chamber slopes down northward, and in its north wall there were several openings: c1, c2, c3 and c4. c2 and c3 seemed to be exploration workings, but c1 and c4 were probably entries into mining galleries. c1 is the most likely connection to the T_N workings, but has not yet been investigated (see Fig. 5.5).

In our 2001 excavations, no underground surveyor was available and it was uncertain how to continue in order to find the connection to the T_N workings.

At the time, it seemed that the most likely connection of T_S with T_N was from Room "b" and for that reason we continued our excavation in that chamber, clearing, as it turned out, further eastwards and actually parallel to any possible connection with T_N. (see Fig. 5.5).

As our efforts progressed, the difficult working conditions under which the early miners had operated again became easy to understand. There was a lack of air and light, and very limited space in which to move. This again suggested that a careful watch of the hanging wall was warranted in order to look for any crack or fault in the rock. At a distance of 4.50 m from the previous shaft T_{1004}, the contours of what we looked for became visible. Soon after that, buckets of sand and small pebbles were being

Fig. 5.10: Shaft T1004.

Fig. 5.11: The "Green Room."

cleared out of the mine through the newly opened shaft T_{1005} (Fig. 5.8). On the inner walls of this shaft there were numerous round and shallow hammer marks, but below the shaft, on its right side, were numerous long and deep "hammer-and-gad" chisel-marks, typical of Egyptian New Kingdom mining.

Careful removal of more of the very hard and compact fill brought to light a beautifully cut narrow tunnel, "b2," approximately 0.25 m in diameter and at least 2 m long. Here we could clearly see how the miners, evidently Egyptians of the New Kingdom, had followed the green vein of copper mineralization. There were chisel marks all around the opening and the walls of this gallery. After removing more of the fill, a supporting pillar was uncovered in the middle of the gallery (Fig. 5.12). The fill was composed of alternating layers of yellowish-brown sand mixed with small pebbles and soft grayish-white sand, mixed with a large quantity of small to medium sized nodules of green copper ore. Charcoal samples were collected from this fill for radiocarbon dating.

We moved south and entered transverse Channel b5, which was oriented more or less north–south. The hanging wall and sidewalls were covered with long, deep chisel marks. Nowhere in the mines of Area T did we feel a stronger Egyptian New Kingdom presence than we felt here.

Lack of oxygen in the heavy underground air again prevented us from continuing our work. We noted a small, round crack in the ceiling rock, but we could not pry it open because of the very difficult conditions. This shaft was marked as Shaft T_{1006}. This shaft seemed oval in shape, its diameter approximately 0.15–0.20 m, slightly widening towards the surface. The main indication that this feature was a shaft was the presence of a large stone stuck in the bottom. Since the stone was larger than the opening itself, the only logical explanation was that the stone had fallen in from the surface before wind-carried sand filled the shaft. The dimensions of this shaft indicated that it was a ventilation shaft.

Fig. 5.12: Support pillar.

THE MINING TECHNOLOGY OF MINE T

INTRODUCTION

Rock breaking methods used in ancient mines can only be attested physically by one or more of three types of evidence, two direct and one indirect. Direct evidence includes tools left behind by the miners, which, when preserved, can tell us the methods used. Besides the tools themselves, tool marks left on the rock *in situ* by the tools used to mine the ore are diagnostic of the tools and methods used for breaking rock.

Indirect evidence is the size and shape of the mine workings. The sizes and shapes of the remaining openings in the rock, combined with knowledge gleaned from the tools and tool marks, can allow technological deductions to be made as to the methods of working used.

ROCK BREAKING

Rock breaking during the Sinai-Arabah Copper Age—Early Phase was by hammer only (there is no evidence of markings left by wedge, spike or gad except in those workings now identified as having been left by intrusive later Egyptian New Kingdom miners) and the tool marks (see Fig. 5.13) are typically a surface of small round overlapping "pock" hammer marks. Some of the galleries are very small in diameter (60 cm), but were still mined by hammer and thus they indicate that the people were small and broke no more volume than they absolutely had to for access.

In Mine T, the earliest mining tools were hafted hammers. In other areas at Timna, where copper ore was excavated by "pit mining," earlier tools ("grooved hammerstones") and mining methods were

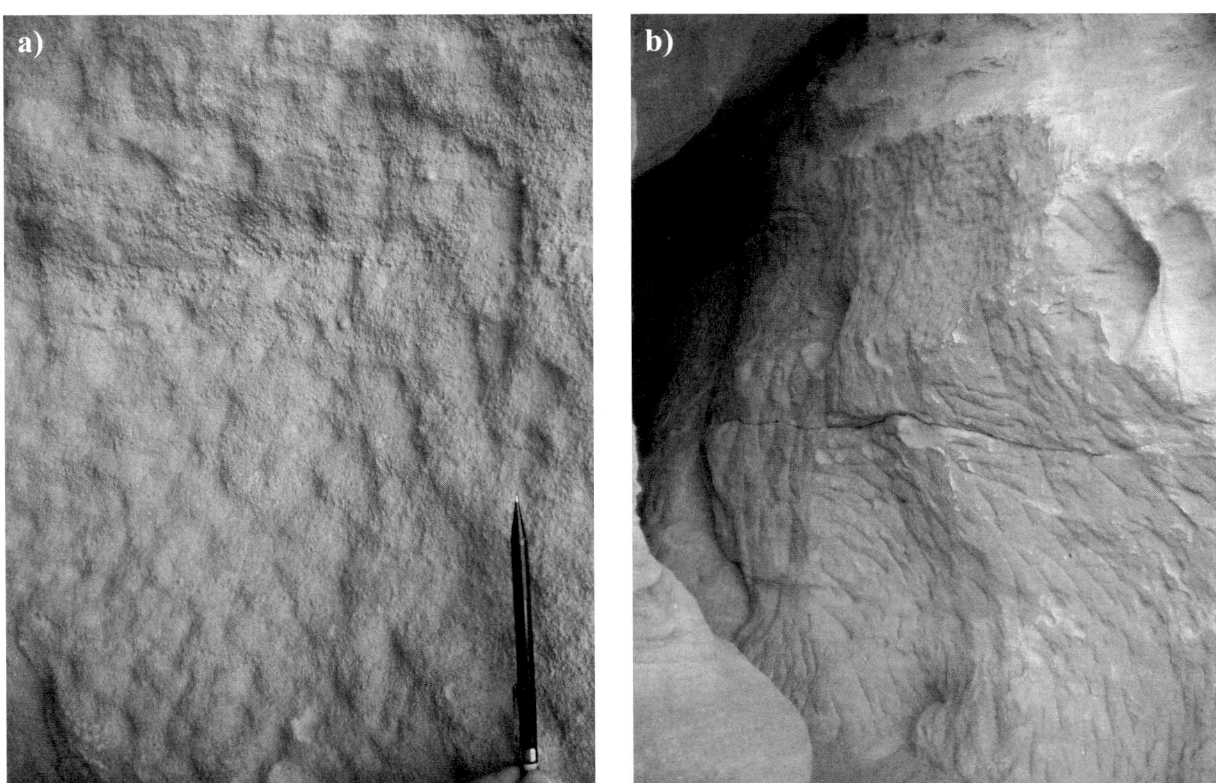

Fig. 5.13: Toolmarks: (a) Phase 1, hammer only; (b) Phase 2, hammer and gad.

attested, but not in Area T. Later on, "hammer-and-gad" methods were used in Area T by the Egyptian New Kingdom miners. There was no evidence of any later periods of mining in Area T, though Roman and Early Islamic mining was established by the Aravah Expedition in the southern as well as northern parts of the Timna mining region. The technology of these workings has not yet been investigated.

MINING TOOLS

Phase 1 Mining Period

During this period, the earliest stage of underground mining at Timna yet identified, dated by Rothenberg to the Sinai-Arabah Copper Age—Early Phase (Chalcolithic), the tool marks show that the shaft sinking and tunneling was done by using hammers alone—no spikes, wedges or gads appear to have been used.

Hafted Hammerstones

In this earliest underground mining phase at Timna, the workers used a unique type of hafted hammer for their mining. These were made from hard stones, mainly local granite. They were discoid in shape with a biconical hole drilled through the middle for a haft. No hafts have been found, but a few complete and many broken hammerstones of this type have been found, both in Mine T and on the surrounding surface (Fig. 5.14).

As far as we know, this type of hammerstone has never been recorded at any other ancient mine anywhere in the world. They are unique to this period of mining at Timna. To the best of our knowledge, at every other ancient mine where hammerstones were used without the use of gads or spikes, the hammerstones have either been unmodified cobbles or have had grooves of various designs etched around them for attachment of a haft. Nowhere else have biconically drilled discoid hammerstones been reported.

These hafted hammerstones were used to break the rock directly by hammering on the rock face. There is no evidence to relate the use of these hammers to any other method; no gads, spikes or wedges appear to have been used with them. Figure 5.15 shows the probable method of working with these hammers.

Fig. 5.14: Tools found in Area T, from left to right: polishing stone ca. 11.5 × 9 × 10 cm pestle/mortar ca. 10 × 7 × 3 cm, half a biconically drilled hammerstone originally roughly 15 cm circular.

These hammers leave tool marks, which consist of round, shallow hollows, about 1.5 to 2 cm in diameter (Fig. 5.13a). These are often overlapping and sometimes cover the entire rock surface. It is extremely difficult to estimate the direction of working from the tool marks of these hammers alone and, in general, the direction of working should be inferred from overall mining considerations rather than from the evidence of tool marks.

Phase 2 Mining Period

This stage of mining has been dated by Rothenberg and the Aravah Expedition to the New Kingdom Egyptian period. Recent revision of the dating of the main smelting sites (Ben-Yosef *et al.* 2012) suggests that this phase also represents early Iron Age activity; however, for convenience of comparison to earlier publications of the Aravah Expedition, we stand by the previous chronological framework and treat this phase as "Egyptian." The mining was done by both the Egyptians and the local Midianites. Mining in this second phase of working was by hammer and spike, wedge or gad. This is attested by the characteristic tool marks left on the walls of the excavations (Fig. 5.13b). These are directional and enable the direction of working to be identified.

Hand Held "Pestle/Mortar" Hammerstones/Mortars

Hand held stones which were both pestles and mortars appear to have been multi-purpose tools, which may have been used at Timna in the Chalcolithic and definitely were used in the Egyptian periods. Stones such as these have been found underground in some of the mines, on the surface near the mines and numerous similar stones have been found in an ore-dressing context at the smelters (Fig. 5.14) They are here designated "pestle/mortar hammerstones."

At the smelting sites these stones were clearly used both as a mortar on which to break up nodules of ore and probably also lumps of slag, and as hammers to do that breaking. They show wear in cup like depressions in the center when they have been used as a mortar and also on their edges when they have been used for hammering. Where mineral dressing may have been done underground, for instance in the openings related to T_1, T_{31} and T_{42}, they may well have been used in a similar manner.

However, we suggest that this type of stone hammer may also have been used in conjunction with a spike or gad (Fig. 5.16). We propose that while the miner held the spike or wedge against the rock to be broken in one hand, he hit the end of it with the hammerstone in the other, with the cup like depression in the middle of the stone being the striking face. In the confined 80 cm diameter galleries, the use of a handle on the hammer when working with a spike or gad would possibly not be favored as the likelihood of hitting a finger would be greater. In addition, the strength of the spikes or wedges, being made of wood, was such that too hard a blow could break them rather than the rock face. For these reasons, the hand held hammerstone would probably have been preferred.

Two fragments of wooden wedges were found in the 1974–1976 excavations in Area S (Conrad and Rothenberg 1980: 178–179; see Chapter 1.1) and it is almost certain that the Egyptians also used wooden mallets with the gads, as has been attested for the excavation of tombs.

No complete mining spikes have been found, but two bronze tips (*ibid.*) were found in Timna, one on a "plate" in the mining area (Site 42) and the other in Smelting Camp Site 2.

Long Metal Tipped Rods

Similarly, it is almost certain that metal tips were used on the end of much longer sticks and used as a jabbing tool to dig longer and smaller prospecting holes. They would also probably have been used in making the initial cut at the start of galleries and in making holes through pillars for ventilation

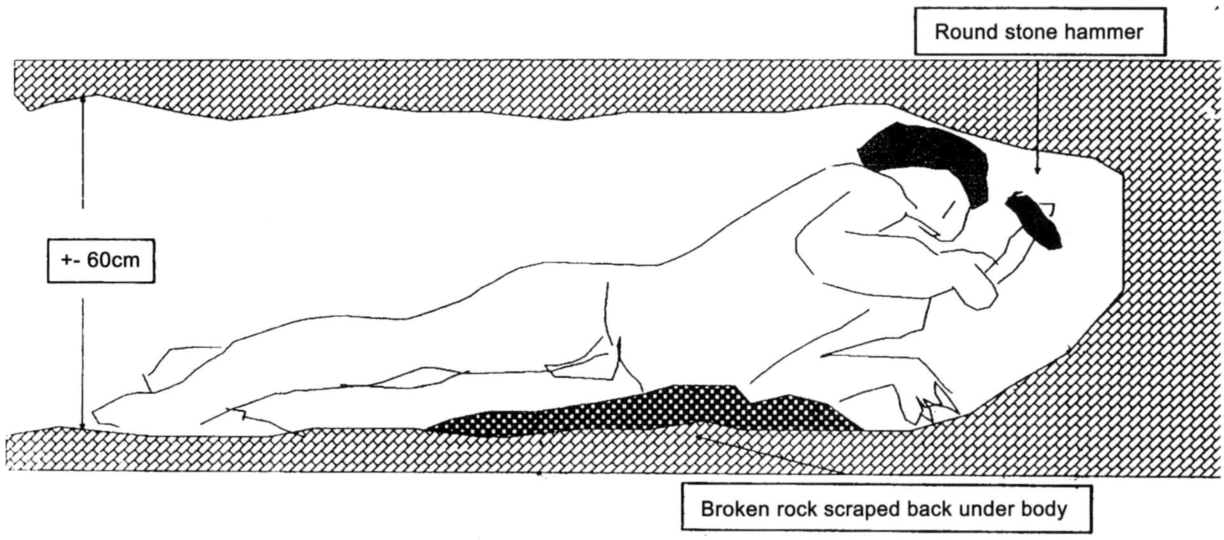

Fig. 5.15 – Miner working with drilled hammerstone.

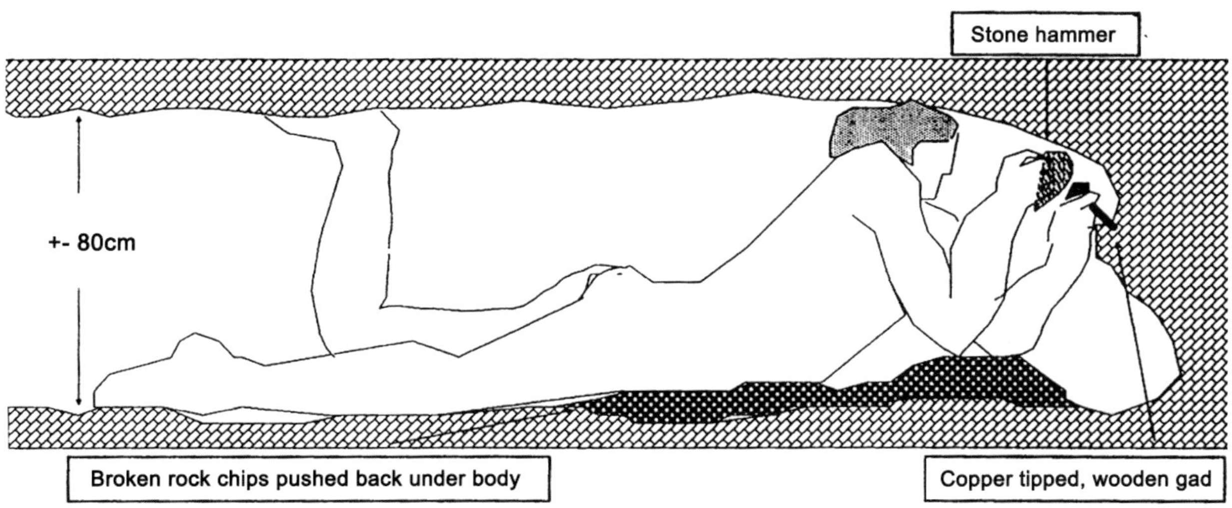

Fig. 5.16: Miner using "pestle/mortar hammerstone."

purposes. Examples of the use of such spikes for making holes through pillars were found in T N. These were also used to make long, very small openings for exploration and/or ventilation purposes ("b3" in T S for example and "ja" in T N). Some of the created holes were akin to modern boreholes.

TOOL MARKS

In Mine T, three types of tool marks have been identified.

1. Hammer Marks (See Fig. 5.13a)

These marks were produced by direct blows of a hammer on the rock face, each hammer mark representing one blow of the hammer. As already mentioned above, it is not possible to judge the direction of working from these tool marks. They are useful only as evidence of the type of breaking technology used in the mine.

There was archaeological evidence in Mine T that this technology was associated with the Sinai-Arabah Copper Age—Early Phase (Chalcolithic) period of mining.

2. Pock-marked Grooves (Fig. 5.17)

Pock-marked grooves were produced by a miner using "hammer and gad" technology. With this type of tool mark, the number of blows of the hammer on the gad can be counted and also the direction of working can be seen from the tool mark. In Fig. 5.16 it can be seen that the worker hammered from top left down towards bottom right, and that it took about 12 blows of the hammer on the gad to produce each groove.

At Timna, this method of working was associated with the Egyptian New Kingdom period of mining. The mining toolkit consisted of mallets, stone hammers and wooden spikes with metal tips, of which numerous stone hammers, two copper points and two wooden wedges have so far been found.

3. Spike Marks

Spike marks were left by the action of a long spike or rod being jabbed at the rock face. The technologies described in 2 and 3 above were associated with the Egyptian New Kingdom mining period.

CLEANING IN MINE T

"Cleaning" is here defined as the removal of the rock, broken by the miners at the face. As Figs. 5.15 and 16 show, initially the miner would break the rock and scrape the resulting rock chips backwards under his body. We assume (from observation of the methods of "modern" miners using similar techniques)

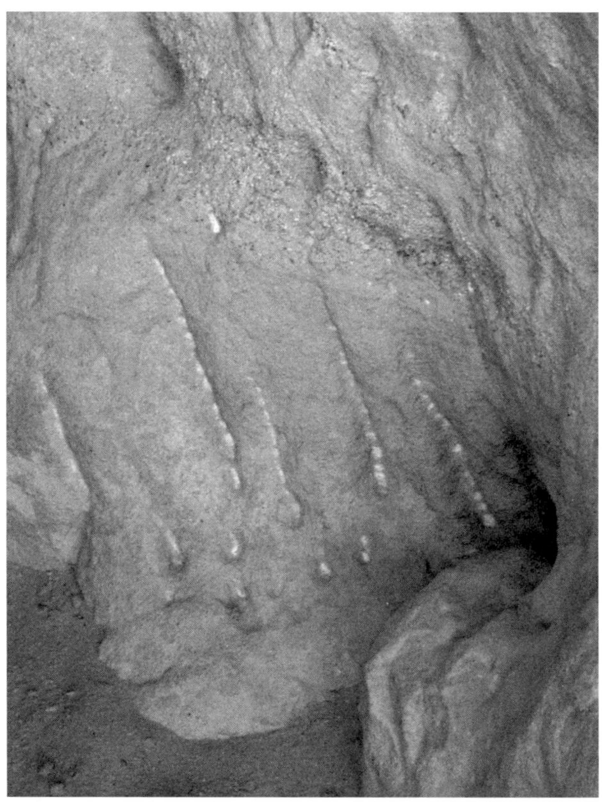

Fig. 5.17: Pock-marked grooves—Phase 2 mining.

that another miner—the "trammer"—would shovel, rake or scrape this loose material into a carrier of some kind, a leather bag, basket, wooden bowl or whatever, and remove it to the place of disposal or ore dressing, dragging the carrier along the floor of the working.

Since the only attested scraping device recovered in the excavation of a mine was a sherd of pottery which had been reused as a scraper in the Roman-Early Islamic mines of the Nahal ʿAmram (Willies 1991), it is not possible at present to give more detail of the cleaning devices used. However, we may assume that throughout the period Mine T was in use, there would have been small scrapers or short handled shovels made from locally available materials, pottery sherds, flat rocks, wood, etc.

TRANSPORT IN MINE T

Tramming is the transport of the broken rock to its place of treatment or disposal. In early mines this would have been limited to the minimum distance needed. Waste was therefore often packed into mined out space and the initial ore dressing was done underground, with the waste again being packed nearby. However, since broken rock takes up a larger volume than it filled *in situ*, there is always some waste rock to be disposed of on the surface.

In Mine T, as in many other mining areas of Timna, there is a conundrum in that there are almost no surface dumps dating to either the Chalcolithic or the Egyptian New Kingdom periods. In addition, the fill cleared by the excavators from the mines did not appear to be packed waste, but windblown and waterborne sand.

What exactly was done with the mined out material? During our surveys we often noticed low dumps around some of the "plates" and it seems most likely that most of the waste dumped on the surface outside the early mines was gradually blown or washed away and very little of it remained *in situ* after several thousand years.

Transport would have to be done manually (by"trammers") using the transport devices mentioned above. To date, it seems that none of the carriers survived in Mine T, which could provide insight into what was being used, either by the Chalcolithic or by the Egyptian New Kingdom miners.

VENTILATION IN MINE T

The Chalcolithic Miners

The workings were not far below the surface, but as the people clearing the infill that had accumulated since the last period of mining could attest (Drenka 2003), their work in the confined spaces of the tunnels—work which does not differ greatly from the work of the original mining—was extremely arduous. Many of the galleries were not much larger than the worker himself, and this means that there was a very restricted flow of air into the work area. In addition, the hard-working miners breathed heavily and breaking the rock—or clearing the infill—polluted the air with dust; the little oxygen there was quickly became depleted.

This required the miners to exit to a nearby opening fairly frequently. The movement of the miners into and out of the gallery would cause some change of air. The "trammers" working behind the rock breaker would help the ventilation a little as they moved in and out of the workings. However, this would only slightly extend the period the breaker could stay in the opening.

For that reason, as we learned from our excavations, the Chalcolithic miners would put in shafts at regular intervals of about 4 to 5 m, in order to improve the airflow. This would of course also have the effect of improving the available light in the mine. We did not find any indication that the Chalcolithic miners used any other method of ventilation.

The Egyptian New Kingdom Miners

The Egyptians of the New Kingdom clearly had a better understanding of the flow of natural ventilation than the Chalcolithic miners. This was attested in other workings, in Area S in particular. However, their working in Mine T, under discussion here, was almost purely of an intrusive nature. They were following the old galleries and extending the mine by additional galleries, following the mineralization and creating new galleries.

The Egyptians, in places, also went down into the floor to a copper horizon at a lower level. All they needed to do for ventilation was to clear the shafts originally excavated by the Chalcolithic miners and this would give them adequate ventilation. New galleries of course needed new shafts and such New Kingdom shafts were indeed found by our excavation in Mine T. There was some evidence that the Egyptians also enlarged some of the Chalcolithic shafts, but whether this was done to improve the ventilation or simply to make travel and lighting easier we could not determine.

LIGHTING IN MINE T

The Chalcolithic Miners

Evidence for the lighting method used by these miners is scanty. However, for reasons of ventilation they sank so many shafts, and the excavations were so close to the surface, that the light underground must have been reasonably good. This could be seen in the portions of the workings cleared by our excavation, as the light where the shafts were all open was relatively good. In addition, evidently they also used clay lamps as a fragment of such a lamp was indeed found in the excavations in Mine T_N in 1976.

The Egyptian New Kingdom Miners

The Egyptian workings, thus far cleared and explored at Timna (see Fig. 5.2: Mines S and T), demonstrated that the Egyptian miners developed galleries and workings far enough from shafts to make artificial lighting essential. Although to date no lamps or other lighting devices dating to this period have been found in the cleared mines, we must assume that the Egyptians of the New Kingdom also used oil lamps in the mines.

SHAFTS IN AREA T

From the 1976 excavations and our subsequent work in 2001–2002, the following general observations can be made about the workings in Mine T. The shafts investigated in Mine T can be divided into two distinct categories:

Chalcolithic Shafts

Chalcolithic shafts had no precise shape in the horizontal plane. Where tool marks were preserved, they were hammer pock marks. Horizontally, the sizes of the shafts also vary greatly depending on their function. These shafts were on average only about 3 m deep, which was all that was required to reach the first mineralized layer which contained sufficient ore of adequate grade to interest these miners, i.e., the ore could be mined "profitably."

The Chalcolithic shafts could be subdivided into two varieties:

1. Shafts with relatively vertical walls and of varying sizes, some very small in diameter. Examples are T_{40}, T_{43}, T_{1000}, T_{1001}, T_{1002}, T_{1006} and T_{1007}. These were probably originally shafts put in for ventilation as well as for light.

2. There were several shafts which connected to the surface as a flight of very rough steps, i.e., they were inclined shafts. Examples are T_1, T_{31} and T $_{42}$. These shafts were also different from the shafts listed in 1), because the gallery into which they connected was a large open working, apparently used for ore dressing.

Careful examination of the tool marks left on the walls of the inclined shafts seemed to indicate—though with this type of tool mark it was almost impossible to be sure—that the cutting of the steps took place subsequent to the sinking of the shaft, i.e., adding steps to what was originally a shaft without steps. It is most unlikely that any shaft was put through to the surface as a vertical raise from underground, though it seems possible that inclined raises were mined from below.

The inclined shafts were probably cut to serve as means for travelling and haulage to the surface in order to minimize the distance the ore had to be dragged through the very small galleries making up the mining network. As mentioned above, some of the underground chambers at the shaft bottoms were used for ore dressing in order to minimize the amount of rock hauled to the surface.

The method of working in all the galleries and shafts of the Chalcolithic period appears to have been breaking the rock by using hafted stone hammers. Many examples of such hammers have been found in the workings and on the surface surrounding the shafts and "plates" in Area T. "Hammer and gad" technology was evidently never used by the Sinai-Arabah Copper Age—Early Phase miners. It is equally most unlikely that any later period miners used "hammer only technology" for mining, and the presence of the typical hammer marks of this period can be used to identify workings as dating to the Sinai-Arabah Copper Age—Early Phase even if they are now part of later mine workings, like sections in Mine T.

Egyptian New Kingdom Shafts

The use of a spike, wedge or gad makes breaking much more efficient, as Hooson (1747) wrote: "A miner that is used to work with these gads will cut the hardest of rocks and as much or more in a solid piece of stone than either by blasting or with picks; and they are very good tools to cut strong and hard ore with." According to the tool marks, the miners of the later mining periods in Timna, in particular the Egyptians of the New Kingdom, indeed used "gad technology."

Mine T has a number of shafts with hammer and gad type tool marks in registers of about 10 cm height on the walls of the shaft. The majority of these "gad" grooves were slanted from right to left, indicating a gad held in the left hand and a hammer in the right. In a few rare cases the grooves went from left to right, probably indicating a left handed worker. In one case there were alternating registers slanting right to left and then left to right, etc. This could indicate an ambidextrous miner, or that two miners, one left handed, took turns in deepening this shaft.

The New Kingdom shafts were all circular or nearly circular and vertical or nearly vertical. In most, but not all, of these shafts there were footholds cut into the sidewalls or, in some rare cases, steps protruding from the sidewalls. Since it is entirely possible in a shaft of this size to travel up and down without the use of footholds, it may be that footholds were only cut where there were workings developed from that shaft in order to ease the travel in the shaft particularly for the miners bringing out the ore. Possibly, shafts with no footholds or steps were purely for exploration.

It has been proposed (Conrad and Rothenberg 1980: 78) that protruding steps should be dated to a different mining period. This seems unlikely since shafts with both varieties connected to the same underground workings. It seems more probable that the reason was geological and that protruding steps were used where the mined rock was hard, with the incised footholds in weaker rocks. At Timna this would mean a predominance of the incised footholds, as indeed was found.

In the archaeologically excavated Mine T, the hammer and gad cut shafts were all later intrusions into the earlier Sinai-Arabah Copper Age—Early Phase (Chalcolithic) workings, dated to the Egyptian New Kingdom.

UNDERGROUND WORKINGS (FIG. 5.5)

The underground workings, accessible from the Sinai-Arabah Copper Age—Early Phase shafts, were more or less horizontal workings and were used to perform a variety of functions. Normally the tunnels created were only as big as they had to be in order to allow the passage of the miner. Only where ore was to be removed would the openings have been made any larger. Their shape, dimension and location, therefore, were to a certain extent dependent on their function and on the mineralization found when they were developed. In the workings excavated in Mine T it was possible to establish that in these Early Phase mines, most galleries, although probably originally just following some mineralization, were used for one or more of four different purposes:

1. Prospecting/exploration
2. Connecting between workings for ease of travel
3. Tunnels developed purely for ventilation
4. Stopes

Prospecting/Exploration Tunnels

These workings have small dimensions in cross-section, and most are relatively short in length. Many are located on the sides of other excavations. The purpose of these workings was clearly to try and trace the extensions of the mineralization. They were normally dug where richer deposits of ore had been found and extracted, as attested by the fact that abundant traces of copper ore can still be seen on the sidewalls of these galleries. Their limited diameter in most cases prevented the miners from entering the galleries. Thus, the length of the Chalcolithic workings seldom goes beyond one meter. In many cases they are only 0.30 to 0.50 m in length. A lot of these workings may have been merely created when the miners followed pockets of ore until they ran out on the boundaries of the workings. The small "stub tunnels" near "c2" and "c3" were examples of this type of working.

The other kinds of exploration tunnels were longer and were probably mined in an attempt to find additional good areas of copper mineralization. Those which have a small diameter, but are large enough for a small person to enter, may date to the Chalcolithic period, though it was not possible to confirm this from tool marks as the modern investigators were too large to enter these exploration tunnels. The end of these tunnels usually narrow down to a point. Gallery "j_a" was an example of this type. Other such exploration tunnels were even smaller and were clearly made by very long mining rods, associated with the Egyptian New Kingdom mining period. Tunnel "b_3," going east near T_{1005}, was an example of such an exploration tunnel.

Connecting Tunnels

Connecting tunnels may have been developed and used to facilitate travel but would have also improved the ventilation and may well have originally been put through for that purpose. Since these tunnels were also later used by the intrusive Egyptian New Kingdom miners; it is possible that some of these tunnels were originally mined, or at least enlarged, by these later miners. These travelling tunnels seldom had well preserved tool marks due to the wear of many people crawling through them. Tunnel "d" is an example of such a tunnel.

Ventilation Tunnels

Ventilation tunnels were often short and had a small diameter of maximum 0.30 m. Some were cut through pillars in places where there was insufficient airflow in the stopes. They enabled air to flow freely through the pillars into the tunnels connecting with the shafts to which the stopes, though in close proximity, do not otherwise connect. A good example of this type of opening is tunnel "g."

Stopes

Stopes are random workings in shape and obviously cut to the height and width required to recover ore from the host rock. They may appear as a widening of the tunnels or as stopes on mineralized rock, probably discovered by exploration tunnels following a horizon of mineralization, and can be defined as "open stopes." These have no preferred direction of working or dimensions, as their limits are entirely dictated by the shape of the ore pockets. A good example of such a stope was room "c," the "Green Room."

The space at the underground entrances to some of the shafts of the Chalcolithic period may have been used as storage rooms for the ore, which had been mined and was to be transported to the surface. We assume that some of the ore was also crushed and sorted in these rooms. However, it seems possible that this was done in the Egyptian period, as was seen in the study of the fill material in the southwestern corner of room "e" at the bottom of shaft T31. This study demonstrated that it was not ordinary sand, but rather a mixture of sand and small lumps of ore, bearing the signs of intentional crushing (It must be noted that according to the tool marks, Room "e" is an Egyptian New Kingdom working). It is important to note that tools used for the crushing of ore were found in these rooms (Conrad and Rothenberg 1980: 170–174), though some of these tools may have been "pestle/mortar hammerstones" which were used later by the Egyptians for "hammer and gad" mining.

Similar fill was found in Room "a" at the bottom of Shaft T1. However, since Egyptian New Kingdom miners intruded into the Sinai-Arabah Copper Age—Early Phase workings, it is quite possible that this underground ore dressing only took place during the Egyptian New Kingdom. Around the bottom of some of the shafts, and of Shaft T40 in particular, there may have been some widening of the working purely to provide underground working space for ore dressing

When the mine was abandoned, most of the horizontal workings as well as the shafts were gradually filled by natural detritus, as clearly shown by the stratigraphy of the fill in the galleries and shafts. The fill was made up of 95% very fine sand, in many cases mixed with small gravel which was carried into the workings by the wind, and the gravel layers were deposited by the rare rainstorms. The latter is clear from the fact that the gravel bearing layers are more cemented and compacted than the pure sand layers, which is due to the water aiding settling and depositing cementing material. The presence of some larger pieces of surficial rocks in the shaft fill is considered to have resulted from the collapse of the material of the shaft collar, the uppermost part of which was sunk through conglomerate. However, some larger sized fragments of rock, found in the fill of the galleries, were different. These did not come from the surface at all, since their composition was different from the conglomerates at surface; they are in fact slabs of rock which fell from the hanging wall and sidewalls of the workings.

GLOSSARY OF TERMS

- Footwall – the floor or rock at the bottom of a mine working
- Hanging wall, Hanging – the rock above a mine opening, the roof of the gallery
- Gad – a spike or wedge with a handle to be used for rock breaking

- Raise, raising – an inclined tunnel worked from the bottom upwards, working from the bottom upwards
- Stope – the underground workings mined solely for the recovery of the ore mineral
- Tramming, trammer – moving the broken ore along horizontally in a tunnel system in a conveyance of some kind, one who trams.
- Winze, winzing – an inclined tunnel worked from the top down
- Workings — The entire system of openings in a mine for the purpose of exploitation (and cf. http://www.rocksandminerals.com/glossary.htm)

REFERENCES

Ben-Yosef, E., Shaar, R., Tauxe, L., Ron, H. 2012. A New Chronological Framework for Iron Age Copper Production in Timna (Israel). *Bulletin of the American Schools of Oriental Research* 367: 31–71.

Ben-Yosef, E., Tauxe, L., Ron, H., Agnon, A., Avner, U., Najjar, M. and Levy, T.E. 2008. A New Approach for Geomagnetic Archaeointensity Research: Insights on Ancient Metallurgy in the Southern Levant. *Journal of Archaeological Science* 35: 2863–2879.

Conrad, H.G. and Rothenberg, B. 1980. *Antikes Kupfer im Timna-Tal*. Bochum.

Drenka, A. 2003. New Excavations in the Chalcolithic Mine T of Timna. Preliminary Report of the Excavations, March - May 2001. *Institute for Archaeometallurgical Studies (IAMS) Newsletter* 23: 21–26.

Hooson, W. 1747. *A Miners Dictionary* (Reprinted by the Inst. of Mining & Metallurgy, 1979). London.

Rothenberg, B. 1980. Die Chronologie des Bergbaubetriebes. In: Conrad, H.G. and Rothenberg, B., eds. *Antikes Kupfer im Timna-Tal*. Bochum: 181–185.

Rothenberg, B. and Glass, J. 1992. Beginnings and Development of Early Metallurgy and the Settlement and Chronology of the Western Arabah, from the Chalcolithic Period to Early Bronze Age IV. *Levant* 24: 141–157.

Rothenberg, B. and Merkel, J.F. 1998. Chalcolithic, 5th Millennium BC, Copper Smelting at Timna—New Radiocarbon Dating Evidence for Timna Site 39. *Institute for Archaeo-Metallurgical Studies (IAMS) Newsletter* 20: 1–3.

Willies, L. 1991. Ancient Copper Mining at Wadi Amram, Israel. *Bulletin of the Peak District Mines Historical Society* 11: 3.

CHAPTER 6

THE INSCRIPTION OF RAMESSESEMPERE IN CONTEXT[1]

Deborah Sweeney

The royal butler Ramessesempere, named on the rock stela at Timna depicting Ramesses III offering to the goddess Hathor, is also known from other monuments in Egypt, which show that he was of Levantine origin, which probably facilitated his contacts with the local miners. As royal butler, he was the king's personal representative, demonstrating the significance the king attached to this expedition—perhaps because mining was resumed after a hiatus related to his raids on the Shasu of Edom, or perhaps because the route to Timna was now incorporated into a network of expedition routes stretching as far as the Arabian spice trade routes.

In P. Harris, Ramesses III claims to have obtained vast amounts of copper from a place named Atika; I suggest this is not Timna but southern Sinai, where gigantic smelting works are known at Bir Nasib, dwarfing the activities at Timna. Nonetheless, given that the Egyptians rarely commemorated their copper-mining expeditions, the presence of the Egyptian shrine and rock inscriptions at Timna indicate significant Egyptian activity there during the New Kingdom.

The royal butler Ramessesempere is the only ancient Egyptian visitor to Timna whose name and title are still known. They are inscribed below a stela engraved on the rock face 20 m above the Hathor shrine at Timna, discovered by A. Nussbaumer in February 1972 (Ventura 1974; Schulman 1976: 117–119; Tallet 2012a: 189–190).

DESCRIPTION OF THE STELA (FIGS. 6.1, 6.2)

On the right stands a goddess, wearing an ankle-length dress and tripartite wig, crowned with a sun-disk and horns, and holding an ankh-sign in her left hand. Although she is not named, she must be Hathor, patroness of natural resources abroad, the only Egyptian goddess worshipped at Timna.

On the left is a king, identified as Ramesses III by his cartouches. He wears a long belted kilt and a nemes headdress with a lappet. In his awkwardly drawn right hand, he presents Hathor with a small round vessel, perhaps a *nw*-pot, traditionally used for wine offerings, to which she was partial. In his left, he presents a bowl, maybe for incense.

The outline of a stela is sketched about them, leaving a large space behind Ramesses, and fading out behind his head. The two figures are exceedingly close, leaving no room for Hathor's scepter or a libation stand (Ventura 1974: 61–62).

Yoav Vakhnin pointed out to me that in this relief, the top of the pharaoh's head is almost the same level as the top of the goddess's sundisk, so that he is half a head taller than she is. This is most unusual.

1 This research was supported by the Israel Science Foundation (grant no 903/09). Many thanks to Itamar Gafni for help in preparing this article, and to Pierre Tallet, Uzi Avner and Tali Erickson-Gini for allowing me to refer to their articles while they were in press.

Fig. 6.1 Relief of Ramesses III offering to the goddess Hathor at Timna (after Tallet 2012b: Pl. 130 and Rothenberg 1972: 132, Fig. 143).

Fig. 6.2 Caption below the relief (after Görg 1989: Pl. 29).

Normally, in Egypt, the pharaoh and the goddess stand eye to eye; the two figures are the same height but the horns of Hathor's headdress give her the greater overall height.

However, at the Egyptian temple at Serabit el-Khadem, in southern Sinai, Hathor is sometimes represented shorter than the king or other gods, although her headdress may make her taller overall (e.g., Sinai Inscriptions 254, 267 and 271). These cases, generated by Egyptians in a temple that was primarily for Egyptian use, with only occasional Semitic monuments (see below), indicate that the phenomenon is rooted in Egyptian art and is not the result of Semitic influence.

Some of these cases are either when Hathor grants blessings to the king (Sinai Inscriptions 83, 89), blesses offerings in her own right (Sinai Inscription 89) or when it was necessary to include Hathor in a horizontal register of decoration (Sinai Inscriptions 83, 124a). In other cases, she is seated and would have been taller than the king if standing (Sinai Inscriptions 94h, 198).

At Timna, perhaps the king was given greater height to accentuate his key role in resuming Egyptian involvement at the site after some years. Alternatively, the artist may have carved the stela outline and the royal figure before realizing that not enough room was left for Hathor.

TEXTS

Before king's face: Cartouche *wsr mꜣꜥ.t rꜥ mrjj jmn* Usimaaatre Meryamun
Between king and Hathor: Cartouche *nb ḫꜥ.w rꜥmss ḥqꜣ jwnw* Lord of regalia, Ramesses ruler of Heliopolis

Band of text below stela: *jr.t ḥjj jn wbꜣ nsw.t rʿmssw[m]prrʿ mꜣʿ* [*ḫrw*] Making an inspection by the royal butler Ramessesempere, justified (following Görg 1989: 176–177).

Various scholars read the text's opening words in different ways: *jj* "coming" (Ventura 1974: 62), *pꜣ jj* "the coming" (Kitchen 1976: 312) and *hꜣjj* "descending/arriving" (Schulman 1976: 118; Tallet 2012a: 190), but Görg's photograph (1989: Pl. 29) clearly supports the reading, *jr.t ḥjj*.

In Egypt, private persons rarely added their names to royal monuments. However, this practice sometimes occurred when the kingship was weaker, such as on the second victory stela of King Kamose, erected in the late Second Intermediate Period, where the power of the Egyptian kings was limited to southern Egypt. It features not only the name, but also the image, of the official who erected the stela (Galán 1998: 426–427). Similar phenomena are known occasionally in Nubia (Galán 1994: 77–78). At mines and quarries abroad, however, officials often combined personal and royal material in the same inscription (Galán 1994, 1998). Far from the constraints of Egyptian society, officials adopted privileges they could not at home.

RAMESSESEMPERE'S CAREER AND BACKGROUND

In a key article, Berlandini-Grenier (1974) assembled Ramessesempere's monuments. They are mostly funerary, probably from his undiscovered tomb (Berlandini-Grenier 1974: 13–14).

A votive stela, probably from Abydos, Cairo JdE 3/7/24/16, reveals more information about his family and background. He is depicted adoring Merneptah's cartouche, worshipping Osiris and Isis, and presenting funerary offerings to his parents (*ibid.*: 7–8).

He also appears with two other names, attested nowhere else: Meriunu (beloved of Heliopolis), maybe where he worked or lived (Giveon 1965: 200), and Ben-Ozen. The latter is a Semitic name. Schneider (1992: 90) suggested it could be related to terms for might, power, permission or obedience. Giveon (1965: 201) proposed "Son of Iden" (Thamudic for "power, authority") and Görg (1989: 178) suggested a connection with the Osnites, a subdivision of the tribe of Gad.

Ramessesempere's father also has a Semitic name, Iupaʿa (Schneider 1992: 53). His mother's name is unfortunately omitted.

Ramessesempere is moreover entitled Siribashani, derived from the toponym Siribashan, also attested in Amarna Letter 201 (Görg 1989: 178). Furthermore, Ramesses II was involved in the Bashan (Berlandini-Grenier 1974: 12) probably during the second decade of his reign (Kitchen 1964: 68–69).

Ben-Ozen may have reached Egypt as a hostage or prisoner of war (Giveon 1965: 202), or as a job-seeker. Or perhaps it was his father Iupaʿa who came to Egypt, possibly bringing his wife from home, maybe marrying in Egypt, with Ramessesempere growing up there. Since Ramessesempere is named after Ramesses II, he probably came to Egypt or entered court service there during Ramesses II's reign (Rowe 1940: 45; Kitchen 1976: 312). It was normal for foreigners who came to Egypt, settled down and became Egyptianized to blend into Egyptian society (Goldwasser 2010: 40). In Egyptian terms, Ramessesempere was an Egyptian of foreign origin, rather than a foreigner (Schulman 1986: 193 n. 2).

Ben-Ozen's representation on his Abydos stela sheds significant light on his background and aspirations. On the one hand, he took Egyptian names, worshipped Egyptian gods and presented Egyptian-style offerings to his late parents, who are also depicted as elite Egyptians. On the other, he also retained his Semitic name and honored his connections to his family and origins. On his other monuments, however, his Semitic connections are omitted.

Ramessesempere is first attested under Merneptah, when he helped deliver coffins to the royal tomb in the Valley of the Kings.

> Year 7 month 4 of summer, day 14. Date of the arrival of the butler Ramessesempere, the scribe Penpamer and the vizier Panehsy to the Valley of the Kings to have the coffins of Pharaoh (life, prosperity and health!) descend to their place (O CGC 25504 r 9–10).

Involvement with the royal burial was a highly responsible mission. Nonetheless, Ramessesempere was only a butler at this point, a relatively minor official (Schulman 1976: 120).

A chief guardian of the Ramesseum named Ramessesempere erected stela BM 796 (Berlandini-Grenier 1974: 15–19). Since this name is rare, Schulman (1976: 125) identifies him with Ben-Ozen and suggests that he held this post before his promotion to royal butler.

Ramessesempere would probably have to have been about 20 years old to function as butler. The interval between Year 7 of Merneptah and the beginning of Ramesses III's reign would have been 20 years (following Hornung *et al.* 2006: 493). Ramessesempere was thus at least 40 when Ramesses III became king, and about 70 if he was still alive when Ramesses III was murdered. The exact date of the Ramessesempere Timna expedition is unknown, since the inscription is undated (Schulman 1976: 124). Since Ramesses III was fighting the Libyans and the Sea Peoples in his fifth, eighth and 11th regnal years, Ramessesempere's expedition probably occurred either before Ramesses III's fifth regnal year (Kitchen 1976: 312), in which case the royal butler came to the Negev in his 40s, or in Ramesses' second decade (Schulman 1976: 124), in which case Ramessesempere was in his 50s or 60s when he came to Timna.

Ramessesempere may have been promoted afterwards, since he uses the prestigious title "Royal Fanbearer" (see Pomorska 1987) on two relief blocks. One was discovered by Berlandini-Grenier (1974: 1–4) at a Cairo antique shop in 1972; the other (Vienne [Isère] Museum no. 1550) has a venerable history in Egyptology: Jean-François Champollion studied it already in 1820 (Gabolde 1990).

RAMESSES III'S ACTIVITIES AT TIMNA

Votive objects of Merneptah (No. 102 and maybe No. 50), Seti II (No. 26) and Tawosret, the last ruler of the 19th Dynasty (No. 41) were found at the Egyptian temple at Timna (Schulman 1988: 145). Tawosret's reign was ended, or followed, by a brief period of civil war, from which Sethnakht, Ramesses III's father, emerged as victor and founded a new dynasty. Avner (2014: 140) observes that Ramessesempere's presence was particularly useful at this point, because Ramesses III was resuming work at Timna after a hiatus. (Perhaps this issue mattered more to the Egyptians than to the local miners at Timna; the break may have been only a few years.)

Somaglino and Tallet (2013) suggest that cartouches of Ramesses III, discovered at Themilat Radadi in the southern Negev, in the Wadi Abu Gada in the Sinai and near the Tayma Oasis in Saudi Arabia, marked stages on a southeastern expedition route branching off to southern Sinai, the southern Negev and the Saudi Arabian incense route. This route may have simplified expedition logistics by allowing expeditions to travel part of the way together, using the same supply network (cf. Tallet 2003: 479). Perhaps Ramessesempere was involved in establishing this framework, although the reference to his "inspection" may indicate that he came to review progress and make recommendations for fine-tuning it. If so, his mission probably occurred later rather than earlier in Ramesses' reign (cf. Somaglino and Tallet 2013: 511).

P. Harris also mentions that Ramesses III campaigned against the Shasu of Seir and Edom, which Grandet (1994b: 245) relates to Egyptian interests in the Timna copper mines. In the wake of this intervention, an Egyptian envoy to Timna might have needed additional diplomatic skills.

EGYPTIAN-CANAANITE COOPERATION IN MINING EXPEDITIONS

For any or all of the above reasons, Ramessesempere may have been chosen for his task because his ethnic background helped him to liase with the local Canaanite miners working with the Egyptians at Timna.

Egyptian-Levantine cooperation in mining expeditions was an old tradition. Over 600 years earlier, at Serabit el-Khadem in the Sinai, groups of Semites joined Egyptian expeditions to the turquoise mines (Černý 1935; Goldwasser 2012: 10–12; Morenz 2011: 225–244). Their leader Khebded was even depicted or named on four of the commemorative expedition stelae erected at the nearby temple of Hathor (Sinai Inscriptions 85, 87, 92, 112).

The groups identified as Canaanites at Serabit were small—ten to 20 men (Černý 1935: 384). However, Goldwasser (2010: 40) notes that during the Middle Kingdom many people from the Levant settled in Egypt and became Egyptian, just like Ramessesempere's family 600 years later. Many other Egyptians of Levantine background could have accompanied the expeditions as miners, soldiers, donkey drivers and interpreters, but were not normally visible in the texts.

One such individual was the expedition leader Ameny. On Sinai Inscription 95 Ameny's mother's name is recorded as Amet "The Canaanite/Syrian"; on his other monuments she has an Egyptian name, Ituneferu. Ameny's partly Near Eastern background made him an ideal choice to lead such an expedition. The same applies to the chamberlain of the treasury, Ptahwer, described as *ꜥꜣm* (Canaanite/Syrian) in Sinai Inscription 414 (Schneider 2003: 72–73). Several other officials were also *ꜥꜣm*: the steward Sanefer (Sinai Inscription 112d), the chief lector priest Werkherephemut (Sinai Inscription 123b), and the chamberlain of Pharaoh *ꜥꜣm* (Sinai Inscription 85; Schneider 2003: 328, 332, 333). A small obelisk from the temple depicts three Canaanite soldiers (Sinai Inscription 163; Bietak 1996: 19) and an Egyptian-style block statue with an alphabetic inscription (Sinai Inscription 346) was erected by the chief miner Naʿam (Goldwasser 2011: 256, 271–272). They too may have been Egyptians of Near Eastern origin.

In the New Kingdom, expedition participants continued to erect personal monuments at the Serabit el-Khadem temple, including two Canaanite participants, Aper-baʿal (Sinai Inscription 423; Gardiner, Peet and Černý 1952: pl. lxxxix; idem 1955: 212) and [Shalem]shema (Giveon 1981).

Similarly, the expedition leader Pa<en>ru, commander of the border fortress of Tjeku, who accompanied the royal butler Neferrenpet to Bir Nasib in Sinai, has a Semitic name (Tallet 2003: 463–464; 2013: 62–63). Tallet (2003: 480) suggests that the royal butler Ashahebsed from Sinai Inscription 252 was of foreign origin. Like Ramessesempere, he had a loyalist name, including a royal name such as Ramesses, or a noun relating to royalty: Ashahebsed means "rich in jubilee festivals." Helck (1958: 273) suggested that loyalist names characterized Egyptianized foreigners in the Egyptian bureaucracy, because they were dependent on the king (cf. Berlandini-Grenier 1974: 10). However, this is not invariably so (Schulman 1986: 193–194).

Most of these individuals are represented by name only. In the New Kingdom, Egyptianized northerners represented in relief, such as Paenru at Bir Nasib, are depicted as elite Egyptians with no foreign characteristics; this choice was also made by Ptahwer, who, by calling himself "leader of multitudes in the land of the others" (Sinai Inscription 54), may have been distancing himself from his Levantine background (Morenz 2011: 77).

During the Middle Kingdom, other options existed. In Sinai Inscription 95 (Gardiner, Peet and Černý 1952, Pl. xxx; idem 1955: 103–104), Ameny is depicted both as a clean-shaven Egyptian and as bearded, the way Egyptian artistic convention represented northerners (Bietak 1996: 19). Ameny held an important post in the Egyptian bureaucracy. The three soldiers of Sinai Inscription 163 (Gardiner, Peet and Černý 1952, Pl. li; idem 1955: 147) were depicted with characteristically Canaanite weapons and hairstyles.

Khebded, the Canaanite leader, was represented on one stela as an Egyptian and on another as a Canaanite dignitary. On Sinai Inscription 87, Khebded is integrated into a group of Egyptian officials, dressed in Egyptian-style but distinguished by his characteristic Near Eastern "mushroom" hairstyle (Goldwasser 2010: 47). On Sinai Inscription 112 he is portrayed riding a donkey—a status symbol in Canaanite culture, although it had ambivalent connotations for Egyptians (Morenz 2011: 228)—and accompanied by attendants. Nonetheless, Khebded was depicted in remarkably positive terms: his image was relatively large; he was named, and so were his attendants. Maybe these were positive gestures towards the Canaanites: foreign leaders rarely shared monuments with Egyptians on relatively equal terms. In official Egyptian iconography, foreigners were normally represented negatively, as enemies smitten by Pharaoh, or, at best, as vassals bringing tribute (O'Connor 2003: 169–171).

ROYAL BUTLERS

In his Timna inscription Ramessesempere commemorated his title of royal butler rather than his Semitic origin. This was normal in inscriptions at mines and quarries, where expedition leaders frequently stressed their closeness to the king. As trusted servants close to the king, royal butlers not only waited on the king at table (Schmitz 1985: 771), but were also sent on important missions, such as administering Egyptian territories in Nubia, leading mining and quarrying expeditions, directing royal building projects, participating in state commissions of enquiry and in military activities (Berlandini-Grenier 1974: 13: Schulman 1976: 123; Schmitz 1985: 771–772).

Schulman (1986: 197) observed that royal butlers were mostly drawn from the middle class rather than from the aristocracy. They were appointed directly by the king, and answered directly to him, rather than following an established route of promotion via the army or the priesthood, which allowed them to acquire their own network of contacts and supporters. Instead, they depended solely on the king.

The royal butlers included a high proportion of Egyptianized foreigners. Just under a quarter of them had foreign names. Another 20% of them had loyalist names, and this group may also have included foreigners (Schulman 1990: 19), although each case must be considered individually (Schulman 1986: 194). These foreigners would be even more loyal to the king, since they would be relative outsiders to Egyptian networks of prestige and influence (Schulman 1990: 19). Sending a royal butler to Timna thus indicated that the king took this expedition very seriously, since he sent a personal representative (Schulman 1986: 199).

THE SCOPE OF EGYPTIAN ACTIVITY AT TIMNA IN THE NEW KINGDOM

In P. Harris, which lists the king's donations to Egypt's temples, Ramesses boasts of sending an expedition that brought huge quantities of copper from a place called Atika.

> I sent my emissaries to the land of Atika, to the great copper mines which are in this place, when their *menesh*-boats were loaded with them, whilst others went overland on their donkeys. It had not been heard of previously, since the time of (ancient) kings. Their mines were found full of copper, loaded up in the tens of thousands to their *menesh*-boats. Setting forth with them to Egypt, arriving safely. The consignment was heaped under the Window of Appearances in the form of very many ingots, like hundreds of thousands, when they were the colour of thrice refined(?) gold. I had everyone see them as a wonder. (P. Harris 78.1–5; Erichsen 1933: 94.6–12; Grandet 1994a: 338–339)

Normally copper was a standard element of offerings or tribute. Its special display here was significant, since Ramesses III had other treasures to offer. P. Harris 78.6 (Erichsen 1933: 95.13–14; Grandet 1994a:

339) relates: "I sent butlers and officials to the land of *mefkat*, to my mother Hathor mistress of *mefkat*," probably to Serabit el-Khadem, since *mefkat* is turquoise (Valbelle and Bonnet 1996: 139; Tallet 2012a: 120, 122–123, 128, etc., *contra* Levene 1997 who views the term more broadly, including also copper ores), and to the land of Punt (P. Harris 77.10–13; Erichsen 1933: 94.13–95.4; Grandet 1994a: 338) perhaps eastern Sudan, Eritrea, northern Ethiopia, or Saudi Arabia (see summary in Fattovich 2012: 1). The Egyptians viewed Punt as the most distant, exotic destination of their expeditions, so Atika was in the leading league.

Since the toponym Atika occurs only once in Egyptian texts, it is difficult to identify. Gebel Atika west of Suez has been suggested (Gauthier 1925: 137–138), although the rocks at that site contain little copper. Nibbi (1995: 71) argued for extending the scope of the toponym Atika 100 kilometres south to Bir Bekheit, the site of a Ramesside copper mine.

Rothenberg originally identified Atika as Timna (Rothenberg 1972: 201), since he dated the extensive smelting Site 30 to the Late Bronze period, and viewed Timna as the Egyptians' main New Kingdom copper-mining center. Recent work by Ben-Yosef *et al.* (2012: 51–52), however, re-dates Site 30 to the late 12th to 9th centuries BCE. Nonetheless, Tali Erickson-Gini's recent excavations at Site 2 (2014: 47–48, 81) show that the Egyptians were active there during the 13th to 11th centuries BCE (cf. Weisgerber 2006: 24.)

Activity at Timna is dwarfed, however, by the scope of copper smelting in southern Sinai during the New Kingdom, and I suggest that this is where Atika should be sought. The central copper smelting area there was Bir Nasib, where the slag heap, accumulating from the 12th Dynasty into the New Kingdom, is estimated at 100,000 tons, enough to produce 5,000 to 5,500 tons of copper (Abdel-Motelib *et al.* 2012: 14). Ramesses III is not commemorated there, but inscriptions at the Bir Nasib smelting site are rare: five Egyptian texts in all (two very brief), five Proto-sinaitic inscriptions and a drawing (Tallet 2012a: 57–70). The Egyptians commemorated turquoise mining because it was prestigious. Copper for workmen's tools and soldiers' weapons was prosaic, less worthy of commemoration.

Given the meager epigraphic evidence from Bir Nasib, the presence at Timna of a shrine with votive offerings including faience and expensive glass items, and also an offering relief, demonstrate that the Egyptians considered this site significant, and Egyptian presence at Timna in the New Kingdom should not be downplayed.

REFERENCES

Abdel-Motelib, A., Bode, M., Hartmann, R., Hartung, U., Hauptmann, A. and Pfeiffer, K. 2012. Archaeometallurgical Expeditions to the Sinai Peninsula and the Eastern Desert of Egypt (2006, 2008). *Metalla (Bochum)* 19.1(2): 3–59.

Avner, U. 2014. Egyptian Timna Reconsidered. In: Tebes, J., ed. *Unearthing the Wilderness: Studies on the History and Archaeology of the Negev and Edom in the Iron Age* (Ancient Near Eastern Supplement Series 45). Leuven: 103–162.

Ben-Yosef, E., Shaar, R., Tauxe, L. and Ron, H. 2012. A New Chronological Framework for Iron Age Copper Production at Timna (Israel). *Bulletin of the American School of Oriental Research* 367: 31–71.

Berlandini-Grenier, J. 1974. Le dignitaire ramesside Ramsès-em-per-Rê. *Bulletin de l'Institut français d'archéologie orientale* 74: 1–19.

Bietak, M. 1996. *Avaris, the Capital of the Hyksos: Recent Excavations at Tell el-Dab'a*. London.

Černý, J. 1935. Semites in Egyptian Mining Expeditions to Sinai. *Archiv Orientalni* 7: 384–389.

Erichsen, W. 1933. *Papyrus Harris I: Hieroglyphische Transkription* (Bibliotheca Aegyptiaca V). Brussels.

Erickson-Gini, T. 2014. Timna Site 2 Revisited. In: Tebes, J., ed. *Unearthing the Wilderness: Studies on the History and Archaeology of the Negev and Edom in the Iron Age* (Ancient Near Eastern Supplement Series 45). Leuven: 47–84.

Fattovich, R. 2012. Egypt's Trade with Punt: New Discoveries on the Red Sea Coast. *British Museum Studies in Ancient Egypt and Sudan* 18: 1–59.

Gabolde, M. 1990. Une lettre inédite de Jean-François Champollion à la Bibliothèque de l'Académie de Lyon. *Bulletin du Cercle lyonnais d'égyptologie Victor Loret* 4: 6–36.

Galán, J.M. 1994. The Stela of Hor in Context. *Studien zur altägyptischen Kultur* 21: 65–79.

Galán, J.M. 1998. Royal Commissioners and Royal Inscriptions. In: Eyre, C.J., ed. *Proceedings of the Seventh International Congress of Egyptologists, Cambridge, 3–9 September 1995* (Orientalia Lovaniensia Analecta 82). Leuven: 419–428.

Gardiner, A.H., Peet T.E. and Černý, J. 1952. *The Inscriptions of Sinai* (2nd edition), Part I: *Introduction and Plates* (Egypt Exploration Society Memoir 36). London.

Gardiner, A.H., Peet T.E. and Černý, J. 1955. *The Inscriptions of Sinai* (2nd edition), Part II: *Translations and Commentary* (Egypt Exploration Society Memoir 45). London.

Gauthier, H. 1925. *Dictionnaire des noms géographiques contenus dans les textes hiéroglyphiques*, Vol. I. Cairo.

Giveon, R. 1965. Two Egyptian Documents Concerning Bashan from the Time of Ramses II. *Rivista degli studi orientali* 40: 197–202.

Giveon, R. 1981. A New Kingdom Stela from Sinai. *Israel Exploration Journal* 31: 168–171.

Goldwasser, O. 2010. How the Alphabet Was Born from Hieroglyphs. *Biblical Archaeology Review* 36(2): 36–50.

Goldwasser, O. 2011. The Advantage of Cultural Periphery. The Invention of the Alphabet in Sinai (circa. 1840 B.C.E.). In: Sela-Sheffy, R. and Toury, G., eds. *Culture Contacts and the Making of Cultures: Papers in Homage to Itamar Even-Zohar*. Tel Aviv: 255–321.

Goldwasser, O. 2012. The Miners Who Invented the Alphabet: A Response to Christopher Rollston. *Journal of Ancient Egyptian Interconnections* 4(3): 9–22.

Görg, M. 1989. *Beiträge zur Zeitgeschichte der Anfänge Israels. Dokumente – Materialen – Notizen* (Ägypten und Altes Testament 2). Wiesbaden.

Grandet, P. 1994a. *Le Papyrus Harris I (BM 9999)*, Vol. 1 (Bibliothèque d'Étude 109/1). Cairo.

Grandet, P. 1994b. *Le Papyrus Harris I (BM 9999)*, Vol. 2 (Bibliothèque d'Étude 109/2). Cairo.

Helck, W. 1958. *Zur Verwaltung des Mittleren und Neuen Reichs* (Probleme der Ägyptologie 3). Leiden and Cologne.

Hornung, E., Krauss, R. and Warburton, D., eds. 2006. *Ancient Egyptian Chronology* (Handbook of Oriental Studies Section 1: The Near and Middle East, 83). Leiden and Boston.

Kitchen, K.A. 1964. New Light on the Asiatic Wars of Ramesses II. *Journal of Egyptian Archaeology* 50: 47–70.

Kitchen, K.A. 1976. Two Notes on Ramesside History. *Oriens Antiquus* 15: 311–315.

Levene, D. 1997. Expedition to Atika. In: Esmael, F.A. and Hawass, Z.A., eds. *Proceedings of the First International Conference on Ancient Egyptian Mining and Metallurgy and Conservation of Metallic Artifacts; Cairo, 10–12 April 1995*. Cairo: 365–378.

Morenz, L.D. 2011. *Die Genese der Alphabetschrift: Ein Markstein Ägyptisch-Kanaanäischer Kulturkontakte.* Würzburg.

Nibbi, A. 1995. A Note on ATIKA/ATTAKA/TAKA and Copper. *Discussions in Egyptology* 32: 67–74.

O'Connor, D. 2003. Egypt's Views of 'Others'. In: Tait, J., ed. *'Never Had the Like Occurred': Egypt's View of Its Past.* London: 155–185.

Pomorska, I. 1987. *Les flabellifères à la droite du roi en Égypte ancienne* (Prace orientalistyczne 34). Warsaw.

Rothenberg, B. 1972. *Timna: Valley of the Biblical Copper Mines*. London.

Rowe, A. 1940. Newly-Identified Monuments in the Egyptian Museum Showing the Deification of the Dead, together with Brief Details of Similar Objects Elsewhere. *Annales du Service des Antiquités de l'Égypte* 40: 1–68.

Schmitz, B. 1985. Truchseß. In: Helck, W. and Westendorf, W., eds. *Lexikon der Ägyptologie* VI. Wiesbaden: cols. 771–772.

Schneider, T. 1992. *Asiatische Personennamen in Ägyptischen Quellen des Neuen Reiches* (Orbis Biblicus et Orientalis 114). Freiburg and Göttingen.

Schneider, T. 2003. *Ausländer in Ägypten während des Mittleren Reiches und der Hyksoszeit. Teil 2: Die ausländische Bevölkerung* (Ägypten und Altes Testament 42). Wiesbaden.

Schulman, A.R. 1976. The Royal Butler Ramessesemperrēᶜ. *Journal of the American Research Center in Egypt* 13: 117–130.

Schulman, A.R. 1986. The Royal Butler Ramessessamiᶜon. *Chronique d'Égypte* 61: 187–202.

Schulman, A.R. 1988. Catalogue of the Egyptian Finds. In: Rothenberg, B., ed. *The Egyptian Mining Temple at Timna* (Researches in the Arabah 1959–1984, I). London: 114–147.

Schulman, A.R. 1990. The Royal Butler Ramessessamiᶜon. An Addendum. *Chronique d'Égypte* 65: 12–20.

Somaglino, C. and Tallet, P. 2013. A Road to the Arabian Peninsula in the Reign of Ramesses III. In: Förster, F. and Riemer, H., eds. *Desert Road Archaeology in Ancient Egypt and Beyond* (Africa Praehistorica 27). Köln: 511–518.

Tallet, P. 2003. Notes sur la zone minière du Sud-Sinaï au Nouvel Empire. *Bulletin de l'Institut français d'archéologie orientale* 103: 459–486.

Tallet, P. 2012a. *La zone minière pharaonique du Sud-Sinaï I. Catalogue complémentaire des inscriptions du Sinaï* (Mémoires publiés par les membres de l'Institut français d'archéologie orientale du Caire 130). Cairo.

Tallet, P. *et al.* 2012b. *La zone minière pharaonique du Sud-Sinaï I. Catalogue complémentaire des inscriptions du Sinaï* (Mémoires publiés par les membres de l'Institut français d'archéologie orientale du Caire 130.2). Cairo.

Valbelle, D. and Bonnet, C. 1996. *Le sanctuaire d'Hathor, maîtresse de la turquoise. Sérabit el-Khadim au Moyen Empire*. Paris.

Ventura, R. 1974. An Egyptian Rock Stela in Timna. *Tel Aviv* 1: 60–63.

Weisgerber, G. 2006. The Mineral Wealth of Ancient Arabia and Its Use: Copper Mining and Smelting at Feinan and Timna—Comparison and Evaluation of Techniques, Production, and Strategies. *Arabian Archaeology and Epigraphy* 17: 1–30.

CHAPTER 7

A PRELIMINARY ARCHAEOMAGNETIC INVESTIGATION OF THE YOTVATA FORTRESS

Ilana Peters, Lisa Tauxe and Erez Ben-Yosef

The hilltop fortress at Yotvata (ʿAin Ghadian), located 15 km northeast of the Timna Valley, has been associated with copper production at Timna. Since the copper smelting and mining sites at Yotvata were dated in accordance with those of Timna, and since the dating of the main phase of copper production at Timna has recently changed to cover ca. 500 years of the Late Bronze and early Iron Ages, it now becomes necessary to re-examine and redate the finds at Yotvata. A preliminary archaeomagnetic study of pottery sherds from the site was undertaken in order to further constrain the time of activity there, based on a distinct difference of ancient geomagnetic field intensities between the Late Bronze and Iron Ages. Our results show a VADM of 108.9 ± 15.9 ZAm^2, which falls below the geomagnetic intensity seen during the early Iron Age and suggests that the fortress was occupied in the Late Bronze Age (13th–early 12th centuries BCE), contemporaneous with the Hathor Shrine (Site 200), Site 2 and Site 3 in the Timna Valley.

INTRODUCTION

Although systematic research and excavations in the Timna Valley have been conducted for over five decades (Ben-Yosef, this volume), there are still many open questions regarding the maintenance of intense copper production there, including the sources of basic supplies such as food and water, that are completely lacking in the valley itself. The nearest water source is 15 km away at Yotvata (ʿAin Ghadian), which has one fortress associated with the copper production industry in the region (Evenari, Shanan and Tadmor 1971; Meshel 1990). However, this fortress's date has alternated between the Late Bronze Age and the Iron Age (Meshel and Sass 1977; Meshel 1990), a change that reflects the difficulty in relying on pottery typologies alone in this peripheral region (cf. Ben-Yosef *et al.* 2012). Moreover, the dating of the fortress was linked to the date of the peak in production in Timna; as this date has recently changed from the Late Bronze to the Iron Age (*ibid.*; Ben-Yosef 2016), the date of the Yotvata Fortress is now also called into question.

To avoid typological uncertainties and dating by proxy, this study aimed at directly constraining the age of pottery sherds from the fortress through archaeomagnetic investigation. Here we report on preliminary results, based on one sherd that yielded reliable magnetic data.

THE SITE AND ITS PREVIOUS DATING

Yotvata is assumed to have been a significant part of the larger system of copper production in this region as it was the closest possible provider of water and wood (Evenari, Shanan and Tadmor 1971)

(Fig. 7.1). This led to a strong association between copper production activities in the Timna Valley and the Yotvata Fortress (Figs. 7.2, 7.3).

Yotvata Fortress was excavated in 1974–1976 by Ze'ev Meshel of the Institute of Archaeology of Tel Aviv University (Meshel and Sass 1977). It is 50 × 70 m and consists of a casement wall on the northwestern and southwestern sides (the other sides are protected by sheer cliffs), with a distinct gate on the southwestern side (Meshel 1990). The wall, built on bedrock, is constructed of stone to a height of 1 m, with an additional course of unbaked clay, indicating mudbrick, and palm fronds serving as a roof. Ceramics were found in the fortress, including Qurrayah Painted Ware (QPW).

Meshel originally dated this fortress to the Iron Age (Meshel and Sass 1977) mostly based on the assumption that Timna's peak period of copper production was during this period. This date had been suggested even earlier by Nelson Glueck, who believed the fortress to be part of the "Israelite Fortress system" in the Negev (Aharoni 1963). In later publications Meshel dated this site to the Late Bronze Age based on the similarities of finds in the fortress and in Timna, especially the presence of Quarrayya Painted Ware pottery, and the assumption of a strong relationship between the water resources at Yotvata and the Egyptian-controlled copper industry (Meshel 1990).

However, the chronology of the main period of copper production in Timna and the southern Arabah has recently been revised (Ben-Yosef *et al.* 2012), and the Egyptian-related copper production in the region (New Kingdom Egypt, 1300–1150 BCE) is now understood to be only the first phase of a several-centuries-long continuous production that peaked in the 10th century BCE (*ibid.*; Ben-Yosef 2016). Moreover, the QPW that was once considered a Late Bronze Age phenomenon, has been found in several Iron Age contexts and cannot be used as a clear indicator for Late Bronze Age activities (Singer-Avitz 2014; Tebes 2007; Ben-Yosef 2016).

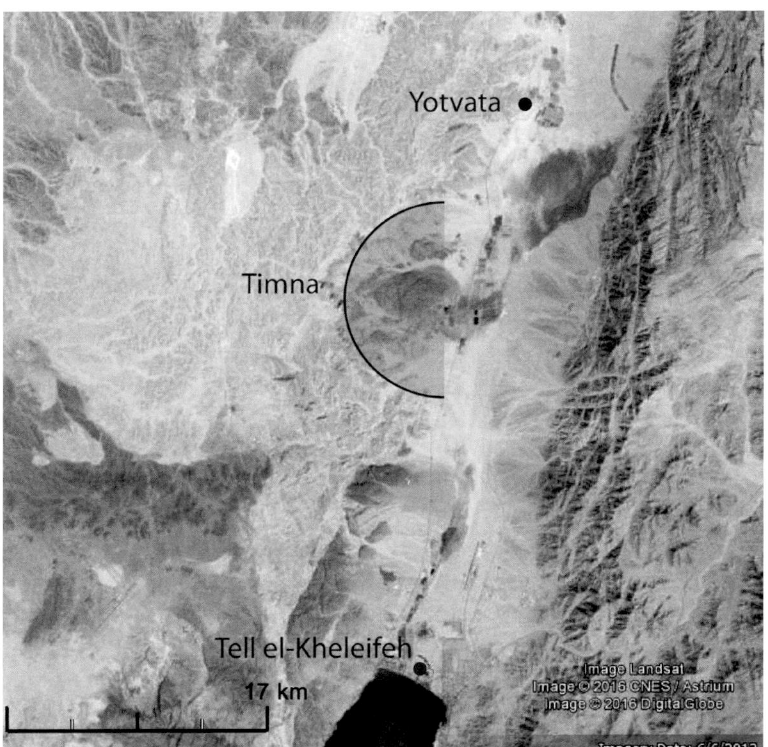

Fig. 7.1: Location map of Timna Valley and Yotvata.

Fig. 7.2: The hilltop fortress at Yotvata next to the modern road (February 2016).

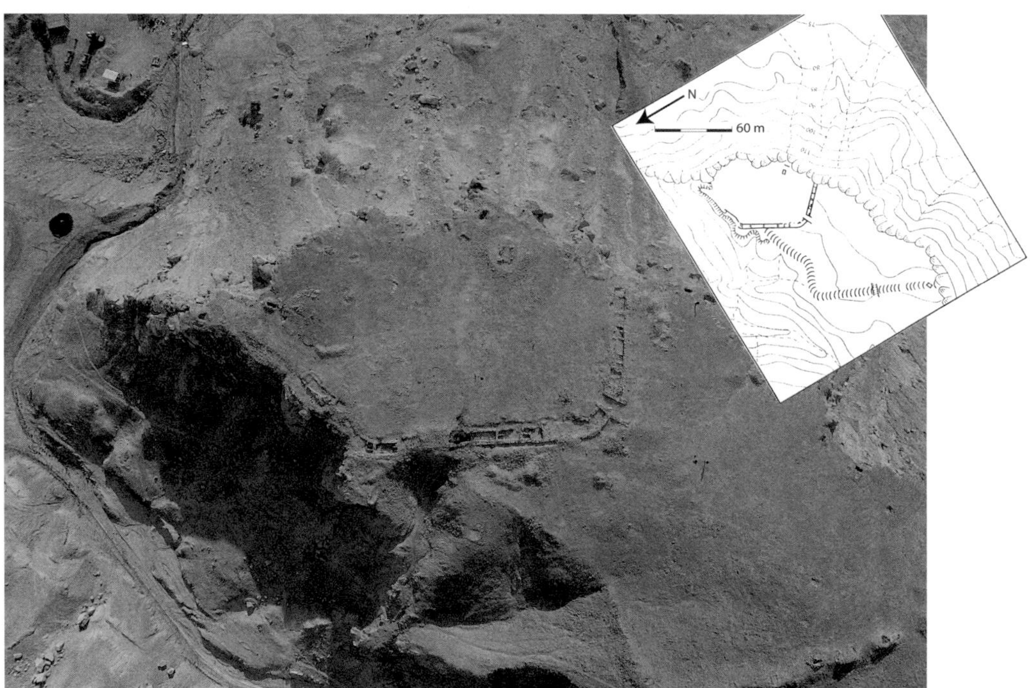

Fig. 7.3: Another view of the Yotvata Fortress (February 2016); inset map after Meshel (1990).

In light of the new chronological framework for Timna and the difficulty with relative dating by pottery typologies, the date of the Yotvata Fortress should be further investigated; here we applied archaeomagnetic experiments to further constrain the date of activities at the site.

METHOD

In this project, we used the changes in the Earth's geomagnetic field to chronologically constrain the manufacturing age of pottery sherds found at the fortress and associated with the main phase of activity there.[1] The Earth's magnetic field rapidly changes over centuries and less as a result of the complicated mechanism of its generation.

Earth's geomagnetic field can be studied based on three attributes: (1) inclination, (2) declination, and (3) intensity, which together represents a geomagnetic field vector specific to a time and place (Sternberg 1997). Inclination and declination are two directional attributes, whereas intensity is not. Additionally, each attribute changes over time in a manner independent of the other two. Therefore, each attribute can outline a range of potential dates for the object and this attribute can be successfully reconstructed by archaeomagnetic experiments. This project concentrated solely on intensity. The other attributes, being directional, are possible to measure only if the object has not moved since its last heating event, for example an intact kiln, which is not the case for the pottery sherds tested during this study (Selkin and Tauxe 2000).

Archaeological material acquires magnetic properties in the following manner: As a heated object cools, the object's ferromagnetic particles' unpaired dipoles will align with the geomagnetic field (Tauxe 2010). Once it is cooled to an ambient temperature, this alignment will not change unless the object is reheated or undergoes chemical weathering or extreme weathering (*ibid*). When the magnetization of the specimen is analyzed, the geomagnetic field at the time of the object's cooling can be determined and then compared to a known curve of regional changes, thus chronologically constraining the object's manufacturing (= heating event) to a limited period of time.

Studies show that many types of heated objects commonly found at archaeological sites yield reliable magnetic data that can be used for archaeointensity dating (Thellier and Thellier 1959). A common material dated by this method is ceramics (e.g., Gallet *et al.* 2009; Valet *et al.* 2003), and likewise in our study we have used pottery sherds to determine when the fortress at Yotvata was active.

Calculating ancient intensity values is based on the principle that the natural remanence magnetization (NRM) acquired by an object during cooling is linearly proportionate to the strength (intensity) of the ambient magnetic field in weak fields, such as the Earth's. The ancient geomagnetic field's intensity can be derived by the following equation (Tauxe 2010):

$$B_{anc} = (M_{NRM})(B_{lab})/(M_{lab})$$

The B_{anc} refers to the ancient geomagnetic field and B_{lab} is a known magnetic field produced in the laboratory and used as part of the experiment which also produces M_{lab}. The M_{lab} is the NRM acquired by the object after the object was heated in a known field (B_{lab}). M_{NRM} is the NRM acquired by an object during the original heating event and is dependent upon the ancient geomagnetic field. In short, two of these elements are produced in an experiment (M_{lab} and B_{lab}), the third measured at the beginning of the experiment (M_{NRM}), and so B_{anc} can be deduced.

The experiments of the current study were conducted at the Paleomagnetic Laboratory of Scripps Institution of Oceanography, University of California, San Diego. The NRM of the object was measured,

1 The site has evidence of earlier activities, including smelting with wind-operated furnaces; see e.g., Avner (2002: 47).

and then the sample was heated at a known magnetic field, cooled, and the sample's newly acquired partial thermal remnant magnetization (pTRM) was then measured. In order to see whether the specimen's ability to acquire NRM changed due to alterations in the crystal lattice during the heating, the specimen was heated in several stages at increasing temperatures and measured in between each heating (Thellier and Thellier 1959). This project used the *IZZI* experimental protocol of the Thellier-Thellier method (Fig. 7.4), in which the specimen was heated to a temperature at a known magnetic field (an in-field), then at a magnetic field of zero (a zero field) which is the *IZ* step (Tauxe and Staudigel 2004). The *ZI* step is heating the object in a zero field, and then heating it in an in-field.

Included in the *IZZI* protocol are several tests applied to determine sample quality, namely the pTRM checks and pTRM tail checks (Tauxe 2010). A pTRM check is conducted in order check whether the specimen's capacity to acquire NRM has changed during the experiment. The samples were heated to a low temperature (T^1) in an in-field, reheated to a higher temperature (T^2) and then reheated to the low temperature (T^1) at the same magnetic field as before. In contrast, a pTRM tail check is to check whether the magnetization has disappeared. In this check, the sample was heated to a temperature (T^1), and then again heated to the previous temperature (T^1) but in a zero field.

Table 7.1 shows the temperatures and magnetic field strength used in this experiment. The results of this project were analyzed via a python program which can be found at http://earthref.org/PmagPy/cookbook/.

The results were analyzed at the sample (= pottery sherd) and specimen (= broken chips of the pottery sherd undergone the experiment and are expected to show same results) level. The selection criteria used in the current study are detailed in Table 7.2.

Cooling rates may also affect our ability to reconstruct the sample's ancient NRM. If the specimen had cooled quickly in ancient times, we can disregard the cooling rate effect. However, slower cooling results in a larger thermal remnant magnetization (Fox and Aitken 1980). This must be taken into account for Yotvata's pottery sherd as it was likely cooled in a kiln over a period of several hours.

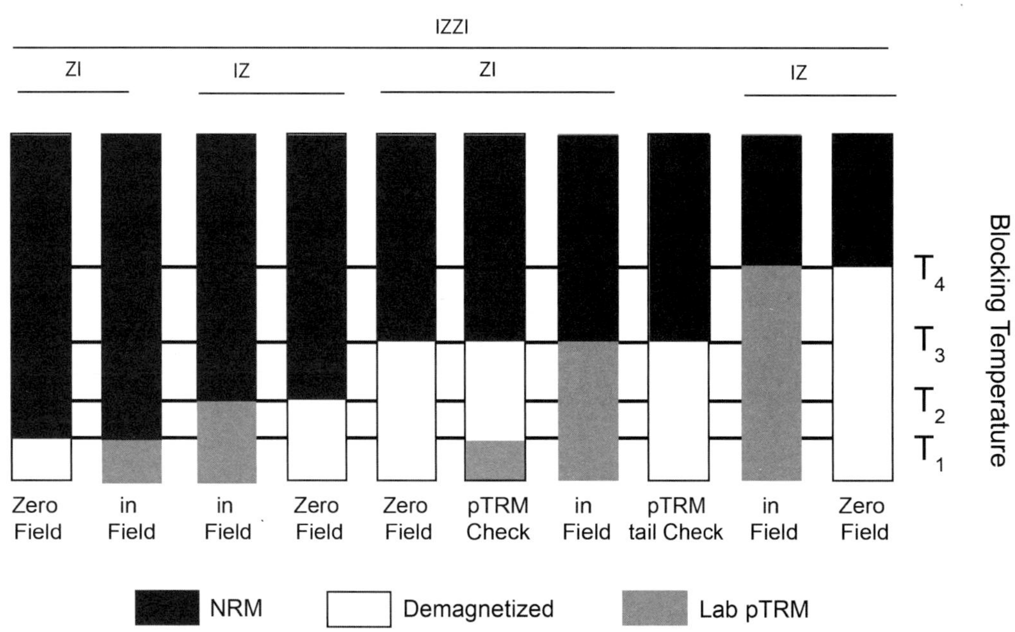

Fig. 7.4: Diagram presenting *IZZI* experimental protocol used in the current experiment (Ben-Yosef *et al.* 2008, and see text for detail).

Table 7.1: Magnetic Field Strength and Temperatures Used for this Project

IZZI Step	*Temperature (°C)*	*Magnetic Field Strength (mT)*
Zero field	100	40
In-field	100	40
In-field	200	40
Zero field	200	0
Zero field	250	0
pTRM check	100	40
In-field	250	40
In-field	300	40
Zero field	300	0
Zero field	330	0
pTRM check	250	40
In-field	330	40
In-field	360	40
Zero field	360	0
Zero field	390	0
pTRM check	330	40
In-field	390	40
In-field	420	40
Zero field	420	0
Zero field	450	0
pTRM check	390	40
In-field	450	40
In-field	480	40
Zero field	480	0
Zero field	500	0
pTRM check	450	40
In-field	500	40
In-field	515	40
Zero field	515	0
Zero field	530	0
pTRM check	500	40
In-field	530	40
In-field	545	40
Zero field	545	0
Zero field	560	0
pTRM check	530	40
In-field	560	40
In-field	580	40
Zero field	580	0

Table 7.2: Specimen Criteria Used in this Project. Sources are as Follows: (1) Kirschvink 1980, (2) Shaar and Tauxe 2013, (3) Tauxe and Staudigel 2004

Criteria Abbreviation	*Criteria Used*	*Definition*	*Failure this Criteria Indicates*	*Source*
int_n	2	Minimum number of data points to be used from the chosen segment of the Arai plot	Not enough data for robust calculation	
n_pTRM	2	Minimum number of pTRM checks to be used from the chosen segment of the Arai plot	Specimen's ability to acquire NRM altered during the experiment	
SCAT	TRUE	Scatter statistic meant to replace DRATS, Z, and MD	The data points in the chosen section in the Arai plot are too scattered: specimen has multi-domain magnetic grains or suffered from alteration	2
FRAC	0.70	Fraction of remanence: vector differences length of selected component divided by total vector differences length	Too little of the experiment's remanence could be used due to scatter or curving in the Arai plot	2
gap_max	0.60	Maximum gap statistic: gap between two data points on an Arai plot	Temperatures chosen were too far apart	2
MAD	7.0	Maximum Angular Deviation	Excessive scatter of Zijderveld plot due, for example, to zig-zagging from pTRM tails	1
DANG	10.0	Deviation of ANGle	Deviation of the best-fit line in the Zijderveld plot caused by specimen carrying multi-component magnetization	3

RESULTS AND CONCLUSIONS

One sample successfully passed all selection criteria and yielded preliminary archaeointensity results (Table 7.3). This was compared to the most updated geointensity curve of the Levant (Fig. 7.5).

The average geomagnetic intensity of the pottery sherd is 108.9 ZAm^2 VADM, with a standard deviation of 15.9 ZAm^2. Comparison to other data from the Levant (Fig. 7.5) suggests that the sherd is dated to the Late Bronze Age (13th–early 12th centuries BCE). While the archaeomagnetic value also allows an Early Islamic date, this has to be discarded for contextual reasons (no evidence of this period at the site, while the main chronological considerations indicate Late Bronze and/or early Iron Age activities, see above).

As mentioned above, cooling rates may also affect our ability to reconstruct the sample's VADM. This must be considered in our analysis of Yotvata's pottery sherd as it was likely cooled in a kiln over a period of several hours. Due to the cooling factor, the VADM can be decreased to a maximum of 3.5% (Genevey, Gallet and Margueron 2003), resulting in a VADM of 105.1 ± 15.9 ZAm^2. This further strengthens our conclusion that it is from the Late Bronze Age and not the Iron Age, as the geomagnetic field had a lower intensity during the Late Bronze Age.

The results from this study correlate with the excavator's more recent publication dating the fortress to the Ramesside period (Meshel 1990), indicating an effort to protect the only oasis in the Timna region during the Late Bronze Age. It is interesting to note that at Timna itself, the copper production during this time had no defensive elements (Yagel, Ben-Yosef and Craddock 2016), thus the oasis is an exception, probably reflecting the importance of this resource in the heart of an extremely arid region.

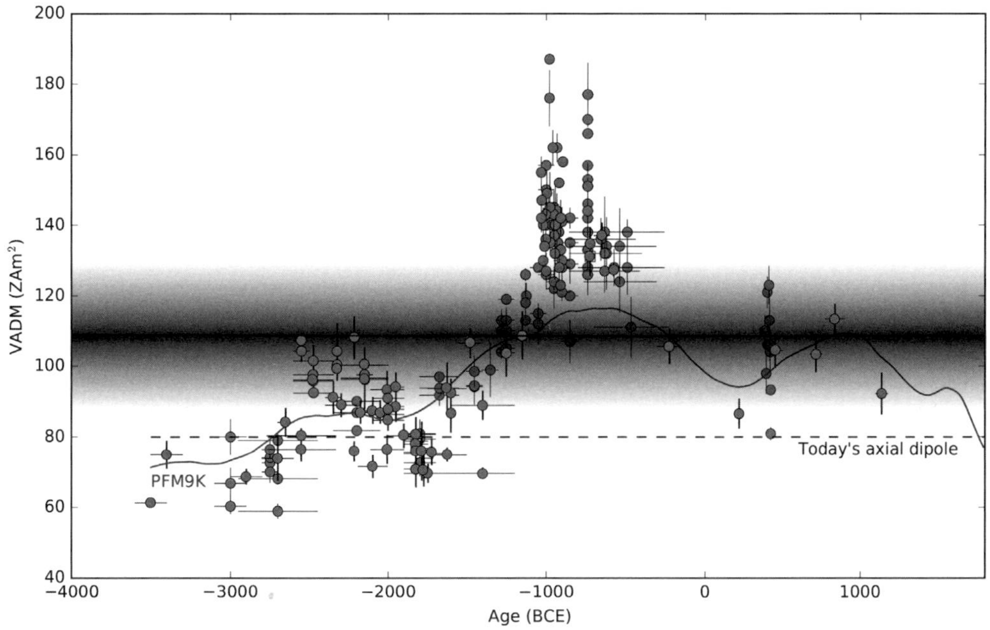

Fig. 7.5. Archaeomagnetic intensity results from a pottery sherd at Yotvata Fortress (this study; the shaded area represents standard deviation of the mean [108.9 ZAm^2]) compared to data from the Levant (after Shaar *et al.* 2016 and references therein). The snaking line at the center is modeled changes by Nilsson *et al.* 2014, and the gray dashed line represents the Earth's current VADM (Finlay *et al.* 2010).

Table 7.3: Successful Specimens and their Results

Specimen	*Passed Criteria*	B_{anc} (μT)	*VADM (Zam^2)*	*VADM (Zam^2) with Cooling Rate Correction*	*Anisotropy Correction*
1	Yes	61.3	120.3	116.1	1.10
2	Yes	49.9	97.9	94.5	1.00

ACKNOWLEDGMENTS

We would like to thank the following individuals and institutions for their help in preparing, measuring and analyzing the samples used in this project: Dr. Ron Shaar and Jason Steindorf at the Scripps Institution of Oceanography, University of California, San Diego and Yael Ebert of the Hebrew University of Jerusalem. This project was supported by the Marie Curie Career Integration Grant (CIG), Grant #33427 (to E.B.-Y.), the Israel Science Foundation (ISF), Grant #1880/17, and the United States-Israel Binational Science Foundation, Grant #20112359 (E.B-Y. and L.T. respectively).

REFERENCES

Aharoni, Y. 1963. Eilat and the Negev. In: Aviram, J., ed. The 18th Congress on Israel Studies (Hebrew). Jerusalem: 54–73.

Avner, U. 2002. *Studies in the Material and Spiritual Culture of the Negev and Sinai Populations, During the 6th–3rd Millennia B.C.* Jerusalem.

Ben-Yosef, E. 2016. Back to Solomon's Era: Results of the First Excavations at Slaves' Hill (Site 34, Timna, Israel). *Bulletin of the American Schools of Oriental Research* 376: 169–198.

Ben-Yosef, E., *et al.* 2008. Application of Copper Slag in Geomagnetic Archaeointensity Research. *Journal of Geophysical Research* 113(B08101).

Ben-Yosef, E., Shaar, R., Tauxe, L. and Ron, H. 2012. A New Chronological Framework for Iron Age Copper Production at Timna (Israel). *Bulletin of the American Schools of Oriental Research* 367: 31–71.

Evenari, M., Shanan, L. and Tadmor, N. 1971. *The Negev, the Challenge of the Desert*. Cambridge.

Finlay, C.C., *et al.* 2010. International Geomagnetic Reference Field: The Eleventh Generation. *Geophysical Journal International* 183(3): 1216–1230.

Fox, J.M.W. and Aitken, M.J. 1980. Cooling-rate Dependence of Thermoremanent Magnetization. *Nature* 283: 462–463.

Gallet, Y., *et al.* 2009. On the Use of Archaeology in Geomagnetism, and Vice-versa: Recent Developments in Archaeomagnetism. *Comptes Rendus Physique* 10: 630–648.

Genevey, A., Gallet, Y. and Margueron, J. 2003. Eight Thousand Years of Geomagnetic Field Intensity Variations in the Eastern Mediterranean. *Journal of Geophysical Research* 108: DOI: 10.1029/2001JB001612.

Kirschvink, J.L. 1980. The Least-squares Line and Plane and the Analysis of Aleomagnetic Data. *Geophysical Journal of the Royal Astronomical Society* 62: 699–718.

Meshel, Z. 1990. *Neve Yotvatah*. Jerusalem (Hebrew).

Meshel, Z. and Sass, B. 1977. Yotvata. *Revue Biblique* 84: 266–270.

Nilsson, A., Holme, R., Korte, M., Suttie, N. and Hill, M. 2014. Reconstructing Holocene Geomagnetic Field Variation: New Methods, Models and Implications. *Geophysical Journal International* 198: 229–248.

Selkin, P.A. and Tauxe, L. 2000. Long-term Variations in Paleointesity. *Philosophical Transactions of the Royal Society of London* Series A 358: 1065–1088.

Shaar, R. and Tauxe, L. 2013. Thellier GUI: An Integrated Tool for Analyzing Paleointensity Data from Thellier-type Data Experiments. *Geochemistry, Geophysics, Geosystems* 14(3): DOI: 10.1002/ggge.20062.

Shaar, R., *et al.* 2016. Large Geomagnetic Field Anomalies Revealed in Bronze to Iron Age Archeomagnetic Data from Tel Megiddo and Tel Hazor, Israel. *Earth and Planetary Science Letters* 442: 173–185.

Singer-Avitz, L. 2014. The Date of the Qurayyah Painted Ware in the Southern Levant. *Antiguo Oriente* 12: 123–148.

Sternberg, R.S. 1997. Archaeomagnetic Dating. In: Taylor, R.E. and Aitken, M.J., eds. *Chronometric Dating in Archaeology*. New York.

Tauxe, L. 2010. *Essentials of Paleomagnetism*. Berkeley.

Tauxe, L. and Staudigel, H. 2004. Strength of the Geomagnetic Field in the Cretaceous Normal Superchron: New Data from Submarine Basaltic Glass of the Troodos Ophiolite. *Geochemistry, Geophysics, Geosystems* 5, Q02H06: DOI: 10.1029/2003GC000635.

Tebes, J.M. 2007. Pottery Makers and Premodern Exchange in the Fringes of Egypt: An Approximation to the Distribution of Iron Age Midianite Pottery. *Buried History* 43: 11–26.

Thellier, E. and Thellier, O. 1959. Sur l'intensité de champ magnétique terrestre dans le passé historique et géologique. *Annales de Geophysique* 15: 285–376.

Yagel, O., Ben-Yosef, E. and Craddock, P. 2016. Late Bronze Age Copper Production in Timna: New Evidence from Site 3. *Levant* 48(1): 33–51.

CHAPTER 8

WHO WAS THE DEITY WORSHIPPED AT THE TENT-SANCTUARY OF TIMNA?

Nissim Amzallag

During the early Iron Age, a Canaanite tent-sanctuary replaced the Hathor temple of the Timna mining area. The shrine retained its metallurgical character but the identity of the deity that replaced Hathor remains obscure. It is suggested here that this deity was no other than YHWH. This opinion is supported by (1) the affinities with metallurgy of the origin, theophany, celestial universe and mode of action of YHWH; (2) biblical indications towards a specific worship of YHWH among the Canaanite metalworkers; (3) indirect evidence regarding a non-Israelite worship of YHWH among the populations involved in copper production in the Arabah Valley; (4) similarities observed between the tent-sanctuary of Timna and the early Israelite cult of YHWH narrated in the Bible. It is concluded here that the metamorphosis of the Hathor temple into a Canaanite tent-sanctuary of YHWH aims towards a "cultural renaissance" accompanied by the renewal of copper production in the Arabah Valley, as well as the emergence of new political entities, including Ancient Israel.

INTRODUCTION

In 1969, the Aravah Expedition led by Beno Rothenberg unearthed a shrine in the midst of the mining area (Site 200) in the Timna Valley. This discovery was especially outstanding due to the richness and the diversity of the findings. Among the five layers Rothenberg described at this site, he identified Strata IV–III (14th–12th centuries BCE) as an Egyptian phase, and Stratum II as a subsequent phase of Canaanite character (Rothenberg 1988: 270–276).

Many of the findings from Strata IV–III (such as the cultic mask, feline figurines and dedications) enabled clear identification of the shrine as a temple dedicated to Hathor (Rothenberg 1988: 276–277). This conclusion is not surprising. A Hathor temple was also built by the Egyptians in the mining areas of Serabit-el Khadim and of Nubia (Valbelle and Bonnet 1996: 8–10). At Serabit, the votive inscriptions evoking the goddess as "the green" (= copper ore), the lady of turquoise (*mafek*), "the golden one", or "the mistress of galena" reveal her close relationship with mineral treasures (Bleeker 1973: 26; Valbelle and Bonnet 1996: 65, 85; Al Ayedi 2007: 23, 26). The bluish glaze that characterize her votive artifacts further suggest her special affinity with copper ore (Pinch 1982: 139–140).

Rothenberg (1972: 150–151, 1999: 171) noticed that significant modifications were carried out at the site during the transition from Strata III to II: the Hathor temple was replaced by a tent-sanctuary comprising a bench and a row of standing stones (*masseboth*), both characteristic of ancient southern Canaanite shrines. Pillars of the Egyptian temple were reused in the tent-sanctuary as standing stones, but only after the representations of Hathor were defaced. This means that the tent-sanctuary was not devoted to the Egyptian goddess, but to an as-yet unknown deity. It is the purpose of this study to investigate the identity of the god worshipped in this shrine.

METALLURGY AND THE TENT-SANCTUARY

The mining and smelting activities did not cease after the Hathor temple was destroyed. A fine analysis of a slag mound from Timna smelting Site 30 recently revealed considerable metallurgical activity during the 11th–10th centuries BCE. These findings led Erez Ben-Yosef *et al.* (2012: 64) to conclude that most of the copper production at Timna occurred in a post-Egyptian context (and see also Ben-Yosef, this volume). It seems, therefore, that the shrine of Site 200 maintained its initial purpose during the Canaanite phase, although the cult of Hathor was no longer practiced there. The following indications support such an assumption:

1. The Canaanite shrine was identified by Rothenberg as a tent-sanctuary due to the large amount of heavy textile found in Stratum II. Rothenberg (1972: 151) noticed that the only two colors of this fabric were red and yellow— the colors of embers and molten metal, expected in a sanctuary devoted to a god of copper smelting.
2. A metallurgical casting workshop with two furnaces has been identified in Locus 109 of Site 200. This workshop was already active during the Egyptian period (Stratum III), and attests to the practice of cultic metallurgy in the Hathor temple (Rothenberg 1988: 192–198). According to Rothenberg (1999: 173), this workshop remained active in the tent-sanctuary (Stratum II).[1]
3. Serpent representations were found in close relation to cultic metallurgy in the shrines of *Bitnah* and *Al Qusais*, in the metallurgical area of *Magan* (Oman) of the end of the second millennium BCE (Benoist 2007: 38). Benoist (*ibid.*: 47–51) noticed the many similarities between these serpent representations (including their association with incense burning) and those seen in the same period in the southern Levant. This suggests that the copper serpent discovered in the *naos* of the tent-sanctuary at Timna (Stratum II) should also be related to cultic metallurgy.

If the shrine's metallurgical vocation was maintained after its transformation into a tent-sanctuary, it may be concluded that the deity worshipped there was also closely related to smelting and metalworking. The identity of the southern Canaanite smelting god remains to ascertained.

THE CANAANITE GOD OF SMELTING

Copper smelting is attested in the Arabah and Negev areas from as early as the fifth millennium BCE (Rothenberg and Glass 1992; Rothenberg, Segal and Khalaily 2004; Shugar 2001: 57–61). Furnace smelting, alloying and metal purification advanced rapidly within the next few centuries (Amzallag 2009a: 502–504; Rowan and Golden 2009: 41–42). This advancement is also evident in the sophisticated Early Bronze Age workshop at Khirbat Hamra Ifdan in the Punon (Feynan) mining area, the largest of its kind in the Ancient Near East. The hundreds of tons of copper that were processed there were certainly the main source of wealth in this desert region (Levy *et al.* 2002). This means that the local god of smelting was probably of central importance during the Early Bronze Age. Copper production was not always extensive during the Bronze Age. However, the huge amount of slag accumulated in the Arabah Valley during antiquity (estimated as 100,000 tons of slag in the Punon area and 10,000 tons at Timna) reveals that metallurgy remained an activity of central economic importance in this region.[2] Accordingly, it is likely that the local god of smelting was not forgotten after the Early Bronze Age.

1 Rothenberg (1988: 273) also noticed another metallurgical workshop at the entrance of the tent-sanctuary in Stratum II (Locus 101). This workshop may have been added during the Canaanite period.

2 For the northern Arabah mining area (Punon), see Bartlett, 1989: 36; Levy *et al.* 2004: 867; Hauptmann 2007. About the same quantity of slag is estimated at the site of Bir Nasib, in the Sinai mining district of Serabit el-Khadim (see Rothenberg 1987: 4). For an estimation of the amount of slag in Timna, see, e.g., Levy and Najjar (2006) and Tebes (2007: 70).

To date this god of smelting remains entirely enigmatic; we do not know of its mention in any text or the existence of any representation. The cults and beliefs of the metalworkers were known to have been secretive (Eliade 1977: 45–53; Kristiansen and Larsson 2005: 52–57; Blakely 2006: 4, 13, 152–156). It is therefore likely that the nature of the southern Canaanite god of smelting remained intentionally enigmatic. This deity apparently had no material representation and its identity remained a secret strictly reserved for metallurgists. It is possible that even mentioning his name was forbidden.

The only famous Canaanite deity whose identity was completely silenced, outside of the Bible, is YHWH. It is explicitly mentioned in the Bible that YHWH is a very ancient Canaanite deity (Ex 3:15; Deut 33:27) whose name and identity remained secret until Moses revealed them to the Israelites (Ex 3:13; 6:3). By promoting a public cult of YHWH, the Israelite theologians apparently violated a traditional taboo. But even after this reform, it seems that YHWH remained an enigmatic and secret deity for the Israelites.[3] Beyond these considerations, some biblical indications enable us to identify YHWH as the traditional god of smelting in southern Canaan:

1. YHWH is assumed to dwell in "mountains of copper" (Zech 6:1–5). He is acknowledged as originating from *Seir* (Judg 5:4), the mountains close to the mining area of Punon.[4] The Book of Isaiah mentions that YHWH intends to give mineral treasures to Cyrus in order to be unambiguously acknowledged by him (Isa 45:3).[5] This means that, as was the case with Hathor in Egypt, YHWH was specifically known for providing mineral resources.
2. In biblical cosmology, the firmament is conceived as a giant metal dome hammered by YHWH. The foundations of the earth were conceived in a similar fashion (Brown 1968; Amzallag 2013: 170–172).[6] This counterintuitive representation is difficult to understand, unless we take into consideration the strong metallurgical affinities of YHWH.
3. The celestial domain of YHWH is envisioned in Ezekiel 1 and 10 as a giant furnace filled with embers and radiant molten metal (ḥašmal) (Driver 1951; Amzallag 2013: 172–175). The Jerusalem Temple, the terrestrial residence of YHWH, which was closely related to his celestial domain (Isa 6:1), is also identified as his "furnace": "Claim of YHWH, whose fire is in Zion, and His furnace in Jerusalem" (Isa 31:9).
4. Volcanism is considered in the Bible as an unavoidable consequence of the presence/revelation of YHWH.[7] This is why the Sinai theophany (as well as many other revelations) is described as a volcanic event (Ex 19:16–19) (Bentor 1990: 336). In Antiquity, the only deity essentially related to volcanoes was the smelting god, due to the strong homology existing between the flowing of lava from a volcano and the release of slag from a furnace (Dieterle 1987). This is another argument confirming the metallurgical background of ancient yahwism (Amzallag 2014).
5. YHWH intervenes on the earth as a divine metallurgist. He is represented as roasting sulfide copper ore, smelting, purifying metals by cupellation and re-melting rusty copper in a furnace (Amzallag 2013: 164–170; 2015a). As revealed in the Book of Isaiah, he is also actively involved in terrestrial

3 This is explicitly attested to in Isaiah 45:15,19; 48:16; Jeremiah 31:33–34; Psalms 25:14; 27:8–11; 31:20.

4 The mentioning of YHWH coming from Paran/Teman (Hab 3:3) and Sinai (Deut 33:2) may also associate him with the mining areas of the southern Arabah and Sinai.

5 "And I [YHWH] will give you [Cyrus] the treasures of darkness, and hidden riches of secret places, that you may know that I am YHWH, who call you by your name, even the God of Israel" (Isa 45:3).

6 This is revealed by the name *rāqîʿa* (firmament), derived from the root *rqʿa* (= to plate metal) also used for the creation of the earth.

7 See Isaiah 13:9; 40:4; 42:14–15; Micah 1:3–4; Joel 3:4; 4:15; Job 9:5; Psalms 97, 104:32, 144:5. See Koenig (1966) for a review of the volcanic theophany of YHWH.

metalworking, wherever it occurs: "Behold, I have created the smith, blow the fire of coals and bring forth a weapon for his work." (Isa 54:16). Cultic metallurgy is therefore explicitly and specifically expected to stimulate his theophany. This conclusion is supported by the identification of kbd-YHWH as the radiance emanating from molten metal (Amzallag 2015b).

6. The emissary (*mal'ak*) of YHWH, the divine being speaking in his name, displays strong affinities with Kotshar, the northern Canaanite smith god (Amzallag 2012). This specific supremacy YHWH holds over the smith god recalls the relationships existing between the mysterious smelters living in mining areas and their clients/representatives—the smiths living in cities.

These observations, when grouped together, leave little doubt that YHWH was traditionally acknowledged in southern Canaan as the god of smelting. We may, therefore, deduce that YHWH was probably the deity worshipped at the tent-sanctuary of Timna.

CULTIC AFFINITIES

The identification of the southern Canaanite god of smelting leads us to look for affinities between the tent-sanctuary at Timna and the biblical cult of YHWH, as it is related in its early stage.

PARALLELS WITH THE ISRAELITE TENT OF MEETING

Rothenberg (1999: 172) assumed a connection between the tent-sanctuary and the biblical *desert tradition of the Tabernacle.* The Tabernacle as described in Exodus 25–28 seems considerably more sophisticated than the Timna tent-sanctuary. However, as Rothenberg already suggested (1972: 184), the Timna sanctuary should be compared with the *'ōhēl mô'ēd* where Moses "encountered" YHWH (Ex 33:7). As shown by Haran (1960), this early sanctuary should be distinguished from the Israelite tabernacle, the conception of which is detailed in Exodus 25–31. In the Bible, the characteristics of this earliest Israelite shrine are totally shrouded, seen only through a pillar of cloud/smoke that appeared shortly after Moses entered the tent (Ex 33:9).[8] At night this pillar of smoke metamorphosed into a pillar of fire, a theme that has already been identified as an expression of the volcanic/metallurgical theophany of YHWH.[9] This means that, in contrast to the tabernacle with theophany in the innermost part, the divine presence here occurs outside the tent (Haran 1960: 55–56). This singularity evokes the activity of a furnace in the entrance of the tent sanctuary from Timna. It is thus likely that mentioning such a pillar evoked the presence of a metallurgical workshop in the "Tent of Meeting." There is a correspondence in structure, therefore, between Moses "encountering" YHWH at the "Tent of Meeting," and the priest-smelters working at the metallurgical workshop at Timna's tent-sanctuary. This parallel is confirmed by the fact that Moses performed a ritual transformation of his rod into a serpent in the name of YHWH (Ex 4:1–5; 7:10). The biblical description of this wonder, the copper nature of the rod and the ability of Aaron and other Egyptians to perform the same act, suggest that such a miracle was no other than the melting of copper, itself symbolized by the serpent (Amzallag 2009b: 395–396, 2013: 161). This "miracle" corresponds exactly to the pouring of molten metal practiced in cultic metallurgy. It clarifies the relationship between serpents, incense (pillar of smoke/cloud) and the liquid element extensively attested in cultic contexts (Benoist 2007).

8 The equivalence between cloud ('ānān) and smoke ('āšān) is revealed in Exodus 19:16, 18.

9 Exodus 13:21, 16:10; Numbers 14:14; Nehemiah 9:12, 19. For the identification of the volcanic nature of the pillar of cloud / fire, see Bentor 1990: 334; Noth 1962: 109; Humphreys 2003: 164–171.

THE COPPER SERPENT

The throne of YHWH is described in Isaiah 6:2 as being surrounded by "burning serpents" (*seraphim*) closely related to the burning of incense with glowing coals (Isa 6:6). Since the Jerusalem Temple was connected in this vision with the celestial throne of YHWH (Isa 6:1), it would seem that serpents and burning incense not only surrounded the celestial throne, but were also a central component of the cult performed at its "terrestrial counterpart." Additional biblical indications confirm the nature of the essential attribute of the serpent in pre-Israelite Yahwism (Amzallag 2016).

A combination of serpents and incense is also characteristic of the Timna tent-sanctuary. The homology between the two is confirmed by the term *seraphim* which is used to describe the burning serpents of the Isaiah vision, as well as the term used in Numbers 21:8–9 to evoke the copper serpent forged by Moses. The Book of Kings (2 Kgs 18:4) confirms that this famous copper serpent (called *Nehushtan*) was present in the Jerusalem Temple and associated with the burning of incense. Its removal from the Temple during Hezekiah's religious reform (2 Kgs 18:4) suggests that it was regarded (by the initiators of this reform) as an obsolete practice belonging to the early phase of the Israelite worship of YHWH.

According to Numbers 21:8, YHWH ordered Moses to forge this cultic copper serpent shortly after the Israelites left *hōr hāhār* (generally identified south of Rekem/Petra). They moved from there towards the Red Sea in order to circumvent the land of Edom (Num 21:4). This means that the Israelite cultic copper serpent was forged in the southern Arabah Valley. This strengthens the homology between the copper serpent of Timna's tent-sanctuary and that of the early Israelite YHWH shrine.

THE EXCLUSIVIST CULT

The main characteristic of the metamorphosis of the Hathor temple into the tent-sanctuary is the disappearance of all the gods and their representations. The defacing of Hathor on the Egyptian pillars, which were then reused in the tent-sanctuary, clearly attests to the exclusivity of the Canaanite smelting god cult at Timna. This evokes the biblical rejection of cultic practices for any gods other than YHWH (Ex 20:2).[10] In the same way that the Egyptian temple was replaced by a tent-sanctuary in which all the other deities were excluded, the Israelites were requested, during the conquest of the "Promised Land," to break down the altars and smash to pieces the pillars and the divine representations in the Canaanite sanctuaries.[11] Hathor is not explicitly mentioned in the Bible. However, she was identified in Canaan as the goddess Ashirat/Asherah.[12] The cult of Hathor was excluded from the tent-sanctuary at Timna, and similarly the prohibition of the Asherah cult became a central motif in the Bible.

ETHNIC CONSIDERATIONS

Two pottery types are encountered during the post-Egyptian period at Timna: the Qurrayah and the Negebite Wares. The former originated in the Hejaz, an area corresponding to the biblical territories of

10 See Schenker (1997). The many gods acknowledged in the Bible (see for example Pss 29:1, 33:6, 82:1, 103:21, 148:2) are regarded as secondary deities to YHWH, who is considered the god of the gods (Deut 10:17).

11 See Exodus 34:12–13; Deuteronomy 7:5, 12:3.

12 The homology between the Canaanite goddess Ashirat and Hathor is revealed by the appellation in Egypt of Hathor as the lady from Gebal. Similarly, the great goddess of Gebal is represented with a typical Hathor headdress (Lipiński 1995: 70). The same is true concerning Ashirat at Ugarit (Mesnil du Buisson 1970: 182–184; Korpel 2001: 139; Stuckey 2002: 133, 135, 148).

Edom and Midian (Rothenberg 1998; Tebes 2007: 85).[13] The latter is found in northern Sinai and in the Negev, the area defined in the Bible as the land of *Amalek* (Rothenberg 1972: 181–182; Tebes 2007: 83–84). This suggests that at least one of these populations was involved in copper production at Timna during the early Iron Age. We must, therefore, evaluate to what extent they acknowledged YHWH at this time.

EDOMITE WORSHIP

Egyptian documents from the Bronze Age evoke the region of *Seir* as the land of *Shossu-Penan*, also termed *Shossu-Srr*. This Shossu population is also named *Shossu-Yah(u)* by the Egyptians, an appellation which suggests that YHWH was traditionally the patron deity of those people involved in copper production at Penan (Weinfeld 1988: 452; Blenkinsopp 2008: 139; Levy 2008a: 257). These findings corroborate the biblical narrative regarding the primeval cult of YHWH among the Kenites, the southern Canaanite metalworkers.[14] They are also confirmed by the biblical account of YHWH's "origin" from Seir (Judg 5:4, Deut 33:2).

Investigation of the cemetery at the copper production area of Wadi Fidan (northern Arabah) reveals an ethnic continuity between the Bronze Age *Shossu* and the Iron Age populations identified with the Qurrayah Ware (Levy, Adams and Muniz 2004: 65–66; Levy 2008a: 253, 2008b: 240–242). Accordingly, the Qurrayah Ware found at Timna from the early Iron Age suggests the presence of a *Shossu*-derived population of metalworkers. This would further support the possibility of a YHWH cult in the tent-sanctuary.

Beyond these considerations, the Bible provides substantial indications for a YHWH cult in Edom. The evidence for this includes: (1) the Kenite/Edomite kinship attested by Esau's genealogy (Gen 36)[15] incorporated the Edomites into the pre-Israelite worship of YHWH which was conducted among the Kenites; (2) YHWH's explicit attribution of his land of origin (*Seir*) to the Edomites (Deut 2:5) is conceivable only if YHWH was worshipped in Edom as a central deity; (3) According to Genesis 27:29, Isaac's blessings, which are related to primogeniture rights, concern mainly a religious authority (necessarily in the YHWH cult). The transfer of these rights from Esau to Jacob reveals that the Edomites were closely related to YHWH even *before* the Israelites; (4) the many similarities between the prophecies concerning Edom and Israel (see Haney 2007: 86–111; Kelley 2009) reveal their parallel status as "people of YHWH," at least in the eyes of the Israelite prophets. For all these reasons, we may conclude that YHWH was worshipped by the population associated with Qurrayah Ware in the Arabah Valley.

AMALEKITE WORSHIP

The abundance of Negebite pottery discovered at Timna led Rothenberg (1999: 170) to identify the miners and/or metalworkers as Amalekites. This conclusion agrees with the biblical account describing Amalek as the son of Timna, the concubine of Elipaz son of Esau (Gen 36:12). This Edomite affiliation of the Amalekites, particularly stressed in the Bible in relation to Timna, provides further evidence for the worship of YHWH among the Timna metalworkers if the identification of the toponym with Manaiyah (the Arabic name for the Timna Valley) is correct.

13 For more about early Iron Age Qurrayah Ware from Timna, see Ben-Yosef *et al.* 2012: 62–63 and Ben-Yosef, this volume.

14 For recent discussions about the Kenite origin of the cult of YHWH, see Blenkinsopp (2008) and Mondriaan, (2011). Weinfeld (1988: 454) identified the Kenites with the *Shossu-Yahu* of the Egyptian documents.

15 See Abramsky (1953: 116, 119). This is confirmed by ethno-historical considerations regarding the populations living in the *Seir* area (Levy 2008b: 243).

The worship of YHWH among the Amalekites is confirmed by a discovery at Kuntillet ʿAjrud (9th–8th century BCE) of a southern Canaanite cult of YHWH, evoked by the explicit description of "YHWH from Teman." The latter description is distinct from the Israelite worship to "YHWH from Samaria."[16] Since Kuntillet ʿAjrud is located in the territory of Amalek, it is likely that this southern Canaanite cult of YHWH also concerned the Amalekites. This conclusion is confirmed in Deuteronomy, where the Amalekites are ashamed of not fearing YHWH because they attacked the Israelites (Deut 25:18). This accusation makes sense only for a people normally expected to fear YHWH. This implies that the Amalekites were regarded by the Israelites as a people who worshipped YHWH as a central deity.

CONCLUSIONS

The present considerations suggest that the tent-sanctuary of Timna was truly devoted to the cult of the Canaanite god of metallurgy, and that this latter god was no other than YHWH. This conclusion is confirmed by the striking similarities found between the worship of YHWH in early Israel, as it is related in the Bible, and the characteristics of the Timna tent-sanctuary. It is also supported by evidence regarding a non-Israelite cult of YHWH among the populations involved in copper production at Timna during the early Iron Age.

In the past, the tent-sanctuary was considered as nothing other than a temporary re-occupation of the Hathor temple destroyed by a natural catastrophe, after the Egyptians left the Timna mining area (Rothenberg 1988: 271, 277–278). In such a context, the metamorphosis of the shrine into a tent sanctuary becomes a circumstantial event of minor historical importance. The recent discovery of the extensive metallurgical activity after the Egyptians left Timna (Ben-Yosef *et al.* 2012), the identification of YHWH as the smelting god worshipped in the tent-sanctuary, the exclusive character of YHWH and the active defacement of Hathor lead us to revisit this opinion. These observations, when combined, suggest rather an active and intentional replacement of the Egyptian temple by a Canaanite sanctuary. These considerations also invite us to integrate this metamorphosis into a large movement of "renaissance" in southern Canaan at the end of the second millennium BCE, coinciding with the social collapse that affected the great regional powers (Mycenae, Hatti, Egypt and Ugarit). In the Arabah Valley, this renaissance is clearly associated with the renewal of extensive mining and copper production (Levy and Najjar 2006; Tebes 2007: 86–87; Levy 2009: 149–150; Levy, Najjar and Ben-Yosef 2014). The present considerations reveal that this movement is also accompanied by the renewal of the worship of the traditional southern Canaanite god of smelting, YHWH, at the expense of foreign cults and practices.

This Canaanite "renaissance" promoted the emergence of new confederations in the southern Levant during the early Iron Age (Midian, Edom, Judah/Israel, Moab and Ammon), all of them considered in the Bible as being affiliated with Abraham, the founding father explicitly blessed by YHWH, through his progeny.[17] This Yahwistic renaissance apparently did not eradicate the cult of the other deities, neither in ancient Israel nor in the neighboring political entities (Stern 2001). Except in Israel, it was apparently not accompanied by transgression of the traditional taboo concerning the mention of YHWH and the secrecy of his cult. Indeed, the combination of exclusion of the other deities with a direct worship of YHWH is currently attested only at the Timna tent-sanctuary and in the early Israelite theology (as it is narrated in

16 See Hadley (1987: 186). For an interpretation of "YHWH of Teman" as a non-Israelite cult of YHWH, see Blenkinsopp 2008: 151; Kelley 2009: 262.

17 YHWH's commitment to Edom, Moab and Ammon is mentioned in Deuteronomy 2:5, 9, 19. The blessing of the Ishmaelite (a term including other southern Canaanite confederations; see Gen 25:13–15) by YHWH is explicitly evoked in Genesis 21:17–18.

the Bible). Both should therefore be considered as radical and uncompromising expressions of a general movement of Yahwistic renaissance occurring in southern Canaan at the beginning of the early Iron Age.

The origin of the Israelite cultic copper serpent from the region of Timna suggests that these two radical developments are interrelated. This assumption is supported on one hand by associating Timna with Amalek (Gen 36:12), and on the other hand by the Amalekite kinship of Ephraim (Judg 5:14), the tribe specifically identified in the Bible with the conquest of the Promised Land. This Amalekite origin is confirmed by the mention of a "Mount of Amalekites" as a prestigious location in the territory of Ephraim (Judg 12:15). It is also supported by the parallel between the *razzia* typically attributed to the Ephraimites, and the pillages regularly performed by the Amalekites.[18] It is also specified that the Ephraimites differ from all other Israelites by their specific accent, a feature which clearly stresses their origin as distinct from that of all the other tribes (Judg 12:5–6). The theology of Ancient Israel seems, therefore, to have evolved in close relation to the transformations which led to the replacement of the Timna Hathor temple by YHWH's tent-sanctuary. It is remarkable that, only three years after the first excavations at Site 200, this point was already unambiguously emphasized by Rothenberg (1972: 184): "Whatever the theological implications one attaches to the Midianite shrine, the discoveries at Timna provide a factual, cultural and historico-geographical background to the early desert part of the Exodus narrative."

REFERENCES

Abramsky, S. 1953. The Kenites. *Eretz-Israel* 3: 116–124 (Hebrew).

al-Ayedi, A.E.R. 2007. The Cult of Hathor as Represented on the Stelae at Serabit el Kadim. *Bulletin of the Egyptian Museum* 4: 23–26.

Amzallag, N. 2009a. From Metallurgy to Bronze Age Civilizations: The Synthetic Theory. *American Journal of Archaeology* 113: 497–519.

Amzallag, N. 2009b. YHWH: The Canaanite God of Metallurgy? *Journal for the Study of the Old Testament* 33: 387–404.

Amzallag, N. 2012. The Identity of the Emissary of YHWH. *Scandinavian Journal of the Old Testament* 26: 123–144.

Amzallag, N. 2013. Copper Metallurgy: A Hidden Fundament of the Theology of Ancient Israel? *Scandinavian Journal of the Old Testament* 27: 155–180.

Amzallag, N. 2014. Some Implication of the Volcanic Theophany of YHWH on His Primeval Identity. *Antiguo Oriente* 12: 11–38.

Amzallag, N. 2015a. Furnace Re-melting as Expression of YHWH's Holiness: Evidence from the Meaning of *qanna* (קנא) in Divine Context. *Journal of Biblical Literature* 134: 233–252.

Amzallag, N. 2015b. The Material Nature of the Divine Radiance and Its Theological Implications. *Scandinavian Journal for the Old Testament* 29: 80–96.

Amzallag, N. 2016 The Serpent as a Symbol of Primeval Yahwism. *Semitica* 58: 208–239.

Bartlett, J.R. 1989. *Edom and the Edomites*. Sheffield.

Benoist, A. 2007. An Iron Age II Snake Cult in the Oman Peninsula: Evidence from Bithnah (Emirate of Fujairah). *Arabian Archaeology and Epigraphy* 18: 34–54.

Bentor, Y. K. 1990. Geological Events in the Bible. *Terra Nova* 1: 326–338.

Ben-Yosef, E., Shaar, R., Tauxe, L. and Ron, H. 2012. A New Chronological Framework for Iron Age Copper Production at Timna (Israel). *Bulletin of the American Schools of Oriental Research* 367: 31–71.

Blakely, S. 2006. *Myth, Ritual and Metallurgy in Ancient Greece and Recent Africa*. Cambridge.

18 Compare Judges 8:1–2; 12:1–5 concerning Ephraim with Deuteronomy 26:17–18 concerning Amalek.

Bleeker, C.J. 1973. *Hathor and Thoth: Two Key Figures of the Ancient Egyptian Religion*. Leiden.

Blenkinsopp, J. 2008. The Midianite-Kenite Hypothesis Revisited and the Origins of Judah. *Journal for the Study of the Old Testament* 33: 131–153.

Brown, J.P. 1968. Cosmological Myth and the Tuna of Gibraltar. *Transactions and Proceedings of the American Philological Association* 99: 37–62.

Dieterle, R.L. 1987. The Metallurgical Code of the Volundarkvida and its Theoretical Import. *History of Religions* 27: 1–31.

Driver, G.R. 1951. Ezekiel's Inaugural Vision. *Vetus Testamentum* 1: 60–62.

Éliade, M. 1977. *Forgerons et alchimistes*. Paris.

Hadley, J.M. 1987. Some Drawings and Inscriptions on Two Pithoi from Kuntillet 'Ajrud. *Vetus Testamentum* 37: 180–213.

Haney, L. 2007. YHWH, the God of Israel... and of Edom? The Relationships in the Oracle to Edom in Jeremiah 49:7–22. In: Goldingay, J., ed. *Uprooting and Planting—Essays on Jeremiah for Leslie Allen*. New York: 79–115.

Hauptmann, A. 2007. *The Archaeometallurgy of Copper—Evidence from Faynan, Jordan*. Berlin.

Haran, M. 1960. The Nature of the ʾOhel Moʿedh in Pentateuchal Sources. *Journal of Semitic Studies* 5: 50–65.

Humphreys, C. 2003. *The Miracles of Exodus*. New York.

Kelley, J. 2009. Towards a New Synthesis of the God of Edom and Yahweh. *Antiguo Oriente* 7: 255–280.

Koenig, J. 1966. Aux origines des théophanies iahvistes. *Revue de l'histoire des religions* 169: 1–36.

Korpel, M.C.A. 2001. Asherah Outside Israel. In: Becking, B., ed. *Only One God?* Sheffield: 127–150.

Kristiansen, K. and Larsson, T.B. 2005. *The Rise of the Bronze Age Society. Travels, Transmissions, Transformations*. Cambridge.

Levy, T.E. 2008a. Ethnic Identity in Biblical Edom, Israel and Midian: Some Insights from Mortuary Contexts in the Lowland of Edom. In: Schloen, D., ed. *Exploring the Longue Durée, Essays in Honor of Lawrence A. Stager*. Winona Lake: 251–261.

Levy, T.A. 2008b. You Shall Make for Yourself No Molten God: Some Thoughts on Archaeology and Edomite Ethnic Identity. In: Dolansky, S., ed. *Sacred History, Sacred Literature. Essays on Ancient Israel, The Bible, and Religion in Honor of R.E. Friedman*. Winona Lake: 239–255.

Levy, T.A. 2009. Pastoral Nomads and Iron Age Metal Production in Ancient Edom. In: Szuchman, J., ed. *Tribes and the State in the Ancient Near East—Cross Disciplinary Perspectives*. Chicago: 147–177.

Levy, T.E., *et al.* 2002. Early Bronze Age Metallurgy: A Newly Discovered Copper Manufactory in Southern Jordan. *Antiquity* 76: 425–437.

Levy, T.E., Adams, R.B. and Muniz, A. 2004. Archaeology and the Shasu Nomads: Recent Excavations in the Jabal Hamrat Fidan, Jordan. In: Propp, W.H.C. and Friedman, R.E., eds. *Le-David Maskil: A Birthday Tribute for David Noel Friedman*. Winona Lake: 63–89.

Levy, T.E., *et al.* 2004. Reassessing the Chronology of Biblical Edom: New Excavations and C^{14} Dates from Khirbat en-Nahas (Jordan). *Antiquity* 302: 865–879.

Levy, T.E. and Najjar, M. 2006. Edom and Copper: The Emergence of Ancient Israelite Rival. *Biblical Archaeology Review* 32: 24–35.

Levy, T.E., Najjar, M. and Ben-Yosef, E. 2014. *New Insights into the Iron Age Archaeology of Edom*, Southern Jordan. Los Angeles.

Lipiński, E. 1995. *Dieux et déesses de l'univers phénicien et punique*. Leuven.

Mesnil du Buisson, R. 1970. *Études sur les dieux phéniciens hérités par l'Empire romain*. Leiden.

Mondriaan, M.E. 2011. Who Were the Kenites? *Old Testament Essays* 24: 414–430.

Noth, M. 1962. *Exodus: A Commentary*. Westminster (English edition).

Pinch, G. 1982. Offerings to Hathor. *Folklore* 93: 138–150.

Rothenberg, B. 1972. *Timna—Valley of the Biblical Copper Mines*. London.

Rothenberg, B. 1987. Pharaonic Copper Mines in South Sinai. *Institute for Archaeo-Metallurgical Studies Newsletters* 10/11: 1–7.

Rothenberg, B. 1988. *The Egyptian Mining Temple at Timna*. London.

Rothenberg, B. 1998. Who Were the "Midianite" Copper Miners of the Arabah? *Metallurgica Antiqua* 8: 175, 197–212.

Rothenberg, B. 1999. Archaeo-Metallurgical Researches in the Southern Arabah 1959–1990. Part 2: Egyptian New Kingdom to Early Islam. *Palestine Exploration Quarterly* 131: 149–175.

Rothenberg, B. and Glass, J. 1992. The Beginning and Development of Early Metallurgy and the Settlement and Chronology of the Western Arabah, from the Chalcolithic Period to the Early Bronze Age IV. *Levant* 24: 141–157.

Rothenberg, B., Segal, I. and Khalaily, H. 2004. Late Neolithic and Chalcolithic Copper Smelting at the Yotvata Oasis (South West Arabah). *Institute for Archaeo-Metallurgical Studies Newsletters* 24: 17–28.

Rowan, Y.M. and Golden, J. 2009. The Chalcolithic Period of the Southern Levant: A Synthetic Review. *Journal of World Prehistory* 22: 1–92.

Schenker, A. 1997. Le monothéisme israélite: un Dieu qui transcende le monde et les dieux. *Biblica* 78: 436–448.

Shugar, A.N. 2001. *Archaeometallurgical Investigation of the Chalcolithic Site of Abu Matar, Israel: A Reassessment of Technology and its Implications for the Ghassulian Culture* (Ph.D. dissertation, University of London). London.

Stern, E. 2001. Pagan Yahwism: The Folk Religion of Ancient Israel. *Biblical Archaeology Review* 27: 20–29.

Stuckey, J.H. 2002. The Great Goddesses of the Levant. *Bulletin of the Canadian Society of Mesopotamian Studies* 37: 127–157.

Tebes, J.M. 2007. A Land Whose Stones Are Iron, and Out of Whose Hills You Can Dig Copper: The Exploitation and Circulation of Copper in the Iron Age Negev and Edom. *DavarLogos* 6: 69–91.

Valbelle, D. and Bonnet, C. 1996. *Le sanctuaire d'Hathor, maîtresse de la turquoise*. Paris.

Weinfeld, M. 1988. The Tradition of Moses and Jethro at the Mountain of God. *Tarbiz* 56: 449–460 (Hebrew).

CHAPTER 9

TRANSGENDERED COPPER MINING IN THE LEVANT

Laura M. Zucconi

In Genesis 36, in the description of Esau's family, the figure of Timna changes gender in the span of a few verses. She is a concubine, a sister, and then the male head of a clan. Timna's transgendered existence also appears in 1 Chronicles 1. In just 15 verses, Timna is a son, a sister, and finally back to a male as a clan leader. The article uses archaeology to help us understand the function of genealogies in the Hebrew Bible, which originated as mental maps of how the various Canaanite tribes related to one another politically and economically. In anthropological study, oral genealogies have an advantage of fluidity with segmented family branches easily adapting to changing social conditions. In terms of political organization, this is known as the heterarchy model and is not normally attributed to written genealogies. The use of multiple genders for one figure in the biblical texts indicates that the advantage of fluidity is not lost by the act of writing down a genealogy.

INTRODUCTION

From the Late Bronze to the Iron I (ca. 1550–1000 BCE), Egypt exploited metallurgical activity in the region of Timna.[1] Recent excavations at Timna have revealed Canaanite mining operations lasting into the 9th century BCE (Ben-Yosef *et al.* 2012). Apart from the archaeological record, Timna appears in the genealogical lists of Genesis 36 and 1 Chronicles 1 as a son, a sister and a concubine.[2] If the biblical name Timna functions as a personification of the region and its inhabitants (Hicks 1962; Williams 1992), then one cannot help but think that the change in gender and marital status of the figure Timna is directly linked to the fortunes of the copper mines in the Timna region. However, the composition of sources in Genesis 36 and 1 Chronicles 1 can post-date the heyday of Timna copper mining which leaves the question as to why they would preserve all three status markers instead of simply referring to Timna as the lowest, a concubine, given the paucity of mining operations for the site from the Iron II through the Hellenistic period. This preservation likely indicates a social fluidity for the region and its inhabitants that is more complex than just a loss of status as the mines were depleted. Timna's social development reflected in the biblical genealogies followed the heterarchy model as it formed alliances with surrounding Canaanites and later Israelites.

Ultimately, the different designations of son, sister, or concubine in original oral genealogies were created at a moment in time when the politico-economic situation warranted it. We cannot pinpoint

1 Nineteenth and 20th Dynasties.

2 This study follows the usage of Timna (ענמת) in the MT. The LXX verses show the same alternating of gender. Timna is not the only name to change genders in the biblical genealogies. Oholibamah, Anah, Maacah, and Ephah also do so. Unfortunately, present archaeological evidence does not allow for a similar analysis of the politico-economic conditions necessitating male and female in the *toledot* for these figures.

that exact time because each designation was recycled as politico-economic situations changed. This study focuses on the meaning of son, sister and concubine in the genealogies with only an outline as to how oral genealogies matched the archaeological record. The later biblical use simply recapitulates these politico-economic situations indicated by further archaeological data giving us insight into how genealogies functioned in Israelite culture.

HETERARCHY MODEL

The heterarchy model allows for flexibility in the ranking of elements depending upon the function of an element at various points in the system (Philip 2005). This model commonly explains the political relationship between various city-states. A similar idea has common acceptance in segmented genealogies, "the relationship between segments for certain political purposes may be different from the relationship between the same segments for social or religious purposes" (Wilson 1977). For this study, economic purposes will be added to the list of functional relationships. A tribe's shifting status in a genealogy may be due in part to whether its economic basis is staple or subsistence finance system. If staple finance underpins the economy, the tribe will control outside access to valued commodities such as copper, whereas under the subsistence system a tribe only controls products for its own use.[3] A male figure in the genealogy represents indigenous control of a staple economy. A female figure such as a sister or daughter indicates a subsistence finance system for the area.[4] Sisters symbolize an allied relationship but with limited access to both political and economic power. Concubines in the genealogy mean a staple finance system under foreign control. The designation of Timna as both male and female in the biblical genealogies may refer to the region's status relative to political and economic positions in the framework of Judahite relations during the composition of the J and P sources as well as Chronicles.

One might foresee a problem with this theory in relation to the Timna region. The area lacked significant occupation when the copper deposits were not being exploited (Rothenberg 1988). Yet, Rothenberg does point out that during these periods the Canaanite groups would retreat to a pastoral semi-nomadic mode occupying the more tenable agricultural areas around Timna (Rothenberg 1972). This retraction from sedentarized copper mining for export still fits with the proposed framework of male/female and marital status markers in the genealogies.

TIMNA'S GENDER SHIFTS

BOOK OF GENESIS

In Genesis 36, Timna is first presented as a concubine of Eliphaz, the son of Esau by Adah, who bore Amalek a clan of Edom.[5] Further on in the same chapter, Timna is listed as a sister of Lotan;[6] both of these are from the P source. By the end of the chapter, Timna is simply designated a clan of Esau,[7] a J source passage.

3 Philip uses the terms "wealth" for an export/profit finance system and "staple" when the system lacks export/profit. This study employs the more commonly accepted terms among economic historians of "staple" and "subsistence" for the respective finance systems.

4 This theory assumes a subsistence finance system would not be controlled by a foreign power.

5 Genesis 36:12.

6 Genesis 36:22.

7 Genesis 36:40.

CHRONICLES

Multiple references to Timna with shifting genders also appear in Chronicles. First, Timna is a son of Eliphaz who happens to immediately precede Amalek in the list.[8] This reference uses the male gender like the J source but with the familial links of P. Shortly thereafter, Timna is designated as the sister of Lotan, the son of Seir, a reiteration of P.[9] By the end of the chapter, Timna appears as the first name in the clan list for Edom[10] which is similar to the J source.

DATING OF TEXTUAL SOURCES

Canaanite and Israelite culture from the Bronze through Iron Ages gave prominence to males. It is reasonable then to work with the idea that males represent a higher status or power position than females. Similarly, the hierarchical culture would impart a greater status to sisters than to concubines.

A simple progression from the oldest source using a masculine designation for Timna to the latest using the feminine designation would show politico-economic decline of the region. The third source's middle dating could act as one point on the linear decline, helping to determine when the shift from masculine to feminine occurred; however, the earliest and latest sources, J and the Chronicler respectively, both refer to Timna as masculine while P uses the feminine. Further complicating the matter is the Chronicler's use of "sister" as well. The simple decline in status cannot be the explanation for the gender shift of Timna. Thus, the dating of the three sources could indicate which relationship such as political, economic or religious held the greatest value for genealogical status in the sources.

At the time of the J source's composition in the 9th century, the region of Timna was at the tail end of intense copper mining controlled by indigenous Canaanites thus designated in the genealogy as a male. From the late 8th to the early 6th centuries, Edom likely controlled Timna. Therefore, from the viewpoint of the Priestly source centered at the Temple in Jerusalem (Hurvitz 1974; Rendsburg 1980; Zevit 1982; Friedman 1987), the region suffered under foreign domination. This situation would warrant a female designation in the genealogy.

The earliest date for Chronicles is 539 BCE since it mentions the Persians while a Greek translation of the text would make its *terminus ad quem* about 200 BCE. If we take the 4th century as a general time period for composition, this would place Timna in an ambiguous politico-economic position. The region was no longer firmly under Persian control but not fully dominated by the Hellenistic kings. The Chronicler's blending of the J and P sources, and consequent changes in Timna's gender, allows the genealogy to adapt to the fluid politico-economic context for the region during the 4th century BCE.

ARCHAEOLOGY

Given the overall fluctuation in settlement patterns for the Levant, let us next trace more specifically the history of Timna through the archaeological record. While outlining the developments in Timna's copper mining, this investigation will match the archaeological record to the functional gender markers in Timna's biblical genealogy.

8 1 Chronicles 1:36.

9 1 Chronicles 1:39.

10 1 Chronicles 1:51. The LXX use of the term prince (ηγεμων) designates a much more settled and centralized political structure than the MT term of clan/chief (אלוף).

CHALCOLITHIC

The Chalcolithic period is the first of three distinct operational phases. Excavations by Rothenberg in 1965 revealed a smelting site with one area of habitation nearby. The surveys of the 1966–1967 seasons documented 38 additional habitation sites consisting of small groups of circular stone enclosures (Rothenberg 1972). This indicates a settlement pattern of pastoral semi-nomads in conjunction with small-scale farming at Yotvata, Biqᶜat Uvda, and Biqᶜat Sayarim. Rothenberg noted that ostrich and gazelle traps at Timna resembled traps used in the Transjordan Plateau as well as similarities between the rock engravings at Site 191 and at Ghassul (Rothenberg 1972).

Although the material remains indicate that a Transjordanian group moved into the Timna Valley, with this as the first significant working of the mines and no occupational layers found indicating another group previously inhabiting the area, it would be reasonable to label them indigenous Canaanites that controlled the Timna mines and its related trade. It is this type of economic system, indigenous control of trade, which would be reflected by a male designation for Timna in the later genealogies.

EARLY BRONZE AGE

The Early Bronze Age saw a shift in settlements away from places like Beer-sheba towards the plains and hill country. Salient features for this new settlement pattern were proximity to natural resources and trade routes (Mazar 1990). Rothenberg's excavations show no significant working of the Timna copper mines during EB I–III but that the region did serve as a caravanserai on the east–west routes moving copper from the mines in the more densely populated southern Sinai (Rothenberg 1972; Mazar 1990). Since Timna did not actively mine and trade its resources but the copper trade did flow through the area, it would be difficult to match a gender or marital status to this time period. A female kinship term such as sister may be used to highlight a subsistence economy, given that the mines were not operational. But, if the focus is on the movement of goods, it could be a male or a concubine in the genealogies depending on the exact role of Timna in a staple finance system based on the copper trade.

MIDDLE BRONZE AGE

Rothenberg did not find evidence of any significant copper mining activity at Timna during the Middle Bronze Age but this time period did see an increase in pastoral-nomadism in the Negev and northern Sinai. Unlike the Early Bronze Age, the economy of these predominantly pastoral-nomadic areas was not linked to the trade routes (Mazar 1990). Although the Middle Bronze experienced an increased urbanization, it does not appear to have significantly affected the pattern of use for local copper resources in Timna from the Early Bronze. Under these conditions, it would be easier to designate Timna with a female status such as sister to indicate a subsistence finance system of local resources being used by the indigenous population. Whereas the trade routes of the Early Bronze could mean any gender marker may apply to Timna, the lack of these routes better pinpoints the term sister as the likely marker. A sister represents a group recognized as related, i.e., an accepted Canaanite tribal entity, but that it does not share to the same degree in the politico-economic power structure as an identified male.

LATE BRONZE AGE TO IRON II

The Late Bronze Age into the Iron II comprises the second significant operational period for Timna. Site 200 and the Hathor shrine attest to Egyptian control from the 14th to 12th centuries. Traditionally, it is understood that Ramesside overseers organized the local labor force (Rothenberg 1972) for copper mining until Canaanite groups took over in the mid-12th century while at the same time the turquoise trade operated out of southern Sinai (Rothenberg 1972). Because of this continuity of operation of the mines, the Late Bronze into Iron II will be treated together rather than as distinct periods.

Eight large mining centers were dated to the Late Bronze through Iron I. Ten camps were connected to the smelting sites with their layout in a semi-circular configuration reminiscent of a Bedouin style encampment (Stager 1985). The camps included not just houses but workshops and storerooms. Not all of them were operational simultaneously. Sites 30 and 34 showed defensive walls and towers whereas Sites 2 and 185 had no fortifications. Rothenberg attributed the fortified sites to an early phase of Egyptian control of the mines that were worked by Canaanites; he initially identified the Canaanite groups as Midianties and Amalekites (Rothenberg 1972). The unfortified sites would then date from a later phase of Egyptian control in which strongly militarized oversight was not necessary.

The idea that the walls of Sites 30 and 34 were fortifications housing slaves stems from Glueck's report in 1935. Yet, the ruined walls may have actually served as "boundaries used to segregate the highly specialized metallurgists from common miners and passerbys" (Ben-Yosef *et al.* 2012). Radio carbon dates from Site 30 show no corresponding dates with the 19th or 20th Dynasties. Instead, Egyptian occupation of these sites may have begun in the late 12th to early 11th centuries down to the 9th century (Ben-Yosef *et al.* 2012). Layer II in Area S coincides with campaigns of Shoshenq I so this dating also parallels development of copper production in Wadi Feynan (Ben-Yosef *et al.* 2012).

Site 2 contained an interesting assemblage of three types of pottery. The first type is normal wheel-made pottery consisting of storage jars, bowls, and jugs akin to normal ware of the Late Bronze–Iron I period in Syria-Palestine. The origin of the Timna normal ware could not be identified. The second form is course hand-made pottery originating from the Timna region with similarities to Negebite Ware. The third type is pink buff ware that originated outside of Syria-Palestine, showing similarity to Edomite or Midianite Ware (Rothenberg 1972).

In Area A of Site 2, Rothenberg excavated a rectangular building with low benches, an altar, five standing stones, a libation bowl, goat bones and all three pottery types. He designated the structure a "Semitic temple" (Rothenberg 1972). Area F contained a destroyed *bamah* near a smelting site (Rothenberg 1972). Found amongst the ruins were Midianite and Negebite Ware,[11] faience, Red Sea shells and ostrich eggs. The floor of the *bamah* had goat horns, copper rings, iron armlets and votives of similar style to those found at other *bamot* as well as the Hathor shrine at Timna. The overlap in votives would indicate a mixed population of Canaanites and Egyptians exploiting the mines at Timna. A rock altar at Site 34 revealed an engraving of a camel with a mounted human which Rothenberg interpreted as an indication of Amalekite occupation (Rothenberg 1972).

The Hathor temple, similar to the one at Serabit el-Khadem, was found at Site 200 with Strata II–V being the significant layers. Stratum V dates to the Chalcolithic period but it is unclear if the area was used as a campsite or for worship. A burial was found under what later became the naos of the Hathor temple. Stratum IV contains the temple to Hathor built by Seti I, which was used throughout the 19th and 20th Dynasties. A break in use occurred during the reign of Seti II, though Ramesses III rebuilt it (marking Stratum III of the site). Stratum II dates to 1150 BCE with the temple area now under the control of the Canaanites. This assumption is based on the installation of low benches, standing stones and the red and yellow remnants of what is thought to be a tent.[12] The

11 It is unclear if he means Midianite and Negebite Ware or the course hand-made and pink buff wares also found in Area A.

12 Although these artifacts indicate a Canaanite rather than Egyptian occupation, it is not clear if the Canaanite group can decisively be labelled Midianite. For discussion of ethnicity, see Killebrew 2005.

images of Hathor were defaced and a copper snake with a gilt head was found. Apart from the defaced Hathor, the destruction of the Egyptian temple likely resulted from an earthquake with the Canaanite installation occurring after the quake.

Rothenberg labels as Midianite and Amalekite the two distinct Canaanite groups represented in the material culture (Rothenberg 1972). He sees the Amalekite group as the first Canaanites in the area who worked in conjunction with the Egyptians and eventually were destroyed. The lack of Midianite Ware at Site 30 reinforces this interpretation for Rothenberg.[13] Site 30 consists of three layers with Layer III dating to the 14th century BCE. Layer II begins with large scale operations under Seti I and Ramesses II with a cessation by the time of Ramesses V's reign in the mid-12th century BCE. Layer I sees a brief renewal of mining activity under Shoshenq I in the mid- to late 10th century (Rothenberg 1988). Although a second distinct Canaanite group worked the Timna mines, including Site 200 and the repurposed Hathor temple, they did not simultaneously occupy Site 30 from the 12th to 10th centuries, hence the gap between Layer II and I.

Labeling as Amalekite the first Canaanite group that was destroyed by the reign of Ramesses V creates an anachronistic problem. Use of the term Amalekite, as well as Midianite, falls out of favor for events after ca. 950 BCE (Ephʿal 1982); according to 1 Chronicles 4:42–43, the Amalekites were not destroyed until about the 8th century BCE. The Esau/Edomite *toledot* combine the Midianites and Amalekites which only further undermines the idea of identifiably distinct pottery types for each clan name. The textual references do not support a shift from Amalekite to Midianite occupation of Timna around 1150 BCE. Rothenberg's assessment of two distinct Canaanite groups occupying Timna may be accurate based on archaeological evidence, but the names Amalekite and Midianite do not accurately reflect these group ethnicities.

Based on the pottery typology of Timna, Rothenberg proposes a "Sinai-Arabah Copper Age" with Early, Middle, and Late phases (Rothenberg 1988). By his 1988 study, he also re-categorizes the pottery into five groups rather than the three groups of his 1972 publication. The first group, with the largest representation at 25%, is labelled Midianite. These are connected to the Hejaz with petrographic analysis showing the same sources as pottery from Qurrayyah (Rothenberg 1988). The second group, designated Local Hand Made, comprises 10% of the pottery. This includes Negebite Ware and the originally labelled Amalekite Ware in 1972. The third group called Local Wheel Made, originally Normal Ware, is contemporary with the Local Hand Made pottery. Egyptian pottery makes up the fourth group. The fifth grouping is a combination of painted Egyptian and Local pottery and is mostly limited to the Hathor temple area. Rothenberg essentially discarded the Midianite and Amalekite distinction since it was an artificial (and rather forced) reconciliation with the Hebrew Bible.

Ben-Yosef reinterprets the pottery assemblages and the strata of Site 200 (Ben-Yosef *et al.* 2012). Stratum III may be 13th century BCE under Ramesses II but a cartouche of Ramesses V can lower the date to mid-12th century BCE. Two Qurrayyah Pottery Ware (QPW) fragments recently found at Site 30, suggesting a 10th century BCE date, challenge Rothenberg's Midianite Ware from the 14th century BCE. The QPW also indicates a "local, semi-nomadic tribal society with possible foreign components" (Ben-Yosef *et al.* 2012). The military campaign of Shoshenq I disrupted metallurgic production in the 10th century BCE. Revival occurred soon after but with a more advanced and efficient system possibly influenced by the Egyptians (Ben-Yosef *et al.* 2012). The abrupt end of copper production in the 9th century BCE was likely due to Hazael of Aram or Judah's campaigns against Edom in the early 8th century.

13 I Chronicles 4:42–43.

But even with these readjustments in periodization and pottery typology, the overall pattern of Egyptian domination of the copper mines worked by Canaanites that eventually gave way to local control by a(nother) Canaanite group still stands as a sound theory for the Late Bronze to the Iron II period. This does match the gender markers of Timna in the biblical genealogies. Timna as a concubine would indicate foreign control, i.e., Egyptians, in a staple finance system. The shift to Canaanite control of the mines by Iron I would then receive a male designation, if they maintained the copper trade. If a decline in trade occurred, then the female role of sister would likely apply.

The Iron II period from the 9th century onward showed no significant settlement pattern for the southern Negev. The area of Timna was likely occupied by nomads who moved on a pastoral-trade circuit dominated by settlements in the northern Negev as well as the Israelite and Judahite monarchies (Mazar 1990). A lack of copper mining in the region lasted until the Nabatean period. Inhabitants living within a subsistence economy would then be designated in genealogies as a sister, much like the situation for Timna in the Middle Bronze period. Timna's location placed it under Edomite control during the 7th century BCE (Zucconi 2007). If the genealogies emphasize the politico-economic relationship, then different designations are possible. From the perspective of Timna as an area fully integrated with a monarchy such as Judah, it could be listed as male or if Timna was viewed as a distinct region dominated by Edom, it may be termed a concubine.

Interestingly, this period of significant inactivity for the copper mines from the 9th century onward corresponds to the composition dates for J and P. The fluid nature of Timna's status in the genealogies in Genesis 36 may reflect the authors' need for adaptation in the face of a changing politico-economic environment for their respective time periods, a sensibility preserved by later redaction and combination of the two sources into one document.

CONCLUSION

As shown, the gender and marital status of genealogical figures are not fixed in the biblical texts. Their switching between masculine, sister and concubine designations reflects a change in the politico-economic status for certain tribal groups. The archaeological record of Timna's copper mines indicates the likely politico-economic relationships influencing gender and marital markers in the biblical genealogies. During the Chalcolithic period, the mines at Timna were worked by local Canaanites warranting a male designation. For the Early Bronze Age, there was no significant mining activity but the area benefited from profitable trade routes, a situation that allows for male as well as sister and/or concubine designations. The Middle Bronze Age still had no mining activity at Timna but the trade routes had also shifted away from the area thus leaving the indigenous Canaanites with a subsistence economy and a sister designation. The Late Bronze Age into Iron I saw a reinvigoration of mining activity under Egyptian domination so a concubine would serve as the genealogical marker. By the Iron IIB, Canaanites had reestablished control of the mines, signaling a shift back to a male.

It was not until the 9th century (Iron IIB), corresponding to the Canaanite controlled mining activity around Timna, that the earliest biblical genealogies began to appear in their written form in the J source, which lists Timna as a male. The P source's composition with both a masculine and feminine Timna correlates to the period in which Judah and Edom vied for control of the area. Similarly, the Chronicler's blended use of masculine and feminine reflects the shifting politico-economic systems under fluctuating Persian and Hellenistic domination. These written sources and their later combination through redaction recapitulate gender and marital markers originating with the oral genealogies for Timna.

REFERENCES

Ben-Yosef, E., Shaar, R., Tauxe, L. and Ron, H. 2012. A New Chronological Framework for Iron Age Copper Production at Timna (Israel). *Bulletin of the American Society of Oriental Research* 367: 31–71.

Eph'al, I. 1982. *The Ancient Arabs: Nomads on the Borders of the Fertile Crescent 9th–5th Centuries B.C.* Jerusalem.

Friedman, R.E. 1987. *Who Wrote the Bible*? New York.

Hicks, R.L. 1962. Timna. In: Buttrick, G.A., ed. *Interpreter's Dictionary of the Bible*. New York: 649.

Hurvitz, A. 1974. The Evidence of Language Dating in the Priestly Code: A Linguistic Study in Technical Idioms and Terminology. *Revue Biblique* 8: 24–56.

Killebrew, A.E. 2005. *Biblical Peoples and Ethnicity: An Archaeological Study of Egyptians, Canaanites, Philistines, and Early Israel, 1300–1100 BCE*. Atlanta.

Mazar, A. 1990. *Archaeology of the Land of the Bible 10,000–586 B.C.E.* New York.

Philip, G. 2005. The Early Bronze Age I–III. In: Adams, R., ed. *Jordan: An Archaeological Reader*. Oakville: 161–211.

Rendsburg, G. 1980. Late Biblical Hebrew and the Date of 'P'. *Journal of Ancient Near East Studies* 12: 65–80.

Rothenberg, B. 1972. *Timna: Valley of the Biblical Copper Mines*. London.

Rothenberg, B. 1988. *The Egyptian Mining Temple at Timna*. London.

Stager, L.E. 1985. The Archaeology of the Family in Ancient Israel. *Bulletin of the American Society of Oriental Research* 260: 1–35.

Williams, D.S. 1992. Timna. In: Freedman, D.N., ed. *Anchor Bible Dictionary*. New York: 553.

Wilson, R.R. 1977. *Genealogy and History in the Biblical World.* New Haven.

Zevit, Z. 1982. Converging Lines of Evidence Bearing on the Date of P. *Zeitschrift für die alttestamentliche Wissenschaft* 94: 481–511.

Zucconi, L. 2007. From the Wilderness of Zin alongside Edom: Edomite Territory in the Eastern Negev during the Eighth–Sixth Centuries B.C.E. In: Malena, S. and Miano, D., eds. *Milk and Honey: Essays on Ancient Israel and the Bible in Appreciation of the Judaic Studies Program at the University of California, San Diego*. Winona Lake: 241–256.

SECTION II

NAHAL ʿAMRAM

CHAPTER 10

ANCIENT COPPER MINES AT NAHAL ᶜAMRAM, SOUTHERN ARABAH

Uzi Avner, Hanan Ginat, Sariel Shalev, Sana Shilstine, Boaz Langford, Amos Frumkin, Rachamim Shem-Tov, Sagi Filin, Reuma Arav, Uri Basson, Omer Shamir, Linda Scott-Cummings

Ancient copper mines at Nahal ᶜAmram were first discovered by IDF soldiers in 1949, then visited by N. Glueck in 1950, surveyed by B. Rothenberg during the 1960s and by L. Willies in 1988. In 2010, the authors began a new, multidisciplinary study supported by the Israel Ministry of Science. The archaeological survey recorded in detail the previously discovered mines but uncovered many others, as well as related sites. The main periods of mining were the Late Bronze and Iron Ages (15th to 10th centuries BCE), the Nabataean-Byzantine (1st to 5th centuries CE) and the Early Islamic (7th to 11th centuries CE). The survey and limited excavations yielded ample finds, while additional scientific studies (geological-geomorphological survey, ground penetrating radar, 3D scanning and a variety of chemical analyses) revealed much new information as to the history and technology of the ancient copper industries in the Arabah Valley.

INTRODUCTION

Nahal ᶜAmram, 10 km north of Eilat, is the third largest ancient copper mining center in the Arabah Valley, after Faynan and the Timna Valley (Fig. 10.1). The area is characterized by complex geology and rich, varying lithology, influenced by the adjacent Syrian-African Fault that created the Arabah Valley (Beyth, Eyal and Garfinkel 2011; Ginat *et al.,* this volume). Copper-bearing sandstone is exposed here, among other rock formations. The region's climate is hyper arid, with annual average rainfall of only 20 mm, against 3600 mm of potential annual evaporation. The closest water sources to the copper mines are Beᵓer Orah, 7 km to the north-northeast, ᶜEin ᶜAvrona, 8 km to the east northeast, and the high water table of Aqaba, 15 km to the south-southeast.

The first report of ancient remains at Nahal ᶜAmram was made in 1949 by soldiers of the Israel Defense Forces who visited the site (Braslavi 1952: 327). It was then visited by N. Glueck (1953: 13–15, 1960: 12–14), who described two mines with shafts and galleries and a large miners' camp.[1] During the 1960s the area was surveyed by Rothenberg (1963: 34–38, 61–62, 1967a: 41–49). He described mines on the escarpments of Har Amir (Site 33) and the large miners' camp (Site 33a) with a slag pile and shallow depressions, or "plates," around the camp. This archaeological complex was first related exclusively to the enterprise of King Solomon (10th century BCE, Rothenberg 1967a: 40–42). Rothenberg also described shafts and mines at the head of Nahal ᶜAmram (Site 38) and dated them to the Roman (1963: 61–64) or Byzantine period (1967a: 147). In 1989, two mining areas were documented in detail by

1 According to Glueck's description, the two mines are those we numbered 35/28 and 36/2–3, following the numbering method of the Israel Archaeological Survey (see map in Figs. 10.2, 10.14).

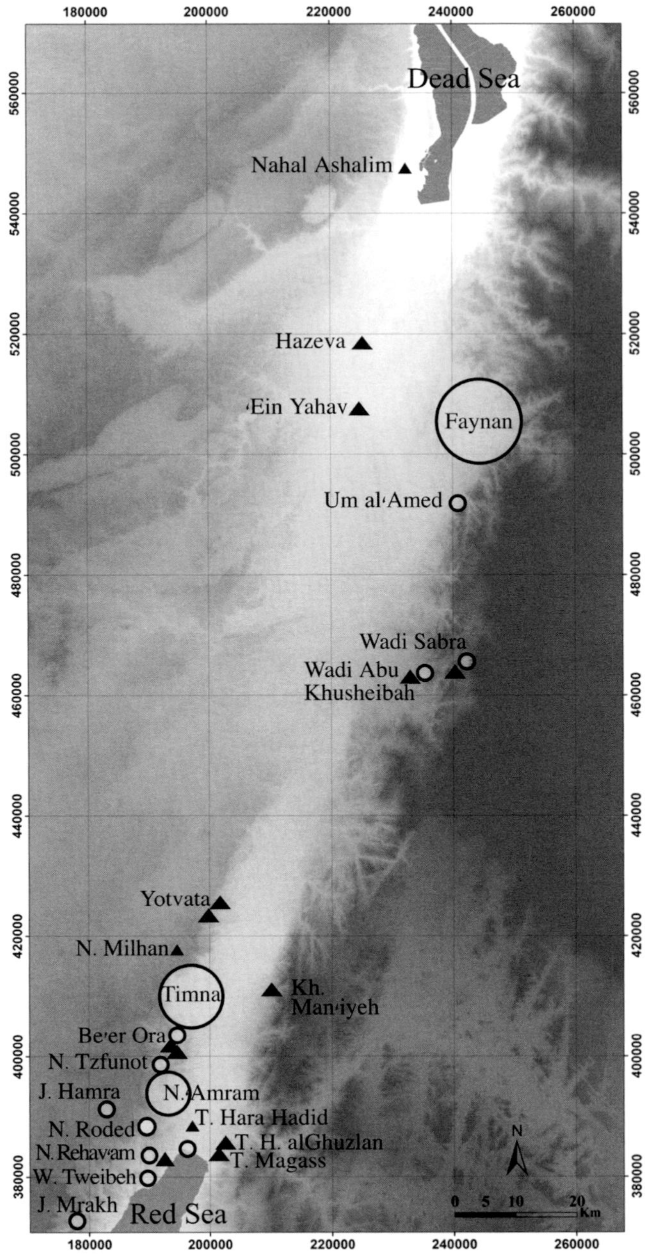

Fig. 10.1: Map of the Arabah Valley. Copper mining areas are indicated by circles and smelting sites by triangles.

L. Willies (1990, 1991), who identified five mining periods: 1. The Late Neolithic to Early Bronze Age (4500–2000 BCE); 2. Narrow shafts and narrow galleries, dated to the Egyptian New Kingdom (14th–12th centuries BCE); 3. Broad galleries and shafts dated to the Roman–Byzantine periods; 4. Similar mining but on a larger scale during the Early Islamic period (7th–8th centuries CE); 5. Some mining or testing in the Mamluk period (13th–15th centuries CE). Rothenberg, in a later publication (1999: 162–166), generally followed Willies' dating.[2]

2 The pottery collections from Rothenberg's and Willies' surveys were examined and dated by M. Gichon of Tel Aviv University.

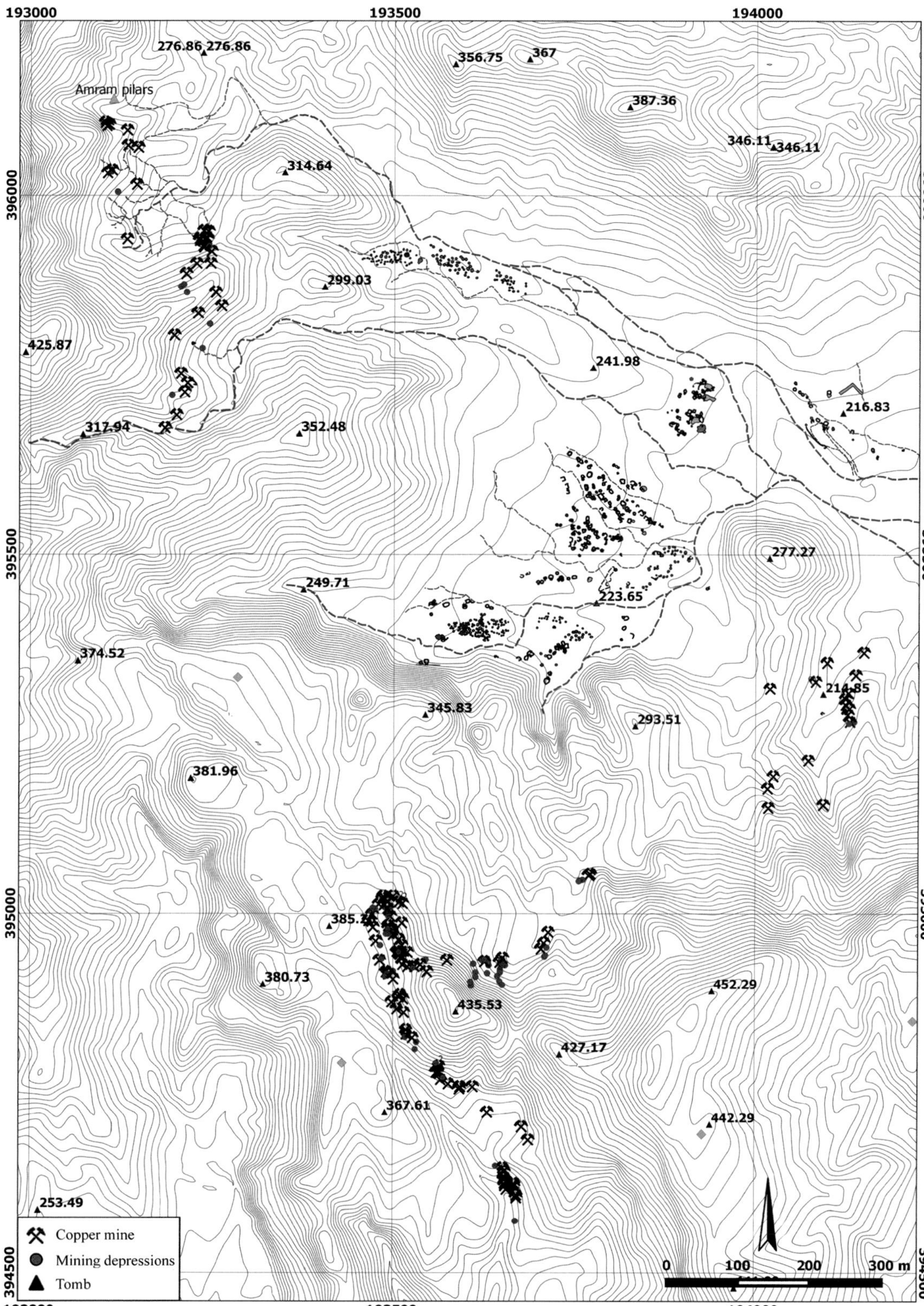

Fig. 10.2: Map showing Nahal ʿAmram and Har Amir, the core area of the survey.

From 2011 to 2013, the authors undertook an interdisciplinary research project, supported by the Israel Ministry of Science and involving several inquiries in the exact science. The study included a meticulous archaeological survey (by Avner and Shem-Tov), detailed mapping of the interior of 47 mines (by Langford), geological and geomorphological studies (by Ginat), a GPR preliminary survey (by Basson), LiDAR scanning of mine dumps and slag piles (by Filin and Arav), chemical analyses of copper ores and slag (by Shilstien and Shalev), ^{14}C dating (by Cummings), and more. Currently, the study of Nahal ʿAmram continues with the support of the Israel Science Foundation (grant No. 60/14).

Following is a brief general description of the research results (for further details, see Arav *et al.*, Ginat *et al.*, Kolska-Horwitz *et al.*, Langford *et al.*, and Shilstein and Shalev, this volume).

ARCHAEOLOGICAL SURVEY AND EXCAVATIONS

Many mines and other sites were recorded during the new survey (Fig. 10.2). Numerous finds were uncovered in limited excavations; here we present only the main results.

THE NEOLITHIC PERIOD

Nine cult sites dated to the Neolithic period were found on the mountains around the copper mines. These sites are part of a much broader phenomenon, with over 300 such sites presently recorded in the Elat Mountains, dated to the seventh and sixth millennia BCE (Avner *et al.* 2014). Rich cultic paraphernalia was found at these sites, including offerings of copper nodules, indicating the interest of the indigenous people in the colorful ore. In fact, copper ore was traded throughout the Near East as early as the Pre-Pottery Neolithic B (eighth–seventh millennia BCE; e.g., Hauptmann 2007: 255–261), and has been found in various uses (e.g., Kingery 1988).

CHALCOLITHIC AND EARLY BRONZE AGE

Exploitation of copper ore for metal production in Nahal ʿAmram probably began early in the Chalcolithic period (ca. 4500 BCE). The northern escarpment of Har Amir was described by Rothenberg (1995: 35, 1999: 163–164) as a Chalcolithic mine, where he apparently referred to open-cast mining. In the new survey, however, we could only identify later mines (see below) and no Chalcolithic artifacts. Currently, evidence for this period at Nahal ʿAmram is limited to four small Chalcolithic–Early Bronze habitations, and some slag with Chalcolithic chemical fingerprints (presently studied by Shilstein and Shalev). Nevertheless, an intensive copper industry has been discovered in two Chalcolithic-Early Bronze I villages in Aqaba, Tell Magass and Tell Ḥujayrat al-Ghuzlan, about a three-hour walk from Nahal ʿAmram (Khalil and Riederer 1998; Hauptman, Khalil and Schmidt-Strecher 2009; Klimscha 2013). A study of the copper ore and artifacts from these sites by Hauptmann, Khalil and Schmidt-Strecker (2009) pointed to the Timna Valley as the source of copper. However, since Nahal ʿAmram is the large copper source closest to Aqaba, we suggested that it could have served as the main source of copper for the Chalcolithic copper industry at these sites. Indeed, a recent pilot study of lead isotopes of copper ores from the Aqaba sites and from the Arabah, by A. Ketelaer and A. Hauptmann, in Bochum (unpublished), supplied the first confirmation to this assumption.[3]

3 The study compared the ratios of lead isotopes Pb 204, 206, 207, 208 in copper ores from the two Chalcolithic villages with those of Nahal ʿAmram, Timna and Faynan. The closest similarity of ratios was found with the ores of Nahal ʿAmram. We are grateful to Hauptmann and Ketelaer for their permission to include the results of their study in this paper.

LATE BRONZE AND IRON AGE

The next period of copper mining and smelting was the Late Bronze and Iron Ages (15th to 10th centuries BCE). Two types of mines are identified in these periods. One was briefly described by Rothenberg (1963: 34–37, 1967a: 41–47), with narrow galleries dug into the sandstone escarpments of Har Amir (erroneously named Givʿat Yocheved by Rothenberg). He also counted 38 narrow shafts that he first interpreted as water holes (Rothenberg 1963: 34–36, Fig. 16; 1967a: 44–46, Fig. 70) as he also did in Timna (1967a: 8–10; 1972: 63–64). Two shafts at Har Amir were considered important support for this interpretation by Rothenberg, but later, in both Timna and Nahal ʿAmram, they were proven to be mining shafts.[4]

During the new survey, 62 narrow mining shafts and 78 horizontal entrances to narrow galleries were recorded (Fig. 10.3). Most shafts are filled today by compact sand and rocks, so only their rims are discernible (Fig. 10.4). A few are still open, connecting to galleries at their bottom or even connecting two different levels of the mines (Fig. 10.5a, b). Both shafts and galleries of Har Amir bear chisel marks (Figs. 10.4, 10.5a) and clearly form one mining system, similar to that of the Timna Valley (Conrad and Rothenberg 1980: 69–167).

A second type of mine from this period is clusters of shallow depressions on alluvial terraces, usually 2–4 m in diameter. To date, 324 such depressions are recorded on the terraces around the miners' camp (Figs. 10.6, 10.8). Initially, the depressions or "plates" were interpreted at Timna and Nahal ʿAmram as ore dressing installations, for crushing and then winnowing the ore (Rothenberg 1963: 38, 1967a: 16), or even for dressing by a flow of water (Rothenberg 1975: 32). Later, however, they were found by Cohen (1976) to be filled-in mining shafts.[5] The "plates" of Nahal ʿAmram are not entirely like those of Timna; they are closer to each other and somewhat deeper. Therefore, it was unclear whether they were covered mining shafts or they were originally open mining depressions dug to capture copper nodules washed and deposited within the alluvium (cf. Ben-Yosef, Levy and Najjar 2009). An attempt to answer this question by means of ground penetrating radar (GPR, by Basson and Shamir, Fig. 7) indicated a possible shaft, but if so, they were actually much wider than the usual shafts, ca. 3 m across, compared to 0.8 m. A dig in one depression (unsurveyed by GPR) was disappointing; it was only 0.5 m deep. Presently it seems that the first option is preferable, but further examinations are still needed, by both GPR survey and excavations. Nevertheless, even now it is clear that these hundreds of depressions add considerable dimension to the copper mining of the Late Bronze and Iron Ages.

Between the clusters of "plates," on a comparatively broad flat area of Nahal ʿAmram, is a large miners' camp previously described by Glueck (1953: 13–14, 1960: 13–14) and by Rothenberg (1963: 34–38, 1967a: 41–53). A ground plan of the camp has now been prepared by a GPS survey, comprising 160 stone-built dwelling structures (hut bases) on an area of 500 × 250 m, divided by small wadis into several "neighborhoods" (Figs. 10.8, 10.9). If an average of five persons were accommodated in one structure, some 800 miners could have occupied the camp during the mining seasons (from late

4 In 1969, Avner first visited the shafts at Har Amir and climbed down to the "water holes" described by Rothenberg. At the bottom of one, galleries extended westward, following a greenish level in the white sandstone bearing the copper nodules (Figs. 10.5a, b). Avner then presented to Rothenberg a series of arguments indicating that the shafts, at both Timna and north ʿAmram, were actually copper mines. In late 1974, the Bochum Mining Museum team, in cooperation with Rothenberg, excavated several shafts in Timna and confirmed that they were indeed mines, not water holes (Conrad and Rothenberg 1980: 69–167).

5 In 1975, E. Cohen excavated a "plate" in the Timna Valley and discovered a shaft in it identical to the so-called water holes (Cohen 1976). From that moment it became clear that the "plates" were actually covered mining shafts. In the following year several "plates" were excavated by Conrad and Rothenberg (1980: 69–72, 127–128; Rothenberg 1978: 55) and fully confirmed Cohen's observation.

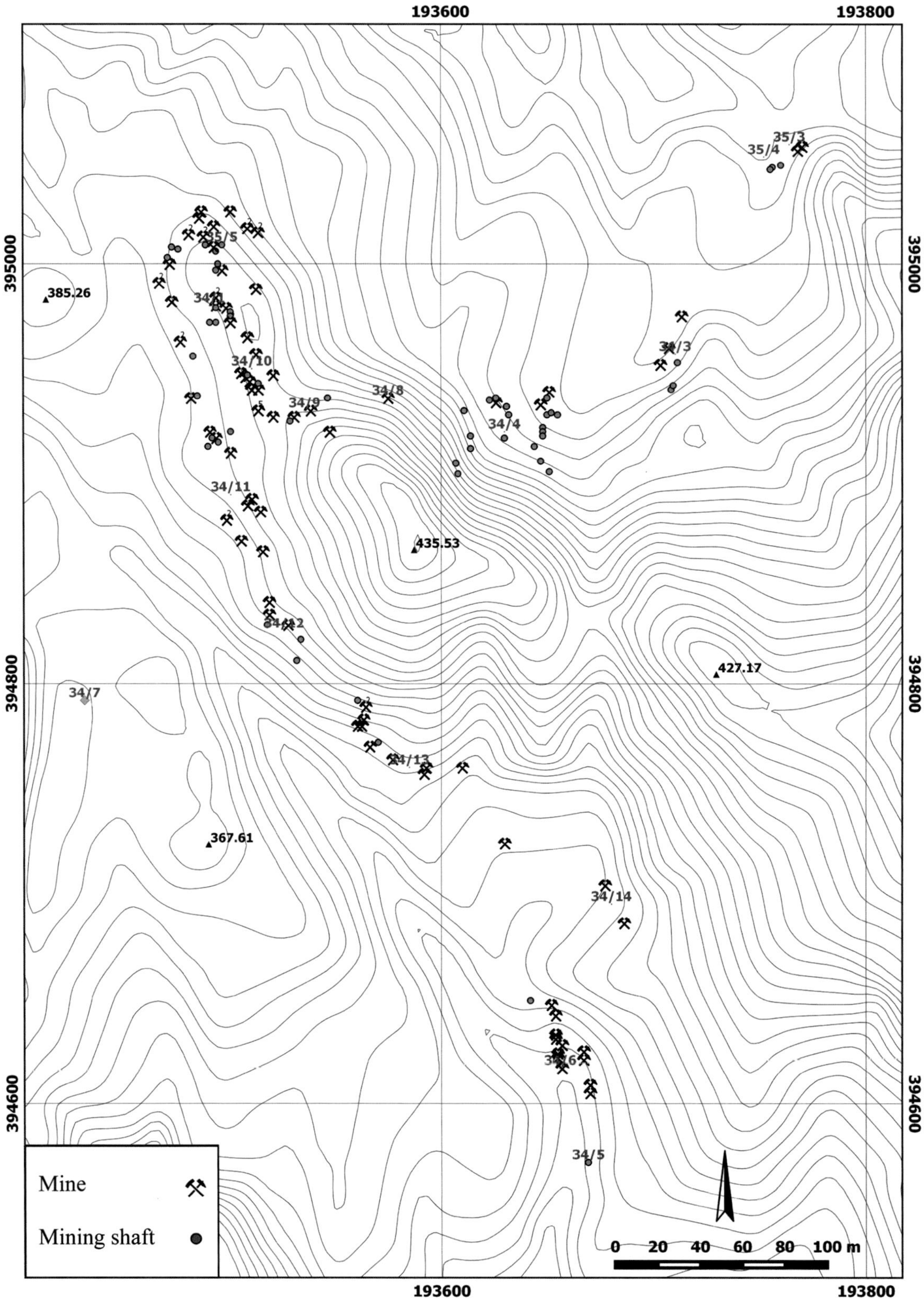

Fig. 10.3: Map of the Har Amir mines (due to close proximity, not all mines are indicated).

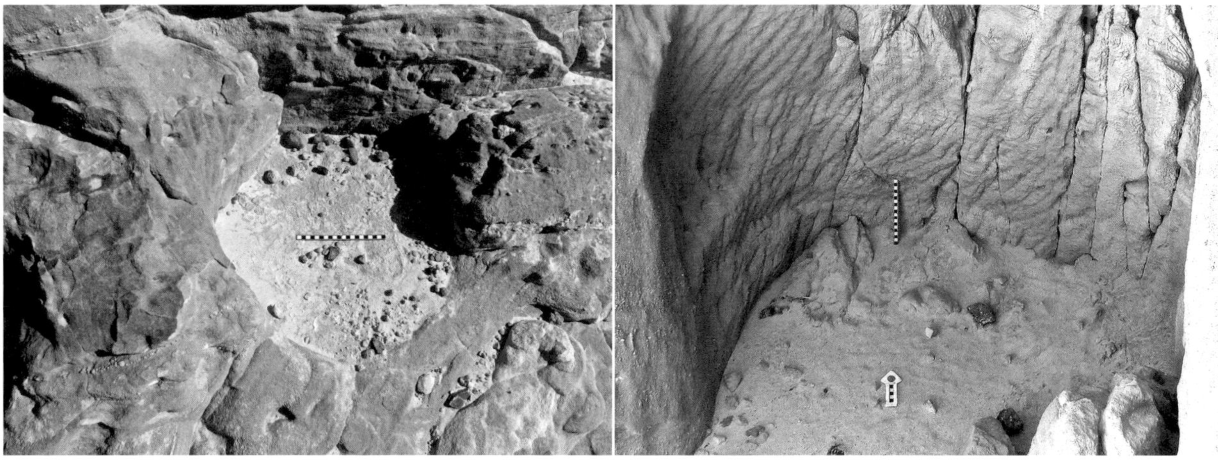

Fig. 10.4: Har Amir Mine 34/4, filled-in narrow shafts, with oblique chisel marks on the shaft's walls.

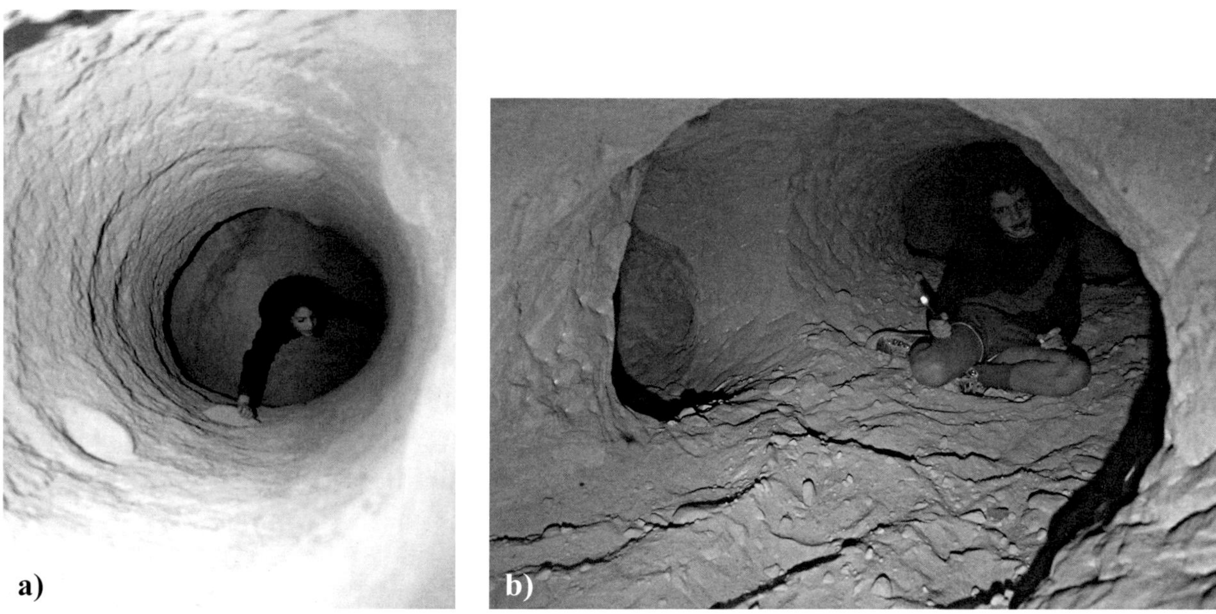

Fig. 10.5: (a) Har Amir Mine 34/9, an open shaft carefully cut by chisel, connecting two levels of the mine; (b) the lower galleries of Mine 34/9.

autumn to spring). Large quantities of stone tools and pottery sherds were previously collected from the camp (by Glueck and by Rothenberg, op. cit.), and artifacts were also taken by visitors during the last six decades. However, in the recent survey we still collected many surface finds in each cluster of structures. These included stone anvils and hammers, some with the remains of crushed copper ore, grinding stones, pottery sherds—mainly of the "regular" wheel-made ware, many fragments of clay furnace linings and of tuyères, and small pieces of furnace slag. The amount of slag is surprisingly small compared to that observed in contemporary camps at Timna. They are found in a thin scatter over most of the camp area, with some denser spots (Fig. 10.8). Due to the small amount of slag, questions arise as to the exact technology of copper smelting in the camp.

The exact time span of the Har Amir mines, the "plates" and the large miners' camp is still unclear. Glueck (1953: 13–14, 1960: 13–14) said the camp was Solomonic, i.e., 10th century BCE, while

Fig. 10.6: A cluster of "plates" at Nahal ʿAmram Site 35/19.

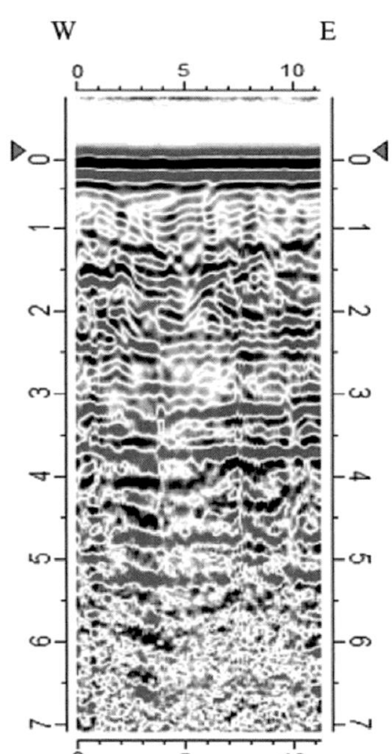

Fig. 10.7: Nahal ʿAmram Site 35/19: a GPR section of a "plate," indicating a possible broad shaft (note the difference between vertical and horizontal scales).

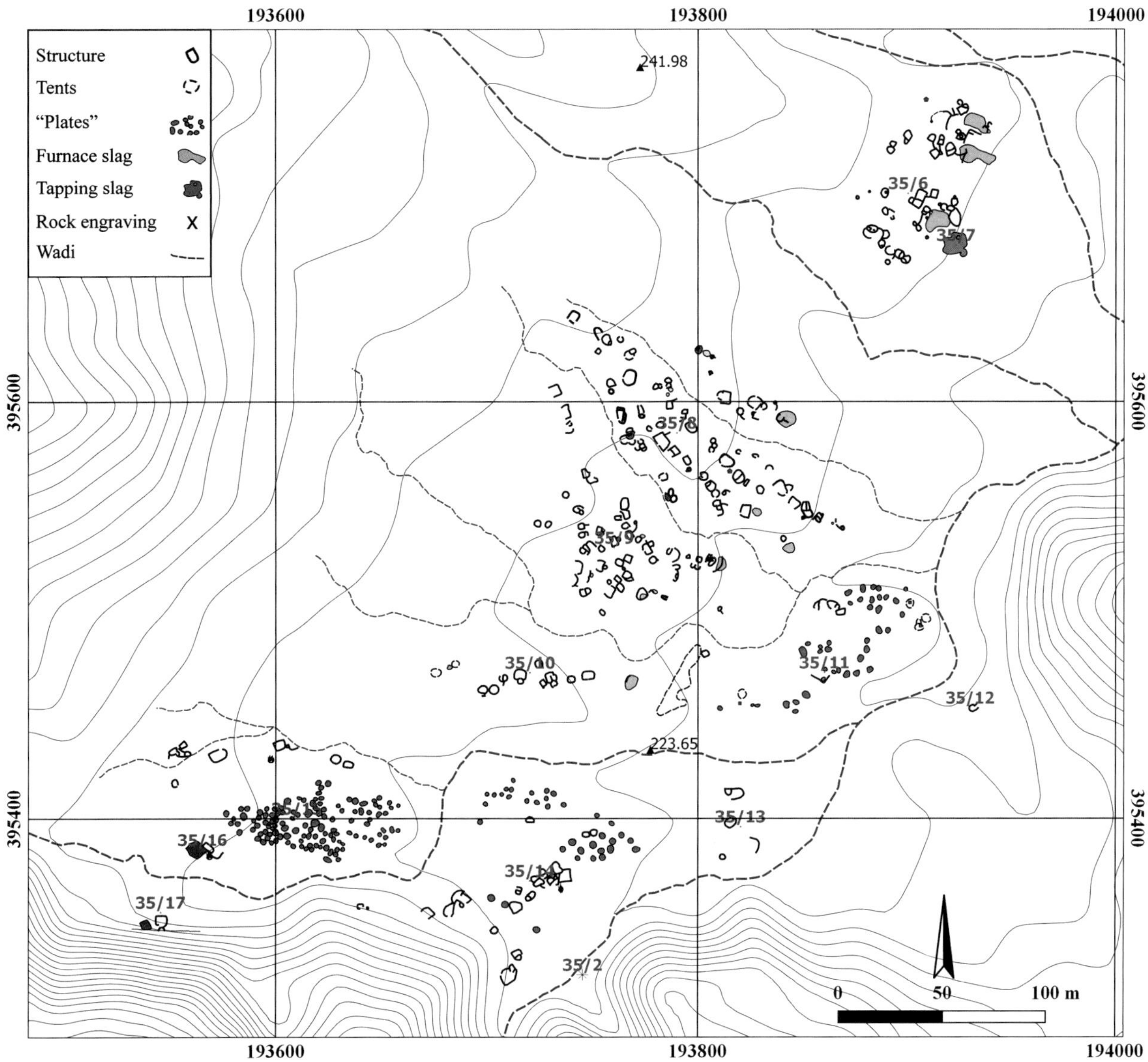

Fig. 10.8: Nahal ʿAmram: a map of the miners' camp.

Rothenberg, in his later publications (1995: 35, 1999: 164) related the complex to the Egyptian New Kingdom, 13th–12th centuries BCE. To date, however, no sign of any Egyptian presence at Nahal ʿAmram has been discovered. In Timna, where many Egyptian artifacts were uncovered in the miners' sanctuary (Rothenberg 1988), a new study demonstrates that the role of Egypt was actually limited (Avner 2014). Moreover, old and new ^{14}C dates from Timna show that the most intensive period of copper production was from the mid-11th to the early 9th centuries BCE, ca. 100–270 years after the end of Egyptian presence in the region (Ben-Yosef *et al.* 2012, Avner *ibid*). This was also the case in the Faynan area, with over 100 ^{14}C dates (Levy, Najjar and Ben-Yosef 2014: 215–222, 129–130, 168–170, 183, 223–228, 814). From Nahal ʿAmram miners' camp, only five ^{14}C date are currently available, from ca. 1420 to 1030 Cal. BCE (Table 10.1: 2–6). Two identical dates, ca. 1220 BCE, fall within the timespan of the Egyptian presence in the region, three others are either earlier or later.

Fig. 10.9: Nahal ʿAmram: the miners' camp, view from the western edge of Har Amir.

Table 10.1: Calibrated ^{14}C Dates from the Nahal ʿAmram Copper Mines and Related Sites (Southern Arabah Valley)*

No.	*Site #* *Site Type*	*Sample #* *Material*	*Depth (Cm)*	*AMS* ^{14}C *BP Date*	*Calibrated Date / Probability (%)*		*~ Mean Cal. Date*
1	Tell Ḥara Ḥadid Smelting site	PRI 4768 Charred twig (unidentified)	20	3322±23	35.2% 33.0% 95.4%	1637–1606 BCE 1584–1454 BCE 1664–1528 BCE	1620/1570 BCE
2	ʿAmram 35/14 Miners' camp	PRI 4764 Charred twig	2	3144±24	68.2% 7.1% 81.0% 7.3%	1446–1401 BCE 1496–1472 BCE 1465–1383 BCE 1340–1317 BCE	1420 BCE
3	ʿAmram 35/14 Miners' camp	PRI 5304 Charred twig	15	2871±24	93.0% 2.4%	1121–974 BCE 957–942 BCE	1030 BCE
4	ʿAmram 35/6 Miners' camp	PRI 4766 Charred twig	10	2984±23	57.3% 12.2% 8.6% 95.4%	261–1192 BCE 1172–1166 BCE 1143–1132 BCE 1276–1132 BCE	1220 BCE
5	ʿAmram 35/6 Miners' camp	PRI 4765 Charred twig	2	2922±23	6.2% 9.8% 52.2% 95.4%	1189–1180 BCE 1159–1145 BCE 1130–1054 BCE 1211–1031 BCE	1120 BCE
6	ʿAmram 35/6 Miners' camp	Pta 4127 Charcoal (Rothenberg 1990a: 72)	?	2920±60	95.4%	1282–932 BCE	1120 BCE

No.	*Site #* *Site Type*	*Sample #* *Material*	*Depth* *(Cm)*	*AMS* *^{14}C BP Date*	*Calibrated Date /* *Probability (%)*		*~ Mean* *Cal. Date*
7	ʿAmram 35/29 Copper mine	PRI 4399 Oil spill	0	1975±22	95.4%	39–71 CE	20/40 CE
8	ʿAmram 35/29 Copper mine	PRI 4352 Wool	0	1782±29	95.4%	138–333 CE	240 CE
9	ʿAmram 35/29 Copper mine	PRI 3036 Fabric	10	1790±15	15.7%	141–197 CE	240 CE
10	ʿAmram 35/29 Copper mine	PRI 4353 Olive pit	10	1724-22	95.4%	251–385 CE	270/340 CE
11	ʿAmram 35/29 Copper mine	PRI 4600 Basket	0	1678±22	2.9% 92.5%	264–274 CE 330–416 CE	360/290 CE
12	ʿAmram 35/29 Copper mine	PRI 4351 Linen	10	1628±22	1.2% 69.1% 3.6% 21.5%	355–365 CE 448–472 CE 448–472 CE 486–535 CE	420 CE
13	ʿAmram 35/32 Copper mine	PRI 5299 Peach pit	20	1776±24	29.7% 38.5% 95.4%	200–260 CE 280–330 CE 140–340 CE	245/310 CE
14	ʿAmram 35/32 Copper mine	PRI 5301 Date pit	15	1758±23	94.8% 0.6%	222–349 CE 371–377 CE	300 CE
15	ʿAmram 35/32 Copper mine	PRI 5300 Walnut shell	25	1712±23	15.3% 52.9% 95.4%	260–280 CE 320–390 CE 250–400 CE	340 CE
16	ʿAmram 35/24 Dwelling cave	PRI 5302 Wood (unidentified)	30	1722±23	95.4%	252–386 CE	260/330 CE
17	ʿAmram 35/24 Mining dump	PRI 3037 Charred twig	5	1800±15	90.6% 4.8%	136–254 CE 301–316 CE	220 CE
18	ʿAmram 36/3 Copper mine	PRI 3038 Twig (unidentified)	0	1765±15	31.9% 36.6% 95.4%	235–260 CE 295–325 CE 220–340 CE	250 CE
19	ʿAmram 35/7 Slag pile	PRI 4166 Charred acacia twig	0	1337±23	88.3% 7.1%	647–740 CE 746–764 CE	670 CE
20	ʿAmram 35/7 Slag pile	PRI 5303 Charred acacia twig	35	1239±23	66.4% 29.0%	687–780 CE 789–872 CE	730/770 CE
21	ʿAmram 35/6 Slag scatter (from 35/7)	PRI 4354 Charred acacia twig	0	1413±22	95.4%	602–659 CE	660 CE
22	ʿAmram 35/7	BM 1163 (Burleigh and Hewson 1979: 350)	0	1240±36	95.4%	681–881 CE	730/770 CE
23	Beʾer Orah Copper furnace	Pta 4117 Charcoal (Rothenberg 1988: 4)	?	1390±50	92.4% 3.0%	560–711 CE 746–764 CE	660 CE
24	Beʾer Orah	Pta 6158 Charcoal (Sharon, Avner and Naḥlieli 1996: 112)	10	1370±20	95.4%	639–676 CE	670 CE

No.	*Site #* *Site Type*	*Sample #* *Material*	*Depth* *(Cm)*	*AMS* *^{14}C BP Date*	*Calibrated Date /* *Probability (%)*		*~ Mean* *Cal. Date*
25	Be'er Orah	Rt 1741 Charcoal (Sharon, Avner and Naḥlieli 1996: 112)	25	1270±55	95.4%	658–881 CE	670 CE
26	Be'er Orah	Pta 6159 Charcoal (Sharon, Avner and Naḥlieli 1996: 112)	15	1210±40	93.7%	687–895 CE	790 CE
27	Be'er Orah	Rt 1742 Charcoal (Sharon, Avner and Naḥlieli 1996: 112)	10	1115!45	3.6% 7.8% 84%	777–793 CE 802–846 CE 854–1017 CE	910 CE
28	Be'er Orah	Rt 1949 Charcoal (Avner and Magness 1998: 57)	10	1150±45	95.4%	770–989 CE	890 CE
29	Be'er Orah	Rt 1740 Charcoal (Sharon, Avner and Naḥlieli 1996: 112)	15	915±50	95.4% 13.5%	1024–1215 CE	1040/1120 CE
30	Be'er Orah	Rt 1825 Charcoal (Sharon, Avner and Naḥlieli 1996: 112)	5	730±55	81.9% 13.5%	1187–1321 CE 1349–1392 CE	1270 CE
31	Timna 39b	Bm1116 Charcoal (Burleigh & Hewson 1979: 349)	?	1945±300	91.9%	595 BCE–640 CE	50 CE
32	Timna 2, Fu 1	Grn4381 (Rothenberg 1990a: 71)	?	1350±50	95.4%	604–770 CE	640 CE
33	Timna 2, Fu Z	BM2242 Charcoal (Rothenberg 1990a: 71)	?	1400±100	90.9% 2.6% 1.9%	421–779 CE 790–830 CE 837–867 CE	650 CE
34	Yotvata (L. 1)	PRI 4167 Charred twig (Chenopodiaceae)	10	2839±23	94.9%	1056–917 BCE	1010 BCE
35	Yotvata (L.2)	PRI 4355 Charred twig (Chenopodiaceae)	15	1278±22	95.4%	673–770 CE	7 0 0 / 7 6 0 CE
36	Yotvata (L.2)	PRI 3263 Charred Acacia twig	10	850±15	95.4%	1161–1223 CE	1180 CE

* PRI samples were prepared for dating by L. Scott Cummings, R. A. Varney, and P. Kováčik at the PaleoResearch Institute, Golden Colorado, and analyzed by AMS at UCIAMS (PRI-3036, 3037, and 30–38 at University of California Irvine) or CAIS (University of Georgia). Calibration based on OxCal 4.2, calibration curve IntCal13 (https://c14.arch.ox.ac.uk/oxcal/OxCal.html). Other dates are referred to in Column 3. Approximate mean calibrated dates are based on the dominant peak in the curve; in cases of two dominant peaks, two mean dates are mentioned.

NABATAEAN TO EARLY ISLAMIC MINES

Mines of later periods, presently generally addressed as Nabataean to Early Islamic (1st to 11th centuries CE), are clustered in two areas. One is located next to the dark red hill of Givʿat Yocheved and another at upper Nahal ʿAmram. In the former (Fig. 10.10), previously unsurveyed, 12 broad horizontal entrances to galleries and five broad shafts were recorded, as well as two large cavities in the sandstone slopes that seem to be the result of opencast mining (Fig. 10.11). Most galleries are now blocked by sand (Fig. 10.12), while shafts are blocked by coarse talus. In open galleries and in partially open broad shafts, long and vertical digging marks are observed, as is also found in the larger mines (Fig. 10.13, and see below), most probably made by large picks held by two hands (cf.

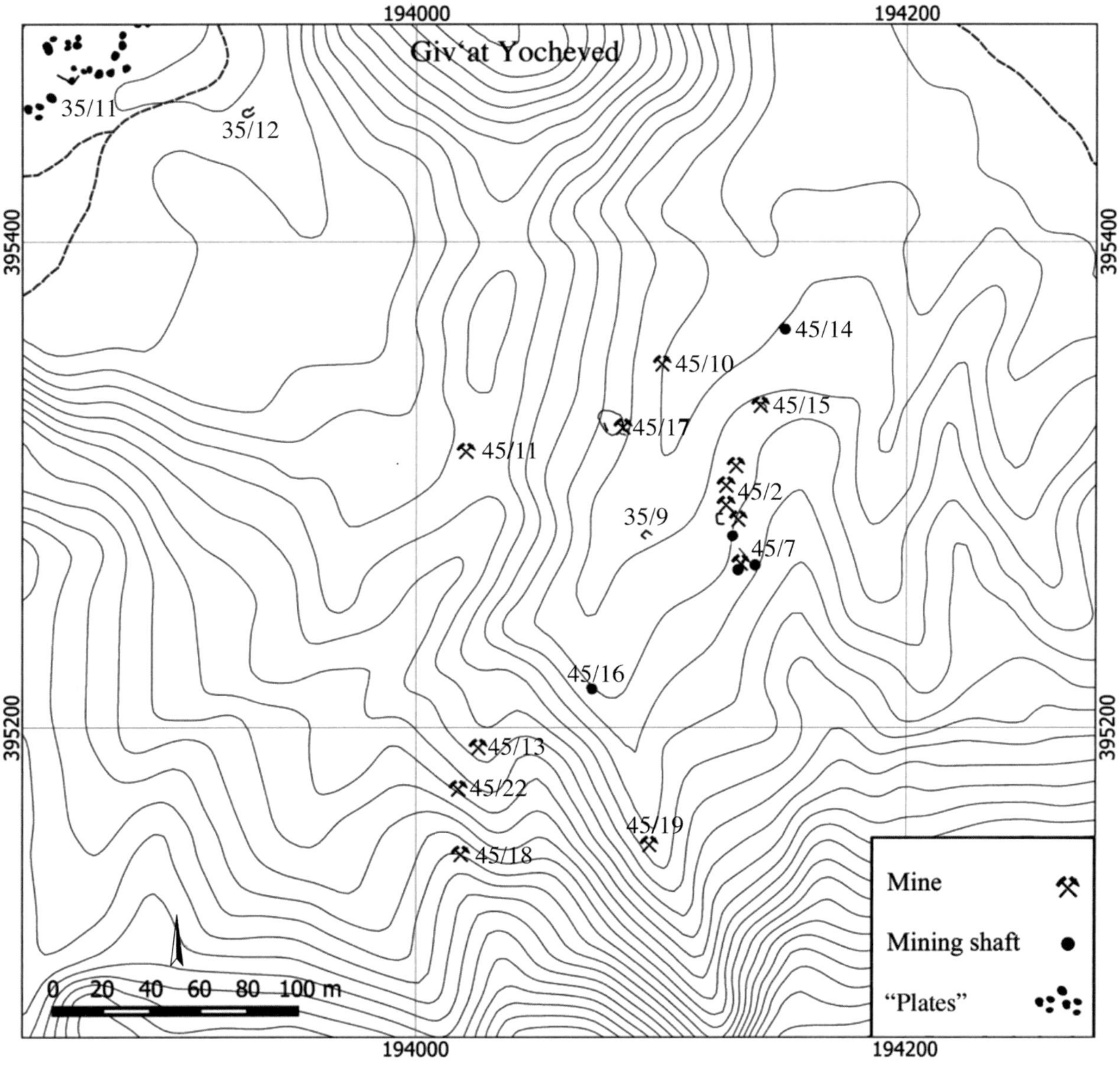

Fig. 10.10: Map of mines near Givʿat Yocheved. The northern cluster is of broad galleries and shafts (Nabataean to Early Islamic); the southern cluster is of narrow galleries (Late Bronze to Iron Ages), aside from Mine 45/16.

Fig. 10.11: Site 45/17 near Givᶜat Yocheved: the cavity in the slope is the result of opencast mining.

Rostovtzeff 1966: Pl. 35).[6] Hence, these mines present technology and strategy different from those of the Har Amir mines.

The largest mining area is that at the head of Nahal ʻAmram (Fig. 10.14). Two of the mines were first described here by Glueck (1953: 13–14, now numbered 35/28 and 36/2–3) and dated by him to the Iron Age. Rothenberg, however, related them to the Roman or Roman-Byzantine periods (1963: 61–62; 1967a: 145–146 respectively). Willies (1990, 1991) surveyed several mines in this area and mapped the largest one (now numbered 35/29). In one gallery he found a mining basket (Willies 1991: 134, Figs. 34, 38).[7] He dated these mines as Roman to Early Islamic, with emphasis on the Early Islamic and with some reuse in the Mamluk period. Rothenberg (1995: 36, 1999: 164–165) generally adopted these dates but pointed at the Byzantine as the most intensive period of mining.

In the new survey we recorded here 29 horizontal entrances to broad galleries and seven broad shafts, up to 11.5 m deep, connecting to galleries and halls, in two clusters on both sides of a steep ridge (Fig. 10.14). The interior of 17 mines could be studied and mapped (Langford *et al.*, this volume); others were blocked by sand or rocks. In some mines the ceiling of an entrance hall collapsed. All mines showed digging marks created by large tools, similar to those of the previous cluster, but in some galleries marks of small picks are also found (Fig. 10.15). Unlike the mines of Har Amir, chisel-marks were not observed in these mines. All galleries contained small niches cut into the walls for oil lamps. In some, these niches were cut very close to the entrance (Fig. 10.16), indicating

6 We thank S. Dar for the reference to this relief.

7 Following the discovery, the basket was held in the IAMS laboratory at University College, London. In 2014 it was returned to Israel and is now stored with organic materials at the Israel Antiquities Authority, Jerusalem.

Fig. 10.12: Site 45/6 near Givᶜat Yocheved: a broad blocked entrance to a mine.

Fig. 10.13: Nahal Tzfunot (north of Har ᶜAmram): a hall in a mine with long, vertical marks of a large digging pick (left wall).

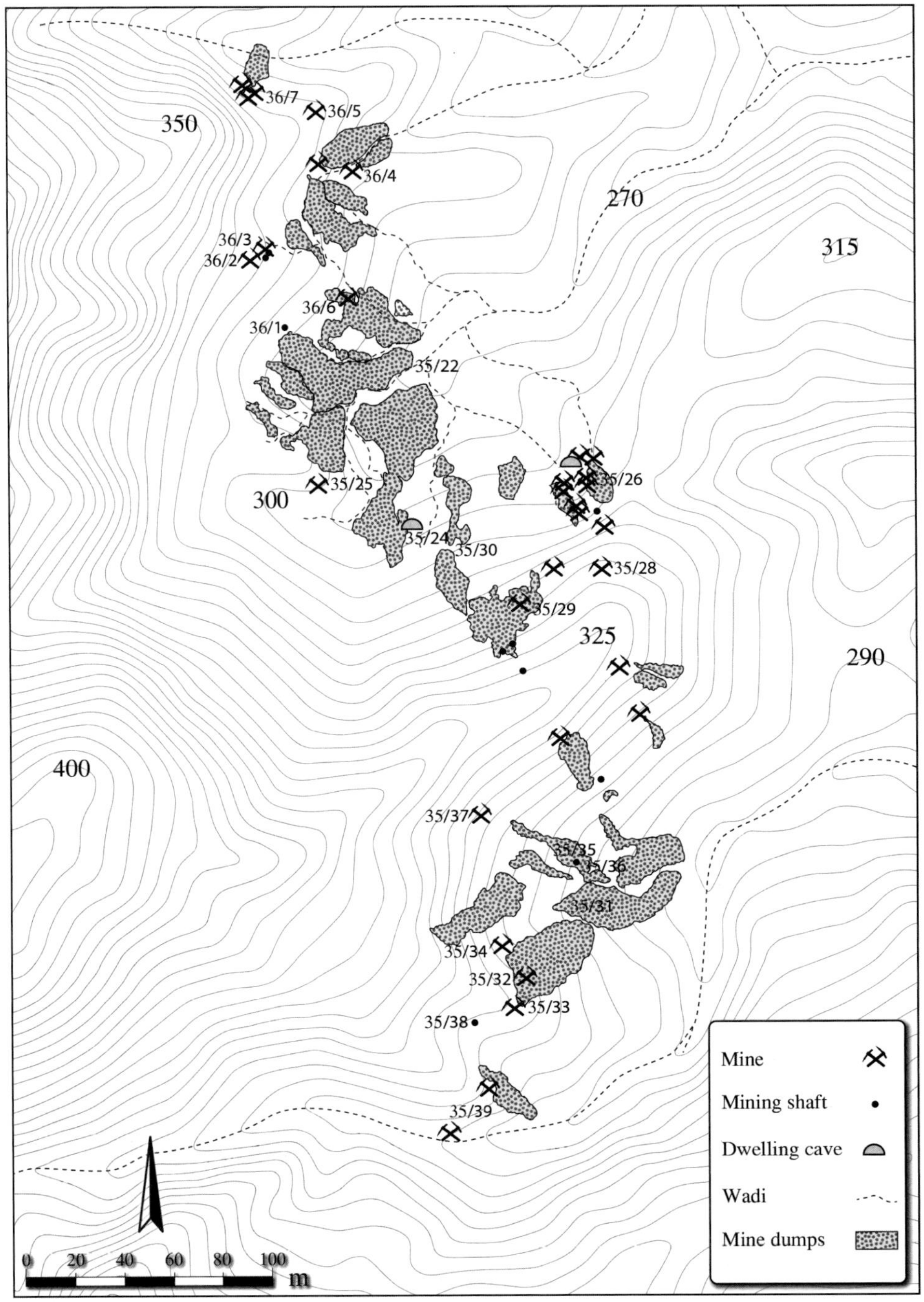

Fig. 10.14: Map of mine clusters at the head of Nahal ʿAmram: the dotted areas indicate the mines' dumps.

night-time work, i.e., two shifts a day. Under two niches in Mine 35/29, oil spills were found. One spill was sampled and oil was extracted (by D. Namdar, the Hebrew University), which now awaits identification.[8] The second sample was analyzed for ^{14}C and dated to the early 1st century CE (Fig. 10.17, Table 10.1: 7).

8 Chemical analysis by Namdar showed that it is not olive oil; other options presently examined are palm or sesame oil.

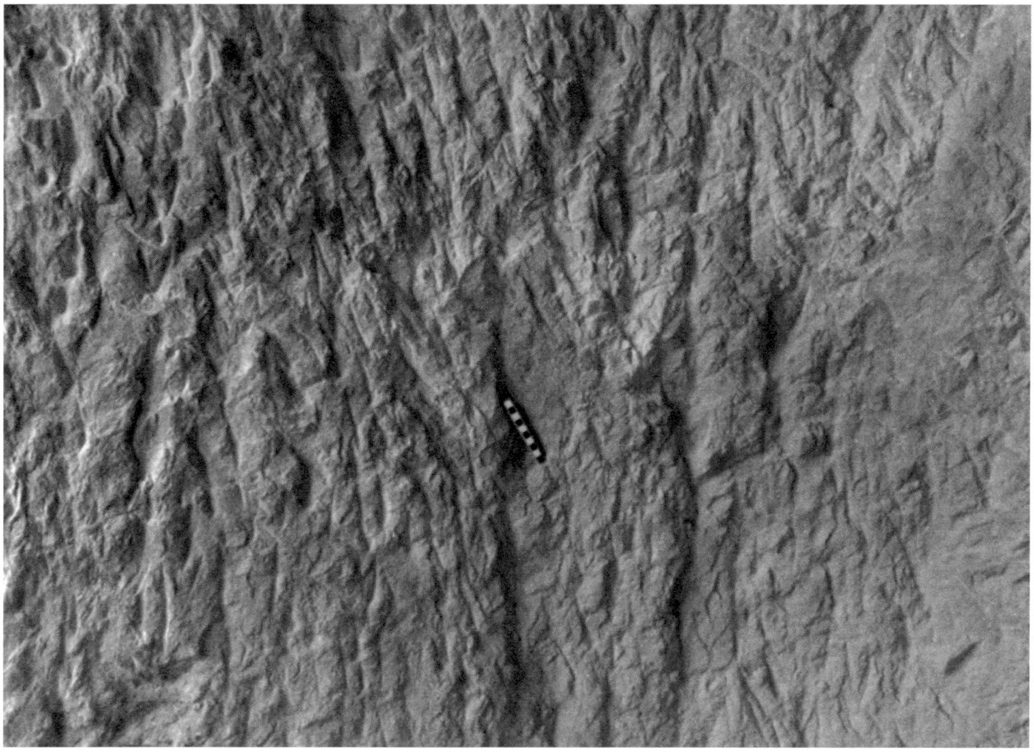

Fig. 10.15: Mine 36/2: marks of small digging picks.

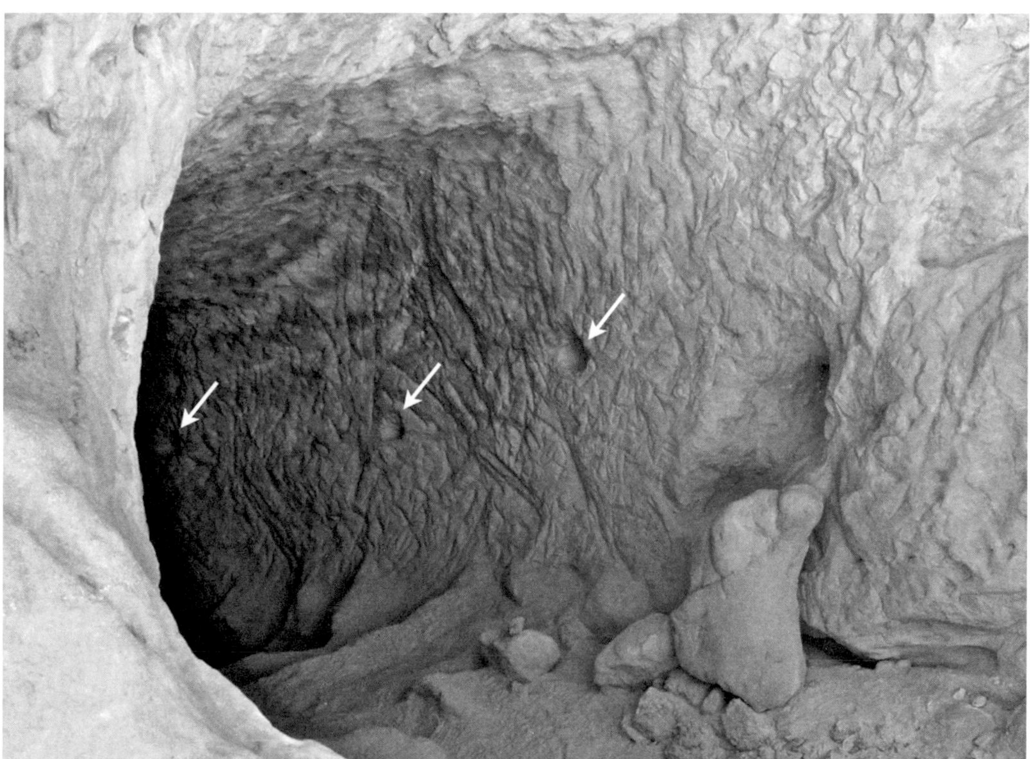

Fig. 10.16: Mine 36/2, small oil lamp niches at the mine entrance, indicating night-time work.

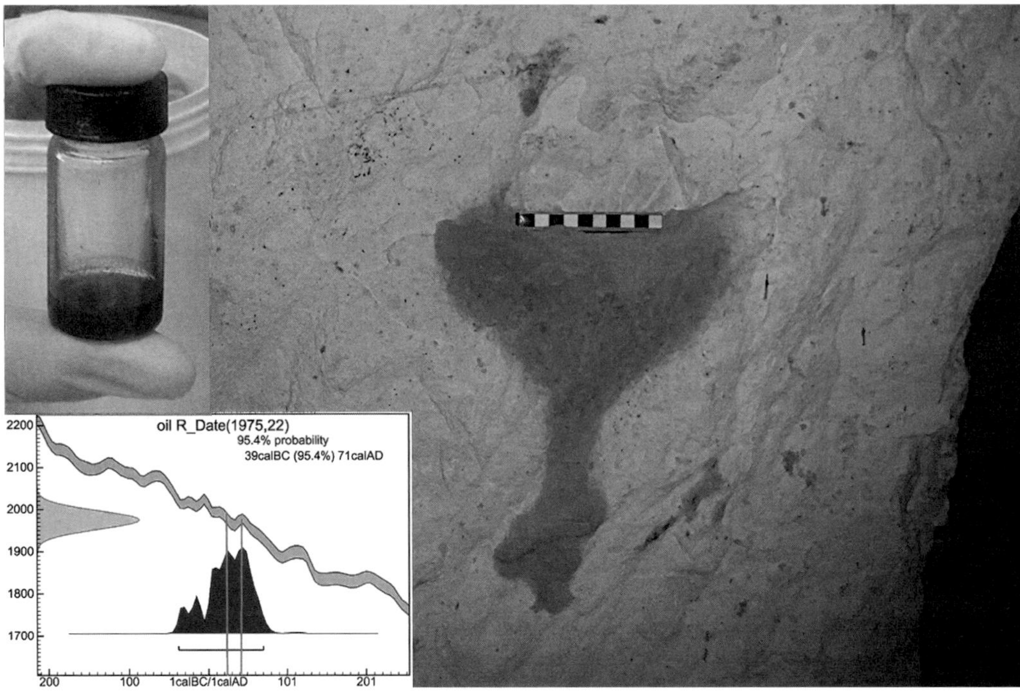

Fig. 10.17: Mine 35/29: oil spill from a lamp niche in the deepest gallery. Oil extracted (by D. Namdar) was radiocarbon dated to ca. 20/40 CE.

In the largest mine (35/29, total length 1100 m) we discovered a lower level of galleries, unnoticed by Willies, with artifacts mixed in the sand. A short dig in this section (2 sq m, 20 cm deep) produced ample finds, including many Early Islamic pottery sherds, from which a jug and a cooking-pot were restored (Fig. 10.18). We also found a Late Nabataean oil lamp and a few fragments dated to the 3rd–4th centuries CE (Fig. 10.19),[9] tridacna shells used as oil lamps, many pieces of textile, leather, goat wool and ropes (Fig. 10.20), many animal bones, (Horwitz *et al.*, this volume), some with green discoloration from copper minerals, numerous fruit remains from a variety of species (see below) and many charcoal and desert plant remains used as firewood. All finds were in the upper 20 cm of the sand fill, with no observed stratigraphy. The basket found by Willis in this mine (Fig. 10.21) is now dated by ^{14}C to ca. 360 or 390 CE (Table 10.1: 11).

One may wonder how cooking activity could have taken place in such a deep, remote, poorly ventilated section of the mine. The answer comes from two shafts seen in the gallery ceiling, blocked by coarse sediment. In the past, however, they obviously supplied enough air to this section. Fluvial sediment blocking galleries and shafts arc common in the mines and provide important information about their history (Ginat *et al.*, this volume). Presently, we have one ^{14}C date, ca. 250 CE, from the bottom of sediment fill of a shaft in Mine 36/2-3 (Table 10.1: 18), which means that a specific flood occurred after this date.

A limited excavation (1 sq m, 25 cm deep) in Mine 35/32 also yielded ample finds, including Early Islamic pottery sherds, smaller amounts of Nabatean–Late Roman sherds, many pieces of fabric, strings and ropes, and numerous fruit remains. This small dig also yielded a basket fragment of the same type illustrated in Fig. 10.21–28 × 18 cm, and the tip of an iron or steel digging tool (Fig. 10.22: 1).

9 The oil lamps were produced in Petra (cf. Grawehr 2006 Nos. 429–433, 500, 502, 504). We thank T. Gini-Erickson for identifying the oil lamp and other pottery sherds, and for the reference to Grawehr.

Fig. 10.18: Mine 35/29: an Early Islamic jug and a cooking-pot, restored from sherds (by T. Levin).

Fig. 10.19: Mine 35/29: Late Nabataean oil lamp and fragments, 4th century CE.

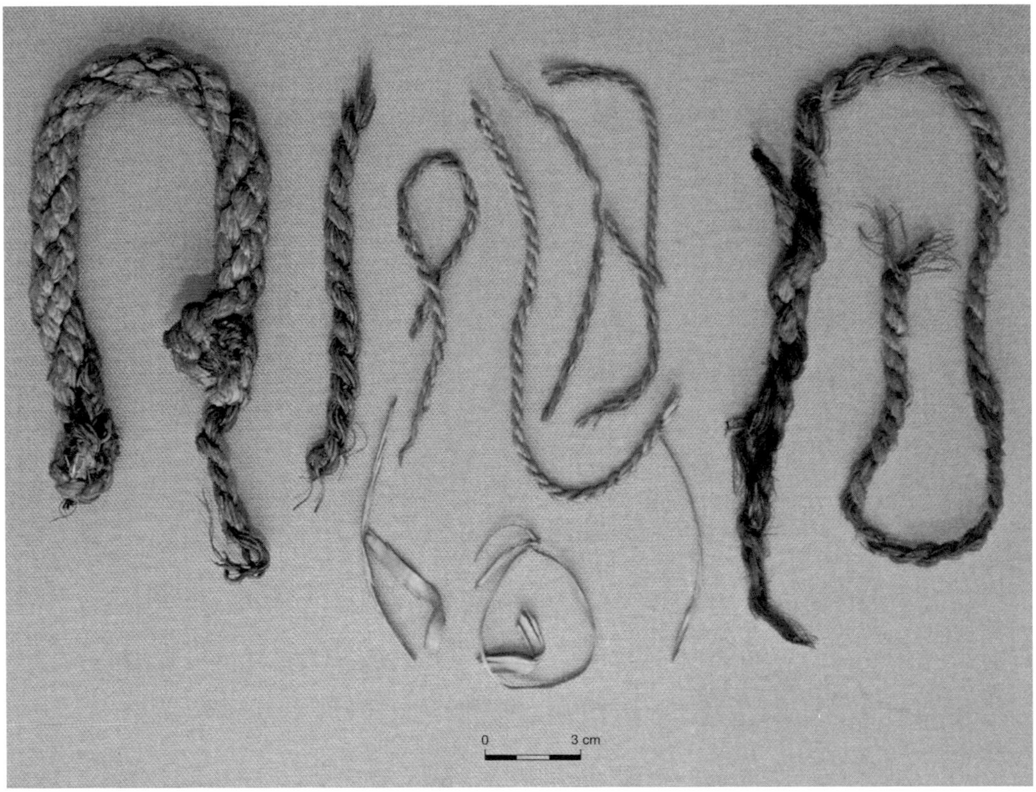

Fig. 10.20: Mine 35/29: ropes made of palm trunk fibers.

Fig. 10.21: Mine 35/29: basket found by Willies, after it was returned from London to the Israel Antiquities Authority.

Another excavation took place in a dwelling cave, ca. 21 sq m, adjacent to opencast Mine 35/24 (Fig. 10.23). Unlike in the mines, some stratigraphy remained. In the upper 30–35 cm, Early Islamic pottery was found together with a few Nabataean–Roman finds, but on the cave's floor only the older artifacts were found, including four Late Roman-early Byzantine coins (Fig. 10.24) dated from 304 to 346 CE. Here, too, many fruit remains[10] and fish bones were found, Red Sea shells and glass beads, many

10 Over 300 fruit remains from the three digs included pits of dates, olives, and peaches; shells of almonds, walnuts, and a pomegranate; grape stalks and others. They were initially identified by Avner, now being studied by M. Daviv and E. Weiss, Bar-Ilan University.

Fig. 10.22: Tips of iron digging tools: 1. from Mine 35.32; 2–4. from dwelling cave at Mine 35/24.

fabric pieces and another fragment of a basket, ca. 28 × 26 cm, of the same type as the previous ones. An important find from the cave's floor was three additional tips of iron digging tools (Fig. 10.22: 2–4). These and the one from Mine 35/32 are the first remains of digging tools found in the Arabah mines.[11]

The fruit remains and animal bones demonstrate a high quality of nutrition available to the miners, indicating that they were not slaves but professionals (Horwitz *et al.*, this volume).

Despite the obvious dominance of Early Islamic pottery, 12 ^{14}C dates currently available from the mines are all earlier, from the early 1st to the early 5th centuries CE (Table 10.1: 7–18). Hence, the immediate understanding from the limited digs is that the same mines, even the same galleries, were exploited over a long period of time, at least from the 1st to the 11th centuries CE, with a possible interruption from the mid-5th century to the mid-7th century. Although we sometimes observe a relative chronology of intersecting galleries in the mines, we are currently unable to reconstruct a map of each mining period separately. Nevertheless, the discovered tips of digging tools may supply another means to identify phases in the mines. These tools seem to be small, similar to modern geologic picks and they are probably responsible for the small digging marks in the mines (Figs. 10.15, 10.16). Since they were found in a Nabataean–Early Byzantine context, they may signify the earlier phase in these mines. Unfortunately, the remains of small

11 Two bronze digging picks were reported from Timna, dated to the 13th–12th centuries BCE (Conrad and Rothenberg 1980: 72–73), but, the same objects were previously published as copper spear-butts (Rothenberg 1972: 105–106). Copper is obviously too soft to dig in sandstone, but no chemical analysis was published to indicate that the tools were indeed of bronze. Hence, the first interpretation of the objects is better accepted.

Fig. 10.23: Opencast Mine 35/24 (center top), with a dump pile below and dwelling cave on the left.

Fig. 10.24: Late Roman–Early Byzantine coins, 304–346 CE, from Nahal ʿAmram Dwelling cave 35/24, L. 4, B 51; L. 5; B. 63.

digging marks are very limited and seem to be obliterated by the later mining with the larger tools; in some locations, one can see long vertical digging marks on top of small marks (Fig. 10.15).

One of the outcomes of this large scale mining was the accumulation of mining dumps on the escarpments and in the wadis below the mines, as mentioned previously by Rothenberg (1967a: 146; 1999: 164) and by Willies (1991: 117, 118). GPS mapping of the dumps enabled calculation of their total area. Presently, they cover 8500 sq m, but their restored area, considering the erosion during the last 900 years (since cessation of mining), is at least 14,060 sq m (Figs. 10.14, 10.25). In the wadis, the

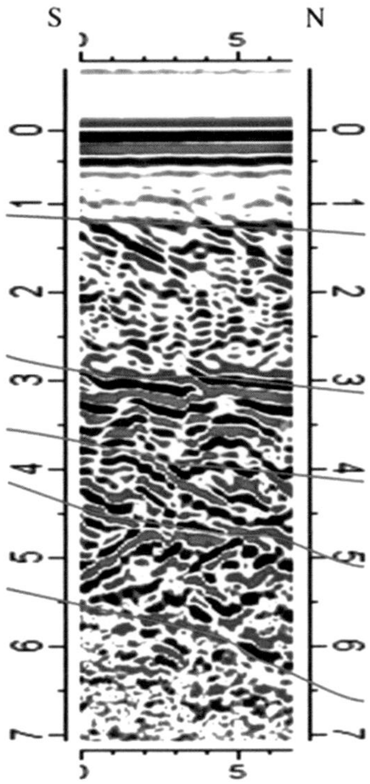

Fig. 10.25: Nahal ʿAmram Site 35/24: a GPR section of a dump. The lines indicate different phases of dumping.

piles are up to 6 m deep (Figs. 10.23, 10.25, and see Arav *et al.*, this volume) but are now extensively intersected by small young wadi channels. On the slopes, the dumps are often eroded (Fig. 10.26) or covered by talus and cannot be recorded. Based on their restored area and varying depths, the volume of the dumps when mining ceased was at least 21,000 cu m or ca. 42,000 tons of sandstone dug out from the mines.[12] In contrast, the calculated volume of the mines around the dumps is only 4520 cu m, ca. 9400 tons. The great difference between the two can be explained in several ways: 1. As mentioned above, some mine entrances and some galleries within mines are presently blocked and cannot be measured. 2. Above some large dumps on the escarpments no mine entrances are visible. This means that some mines remain unknown to us, covered by talus. 3. In some places opencast mines are observed (Figs. 10.11, 10.23), with no preserved digging marks, but the rock face seems unnatural (cf. Willies 1991: 115, 121, 135). This mining technique could have created large dumps with no galleries.

Similar to the mines, the dumps also show long-term accumulation. The vast majority of pottery sherds collected from their surface was Early Islamic, however, the only charcoal from a dump dated by ^{14}C was ca. 220 CE (Table 10.1: 17). GPR sections of the dumps (Fig. 10.25) show that they accumulated in several phases, in different gradients, with occasional leveling.

12 The erosion of dumps during the 1000 years of mining was not included in the above estimation, hence, the true amount of eroded dumps over 2000 years should be much higher. Also, the specific weight of the dumps, 2.1 gr/cu cm, was measured from the top of a pile. However, as seen in the GPR section (Fig. 10.25), the lower levels are more compressed, so the general dump weight may be higher than that calculated here. A laser scanning of three dump piles before and after the November 2012 rain storm, by Arav *et al.* (this volume), showed that ca. 0.4% of the dumps' volume was washed away. This may provide a general idea as to the loss of dump material during almost a millennium of mining and during the last millennium, since the end of the mining activity.

Fig. 10.26: Site 45/6, an eroded dump on a slope. The pillar is the remains of an eroded layer of a dump; the rock above the pillar is from a talus cover; a 1 m layer of dump still lies beneath the pillar.

In light of the large scale mining in the later periods, the question arises: Where was the copper ore smelted? Within the earlier miners' camp there is one late small pile of broken slag "cakes" (Fig. 10.27), dated now by four ^{14}C dates from ca. 660 to 770 CE (Table 10.1: 19–22).[13] The slag pile is 102 sq m, and two probes excavated in it showed that its depth is 20–40 cm. The total weight of the pile, based on the excavated slag, is ca, 30 tons, theoretically representing a production of only three tons of copper. This calculation follows Hauptmann (2007: 253), presenting a ratio of ca. 1:10 copper to slag. This ratio may be true for earlier slag; however, for the later periods it should be taken with reservation. Almost all Early Islamic smelting sites with slag "cakes" along the Arabah, are located next to older smelting sites, containing small furnace slag. This may mean that the older slag, containing 30%–50% of iron, could have been reused as flux by the later smiths. If slag could be reused as flux (after crashing), the estimated amount of produced copper should be much higher.

In any case, the slag pile at Nahal ʿAmran is indeed small, but several larger smelting camps from this period are known in the southern Arabah. The primary one is near Beʾer Orah, close to a water source (Fig. 10.28). A 3D laser scan, by Arav *et al.* (this volume) indicated ca. 3000 tons of slag.[14] For other nearby and contemporary smelting sites, mainly Beʾer Orah South, Tell Ḥara Ḥadid and near Umm Rashrash[15] (see the sites' location in Fig. 10.1), about the same amount combined is estimated. Therefore, the current conjectured total amount of copper produced in the later periods is at least 600 tons, or much more.

13 Geomagnetic dating of this slag pile was somewhat later than the ^{14}C dates, around 900 CE (Ben Yosef *et al.* 2008, Fig. 13).

14 The specific weight of slag "cakes" from Be'er Orah was measured at 4.1 gr/cu cm.

15 Two large piles of slag were mentioned by Glueck (1935: 47–48) on a hill above Um Rashrash. Unfortunately, they were erased during the early construction stage of modern Eilat.

Fig. 10.27: Nahal ʿAmram Site 35/7, from the south: a small pile of fragments of tapped slag "cakes," with two excavated probes.

Fig. 10.28: Beʾer Orah: large slag piles, mainly Early Islamic.

HISTORICAL NOTES

Although Chalcolithic-Early Bronze mining at Nahal ʿAmram is elusive, the area seems to have served as the main source of copper ore for the intensive industry of the two Chalcolithic villages at Aqaba: Tell Magass and Tell Ḥujayrat al-Ghuzlan. As mentioned above, the first support for this assumption is based on the recent pilot study of lead isotopes of copper ores from these villages and from the largest copper ore sources along the Arabah Valley.

As for the Late Bronze and Iron Ages, for many years King Solomon was considered the historical figure behind the establishment of the Arabah copper industry (Phythian-Adams 1933; Glueck, e.g., 1938, 1942, 1970: Chapter 4). Rothenberg (1963: 42, 64) first followed this historical view, but later (e.g., 1967a: 51–53; 1972: 180–181; 1988: 270–278; 1999 *passim*) dated the complex to an earlier period and attributed it to the Egyptian initiative of the 13th–12th centuries BCE. Nevertheless, as mentioned above, the role of the Egyptians in Timna now seems limited, and besides Timna, no sign of the Egyptian New Kingdom was found at other mines along the Arabah Valley. Based on ample ^{14}C dates from Faynan and Timna, the climax of copper production was the 11th–9th centuries BCE. Of five ^{14}C dates presently available from Nahal ʿAmram, from these periods, two fall within the time of Egyptian presence at Timna, but there is no indication of any Egyptian technology or control of work. Hence, the question remains regarding who stood behind the organization of this large-scale operation? Apparently, this time span permits the reintroduction of King Solomon into the scenario. However, unlike at Negev Highland sites, no Judean pottery was found at Nahal ʿAmram or in Timna. Unavoidably, the only candidates that could have operated this industry were the indigenous desert tribes, the *"Shasu"* (cf. Levy, Adams and Muniz 2004; Levy, Najjar and Ben-Yosef 2014: 986–994; Levy *et al.* 2005; Avner 2014).

Another intriguing period at Nahal ʿAmram spans the 1st to 5th centuries CE. The later part of this period was included in the historical overview by Rothenberg and Willies (see above). However, among the tens of kilograms of pottery sherds collected during the new survey from the mines and dumps, as well as from the excavations, we recognized only very few Late Roman or Byzantine sherds. In Rothenberg's survey of the Arabah and the Elat Mountains (1963, 1967b, 1971), 44 sites out of a total of 226 were identified as Byzantine, based on surface collection of pottery. Today, however, based on a reexamination of the pottery, excavations and ^{14}C dates, all of these 44 sites are dated to the Early Islamic period (Avner and Magness 1998; R. Avner 1998 Rapuano 2013), except the Late Roman fortress of Yotvata.[16] This concurs with our recent observation of the large amount of pottery from Nahal ʿAmram and the new ^{14}C dates.[17]

Nevertheless, 12 ^{14}C dates from the ʿAmram mines (Table 10.1: 7–18) shed new light on the copper mining during the 1st to 5th centuries CE. This time-span began during the peak days of the Nabataean kingdom, the time of King Aretas IV, and continued through the Late Roman and Byzantine periods. Until recently, the only clue for a Nabataean presence at Nahal ʿAmram was a large and fine rock stella of an "eye idol" on the cliff of Givʿat Yocheved (Fig. 10.29, and see Avner 2000: 109), but now this stella is no longer in isolation.

In 106 CE the Nabataean kingdom was annexed to the Roman empire (Bowersock 1983: 76–109), but 25 years later (131–132 CE), the Bar-Kochba Revolt broke out and forced the Roman army to abandon the southern Negev and other regions (Eck 1999). Therefore, it seems that only the Nabataeans could

16 Meshel 1989; Roll 1989; Davies and Magness 2008, 2011.

17 Following our own examination, K. Damgaard (Copenhagen University) carefully examined all the pottery from the mines and confirmed the initial observation. His thorough study (2011) of the Early Islamic pottery from Aqaba validates this conclusion.

Fig. 10.29: Givʿat Yocheved, Site 45/20: a rock-stela of an "eye idol," representing a Nabataean goddess.

have continued to operate and run the copper mines. They even maintained a variety of organizational institutions, without a king or a government to instruct them. Only from Diocletian's Reform, at the very end of the 3rd century CE, can one see a resumption of Roman control of the southern Negev (Erickson-Gini 2010: 65–74: 98; Avner 2015, with references). However, even then there are no indications of Roman or Byzantine control over the copper mines. The very few Late Roman-Byzantine pottery sherds found to date in the survey and excavations, and the four coins of the early 4th century from the dwelling cave (Fig. 10.24), cannot indicate yet a Roman or Byzantine control of the mines. So, once again the industry seems to be in the hands of the local population, this time the Nabataeans. The dimension of the Nabataean mines, although difficult to detect today, is well reflected in the nearby Nabataean–Roman–Byzantine Aila (ancient Eilat, in modern Aqaba), where numerous remains of the copper industry were uncovered, including over 500 copper objects, in comparatively small excavated areas (Parker 2006: 228; Power 2012: 29). Still missing are the remains of Nabataean copper smelting, of which we have only clues: a small number of Nabataean pottery sherds in the large smelting site of Beʾer Orah, additional Nabataean sherds at a small smelting site near Yotvata, a geomagnetic date of the 1st century BCE–CE from the same site (Ben Yosef *et al.* 2008, Fig. 13), and a ^{14}C mid-1st century CE date from the Chalcolithic smelting furnace of Timna 39b (Table 10.1: 31, Rothenberg 1990b).[18]

18 Paleomagnetic dating of the slag from the Timna 39b furnace, by Ben-Yosef *et al.* (2008: 2872–2874) confirmed its Chalcolithic date, but also pointed to a later, secondary use of the furnace. This may explain the late ^{14}C date from the furnace, as suggested by Rothenberg (1990b).

It is interesting to note that from the Timna Valley only little is known of Nabataean–Roman activity (one mine, two Greek inscriptions with Nabataean names, and some pottery). So it seems that Nahal ʿAmram gained preference in these periods.

The Early Islamic (7th to 11th centuries CE) was a major mining period at Nahal ʿAmram, when the mines reached their maximum dimensions. This intensive mining fits well with what we know of the period in the southern Arabah Valley and the vast development in the region: the construction of a series of villages, large farms based on *qanat* (underground water tunnels), gold production, and more (Gilat *et al.* 1993; Avner and Magness 1998; Avner 2016). All these created a lively hinterland for the town of Ayla,[19] an important administrative and commercial center, pilgrim station and even a focal point of Islamic scholarship (Cobb 1995; Whitcomb 1988, 1994, 1998; Damgaard 2011).

Compared with the extensive Early Islamic mining activity at Nahal ʿAmram, the situation at Timna is puzzling. Early Islamic presence there is known only on top of older smelting camps, mainly at Site 2, but no contemporary mines were ever identified.[20] Nor has early Islamic mining been found in the Faynan area, but additional mines are known at Nahal Reḥavʿam, near Eilat and at Jebel Mraḥ, in eastern Sinai, 18 km west-southwest of Eilat (unpublished).

Little is known of the Mamluk period (13th–15th century CE) in the Eilat region, aside from the Ayla Fortress and the remains along Darb al-Hajj. Willies (1991: 136) and Rothenberg (1999: 164) suggested some Mamluk mining at Nahal ʿAmram, which finds some support in two glazed pottery sherds collected in the recent survey. Two buildings with adjacent installations at Nahal ʿAmram (Site 45/23) and northeast Timna Valley, and one ^{14}C sample from Beʾer Orah, ca. 1270 CE (Table 10.1: 30) are probably Mamluk.

CONCLUDING NOTE

The preliminary results of the renewed research at Nahal ʿAmram demonstrate its potential for studying the past of the ʿArabah Valley and the role of copper in the history, technology, economy and sociology of desert populations. Obviously, to reach a better understanding of these themes, additional excavations are needed, as well as numerous chemical analyses and series of ^{14}C dates.

REFERENCES

Avner, R. 1998. Elat-Elot: An Early Islamic Village. *ʿAtiqot* 36: 21*–40*.

Avner, U. 2000. Nabatean Standing Stones and Their Interpretation. *ARAM* 11–12: 95–120.

Avner, U. 2014. Egyptian Timna—Reconsidered. In: Tebes, J., ed. *Unearthing the Wilderness: Studies on the History and Archaeology of the Negev and Edom in the Iron Age*. Leuven: 103–163.

Avner, U. 2015. The Nabataeans in Sinai. *ARAM* 27: 389–421.

Avner, U. 2016. Desert Farming in the Southern ʻArabah Valley, Israel, 2nd Century B.C. to 11th Century A.D. In: Davies, A., Retamero F. and Schjellerup I., eds. *Making Agro-Pastoral Landscapes in Pre-Industrial Societies: Choices, Stability and Change*. Oxbow: 19–36.

19 Aila refers to the Nabataean–Byzantine town; Ayla refers to the Early Islamic town (see the difference in the publication of Parker, Whitcomb and others).

20 In Timna Site 2, four furnaces of the same types were excavated by Rothenberg (1972: 71–78, 1990a: 13–44), next to piles of slag "cake" fragments. They were identified as New Kingdom Egyptian furnaces but actually two of them were directly dated by ^{14}C as Early Islamic (Table 10.1: 31, 32), much like Be'er Orah and Yotvata, where slag "cakes" are dated to this period (Table 10.1: 23–30, 33–35 and see discussion in Avner 2014: 131).

Avner, U. and Magness, J. 1998. Early Islamic Settlement in the Southern Negev. *Bulletin of the American Schools of Oriental Research* 310: 39–57.

Avner, U., *et al.* 2014. Survey of Neolithic Cult Sites in the Eilat Mountains, Israel. *Journal of the Israel Prehistoric Society* 44: 101–116.

Ben-Yosef, E., Levy, T. and Najjar, M. 2009. New Iron Age Copper-Mine Field Discovered in Southern Jordan. *Near Eastern Archaeology* 72: 98–101.

Ben-Yosef, E., Sha'ar, R., Tauxe, L. and Ron, H. 2012. A New Chronological Framework for Iron Age Copper Production at Timna (Israel). *Bulletin of the American Schools of Oriental Research* 367: 31–71.

Ben-Yosef, E., *et al.* 2008. A New Approach for Geomagnetic Archaeointensity Research: Insights on Ancient Metallurgy in the Southern Levant. *Journal of Archaeological Sciences* 35: 2863–2879.

Beyth, M., Eyal, M. and Garfinkel, Z. 2011. The Geology of the Eilat Sheet. *GSI Report* 22/2011.

Bowersock, G.W. 1983. *Roman Arabia*. London.

Braslavi, Y. 1952. *Knowing the Country,* Vol. IV. Tel Aviv (Hebrew).

Burleigh, R. and Hewson, A. 1979. British Museum Natural Radiocarbon Measurements XI. *Radiocarbon* 21(3): 339–352.

Cobb, P. 1995. Scholars and Society at Early Islamic Ayla. *Journal of the Economic and Social History of the Orient* 38: 417–428.

Cohen, E. 1976. The Mystery of "Plates" at Timna Valley. *Israel Land and Nature* 19: 15–17 (Hebrew).

Conrad, H. and Rothenberg, B. 1980. *Antikes Kupfer im Timna-Tal*. Bochum.

Damgaard, K. 2011. *Modeling Mercantilism: An Archaeological Analysis of Red Sea Trade in the Early Islamic Period (ca. 650–1100 CE)* (Ph.D. dissertation, Copenhagen University). Copenhagen.

Davies, G. and Magness, J. 2008. The Roman Fortress at Yotvata 2007. *Israel Exploration Journal* 53: 103–114.

Davies, G. and Magness, J. 2011. The Roman Fort at Yotvata: A Foundation Under Valens? *Journal of Roman Archaeology* 24: 469–480.

Eck, W. 1999. The Bar Kochva Revolt: The Roman Point of View. *Journal of Roman Studies* 89: 76–89.

Erickson-Gini, T. 2010. *Nabataean Settlement and Self-Organized Economy in the Central Negev: Crisis and Renewal* (BAR International Series 2954). Oxford.

Gilat, A., Shirav, M., Bogoch, R., Halicz, L. and Avner, U. 1993. Significance of Gold Exploitation in the Early Islamic Period, Israel. *Journal of Archaeological Science* 20: 429–437.

Glueck, N. 1935. *Explorations in Eastern Palestine, II* (Annual of the American Schools of Oriental Research 14–15). New Haven.

Glueck, N. 1938. The Topography and History of Ezion-Geber and Elath. *Bulletin of the American Schools of Oriental Research* 72: 2–13.

Glueck, N. 1942. The Excavations of Solomon's Seaport: Ezion-Geber. *The Smithsonian Report for 1941*: 453–476.

Glueck, N. 1953. Exploration in Western Palestine. *Bulletin of the American Schools of Oriental Research* 131: 6–15.

Glueck, N. 1960. Archaeological Exploration of the Negev in 1959. *Bulletin of the American Schools of Oriental Research* 159: 3–14.

Glueck, N. 1970. *The Other Side of the Jordan*. Cambridge.

Grawehr, M. 2006. Petra, *Ez Zantur III: Die Lampen der Grabungen auf ez-Zantur, in Petra*. Mainz am Rhein.

Hauptmann, A. 2007. *Copper: Evidence from Faynan, Jordan*. Berlin.

Hauptmann, A., Khalil, L. and Schmidt-Strecker, S. 2009. Evidence of Late Chalcolithic/Early Bronze Age I Copper Production from Timna Ores at Tell al-Magass, 'Aqaba. In: Khalil, L. and Schmidt, C., eds. *Prehistoric 'Aqaba I*. Berlin.

Khalil, L. and Riederer, J. 1998. Examination of Copper Metallurgical Remains from a Chalcolithic Site at el-Magass, Jordan. *Damaszener Mitteilungen* 10: 1–9.

Kingery, D. 1988. Nahal Hemar Cave: Plaster Beads. *'Atiqot* 18: 45–46.

Klimscha, F. 2013. Innovations in Chalcolithic Metallurgy in the Southern Levant during the 5th and 4th Millennum BC: Copper-Production at Tell Ḥujayrat al-Ghuzlan and Tell al-Magass, ʻAqaba Area, Jordan. In: Burmeister, S., Hansen, S., Kunst, M. and Müller-Scheeßel, N., eds. *Metal Matters: Innovative Technologies and Social Changes in Prehistory and Antiquity*. Rahden.

Levy, T., Adams, R.B. and Muniz, A. 2004. Archaeology of the Shasu Nomads—Recent Excavations in the Jabal Humrit Fidan, Jordan. In: Freedman, R.E. and Propp, W., eds. *Le-David Maskil: A Birthday Tribute to David Noel Friedman.* Winona Lake: 63–89.

Levy, T.E., Najjar, M. and Ben-Yosef, E. 2014. *New Insight into the Iron Age Archaeology of Edom, Southern Jordan,* Vols. 1, 2. Los Angeles.

Levy, T., *et al*. 2005. Lowland Edom and the High and Low Chronologies. In: Levy, T. and Higham, T., eds. *The Bible and Radiocarbon Dating: Archaeology, Text and Science*. London: 129–163.

Meshel, Z. 1989. A Fort at Yotvata from the Time of Diocletian. *Israel Exploration Journal* 39: 228–238.

Parker, T. 2006. Roman Aila and the Wadi ʻArabah: An Economic Relationship. In: Bienkowski, P., Galor, K., eds. *Crossing the Rift: Routes Settlement Pattern Interaction in the Wadi ʻArabah*. Oxford: 223–230.

Phythian-Adams, C. 1933. Israel in the ʻArabah. *Palestine Exploration Fund Quarterly Statement* 66: 137–145.

Power, T. 2012. *The Red Sea from Byzantium to the Caliphate: AD 500–1000*. Cairo, New York.

Rapuano, Y. 2013. An Early Islamic Settlement and Possible Open-Air Mosque at Eilat. *'Atiqot* 71: 129–165.

Roll, I. 1989. A Latin Imperial Inscription from the Time of Diocletian Found at Yotvata. *Israel Exploration Journal* 39: 239–260.

Rostovtzeff, M. 1966. *The Social and Economic History of the Roman Empire,* Vol. 1. Oxford.

Rothenberg, B. 1963. Ancient Copper Industries in the Western ʻArabah. *Palestine Exploration Fund Quarterly* 95: 5–64.

Rothenberg, B. 1967a. *Negev*. Ramat Gan (Hebrew).

Rothenberg, B. 1967b. Archaeological Sites in the Southern ʻArabah and the Eilat Mountains. In: Ron, Z., ed. *Survey of the Eilot District*. Eilat: 283–341 (Hebrew).

Rothenberg, B. 1971. The ʻArabah in Roman and Byzantine Times in the Light of New Research. In: Applebaum, S., ed. *Roman Frontier Studies, 1967: The Proceedings of the Seventh International Congress Held at Tel Aviv*. Tel Aviv: 211–223.

Rothenberg, B. 1972. *Timna: Valley of Biblical Copper Mines*. London.

Rothenberg, B. 1975. Timna Valley. *Hadashot Arkheologiyot* 54–55: 31–34.

Rothenberg, B. 1978. Timna Mines. *Hadashot Arkheologiyot* 65–66: 55–57.

Rothenberg, B., ed. 1988. *Researches in the Araba 1959–1984, Vol. 1, The Egyptian Mining Temple at Timna* (Metal in History 2). London.

Rothenberg, B., ed. 1990a. *The Ancient Metallurgy of Copper: Researches in the ʻArabah 1959–1984,* Vol. 2. London.

Rothenberg, B. 1990b. The Chalcolithic Copper Smelting Furnace in the Timna Valley—Its Discovery and the Strange Argument Surrounding Its Dating. *Institute for Archaeo-Metallurgical Studies* 15/16: 9–12.

Rothenberg, B. 1995. A Summary of Thirty Years of Archaeological and Archaeometallurgical Field and Laboratory Research Concerning the Timna Valley and Its Surroundings. In: Aviram, Y., ed. *Eilat: Studies in the Archaeology, History and Geography of Eilat and the ʻArabah*. Jerusalem: 1–45 (Hebrew).

Rothenberg, B. 1999b. Archaeo-Mettallurgical Researches in the Southern ʻArabah 1959–1990, Part 2: Egyptian New Kingdom (Ramesside) to Early Islam. *Palestine Exploration Fund Quarterly* 131: 149–175.

Sharon, M., Avner, U. and Naḥlieli, D. 1996. An Early Muslim Mosque Near Beʾer Orah in the Southern Negev: Possible Evidence for an Early Eastern Qiblah? *ʻAtiqot* 30: 107–114.

Whitcomb, D. 1988. *ʻAqaba: ʻPort of Palestine on the China Seaʼ*. Amman.

Whitcomb, D. 1994. *Ayla: Art and Industry in the Islamic port of Aqaba*. Chicago.

Whitcomb, D. 1998. Out of Arabia: Early Islamic Aqaba in its Regional Context. In: Gayraud, R.P., ed. *Colloque international d'archéologie islamique*. Cairo: 403–318.

Willies L. 1990 "Exploring the Ancient Copper Mines of the Wadi 'Amram (Southern 'Araba)," *Institute for Archaeo-Metallurgical Studies* 15–16: 12–15.

Willies, L. 1991. Ancient Copper Mining at Wadi 'Amram, Israel: An Archaeological Survey. *Bulletin of the Peak District Mines Historical Society* 11: 109–131.

CHAPTER 11

VOLUME AND MASS ESTIMATION OF MINE DUMPS AND SLAG PILES USING HIGH-RESOLUTION TERRESTRIAL LASER SCANS

Reuma Arav, Sagi Filin and Uzi Avner

During the new archaeological study of the copper mining center at Nahal ʿAmram in the southern Arabah Valley, mounds of mining dumps at the headwaters of Nahal ʿAmram and piles of copper slag at Beʾer Orah were scanned using a high resolution terrestrial laser scanner (LiDAR). The 3D scanning was undertaken in an attempt to estimate the volume and mass of each in order to evaluate the intensity of copper mining and production during the 1st to 11th centuries CE. The method, challenges and results are discussed in this article.

INTRODUCTION

In the early 1950s, Glueck (1953: 13–15; 1960: 12–14) and in the 1960s Rothenberg (1963: 34–38, 61–62; 1967: 41–49) surveyed copper mines of various types and periods at Nahal ʿAmram. In the late 1980s Willies (1990, 1991) mapped two large mines in the area. A new detailed survey (2010–2013, Avner *et al.* this volume) recorded and mapped many more mines and other related sites. The most intensive mining was identified during three periods: the Late Bronze–Iron Age (13th to 9th centuries BCE), the Late Nabataean (1st to 5th centuries CE) and the Early Islamic (7th to 11th centuries CE). The latter two are the focus of this paper.

The mining dumps were inside a cluster of mines in headwaters of Nahal ʿAmram (Fig. 11.1a). Mapping the dumps by means of an accurate GPS indicated a total of 8500 sq m for the dump area presently preserved (reduced from approx. 14,060 sq m, before being intersected by floods), while a GPR survey showed that the piles were up to 6 m deep (Avner *et al.*, this volume, Figs. 11.23, 11.25). The area documented in this study was limited to the higher dump piles and the escarpment on their east (ca. 17,000 sq m), and calculation of the piles volume there may help to estimate the total volume and mass, which can be compared with the volume of the mapped mines.

Beʾer Orah is a water source 12 km north of Nahal ʿAmram, surrounded by ancient sites of several types and periods. One site is a copper smelting camp (Fig. 11.1b), the largest of its kind in the southern Arabah, dated mainly to the Early Islamic period (Rothenberg 1999: 166–168; Avner *et al.* this volume, Fig. 11.27). Smaller smelting camps of this period are known in the southern Arabah, where copper ore from the Arabah mines was brought for smelting; estimating the volume and mass of slag piles at the major site (Beʾer Orah) can assist in assessing the total mass of copper produced in the southern Arabah during the 1st to 11th centuries CE.

Reliable estimations of volume and mass challenge the commonly used surveying practices as both the dumps in Nahal ʿAmram and slag piles in Beʾer Orah bear no simple shape and are embedded within complex topographies. To handle the challenges, terrestrial laser scanning campaigns were carried

Fig. 11.1: The studied sites, featuring: (a) the dump piles in Nahal ʿAmram (white lines delineate the piles' boundaries); (b) the slag piles in Beʾer Orah.

out to document and characterize these features and to estimate their volume and mass. Terrestrial laser scanning can generate an accurate documentation of scenes and offer dense and realistic three-dimensional data. As demonstrated below, it facilitates a detailed description of surface and objects, irrespective of their shape complexity. We describe the technology and related processes leading to the depiction of the complete scene, its characterization, and an approximation of the derived products.

METHODOLOGY

The Nahal ʿAmram and Beʾer Orah sites cover relatively wide areas and feature complex topography and entities (cf. Fig. 11.1). Therefore, use of conventional documentation practices may fall short of offering the levels of detail and accuracy necessary to characterize them. Using land surveying practices, e.g., total stations or GPS receivers, may depict the basic physical characteristics of objects and their surroundings but they are subjective and represent the three dimensional space in a limited way (Ergincan *et al.* 2010). Use of photogrammetric techniques provide an affordable solution which may depict small scale areas in detail (e.g., Lambers and Remondino 2007; Yilmaz *et al.*, 2007; Nadel *et al.* 2008; Al-kheder, Al-Shawabkeh and Haala 2009; Lerma *et al.* 2010), but in the current settings involves the handling of a large amount of images, reconstruction of complex topographies, and requires expert specialists and a considerable investment of time (Al-kheder, Al-Shawabkeh and Haala 2009; Lerma *et al.* 2010). Instead, the use of terrestrial laser scanning technology, directly measuring a dense and accurate set of three-dimensional points and facilitating a detailed surface and object description irrespective of shape complexity, is chosen here. Generally, terrestrial laser scanners (Fig. 11.2a) employ the principle of emission of collimated and coherent energy pulses and measurement of the round-trip travel time from the scanner to the illuminated object. Translated into range and adding longitudinal and latitudinal beam diversion mechanisms, a panoramic coverage of the surveyed scene can be obtained (Fig. 11.2b). Laser scanners do not rely on external illumination source and can operate day and night. However, as the measured range provides no radiometric content, cameras are incorporated into the scanners to supplement the three-dimensional positional data by color information. Fusion of the color content and the laser points cloud yields a three-dimensional photorealistic depiction of the scanned environment (Fig. 11.2c).

A complete three-dimensional site documentation requires scanning a scene from several stations to guarantee complete coverage of the region of interest, to obtain a sufficient level of detail and to limit occlusions. Data acquired from each scan is measured in the scanner's reference coordinate system, requiring registration of the individual scans into a common reference frame. Two approaches are usually utilized for registering the individual scans: 1. enlistment of designated reflector targets which can then be automatically recognized (Fig. 11.2d); 2. identification of common natural targets. In most cases reflector targets are deployed as tie elements within a scanning campaign whereas natural targets are used as tie elements between scanning campaigns.

In our study, scanning was performed using the Leica ScanStation C10 terrestrial scanner (Fig. 11.2a) whose accuracy is ±4 mm in range measurements and ±12" in angle measurement. In both sites the vertical and horizontal angular scan resolution was ~0.057°, spanning 360° horizontally and 90° vertically, and yielding nearly 14 million points. Although each scan can cover up to 150 m in radius, occlusions usually limited the range.

CASE STUDIES

NAHAL ʿAMRAM

The Nahal ʿAmram site was scanned twice, first in February 2012 and again in March 2013 following an extreme rain storm that affected the region in November 2012. Due to the topographic complexity of the site and the limited field of view, scans from six positions were required to generate a complete three-dimensional imaging. A total area of ~130 × 130 sq m (including the escarpment to the east) was

a)

b)

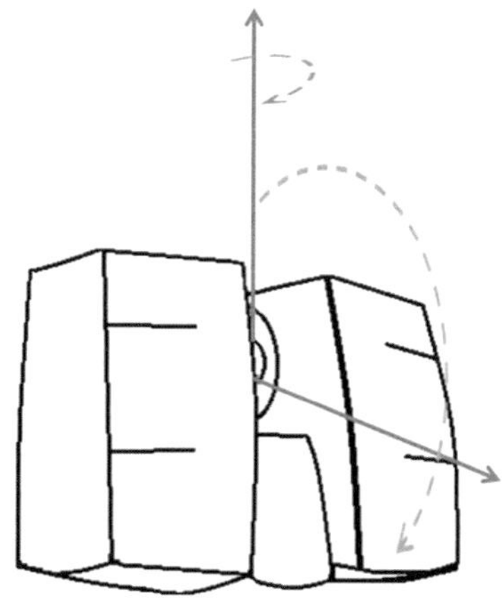

c)

d)

Fig. 11.2: Terrestrial laser scanning technology: (a) The Leica ScanStation C10; (b) diagram of the scanning principle featuring the emitted beam and the longitudinal and latitudinal rotations of the scanner; (c) a photorealistic three-dimensional rendering of a scanned site; (d) reflectors used for registration positioned within the scanned scene.

measured by ~42 million laser points with an average ground resolution of 1 point per 1 sq cm (Fig. 11.3). For co-registration of the individual scans in each campaign, target reflectors were used, yielding registration accuracy of ~4 mm.

There are two main challenges in detecting and estimating the volume of the mine dumps from the point cloud data. First is the lack of clear markers that separate the dumps from the surrounding topography; second is invisibility of the buried valley floor from which the actual dump height can be measured.

Using the color content which documents the change in hues between the soil surface and the dump, and analyzing the point cloud derived topography, the dumps were determined and isolated. To overcome the invisibility of the underlying surface topography, the pile edges, which feature the valley topography, were triangulated. This enabled us to approximate the ground surface underneath the pile. Measurement of the pile depth with respect to the approximated surface shows up to 6 m discrepancies, which is in accordance with the GPR survey that was performed as part of this project (Avner *et al.*, this volume).

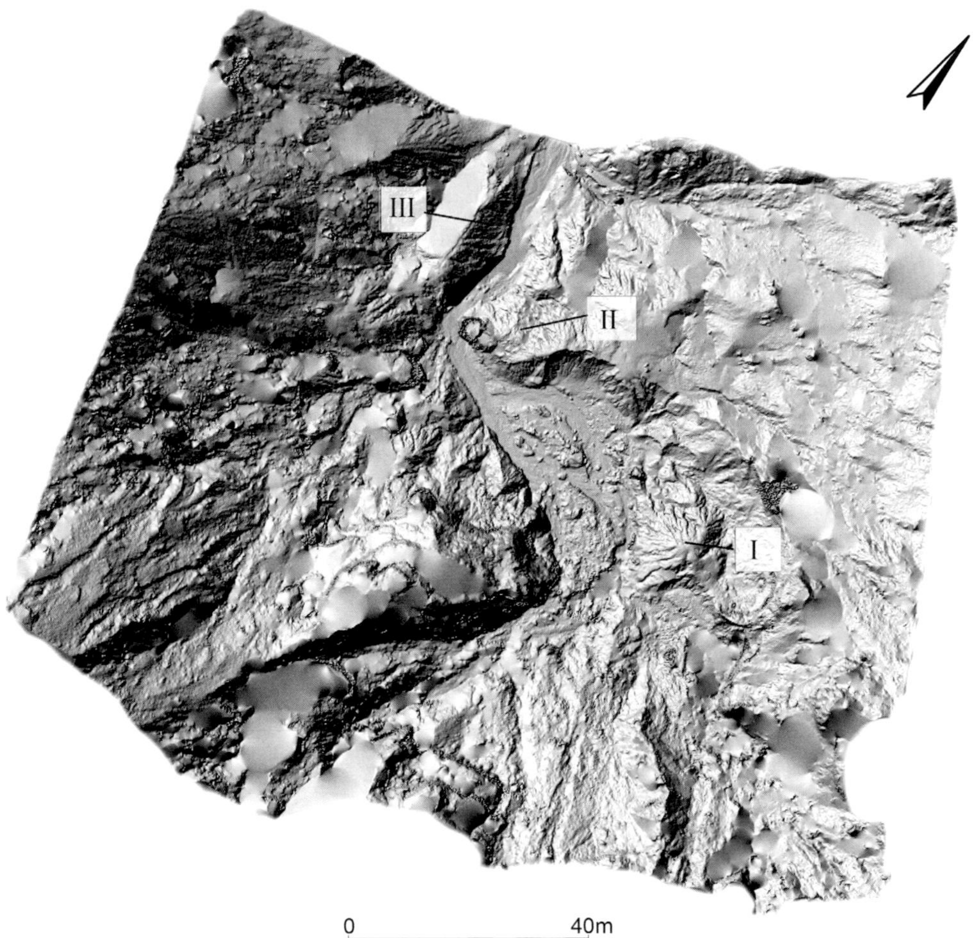

Fig. 11.3: Shaded relief maps of the Nahal ʿAmram site generated from the three-dimensional point cloud. The three piles which were studied are marked on the scan.

Following the isolation of the dump (Fig. 11.5), volume was computed by polyhedra volume summation. Polyhedra were formed by triangulation of the laser points that capture the dump surface topography as the polyhedra heads, and the approximated valley floor surface as their base level. Each polyhedron volume was computed by its subdivision into three tetrahedra (Fig. 11.4), whose volume is computed by:

1. $$v_T = \frac{(p_1 - p_2)\cdot((p_2 - p_4)\times(p_4 - p_3))}{6}$$

The pile volume was then computed by:

2. $$V_d = \sum_{j=1,n}\sum_{i=1,3} V_{T(i,j)}$$

with n= number of polyhedra.

The measurements based on the 3D scan are presented in Table 11.1.

EVALUATION OF CHANGES DUE TO THE STORM EVENT

The rainstorm of November 18, 2012 was rare in its intensity—approximately 40 mm of rainfall in a short time (Ginat *et al.,* this volume). The rain's effect on the dumps was analyzed by comparing the scans before and after the storm. Registration of the scans of the two campaigns was performed by identification of natural objects which appeared in both epochs and were not affected by the storm. The registration accuracy of the two campaigns is ±5 mm.

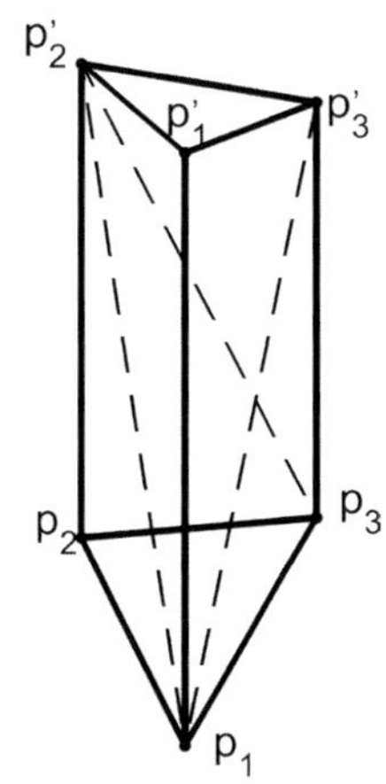

Fig. 11.4: Schematic diagram of a polyhedron's partition into three tetrahedra for volume computation.

TABLE 11.1: MEASURED AREA AND VOLUME OF THE NAHAL ʿAMRAM DUMPS

Pile No.	*Plain Area (sq m)*	*Surface Area (sq m)*	*Volume (cu m)*	*Weight (Tons)*
I	300	870	500	1000
II	1000	1820	3020	6040
III	400	1400	1130	2230
Total	*1700*	*4090*	*4650*	*9270*

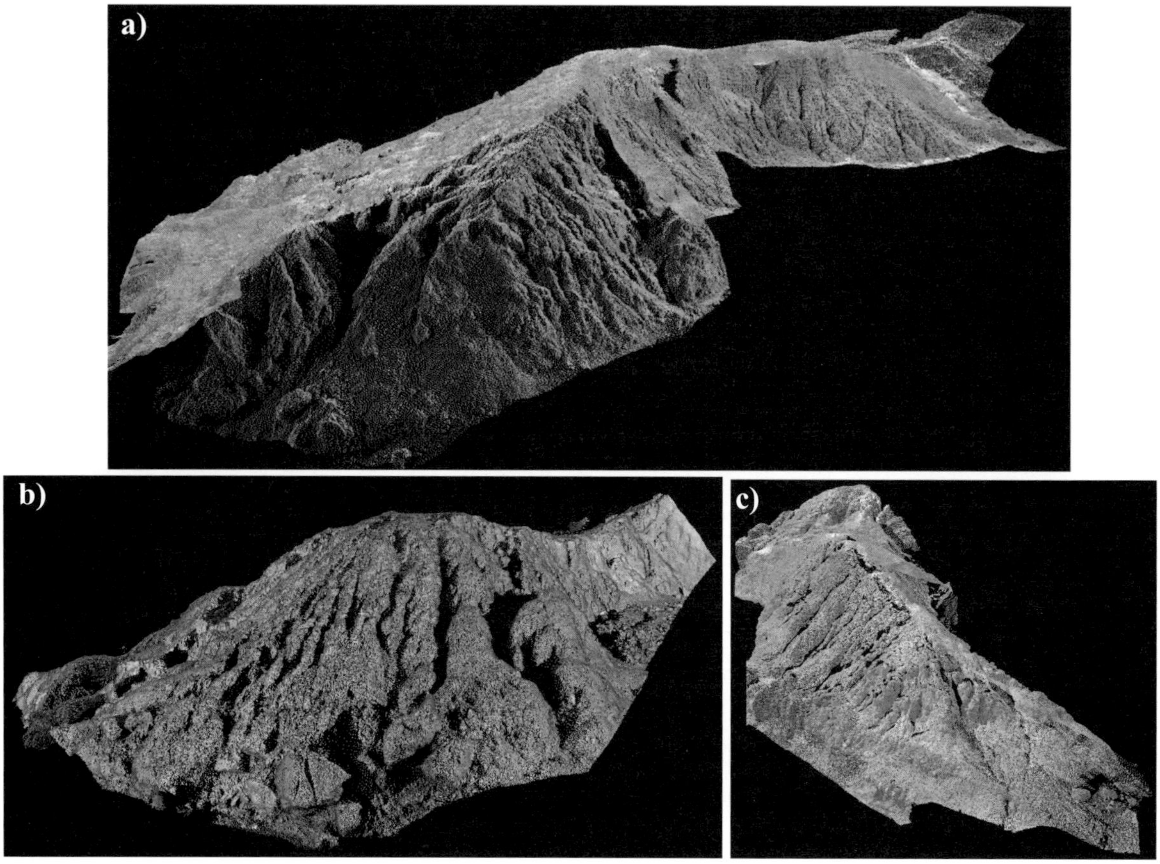

Fig. 11.5: Isolated dump piles: (a) pile II; (b) pile I; (c) pile III in Fig. 11.3.

Before the rainstorm, the dump piles were brownish-pink, but following the rain, they turned white (Fig. 11.6 and cf. Fig. 11.1) due to removal of the patinated surface. In addition, some of the small rills on the piles' slopes deepened, up to 30 cm (on the eastern slope of Pile II, Fig. 11.3). Evaluating the height differences within the study area shows sediment deposition of up to 50 cm along the wadi's channel and on a lesser scale along its tributaries. Analyzing height differences on dump faces shows that those that exceed the registration and scanner's level of accuracy can only be traced by changes along the dump edges. Other changes may be within the ±4 mm resolution of the registration, so the amount of dump washed away cannot be measured with accuracy. Assuming removal of 4 mm on average during this rain event, and given that the scanned surface areas were 4090 sq m, we may estimate that about 16 cu m were removed, i.e., ca. 32 tons, or 0.35% of the piles' volume and weight. Considering that such a rain event occurs once in 50 years, this result provides a possible tool to estimate the amount of dump eroded during the last millennium, since the cessation of mining, or even during the last two millennia since the beginning of mining in this cluster of mines.

BEʾER ORAH

The slag piles at Beʾer Orah were scanned in March 2013. Beʾer Orah is located in a relatively flat topography with surrounding hills. Three scans were sufficient to secure complete coverage, with the first offering an overview of the site (Fig. 11.7) and the rest adding the necessary details for complete three-dimensional documentation. The average ground resolution was ca. 1 point per 1 sq cm. Co-

Fig. 11.6: Nahal ᶜAmram dump piles following rain storm.

registration of the individual scans yielded accuracy of ±3 mm. With each scan covering ca. 150 m in radius and generating ca. 5 million points, an overall area of 215 × 320 sq m was covered by ca. 16.5 million points, in which the slag piles area was 120 × 50 sq m.

To separate the slag piles from the ground, a ground-to-object separating filter was applied. The morphological filter is built on the premise that a large height difference between nearby points reflect off-terrain entities rather than a steep topography. Therefore, points that conform to their surrounding are categorized as terrain whereas ones that do not meet this criterion are categorized as off-terrain. Following their separation, the pile's volume was computed in a similar manner to the dumps in Nahal ᶜAmram (Equations 1 and 2), resulting in 365 cu m. However, excavations at the site (Sharon, Avner and Nahlieli 1996) revealed that the slag base level is up to 25 cm below surface. Thus, the volume computation was modified to ca. 700 cu m. The average volumetric mass density was measured as ρ = 4.1 kg/cu m, suggesting 1500 tons of slag deposited on ground level and a total of ca. 3000 tons to the base level. Assuming a copper to slag ratio of 1:10 (cf. Hauptmann 2007: 253), we suggest that ca. 300 tons of copper were extracted at Beʾer Orah.[1]

CONCLUDING NOTES

The dumps currently cover 8500 sq m, but originally covered at least 14,060 sq m. The measurements based on the 3D scanning presented in this article pave the way for a future general estimation of the dumps' volume and weight loss over the past 2000 years.

1 cf. Rothenberg's (1995: 37) estimation—5000 tons of slag and 300–500 tons of copper.

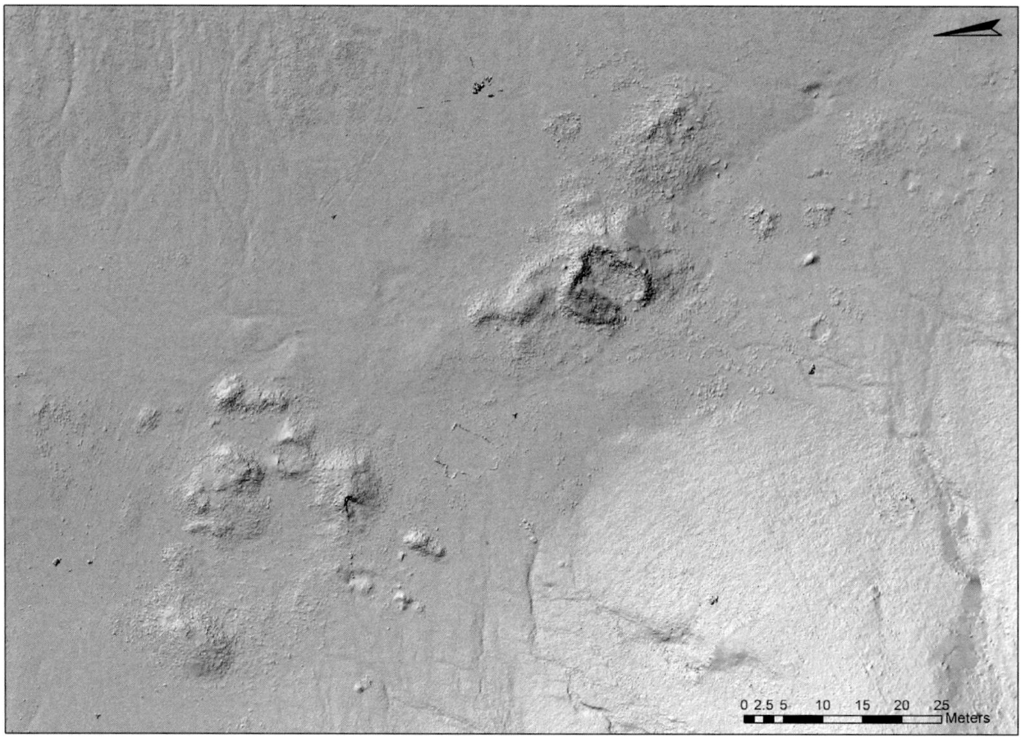

Fig. 11.7: Shaded relief map of Beʾer Orah generated from the three-dimensional point cloud.

The results from Beʾer Orah can now be applied to the contemporary smelting sites in the Eilat region: Beʾer Orah South, Naḥal ʿAmram, Tell Ḥara Ḥadid and near Umm Rashrash (in present day Eilat), as well as a few smaller sites. Approximately similar total amount as at Beʾer Orah is estimated for them, so the presently conjectured total amount of copper produced during the first millennium CE is at least 600 tons (see Avner *et al.* this volume, n. 12).

Amendment of the data from both mining dumps and slag piles is expected as the research progresses.

REFERENCES

Al-Kheder, S., Al-Shawabkeh, Y. and Haala, N. 2009. Developing a Documentation System for Desert Palaces in Jordan Using 3D Laser Scanning and Digital Photogrammetry. *Journal of Archaeological Science* 36(2): 537–546.

Ergincan, F., Çabuk, A., Avdan, U. and Tün, M. 2010. Advanced Technologies for Archaeological Documentation: Patara Case. *Scientific Research and Essays* 5(18): 2615–2629.

Glueck, N. 1953. Exploration in Western Palestine. *Bulletin of the American Schools of Oriental Research* 131: 6–15.

Glueck, N. 1960. Archaeological Exploration of the Negev in 1959. *Bulletin of the American Schools of Oriental Research* 159: 3–14.

Hauptmann, A. 2007. *Archaeo-Metallurgy of Copper: Evidence from Faynan, Jordan*. Berlin.

Lambers, K. and Remondino, F. 2007. Optical 3D Measurement Techniques in Archaeology: Recent Developments and Applications. In: Posluschny, A., Lambers, K. and Herzog, I., eds. *"Layers of Perception": Proceedings of the 35th International Conference on Computer Applications and Quantitative Methods in Archaeology (CAA)*. Berlin: 27–35.

Lerma, J.L., Navarro, S., Cabrelles, M. and Villaverde, V. 2010. Terrestrial Laser Scanning and Close Range Photogrammetry for 3D Archaeological Documentation: The Upper Paleolithic Cave of Parpalló as a Case Study. *Journal of Archaeological Science* 37(3): 499–507.

Nadel, D., *et al.* 2008. The Raqefet Cave 2008 Excavation Season. *Journal of the Israel Prehistoric Society* 39: 21–61.

Rothenberg, B. 1963. Ancient Copper Industries in the Western ʿArabah. *Palestine Exploration Fund Quarterly* 95: 5–64.

Rothenberg, B. 1967. *Negev*. Ramat Gan (Hebrew).

Rothenberg, B. 1995. A Summary of Thirty Years of Archaeological and Archaeometallurgical Field and Laboratory Research Concerning the Timna Valley and its Surroundings. In: Aviram, Y., ed. *Eilat: Studies in the Archaeology, History and Geography of Eilat and the Arava*. Jerusalem: 1–45 (Hebrew).

Rothenberg, B. 1999. Archaeo-Mettallurgical Researches in the Southern Arabah 1959–1990, Part 2: Egyptian New Kingdom (Ramesside) to Early Islam. *Palestine Exploration Quarterly* 131: 68–89.

Sharon, M., Avner, U. and Nahlieli, D. 1996. An Early Muslim Mosque near Be'er Orah in the Southern Negev: Possible Evidence for an Early Eastern Qiblah? *ʿAtiqot* 30: 107–114.

Willies L. 1990. *Exploring the Ancient Copper Mines of Wadi ʿAmram (Southern ʿAraba)*. Institute for Archaeo-Metallurgical Studies 15–16: 12–15.

Willies, L. 1991. Ancient Copper Mining at Wadi ʿAmram, Israel: An Archaeological Survey. *Bulletin of the Peak District Mines Historical Society* 11: 109–131.

CHAPTER 12

EVIDENCE OF PAST FLOOD INTENSITIES IN THE NAHAL ʿAMRAM COPPER MINES

Hanan Ginat, Dagan Meeshly, Uzi Avner and Boaz Langford

Nahal ʿAmram was an important center of copper mining throughout history. In the mines with broad galleries and shafts two different effects of past floods are found: incisions in gallery floors and accumulation of fluvial sediments. Elongated signs of digging picks discovered on some sediments indicate that workers needed to reopen the mines before continuing exploitation. A high intensity rain event occurred in the ʿAmram area in November, 2012, with a downpour of over 40 mm in less than one hour. Floods created by the rain penetrated some mines and supplied a live example of the effect of similar events in the past. Studying the fluvial sediments may supply vital information about the mines' history and paleo-environment. This paper presents preliminary results of the study of alluvial terraces in the ancient copper mines.

INTRODUCTION

Nahal ʿAmram is a small erosive valley located on the southwestern margin of the Arabah Valley, 10 km north of Eilat. The area drains eastward to the Avrona Salt-flat. Copper mines were first discovered on its white sandstone escarpments in the early 1950s, but many more were recorded during a recent study (Avner *et al*; Langford *et al.* this volume). There is a unique phenomenon of fluvial terraces, up to 155 cm high, inside the mines, created by floodwaters running across the escarpments during intensive rain events. Several layers are exposed in the terraces' cutoff, including thin sheets of sand, silt and clay and thicker strata of sandstone and limestone pebbles. Some of the shafts within or outside the mines are also filled with the fluvial sediments. The source of these sediments is obviously the colluvium on the escarpments above the mines. Small plant remains found in this sediment were ^{14}C dated to ca. 250 CE (see below).

GEOLOGICAL BACKGROUND

The various igneous and sedimentary rock formations exposed at Nahal ʿAmram, Har ʿAmram, and Har Amir are evidence of the complex geological history of this region (Fig. 12.1).

The igneous Har ʿAmram, an intrusive, young Proterozoic (~600 ma) complex, is one of the northernmost outcrops of the Arab-Nubian shield in the Sinai microplate. Har ʿAmram, defined as the "ʿAmram Massif," consists of well exposed magmatic rocks (granite, rhyolite, monzonite, quartz-syenites and various types of mafic and felsic dykes). The ʿAmram lithological sequence evolved through three magmatic cycles (Mushkin *et al.* 2003; Mushkin *et al.* 2005), later overlain unconformably by Cambrian sandstone (Garfunkel 1980).

In Nahal ʿAmram and on the escarpment of Har Amir, red and purple Lower Paleozoic sandstones (the Netfim and Shehoret formations) are exposed, up to 100 m. Copper-bearing Cambrian marine

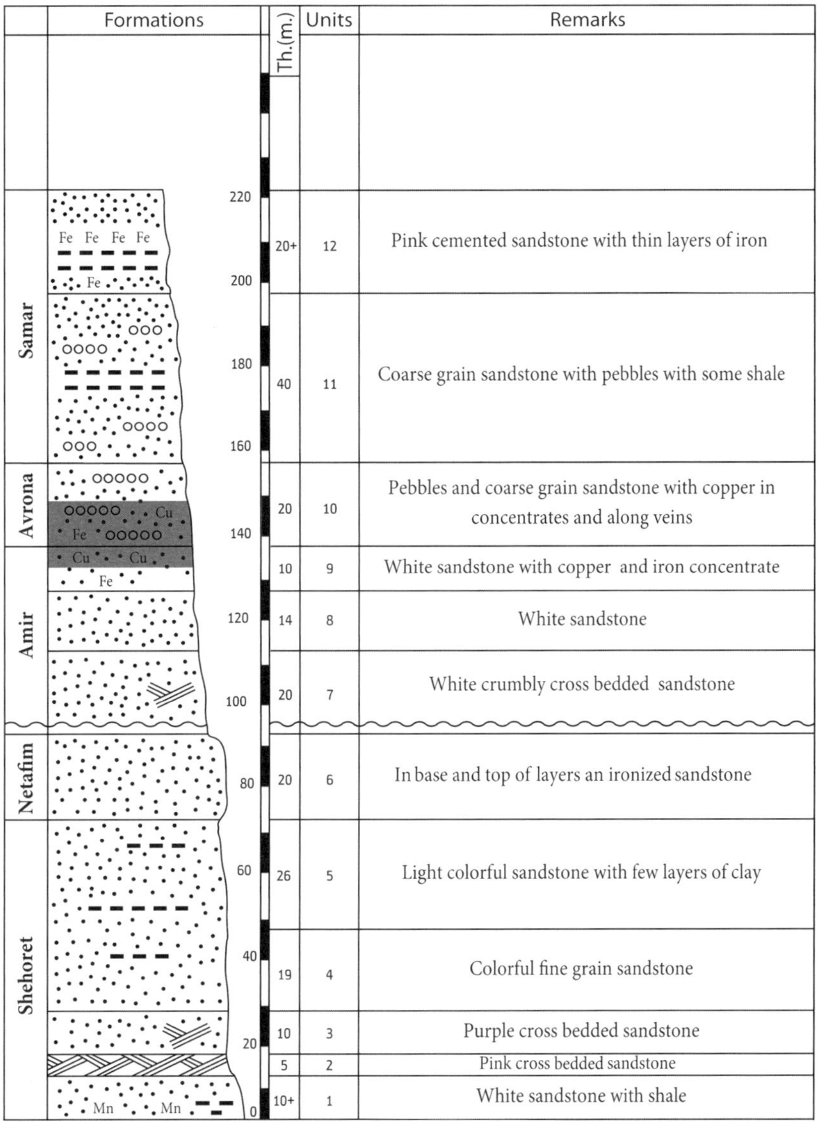

Fig. 12.1: Composite columnar section of the Har Amir Nubian sandstones. Gray fill indicates copper mineralization (in Amir and Avrona formations).

Dolomite, exposed on Har Timna, is absent at Nahal ʿAmram (Beyth *et al.* this volume). The lower part of the ʿAmram Cliffs, on the west, is formed by Lower Cretaceous sandstones of the Kurnub Group, 200 m thick, crowned by 100 m of limestone and dolostone of the Upper Cretaceous Judea Group (Fig. 12.2). The latter, with steeply declined layers, also forms the eastern side of Har Amir.

The major structural feature dominating the region is the Dead Sea Transform (DST), an active plate boundary that separates the Arabian plate from the Sinai sub-plate. Tectonic activity here includes left-lateral faulting and rifting (Bartov 1974; Garfunkel 1980, 1981; Eyal *et al.* 1981; Ginat, Enzel and Avni 1998).

Today the DST is a morpho-tectonic depression 10–20 km wide, partially filled with alluvial sediments, up to 1200 m deep. In the Arabah Valley the DST is composed of a series of *en-echelon*

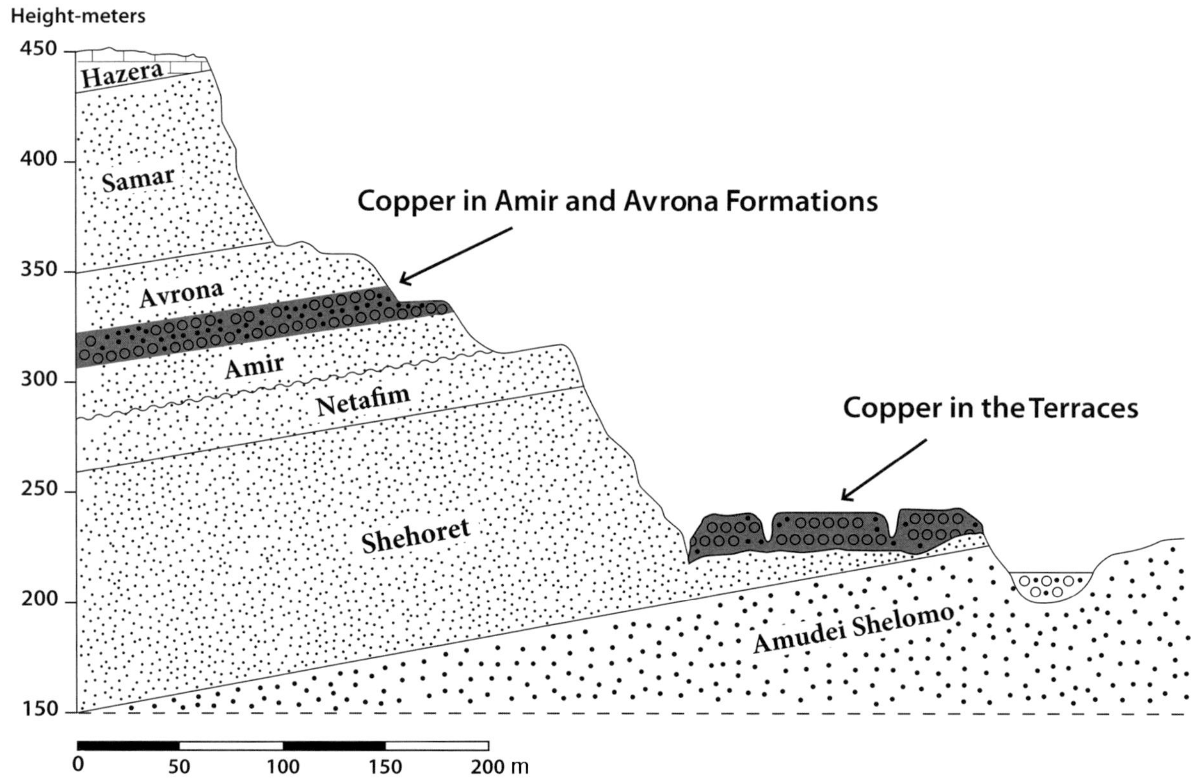

Fig. 12.2: Geological cross-section of Har Amir northern slope, the gray fill indicates copper mineralization (in Amir and Avrona Formations and in the alluvial terrace).

strike-slip faults, which form structural and topographical basins (Garfunkel 1981). The Avrona playa, the erosion-base for Nahal ʿAmram, is a deep basin in the southern Arabah Valley, filled by young sediments more than 200 m thick (Frieslander 2000).

The Har ʿAmram structure is limited by young northeastern striking, normal and sinistral faults and by east–west dextral faults. Har Amir is a north–northeast syncline also affected by the DST (Beyth, Eyal and Garfinkel 2011).

The copper ore at Nahal ʿAmram occurs in white sandstone layers of the Amir and ʿAvrona formations (Kurnub Group). It is mainly fine-grained sandstone, with some coarser sand and with a Kalonitic matrix (Weissbrod 1970; Weillies 1991). The copper ore occurs in the upper eight m of the Amir Formation and in the lower 20 m of the Avrona Formation (Figs. 12.2, 12.3). It is found in green nodules, usually from a few mm to several cm across, and also as "strings." The minerals are carbonates, mainly malachite ($Cu_2CO_3(OH)_2$) and related minerals, such as atacamite ($Cu_2Cl(OH)_3$). The source of copper here is unclear as yet, though probably results from thermal activity created by the development of the Dead Sea Transform (Segev 1986).

CLIMATE AND HYDROLOGICAL BACKGROUND

The southern Arabah Valley is the driest and hottest area in the entire Negev region and southern Jordan, classified as extreme hyper-arid (Kafle and Bruins 2009). The mean annual rainfall in the area (measured in Eilat) is 20 mm. Rainfall varies considerably not only from year to year, but also from

Fig. 12.3: Mine 35/29, alluvial sediments inside galleries, reaching the ceiling.

place to place. The so-called spottiness of rainfall is characteristic of the area (Sharon 1972). The climate is characterized by intense radiation, high air temperatures (annual average > 23 °C; summer mid-day temperature over 40^0) and high potential evaporation of 3,600 mm/yr (Goldreich and Karni 2001).

Flood events in the southern Negev are mainly caused by the Red Sea Trough synoptic system, while a secondary source is the Mediterranean depression, approaching Syria and Israel from the west, accompanied by a distinct upper-level trough (Kahana *et al.* 2002).

Intense rain events, amounting to tens of mm in a short time, occur in the Eilat region once in 5–10 years. The most recent event occurred on Nov. 18, 2012 (see below) while the previous intense rainfall was recorded on March 2, 2006 (Givᶜati *et al.* 2012). Such rain events cause torrential floods, mostly recorded by slake water deposits found in outer spaces above or beside the present channels (Greenbaum, Schick and Baker 2000). These are mostly very fine deposits which also record the ancient flood levels. From the Eilat region the only recorded past mega-floods were from Nahal Netafim, with a catchment of 35 sq km, reconstructed as 200–600 m^3/s, and dated 33,000 to 29,000 years BP (Grodeck, Lekah and Shick 2003). Until recently, no past flood event was observed in the Nahal ᶜAmram area.

ALLUVIAL TERRACES IN THE COPPER MINES

Over 200 copper mines were recorded at Nahal ᶜAmram (Avner *et al.*; Langford *et al.* this volume); in some of them fluvial terraces were found, up to 155 cm high. Layers of different nature can be identified in the fluvial deposits:

1. Thin lamination of sand, silt and clay, mainly on the upper part of the terraces (Figs. 12.3, 12.4).
2. Fine sand with some angular sandstone pebbles.
3. Coarse clastic materials, mostly sandstone pebbles.
4. In some terraces, angular limestone rocks, 2–20 cm across. Following is the description of some of the sediments in the mines:

Fig. 12.4: Mine 35/32, fluvial sediments in a hall, 1.5 m high. The lower part contains sandstone pebbles and grains, the upper part is finely laminated. The sediment was incised by later floods.

Mine 36/2–3, on the western escarpment of the head of Nahal ʿAmram, consists of two declined broad galleries, 32 and 21 m long (entrances are 2.4 × 1.8 m and 3.9 × 1 m, Fig. 12.5). In the upper, steeper gallery the miners cut steps to ease the way up and down, but now they are partly disconnected as a result of past floods that penetrated the mine (Fig. 12.5). Eight m from the entrance of the lower gallery, a shaft can be seen in the ceiling, ca. 1 m in diameter, sealed by mixed sediments (Fig. 12.6). The shaft did not penetrate the gallery from the escarpment above it, but from an unseen gallery east of the upper entrance. Remains of sediments on the gallery wall next to the shaft indicate that it was filled through the entrance. Due to a collapse of part of the fill, seven levels of sediments are visible, indicating more than seven floods that filled in the shaft. In the second layer (Unit 2 in Fig. 12.5), small fragments of plant twigs were found, dated by ^{14}C to ca. 250 CE (Avner *et al.* this volume, Table 10.1: 18). This implies that the shaft and mine were constructed prior to this date, while the floods that filled in the shaft occurred sometime later then the mid-3rd century CE.

In the same gallery, 12 m from the entrance, fluvial deposit 2.5 m long remained on its western side, which originally blocked the entire gallery's height (Fig. 12.6). The lower part of the sediment is coarse, indicating a high intensity flood, while the upper part is finely laminated sand, silt and clay, indicating a gradual and prolonged sedimentation process when the mine was out of use. The source of these sediments must have been the second order tributary which presently runs 2 m below the lower gallery's threshold. Based on observations in Timna Valley (Shlomi, Holzer and Ginat 2011) it is assumed that the tributary was higher in the past, closer in level to the mine entrance, while incision continued with time in the sandstone bottom.

On the sediment face, vertical marks of digging picks are observed, ca. 20 cm long and 3–4 cm wide (Fig. 12.6), representing re-exploitation of the mine in a later period. Since the gallery continued much beyond the sediment, it should be assumed that it accumulated in the lowest point of the mine at the time. However, during the new mining period, after the gap, most of the sediment was re-excavated and mining continued further down.

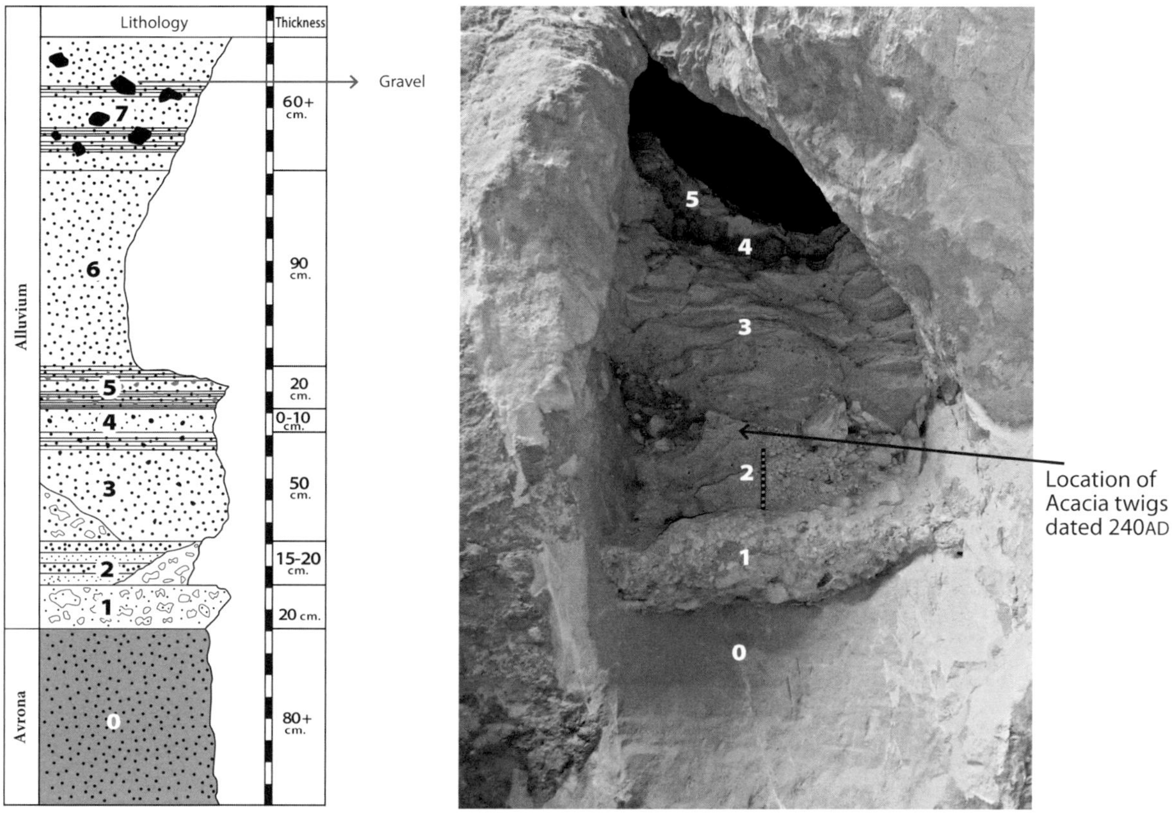

Fig. 12.5: Mine 36/2, fluvial fill in the shaft: 0. Avrona sandstone, 1. Coarse and angular gravel of sand stone 2. Light coarse gravel of sediment with fine sediments X- location of Acacia twig dated to ca. 240 CE, 3. Mostly sand and silt grains, 4. Thin layers of fine sediments, 5. Thin sandy layers, 6. Mostly alluvial sand, 7. Thin layers with silt, sand and angular pebbles of limestone (scale- 20 cm).

Based on the sediment situation, the pottery sherds and the ^{14}C date from this mine and the neighboring ones (Avner *et al.* this volume, Table 10.1), the following sequence of events in the mine is suggested: In the 1st to 3rd centuries CE, the lower gallery was only ca. 15 m long and the shaft connecting to it from an upper gallery (unseen today) was opened. The period was the Late Roman and Byzantine, but the mine was actually operated by Nabataeans (Avner *et al.* this volume). Mining ceased probably in the 5th century CE, and sediments gradually filled in the gallery up to the ceiling level and the lower part of the shaft. The digging marks in the sediment can be related to the Early Islamic period, 7th to 11th centuries CE, during which the gallery was reopened and extended to a length which is currently impossible to measure. At the end of this period, mining ceased and floods filled in the lower section of this gallery. As for the upper gallery, it seems that incision of the staircase occurred by floods after the 11th century CE, which also filled its lower part. Further information and better dating of the mine phases require excavation.

Mine 35/29 is the largest and most complex recorded at Nahal ʿAmram, consisting of a main inclining level of galleries and halls, with additional upper and lower levels, totaling 1100 m in length. Five entrances, on both sides of the ridge, lead into the mine (see details and map in Langford *et al.* this volume; Fig. 12.9). In the lowest gallery many artifacts were found in the survey and during a short excavation. Most are dated to the Early Islamic period, but six ^{14}C dates were from the early 1st to early 5th centuries CE (Avner *et al.* this volume, Table 10.1: 7–12).

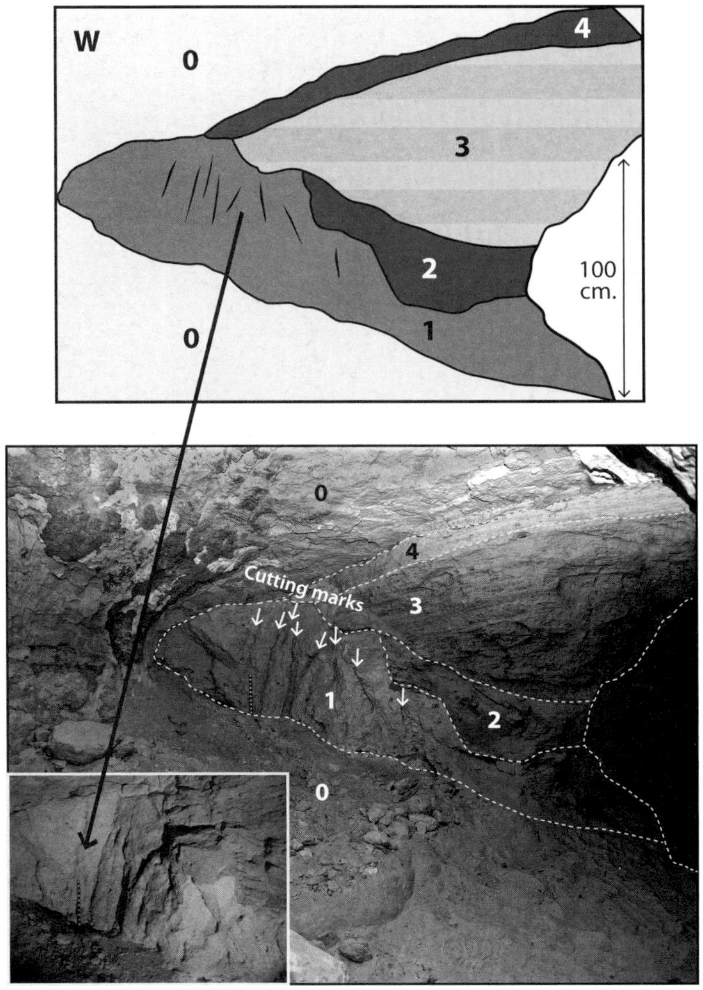

Fig. 12.6: Mine 36/2, fluvial sediment remains with four observed layers: 0. Avrona sandstone, 1. Coarse sand, 2. Coarse sand with silt and clay, 3. Layers of alluvial sand, 4. Fine sediments, mostly clay.

Several types of alluvial terraces were found in the mine. High intensity floods washed in coarse sediments, while in the later phases of sedimentation, sand, silt and clay were deposited by low intensity floods. The remains of one terrace, hanging between two levels of galleries, presents four different layers of sediments of different nature (Fig. 12.7). However, the sandstone bottom of this gallery is deeply incised by extreme floods that filled in deeper parts of the mine. Some galleries were filled in by sediments up to the ceiling level, but then re-excavated by miners either soon after the flood or during a later period of mining.

Alluvial terraces are exposed in other mines as well. Some terraces consist of one finely laminated thick layer, representing many moderate floods. In some mines the fine sediments seem quite recent, long after the cessation of mining. Shafts were also filled in, partly or totally by coarse sediments (Fig. 12.8).

FLOOD IN THE MINES

A strong rain event created by the Red Sea Trough occurred at Nahal ʿAmram on November 18, 2012. No rain gauge was set near the mine, but 5 km to the east, 27 mm of concentrated rain were measured

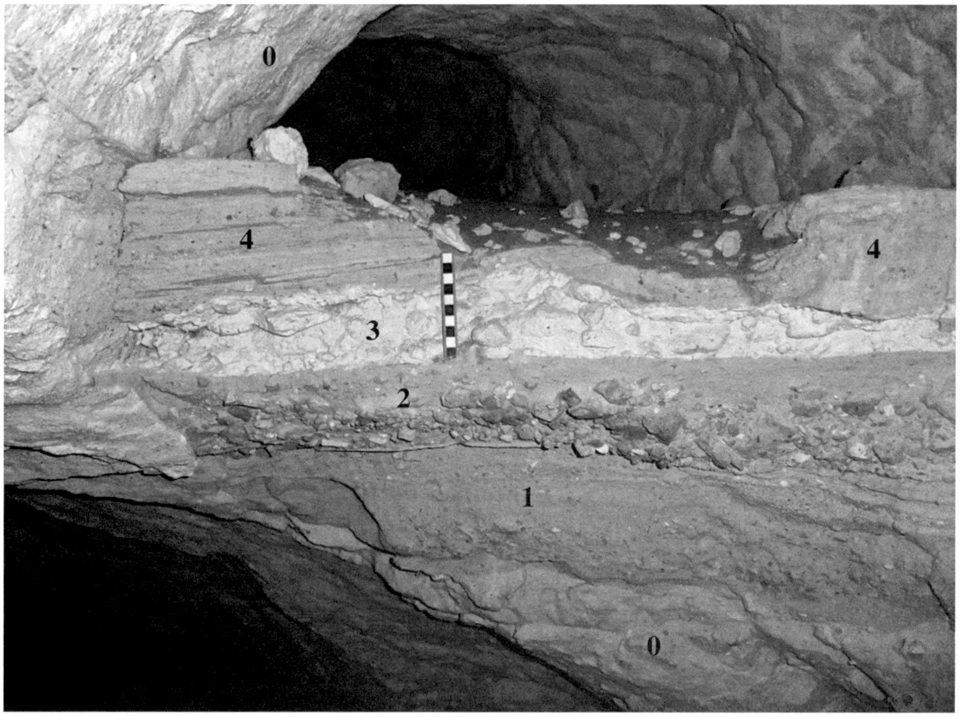

Fig. 12.7: Mine 35/29, different sediment types: 0. Avrona sandstone; 1. Fine sand, silt and clay with fine gravel and few copper concretions; 2. Limestone and sandstone coarse gravel; 3. Light sandstone gravel; and 4. Layers of fine sand, silt and clay.

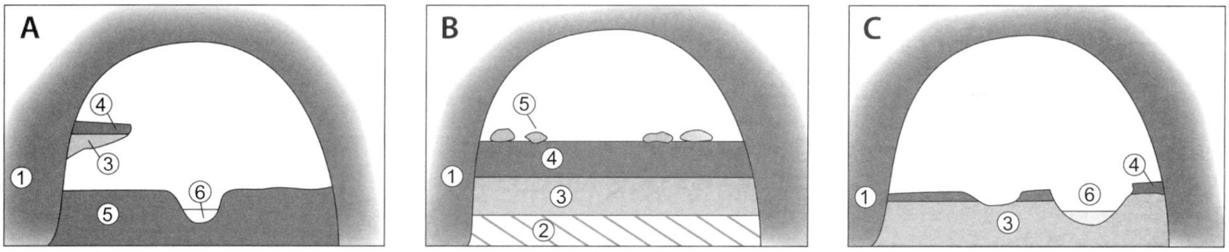

Fig. 12.8: Different types of alluvial sediments in the galleries: 1. Avrona sandstone; 2. Coarse sandstone gravel; 3. Thin lamination of sand, silt and clay; 4. Fine sand with some angular sandstone pebbles; 5. Coarse clastic materials, mostly limestone pebbles, 2–20 cm across; and 6. Fine sediments from recent floods.

within ca. 30 minutes. A hydrologic station at Nahal ʿAmram recorded a 1.4 m high flood, accumulated from a catchment of only 4 sq km that ran for only 20 minutes (Givʿati 2012). Based on these data, it is estimated that 40 mm of rain fell on Nahal ʿAmram in a short period of time and caused the high intensity flood, calculated as 20 cu m per second on each sq km! The probability of a flood in this capacity is 1% (i.e., once in 50 years, following the hydrology mode—*ibid.*). The event provided us with the unique opportunity to examine the flood effect within the mines.

The catchment area above Mine 36/2-3 is 386 sq m (Fig. 12.9). Assuming 40 mm of rain and 20% conductivity, then ca. 3.1 cu m flowed into the mine during this short event. As a result, a 20 m long incision was created in the declining bottom of the lower gallery, 20–80 cm deep. The water cut the old, compact clastic sediment, but not the sandstone underneath. At least two cubic meters of sediment

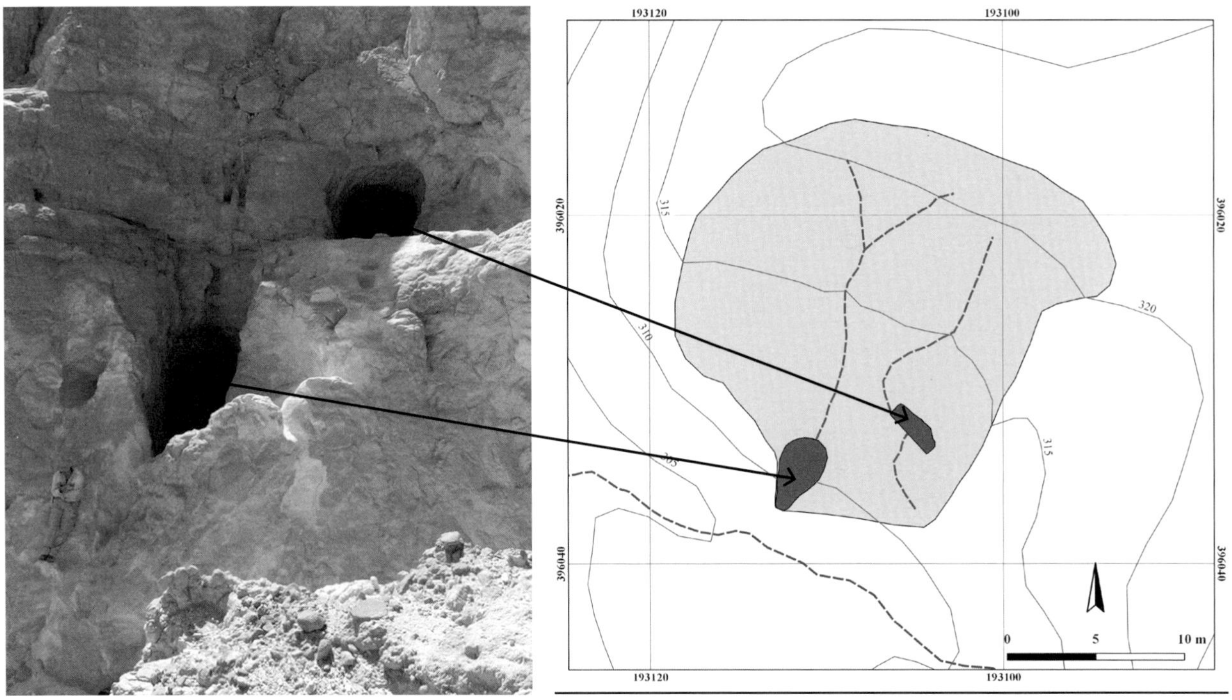

Fig. 12.9: Left: The two mine entrances. Right: The catchment area above Mine 36/2, 386 sq m.

were transported from the floor into the current base of the mine. Here, therefore, we find a small-scale current example for an incising flood. It should be mentioned again that today, the nearby tributary is much lower than the gallery's threshold and cannot feed the mine with floods. Therefore, the estimated amount of water that penetrated the mine could only have come from the escarpment above the mine.

In the largest mine (35/29), the water flow reached a level of 80 cm, and a new layer of fine sediments, mostly silt and clay, accumulated in the lower galleries. This may demonstrate that some floods in the past, which washed in coarser, clastic sediments or caused incision in galleries, were even stronger than the recent one.

Many nodules and chunks of copper ore were found in the channels of Nahal ʿAmram following the flood. These measured up to 20 cm across, 3.5 kg in weight, and contained 45–55% copper (measured by a handheld XRF). It would be logical to assume that such nodules and chunks were also intensively collected by the ancient miners after floods.

CONCLUSIONS

The new study of Nahal ʿAmram demonstrates that the area was an important copper mining center in the past, more than previously assumed. The mines with broad galleries and shafts discussed above form a type which is different from earlier ones. Based on a large amount of pottery, coins and the presently available ^{14}C dates, they are dated from the 1st to 11th centuries CE.

Following are preliminary, general observations of the fluvial sediments:

Most sediment in the mines consist of fine grains deposited in thin layers. The rain event of November 2012, although rare in its intensity, also washed into the mines only fine grained sand, silt and clay. Nevertheless, it was strong enough to cut through older accumulations on a gallery floor. This means that most floods that penetrated the mines in the past were of the scale of the recent one, or less. However, clastic sediments, with sandstone and limestones gravel, represent floods of higher intensity.

Study of the fluvial sediments inside the mines may contribute important information about their history, as well as about the paleo-environment. Unfortunately, we currently have no reliable means to date the sediments (an attempt by N. Porat of the Israel Geological Survey to use OSL dating was unsuccessful). The dating of sediment layers depends on finding organic remains for ^{14}C dating, either accidentally or by careful excavations of sediments. Much work on the mines' sediments is still required.

ACKNOWLEDGMENTS

We would like to thank Rachamim Shem-Tov for his help in measuring the fluvial deposits in the caves, to thank Rina Feldman-Avner for the English corrections and Danny Schiffer for the graphic illustrations. Photographs were taken by Ginat, Avner and Langford. The survey of Nahal ʿAmram was supported by the Israel Ministry of Science; the current stage of research is supported by the Israel Science Foundation (Grant No. 60/14).

REFERENCES

Bartov, Y. 1974. *A Structural and Paleogeographical Study of the Central Sinai Faults and Domes* (Ph.D. dissertation, The Hebrew University). Jerusalem (Hebrew with English abstract).

Beyth, M., Eyal, M. and Garfinkel, Z. 2011. The Geology of the Eilat Sheet *(Geological Survey of Israel Report* 22/2011). Jerusalem.

Eyal, M., Eyal, Y., Bartov, Y. and Steinitz, G. 1981. Tectonic Development of the Western Margin of the Gulf of Eilat (Aqaba) rift. *Tectonophysics* 80: 39–66.

Frieslander, U. 2000. *The Structure of the Dead Sea Transform Emphasizing the Arava Using New Geophysical Data* (Ph.D. dissertation, The Hebrew University). Jerusalem (Hebrew with English abstract).

Garfunkel, Z. 1980. Contribution to the Geology of the Precambrian of the Elat Area. *Israel Journal of Earth Sciences* 29: 25–40.

Garfunkel, Z. 1981. Internal Structure of the Dead Sea Leaky Transform (Rift) in Relation to Plate Kinematics. *Tectonophysics* 80: 81–108.

Ginat, H., Enzel, Y. and Avni, Y. 1998. Translocated Plio-Pleistocene Drainage Systems Along the Arava Fault of the Dead Sea Transform. *Tectonophysics* 284: 151–160.

Giv'ati, A. 2012. *Primary Report of the 18.11.2012 Rain and Floods in Amram Region* (Report of the Israel Water Authority). (Hebrew).

Goldreich, Y. and Karni, O. 2001. Climate and Precipitation Regime in the 'Arava Valley. *Israel Journal of Earth Sciences* 50: 53–59.

Greenbaum, N., Schick, A. and Baker, V. 2000. The Palaeoflood Record of a Hyper-Arid Catchment, Nahal Zin, Negev Desert, Israel. *Earth Surface Processes and Landforms* 25: 951–971.

Grodeck, T., Lekah, J. and Shick, A. 2003. Floods Convenyance on Urbanizing Alluvial Fans as an Alternative for Total Protection Systems. *Horizons in Geography* 57–58: 70–88 (Hebrew).

Kafle, H.K. and Bruins, H.J. 2009. Climatic Trends in Israel 1970–2002: Warmer and Increasing Aridity Inland. *Climatic Change* 96: 63–77.

Kahana, R., Ziv, B., Enzel, Y. and Dayan, U. 2002. Synoptic Climatology of Major Floods in the Negev Desert, Israel. *International Journal of Climatology* 22: 867–882.

Mushkin, A., Navon, O., Halicz, L., Hartmann, G. and Stein, M. 2003. The Petrogenesis of A-Type Magmas from the ʿAmram Massif, Southern Israel. *Journal of Petrology* 44: 815–832.

Mushkin, A., *et al*. 2005. Geology and Geochronology of the ʿAmram Massif, Southern Negev Desert, Israel. *Israel Journal of Earth Sciences* 48: 179–193.

Segev, A. 1986. Lithofacies Relations and Mineralization Occurrences in the Timna Formation, Timna Valley (Ph.D. dissertation, The Hebrew University). Jerusalem (Hebrew with English abstract).

Sharon, D. 1972. The Spottiness of Rainfall in the Desert area. *Journal of Hydrology* 17: 161–175.

Shlomi, Y., Holzer, Y. and Ginat, H. 2011. Erosion Rates in an Ancient Copper Mine in Timna Valley. *Abstracts for the Annual Geological Conference, Mitzpeh Ramon* 12.

Weissbrod, T. 1970. The Stratigraphy of the Nubian Sandstone in Southern Israel. *Geological Survey of Israel Report* OD/2/70.

Willies, L. 1991. Ancient Copper Mining at Wadi ʻAmram, Israel: An Archaeological Survey. *Bulletin of the Peak District Mines Historical Society* 11: 109–131.

CHAPTER 13

MINERS' MEALS AT THE COPPER MINES OF NAHAL ʿAMRAM, SOUTHERN ISRAEL

Liora Kolska Horwitz, Uzi Avner and Omri Lernau

Here we describe an assemblage of faunal remains retrieved from two copper mines in Nahal ʿAmram in the Arabah Valley, dating to the Early Islamic and/or Roman periods. The animal remains comprise bones of domestic caprines, camel as well as fish bones representing six different Families, four of which only inhabit the Red Sea. The mammal bones were disarticulated, such that it is highly unlikely that they represent natural mortalities. Moreover, the bones show characteristic butchery damage resulting from carcass dismemberment and filleting, long bone breakage resulting from marrow extraction as well as human gnawing. Most of the mammal bones represent meat-rich limb bones. Consequently, they have been interpreted as food remains discarded by miners. This is corroborated by their association in the excavated area with cooking-pots, firewood, charcoal and remains of fruit. Clearly, miners in Nahal ʿAmram enjoyed a rich and varied diet, atypical of slaves or poor people.

We compare the Nahal ʿAmram findings with published data on food remains from mines and mining camps elsewhere in the world, including the copper mines of Faynan (Jordan) and the neighboring mines in Timna. We conclude that the mines were probably provisioned by settlements in the surrounding area.

INTRODUCTION

During renewed investigations of copper mines and related sites situated around Nahal ʿAmram, ca. 10 km north of the city of Eilat (Avner *et al.*, this volume), artifacts were found inside several mines. Mine 35/29, the largest known to date in the Arabah Valley (total length 1100 m, Langford *et al.*, this volume: Fig. 14.9) was first surveyed by Willies (1990, 1991), but the new survey discovered a lower level of galleries, rich with artifacts and organic remains. A short excavation in one spot, measuring 4 × 1.5 m, uncovered many finds in an unstratified soft sand fill, including Roman (Late Nabataean) oil lamps, Early Islamic pottery, botanical remains and animal bones among others. The majority of finds date to the Early Islamic period (7th to 11th centuries CE). However, five ^{14}C dates on organic remains gave ages spanning the Roman period (mid-1st to early 5th centuries CE) (see Avner *et al.*, this volume for details).

Though the excavated location in Mine 35/29 is in its deepest and most remote gallery, and presently has very poor ventilation, it probably served as the miners' kitchen. This suggestion is based on the finds, which comprise a rich assemblage of mammal and fish bones, the remains of a wide variety of fruit, cooking-pots, firewood and charcoal. Two vertical shafts observed in the gallery's ceiling, and currently blocked by coarse fluvial sediment, supplied ventilation to this gallery in the past.

During the survey of nearby Mine 35/32, several animal bones and fruit remains were discovered and collected in a localized area within a gallery ca. 12 m from the entrance. This spot still awaits excavation.

The following is a description of the animal bones currently retrieved from the two mines. As at present we do not have ^{14}C dates directly measured on the faunal remains, we have to consider them as associated either with the Roman or Early Islamic periods, or spanning both. The remains are discussed in relation to fauna retrieved from other mines and associated settlements in the Arabah as well as those found elsewhere in the world. Specific reference is made as to whether the Nahal ʿAmram mines were provisioned by outside suppliers or were self-sufficient.

THE FAUNAL REMAINS

The faunal assemblage comprises a total of 244 archaeozoological and archaeomalacological remains—232 from Mine 35/29 and only 12 from Mine 35/32. They represent the remains of sheep/goat, goat, camel, a species of mollusk and fish (Tables 13.1–2).

SPECIES

The animal species identified in the Nahal ʿAmram faunal assemblages were relatively limited, although this may relate to the small size of the sample. Two species of terrestrial mammal (NISP = 195) comprised 81.0% of the identified remains, while fish (NISP = 43) and mollusks (NISP = 3) made up the remaining 18.0% and 1.0% respectively (Table 13.1). Only two mammalian species were identified—the dromedary (*Camelus dromedarius*) and domestic goat (*Capra hircus*), while the largest proportion of remains belonged to unidentified caprines, probably domestic sheep or goat (*Ovis aries* or *Capra hircus*). At least six fish Families were identified in addition to three *Tridacna* shells (Tables 13.1, 2).

Camel

Camel is represented by 23 bones (11.8% of identified mammals), all from Mine 35/29 (Table 13.1). With the exception of one complete posterior first phalanx (toe bone), only fragments of trunk elements are represented—the centrum and spines of vertebrae, rib shaft fragments and ossified cartilage of the sternum (Table 13.3). Following morphological and biometric criteria of first phalanges, outlined in Studer and Schneider (2008), the camel phalanx was identified as belonging to a dromedary (*Camelus dromedarius*) (Fig. 13.1).

Caprines

The majority of animal remains represent domestic caprines (either sheep or goat) (NISP = 159; 81.5% of identified mammals). In addition, a few caprine remains could be identified as domestic goat, *Capra hircus* (NISP = 13; 6.7% of identified mammals) (Table 13.1), based on their morphology (Boessneck 1969; Halstead, Collins and Isaakidou 2002). They include an adult ram horncore that had been severed from the skull at its base. Part of the horn sheath still adheres to its surface (Fig. 13.2), attesting to the excellent bone preservation at the site.

On the basis of tooth eruption and attrition (Payne 1973), the combined caprine-goat assemblage includes at least one mature adult goat, perhaps the ram, aged 3.5–6 years (Payne wear stages for lower M2 and M3 were 9 and 10 respectively), and at least two immature animals—one aged ca. 1 month old and another slightly older animal aged ca. 6 months (Payne wear stages for the lower deciduous m3's are 0 and 2 respectively). Pathologies are evident on some of the teeth. Two very worn, paired upper third molars exhibit enamel hypoplasia. These lesions form due to trauma, disease or other kinds of stress during tooth development (Baker and Brothwell 1980). In addition, both the very attrited second and third lower molars of an adult goat exhibit deposits of dental calculus (calcified dental plaque; *ibid.*).

Table 13.1: Faunal Remains by Location (NISP Counts and Relative Frequencies)

Taxa	*Mine 35/32*		*Mine 35/29*		*Components of Assemblage*
	NISP	*%*	*NISP*	*%*	*%*
Camel			23	11.8	
Goat			13	6.7	
Sheep/Goat	3		159	81.5	
Mammal Total	3		195	100	81.0
Fish Total	9		34		18.0
Mollusk Total			3		1.0
Faunal Total	12		232		100

Table 13.2: Identification, Skeletal Element Distribution and Origin of Fish Remains

Family	*Genus*	*Species*	*Element*	*N*	*Origin*
Scaridae (Parrotfish)			First vertebra	1	
			Abdominal vertebra	1	
			Caudal vertebra	7	
			Hyomandibular	1	
			Tail bone	1	
	Scarus	*sp.*	Basioccipital	1	Red Sea
	Scarus	*gibbus*	Dentary	3	Red Sea
Serranidae (Groupers)			Caudal vertebra	1	Red or Mediterranean Sea
			Dentary	1	
			Opercular	1	
			Preopercular	3	
Carangidae (Jacks, Pompanos)			Opercular	1	
			Quadrate	1	Red or Mediterranean Sea
Lethrinidae (Emperors)			Basipterygium	1	Red Sea
	Lethrinus	*sp.*	Premaxilla	1	
Pomacanthidae (Angel fish)			Caudal vertebra	1	Red Sea
Priacanthidae (Big eyes)	*Priacanthus*	*hamrur*	Third vertebra	1	Red Sea
Unidentified fish			Abdominal vertebra	1	
			Caudal vertebra	5	
			Interopercular	1	
			Quadrate	1	
			Scale	4	
			Tail bone	1	
			Vertebra	1	
			Unknown	2	
Total Bones				*43*	

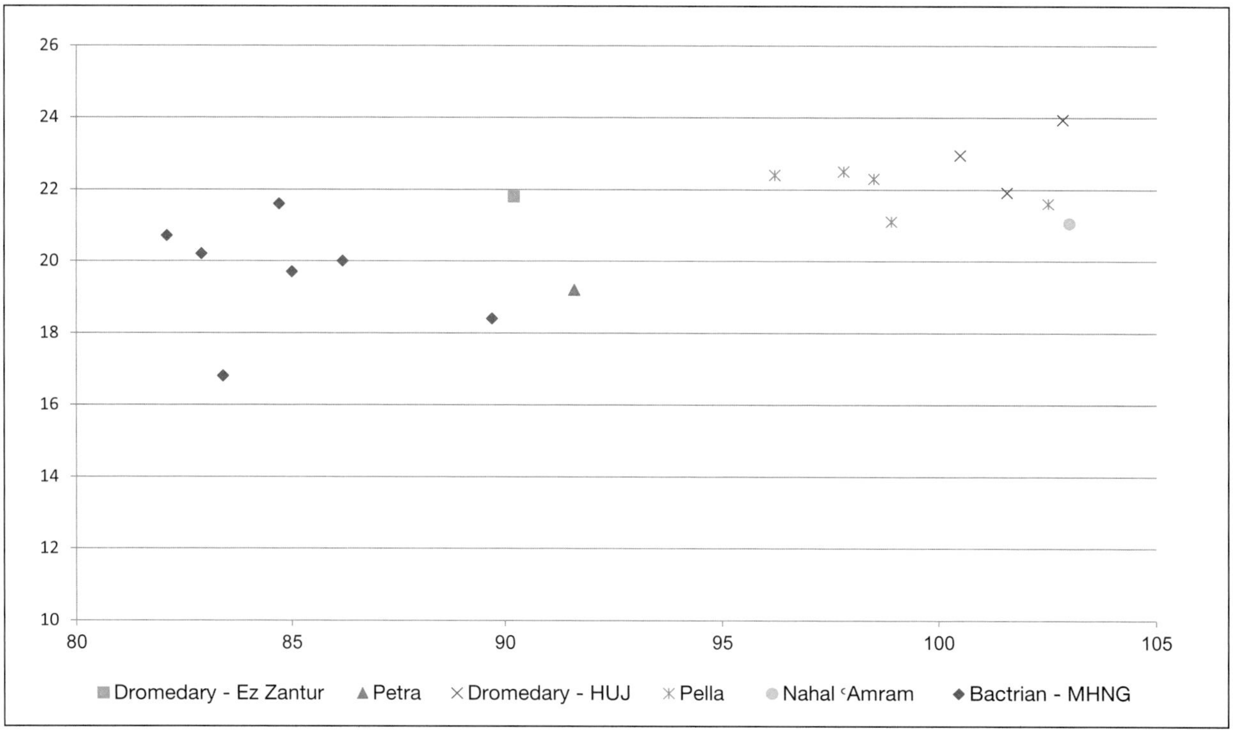

Fig. 13.1: Scatterplot showing camel first phalanx measurements (GL = Greatest length; SD= minimum shaft width) used to determine species attribution. Data for HUJ specimens—taken by LKH; data for specimens from MHNG, Pella and Petra (ez-Zantur) from Studer and Schneider (2008: Table 1).

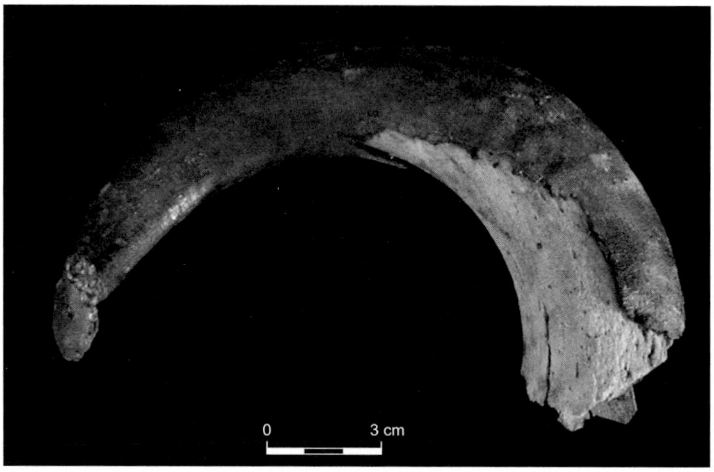

Fig. 13.2: Goat horncore from Nahal ʿAmram with preserved horn sheath. Note the severed distal end where the horncore has been detached from the skull.

None of the Nahal ʿAmram faunal remains were found in articulation, emphasizing that they do not represent natural mortalities, i.e., remains of animals that accidentally entered the mine and died there. As illustrated in Tables 13.3 and 13.4, all skeletal elements of the caprine carcass are represented in the assemblage. There is however, an apparent bias in skeletal elements represented with selection of meat-rich upper fore and hindlimb elements as well as meaty trunk parts. Very few cranial remains,

Table 13.3: Skeletal Element Representation of Mammalian Remains (NISP counts)

Skeletal Element/Species	*Camel*	*Goat*	*Sheep/Goat*
Horncore	-	1	-
Cranium-occipital	-	-	1
Unid. cranial bones	-	-	6
Maxilla	-	-	2
Pre-maxilla	-	-	2
Mandible	-	3	-
Mandible fragments	-	-	2
Cervical vertebral body*	-	-	12
Thoracic vertebral body*	-	-	2
Lumbar vertebral body*	1	-	14
Caudal vertebra whole	1	-	-
Thoracic spine	1	-	4
Unid. Vertebral spine	2	-	-
Unid. vertebral body	8	-	2
Unid. vertebral epiphysis	-	-	3
Sacrum	1	-	2
Rib prox.	-	-	15
Rib shaft	3	-	37
Rib dist.	1	-	-
Sternum cartilage	4	-	-
Scapula dist.	-	2	4
Scapula blade	-	-	2
Humerus prox.	-	-	3
Humerus dist.	-	1	-
Radius prox.	-	1	3
Ulna prox.	-	-	3
Metacarpal prox.	-	-	2
Pelvis ilium	-	-	10
Femur prox.	-	-	3
Femur shaft	-	-	4
Femur dist.	-	1	5
Tibia prox.	-	-	3
Tibia shaft	-	-	6
Tibia dist.	-	-	1
Patella	-	-	1
Astragalus	-	-	1
Metatarsal prox.	-	-	2
Metatarsal shaft	-	-	1
Metatarsal dist.	-	1	-
1st Phalange	1	1	-
2nd Phalange	-	2	-
Hoof sheath	-	-	1
Total NISP	*23*	*13*	*159*

* These vertebrae are all halved through their centrum; Unid. = unidentified; dist. = distal; prox. = proximal.

Table 13.4: Bodypart Breakdown for Main Species

Body Parts	*Camel*		*Caprines*		*Key*
	NISP	*%*	*NISP*	*%*	
Cranial	-	-	17	9.8	Skull, antler/horn, maxilla, mandible and loose teeth
Trunk	22	96	91	53.0	Atlas, axis, cervical, thoracic, lumbar and caudal vertebrae, ribs
Upper forelimb	-	-	19	11.0	Scapula, humerus, radius, ulna
Lower forelimb	-	-	2	1.1	Metacarpal, carpals
Upper hindlimb	-	-	34	19.7	Pelvis, femur, tibia, patella
Lower hindlimb	-	-	5	3.0	Metatarsal, calcaneum, astragalus, tarsals
Foot	1	4	4	2.3	1st, 2nd and 3rd phalanges
Total	*23*	*100*	*172*	*100*	

mandibles, lower fore/hindlimb and foot elements are represented, these being the least meaty parts of the carcass and are usually discarded at the slaughter location. Hence, the body part distribution supports the interpretation of this assemblage as representing choice food remains. This contention is corroborated by the butchery damage and evidence of human gnawing evident on a high proportion of the caprine bones (see below).

Fish

A total of 43 fish bones were recovered from two mines, of which 27 could be attributed to six different fish Families (Table 13.2). In Mine 35/32 only nine fish bones were recovered, of which three were identified to Family and six bones could not be identified at all. The identified material represented two bones of Scaridae and one of Carangidae. Mine 35/29 yielded a larger fish bone assemblage comprising 34 bones. Of these 13 were identified as belonging to Scaridae, six as Serranidae, two Lethrinidae, and one each belonging to Carangidae, Pomacanthidae and Priacanthidae respectively. Sixteen of the fish remains could not be identified to Family.

With respect to the entire identified fish assemblage, three of the Families inhabit the Red Sea and are not found in the Mediterranean: Pomacanthidae (Angel fish), one vertebra; Priacanthidae (Bigeyes), one vertebra, which belonged to the species *Priacanthus hamrur*, Moontail Bullseye; and Lethrinidae (Emperors) represented by a cranial bone. Scaridae (parrotfish) were represented by 15 vertebrae and head bones, of which three could be further identified to the species *Scarus gibbus*, the heavybeaked parrotfish. Although one representative of Scaridae is found in the Mediterranean Sea, *Sparisoma cretense,* it is easily differentiated from the Red Sea Scarids. On the basis of morphology, the Scarid bones from Nahal ʿAmram certainly do not belong to *S. cretense,* so that they clearly originate from the Red Sea even without further identification to genus/species. Eight bones were identified as belonging to the Families Carangidae (Jacks and Pompanos, two head bones) and Serranidae (Groupers, five head bones and one vertebra). These two families have representatives both in the Red Sea and in the Mediterranean. Taxonomic identification of these bones to the level of species was not attained and therefore the fish could originate in either of the seas. Given that all other identified fish and the clam shells originated in the Red Sea, it is assumed that the Serranidae and Carangidae are also derived from this source. Moreover, the Red Sea is much closer to Nahal ʿAmram than the Mediterranean (13 km vs 220 km). The minimum number of fish (MNI) calculated for the assemblages from both mines is at least seven.

The skeletal element breakdown for the combined fish sample from both mines is given in Table 13.2. There are both head bones and vertebrae represented. The Scaridae (parrotfish) are represented by head, trunk and tail elements; Lethrinidae by a jaw bone; Priacanthidae by a first vertebra, i.e., that is close to the head; Pomacanthidae by a tail vertebra and Serranidae by elements of the head. It is most probable that the Nahal ʿAmram fish were processed, (salted/dried/smoked) at the coast for transport inland. However, at least for some Families, complete or relatively complete fish were brought to the mine, as attested to by the presence of scales and cranial elements.

Mollusks

Three valves of the elongated giant clam (*Tridacna maximus*) recovered from Mine 35/29 (and four more found in other mines), all originated from the Red Sea. The valves are large and thick with a relatively deep inner depression. No modifications are visible on the shells though one of them bares a dark organic stain in the inner depression, probably indicating use as an oil lamp. This interpretation is supported by the numerous shell-shaped oil lamps of ceramics and stone which are found at Near Eastern sites (e.g., Douglas 2001). Other uses are also suggested for the *Tridacna* shells: cosmetics and incense bowls, offering dishes, libation receptacles, musical instruments or they served as ritual offerings in themselves (e.g., Reese 1988: 264–65; Gensheimer 1984). At the Nahal ʿAmram mine, the shells may have served as food items (cf. Mienis 1994) and were then used as oil lamps.

BONE MODIFICATIONS

Modifications were observed on numerous faunal remains in the Nahal ʿAmram assemblage and could be attributed to both abiotic agents (discoloration resulting from proximity to copper and possibly also iron oxides in the sediment) and biotic agents (animal derived damage—gnawing; and human related damage— butchery, gnawing, and burning - Lyman 1994) (Table 13.4).

Discoloration

Many of the animal bones recovered from the gallery bore a blue/green hue and/or a dark brown color. These discolorations are evident on remains of all species: on 12 of the 23 camel bones (52.1%); on 117 out of 172 caprine-goat bones (62%), and on eight of the 43 fish bones (18.6%) (Table 13.5). Green-blue discoloration is known in the literature as either due to (a) direct contact with a copper or brass

Table 13.5: Discoloration to Bones (Numbers are NISP's)

	Camel (Total NISP=23)		*Caprine-Goat (Total NISP =172)*	
Damage Category	*Discoloration*		*Discoloration*	
Skeletal Element	*Dark Brown*	*Green*	*Dark Brown*	*Green*
Cranial	-	-	2	7
Trunk	12	7	22	38
Forelimb	-	-	3	9
Hindlimb	-	-	13	23
Feet	-	-	-	-
NISP	12	7	40	77
% Discolored of Total NISP	*52.1*	*30.4*	*23.2*	*44.7*

object, (b) contact with copper enriched soil, (c) presence of a copper salt deposit on the bone surface, (d) heat induced transformation or, (e) intentional coloration using a dye, such as woad, as known from the Roman period (Morris 1981; Reiche, Vignaud and Menu 2000; Ferrand *et al.* 2014). In the case of Nahal ʿAmram, the blue/green discoloration is undoubtedly due to impregnation by copper carbonate solutions present in the sediment given their discovery inside a copper mine. The dark brown bones appear to be the result of permeation of iron oxides and only occasionally due to burning (Dutton and Fasham 1994: 269; Pyatt *et al.* 2005; Dupras and Schultz 2013). Geochemical analyses are underway to confirm this.

Only two bones exhibit convincing evidence of burning. One is a right distal goat femur with a fused distal epiphysis and a shaft severed 1/3 up from the distal end. The shaft rim was stained black with a patch of blackened and calcined bone on the adjacent shaft area. The medullary cavity was filled with a dark colored material that may be soot, suggesting that it may have been used as a lamp of some sort. The second is a caprine occipital fragment with a scorched red brown-gray patch on one condyle.

BUTCHERY

Table 13.6 presents the data on butchery for the Nahal ʿAmram faunal assemblage. A total of 52.1% of all camel remains and 64.5% of caprine bones exhibit butchery modifications. Most commonly modified bones are vertebral bodies that have been severed in two, usually through the middle of the centrum. This butchery pattern was observed on nine camel and 32 caprine vertebrae (Fig. 13.3). The second most common damage is observed on ribs that had either the distal, proximal or both ends severed. This pattern was found on three camel and 40 caprine ribs. The damage seen on the trunk elements indicates that during butchery, carcasses were probably divided in half through the vertebral column (by splitting the vertebral bodies), following which rib-racks were removed by chopping through the proximal rib ends, severing them from the vertebrae. A further modification ensued whereby the rib-racks were subdivided, resulting in sections comprising short pieces of ribs. This is corroborated by measurements taken of 37 such rib shafts which shows them to be of standardized lengths; 20 smaller pieces had a mean length of 40.2 mm (range 22.2–50.4 mm, SD 6.9), while a further 17 larger pieces had a mean length of 66.4 mm (range 51.1–93.1 mm, SD 11.7). In addition to severing of the trunk elements, light cut marks associated with defleshing were observed on the bodies of five vertebrae and on five rib shafts; all had also been severed.

TABLE 13.6: BUTCHERY DAMAGE ON CAMEL AND CAPRINE-GOAT BONES (NUMBERS ARE NISP'S)

Skeletal Element	*Camel (Total NISP=23)*		*Caprine-Goat (Total NISP =172)*	
	Severed	*Cut Marks*	*Severed*	*Cut Marks*
Cranial	-	-	8	-
Trunk	12	-	72	10
Upper forelimb	-	-	8	4
Lower forelimb	-	-	-	-
Upper hindlimb	-	-	23	10
Lower hindlimb	-	-	-	1
Feet	-	-	-	-
NISP	12	-	111	25
% Butchered of Total NISP	*52.1*		*64.5*	*14.5*

Fig. 13.3: Two caprine vertebrae from Nahal ᶜAmram that have been severed at right angles to the axis of the bone (vertically) through the centrum.

Caprine-goat upper fore- and hindlimbs also exhibit butchery damage. Parts of eight forelimb elements (six scapulae, two humeri) and 23 hindlimbs (ten pelves, nine femora, four tibiae) had been severed, forming smaller "parcels" of bone and meat. In addition, light cut marks were observed on one upper forelimb (a humerus shaft) and on four upper hindlimbs (two on a pubis shaft, one on an acetabulum, one on a tibia shaft). Deep chop marks were found on three upper forelimbs (one on a scapula neck, one on a radius shaft and one on a proximal ulna), as well as on six upper hindlimbs (one on a pubis, three on ilium shafts, two on femur shafts). Some chop marks possibly resulted from failed attempts to sever the bones, while the lighter cut marks are all associated with filleting, i.e., meat removal. A single metatarsal exhibited short parallel cut marks adjacent to the proximal end, characteristic of skinning (Fig. 13.4).

The extensive butchery damage evident on the majority of bones in the Nahal ᶜAmram assemblage demonstrates that small joints of meat with their associated bones are represented here, i.e., portions of vertebral column, rib-racks or meat-rich limb bones. This clearly supports the interpretation of these faunal remains as representing leftovers from a meal/s.

HUMAN GNAWING AND BONE FRACTURING

Several studies have focused on how to distinguish gnaw marks made by carnivores and those made by humans and other primates, resulting in a corpus of diagnostic marks for human gnawing (e.g., Laroulandie 2000; Landt 2007; Martinez 2009; Fernández-Jalvo and Andrews 2011). The most frequent and characteristic modification to bones resulting from human chewing is that the ends of thin bones, such as ribs, are bent relative to the long axis of the bone and the damaged area is frayed or exhibits a roughened surface with parallel grooves and fibrous texture, i.e., peeling (as defined by White [1992: 140]: "when fresh bone is fractured and peeled apart similar to bending a small fresh twig from a tree branch between two hands"). Fernandez-Jalvo and Andrews (2011: 121) noted that "Some of our European experimenters commented that this is the easiest way to break the bone and suck the marrow."

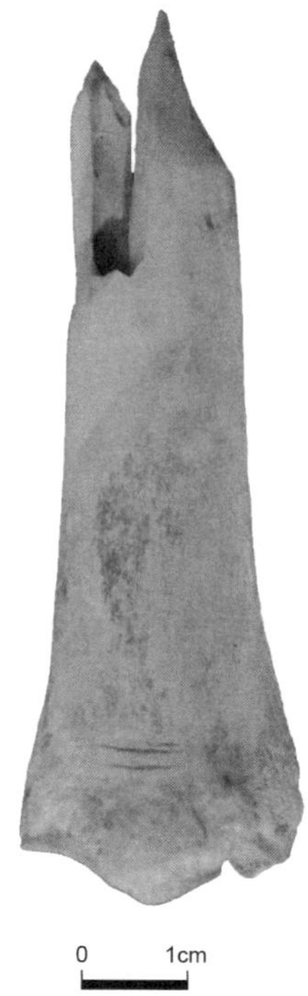

Fig. 13.4: Caprine proximal metatarsal from Nahal ʿAmram with short, parallel cut marks.

A second criterion of human gnawing is the presence of relatively superficial surface markings. Human gnawing leaves puncture marks that are generally triangular in shape, dispersed and small (<4 mm). Fernández-Jalvo and Andrews (2011) noted these as present but rare, while the more common features were crenulated or ragged bone edges and shallow, linear marks on the bone surface—transverse or oblique to the long axis of the bone, probably caused by the incisors. Similarly, Binford (1981), who documented human bone gnawing by the Nunamiut Eskimo, and Gifford-Gonzalez (1989), who studied the Dassanetch people, both remarked that the extent of damage and nature of the marks left by people following bone gnawing, were more superficial, far less developed (less intense) as well as lower in number, than those resulting from gnawing by large carnivores.

Studer and Pillonel (2007) studied human damage to bones of an immature wild boar consumed by the Yali people of Papua, New Guinea and found such damage on 57% of all examined ribs. The damage was exclusively on the distal end of the ribs, at a distance ranging from 2 mm to 2 cm from the bone's border, and took the form of ragged edges, crushing of the bone's periphery, shallow pits and cracks along the edges. The young age of the animal was said to be the primary factor responsible for the high frequency of damage to the distal ribs, since this end is cartilaginous and soft in an immature animal. A similar age-related damage pattern was documented by Brain (1969) on bones recovered following consumption of a domestic goat by a community of Khoi in Namibia. Brain's study demonstrated that considerable damage was inflicted during consumption on unfused limb elements (proximal and distal femur; distal metapodials), but in his study surprisingly few distal rib ends were damaged.

In the Nahal ʿAmram assemblage, human gnawing was clearly identified on 19 caprine-goat bones, which is 11% of the caprine-goat sample as a whole. On ten severed sections of rib corpus, the damage from gnawing took the form of peeling, crushed edges, presence of light gouge marks and striations adjacent to the bone's edge (a 2–3 cm distance at most), often on more than one end (Fig. 13.5). This

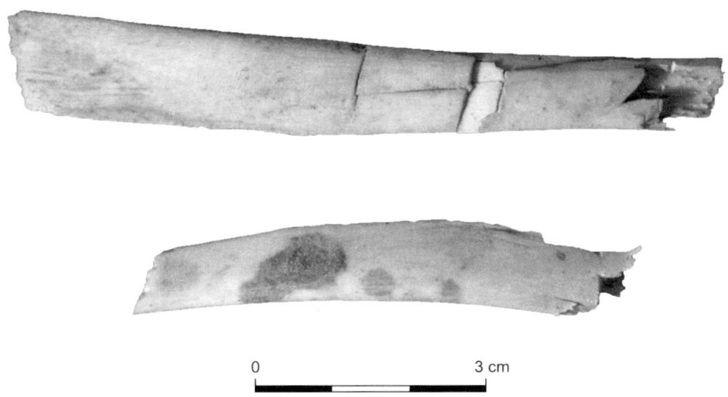

Fig. 13.5: Two caprine rib shafts from Nahal ʿAmram exhibiting typical edge "peeling" damage resulting from human gnawing.

is a relatively high frequency of gnawing damage—19.2% of all caprine ribs in the assemblage—and recalls the results of the study undertaken in Papua, New Guinea. The same pattern of damage was also observed on three thoracic vertebrae spine tips adjacent to the edges of two long bone shafts (humerus and metatarsal) and adjacent to the border of a severed mandible corpus. Puncture holes, the result of a bite penetrating the outer table of the bone, was observed on a proximal ulna, a scapula blade (adjacent to the edge) and on the neck of a scapula close to the distal epiphysis. The distinctive gnaw damage, combined with the pattern of butchery, leave no doubt that the faunal remains represent the debris of the miners' meals and are indicative of the intensity of animal food exploitation.

A unique find in the Nahal ʿAmram assemblage was a caprine metatarsal midshaft fragment, without epiphyseal ends, but with a wooden twig inserted through the medullary cavity so that it protrudes from both shaft ends (Fig. 13.6). The bones' ends are oblique but jagged and the shaft is stained light green, with a deep chop mark adjacent to one end. Given the tight fit of the twig inside the bone shaft, it is apparent that it was not inserted accidentally but perhaps was used as an *ad hoc* tool to extract bone marrow. Similarly, green (spiral) fractures were evident on 19 caprine-goat bones (11% of the assemblage) that comprise 18 long bone shafts: one humerus, four radii, two ulnae, four femora, six tibiae, a metatarsal and a mandible corpus. Percussion notches are evident on two of the six fractured tibia shafts. The fracturing of bones to extract marrow is yet another indication of intensive exploitation of animal foods in the mine.

DISCUSSION

In recent years, there has been increased interest worldwide in faunal remains recovered from mines. They fall into two categories; animal bones used as mining tools, and more commonly, food debris, recovered from within mines and from associated miners' camps.

BONE TOOLS

Examples of bone tools recovered from ancient mines are cattle scapulae from a Bronze Age metallurgical work camp at the Ross Island mine (Ireland) or from flint mines on the continent (O'Brien 1995: 26). These items were probably used as scoops or shovels. Antler picks, usually on naturally shed antlers, were another common mining tool in antiquity, such as those discovered in the prehistoric flint mines in Spiennes (Belgium) (dated to 4400/4200 to 3350/2900 cal BC) (Collet, Hauzeur and Lech 2008: Fig. 13), the Neolithic

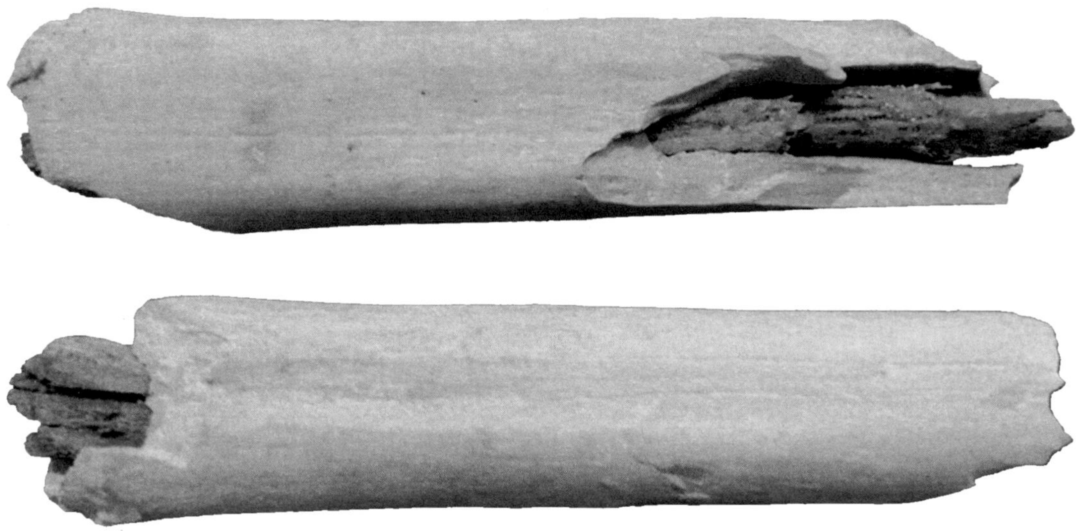

Fig. 13.6: Caprine metatarsal shaft fragment from Nahal ᶜAmram with a small wooden stick inserted into the medullary cavity.

flint mine at Grimes Graves (UK) (Clutton-Brock 1984: 25) or in the Bronze Age (1880–1630 BCE) copper mine at Ecton, Staffordshire (UK) (Barnatt and Thomas 1998). In Peru, picks and scrapers made of deer antlers as well as from the iliac bone (part of the pelvis) of deer and llamas were used in mining, while on the coast, whale bones were used for chisels (Petersen 2010). Some of these bones/antlers may have been hafted or attached with a cord to wooden handles, as reported by Peterson (2010: 37) for a Chavin period (1200 BCE–200 CE) antler pick from the site of Las Aldas. Excavations of a Bronze Age smelting site in the Pyllau Valley, part of the Great Orme mining complex located in Wales (UK), have produced thousands of bone artifacts—predominantly long bones (tibia, ulna and ribs). Lewis (1998: 49) noted that the majority of these bones served as mining tools as evidenced by the pattern of their wear and fracturing that is consistent with their use as scrapers, chisels and gouges, and that only a small portion represented food remains.

No bone artifacts that may have served as mining tools have been recognized in either the Nahal ᶜAmram copper mines or from mines in the neighboring Timna Valley. One explanation may be the late date of much of the mining in this area and the widespread use of iron implements from the Iron Age onwards. Indeed, most of the digging marks in the mines are from large picks, apparently of iron.

FOOD REMAINS

Food Remains from Inside Mines

The location of food residue, cooking-pots, firewood and charcoal within the Nahal ᶜAmram mines demonstrates that the miners camped and worked inside the gallery. This, however, is not a unique find. A parallel example is from the Chehr Abad salt mine in northwestern Iran, which dates from the 4th century BCE to the 4th century CE (Ramaroli *et al.* 2010). In addition to remains of eight mummified miners, remains of tools, textiles, cultivated plants and animal bones were found there. Sheep and goat were consumed in the salt mine, while the consumption of beef is assumed due to the presence of a container manufactured from cattle horn at the site. It is also proposed that pig, a species found at other sites in the region, may have been consumed by the miners, though its remains were not found in the mine (Aali *et al.* 2012).

Another example relates to remains from the pre-Hispanic (Nasca and Wari cultures) haematite mine of Mina Primavera (Nasca region, Peru). The mine functioned in the first few centuries of the Common Era, and later between ca. 900–1500 CE. Excavations inside the mines have revealed the presence of domestic refuse—textiles, stone and shell beads, flaked stone, musical instruments, rope, pottery sherds, fragments of the sea shell *Spondylus princeps*, plant remains and a few faunal remains. These archaeological finds have been interpreted as indications that the miners camped inside the mine, bringing with them tools, food and perhaps ritual offerings (Vaughn *et al.* 2007, 2013). However, to date, the only fauna found represent small mammals that may have died naturally after the mine had been worked (Vaughn *et al.* 2013: 170–171).

The most recent documentation of food consumed and even cooked inside mines is Cornish Pasty—traditionally comprising diced meat, potato, turnip and onion wrapped in a pastry—which was invented by Cornish miners to be consumed underground, when it was not possible to surface for lunch (website of the Cornish Mining Organization http://www.cornish-mining.org.uk/delving-deeper/food). Apparently, some mines had stoves down the mineshafts to cook the raw pasties. During the 19th century, Cornish miners who migrated overseas for work introduced the pasties to other parts of the world— Australia, Mexico, North America (Shoemaker 2001).

Food Remains from Mining Settlements and Their Source

Remains of food consumed by miners have primarily been recovered from settlements adjacent to mines and have been reported from mining localities worldwide. They represent residues of meals cooked and consumed by those who worked in the mine, or were associated in some manner with them. In all instances, the majority of foods were imported to the mine from neighboring settlements since mining settlements were not self-sufficient. A few examples are detailed below.

In the Bronze Age metallurgical work camp at Beaker Hut site on Ross Island (Ireland), remains of animal bones, mostly cattle, pig and some sheep, were recovered during excavations in the settlement, that represent refuse of food animals brought to the camp from neighboring farms on the Irish mainland (O'Brien 2004: 473). From the Late Bronze Age copper mining area of Mauken, in the Tyrol (Austria), Schibler *et al.* (2011) describe a rich and varied botanical and faunal repertoire. Here, the food remains were recovered from a midden associated with a work camp for ore processing adjacent to the mine adits (Mauk D) and from a contemporaneous smelting site (Mauk A). They largely represent imports of domestic animals—cattle, pig and sheep/goat—supplied by farms in the surrounding region. The authors posit that choice cuts of meat, possibly as deboned or preserved cuts (e.g., ham, bacon) that were ready to eat or uncured for cooking, were brought to the site rather than whole carcasses. Some evidence points to limited production of dairy products on-site (on a seasonal basis), as well as hunting, though the researchers concluded that the latter was of little caloric importance.

Likewise, mines and quarries in later periods primarily relied on imported provisions. At the Roman marble quarries of Mons Porphyrites (Gebel Abu Dukhan) and Mons Claudianus (Gebel Abu Hamr), located in the Eastern Desert of Egypt, food was imported from the Nile Valley (cereals, lentils, dates, olives, wine, donkey, pig, camel, goat, sheep, horse and chicken), the Red Sea coast (shellfish and fish) and the Eastern Desert (game and snails), while exotic items (artichoke, citron and pepper), apparently imported from India, were also found (Hamilton-Dyer 2001, 2007; Van Der Veen and Tabinor 2007). The presence of vegetable seeds (of cabbage, leaf beet, lettuce, mint, basil) indicates that some plants were cultivated locally. In the specularite mines at Doornfontein and Blinkklipkop (Northern Cape Province, South Africa), a wide range of domestic and wild fauna, all representing food remains, were recovered from archaeological deposits dating to ca. 1200 BP (Thackeray, Thackeray and Beaumont 1983). The source of these animals was not discussed in the publication, but it is likely that they were hunted locally and either brought to the locality by the miners or bartered from settlements near the mines. Claughton and Smart (2008) reported that during the later Medieval period (ca. late 1290s), the majority of the workforce employed at the silver mine at Combe Martin (Devon, UK) were immigrants who had no access to agricultural holdings, and as such were largely reliant on imported foodstuffs. With respect to foodstuffs, irrespective of temporal or spatial differences, it is evident that mines are not self-sufficient entities but consumer sites dependent on provisioning from outside producers.

In the Timna copper mining area, 15km north of Nahal ʿAmram, faunal remains representing consumption refuse were recovered from two Late Bronze Age/early Iron Age sites. At Site 2, a smelting camp, remains of goat, cattle, camel, donkey and fragments of ostrich eggshell, were recovered (Rothenberg 1972: 78; Hakker-Orion 1984). Fauna from more recent excavations at the site have yielded goat remains (but not sheep), and surprisingly—domestic pig (Bar-Oz and Erickson-Gini in press). Some of the bones are burnt and a few exhibit butchery damage. Remains from Site 30, another smelting camp, were predominantly those of food animals—domestic sheep and goats, with scanty remainders of cattle, camels, equids and carnivores. In addition, two Mediterranean Sea fish species and another freshwater fish species, probably from the Nile River, were identified in the assemblage (Lernau 1988b; Grigson 2012). Recently, Sapir-Hen and Ben-Yosef (2013) presented an archaeozoological analysis of remains from renewed excavations at the early Iron Age sites Timna 30 and 34, with domestic sheep and goats dominating the assemblages, accompanied by remains of donkey. Red Sea shells and scanty remains of cattle and ostrich eggshells and fish complete the faunal spectrum. The fish mostly originate from the Mediterranean Sea, but also from the Nile River (Nile perch) and several other freshwater species that may be from the Nile or the Levant. The absence of Red Sea fish differentiates these Iron Age assemblages from that recovered at Nahal ʿAmram.

The third locality in the Timna Valley is the miners' sanctuary, Site 200 (dated to the 13th–10th centuries BCE). Animal remains recovered during excavations were restricted to remains of domestic sheep and goat, in addition to numerous ostrich eggshells (Avner 2014: Fig. 7) and a small collection of fish bones imported from the Nile, from the Mediterranean, and possibly also from the Red Sea (Lernau 1988a, b). It should be borne in mind that these remains represent food offerings/sacrifices or remains of ritual meals offered to the gods. As such, they may represent intentionally selected taxa and not the full spectrum of animals consumed by the miners.

At the Iron Age copper production site of Khirbat en-Nahas, Faynan, Jordan (Muniz and Levy 2014) caprines, cattle, donkey and camel were found, in addition to scanty remains of wild mammals (gazelle, hare), birds, mollusks and reptiles. Fish families identified are Sparidae, Scaridae, Sciaenidae and Labridae which could originate either in the Red or Mediterranean Seas. Both donkey and camel appear to have been exploited in the Faynan/ Arabah mining settlements, as well as consumed by the miners. Based on age and skeletal element profiles, the researchers suggest that the Iron Age mining settlement was self-sufficient since domestic animals, which formed the mainstay of the economy, were maintained (produced) and exploited (consumed) locally (Muniz and Levy 2014: 659).

In terms of mammalian faunal diversity these archaeozoological mining assemblages in the Faynan/ Arabah mining settlements resemble each other, but the source/s of the animals consumed at all these sites is unclear. If animals were kept by the miners, then a local infrastructure for husbandry and/or farming would have been required, including access to a permanent water supply, but little or no evidence is found for these features. There is a permanent spring at Khirbat en-Nahas (Faynan), and Levy, Ben-

Yosef and Najjar (2012; Fig. 21.1) illustrate agricultural fields adjacent to the Iron Age mines but do not discuss them further, while Knabb (2008) emphasizes the paucity of evidence for agriculture at the site. Barker (2002) also identified a large field system in Faynan and noted that local agricultural production was probably aimed at feeding the laborers associated with the Roman and Byzantine period copper mine Phaeno. No permanent water source exists in either Timna or Nahal ʿAmram. However, water sources and large farms were situated in the Arabah Valley. The Yotvata oasis, 10 km north of Timna, supplied plenty of water and farming took place all through the first millennium CE. At Evrona, 7 km southeast of Nahal ʿAmram, a large farm was cultivated during the 7th–11th centuries CE, based on *qanats*. Both farms produced a wide variety of crops while herding was practiced all along the Arabah in all seasons. These farms supplied food for the population of Aila, the nearby villages, the copper industry, trade caravans and pilgrims (Avner and Magness 1998; Horwitz 1998; Avner 2016; Kishon *et al.* in press).

Comparison of species representation demonstrates the range of animals at Nahal ʿAmram to be far narrower than that found in the nearby Early Islamic settlements noted above. Noteworthy is the absence of equids, wild ungulates, carnivores, birds, etc. The mine sample contains only the most common food taxa found in the settlements—caprines and fish (and to a lesser extent, camel). It seems unlikely that the small size of the Nahal ʿAmram faunal sample is responsible for the limited range of species (Grayson 1984). Rather, the specialized activities carried out at the mines have dictated the composition of the archaeozoological sample.

Fish and mollusks were evidently traded from some distance (Nile River, Mediterranean coast as well as the Red Sea) to both the inland settlements and the mines, with all species found at Nahal ʿAmram also occurring in the Early Islamic settlements (Mienis 1994; Lernau 1998). The extent of fish consumption is difficult to gauge, but was probably extensive given the quantity of fish recovered from the Early Islamic settlements.

In sum, it seems that the nutritional status of miners at Nahal ʿAmram was high. They consumed protein from three sources (meat, fish, nuts) as well as several different species of fruit whose remains were found intermingled with the animal bones. These elements are indicative of a rich and varied diet typical of wealthy people, and not of slaves or the poor, offering a clue as to the identity of the miners in the Roman and/or Early Islamic periods. Most importantly, given what has been documented at other mines, provisioning of the Nahal ʿAmram mines from surrounding agrarian settlements undoubtedly took place.

ACKNOWLEDGMENTS

Research at Nahal ʿAmram was supported by a grant from the Israel Science Foundation (Grant No. 60/14); comparative camel remains were measured in the National Natural History Collections of the Hebrew University of Jerusalem.

REFERENCES

Aali, A., *et al*. 2012. Ancient Salt Mining and Salt Men: The Interdisciplinary Chehrabad Douzlakh Project In North-Western Iran. *Antiquity* (Online Gallery) 86(333). http://antiquity.ac.uk/projgall/aali333/

Avner, U. 2014. Egyptian Timna Reconsidered. In: Tebes, J., ed. *Unearthing the Wilderness: Studies on the History and Archaeology of the Negev and Edom in the Iron Age*. Leuven: 103–163.

Avner, U. 2016. Desert Farming in the Southern 'Arabah Valley, Israel, 2nd Century B.C. to 11th Century A.D. In: Davies, A., Retamero, F. and Schjellerup, I., eds. *Making Agro-Pastoral Landscapes in Pre-industrial Societies: Choices, Stability and Change*. Oxford: 19–36.

Avner, U. and Magness, J. 1998. Early Islamic Settlement in the Southern Negev. *Bulletin of the American Schools of Oriental Research* 310: 39–57.

Baker, J. and Brothwell, D. 1980. *Animal Diseases in Archaeology*. London.

Barker, G.A. 2002. A Tale of Two Deserts: Contrasting Desertification Histories on Rome's Desert Frontiers. *World Archaeology* 33: 488–507.

Barnatt, J. and Thomas, G.H. 1998. Prehistoric Mining at Ecton, Staffordshire: A Dated Antler Tool and Its Context. *Mining History* 13 (5): 72–78.

Bar-Oz, G. and Erickson-Gini, T. In press. Late Bronze Faunal Remains From Timna, Site 2. In: Bar-Oz, G. and Horwitz, L.K., eds. *Discovering Noah's Ark: Zooarchaeology of the Holyland* (IAA Reports). Jerusalem.

Beaumont, P. B. 1973. The Ancient Pigment Mines of Southern Africa. *South African Journal of Science* 69: 140–146.

Binford, L. 1981. *Bones: Ancient Men and Modern Myths*. New York.

Boessneck, J. 1969. Osteological Differences between Sheep (*Ovis aries*) and Goat (*Capra hircus*). In: Brothwell, D.R. and Higgs, E.S., eds. *Science in Archaeology*. London: 331–358.

Brain C.K. 1969. The Contribution of Namib Desert Hottentots to the Understanding of Australopithecine Bone Assemblages. *Scientific Papers of the Namib Desert Research Station* 39: 13–22.

Claughton, P. and Smart, C. 2008. The Bere Ferrers Project and Discovery of a Roman Fort at Calstock. Unpublished Online Research Report. http://people.exeter.ac.uk/pfclaugh/mhinf/Roman%20fort%20-%20Tamar%2030.pdf

Clutton-Brock, J. 1984. *Excavations at Grimes Graves Norfolk 1972–1976 Fascicule 1: Neolithic Antler Picks From Grimes Graves, Norfolk, and Durrington Walls, Wiltshire: A Biometrical Analysis*. London.

Collet, H., Hauzeur, A. and Lech, J. 2008. The Prehistoric Flint Mining Complex at Spiennes (Belgium) on the Occasion of its Discovery 140 Years Ago. In: Allard, P., Bostyn, F., Giligny, F. and Lech, J., eds. *Flint Mining in Prehistoric Europe Interpreting the Archaeological Records* (BAR International Series 1891). Oxford: 41–77.

Douglas, D.A. 2001. *The Development and Demise of the Early Bronze Age IV, Near Eastern Oil-Lamp* (PhD dissertation, Commonwealth Open University). London.

Dupras T.L. and Schultz J.L. 2013. Taphonomic Bone Staining and Color Changes in Forensic Contexts. In: Pokines, J. and Symes, S.A., eds. *Manual of Forensic Taphonomy*. Boca Raton. 315–340.

Dutton, L.A. and Fasham, P.J. 1994. Prehistoric Copper Mining on the Great Orme, Llandudno, Gwynedd. *Proceedings of the Prehistoric Society* 60: 245–286.

Fernández-Jalvo, Y. and Andrews, P. 2011. When Humans Chew Bones. *Journal of Human Evolution* 60: 117–123.

Ferrand, J., *et al*. 2014. On the Origin of the Green Colour of Archaeological Bone Artefacts of the Gallo-Roman Period. *Archaeometry* 56(6): 1024–1040.

Gensheimer, T.R. 1984. The Role of Shell in Mesopotamia: Evidence for Trade Exchange with Oman and the Indus Valley. *Paléorient* 10: 65–73

Gifford-Gonzalez, D. 1989. Modern Analogues: Developing an Interpretive Framework. In: Bonnichsen, R. and Sorg, M.H., eds. *Bone Modification*. Orono, ME: 43–52.

Grayson, D.K. 1984. *Quantitative Zooarchaeology: Topics in the Analysis of Archaeological Faunas*. New York.

Grigson, C. 2012. Camels, Copper and Donkeys in the Early Iron Age of the Southern Levant: Timna Revisited. *Levant* 44(1): 82–100.

Hakker-Orion, D. 1984. The Role of the Camel in Israel's Early History. In: Clutton-Brock, J. and Grigson, C., eds. *Animals and Archaeology 3. Early Herders and their Flocks* (BAR International Series 202). Oxford: 207–12.

Halstead, P., Collins, P. and Isaakidou, V. 2002. Sorting the Sheep from the Goats: Morphological Distinctions between the Mandibles and Mandibular Teeth of Adult *Ovis* and *Capra. Journal of Archaeological Science* 29: 545–553.

Hamilton-Dyer, S. 2001. The Faunal Remains. In: Maxfield, V. A. and Peacock, D.P.S., eds. *Survey and Excavation Mons Claudianus Vol. II: Excavations I*. Cairo: 251–312.

Hamilton-Dyer, S. 2007. Faunal Remains. In: Peacock, D.P.S. and Maxfield, V.A., eds. *The Roman Imperial Quarries, Survey and Excavation at Mons Porphyrites 1994–1998, Vol. 2: The Excavations* (Egypt Exploration Society). London: 143–175.

Horwitz, L.K. 1998. Animal Exploitation during the Early Islamic Period in the Negev: The Fauna from Elat-Elot. *ʿAtiqot* 36: 27–38.

Kishon, V., Hellwing, S., Meshel, Z. and Ayalon, E. In press. Early Islamic Faunal Remains from Yotvata. In: Bar-Oz, G. and. Horwitz, L.K., eds. *Discovering Noah's Ark: Zooarchaeology of the Holyland* (IAA Reports). Jerusalem.

Knabb, K.A. 2008. *Understanding the Role of Production and Craft Specialization in Ancient Socio-economic Systems: Toward the Integration of Spatial Analysis, 3D Modeling and Virtual Reality in Archaeology* (MA thesis, University of California). San Diego.

Landt, M.J. 2007. Tooth Marks and Human Consumption: Ethnoarchaeological Mastication Research among Foragers of the Central African Republic. *Journal of Archaeological Science* 34: 1629–1640.

Laroulandie, V. 2000. *Taphonomie et archéozoologie des aiseaux en grotte. Applications aux sites paléolithiques du Bois-Ragot (Vienne), de Combe Saunière (Dordogne) et de la Vache (Ariège)* (PhD dissertation, Université Bordeaux-I). France.

Lernau, H. 1988a. Mammalian Remains. In: Rothenberg, B., ed. *Researches in the 'Arabah 1959–1984, Vol. 1, The Egyptian Mining Temple at Timna*. London: 246–253.

Lernau, H. 1988b. Fish Remains. In: Rothenberg B., ed. *Researches in the Arabah 1959–1984, Vol. 1: The Egyptian Mining Temple at Timna*. London: 241–246.

Lernau, O. 1998. Fish Remains at Eilat-Elot. '*Atiqot* 36: 41–46 (Hebrew).

Levy, T.E., Ben-Yosef, E. and Najjar, M. 2012. Khirbat en-Nahas. In: Kassianidou, V. and Papasavvas, G., eds. *Eastern Mediterranean Metallurgy and Metal Work in the Second Millennium BC*. Oxford: 197–214.

Lewis, C.A. 1998. The Bronze Age Mines of the Great Orme and Other Sites in the British Isles and Ireland. In: Mordant, C., Pernot, M. and Rychner, V., eds. *L'atelier du bronzier en Europe du XX au VIII siecle avant notre ère* (Actes du colloque international Bronze '96, Neuchâtel et Dijon. Tome II: Du minerai, du métal à l'objet). Paris: 45–58.

Lyman, R.L. 1994. *Vertebrate Taphonomy*. Cambridge.

Martínez, G., 2009. Human Chewing Bone Surface Modification and Processing of Small and Medium Prey amongst the Nukak (Foragers of the Colombian Amazon). *Journal of Taphonomy* 7: 1–20.

Mienis, H.K. 1994. The Mollusks of the Excavations of the Early Arabic Site of Yotvata, Arava Valley, Israel. *Soosiana* 21/22: 22–28.

Morris, A.G. 1981. Copper Discolouration of Bone and the Incidence of Copper Artefacts with Human Burials in South Africa. *The South African Archaeological Bulletin* 36(133): 36–42.

O'Brien, W. 1995. Ross Island: The Beginning. *Archaeology Ireland* 9(1): 24–27.

O'Brien, W. 2004. *Ross Island: Mining, Metal, and Society in Early Ireland*. Galway.

Payne, S. 1973. Kill-Off Patterns in Sheep and Goats: The Mandibles from Aşvan Kale. *Anatolian Studies* 23: 281–303.

Petersen, G. 2010. *Mining and Metallurgy in Ancient Perú. Translated from Spanish by W.E. Brooks* (The Geological Society of America, Special Paper 467). Reston, VA.

Pyatt, F.B., Pyatt, A.J., Walker, C., Sheen, T. and Grattan, J.P. 2005. Environmental Toxicology: Heavy Metal Content of Skeletons from an Ancient Metalliferrous Polluted Area of Southern Jordan with Particular Reference to Bioaccumulation and Human Health. *Ecotoxicology and Environmental Safety* 60: 295–300.

Ramaroli, V., *et al.* 2010. The Chehr Abad "Salt Men" and the Isotopic Ecology of Humans in Ancient Iran. *American Journal of Physical Anthropology* 143: 343–354.

Reese, D.A. 1988. Recent Invertebrates as Votive Gifts. In: Rothenberg, B., ed. *Researches in the Arabah 1959–1984, Vol. 1: The Egyptian Mining Temple at Timna*. London: 260–265.

Reiche, I., Vignaud, C. and Menu, M. 2000. Heat Induced Transformation of Fossil Mastodon Ivory into Turquoise 'Odontolite'. Structural and Elemental Characterization. *Solid State Sciences* 2: 625–636.

Rothenberg, B. 1972. *Timna*. London.

Sapir-Hen, L. and Ben-Yosef, E. 2013. Food and Culture in Smelting Sites: A View from Timna. Paper Presented at Timna Park International Conference "Mining for Copper: Environment, Culture and Copper in Antiquity" In Memory of Professor Beno Rothenberg, 22–25 April 2013, Timna. Abstract: 34.

Schibler, J., *et al*. 2011. Miners and Mining in the Late Bronze Age: A Multidisciplinary Study from Austria. *Antiquity* 85(330): 1259–1278.

Shoemaker, L.C. 2001. Passionate for the Pasty: the Cornish Pasty in Michigan's Upper Peninsula. In: Walker, H., ed. *Food and the Memory: Proceedings of the Oxford Symposium on Food and Cookery, 2000*. Oxford: 245–253.

Studer, J. and Pillonel, D. 2007. Traditional Pig Butchery by the Yali People of West Papua (Irian Jaya): An Ethnographic and Archaeozoological Example. In: Albarella, U., Dobney, K., Ervynck, A. and Rowley-Conwy, P., eds. *Pigs and Humans. 10,000 Years of Interaction*. Oxford: 308–329.

Studer, J. and Schneider, A. 2008. Camel Use in the Petra Region, Jordan: 1st century BC to 4th century AD. In: Vila, E., Gourichon, L., Choyke, A. M. and Buitenhuis, H., eds. *Archaeozoology of the Near East VIII* (Travaux de la Maison de l'Orient et de la Méditerranée N° 49). Lyon: 581–596.

Thackeray, A.I., Thackeray, J.F. and Beaumont, P.B. 1983. Excavations at the Blinkklipkop Specularite Mine near Postmasburg, Northern Cape. *South African Archaeological Bulletin* 38(137): 17–25.

Van der Veen, M. and Tabinor, H. 2007. Food, Fodder and Fuel at Mons Porphyrites: The Botanical Evidence. In: Maxfield, V.A. and Peacock, D.P.S., eds. *Survey and Excavation at Mons Porphyrites 1994–1998, Vol. 2: The Excavations*. London: 83–142.

Vaughn, K.J., Linares Grados, M., Eerkens, J.W. and Edwards, M.J. 2007. Hematite Mining in the Ancient Americas: Mina Primavera, A 2,000 Year Old Peruvian Mine. *JOM: The Journal of the Minerals, Metals and Materials Society* 59(21–25): 16–20.

Vaughn, K.J., Van Gijseghem, H., Whalen, V.H., Eerkens, J.W. and Linares Grados, M. 2013. The Organization of Mining in Nasca during the Early Intermediate Period: Recent Evidence from Mina Primavera. In: Tripcevich N. and Vaughn K.J., eds. *Mining and Quarrying in the Ancient Andes*. New York: 157–182.

White, T.D. 1992. *Prehistoric Cannibalism at Mancos 5MTUMR-2346*. Princeton.

Willies, L. 1990. Exploring the Ancient Copper Mines of the Wadi ʿAmram (South Arabah). *Institute for Archaeo-Metallurgical Studies Newsletter* 15–16: 12–15.

Willies, L. 1991. Ancient Copper Mining at Wadi Amram Israel: An Archaeological Survey. *Bulletin of the Peak District Mines Historical Society* 11: 109–131.

CHAPTER 14

NAHAL ᶜAMRAM, SOUTHERN ARABAH VALLEY: A SURVEY OF UNDERGROUND COPPER MINES

Boaz Langford, Amos Frumkin, Uzi Avner and Hanan Ginat

Nahal ᶜAmram is an erosive valley on the southwestern side of the Arabah Valley, exposing a wide range of sandstone and magmatic rocks. Several copper mines have been explored here since the 1950s, but in a recent study, many more mines were discovered and charted. Mapping the interior of the mines was undertaken to allow analysis of their plans and extent, the use of digging tools, methods and strategies of mining, periods of exploitation and temporal changes. This paper presents the preliminary mapping and survey results of the Nahal ᶜAmram major mines.

INTRODUCTION

Nahal ᶜAmram is situated in the southwestern Arabah, ca. 10 km north of Eilat and 12 km south of the Timna Valley (Avner *et al.* this volume, Fig. 10.1). The area is characterized by complex geology and lithology, with igneous rocks and sandstone layers on the surrounding cliffs, crowned by limestone, up to 648 m ASL, 350 m above the wadi level (see also Ginat *et al.* this volume). Copper bearing layers of white sandstone (Weissbrod 1970) are exposed in several locations at Nahal ᶜAmram and vicinity,[1] in which mines were developed in the past using various techniques.

Previous surveys by N. Glueck (1953: 13–14) and by B. Rothenberg (1963: 33–38; 1967: 41–50, 145–148), exposed several copper mines of various types. Initial mapping of the largest mine was carried out by L. Willies (1990, 1991), but the recent detailed survey has disclosed many more mines (Avner *et al.*, this volume). General surface mapping of the sites was performed by means of accurate GPS (Trimble GeoXT 2005) and GIS (by R. Shem-Tov, see Avner *et al.,* this volume, Figs. 10.2, 10.3, 10.10, 10.14). The interior of 47 mines was mapped by the ICRC (Israel Cave Research Center), recording and analyzing characteristics of the mines' interiors, including dimensions, engineering issues, tools and mining strategies, artifacts. The survey of Nahal ᶜAmram was supported by the Israel Ministry of Science; the current stage of research is supported by the Israel Science Foundation (Grant No. 60/14).

CHARTING THE MINES

Systematic cave studies have been carried out in Israel for several decades (Shivtiel 2009). The recorded information shows that the Eilat area is relatively poor in natural caves but rich in artificial cavities—most of them the ancient copper mines of the Timna Valley, Nahal ᶜAmram and some other locations.

---

1 Mining sites found in Nahal Zefunot (north of Har ᶜAmram) and in Nahal Rehavᶜam (near Eilat) are not presented here.

Measuring and charting is essential for the study of the mines. Inside, polygons are measured by compass and a hand-held laser rangefinder/clinometer. The polygon serves as a baseline, from which additional measurements are taken to walls, ceilings, floors and other elements, to create the general plan and draw sections (see, e.g., Frumkin 1984; Dasher 1994). The map and sections provide the basic information as to the length, width, height, gradient and curves of each gallery or hall, in addition to photographs and descriptive text.

To date, the ICRC has measured and documented 47 underground mines longer than 5 m in the Nahal ᶜAmram region (Table 14.1): 22 at the escarpments of Har Amir, eight at Givᶜat Yocheved and 17 at upper Nahal ᶜAmram. Presently, the total measured length of the mapped mines is 2394 m, from which at least 5000 cu m of rock (*over* 10,000 tons) were mined out. These figures, however, are only partial, since many galleries and entrances are presently blocked by sediment. An additional large mine was mapped in Nahal Tzfunot, north of Nahal ᶜAmram. Following is a description of the major mine clusters and their characteristics.

HAR AMIR

The mines of Har Amir are clustered on two faces of the mountain (Avner *et al.*, this volume, Fig. 10.3), the northern and the southeastern, where the Amir and ᶜAvrona sandstone formations are exposed (Fig. 14.1). Here, 62 narrow mining shafts and 78 horizontal gallery entrances were recorded.

Today most shafts are filled with sediment, so only their rims are discernible; a few are still open and can be studied without excavation. At Site 34/4, one shaft in a cluster of 20 is cylindrical, 0.8 m in diameter and 6.5 m deep, with small protruding and recessed steps to ease climbing up and down (Fig. 14.2). The

Fig. 14.1: Har Amir Mine 35/5, entrances to galleries on the Avrona Formation cliff.

base is filled with sand, so it is presently unknown whether galleries extended from the bottom of the shaft. A short horizontal gallery extends from the shaft at mid-depth. Mine 34/9 has a similar open shaft, ca. 4 m deep, connecting the upper level of the mine with the lower galleries (Avner *et al.,* this volume, Fig. 10.5a, b).

The galleries of Har Amir are narrow, up to 1 m in diameter, but sometimes only 0.45 m, hardly passable (Fig. 14.3). They are usually quite short, up to 20 m long and inclined, following the dip of the sandstone layers. All shafts and galleries on Har Amir were cut by means of chisels and hammers.

Table 14.1: Summation of the Mines' Data According to Location and Period

Location	*Mine No.*	*Survey Map No.*	*Altitude (ASL)*	*Total Length in m*	*General Volume m^3*	*Period of Activity*
Har Amir	4	34/1	370	207	269.1	LB-IA
	5	34/9	365	6.71	4.83	
	6	34/9	360	8.64	6.32	
	7	34/8	360	4	3.24	
	8	34/4	355	9.6	8.64	
	9	35/5	370	34	21.7	
	10	35/5	366	6.5	2.86	
	11	35/5	360	6	3.6	
	12	35/5	365	6.8	2.14	
	13	34/11	368	9	6.48	
	14	34/11	368	13.5	4.86	
	15	34/11	364	14.5	11.78	
	16	35/5	374	8	3.2	
	17	35/5	360	11.5	4.6	
	18	34/1	365	45	11.25	
	19	34/11	360	3	1.5	
	41	34/6	299	12	5.76	
	42	34/6	309	27	34.02	
	43	34/6	316	19	22.8	
	44	34/13	348	5	2.2	
	45	34/13	341	5	1.01	
	46	34/13	344	13	5.72	
	22	45/13	239	5.6	3.8	
Givᶜat Yocheved	23	45/18	250	3.7	1.44	
	24	45/22	254	30.2	21.74	
	25	-	266	4.25	2.82	
	26	45/19	257	5.7	5.7	
	27	45/16	232	5.5	2.69	
	Total			*529.7*	*475.8*	
Givᶜat Yocheved	20	45/1	215	18.5	51.06	Nab-EI
	21	45/3	228	4	10.8	
	Total			*22.5*	*61.86*	

Location	*Mine No.*	*Survey Map No.*	*Altitude (ASL)*	*Total Length in m*	*General Volume m^3*	*Period of Activity*
Nahal ʿAmram Basins A,B	1	35/28	300	111.5	589.5	Nab-EI
	2	35/29	300	1100	2640.9	
	3	36/1	309	7.4	4.73	
	28	36/3	322	53	148.4	
	29	36/5	279	14.5	34.69	
	30	35/24	295	9.1	28.66	
	31	35/26	302	6.4	10.56	
	32	35/26	299	8.5	57.12	
	33	35/26	301	7	14.1	
	34	35/26	302	8	6.75	
	35	35/32	306	437	845.3	
	36	35/38	314	3.5	11.14	
	37	35/39	318	4	3.2	
	38	35/25	297	5.5	7.7	
	39	36/6	294	8.2	6.99	
	40	35/39	318	69	111.78	
	47	36/7	290	27	27	
	Total			*1859.6*	*4548.52*	

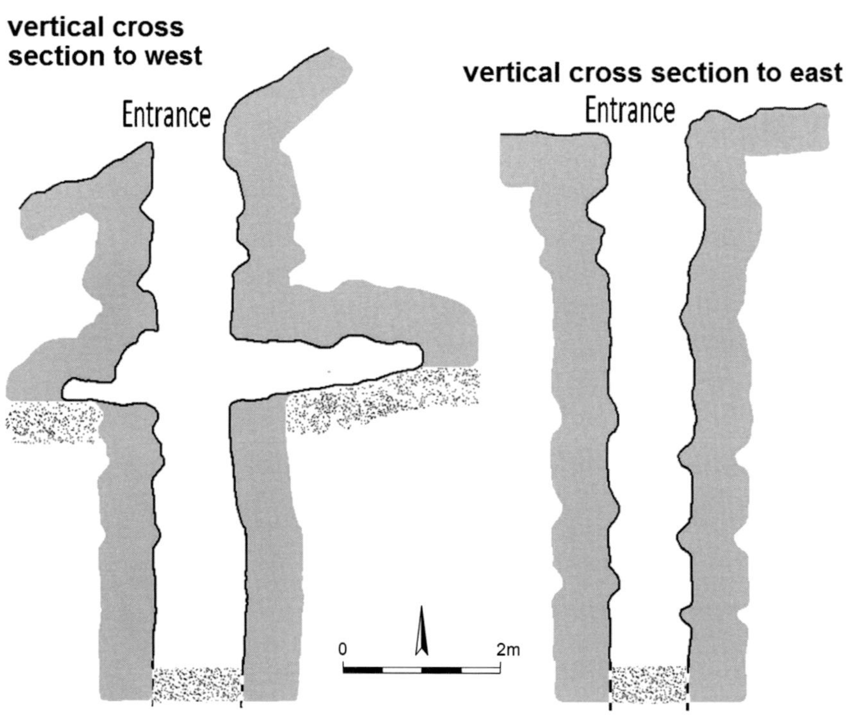

Fig. 14.2: Har Amir Mine Site 34/4, sections of an open shaft, exemplifying the nature of the blocked ones.

Fig. 14.3: Har Amir Mine 34/1, an example of narrow galleries.

They present a uniform style, currently generally dated to the 15th to 10th centuries BCE, based on a small number of pottery sherds and on ^{14}C dates from the nearby miners' camp (Avner *et al.*, this volume, Table 10.1: 2–4). More precise dating requires excavations in selected mines. The few open shafts represent the mining strategy of the Late Bronze–Iron Age. Each shaft served first as a "testing drill." If greenish levels were exposed during the dig, galleries would follow these layers to capture copper nodules contained in them. The shafts are usually only a few meters apart from each other, to allow ventilation in the galleries connecting the shafts, sometimes in two or three levels. Other mines include galleries only, dug directly from the sandstone cliff face (Fig. 14.4). The use of bronze chisels, which became available around 2000 BCE, enabled the miners to dig the narrow shafts and galleries.

GIVᶜAT YOCHEVED

The mines next to Givᶜat Yocheved comprise two main clusters (Avner *et al.,* this volume, Fig. 10.10). In the southern one, narrow galleries were dug into the cliff faces (Fig. 14.4), with some narrow shafts. Both were made by chisels, similar to those at Har Amir, and they actually belong to the same complex. Some entrances are accessible today only by means of rappelling, implying that much erosion affected the cliffs during the last 3000 years. Two shafts truncated by torrential runoff (during heavy, concentrated rainfall events) support this view.

The northern cluster comprises broad, semi-rounded galleries, up to 2.2 m wide, and broad shafts, 1 m or more in diameter. Both are commonly blocked today by sediments (Fig. 14.5), some barely discernible. In both shafts and galleries, marks of large digging picks were preserved. These indicate a very different technology and strategy of mining, later than that of the Har Amir mines.

Fig. 14.4: Givᶜat Yocheved Mine 34/6, a narrow gallery dug into the cliff face.

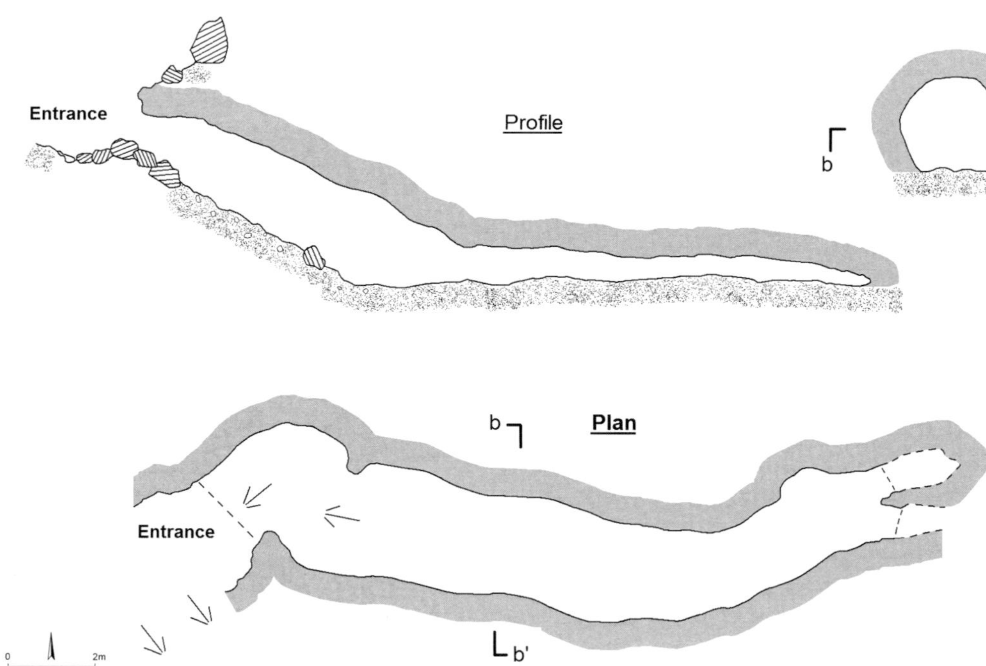

Fig. 14.5: Givᶜat Yocheved Mine 45/1, plan and profile of a wide, semi-rounded gallery.

NAHAL ʿAMRAM, BASINS A AND B

Two clusters of mines were recorded in two adjacent basins at the mouth of Nahal ʿAmram (Avner *et al.*, this volume, Fig. 10.14), with 34 horizontal entrances leading into inclining galleries and eight broad mining shafts. Large mining dumps were mapped on the slopes and in the wadis below the mines, indicating intensive mining (Filin *et al.*, this volume). The mines are complex, consisting of broad galleries with a semi rounded or a square section, broad halls, and shafts connecting different levels of galleries. There are also narrow galleries, some of which are truncated by broader ones and are therefore attributed to earlier mining periods. Digging marks in the mine are mostly of large picks, as well as some small picks, but no chisel marks are present. Open mines are also observed in these clusters (Avner *et al.*, this volume, Fig. 10.24).

Following is the description of three mines:

Mine 35/28 (Fig. 14.6) is dug into the ridge separating the two basins. Two entrances lead into a central hall, 17 × 16 m and up to 2.3 m high, with a mass of rocks collapsed from the ceiling. Seven inclining galleries spread from the hall eastward, in two levels. Some galleries are filled by sand, so their length is unknown. The currently known volume of the mine is ca. 590 cu m (ca. 1360 tons of sandstone dug-out). Tool marks in the mine indicate the use of large picks; a few Early Islamic pottery sherds were collected inside and near the northern entrance.

Mine 35/32 consists of several galleries and a large hall, currently totaling 437 m in length (Fig. 14.7). Six shafts connect galleries of different levels, mostly blocked today by sediments. The hall size is 11 × 8 m, originally ca. 3 m high but now filled in by fine fluvial sediments, 155 cm thick, accumulated by moderate floods. Part of the sediment mass, however, was later truncated by vigorous floods which filled the deeper parts of the mine, presently inaccessible. During the survey, fruit remains were found

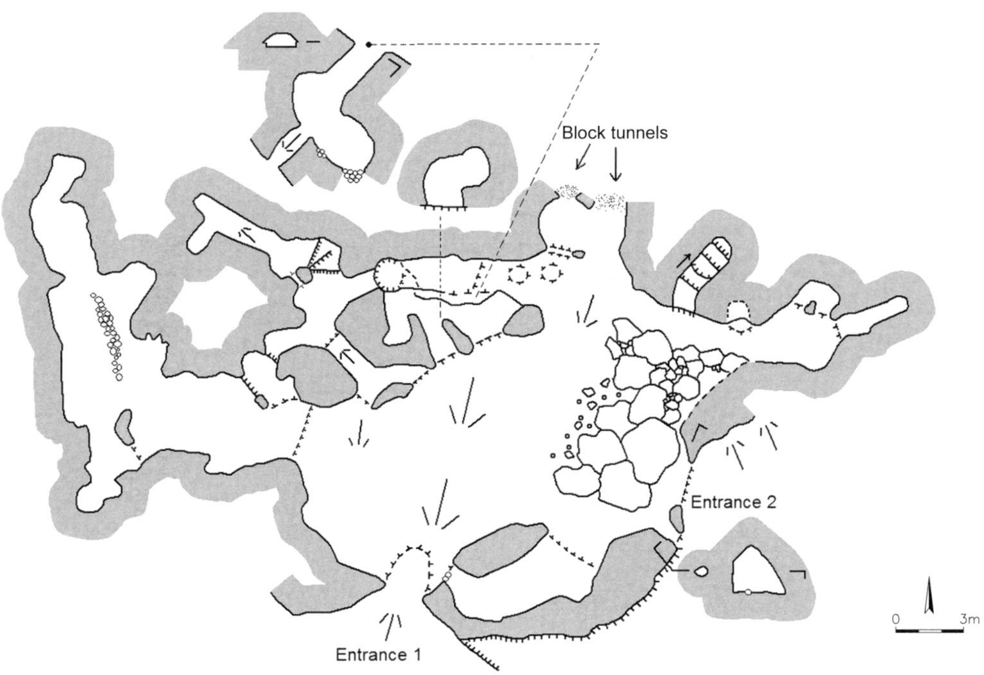

Fig. 14.6: Mouth of Nahal ʿAmram, plan of Mine 35/28.

in the main gallery, including olive, date and peach pits, as well as an almost complete Early Islamic jar (Fig. 14.8). The currently measured volume of the mine is 845 cu m (ca. 2000 tons of dug-out sandstone).

Mine 35/29 is the largest known at Nahal ᶜAmram and probably the most complex in the entire Arabah Valley. It was first mapped and described in detail by Willies, who also found a mining basket

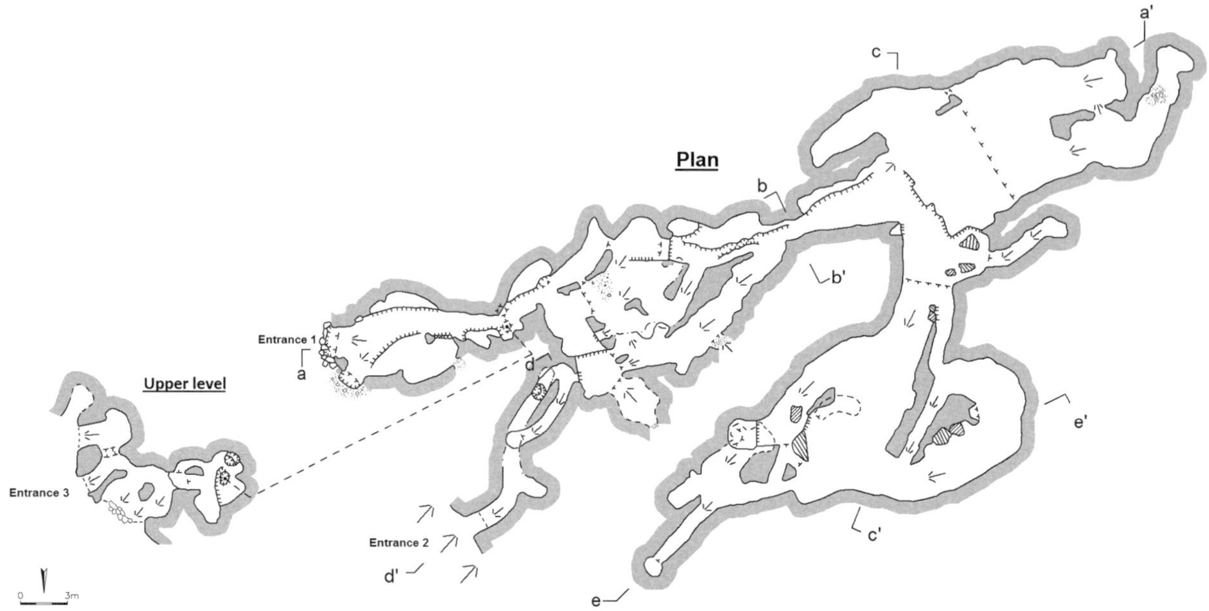

Fig. 14.7: Mouth of Nahal ᶜAmram, plan of Mine 35/32.

Fig. 14.8: Mine 35/32, an Early Islamic jar, as found.

made of palm leaves inside (Willies 1991, Figs. 34, 38). The new survey, with better measuring means, produced a more accurate map and sections of the mine (Fig. 14.9) and discovered a new, lower level of galleries rich with finds. Five entrances and one broad shaft lead into the mine on both sides of the ridge separating the two clusters of mines. The main entrance, on the western escarpment, leads into Hall A,

Fig. 14.9: Plan of Mine 35/29, the largest in the southern Arabah Valley. Main halls are marked by capital letters.

ca. 13 × 10 m in size and up to 3 m high, into which an 11.5 m deep shaft was dug. From this hall, a short gallery leads to the larger Hall B, 20 × 30 m, currently 1.5 m high but the floor is covered by thick debris. From this hall, a complex of galleries spread eastward forming the inclined main level of the mine. In addition, lower and higher galleries were dug in various sections, interconnected by shafts. The galleries are either semi-rounded or square in section; some galleries truncate older ones. All tool marks observed in the mine are of large picks, while in the older, narrow galleries no tool marks remained. The total measured length of the mine is 1100 m and its estimated volume is at least 2640 cu m, ca. 5300 tons of sandstone dug out from the measurable sections of this mine.

The newly discovered level, the deepest in the mine (Section K), is a steeply inclined gallery with two shafts visible in the ceiling of its upper side, blocked by coarse sediments. Artifacts were observed here during the survey while a short excavation yielded ample finds mixed with the sand. These included many fabric pieces, ropes, wool and leather, and many remains of dates, olives, peaches, almonds, walnut and pomegranate, firewood and charcoal from at least 19 plant species (Avner *et al.*, this volume), and many bones of sheep/goat and fish (Horwitz and Avner, this volume). A Late Nabataean oil lamp, dated to the 4th century CE, and Tridacna shells from the Red Sea, also used as oil lamps, were found as well. Two small niches cut into the gallery wall for oil lamps showed stains of oil (Avner *et al.,* this volume, Fig. 10.17), from which oil was successfully extracted for chemical analysis and ^{14}C dating. The vast majority of pottery sherds were Early Islamic (7th to 11th centuries CE), however, six ^{14}C dates from organic materials from this mine were from the 1st to 5th centuries CE (Avner *et al.,* this volume, Table 10.1: 7–12).

PRELIMINARY CONCLUSIONS

Based on the pottery collected, other finds and the ^{14}C dates, the general history of mining at Nahal ᶜAmram can currently be described, keeping in mind that much work is yet to be carried out before defining the exact time-span of each mining period.

According to small amounts of Chalcolithic–Early Bronze pottery sherds in habitation sites and in the miners' camp, some mining took place during these periods. However, we could not attribute any mine to these periods. Also, we could not confirm Rothenberg's identification of Har Amirᶜs northern escarpments as a Chalcolithic mine (Rothenberg 1999: 163–164).

The next period is generally termed here Late Bronze–Iron Age, the 15th–10th centuries BCE, but more precise dating is still needed. The 62 narrow shafts and 78 gallery entrances can safely be related to this period, as well as 324 shallow depressions on alluvial terraces, which most probably served as open mining pits (Avner *et al.*, this volume). Unlike in the Timna Valley, no sign was found here of New Kingdom Egyptian intervention in copper production. Rather, it seems that the entire plant was operated by the local desert population.

The next period of interest is the 1st to 5th centuries CE, with complex mines based on broad galleries and shafts. The period is defined as Roman–Byzantine, however, almost no Roman or Byzantine pottery was found in the mines while some pottery of the period is Late Nabataean, 3rd–4th centuries CE, produced in Petra (Avner *et al.,* this volume). The finds and ^{14}C dates indicate that mining began in the kingdom time of the Nabataeans and continued, by Nabataeans, after their kingdom ceased to exist.

The period of most intensive mining was the Early Islamic (7th to 11th centuries BCE). The pottery collected from the mines in this period is richer by far than from any other period. It generally seems that Nahal ᶜAmram mines reflect an almost continuous tradition of copper mining throughout the first millennium CE. In light of the enormous scale of mining during these periods at Nahal ᶜAmram, it is interesting to note that very little is known of Late Nabataean mining in Timna Valley; and no mining is known there during the Early Islamic period.

To date, the estimated volumes of dug-out sandstone from the mines and dumps is very preliminary. However, it seems that the volume from the dumps far exceeds that from the mines, although part of the dumps was washed away in the past by floods. Explanations of this discrepancy may be offered (Avner *et al.*, this volume).

The copper mines of Nahal ʿAmram are much larger and more complex than was previously known, but there is still much to learn from them. Efforts must be concentrated in identifying strategies and technologies of mining in higher resolution and relate them to corresponding periods, based on all methods of dating.

ACKNOWLEDGMENTS

We would like to thank the I.C.R.C volunteers: M. Ullman, E. Cohen, Sh. Yaaran, A. Ginat, D. Oren, Sh. Ravek and L. Enmar for their assistance in the research. Special thanks to V. Buslov who helped us with the cartographic processing of the maps. This research was supported by the Israel Ministry of Science and the Israel Science Foundation (Grant No. 60/14).

REFERENCES

Dasher, G.R. 1994. *On Station: a Complete Handbook for Surveying and Mapping Caves.* Huntsville, Alabama.

Frumkin, A. 1984. Mapping Caves. *Niqrot Zurim 10*: 113–124 (Hebrew).

Glueck, N. 1953. Exploration in Western Palestine. *Bulletin of the American Schools of Oriental Research* 131: 6–15.

Rothenberg, B. 1963. Ancient Copper Industries in the Western ʿArabah. *Palestine Exploration Quarterly* 94: 5–71.

Rothenberg, B. 1967. *Negev: Archaeology in the Negev and the ʿArabah.* Ramat Gan (Hebrew).

Rothenberg, B. 1999. Archaeo-Mettallurgical Researches in the Southernʿ Arabah 1959–1990, Part 2: Egyptian New Kingdom (Ramesside) to Early Islam. *Palestine Exploration Quarterly* 131: 68–89.

Shivtiel, Y. 2009. The History of the Research of Israel Caves. *Ariel* 188: 6–39 (Hebrew).

Weissbrod, T. 1970. The Stratigraphy of the Nubian Sandstone in Southern Israel. *Geological Survey of Israel Report* OD/2/70.

Willies, L. 1990. Exploring the Ancient Copper Mines of the Wadi ʿAmram (Southern ʿAraba). *Institute for Archaeo-Metallurgical Studies* 15–16: 12–15.

Willies, L. 1991. Ancient Copper Mining at Wadiʿ Amram, Israel: An Archaeological Survey. *Bulletin of the Peak District Mines Historical Society* 11: 109–131.

CHAPTER 15

A PRELIMINARY STUDY OF COPPER SLAG IN THE SOUTHERN ARABAH VALLEY

Sana Shilstein and Sariel Shalev

As part of the archaeological study of Nahal ʿAmram (see Avner et al.*, this volume) copper slag samples from a miners' camp were chemically analyzed by X-ray fluorescence (XRF) in an attempt to determine the relative technological level of each of its periods. The results were then compared with those from other smelting sites in the southern Arabah Valley. The method provides a tool to distinguish between slag of various technological phases and periods.*

INTRODUCTION

Ancient extractive metallurgy in the Arabah Valley has been studied since the early 1960s, with the pioneering work of the late B. Rothenberg, in cooperation with Tylecote, Lupu, Bachmann, Bamberger, Merkel and others.[1] Naturally, some mistakes in research were made along the way. For example, the tapping furnaces at Timna Site 2, a key site for the study of Arabah metallurgy, had been related to the New Kingdom (NK) Egyptian conquest of the region (13th–12th centuries BCE), but were later found to be Early Islamic. Accordingly, the large tapping slag "cakes" produced by these furnaces were also dated to this period.[2] The work of Levy *et al.* in the same region also contributed much new data with ample ^{14}C dates (e.g., Levy *et al.* 2002, 2007, 2008, 2012). Ben-Yosef re-dated slag deposits along the Arabah, using paleomagnetism and new ^{14}C dates from Timna Sites 30 and 34 (i.e., Ben-Yosef, Tauxe and Levy 2010; Ben-Yosef *et al.* 2008, 2012).

A major obstacle in studying the chemical composition of slag is the great heterogeneity in every chunk, even in the small ones (XRF lines of each element may differ up to 100 times). Therefore, the method adopted here includes multiple XRF readings from many slag pieces and from different spots on each, then calculating the percentage of various ratios of Fe+Mn and Cu. The Fe and Mn in the slag originate mainly from fluxing agents and the results reflect the relative level of pyrotechnology of slag from various sites. As the technology advanced, this ratio increased in favor of Fe and Mn in a higher percentage of readings and the level of homogeneity increased, indicating a better use of fluxes, better control of the furnace temperature and better extraction of copper from the smelted ore (Bachmann 1980; Ben-Yosef 2010).

---

1 See, e.g., Rothenberg 1962, 1972, 1985, 1990; Bachmann 1980; Merkel 1983, 1996.

2 Four furnaces were excavated by Rothenberg (1972: 67–111; 1983) at Site 2, next to the piles of larger tapping slag. He attributed them to the New Kingdom Egyptian activity at the site (13th–12th centuries BCE), but ^{14}C dates from two of them were of the Early Islamic period (Rothenberg 1990: 71, n. 23). Their Early Islamic date was further supported by a recent technological study (Ben-Yosef 2010: 671–672, 695). Four similar furnaces at Beʾer Orah and Nahal ʿAmram were also proven to be Early Islamic (7th–11th centuries CE, see Avner *et al.*, this volume, Table 10.1: 9–19, 32–36).

In the present study we analyzed large numbers of slag samples from several sites in Nahal ʿAmram and from other sites in the southern Arabah (Fig. 15.1). The sites were preliminarily dated based on surface finds of flint and pottery sherds, on ^{14}C dates, on the slag morphology, and on the surveys and excavations of Rothenberg and others. The results of these analyses are in agreement with the dating of most sites.

METHODOLOGY

At least 20 slag samples were analyzed from each site and sorted to different types by shape, size and color and in some cases also by the presence or absence of magnetic properties. Each sample was then broken mechanically into pieces of ca. 1 cm and measured by ED XRF, using a Benchtop EX–Calibur (Xenemetrix). The intensities (I) of Ca, Mn, Fe, Cu, Pb, Sr, Ba were measured at 45 kV, 0.25 mA (Mo filtering), for 60 seconds, with a primary beam of 2 mm in diameter on slag pieces or from powder produced by mechanical crushing. In an attempt to overcome the effect of heterogeneity, two to three readings were taken from different points on each fragment of a sample, which translates into tens of measurements from each piece of slag and hundreds from each site. Readings were taken from both the

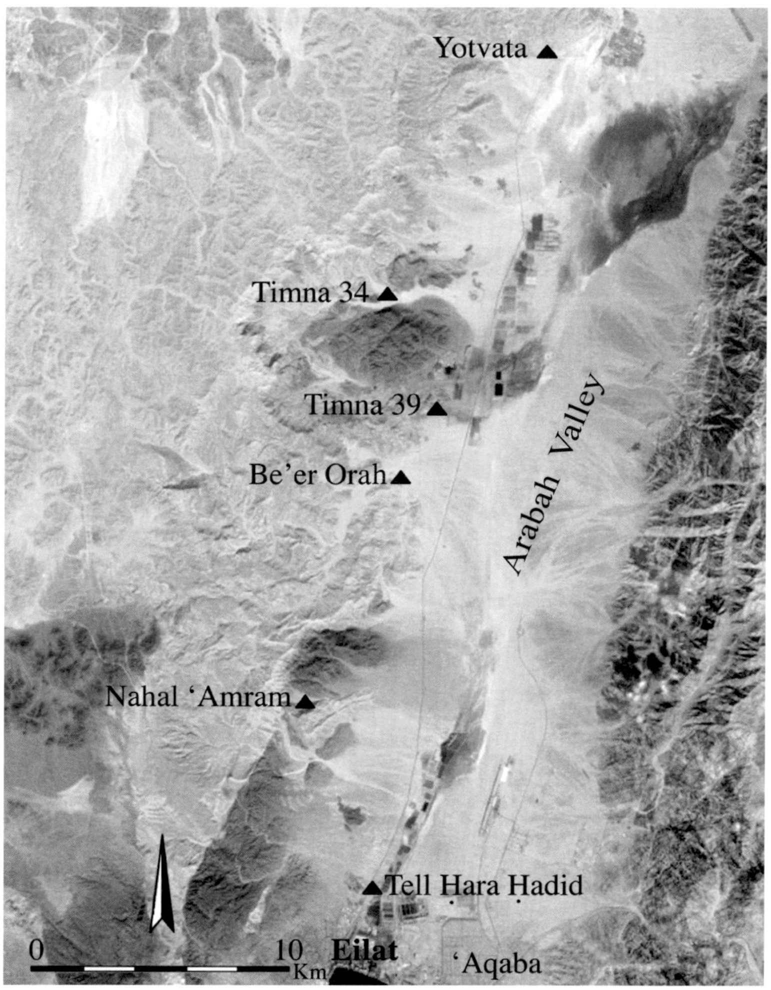

Fig. 15.1: Map of the Arabah Valley with the main copper production sites (triangles). Location of sites is approximate (after U. Avner).

surface and the interior of the slag. Since the XRF beam penetrates only several tens of microns and surface corrosion may bias the elements reading, the rough surface was polished first. The analyses were based on three criteria: the average content of copper in the slag, the ratio of intensities of Fe+Mn and Cu, and the heterogeneity of these ratios.

The average intensities of all the detected elements (Fig. 15.2) were calculated separately for each sample and site (see Table 15.1). The precision of elements with measured concentration of up to 10% is 5–10% and was calculated by calibrating the results of intensities by comparison to specially prepared standards of sea sand mixed with definite amounts of each of the above elements (Fe, Mn and Cu). For elements like Pb, Sr and Ba, their measured intensities could be used only qualitatively or semi-quantitatively due to low and insufficient statistical precision. The measured results for Ca are strongly influenced by absorption of other elements and therefore the real precision is relatively poor even when Ca intensity reading is high. Thus, data about these components in the slag could be used only as an additional qualitative indication in clustering chemically different groups of samples.

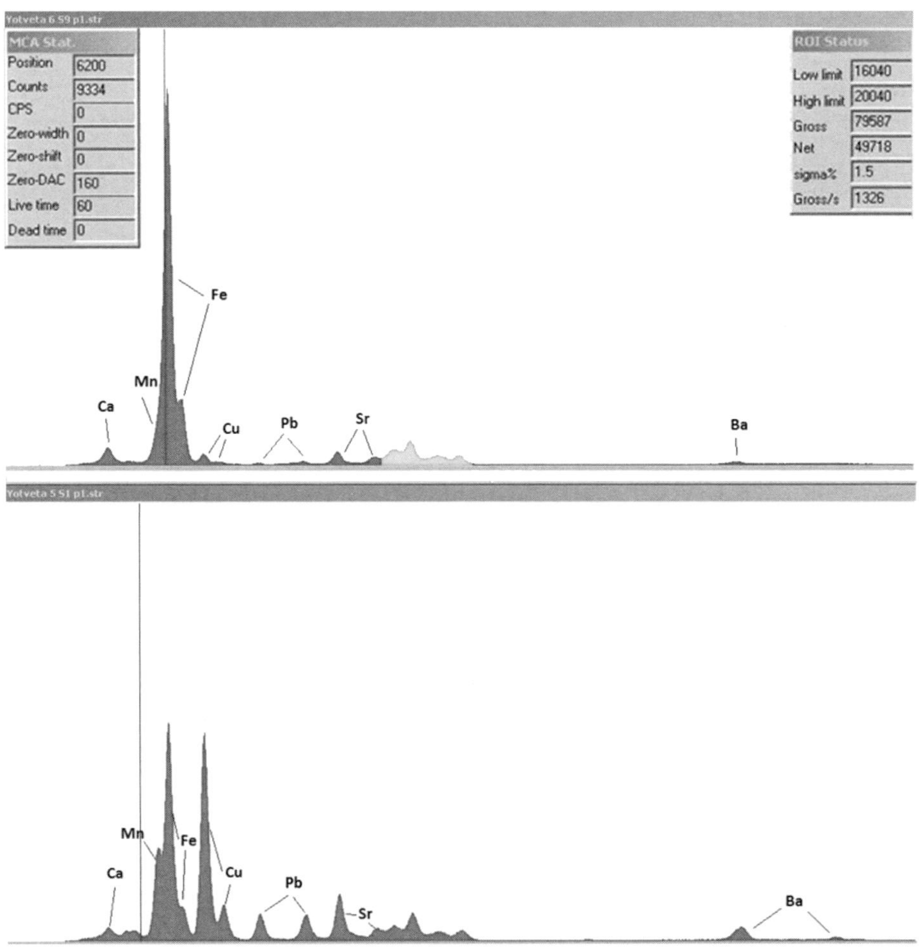

Fig. 15.2: Top image–spectrum of slag with low Cu content (rather later). [I(Fe)+I(Mn)]/I(Cu)≈50. Marked part of spectrum originates from primary radiation of X-ray tube. Bottom image–spectrum of slag with high Cu content (rather older). [I(Fe)+I(Mn)]/I(Cu)≈1.

Table 15.1: Average Intensities of Slag Samples from Different Sites (SD = Standard Deviation)

Site	*I(Ca)*	*I(Fe+Mn)*	*I(Mn)/I(Fe)*	*I(Cu)*	*I(Fe+Mn)/I(Cu)*	*I(Pb)*	*I(Sr)*	*I(Ba)*	*Comments*
ʿAmram 35/7 I	8	250	0.11	6	40	0	5.6	3.3	Add. Zn
SD	2	70	0.07	4		3	2		
ʿAmram 35/6	6	160	0	60	2.6	0	5.5	2	Add. Zn
SD	2	60		50		3	2		
ʿAmram 35/7 II	8.7	270	0	10	27	0	6	3.2	Add. Zn
SD	2	40		7			1	0.6	
Yotvata I	7	210	0.3	55	4	5	11	3.5	
SD	3	50	0.15	30		5	8	4	
Yotvata II	10	260	0.14	3	87	1	8	2.3	
SD	2	100	0.01	1.5		0.3	1	0.8	
Timna 34 I	7.5	330	0	11	30	0	5	0	Add. Zn
SD	1.3	70			7			1	
Timna 34 II	7	320	0	12	27	3	6	0	Add. Zn
SD	4	120		12			2.5	3	
Timna 34 III	6	340	0.08	19	18	2	4.5	2	Add. Zn
SD	1.5	70	0.02	14		1.5	1	0.6	
Timna 39a	8.5	250	0	31	8	2.5	9	0	
SD	2	100		25		1.6	5		
Timna 39b	8.5	230	0	77	3	4.5	5.5	0	
SD	2.7	60		90			2.2	2.5	
Beʾer Orah I	8.5	210	0.28	120	1.7	3	6.5	0	
SD	4	100	0.09	120			2	3	
Beʾer Orah II	10	210	0.16	22	9.5	2	9	0	
SD	3	60	0.07	10			1	2	
Beʾer Orah III	10	210	0.08	24	8.7	2.5	10	0	
SD	2	40	0.05	10			1	3	
Beʾer Orah IV	14	190	0.22	16	12	2	14	0	
SD	2	70	0.06	15			1	3	
Hara Hadid I	7	170	0.05	11	15	1.2	5	0	Add. Zn
SD	2	30	0.01	6			0.6	3	
Hara Hadid II	6.5	190	0.28	110	1.7	1.5	6.5	3	Add. Zn
SD	3	60	0.1	60		1.5	3	2.5	

In order to reliably determine the actual contents of the major components in the slag, i.e., Cu and fluxing elements such as Fe and Mn, some technical limitations have to be taken into consideration while using XRF as an analytical method. For an expected high concentration level of major elements (20%–30% or higher), the intensity dependence on the concentrations is significantly non-linear. In addition, the intensities are influenced by the absorption of Cu-Kα fluorescent radiation by Fe or Mn atoms. Therefore, instead of the usual procedure of calibrating these dependencies for even two elements (e.g., Fe and Cu) by measuring samples of known definite concentrations as standards, we used a different procedure. For several samples with the highest intensities of Fe and Cu lines, we measured their intensities in a highly diluted state, by mixing the powdered sample with sand of definite and low concentration of Fe (1:10). By this procedure, we obtained the precise concentrations of those samples. Later, we used these results for calculating the concentrations of all other samples by plotting their measured intensities on a linear interpolation line. By using different dilution ratios for the same powdered sample, we were able to determine the real corrections for calculating the Fe (Mn) and Cu concentrations. The results of the described determination of averaged concentrations in different groups of samples from different sites are presented in Table 15.2.

In order to evaluate the level of copper smelting technology and check its correlation to different sites and periods, the XRF intensity lines of Fe and Mn (major fluxing elements) were compared to the intensity line of Cu that remained in the slag during smelting production. The typical Fe+Mn content in the slags is about 20%. Therefore, a high ratio of [I(Fe)+I(Mn)]/I(Cu) corresponds to low concentration of Cu in the slag. For instance, the ratio of 100 corresponds to 0.1% Cu in the slag while the ratio of 0.1 corresponds to 10% Cu. We present below this ratio distribution of all measured points for each of the studied sites (Figs. 15.4–15.9). Our main preliminary working hypothesis is that clear differences in the above parameters (i.e., the changes in relative amount of copper left in the slag) represent a different industrial process and have direct connection to different sites and periods. The ratios of [I(Fe)+I(Mn)]/ I(Cu) are presented in logarithmic scale in the figures below.

In this article we focus on two major types of slag. The earlier (Chalcolithic to Iron Age) and the later, larger tapping slag (Nabataean to Early Islamic). The results will serve later as a base for finer distinctions of industries within each general group.

TABLE 15.2: AVERAGE CONCENTRATION OF FE, MN AND CU CALCULATED FOR SLAG FROM DIFFERENT SITES (THE CU CONCENTRATIONS FOR ANCIENT SITES ARE SHOWN IN BOLD)

Site	*%Cu*	*%Fe*	*%Mn*
ʿAmram 35/7 I	0.2	16	5
ʿAmram 35/6	**2**	13	0
ʿAmram 35/7 II	0.3	21	0
Beʾer Orah I	**3**	13	7
Beʾer Orah II	0.7	11	8
Beʾer Orah III	0.7	14	3
Beʾer Orah IV	0.5	13	7
Yotvata I	**1.5**	11	12
Yotvata II	0.1	20	7
Tell Ḥara Ḥadid I	0.2	14	2
Tell Ḥara Ḥadid II	**2.5**	10	11
Timna 34 I	0.2	30	0
Timna 34 II	0.25	24	0
Timna 34 III	0.4	30	5
Timna 39a	**0.6**	20	0
Timna 39b	**1.5**	17	0

THE SITES AND ANALYSES

NAHAL ʿAMRAM SITES 35/6 AND 35/7

A large miners' camp, 250 × 500 m, is located in the heart of the mining area (Avner *at al.,* this volume, Figs. 10.2, 10.8, 10.9), dated to the Late Bronze–Iron Age, (15th to 9th centuries BCE). The camp is represented here by one "neighborhood," numbered 35/6, from which three ^{14}C dates are currently available; they are ca. 1120 BCE, 1120 BCE and 1220 BCE (Avner *et al.*, this volume, Table 10.1: 4–6). Small furnace slag, usually smaller than 15 mm and containing tiny copper inclusions, is loosely scattered on most of the camp's surface, with some areas of a denser scatter. Generally, the amount of slag in the entire camp is very small compared to that in the contemporaneous camps in the Timna Valley.

Samples from Miners' Camp 35 are divided by finding areas and size accordingly: 35/7 I and 35/7 II are from a later pile of fragments of tapped slag "cakes" (Fig. 15.3), ca. 10 m across, next to Site 35/6, with some small slag which are the by-product of the tapping process. Four ^{14}C dates from this pile are between ca. 670 and. 780 CE, the Early Islamic period (Avner *et al.,* this volume, Table 10.1: 19–22).[3] Following the smelting, tapping-out of the slag from the furnace and cooling down, the slag "cake" was broken into pieces 5–20 cm across to make sure no copper was tapped out of the furnace together with the slag. The slag "cakes" show much higher homogeneity than that of the small furnace slag, much lower Cu content and a much higher ratio of (Fe+Mn)/Cu (circles in Fig. 15.4). Obviously, these slag represent a better technology, including intensive fluxing and a better control of furnace temperature.

The small slag of sample 35/7 II are solid, similar to the "cake" fragments, showing an analogous distribution of (Fe+Mn)/Cu ratios. These may indicate that some of the slag was not tapped out from the furnace, but rather taken out with the copper metal created on the furnace bottom, and then separated from it after cooling. Both samples of 35/7 were found to be more magnetic than those of 35/6.

The measured distributions of intensities of the fluorescence line of fluxing elements in all samples which were compared with the intensity of the Cu line for Sites 35/7 I and 35/7 II (Fig. 15.4) are similar. The most common ratio [I(Fe)+I(Mn)/I(Cu)] is about 50–60. For our instrument, the sensitivity to Cu is about 2 times more than for Fe. It means that for a realistic concentration of Fe (20%–30%), the Cu concentration is about 0.25%. On the graph (Fig. 15.4) some trend or shift to the left for Sample 35/7 II is visible, which means that the slag samples of this site probably contain a little bit more Cu than those from Samples 35/7 I. Slag samples from 35/7 II contain only Fe, while samples from Site 35/7 I also contain Mn. However, comparison between 35/7 I and 35/7 II (combined) and 35/6, shows a stronger difference than between the first two. The distribution presented by squares is much wider (logarithmic scale!), the most common ratio [I(Fe)+I(Mn)/I(Cu)] is about 10. In other words, the slag from 35/6 is much more heterogeneous than that from 35/7, containing much more Cu (Table 15.2) and representing more primitive technology than that of 35/7 I, II. (cf. Ben-Yosef 2010: Chapter 8).

In order to distinguish between older and more recent sites, a value I(Fe)+I(Mn)]/I(Cu)=3 was chosen as a cut-off. Thus, if the calculated ratio of slag from a given site is significantly greater than 3 (as for 35/6), the site may be identified as older. If the ratio is smaller than 3 (as for Site 35/7), the site may be identified as more recent. The difference should also be observed with regard to the existence

3 Sample No. 21 in the ^{14}C table has been retrieved from a small slag in Site 35/6, but obviously originated from the Early Islamic pile of slag.

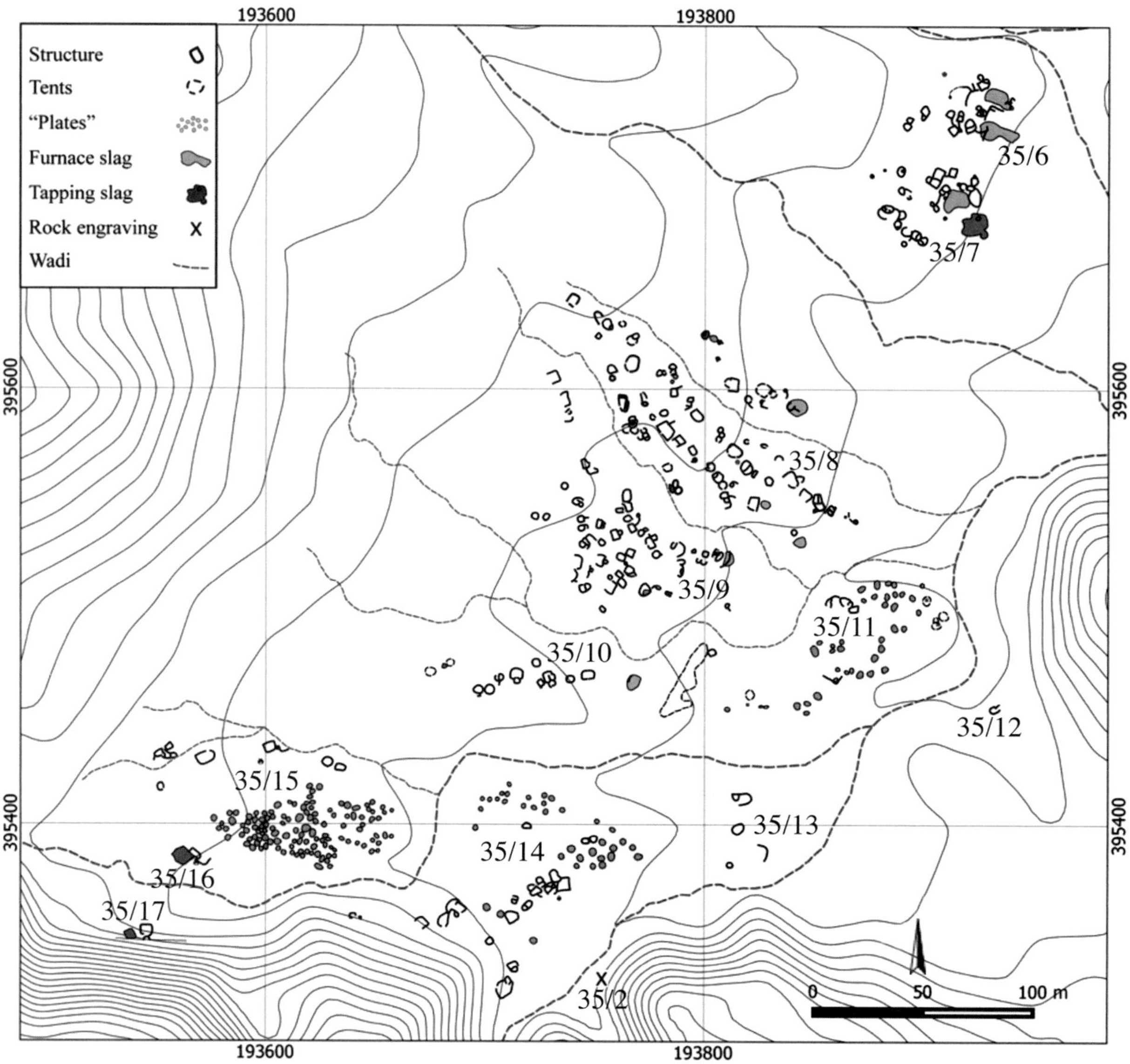

Fig. 15.3: Nahal ʿAmram, a map of the miners' camp.

or absence of a "tail" on the distributions of [I(Fe)+I(Mn)]/I(Cu) less than about 10. Data for other elements will be discussed below but we wish to note here that some amount of Zn occurs in these slag samples. Zn concentration is several times smaller than Cu, but Zn occurs consistently in all slag from Nahal ʿAmram. In summary, the results present clear chemical signatures of two different smelting technologies in ʿAmram, probably representing two different periods, and both periods have Zn in the slag samples.

For one slag sample from Site 35/6 with a relatively high intensity Cu line and minimal intensity of Fe line, we used a full procedure of powdering, dilution in various ratios up to about 20, and, as a result, the precise values for Cu and Fe concentrations were determined: 25% Fe and 5.4% Cu. This sample also contained some Cu inclusions which were measured separately.

YOTVATA

A low pile of fragments of larger tapping slag "cakes" ca. 6 m across was found some 200 m west of the Late Roman fortress. Rothenberg (1967: 292) dated the site to the Byzantine period; however, it contains distinctive Nabataean pottery sherds (1st century BCE to 4th century CE). Adjacent to this pile is a larger area scattered with small pieces of slag, unnoticed by Rothenberg and dated only by archaeomagntism to the Iron Age (Ben-Yosef *et al.* 2008).

Fig. 15.5 show a clear difference between the two types of slag. The small ones (Yotvata I) are with lower I (Fe)+I (Mn)]/I (Cu) ratios and high inhomogeneity, while in the slag "cakes" (Yotvata II)

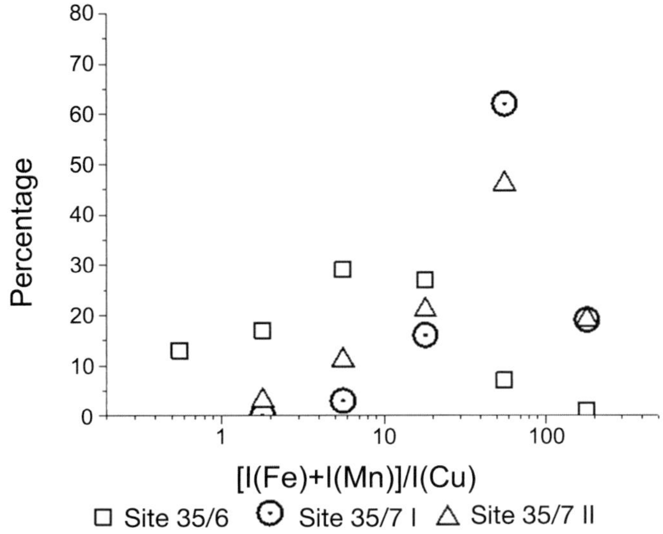

Fig. 15.4: Resulting distributions for the Nahal ʿAmram sites (part of samples with [I(Fe)+I(Mn)]/I(Cu)<3 is equal 0 for 35/7 I, 3% for 35/7 II and 34% for 36/6).

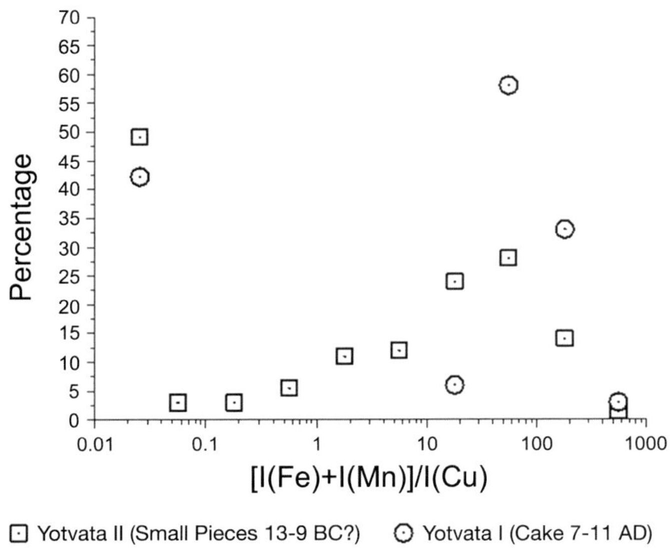

Fig. 15.5: Resulting distributions for Yotvata sites (part of samples with [I(Fe)+I(Mn)]/I(Cu)<3 is equal 0 for Yotvata II and 23% for Yotvata I).

the ratios are much higher in favor of Fe+Mn, with much higher homogeneity (in 57% of the readings, I (Fe)+I (Mn) is 70 times higher than I (Cu); in 34% of the readings it is 110 times higher). Therefore, we can see that the slag "cakes" represent better technology with high efficiency of copper extraction from the ore and maybe the use of a lower grade ore as well. It is interesting to note that both types of slag contain Mn, a point discussed below.

TIMNA SITE 34

This smelting camp on a large flat hilltop has large areas covered by low piles of small but solid slag. Recent excavations at the site by Ben-Yosef (2016) yielded ^{14}C dates from the smelting refuse and the gatehouse, of the late 11th–10th century BCE.

Three samples were taken from the site: Samples 34 I and II are of the small, solid slag, probably broken off from slag "plates" created by tapping, but smaller and thinner than the later "cakes."[4] Indeed, similar slag from nearby Site 30 were suggested by Ben Yosef *et al.* (2012: 48) to be tapping slag, also dated to the 10th century BCE. Sample 34 III is of smaller slag pieces, visually similar to those of Nahal ʿAmram 35/6, Yotvata II and to other sites in the Timna Valley; they were collected next to a small "high-place" on the northern end of the site (Rothenberg 1967: 27–28). Samples I and II are quite close in distribution of [I (Fe)+I (Mn)]/I (Cu) ratios (Fig. 15.6), while Sample III shows wider distribution, i.e., lower ratios of (Fe+Mn)/Cu and a higher degree of heterogeneity. The three samples on the graph in Fig. 15.6 are nevertheless similar to those typical of later dated sites (like ʿAmram 35/7).

TIMNA SITES 39A AND 39B

Site 39a is a habitation site dated to the fifth millennium BCE, with some small furnace slag (Rothenberg and Merkel 1998a). Site 39b is a furnace located 130 m to the north of 39a, with small furnace slag scattered around, dated by Rothenberg (1990: 79–82) to the Chalcolithic period (fifth–fourth millennia BCE).

The Fe/Cu distribution ratios of Sites 39a and 39b are similar (Fig. 15.7), indicating the possible connection between the two, despite the distance and the questionable dating of 39b. Both slags show a high level of inhomogeneity, yielding additional support for their early date.

BEʾER ORAH

This is a group of four different smelting sites in the vicinity of a water source. Beʾer Orah I is an unusual copper mine, on a limestone hill, with Chalcolithic–Early Bronze surface finds (fifth–third millennium BCE) and a dump of ash and charcoal and small furnace slag. The dump, however, yielded four ^{14}C dates of the M B II, around 1800 BCE (Avner 1993; Segal and Carmi 1996: 98; 2004: 146), a period that is considered to be totally absent in the desert. The ratio distribution (Fig. 15.8) is somewhat similar to those of other early sites with furnace slag, though with a lower presence of copper and a lower level of heterogeneity, i.e., a somewhat better technology.

Beʾer Orah II is the largest smelting camp in the southern Arabah, 18 km north of Eilat, ca. 900 m northeast of the water well bearing this name, The site contains piles of slag "cake" fragments, mainly dated to the Early Islamic period by pottery and ^{14}C dates; (Avner *et al.*, this volume, Fig. 10.28, Table 10.1: 23–30; Avner and Magness 1998; Rothenberg 1999). However, some Nabataean pottery sherds (1st century BCE to 4th century CE) are also found among the slag.

4 See Rothenberg 1990c: Fig. 80, which is ca. 25 cm across, although in the caption it is termed "cake."

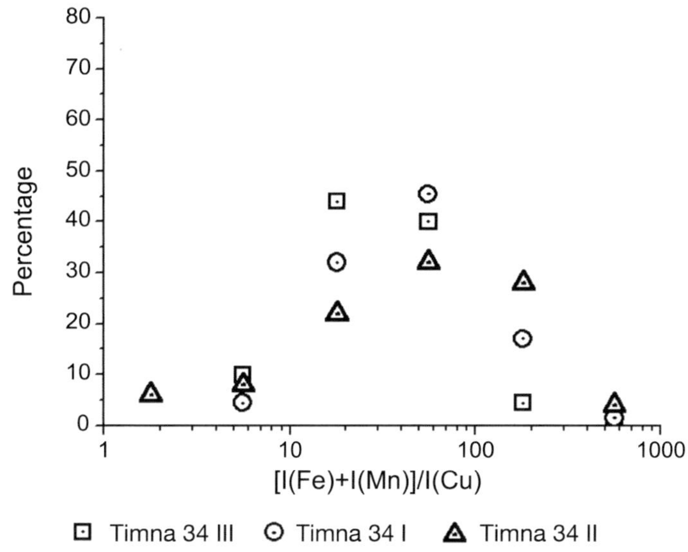

Fig. 15.6: Resulting distributions for the Timna 34 site (part of samples with [I(Fe)+I(Mn)]/I(Cu)<3 is about 0 for all three samples).

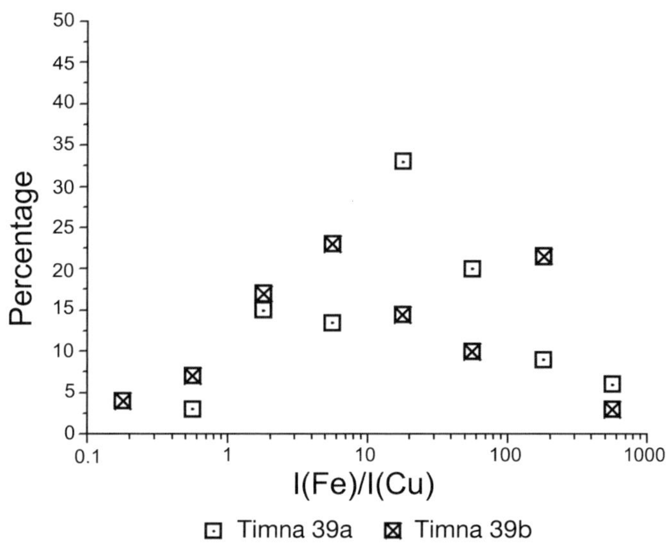

Fig. 15.7: Resulting distributions for Timna 39 sites (part of samples with [I(Fe)+I(Mn)]/I(Cu)<3 is about 30% for both sites).

Be᾽er Orah III, 1 km south of the Be᾽er Orah well, contains smaller piles of the same type of slag, dated according to the pottery sherds as Early Islamic. Be᾽er Orah IV is a small pile of slag "cake" fragments, with no pottery, and therefore now generally dated as Nabataean to Early Islamic.

Results of Be᾽er Orah II–IV are generally similar (Fig. 15.8, circles and triangles), with a low content of copper and comparatively high content of Mn, which is almost unique to these sites in the southern Arabah Valley. Although most of the evidence (mainly ^{14}C dates from Be᾽er Orah II) points to the early Islamic period, we do not exclude the possibility that some of the tapping slag "cakes" are actually Nabataean.

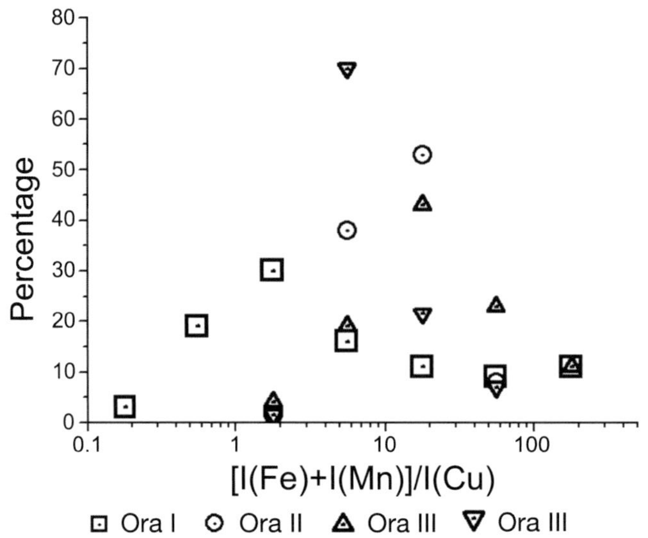

Fig. 15.8: Resulting distributions for Beer Orah sites (part of samples with [I(Fe)+I(Mn)]/I(Cu)<3 is about 50% for Beʾer Orah I and about 0 for other sites).

TELL ḤARA ḤADID

Smelting Site I has a large pile of tapping slag "cakes" (Frank 1934a): 242–243) similar to those of Beʾer Orah. It was dated by Rothenberg (1967: 286) as Byzantine; however, pottery sherds from the site collected by Avner are Early Islamic. The hilltop Site II (just above Site I) was scattered with small furnace slag and a few Chalcolithic–Early Bronze pottery sherds (now mostly erased by modern activity).

The difference in [I(Fe)+I(Mn)]/I(Cu) ratios between the two types of slag (Fig. 15.9) is clear again and strikingly similar to the case of Yotvata and Nahal ʿAmram—low ratios and high heterogeneity for the small slag from the hilltop. A particularly interesting facet is the 11% Mn in this slag type. The slag "cakes," on the other hand, show high [I (Fe)+I (Mn)]/I (Cu) ratios and high homogeneity.

ADDITIONAL COMMENTS

In most of the above analyzed sites we observed, as expected, the abundant use of iron oxide as a fluxing substance, but in some cases, we saw an occurrence of manganese oxide (in about half of all sites, the ratio Mn/Fe is equal from 0 to 1). Due to the relatively small amount of slag analyzed in this pilot project, we could not find any correlation between the ratio of Mn/Fe and the age of the slag or the site. For half of the sites, Zn impurity exists in the slag. Zn content is not correlated with the Cu content and therefore could represent traces of a mineral and not necessarily metal production. The level of concentration is up to about 10% of the Cu content.

The variation of relative amount of Ca, Pb, Sr and Ba in the slag could be an additional indication of differences between the sites and the use of different kinds of local minerals. For instance, the slag samples from sites in ʿAmram do not contain Pb, and slag samples from Orah and Timna 39 do not contain Ba. Surely, these aspects need more study in the future and a much larger sample base. The level of concentration detected in the slags here is up to 15–20% for Ca, up to 0.05% for Sr, and less than 0.5% for Pb and Ba.

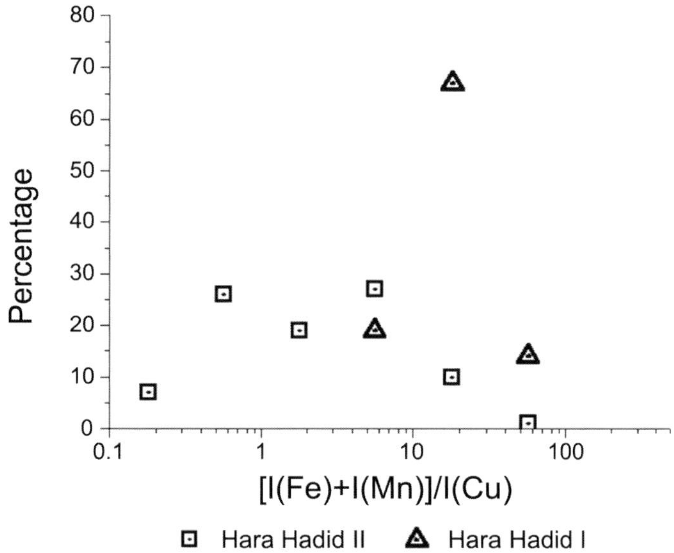

Fig. 15.9: Resulting distributions for Hara Hadid sites (part of samples with [I(Fe)+I(Mn)]/I(Cu)<3 is about 50% for for Tell Hara Hadid and about 0 for Hara Hadid).

Using the technique described above, we calculated the mean value of Fe concentrations as 25% in maximum intensities of Fe fluorescent lines, and the mean value of Cu concentration as 2.5% for maximum intensity of Cu line, as in Hara Hadid II. Concentrations of Cu, Fe and Mn for the sites discussed here are shown in Table 15.2. For several samples with high Cu content, concentrations of up to about 10% were calculated. The typical content levels are 10–20% Fe+Mn and 0.5–2.5% Cu for the "older" sites, and up to 25% Fe+Mn and 0.1–0.3% Cu for sites of "later" periods.

The data in Table 15.2 show that a distinction between "older" and "younger" sites by averaging the Cu content in the slag could be problematic and less clear than the proposed comparison and distinction by using the distribution of relative intensities of Fe (Mn) and Cu fluorescence lines as described in detail in Ben-Yosef, Tauxe, Levy 2010, as well as in this paper.

CONCLUSIONS

1. A clear difference is apparent between the earlier furnace small slag of the Chalcolithic to Iron Ages, and the later tapped slag "cake" fragments of the Nabataean to Early Islamic periods in all the studied sites. The former presents comparatively high content of Cu (typically 1%–3%), rather low ratio of (Fe+Mn)/Cu (6–30) and a high level of heterogeneity. The latter presents low content of Cu (0.1%–0.7%), high ratios of (Fe+Mn)/ Cu (20–300) and a much higher level of homogeneity. This is applicable to all the sampled slags from Yotvata, Beʾer Orah, Nahal ʿAmram and Tell Hara Hadid.
2. Only one site is thus far outstanding in its slag composition—Timna 34, now safely dated to the 10th century BCE (Ben Yosef 2016). This slag contains very little Cu compared to other early sites (0.2%–0.4%, Table 15.2) and high Fe (24%–30%), and its matrix is more solid. Therefore, the Timna 34 slag could very well represent an advanced technology, and probably resulted from some tapping (cf. Ben-Yosef *et al.* 2012: 48).

3. The occurrence next to each other of both types of slag does not seem to be accidental. It could attest to a possible use of "older" slags as a fluxing agent in a "later" smelting process. This possibility finds support in the presence of Mn where both types are found adjacent to each other, in Yotvata, Nahal ʿAmram, Beʾer Orah and Tell Hara Hadid. Also, Zn is found in both types of slag at Timna 34, Nahal ʿAmram and Tell Hara Hadid, but it is absent in both types at Beʾer Orah and Yotvata. Therefore, the exploitation of "older" slag for fluxing by later smelters may explain why the latter located themselves next to earlier smelting sites.
4. The method adopted here proves more valuable than previous approaches in identifying technological phases (see, e.g., Rothenberg 1990: 69–70). However, much work is needed in order to reach a finer distinction, with more samples from a greater number of sites, many more XRF readings for each sample, more complex calculations and ^{14}C dates. Despite some five decades of study of metallurgy in the southern Arabah Valley, the specific technology of each phase or period is still unclear. It is expected, however, that as this study advances, a more detailed history of smelting technology in the Arabah region will come to light.

ACKNOWLEDGMENTS

This research was conducted with the help of Israel Science Foundation grant no. 60.14.

REFERENCES

Avner, U. 1993. Be'er Ora: Copper Production. *Excavations and Surveys in Israel* 12: 115.

Avner, U. and Magness, J. 1998. Early Islamic Settlement in the Southern Negev. *Bulletin of the American Schools of Oriental Research* 310: 39–57.

Bachmann, H.G. 1980. Early Copper Smelting Techniques in Sinai and in the Negev as Deduced from Slag Investigations. In: Craddock P.T., ed. *Scientific Studies in Early Mining and Extractive Metallurgy* (British Museum Occasional Paper 20). London: 103–134.

Ben-Yosef, E., *et al.* 2008. A New Approach for Geomagnetic Archaeointensity Research: Insights on Ancient Metallurgy in the Southern Levant. *Journal of Archaeological Science* 35: 2863–2879.

Ben-Yosef, E. 2010. *Technology and Social Processes: Oscillations in Iron Age Copper Production and Power in Southern Jordan* (Ph.D. dissertation, University of California). San Diego.

Ben-Yosef, E., Tauxe, L., Levy, T.E. 2010. Archaeomagnetic Dating of Copper Smelting Site F2 in Timna Valley (Israel) and Its Implication on Modeling Ancient Technological Developments. *Archaeometry* 52: 1110–1121.

Ben-Yosef, E., Shaar, R., Tauxe, L. and Ron, H., 2012. A New Chronological Framework for Iron Age Copper Production in Timna (Israel). *Bulletin of the American Schools of Oriental Research* 366: 1–41.

Ben-Yosef, E. 2016. Back to Solomon's Era: Results of the First Excavations at Slaves' Hill (Site 34, Timna, Israel). *Bulletin of the American Schools of Oriental Research* 376: 169–198.

Frank, F. 1934. Aus der Araba I: Reiseberichte. *Zeitschrift des Deutschen Palästina-Vereins* 57: 191–280.

Frank, F. 1934a. Aus der 'Arabah, I: Tell el-Chlefi. *Zeitschrift des deutschen Palästina-Vereins* 57: 243–245.

Levy, T.E., *et al.* 2002. Early Bronze Age Metallurgy: A Newly Discovered Copper Manufactory in Southern Jordan. *Antiquity* 76(292): 425–437.

Levy, T.E., Najjar, M. and Higham, T. 2007. Iron Age Complex Societies, Radiocarbon Dates and Edom: Working with the Data and Debates. *Antiguo Oriente* 5: 13–34.

Levy, T.E., *et al.* 2008. High Precision Radiocarbon Dating and Historical Biblical Archaeology in Southern Jordan. *Proceedings of the National Academy of Science* 105: 16460–16465.

Levy, T.E., *et al.* 2012. Cyber-archaeology in the Holy Land-the Future of the Past. *Washington, DC: Biblical Archaeology Society eBook.*

Merkel, J. 1983. A Laboratory Reconstruction of Late Bronze-Early Iron Age Copper Smelting in the Arabah. In: Sawyer, J.F.A. and Clines, D.J.A., eds. *Moab and Edom.* Sheffield: 125–128.

Merkel, J. 1996. Simulating Experiments on Egyptian New Kingdom Copper Smelting. In: Feisal, A.E., ed. *Proceedings of the First International Conference on Ancient Egyptian Mining and Metallurgy and Conservation of Metallic Artifacts*. Cairo: 79–87.

Rothenberg, B. 1962. Ancient Copper Industries in the Western Arabah: An archaeological survey of the Arabah, Part 1. *Palestine Exploration Quarterly* 94: 5–71.

Rothenberg, B. 1967. Archaeological Sites in the Southern 'Araba and the Eilat Mountain. In: Ron, Z., ed. *Survey of the Eiloth Region*, Vol. 2. Eilat: 283–331.

Rothenberg, B. 1972. *Timna: Valley of the Biblical Copper Mines*. London.

Rothenberg, B. 1983. A Rock-Cut Copper Smelting Furnace in the Timna Valley. *Historical Metallurgy* 17: 116–119.

Rothenberg, B. 1985. Copper Smelting Furnaces in the Arabah, Israel: The Archaeological Evidence. In: Craddock, P.T. and Hughes, M.J., eds. *Furnaces and Smelting Technology in Antiquity.* London: 123–150.

Rothenberg, B., ed. 1990. *The Ancient Metallurgy of Copper, Vol. 2: Researches in the Arabah 1959–1984*. London.

Rothenberg, B. 1999. Archaeo-Metallurgical Researches in the Southern Arabah 1959–1990, Part 2: Egyptian New Kingdom (Ramesside) to Early Islam. *Palestine Exploration Quarterly* 131: 149–175.

Rothenberg, B. and Merkel, J. 1998a. Chalcolithic, 5th Millennium BC, Copper Smelting at Timna. *Institute for Archaeo-Metallurgical Studies* 20: 1–3.

Rothenberg, B. and Merkel, J. 1998b. Chalcolithic, 5th Millennium BC, Copper Smelting at Timna - New Radiocarbon Dating Evidence for Timna Site 39. *Institute for Archaeo-Metallurgical Studies* 20: 1–3.

Segal, D. and Carmi, I. 1996. Rehovot Radiocarbon Date List V. *ʿAtiqot* 29: 79–106.

SECTION III

FAYNAN, THE NEGEV AND BEYOND

CHAPTER 16

INTENSIVE SURVEYS, LARGE-SCALE EXCAVATION STRATEGIES AND IRON AGE INDUSTRIAL METALLURGY IN FAYNAN, JORDAN: FAIRY TALES DON'T COME TRUE

Thomas E. Levy, Erez Ben-Yosef and Mohammad Najjar

Beno Rothenberg pioneered regional studies of ancient metallurgy through his research and publications in the Timna Valley (Israel), Rio Tinto (Spain) and other regions in the Old World. From 1983 to 1993, Andreas Hauptmann and researchers from the Deutsches Bergbau-Museum (DBM) were influenced by Rothenberg to carry out a similar pioneer regional study in the copper ore rich Faynan region in Jordan's northern Nahal Arabah. The DBM project focused on the history of copper technology from the beginning of ore exploitation in the Pre-Pottery Neolithic (PPNB) to the Islamic period by surveying and small-scale sampling of sites specifically related to metallurgy. In 1997, the University of California, San Diego and the Department of Antiquities of Jordan embarked on an anthropological archaeology project that looked at the same "deep time" trajectory in Faynan but with a different focus. The new project continues today. It aims to understand the social, cultural and historical dimensions of ancient metallurgy in Faynan. By recording data through cyber-archaeology, and by applying in-depth 100% pedestrian surveys coupled with large-scale horizontal excavations at the main Faynan production sites to document archaeological remains, new insights were reached on the role of metallurgy in south Levantine culture and its impact on understanding social change. This paper tabulates the surveys and excavations carried out by the ELRAP team and summarizes some of the conclusions concerning the presence of industrial scale copper production in the 10th and 9th centuries BCE (Iron Age) and outlines the implications for local complex societies in this part of the southern Levant during those centuries.

INTRODUCTION

Since 1997, the University of California, San Diego Levantine and Cyber-Archaeology Laboratory's Edom Lowland Regional Archaeology Project (ELRAP) has carried out a social archaeology "deep-time" study of ancient mining and metallurgy in Jordan's copper ore Faynan district, approximately 50 km south of the Dead Sea (Fig. 16.1). In our work, technology is used as a lens for examining the social evolution of cultures in the southern Levant from the Pre-Pottery Neolithic to Medieval Islamic times. The earliest historical reference to Faynan is from the time of the Romans, when Christians were especially persecuted. Christians were condemned to work in the Faynan mines "damnatio ad metalla" as recorded by Eusebius, Bishop of Caesarea, in the late 3rd/4th centuries CE (Freeman-Grenville, Chapman III and Taylor 2003; Najjar and Levy 2011; Mattingly 2011).

The earliest mapping of the copper region was carried out by Jean Baptiste Bourguignon D'Anville in 1732. Although he never seems to have left the city of Paris, he was a great cartographer who relied on data from other researchers and explorers (D'Anville 1732). Starting in the 19th century, explorers, the military, biblical scholars, archaeologists and geologists have been drawn to the Faynan region for a variety of reasons (Robinson 1867; Kitchener 1884; Musil 1907; Frank 1934; Glueck 1935, 1940; Kind 1966; MacDonald and Khairieh 1992). However, the first large-scale science-based research that focused on the archaeometallurgical record in Faynan was carried out by the Deutches Bergbau Museum (DBM) under the direction of Andreas Hauptmann (1987; 2007). The Council for British Research in the Levant (CBRL) carried out major projects on desertification (Barker, Gilbertson and Mattingly 2007) and research concerning the earliest phases of the Neolithic period in this desert region (Finlayson and Mithen 2007; Finlayson *et al.* 2011). Other Neolithic projects in the Faynan area were carried out by Mohammad Najjar and Alan Simmons (Simmons and Najjar 2006). These late prehistory projects had little to do with Faynan being a copper ore resource zone. While the Wadi Faynan Landscape Survey project, led by Graeme Barker, David Gilbertson and David Mattingly, focused on pedestrian surveys aimed at contextualizing desertification and ancient copper pollution, little excavation work was done (Grattan *et al.* 2002; Grattan *et al.* 2005; Grattan, Gilbertson and Hunt 2007). Similarly, the

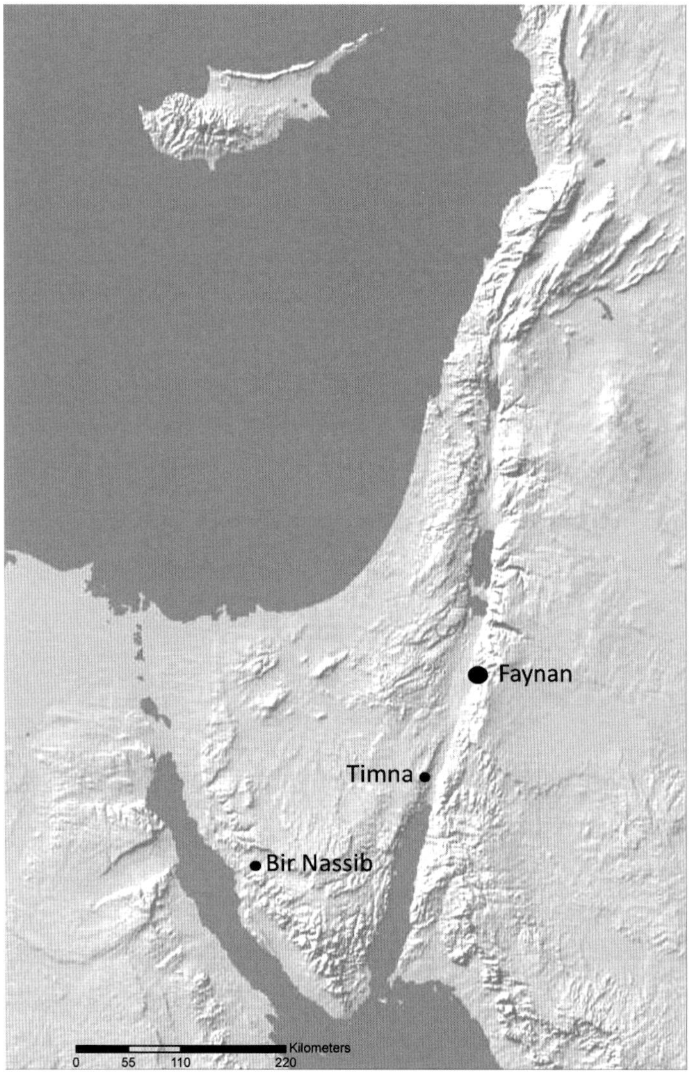

Fig. 16.1: The main copper ore resources zones in the southern Levant.

pioneering work of the DBM, while providing one of the most in-depth studies of changing systems of ore extraction and copper production with hard analytical data, did not conduct large-scale excavation projects. Instead, surveys were aimed specifically at locating metallurgical sites and, when needed, small soundings were carried out at sample sites. Thus, ELRAP expeditions differed from previous research in that they focused on large-scale excavations of copper production sites and other types of sites to help flesh out the social dynamics of the societies that lived and worked in the Faynan regions. More recent research in Faynan by the ELRAP team and others have returned to the issue of ancient pollution and the movement of human populations throughout the late Holocene (Grattan, Gilbertson and Hunt 2007, Adams *et al.* 2010, Beherec *et al.* 2015, Knabb *et al.* 2016). This paper will provide a snapshot of some of the results of ELRAP's expeditions, co-directed by M. Najjar and T.E. Levy.

THE FAYNAN COPPER ORE RESOURCE ZONE

The copper ore bearing deposits in Faynan represent the largest source of copper in the southern Levant and provide a unique opportunity to study the role of ancient mining and metallurgy on the evolution of societies from the Neolithic period (ca. 8500 BCE) to Medieval Islamic times. In terms of the regional scale of copper deposits in the Arabah Valley that separates modern Jordan and Israel, Faynan is the 'Big Brother' to Timna (Levy, Najjar and Ben-Yosef 2014; Ben-Yosef 2016). The copper ore district of Faynan encompasses an area of over 400 sq km. While there are a number of ancient mining regions around the world—Rio Tinto in Spain, the Central Taurus Mountains in Turkey, Timna in Israel, many locales in Saudi Arabia, the Troodos Mountains in Cyprus, the Tonglu Mountains in China, Arisman on the Central Iranian Plateau, and other areas—perhaps the best preserved and longest exploited region is Faynan in Jordan. The Faynan district is shaped like a "box canyon" oriented towards the northeast. East–west faults exposed in the main wadi system of the Wadi Fidan–Faynan–Dana and smaller systems are home to the main copper ore bodies in the region. According to Hauptmann (2007: 43) change in the flow regime in the district during the third millennium BCE may have led to the exposure of copper mineralization first in the Burj Formation (Dolomite–Limestone–Shale [DLS] unit), providing an opportunity for intensive mining activities during the Early Bronze Age (ca. 3600 to 2000 BCE). The Cambrian Burj/DLS unit is equivalent to the Timna Formation in Israel. Higher up in the geological stratigraphy of Faynan is the Massive Brown Sandstone (MBS) unit whose copper mineralization is equivalent to that found in the Amir and Evrona Formations in the Timna area. The MBS unit contains oxidic and sulphidic copper ores particularly in the secondary wadi systems of Qalb Ratiye and Wadi Abiad north of the main Faynan drainage. These copper ores were the focus of the earliest mining activities in Faynan (*ibid.*: 67). There are three major contrasting phytogeographic climatic zones that stretch across a distance of less than 25 km from the edge of the Transjordanian Plateau and Shara Mountains in the east where a narrow band of the Mediterranean region is found, followed by a narrow band of semi-arid Irano-Turanian environment along the plateau edge, to the main Saharo-Arabian Desert environmental zone where average annual rainfall is less than 90 mm, forming the main climatic zone of Faynan. The main wadis and associated interfluves are characterized by trees such as *Acacia raddiana*, *Calotropis procera*, and *Tamrix nilotica* that, in different archaeological periods, served as important sources of charcoal for smelting. In various part of the Faynan region there are pockets of oasis environments characterized by perennial fresh water springs where date palms (Phoenix dactylifera) are often found with *Juncus arabicus*, and trees such as *Moringa peregrine* (cf. Ain Fidan, Ain Ghuwayba, Ain Ghweir). Throughout history these oases have provided rich garden ecosystems for the occupants of Faynan and other parts of the south Levantine deserts (Danin 1983, 1998).

CONTEXTUALIZING RECENT ARCHAEOLOGICAL RESEARCH IN FAYNAN

As noted above, the first systematic surveys in the Faynan region were carried out by the DBM under the direction of Andreas Hauptmann from 1983 to 1993 with the specific aim of identifying sites related to all periods of copper production. Approximately 160 sites were identified (Table 16.1; Hauptmann 2007), the majority of which are ancient mines. All of the most important ancient copper production sites were recorded and often sampled. Using mostly geological methodologies, the DBM established the basic stratigraphy, radiocarbon dating, and archaeometallurgical history of the region. This work has become a standard for regional archaeometallurgical investigations around the world (Roberts and Thornton 2014). The DBM framework established base-line archaeometallurgical data and the history of copper production for Faynan, but it did not present a social archaeological framework for understanding ancient craft production in all its socio-economic dimensions. On the other hand, the CBRL project in Faynan was designed as a diachronic and global comparative study of processes of desertification and ancient pollution. Intensive surveys in the main Faynan Valley were aimed at studying changes in ancient landscape use in a desert environment through the dating of well over 1,000 ancient agricultural terraces. Excavation of sites was not a priority, with the exception of a small, detailed excavation and microarchaeology, paleoenvironmental and ancient pollution study of a terrace system near the largest site in the valley—Khirbat Faynan (Barker *et al.* 1997; Barker 2001; Barker, Gilbertson and Mattingly 2007; Hunt *et al.* 2007). Important early Neolithic research has been carried out at the Pre-Pottery Neolithic A site of WF 16 (Finlayson *et al.* 2007, 2011), at Ghwair I (Simmons and Najjar 2006) and the Pre-Pottery Neolithic site of Tell Tifdan (Twiss 2007) dating to the Pre-Pottery Neolithic B. Like the DBM research, the CBRL Faynan work is an especially important regional geomorphological study that helps establish a framework for studying paleoenvironmental and landscape change in Faynan. As shown in Fig. 16.2, there have been 12 major survey projects in and around the Faynan district. What distinguishes ELRAP work from these earlier investigations is the combination of

Table 16.1: Regional Archaeological and Archaeometallurgical Surveys in the Faynan District

Survey Project	*Researchers*	*Number of Sites*
Deutsches Bergbau Museum Survey 1983-1992	Hauptmann	160
Southern Ghors and NE ‘Arabah 1985-1986	MacDonald	240
Tafila – Busayra Archaeological Survey 1999-2001	MacDonald	290
Wadi Faynan Landscape Survey (CBRL)	Barker, Gilbert, Mattingly, Grattan	N.a.
Barqa Landscape Survey, 2010 (Hamilton U - Abersy	Adams, Grattan	N.a
Total		*690*
Wadi al-Jariya Survey 2002 (Jabal Hamrat Fidan, UCSD – DOAJ), 2007 (ELRAP, UCSD DOAJ)	Levy, Najjar, Adams, Knabb	218
Wadi Fidan Survey (Jabal Hamrat Fidan, UCSD - DOAJ)	Levy, Najjar, Adams	162
Faynan – Busayra Regional Survey (UCSD – DOAJ)	Ben-Yosef, Najjar, Levy	116
Lowlands to Highlands of Edom Project (UCSD – DOAJ)	Smith, Najjar, Levy	48
Wadi Feid Expedition, 2009 (UCSD – DOAJ)	Knabb, Najjar, Levy	123
Buweirda Springs Survey, 2011 (UCSD-DOAJ)	Knabb, Najjar, Levy	18
Total		*685*

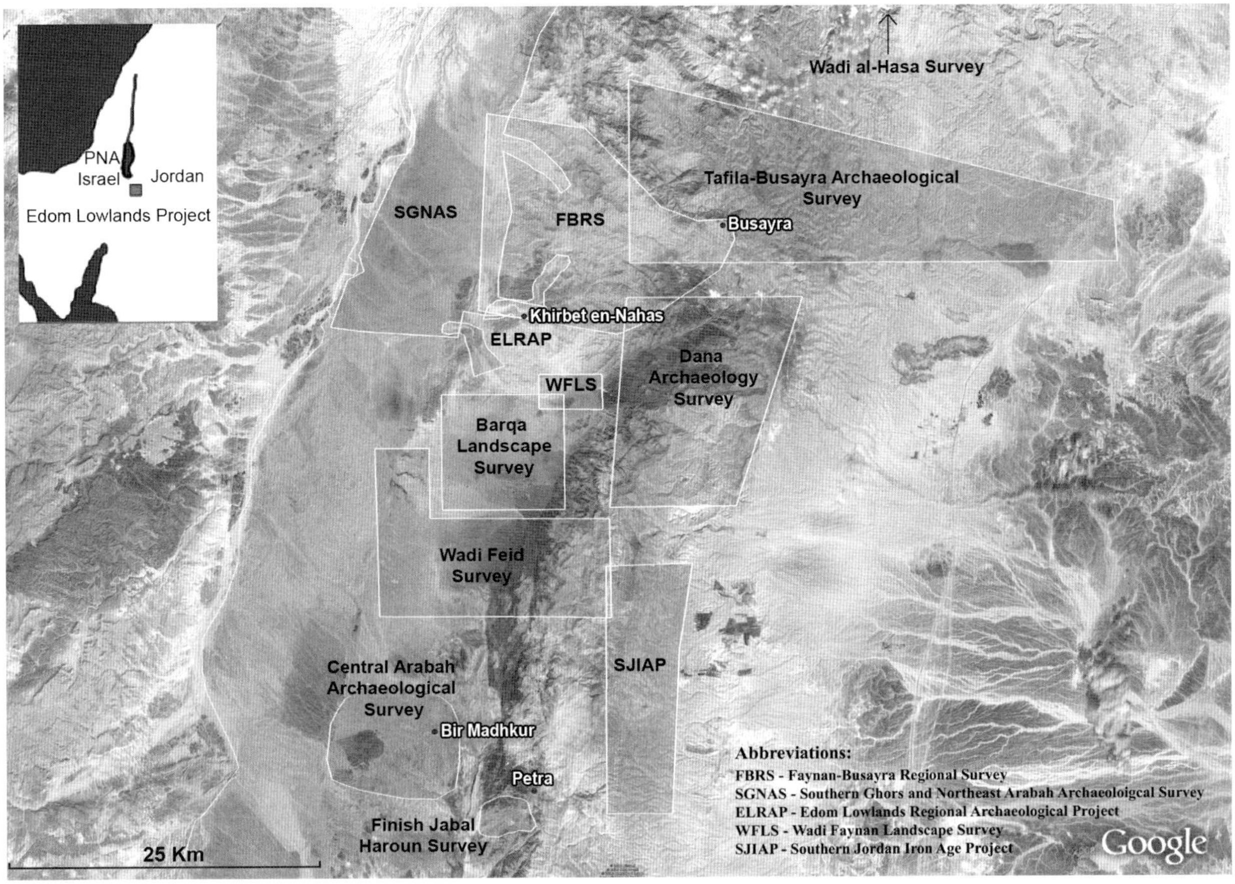

Fig. 16.2: Major archaeological surveys carried out in the Faynan district, Jordan.

intensive archaeological surveys coupled with large-scale excavations aimed at teasing out the socio-economic dimensions of the people who practiced ancient mining and copper production in Faynan from the Neolithic to Medieval Islamic times (Tables 16.1 and 16.2; Fig. 16.2). According to Costin (1991), explanations for the organization of craft production are keys to understanding the political, social and economic nature of cultural systems. For ELRAP researchers, some of the major goals have been to understand the degree of elite control and sponsorship of any given production system, the scale and intensity of production, the nature of workshop organization, the concentration of production, its scale and finally its intensity during each production period. In the following discussion, some of the highlights of the ELRAP excavation work related to ancient mining and metallurgy and their social implications in Faynan are presented. To help parse out the socio-economic underpinnings of the ancient metallurgical systems through time in the Faynan region, our various projects have applied the *chaîne opératoire* concept of anthropological discourse (Leroi-Gourhan 1943; Dobres 1999a; 1999b; Bar-Yosef and Van Peer 2009). By identifying the "operational sequence" that underlies technological processes and the behavioral activities that are associated with them, it is possible to move from the static archaeological record of ancient production to the dynamic socio-economy and culture of those who worked in the mines and copper production centers of Faynan. One example of the application of the *chaîne opératoire* method is presented below for the Iron Age site of Khirbat en-Nahas.

Table 16.2: UC San Diego – Department of Antiquities of Jordan Edom Lowlands Regional Archaeology Project (ELRAP) Excavations, 1997–2012

Excavation	*Period*	*Function*	*Project Year*
Tell Tifdan	PPNB	Village	1999
Wadi Fidan 61	PPN–PN Iron IIA	Village Tombs	2012
Wadi Fidan 51	Late Neolithic–Early Chalcolithic	Village (small)	1999
Wadi Fidan 4	EB I Iron IIA	Village workshop Tombs	1997, 2003
Khirbat Hamra Ifdan	EB II–IV Iron Age	Copper Production Manufactory; Smelting	1999, 2000, 2007, 2011
Wadi Fidan 40	Iron I–II	Cemetery	1997, 2003, 2004, 2009
Rujm Hamra Ifdan	Iron II	Watchtower	2007
Khirbat en-Nahas	Iron I–IIA	Smelting site	2002, 2006, 2009
Khirbat al-Jariya	Iron I–IIA	Smelting site	2006, 2014
Ras-al Miyah	Iron IIC	Fortress complex	2009
Khirbat al-Ghuweiba	Iron I–IIA, Roman/Nabatean	Small scale smelting village	2009
Jabal al-Jariya Mines	Iron IIB/C?	Pit mines	2007
Khirbat Faynan	EB IA–II, Roman/Nabatean	Settlement center	2011, 2012
Khirbat Nuqayb al-Asaymir	Islamic (Middle)	Smelting site	2002, 2011, 2012
L2HL Khirbat al-Malayqtah, al-Kur, al-Iraq, Tawilan	Iron IIB	Settlements	2006, 2007, 2010, 2011

THE KHIRBAT EN-NAHAS CASE STUDY

Khirbat en-Nahas (KEN; 30.661394N, 35.392700E) provides the chronological, socio-economic, technological and historical anchor for assessing the Iron Age archaeometallurgy of the Faynan district. As noted above, KEN is the largest Iron Age production site (ca. 10 ha) in this copper ore region. It is covered with thick deposits of ancient black "slag mounds"—the industrial waste from copper smelting. Prior to our investigations, scholars debated the date and significance of this site. However, since 2002 and the large scale UCSD–DOAJ excavations at the site, coupled with a program of high precision radiocarbon dating, there is no doubt about the Iron Age date of this massive site. Seven distinct excavation areas have been opened by our team, providing a rich sample of the range of construction and metallurgical activities carried out here. The excavations at KEN have produced superb archaeometallurgical, ground stone, ceramic and other material culture collections for the study of mining and metallurgy during this period. Here a brief summary of these areas is presented from the 2002, 2006 and 2009 expeditions. Critical to understanding Iron Age copper production at this site is the fact that after the 9th century BCE, copper production ceased at this site and its immediate surroundings.

In a recent paper, we briefly summarized the seven main excavation areas and their dating at KEN (Levy, Ben-Yosef and Najjar 2012). Here we refer to the functional attribution of the two main excavation areas and their dates to aid in socio-economic interpretation of the site during the two main occupation phases—the 10th and 9th centuries BCE (see also the final publication of KEN and the ELRAP Iron Age research in Faynan [Levy, Najjar and Ben-Yosef 2014]).

Fig. 16.3: Aerial view of the center of Khirbat en-Nahas, Jordan, during ELRAP's excavations.

AREA A

Two main construction phases have been identified in the gatehouse (Area A) that is incorporated into the main fortification wall of the fortress. Radiocarbon dates place the initial construction squarely in the early to mid-10th century BCE. In the 9th century BCE, the gatehouse was "decommissioned" and reconfigured as a large building with no military function. In this last phase of occupation, the gatehouse was used for metallurgical activities. The four-chamber plan of the gatehouse at KEN is similar to other 10th century BCE examples at south Levantine sites as Ashdod 10a, Beer-sheba V, Khirbat Qeiyafa (Garfinkel *et al.* 2012), and Megiddo VA–IVB (Gate 1567) (Levy, Najjar and Ben-Yosef 2014). We intentionally left one of the chambers of this gatehouse unexcavated for future researchers. Interested readers can access the basic radiocarbon dating framework for KEN published previously, and the full compendium with illustrated contextual information (*ibid.*).

AREA M

One of the numerous "slag mounds" at Khirbat en-Nahas was sampled by our team from surface to virgin soil, together with a structure complex that was partially covered by the top layers of the mound. More than six meters of copper-production-related deposits, including interchanging horizons of various types of slag (fine-crushed, tap, and furnace in different sizes), clayey and ashy sediments mixed with tuyère and furnace fragments, were exposed. The Area M slag mound has been an important proxy for measuring changes in both copper production and changes in the intensity of the earth's magnetic field from the end of the Late Bronze Age through the Iron Age (Ben-Yosef *et al.* 2008; Levy *et al.* 2008). For our purposes here, it is important to emphasize that the Area M excavations document two major phases of industrial copper production at Khirbat en-Nahas that occurred during the 10th and 9th centuries BCE. These technological data, associated with local Edomite pottery assemblages (Smith and Levy 2008), provide another source for demonstrating the presence of a complex south Levantine society (archaic state/paramount chiefdom) which controlled copper production during both of these centuries. The Area M slag mound excavation also provides evidence for a disruption and transition in copper production that dates to the end of the 10th century BCE and corresponds to the Egyptian campaign of Shishak (or Shoshenq

TABLE 16.3: COMPARATIVE STRATIGRAPHY AND CHRONOLOGY OF KHIRBAT EN-NAHAS BASED ON HIGH PRECISION RADIOCARBON DATING (LEVY, NAJJAR AND BEN-YOSEF 2014)*

St.	^{14}C	*KEN-A Phases and Description (2002,2006)*	*KEN-M Phases and Description (2002,2006)*	*KEN-F Phases and Description (2006)*	*KEN-S Phases and Description (2002)*	*KEN-T Phases and Description (2006)*	*KEN-R Phases and Description (2006,2009)*	*KEN-W Phases and Description (2009)*
I		A1a, A1b Post-abandonment	- Post-abandonment	F1a F1b Post-abandonment	S1a S1b Post-abandonment	T1a Post-abandonment	R1a, R1b Post-abandonment	W1A Collapse
II	9th century BCE	A2a Ephemeral metal production outside gate	M1; 1a Smelting; (1a after building (?abandonment	(Unoccupied)	S1 Abandoned possible corral	(Unoccupied)		W1B Post-abandonment ((Unoccupied
	9th century BCE	A2b Intensive metal production	M2a Intensive smelting; late phases in building	F2a Addition of small architectural units	S2a Minor extensions of main building		R3 Courtyard copper production complex	W2 Structure complex; ?storage areas; cultic Undated
III	Late 10th–Early 9th century BCE	A3a Decommissioning ?Residential	M2b First phase of building; intensive smelting	F2b Main building with basins, recycling	S2b Main building –casting/slag processing	T1bT2a 2nd occupation phase floor	R2a Some additions to structure	
IV	10th century BCE	A3b Building of gatehouse	(M3 (IV-V Industrial horizon; intense copper production	F2c Crushed slag below building	- Unoccupied	T2b Main building phase, residential	R2b Main building phase, residential	
V	Early 10th century BCE	A4a Crushed slag/ below gate			S3 Sparse crushed slag	T3 Pre-building smelting	R3a, R3b Crushed slag below building	W3 Metallurgical debris
VI	Late 12–11th century BCE	- Unoccupied	M4 Copper production	- Unoccupied	S4 Small scale metal prod and domestic activities	- Not excavated	- Not excavated	
VII	13–12th century BCE		M5a, M5b Installations; small scale production		S5 Unoccupied			Not excavated
-	-	A4b Virgin sediment	- Virgin sediment	F3 Virgin sediment	S5 Virgin sediment			

* Color key: white = unoccupied / post-abandonment contexts; orange = mainly smelting and other pyrotechnological activities; green = mainly architectural remains (buildings); blue = unexcavated contexts; pink = virgin soil.

I known from the Hebrew Bible [I Kings 14:25; Chronicles 12:1–12]) (Levy *et al.* 2008). Interestingly, at about the same time the sites of Khirbat al-Jariya and probably also the site of Khirbat al-Ghuweiba were abandoned (Ben-Yosef *et al.* 2010). Taken together, the Khirbat en-Nahas excavations in the fortress gatehouse, the industrial slag copper slag mound and other areas at the site demonstrate conclusively the presence of a local kingdom-level society at Faynan during both the 10th and 9th centuries BCE. In the final two-volume monograph publication of our Iron Age research in Faynan (Levy, Najjar and Ben-Yosef 2014), a total of 108 accelerator mass spectrometry (AMS) dates processed from KEN are analyzed for the seven different excavation areas on the site (*ibid.*). This final publication adds more evidence for dating the fortress construction and the first Iron Age industrial-scale copper production at KEN and the Faynan region to the 10th century BCE. In Table 16.3 we summarize the results of all the stratified excavations at Khirbat en-Nahas, correlating the stratigraphy of the different excavation areas based on high precision radiocarbon dating. Accordingly, Levy *et al.* (*ibid.*; and Table 16.3 presented here) provide the benchmark for understanding biblical Edom and the tempo of the emergence of industrial-scale copper production, Iron Age social organization in this part of the southern Levant, and the role of a local Edomite kingdom or paramount chiefdom in the history of the southern Levant during the 10th and 9th centuries BCE.

CONCLUSION

When we began our deep-time study of the role of technology (ancient mining and metallurgy) on social evolution in the southern Levant, we had no idea that by the time we reached the Iron Age, we would enter the complex and contentious world of biblical archaeology. From an anthropological perspective, our decade of Iron Age fieldwork and research culminated in a definitive database for large-scale stratified Iron Age excavations for the region of Edom, which helps to explain the rapid culture change in Edom during the 10th and 9th centuries BCE. After excavating at Khirbat en-Nahas in 2002 and 2006, we produced a number of publications that demonstrated the scale of Iron Age copper production during these centuries using a chronology based on high precision radiocarbon dates from secure stratified contexts (Higham *et al.* 2005; Levy, Najjar and Higham 2005; Levy *et al.* 2008). Using the social organization of copper production as a proxy, the data showed that during the 10th century BCE local Edomite ceramic assemblages were intertwined with both the construction of monumental architecture (a fortress) and industrial scale copper production belonging to a local Levantine kingdom or chiefly confederacy (Levy, Ben-Yosef, and Najjar 2014). Previous researchers such as Crystal Bennett and Piotr Bienkowski (Bennett 1966; Bennett and Bienkowski 1995; Bienkowski 2002; Porter 2004) had argued that the Iron Age began late in Edom, most likely during the 7th century BCE, fitting into a biblical minimalist interpretation of the Hebrew Bible. By demonstrating the presence of complex societies in Edom during the 10th century, ELRAP negated these early views. Consequently, advocates of the "low chronology" were confounded by the new data that necessitated acknowledging a completely new understanding of Edom with a much longer chronology and a greater level of social complexity than previously assumed. Israel Finkelstein and Neil Silberman (Finkelstein and Silberman 2006) ignored new data from Khirbat en-Nahas (Higham *et al.* 2005; Levy *et al.* 2005) and suggested:

> Another important source of copper is the area of Wadi Feinan on the eastern margin of the Arabah valley.... Recent studies by German, American, and Jordanian scholars have revealed evidence for continuous activity in the Iron Age, with one of the intense periods of mining and production dated to the late eighth and seventh centuries BCE. Like all other lucrative economic activities in the region, this industry was carried out under Assyrian auspices. (Finkelstein and Silberman 2006: 174–175)

Simply put, there is no evidence for industrial scale copper production in Faynan during the late 8th or 7th centuries BCE and no evidence of an Assyrian presence in Faynan during these periods or the preceding 10th and 9th centuries BCE. Finkelstein's error is based not only on ignoring the hard archaeological stratigraphy and georeferenced archaeometallurgical data that was published at the time for the KEN excavations, but most importantly, his uncompromising commitment to his "low chronology" for the Iron Age of the southern Levant that erases all evidence for the presence of local complex archaic states in the region during the 10th century BCE (Finkelstein 2005a). The problems with the low chronology were debated at the time (Finkelstein 2005a; Levy and Higham 2005; Levy *et al.* 2005; Mazar 2005; Mazar *et al.* 2005) and continue to be discussed (Finkelstein and Piasetzky 2011), however, most scholars support the traditional high or modified high chronology (Levy and Higham 2005; Mazar and Bronk Ramsey 2010; Mazar 2011). Regardless of how one aligns to this chronological debate, the large-scale excavations at Khirbat en-Nahas demonstrate that copper production was carried out by local south Levantine complex societies capable of building monumental defensive works and organizing intricate systems of ore procurement, labor distribution, complex technological organization, and exchange networks (cf. Ben-Yosef 2010) that contradict Finkelstein's minimalist reconstructions of social complexity during the 10th century BCE not only in Transjordan but in the southern Levant as a whole.

In the case of Iron Age Edom, Finkelstein's approach is seen most clearly in one of his earlier critiques of our work where he uses a method that runs counter to standard archaeological practice:

> ... [A] close look at the aerial photo and plan of Khirbat en-Nahas (Levy *et al.* 2004: 868–869) seems to show that the fort was built on top of the site, cutting into the piles of copper industry waste. Second and more important, the Khirbet en-Nahs fort is almost identical in layout and plan of the gate to two other forts in the south—Ein Hazeva, which has already been mentioned, and Tell el-Kheleifeh. But these forts date to the late 8th and 7th century BCE...; they seem to have been built by the Assyrians to protect their interests along the main Arabian trade routes in the desert. Moreover, there is no evidence for a similar fort in the Negev, Transjordan or west of the Jordan —in fact, in the entire Levant—prior to the Assyrian takeover. (Finkelstein 2005b)

Granted, Finkelstein wrote this interpretation more than a decade ago. In point of fact, our chapter presented here was written 4 years ago and we are only now able to revise our remarks. However, as recently as 2014 (Finkelstein 2014: 130) argues that Khirbat en-Nahas is typical of his "Phase C," large solid-wall fortresses that were part of "a system of Assyrian strongholds in the south, which also included En Hazeva and Khirbet en-Nahas (Fig. 6)." By 2016 (Finkelstein 2016), Finkelstein still adheres to the incorrect interpretation of an "Assyrain fortress" at Khirbat en-Nahas and then presents a spurious model of the ^{14}C dates from the site that simply adds the number of dates for each 50 year interval (i.e., 1100–1050 BCE, 1050–1000 BCE, 1000–950 BCE, 950–900 BCE, 900–850 BCE, etc.) and argues that the number of dates per half century interval equals the periods when copper production was most intense. This approach is erroneous as we simply excavated more area from the easily accessible 9th century BCE levels than the deeper 10th century levels. This model ignores the stratigraphy and archaeological contexts of the large ^{14}C date sequence at Khirbat en-Nahass.

Examining an aerial photo to make chronological observations on a site is unheard of for archaeological practice. As for Ein Hazeva, radiocarbon dates from the earliest layers at the site show that it was occupied in the Iron I, with at least the core of the massive fortress dated to the 10th century BCE (Y. Israeli and T. Erickson-Gini, personal communication) and the recording of the stratigraphy at Tell el-Kheleifeh is acknowledged to be poor, making it difficult to determine the occupational history of the site (Pratico 1993a, b). While all of Finkelstein's attempts to re-date the stratigraphy and history of Khirbat en-Nahas have been addressed with in-depth discussions of the stratigraphy, radiocarbon dating, and material culture from the site (Levy *et al.* 2005; Levy and Najjar 2006; Levy *et al.* 2008), as noted above it is his recent paper concerning Edom (Finkelstein 2014), where he reiterates his arguments for "down-dating" the fortress to the late 8th century BCE—in his opinion Tell el-Kheleifeh has a) no late Iron I or early Iron IIA finds to use it to cross-date KEN, b) organic samples found under the gate of the Khirbat en-Nahas fortress were radiocarbon dated to the 9th century, making it "impossible to date the construction of the gate to the 10th century"; and c) the pottery from Khirbat en-Nahas "belongs mainly to the Iron IIB–C."

Without repeating ourselves, his presentation of the suite of KEN radiocarbon dates is selective and does not treat the entire assemblage of dates published previous to the final publication or in the final publication (Levy, Najjar and Ben-Yosef 2014). With regard to the pottery from KEN, Finkelstein actually published his opinion on the assemblage (Finkelstein and Singer-Avitz 2009) before our paper appeared. Our paper presents the first detailed data for scholars to work with and demonstrates the local development of Edomite pottery from the late 11th to 9th centuries BCE, with a strong 10th century BCE representation (Smith and Levy 2008).

The final publication of the excavations at Khibat en-Nahas (Levy, Najjar and Ben-Yosef 2014) provide a model for Iron Age excavation reporting in the southern Levant. Each excavation area of the site is presented with a high precision radiocarbon dating sequence and Bayesian model to date the area and connect it to all the other excavations at the site. These results show conclusive evidence of local, non-Mesopotamian, construction and industrial scale copper production in the 10th century BCE. However, it is the Assyrian kingdom that Finkelstein would like readers to believe carried out industrial copper production in Faynan during late 8th and 7th centuries BCE, or—in other cases—be responsible for massive construction operations in relation to developing trade routes (and cf. Ben-Yosef, Najjar and Levy 2014 for contradicting this paradigm). Given Finkelstein's (2016) attempt to work with the ^{14}C dating sequence at Khirbat en-Nahas, in the future he should include all the dates published in the final publication (Levy *et al* 2014) as his recent (*ibid*: 26) model ignores this. Ideally, Finkelstein's "Assyrian fortress" model should be retracted and our debate can move on to issues concerning local complex societies in Edom and Israel during the 10th and 9th centuries BCE, how they evolved, were maintained and ultimately—collapsed.

REFERENCES

Adams, R.B., *et al.* 2010. Report on the First Season of the Barqā Landscape Survey, South-West Jordan. *Annual of the Department of Antiquities of Jordan* 54: 95–120.

Barker, G.W. 2001. A Tale of Two Deserts: Contrasting Desertification Histories on Rome's Desert Frontiers. *World Archaeology* 33(3): 488–507.

Barker, G.W., *et al.* 1997. The Wadi Faynan Project, Southern Jordan: A Preliminary Report on Geomorphology and Landscape Archaeology. *Levant* 29: 19–40.

Barker, G., Gilbertson, D. and Mattingly, D., eds. 2007. *Archaeology and Desertification: The Wadi Faynan Landscape Survey, Southern Jordan*. Oxford.

Bar-Yosef, O. and Van Peer, P. 2009. The Chaine Operatoire Approach in Middle Paleolithic Archaeology. *Current Anthropology* 50(1): 103–131.

Beherec, M.A., *et al.* 2016. Iron Age Nomads and Their Relation to Copper Smelting in Faynan (Jordan): Trace Metal and Pb and Sr Isotopic Measurements from the Wadi Fidan 40 Cemetery. *Journal of Archaeological Science* 65: 70–83.

Bennett, C.M. 1966. Umm el-Biyara. *Revue Biblique* 73(3): 400–401.

Bennett, C.M. and Bienkowski, P. 1995. *Excavations at Tawilan in Southern Jordan* (British Academy Monograph in Archaeology 8*)*. Oxford.

Ben-Yosef, E. 2010. *Technology and Social Process: Oscillations in Iron Age Copper Production and Power in Southern Jordan* (Ph.D. dissertation, Univeristy of California). San Diego.

Ben-Yosef, E., Najjar, M. and Levy, T.E. 2014. Local Iron Age Trade Routes in Northern Edom: From the Faynan Copper Ore District to the Highlands. In: Levy, T.E., Najjar, M. and Ben-Yosef, E., eds. *New Insights into the Iron Age Archaeology of Edom, Southern Jordan*, Vol. 2. Los Angeles: 493–575.

Ben-Yosef, E. 2016. Back to Solomon's Era: Results of the First Excavations at "Slaves' Hill" (Site 34, Timna, Israel). *Bulletin of the American Schools of Oriental Research* 376: 169–198.

Ben-Yosef, E., Levy, T.E., Higham, T., Najjar, M. and Tauxe, L. 2010. The Beginning of Iron Age Copper Production in the Southern Levant: New Evidence from Khirbat al-Jariya, Faynan. *Antiquity* 84(325): 724–746.

Ben-Yosef, E., *et al.* 2008. A New Approach for Geomagnetic Archaeointensity Research: Insights on Ancient Metallurgy in the Southern Levant. *Journal of Archaeological Science* 35(11): 2863–2879.

Bienkowski, P. 2002. *Busayra: Excavations by Crystal-M. Bennett, 1971–1980* (British Academy Monographs in Archaeology 13). Oxford.

Costin, C.L. 1991. Craft Specialization: Issues in Defining, Documenting, and Explaining the Organization of Production. *Archaeological Method and Theory* 3: 1–56.

Danin, A. 1983. *Desert Vegetation of Israel and Sinai*. Jerusalem.

Danin, A. 1998. Man and the Natural Environment. In: Levy, T.E., ed. *The Archaeology of Society in the Holy Land.* New York: 24–39.

D'Anville, J.B.B. 1732. *Patriarcharus Hierosolymitanus*. Paris.

Dobres, M.A. 1999a. Of Great and Small *Chaines* of Being: Toward Understanding the Sensual and Social Embodiment of Prehistoric Technology. In: Sillar, B. and Boyd, B., eds. *Embedded Technologies: Reworking Technological Studies in Archaeology*. Lampeter: 1–19.

Dobres, M.A. 1999b. Technology's Links and Chaines: The Processual Unfolding of Technique and Technician. In: Dobres, M.A. and Hoffman, C.R., eds. *The Social Dynamics of Technology: Practice, Politics, and World Views*. Washington: 124–146.

Finkelstein, I. 2005a. A Low Chronology Update: Archaeology, History and Bible. In: Levy, T.E. and Higham, T., eds. *The Bible and Radiocarbon Dating: Archaeology, Text and Science.* London: 31–42.

Finkelstein, I. 2005b. Khirbet en-Nahas, Edom and Biblical History. *Tel Aviv* 32(1): 119–125.

Finkelstein, I. 2013. *The Forgotten Kingdom: The Archaeology and History of Northern Israel*. Atlanta.

Finkelstein, I. 2014. The Archaeology of Tell el-Kheleifeh and the History of Ezion-geber/Elath. *Semitica* 56: 105–136.

Finkelstein, I. 2016. The Southern Steppe in the Early Phases of the Iron Age: A Framework for Territorial History (in Hebrew). In: Patrich, J., Peleg-Barkat, O. and Ben-Yosef, E., eds. *Arise, Walk Through the Land: Studies in the Archaeology and History of the Land of Israel in Memory of Yizhar Hirschfeld on the Tenth Anniversary of his Demise*. Jerusalem: 17–28.

Finkelstein, I. and Piasetzky, E. 2011. The Iron Age Chronological Debate: Is the Gap Narrowing? *Near Eastern Archaeology* 74: 50–54.

Finkelstein, I. and Silberman, N.A. 2006. *David and Solomon—In Search of the Bible's Sacred Kings and the Roots of Western Tradition*. New York.

Finkelstein, I. and Singer-Avitz, L. 2009. The Pottery of Khirbet en-Nahas: A Rejoinder. *Palestine Exploration Quarterly* 141(3): 207–218.

Finlayson, B. and Mithen, S., eds. 2007. *The Early Prehistory of Wadi Faynan, Southern Jordan: Archaeological Survey of the Wadis Faynan, Ghuwayr and al-Bustan and Evaluation of the Pre-Pottery Neolithic A Site of WF 16* (Wadi Faynan Series Volume 1; Levant Supplementary Series Volume 4). Oxford.

Finlayson, B., *et al.* 2011. Architecture, Sedentism, and Social Complexity at Pre-Pottery Neolithic A WF16, Southern Jordan. *Proceedings of the National Academy of Sciences* 108(20): 8183–8188.

Frank, F. 1934. Aus der Araba I: Reiseberichte. *Zeitschrift des Deutschen Palastina-Vereins* 57: 191–280.

Freeman-Grenville, G.S.P., Chapman III, R.L. and Taylor, J.E., eds. 2003. *Palestine in the Fourth Century A.D.: The Onomasticon by Eusebius of Caesarea with Jerome's Latin Translation and Expansion in Parallel from the Edition of E. Klostermann*. Jerusalem.

Garfinkel, Y., Streit, K., Ganor, S. and Hasel, M.G. 2012. State Formation in Judah: Biblical Tradition, Modern Historical Theories, and Radiometric Dates at Khirbet Qeiyafa. *Radiocarbon* 54: 359–369.

Glueck, N. 1935. *Explorations in Eastern Palestine, II* (Annual of the American Schools of Oriental Research 15).

Glueck, N. 1940. *The Other Side of the Jordan*. New Haven.

Grattan, J., *et al.* 2005. Analyses of Patterns of Copper and Lead Mineralization in Human Skeletons Excavated from an Ancient Mining and Smelting Centre in the Jordanian Desert: A Reconnaissance Study. *Mineralogical Magazine* 69: 653–666.

Grattan, J., *et al.* 2002. 'Death... More Desirable than Life'? The Human Skeletal Record and Toxicological Implications of Ancient Copper Mining and Smelting in Wadi Faynan, Southwestern Jordan. *Toxicology and Industrial Health* 18(6): 297–307.

Grattan, J. P., Gilbertson, D.D. and Hunt, C.O. 2007. The Local and Global Dimensions of Metalliferous Pollution Derived from a Reconstruction of an Eight Thousand Year Record of Copper Smelting and Mining at a Desert-Mountain Frontier in Southern Jordan. *Journal of Archaeological Science* 34(1): 83–110.

Hauptmann, A. 1987. The Earliest Periods of Copper Metallurgy in Feinan, Jordan. In: Hauptmann, A., Pernick, E. and Wagner, G.A., eds. *Old World Archaeometallurgy: Proceedings of the International Symposium "Old World Metallurgy," Heidelberg 1987*. Bochum.

Hauptmann, A. 2007. *The Archaeo-Metallurgy of Copper: Evidence from Faynan, Jordan*. New York.

Higham, T., *et al.* 2005. Radiocarbon Dating of the Khirbat-en Nahas Site (Jordan) and Bayesian Modeling of the Results. In: Levy, T.E. and Higham, T., eds. *The Bible and Radiocarbon Dating—Archaeology, Text and Science*. London: 164–178.

Hunt, C., David, O., Gilbertson, D. and El-Rishi, H.A. 2007. An 8000-Year History of Landscape, Climate, and Copper Exploitation in the Middle East: The Wadi Faynan and the Wadi Dana National Reserve in Southern Jordan. *Journal of Archaeological Science* 34(8): 1306–1338.

Kind, H.D. 1966. Antike Kupfergewinnung zwischen Rotem und Totem Meer. *Zeitschrift des Deutschen Palästina-Vereins* 81: 56–73.

Kitchener, H.H. 1884. Major Kitchener's Report. *Palestine Exploration Fund, Quarterly Statement* 16(4): 202–21.

Knabb, K.A., *et al.* 2016. Environmental Impacts of Ancient Copper Mining and Metallurgy: Multi-proxy Investigation of Human-landscape Dynamics in the Faynan Valley, Southern Jordan. *Journal of Archaeological Science* 74: 85–101.

Leroi-Gourhan, A. 1943. *Evolution et techniques: l'homme et la matière*. Paris.

Levy, T.E., *et al.* 2004. Reassessing the Chronology of Biblical Edom: New Excavations and 14C Dates from Khirbat en-Nahas (Jordan). *Antiquity* 78: 865–879.

Levy, T.E., Ben-Yosef, E. and Najjar, M. 2012. New Perspectives on Iron Age Copper Production and Society in the Faynan Region, Jordan. In: Kassianidou, V. and Papasavvas, G., eds. *Eastern Mediterranean Metallurgy and Metalwork in the 2nd Millennium BC*. Oxford: 197–214.

Levy, T.E. and Higham, T. 2005a. Radiocarbon Dating and the Iron Age of the Southern Levant: Problems and Potentials for the Oxford Conference. In: Levy, T.E. and Higham T., eds. *The Bible and Radiocarbon Dating: Archaeology, Text and Science*. London: 3–14.

Levy, T.E. and Higham, T., eds. 2005b. *The Bible and Radiocarbon Dating - Archaeology, Text and Science*. London.

Levy, T.E. and Higham, T. 2005c. Preface. In: Levy, T.E. and Higham, T., eds. *The Bible and Radiocarbon Dating: Archaeology, Text and Science*. London.

Levy, T.E., *et al.* 2008. High-Precision Radiocarbon Dating and Historical Biblical Archaeology in Southern Jordan. *Proceedings of the National Academy of Sciences* 105(43): 16460–16465.

Levy, T.E. and Najjar, M. 2006. Some Thoughts on Khirbat en-Nahas, Edom, Biblical History and Anthropology—A Response to Israel Finkelstein. *Tel Aviv* 33: 107–122.

Levy, T.E., Najjar, M. and Ben-Yosef, E., eds. 2014. *New Insights into the Iron Age Archaeology of Edom, Southern Jordan—Surveys, Excavations and Research from the Edom Lowlands Regional Archaeology Project (ELRAP).* Los Angeles.

Levy, T.E, Najjar, M. and Higham, T. 2005. *How Many Fortresses Do You Need to Write a Preliminary Report? Or Response to Edom and the Early Iron Age: Review of a Recent Publication in Antiquity*. Manchester.

Levy, T.E., *et al.* 2014. Excavations at Khirbat en-Nahas 2002–2009: Unearthing an Iron Age Copper Production Center in the Lowlands of Edom (Southern Jordan). In: Levy, T.E., Najjar, M. and Ben-Yosef, E., eds. *New Insights into the Iron Age Archaeology of Edom, Southern Jordan: Surveys, Excavations and Research from the Edom Lowlands Regional Archaeology Project (ELRAP).* Los Angeles.

Levy, T.E., *et al.* 2005. Lowland Edom and the High and Low Chronologies: Edomite State Formation, the Bible and Recent Archaeological Research in Southern Jordan. In: Levy, T.E. and Higham, T., eds. *The Bible and Radiocarbon Dating—Archaeology, Text and Science*. London: 129–163.

MacDonald, B. and Khairieh, A. 1992. *The Southern Ghors and Northeast 'Arabah Archaeological Survey 1985–1986, Southern Jordan* (Sheffield Archaeological Monographs 5). Sheffield.

Mattingly, D.J. 2011. *Imperialism, Power, and Identity: Experiencing the Roman Empire*. Princeton.

Mazar, A. 2005. The Debate over the Chronology of the Iron Age in the Southern Levant: Its History, the Current Situation, and a Suggested Resolution. In: Levy, T.E. and Higham, T., eds. *The Bible and Radiocarbon Dating—Archaeology, Text and Science*. London: 15–30.

Mazar, A. 2011. The Iron Age Chronoloical Debate: Is the Gap Narrowing? Another Viewpoint. *Near Eastern Archaeology* 74(2): 105–111.

Mazar, A., Bruins, H.J., Panitz-Cohen, N. and Van der Plicht, J. 2005. Ladder of Time at Tel Rehov: Stratigraphy, Archaeological Context, Pottery and Radiocarbon Dates. In: Levy, T.E. and Higham, T., eds. *The Bible and Radiocarbon Dating: Archaeology, Text and Science*. London: 193–255.

Mazar, A. and Bronk-Ramsey, C. 2010. A Response to Finkelstein and Piasetzky's Criticism and "New Perspective". *Radiocarbon* 52: 1681–1688.

Musil, A. 1907. *Arabia Petraea. I. Moab; II. Edom: Topographischer Reisebericht*. Vienna.

Najjar, M. and Levy, T.E. 2011. Condemned to the Mines. *Biblical Archaeology Review* 37(6): 30–39, 71.

Porter, B.W. 2004. Authority, Polity, and Tenuous Elites in Iron Age Edom (Jordan). *Oxford Journal of Archaeology* 23(4): 373–395.

Pratico, G.D. 1993a. *Nelson Glueck's 1938–1940 Excavations at Tell el-Kheleifeh: A Reappraisal* (American Schools of Oriental Research Archaeological Reports 3). Atlanta.

Pratico, G.D. 1993b. Kheleifeh, Tell-El. In: Stern, E., ed. *The Encyclopedia of Archaeological Excavations in the Holy Land*. Jerusalem.

Roberts, B.W. and Thornton, C. 2014. *Archaeometallurgy in Global Perspective: Methods and Syntheses*. New York.

Robinson, E. 1867. *Biblical Researches in Palestine and Adjacent Regions* (3rd ed. 3 vols., 1970). London.

Simmons, A.H. and Najjar, M. 2006. Ghwair I: A Small, Complex Neolithic Community in Southern Jordan. *Journal of Field Archaeology* 31(1): 77–95.

Smith, N.G. and Levy, T.E. 2008. The Iron Age Pottery from Khirbat en-Nahas, Jordan: A Preliminary Study. *Bulletin of the American Schools of Oriental Research* 352: 1–51.

Twiss, K.C. 2007. The Zooarchaeology of Tel Tifdan (Wadi Fidan 001), Southern Jordan. *Paléoorient* 33(2): 127–145.

CHAPTER 17

KEY FEATURES FOR DEDUCING TECHNOLOGICAL INNOVATIONS AND ORGANIZATIONAL STRUCTURES IN THE BRONZE AGE MINING DISTRICT OF FAYNAN, JORDAN

Ingolf Löffler

The production and use of raw materials have been related to land use regulation and the ecologic, economic and social stages of each period. A study of the economic structures of the Faynan mining district—which takes into account production and manufacturing sites as well as commodities produced—must evaluate ancient technology in terms of standardization, centralization and scale. These in turn are directly related to overarching questions concerning political, social and cultural aspects in the procurement and use of raw materials and can be useful tools for relating raw materials and their products to larger, primary questions concerning the structure and organization of the Bronze Age mining district.

The first metallurgical phase in the Chalcolithic and EB I periods are evidenced only to a very limited extent at Faynan itself. In contrast, the younger phase of the developed Early Bronze Age shows a fundamental change of development in settlement pattern and metallurgy in terms of centralization, standardization and scale. The available evidence suggests on one hand craft specialization and on the other a complex social organization of mining, trade and distribution of copper.

INTRODUCTION

Research on the development of historical and technical innovations in mining and the associated social aspects in the Faynan district is based on archaeological investigation and archaeometallurgical studies in copper metallurgy (Hauptmann 2007). These two analytical research avenues answer questions concerning the factors influencing the development and innovation of extractive metallurgy at the Faynan mining district in context with the economy, spatial distribution and settlement progression in more than single utilization phases. The settlement pattern and archaeometallurgical installations in the Faynan district are unique because the ancient mining sites managed to survive and were not impacted by more recent mining activities. The metallurgical remains include furnace fragments, crucible fragments, casting molds, prills and lumps of copper, ingots, copper tools, slag, ores and mining tools.

The copper district of Faynan is located 80 km south of the Dead Sea (Fig. 17.1), halfway between the Dead Sea and the Red Sea, west of the Jordanian Plateau. The ancient ore district (Fig. 17.2) is one of the largest copper producers in the southern Levant. Its long-lasting exploitation has spanned from the Pre-Pottery Neolithic A (Finlayson and Mithen 2007) until the Islamic period.

The exploitation of the ore district took place in various stages and especially during the late Chalcolithic–Early Bronze Age (mid-/end of fourth and third millennium BCE), the early Iron Age (ca. 1200–800 BCE) and Nabatean-Roman times (100 BCE–400 CE).

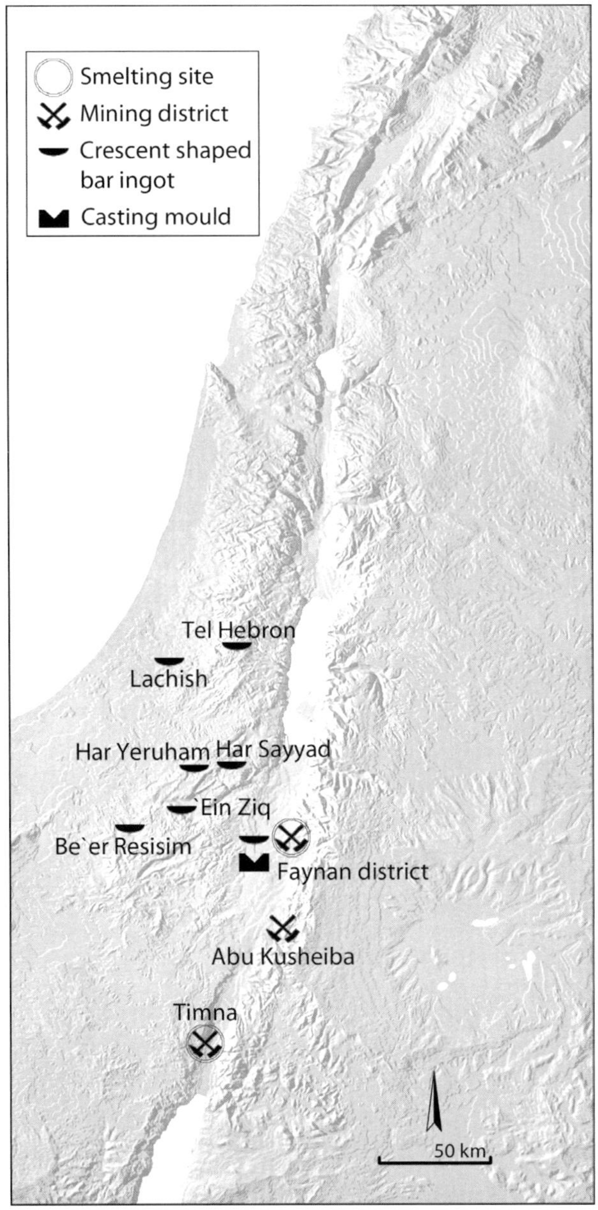

Fig. 17.1: Distribution map of crescent-shaped bar ingots and mining areas in the southern Levant.

The fundamental investigations for this article were undertaken by the Deutsches Bergbau-Museum Bochum between 1983–1993 under the administration of Prof. Dr. A. Hauptmann in collaboration with Prof. Dr. G. Weisgerber and other universities and institutions (Adams and Genz 1995; Levy 1995; Engel and Frey 1996; Weisgerber 1996, 2002, 2006; Hauptmann 2000, 2007; Levy *et al.* 2002, 2005; Barker, Gilbertson and Mattingly 2007; Ben-Yosef 2010; Ben-Yosef, Levy and Najjar 2009; Ben-Yosef *et al.* 2010; Ben-Yosef *et al.* 2016).

This study presents an overview of the changing technological innovations of mineral exploitation from the Chalcolithic through the EBA IV by looking at settlement and mining activities as well as the production of copper in relation to standardization, centralization and scale.

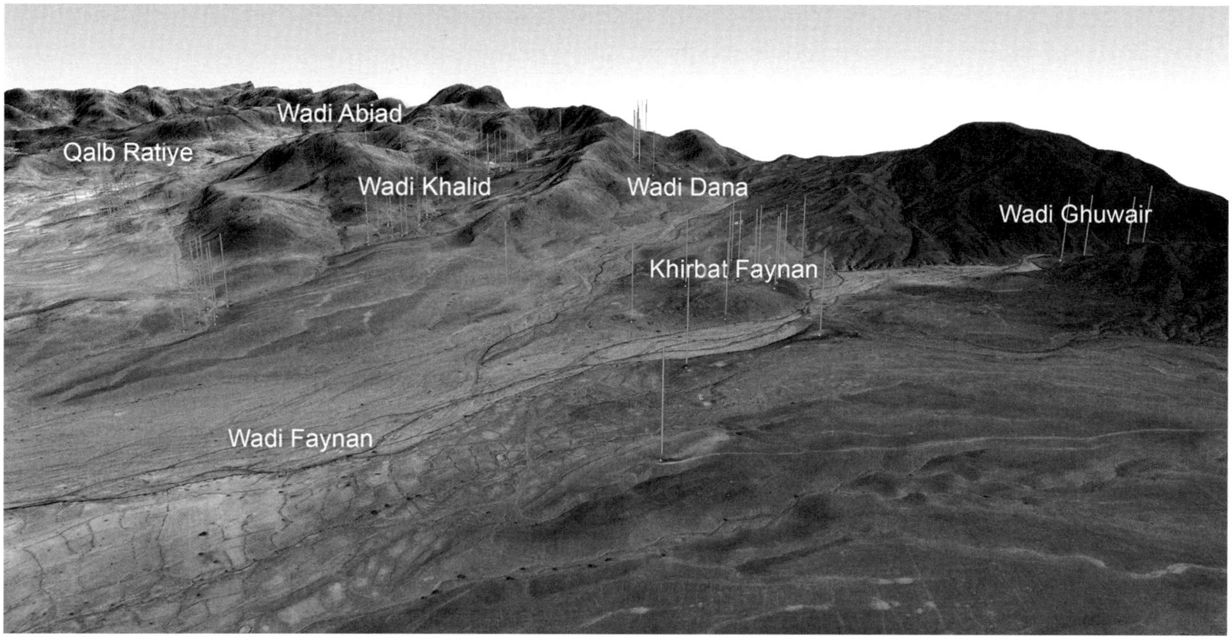

Fig. 17.2: 3D perspective overview of the Byzantine ruins at Kirbet Faynan (center) and of the mining and smelting sites in its vicinity (light gray lines) (Hauptmann 2007: Fig. 5.4a).

THE CHALCOLITHIC/EB I METALLURGY

In the southern Levant, Chalcolithic metallurgy was connected to the emergence of the first chiefdom organizations which seem to have been controlled by social elites (Levy 1995; Levy *et al.* 2002: 427). They lived in small nucleated villages, situated in a two-tier settlement hierarchy of numerous small villages and larger regional village centers. With the regression of Chalcolithic society during the early fourth millennium BCE, the common pattern of nucleated villages continued into the EB I (3600–3300 BCE) period. In such a small village community the earliest evidence for small scale metallurgy at Faynan (Hauptmann *et al.* 1992) comes from the site of Wadi Fidan 4 (3400/3500 BCE) (Adams and Genz 1995; Adams 2002: 22–23). The village community of Wadi Fidan 4 was only partly specialized in the production of copper by extractive metallurgical processes.

The majority of the crucible-based productions of copper and copper beads originated from several copper mineral deposits from the Faynan region (Hauptmann 2007: 158–163). Just a few kilograms of crushed slag, ash, copper ores, a small number of ca. 30 crucible fragments and copper prills were discovered mainly in a limited corner of the settlement (Hauptmann *et al.* 1993).

The copper and slags were produced under "primitive" conditions (Craddock, Meeks and Timberlake 2007) and of low scale. No casting molds were excavated at Wadi Fidan 4 that might prove the production of copper ingots or final products (Hauptmann 2007; Hauptmann and Löffler 2013). The evidence of stone mining hammers and raw material for stone tools and other finds reveals that the occupants were also involved in mining, trading of ores and copper production activities as a supplement to general subsistence activities (Adams 2002: 23).

At the well-structured EB I settlement complex Wadi Faynan 100, several ores, crucible fragments, casting molds and crushed slags were found as well (Hauptmann 2007: 110). The range of production is unknown because of the limited excavations.

The focus at the settlement was on craft as well as food production. The latter included cereal cultivation, tree-crop agriculture and animal husbandry as can be demonstrated by the evidence of field systems, and agricultural and pastoral sites (Barker, Gilbertson and Mattingly 2007: 236–270).

During the Chalcolithic/EB I periods, over 100 small mines were dug. Following the outcrops of mineralizations, these small mines were opened in the solid rock of the Massive-Braun-Sandstone (MBS, Fig. 17.3) with a lateral enlargement of only a few meters. The mining constructions are simple pits, with an initial development to room-and-pillar constructions (Weisgerber 2006: 5). The mining was performed right at the ore deposit, but the main activities of smelting and casting were carried out in villages elsewhere.

The mined ores were also probably transported by pastoral nomads (Anfinset 2004) to regional centers of cultural and regional gravity, such as the Beer-sheba Valley and the Aqaba region (Hauptmann 2000, 2007). Tell Abu Matar, in the Beer-sheba Valley, is one of the major centers for copper smelting and production of artifacts dating to the late fifth and fourth millennium BCE. It is located some 100 km north of Faynan and produced—in contrast to Wadi Fidan 4—tons of copper slag which constitutes evidence for extensive metal production (Shugar 2000) using ores from the Faynan copper district (Hauptmann 1989).

The main characteristics of this period are the concentration of mining in Qalb Ratiye, Wadi Abiad, in the upper part of the Wadi Khalid, the centralization of settlement and development of craft—as well as food-production activities.

THE CHANGE IN METALLURGY AT FAYNAN DURING THE EB II

The decreasing number of Levantine settlements in EB II (Milevski 2009: 133) resulted in the centralization of goods distribution, which influenced developments in the Faynan mining district. The population growth and a reduction in the actual numbers of sites suggest that larger sites began to

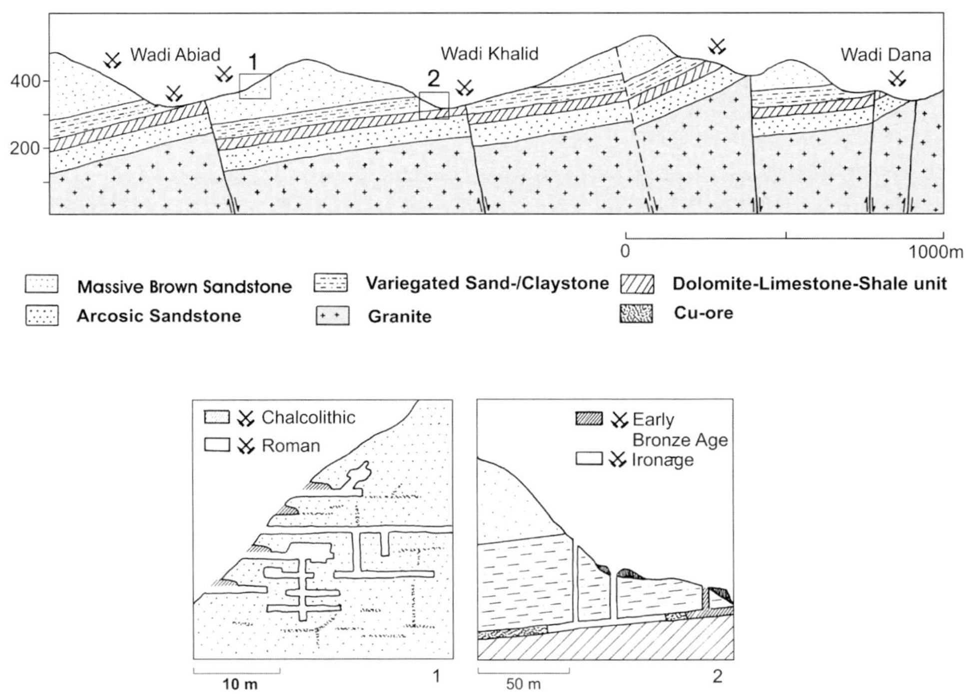

Fig. 17.3: Schematic illustration of the mining development in relation to the different geological sections (after Hauptmann 2007: Fig. 5.46).

play an increasingly important role (Adams 2002: 23). Walled towns were built during this period and societal changes took place, with increased centralization and control, as well as the development of large-scale building projects (Amiran and Gophna 1989).

These external political forces seem to have played a major role in altering and enlarging the small-scale metal production which was practiced in the Chalcolithic/EB I periods at Faynan to a more organized operation (Dever 1995).

During this time the first evidence for major changes in centralization, scale and standardization of mining and copper production can be observed (Hauptmann and Rohden 1988; Hauptmann *et al.* 1992). Adits were dug in the higher and terraced formations of Wadi Dana and Wadi Khalid. There are 66 documented mines in Wadi Khalid (Weisgerber 2006); this suggests a dramatic increase in the scale of mining activities.

The second type of copper mineralization in the Dolomite-Limestone-Shale (DLS, Fig. 17.3) was exploited for the first time in underground mines; the ores mined in the fourth millennium were totally ignored.

Besides the fact that the new ore deposits (which were much more difficult to mine) were more plentiful (Hauptmann 2000, 2007), the main reason for their exploitation may have been their higher quality due to lead impurities that helped in the casting of high-end copper ingots like those from Khirbat Hamra Ifdan (KHI) (Fig. 17.4). In addition, the new copper with higher lead concentration is easier to cast in contrast to other copper sources with fewer trace elements.

Another development in the smelting process was smelting ores in closer proximity to the mines. This was associated with moving from small volume crucible smelting to larger standardized smelting furnaces. These furnaces were large-scale operations with multiple installations constructed and rebuilt close to one another, positioned on the windward side of the high hills to use wind power for smelting (Hauptmann 2007: 104–108).

In order to organize the labor and resources needed, a degree of control centralization as well as manipulation of the labor base would have been essential. The transition from small-scale specialization in the Chalcolithic and EB I to this highly-sophisticated and extensive production process that took place in Faynan (Adams 2002: 24) was based on a growing demand and a developed superior organizational structure.

Fig. 17.4: Reconstructions of the casting mold and crescent-shaped bar ingots (length 22 cm). This type of ingot was traded during the EB III/IV (photo I. Löffler, Deutsches Bergbau-Museum).

At this time a new relationship evolved between the town centers and their hinterlands, which supported the concept of urbanized and rural space (Joffe 1993: 73). The combined evidence from the increased production specializations and technology developments suggests changing consumption patterns and the production of articles for trade (Adams 2002: 23).

THE COPPER PRODUCTION AT FAYNAN DURING THE EB III–IV

The production level of metal increased dramatically in the middle of the third millennium BCE in order to answer a growing demand for copper in the Eastern Mediterranean and beyond (Hauptmann 2007; Hauptmann and Löffler 2013). Similar activities are observed in Anatolia, Central Asia, the Iranian Plateau, the Mediterranean Basin and Central Europe (Amzallag 2009).

Analyses of ores found in the settlements of Barqua el-Hetiye and KHI (Hauptmann 2000, 2007), and at the surrounding 13 Early Bronze Age smelting sites, as well as the chemical composition of Early Bronze Age slags, clearly indicate that the raw ore source at the Faynan mining area had changed at the beginning of the third millennium BCE. While copper ores from the MBS were smelted in the fourth millennium BCE, mixed copper-manganese ores from the DLS were utilized exclusively to produce metal in EB II–IV. The center of the DLS mines was in the lower course of Wadi Khalid and Wadi Dana (Fig. 17.2) in the eastern part of the Faynan copper district, 1–2 km away from the major Early Bronze Age slag field at the Khirbe of Faynan (Hauptmann 2000, 2007).

One thousand tons of slag are located in the area around the Khirbe of Faynan, as well as in the lower Khalid, Dana and Ghuwair Wadis. An additional ca. 500 tons were discovered at 13 contemporary EB II–IV smelting sites (Hauptmann 2007). At these Early Bronze Age smelting sites an extraordinary feature can be observed that offers not only a glimpse into the metal smelting organization but also into the organization of the mining activities.

All the Early Bronze Age smelting sites in the Faynan mining area are positioned on slope edges and hills that are exposed to strong blowing seasonal winds from the Jordanian Plateau as well as from the Nahal Arabah. These smelting sites have been identified as the earliest smelting furnaces known so far (Hauptmann 2000, 2007, Hauptmann and Löffler 2013). Weisgerber (2006: 12–13) argues that is why the furnaces are so widely distributed on suitable summits, and that uncentralized production in the hands of several producers can be assumed. In any case smelting did not happen in one centralized place but was rather concentrated on the slope edges and several hilltops close to the different mining areas.

DEVELOPMENT OF SETTLEMENTS IN RELATION TO COPPER PRODUCTION

The first stage of craft separation can be seen at the site of Wadi Fidan 4 (3400/3500 BCE). The settlement was surrounded by smaller satellite sites. Explicit investigations of its settlement architecture indicate that copper production at the site was relatively small scale (Adams 2002; Barker, Gilbertson and Mattingly 2007: 269). The features demonstrate that part-time metalworkers carried out their activities in restricted zones of what was otherwise a domestic agricultural village.

Few settlements in the mining area reveal significant data on this key period with respect to relations between Faynan and the western/northwestern Levant and beyond during the later Early Bronze Age (ca. 3000–2000 BCE). The first is the Wadi Faynan 100 EB I settlement complex (3600–3000 BCE). The magnitude of the site is much greater than that of Wadi Fidan 4. The densest surface material covers an area of ca. 11 ha in connection with a substantial boundary wall and a cluster of house-and-courtyard complexes (Wright *et al.* 1998; Barker, Gilbertson and Mattingly 2007: 268). The focus at the settlement was on craft activities as well as on food-production. The latter included cereal cultivation, tree-crop agriculture and animal husbandry, as demonstrated by the evidence of field systems, agricultural and pastoral sites (Barker, Gilbertson and Mattingly 2007: 236–270). The key question that arises is who organized the complex arable, pastoral and metallurgical activities? The implication of all these facts is that there were individuals or groups at Wadi Faynan 100 who controlled both the metal-extraction process as well as other self-sufficient activities (*ibid.*: 268).

The settlement of Barqa el-Hetiye is slightly younger (dating to the beginning of the third millennium, 3080–2875 calibrated BCE, i.e., EB II), and is comprised of only a few rooms (Fritz 1994; Adams 2003). This settlement may also have played a role as a local trading center strategically positioned near trade routes crossing the Nahal Arabah rift valley (Adams 2003). The craftsmen here were involved in metal production and processing (Adams 2000; Levy *et al.* 2002). Similar ingot mold forms at Barqa el-Hetiye and KHI show a possible continuity of trade and production from EB II to III. Only further excavation can verify this hypothesis.

Khirbat Hamra Ifdan (*ibid.*; Hauptmann 2007) is located at a perennial water source and was probably a checkpoint of great strategic importance. Its main occupation phase is the EBIII/IV (Ben-Yosef *et al.* 2016). The excavations there revealed a slice of a solid structure containing important quantities of copper production debris. This key settlement represents the largest metal workshop of this period in the entire Near and Middle East. It consists of 80 rooms, in which more than 600 crucible fragments, more than 800 fragments of casting molds, copper lumps and prills, slag, ingots and final objects were found.

The archaeometallurgical evidence from this site suggests that the trend of specialization and an increased scale of production continued until the end of the main phase of the Early Bronze Age and further into EB IV (Adams 2002: 24; Hauptmann 2007: 240; Ben-Yosef *et al.* 2016).

The corresponding evidence for the final production of copper objects and ingots from this site suggests that this was a highly organized and specialized operation for the exclusive beneficiation of copper which was prepared for export in the shape of both ingots and finished tools (Hauptmann 2007: 242). This evidence for such an industrialized production scale has led to the suggestion that this site be termed a manufactory installation (Adams 1999, 2002; Levy *et al.* 2002).

The excavation of ingots and casting molds at Barqua el-Hetiye (EB II) and KHI (EB III–IV) confirms that these ingots originated in the Faynan district (Adams 2002: 25; Hauptmann 2007). The chemical composition and lead isotope ratios of the KHI ingots and the other analyzed Levantine ingots (Fig. 17.1) are identical to those from the DLS ore and the raw copper production in Faynan (*ibid.*).

In contrast to the technological developments during EB IV (ca. 2500–2000 BCE, Regev *et al.* 2012), this period is characterized by a process of de-urbanization and social dissolution, and only a few centers (which were founded in the EB II) continued to exist. As a result, the EB IV has been defined in past research as a period of ruralism (Dever 1995: 295).

DISCUSSION

The questions regarding the archaeological evidence are: Who coordinated the enormous scale of copper production? Was there a central administration that controlled the production and organized the trade?

More than 100 crescent-shaped bar ingots of the type found at Barqua el-Hetiye and KHI (Levy *et al.* 2002; Hauptmann *et al.* 2015) were discovered in many later Early Bronze Age Levantine sites. They were found in EB IV settlements (Tufnell 1958; Gophna 1992) in the southern Levant and the Dead Sea area, e.g., Beʾer Resisim, En Ziq (Segal, Halicz and Cohen 1999), Har Yeruham and Hebron (Dever and Tadmor 1976).

The manner in which these Levantine sites were dated to the EB IV, the shallowness of the stratigraphy and the fact that most of these finds come from "ingot hoards," have placed the exact dating of these ingots in doubt; some have even suggested that they should be redated to the EB III (Adams 2002: 25). However, the evidence for contact between the Faynan area and western Palestine is not only demonstrated by the metallurgical finds, but also by the ceramic similarities between KHI and the sites of ʿEn Ziq and Beʾer Resisim (Goren 1996; Adams 1999). This evidence from these sites seems to support direct links between the Faynan area and the Negev.

In addition to the standardization of smelting, melting and casting that is evident at the metallurgical center of KHI, there was also a standardization of ingots. These standardized ingots are easy to cut and transport and can be used as a means of payment.

The standardized, crescent-shaped copper ingots (Dever and Teadmore 1976) were established as a "trade mark" in quality and form (Fig. 17.5), in contrast to other contemporaneous ingots and casting molds (Müller-Karpe 1994: 131–143; Weisgerber and Yule 2003) like the associated finds of pin casting and axe casting molds of the KHI-workshop. The demand for different types of standardized ingots indicates different weight systems. The ingot standardization is the best indicator of a central administration and control of the copper production and trading systems.

Fig. 17.5: The polished section of a crescent-shape bar ingot found in the settlement of Khirbat Hamra Ifdan, Stratum III (EB III). The object shows a low porosity and is dotted by only tiny flaws. Gritted inclusions of copper sulfides are visible. The section shows a cooling rim following the incision of the casting mold. The copper contains no slag inclusions, is low in As and contains ca. 1 wt.-% Pb. In comparison with Late Bronze Age oxhide ingots from the Uluburun shipwreck, the KHI copper is of much higher quality (Hauptmann, Maddin and Prange 2002; Hauptmann 2007: 242, Fig. 7.16).

CONCLUSION

The development of the structured agro-pastoral system in the EB I not only created the impetus for copper production at Wadi Faynan 100, but was responsible for the development of the entire mining area as well as for large-scale and multi-phased copper manufacture and production. For more detailed information on subsistence strategies we need to further investigate food-production activities of the agro-pastoral system in EB II–III.

The first metallurgical phase in the Chalcolithic and EB I periods are evidenced only to a very limited extent at Faynan itself. However, provenance studies (utilizing analytical measurements of trace elements and lead-isotopic relationships) of metal artifacts from other sites have demonstrated that they originated in the Faynan district. In contrast, the younger phase of the developed Early Bronze Age shows a fundamental developmental change in settlement pattern and metallurgy (Hauptmann 2000, 2007; Hauptmann and Löffler 2013) in terms of centralization, standardization and scale. Barker, Gilbertson and Mattingly (2007: 269), however, argue that the structured agro-pastoral system continued into the EB II (3000–2700 BC) on a reduced scale.

The technological and cultural advancements in the Faynan mining area demonstrate the centralization of mining and smelting activities at suitable places in relation to settlement development, infrastructure and natural resources. The evidence for centralization, standardization and large-scale copper production point on the one hand to craft specialization and on the other to a complex social organization of copper mining, trade and distribution.

It is not clear whether southern Levantine fortified urban centers directly controlled mining and metal production at Faynan. The EB II–III urban centers may have controlled production and distribution of the produced metals, or they may have only supervised the distribution. Similarly, they may have controlled the transportation and distribution routes alongside an indirect command over the copper production (Milevski 2009: 135).

The Early Bronze Age south Levantine urban centers indeed denote a certain concentration of commodities. Administration of copper exchanges would have been economically profitable. Unfortunately, there are no administrative records explaining the mechanisms responsible for the circulation of these commodities (Milevski 2009: 139), and we have no information regarding the EB I elites or rulers and their administrations in the EB II–III in these centers (Milevski 2009: 136–137).

Looking at the long time-span and intermittent exploitation of the Faynan mining district, a holistic approach is necessary in order to understand regional and superior adjustments. Knowledge of the organization of technological and manual processes, as well as of the social environment involved in mining and metalworking, are the basis for any attempt to understand and construe the interaction between metallurgy and cultural historical evolutions.

REFERENCES

Adams, R. 1999. *The Development of Copper Metallurgy during the Early Bronze Age of the Southern Levant: Evidence from the Feinan Region, Southern Jordan* (Ph.D. dissertation, University of Sheffield). Sheffield.

Adams, R. 2000. The EBA III-IV Transition in Southern Jordan: Evidence from Khirbet Hamra Ifdan. In: Philip, G. and Baird, D., eds. *Breaking with the Past: Ceramics and Change in the Early Bronze Age of the Southern Levant* (Levantine Archaeology 2). Sheffield: 379–401.

Adams, R. 2002. From Farms to Factories: The Development of Copper Production at Faynan, Southern Jordan, during the Early Bronze Age. In: Ottaway, B.S. and Wagner, E.C., eds. *Metals and Society* (British Archaeological Reports, International Series). Oxford: 21–32.

Adams, R. 2003. External Influences at Faynan during the Early Bronze Age: A Re-Analysis of Building I at Barqa el-Hetiye, Jordan. *Palestine Exploration Quarterly* 135, 1: 6–21.

Adams, R. and Genz, H. 1995. Excavations at Wadi Fidan 4: A Chalcolithic Village Complex in the Copper Ore District of Feinan, Southern Jordan. *Palestine Exploration Quarterly* 127: 8–20.

Amiran, R. and Gophna, R. 1989. Urban Canaan in the EB II and EB III Periods. In: de Miroschedji, P., ed. *L'urbanisation de la Palestine à l'âge du bronze ancien. Bilan et perspectives des recherches actuelles* (British Archaeological Reports International Series 527). Oxford: 109–116.

Amzallag, N. 2009. From Metallurgy to Bronze Age Civilizations: The Synthetic Theory. *American Journal of Archaeology* 113: 479–519.

Anfinset, N. 2004. Metallurgy and Pastoral Nomads in the Chalcolithic Period of the Southern Levant. In: Oestigaard, T., Anfinset, N. and Saetersdal, T., eds. *Combining the Past and the Present* (BAR International Series 1210). Oxford: 61–73.

Barker, G.W., Gilbertson, D.D. and Mattingly, D.J., eds. 2007. *Archaeology and Desertification: The Wadi Faynan Landscape Survey, Jordan*. Oxford.

Ben-Yosef, E. 2010. *Technology and Social Processes: Oscillations in Iron Age Copper Production and Power in Southern Jordan* (Ph.D. dissertation, University of California). San Diego.

Ben-Yosef, E. *et al*. 2016. Early Bronze Age Copper Production Systems in the Northern Arabah Valley: New Insights from Archaeomagnetic Study of Slag Deposits in Jordan and Israel. *Journal of Archaeological Science* 72: 71–84.

Ben-Yosef, E., Levy, T.E., Higham, T. and Tauxe, L. 2010. The Beginning of Iron Age Copper Production in the Southern Levant: New Evidence from Khirbat al-Jariya, Faynan, Jordan. *Antiquity* 84: 724–746.

Ben-Yosef, E., Levy, T.E. and Najjar, M. 2009. New Iron Age Copper-Mine Fields Discovered in Southern Jordan. *Near Eastern Archaeology* 72(2): 98–101.

Craddock, P.T., Meeks, N. and Timberlake, S. 2007. On the Edge of Success: the Scientific Examination of the Products of the Early Mines Research Group Smelting Experiments. In: LaNiece, S., Hooks, D. and Craddock, P., eds. *Metals and Mines* (Studies in Archaeometallurgy). London: 37–45.

Dever, W. 1995. Social Structure in the Early Bronze IV Period in Palestine. In: Levy, T.E., ed. *The Archaeology of Society in the Holy Land. Facts on File*. New York: 282–296.

Dever, W.G. and Tadmor, M. 1976. A Copper Hoard of the Middle Bronze Age I. *Israel Exploration Journal* 26: 163–169.

Engel, Th. and Frey, W. 1996. Fuel Resources for Copper Smelting in Antiquity in Selected Woodlands in the Edom Highlands to the Wadi Araba/Jordan. *Flora* 191: 29–39.

Finlayson, B. and Mithen, S., eds. 2007. *The Early Prehistory of Wadi Faynan, Southern Jordan: Archaeological Survey of Wadis Faynan, Ghuwayr and al-Bustan and Evaluation of the Pre-Pottery Neolithic A Site of WF16*. Oxford.

Fritz, V. 1994. Vorbericht über die Grabungen in Barqa el-Hetiye im Gebiet von Fenan, Wadi el-Araba (Jordanien) 1990. *Zeitschrift des Deutschen Palästina-Vereins* 110(2): 125–150.

Gophna, R. 1992. The Intermediate Bronze Age. In: Ben-Tor, A., ed. *The Archaeology of Ancient Israel*. New Haven/Tel Aviv: 126–158.

Goren, Y. 1996. The Southern Levant in the Bronze Age IV: The Petrographic Perspective. *Bulletin of the American Schools of Oriental Research* 303: 33–72.

Hauptmann, A. 1989. The Earliest Periods of Copper Metallurgy in Feinan/Jordan. In: Hauptmann, A., Pernicka, E. and Wagner, G.A., eds. *Archäometallurgie der Alten Welt/Old World Archaeometallurgy* (Der Anschnitt, Beiheft 7). Bochum: 119–135.

Hauptmann, A. 2000. *Zur frühen Metallurgie des Kupfers in Fenan/Jordanien* (Der Anschnitt, Beiheft 11). Bochum.

Hauptmann, A. 2007. Early Metallurgy of Copper. Evidence from Faynan, Jordan. In: Wagner, G.A. and Herrmann, B., eds. *Natural Science in Archaeology*. Heidelberg/Berlin/New York.

Hauptmann, A., Begemann, F., Heitkemper, E., Pernicka, E. and Schmitt-Strecker, S. 1992. Early Copper Produced at Feinan, Wadi Araba, Jordan: The Composition of Ores and Copper. *Archeomaterials* 6(1): 1–33.

Hauptmann, A. and Löffler, I. 2013. Technological Innovations and Organisational Structures of Prehistoric Mining and Metal Production–Examples from Faynan, Jordan. In: Burmeister, St., Hansen, Sv., Kunst, M. and Müller-Scheeßel, N., eds. *Metal Matters. Innovative Technologies and Social Change in Prehistory and Antiquity. Menschen–Kulturen–Traditionen.* Rahden/Westfalen.

Hauptmann, A., Maddin, R. and Prange, M. 2002. On the Texture and Composition of Copper and Tin Ingots Excavated from the Shipwreck of Uluburun. *Bulletin of the American Schools of Oriental Research* 328: 1–30.

Hauptmann, A., Pernicka, E., Lutz, J. and Yalcin, Ü. 1993. Zur Technologie der frühesten Verhüttung von Kupfererzen im östlichen Mittelmeerraum. In: Frangipane, M., Hauptmann, H., Liverani, M., Matthiae, P. and Mellink, M., eds. *Between the Rivers and over the Mountains: Archaeologica Anatolica et Mesopotamica. Alba Palmieri Dedicata*. Rome: 541–572.

Hauptmann, A. and Rohden, C. 1988. Archäometallurgische Untersuchungen zur Kupferverhüttung der frühen Bronzezeit in Fenan, Wadi Arabah, Jordanien. In: *Archäometallurgie von Kupfer und Eisen* (Jahrbuch des Römisch-Germanischen Zentralmuseums Mainz 35). Mainz: 510–516.

Hauptmann, A., Schmitt-Strecker, S., Levy, T.E. and Begemann, F. 2015. On Early Bronze Age Copper Bar Ingots from the Southern Levant. *Bulletin of the American Schools of Oriental Research* 373: 1–24.

Joffe, A.H. 1993. *Settlement and Society in the Early Bronze Age I and II, Southern Levant: Complementarity and Contradiction in a Small-Scale Complex Society* (Monographs in Mediterranean Archaeology 4). Sheffield.

Levy, T.E. 1995. Cult, Metallurgy and Rank Societies—Chalcolithic Period (ca. 4500–3500 BCE). In: Levy, T.E., ed. *The Archaeology of Society in the Holy Land.* New York: 226–243.

Levy, T.E., *et al.* 2002. Early Bronze Age Metallurgy: A Newly Discovered Copper Manufactory in Southern Jordan. *Antiquity* 76: 425–437.

Levy, T.E., *et al.* 2005. Iron Age Burial in the Lowlands of Edom: The 2004 Excavations at Wadi Fidan 40, Jordan. *Annual of the Department of Antiquities Jordan* 49: 443–487.

Milevski, I. 2009. Local Exchange in the Southern Levant during the Early Bronze Age: A Political Economy Viewpoint. *Antiguo Oriente: Cuadernos del Centro de Estudios de Historia del Antiguo Oriente* 7: 125–159.

Müller-Karpe, A. 1994. Anatolisches Metallhandwerk. *Offa-Bücher*, Bd. 75. Neumünster.

Regev, J., de Miroschedji, P., Greenberg, R., Braun, E., Greenhut, Z. and Boaretto, E. 2012. Chronology of the Early Bronze Age in the southern Levant: new analysis for a high chronology. *Radiocarbon* 54 (3–4): 525–566.

Segal, I., Halicz, L. and Cohen, R. 1999. A Study of Ingots and Metallurgical Remains from `Ein Ziq and Beer Resisim, Central Negev, Israel. In: Young, S.M.M., Pollard, A.M., Budd, P. and Ixer, R.A., eds. *Metals in Antiquity* (BAR International Series 792). Oxford: 179–186.

Shugar, A. 2000. *Archaeometallurgical Investigation of the Chalcolithic Site of Abu Matar, Israel: A Re-Assessment of Technology and its Implications for the Ghassulian Culture* (Ph.D. dissertation, University of London). London.

Tufnell, O. 1958. *Lachish IV (Tell Ed-Duweir), The Bronze Age*. Oxford.

Weisgerber, G. 1996. Montanarchäologie—Mehr als Technikgeschichte: Das Beispiel Fenan (Jordanien). In: Kroker, W., ed., *Montantechnologie an historischen Schnittstellen* (Bochum: Die Technikgeschichte als Vorbild moderner Technik 20). Bochum: 19–34.

Weisgerber, G. 2002. Spatial Organisation of Mining and Smelting at Feinan, Jordan: Mining Archaeology beyond the History of Technology. In: Craddock, P.T. and Lang, J., eds. *Mining and Metal Production through the Ages.* London: 76–89.

Weisgerber, G. 2006. The Mineral Wealth of Ancient Arabia and its Use. Copper Mining and Smelting at Feinan and Timna–Comparison and Evaluation of Techniques, Production, and Strategies. *Arabian Archaeology and Epigraphy* 17: 1–30.

Weisgerber, G. and Yule, P. 2003. Al-Aqir near Bahlā'—An Early Bronze Age Dam site with Planoconvex 'Copper' Ingots. *Arabian Archaeology and Epigraphy* 14: 24–53.

Wright, K., *et al.* 1998. The Wadi Faynan Fourth and Third Millennia Project, 1997: Report on the First Season of Test Excavations at Wadi Faynan 100. *Levant* 30: 33–60.

CHAPTER 18

COPPER TRADE AND THE SETTLEMENT RISE IN THE SOUTH LEVANTINE DESERTS IN THE EB IV

Moti Haiman

About 1500 EB IV sites have been surveyed in the Negev and Sinai Deserts. These sites can be divided into two groups. The first group is made up of about seven large sites, such as Beʾer Resisim, Mashabe Sade and ʿEn Ziq, each containing up to 250 structures, located near water sources and defined as permanent settlements. Excavations have demonstrated that copper processing and trade constituted the main economic activity. This is based on the abundance of copper ingots and an unusual amount of stone tools, which were probably used for copper processing. At these sites no evidence has been found of subsistence activity, such as sickle blades or animal pens. The second group consists of small, poorly built sites located far from water sources, leading to the conclusion that they were temporary. The enormous number of animal pens found at these sites indicates that pastoralism was their dominant economic activity. Sickle blades found at several sites implies that the population was also engaged in seasonal agriculture. In addition, rich findings of cairns, stone piles and stone lines—evidence of traditional desert population burials and cult features—have also been found. The settlement rise in the deserts of the southern Levant is a result of circumstances that enabled Asiatics to transport copper to Egypt. The permanent settlements reflect copper-specialized groups transporting standard ingots from the Faynan copper mines to Egypt. Following the establishment of those sites there was a rise in semi-nomadic activity throughout the area south of the permanent sites. The settlement pattern reflects a complex system consisting of copper specialists and semi-nomads. However, the orchestrators of the system remain enigmatic. Based on the examination of several cultural characteristics both at the permanent and the temporary sites, the possibility to locate the system's orchestrators in one of the Arabian Desert oases should not be excluded.

INTRODUCTION

The south Levantine deserts include the Sinai Peninsula and the Negev, spanning an area of 70,000 sq km, and constitute part of the Arabian Desert (Fig. 18.1). The central Negev and Sinai, the focus of the present article, are hyper arid zones, where the environmental conditions are extremely harsh. These arid desert regions have few water sources, and an annual rainfall of less than 100 mm—below the minimum necessary for grain growth. These poor conditions allow for the sustenance of small herds only—and thus cannot serve as the sole source of income of even a small family. Only in the peripheral areas of that desert are there minimal conditions for permanent settlement: the hill area of biblical Edom north of Petra, the Beer-sheba Valley and the northern and southern Sinai. Despite the harsh environmental conditions, this region became the domain of the ca. 1500 EB IV sites known to date (Haiman 2009: 38).

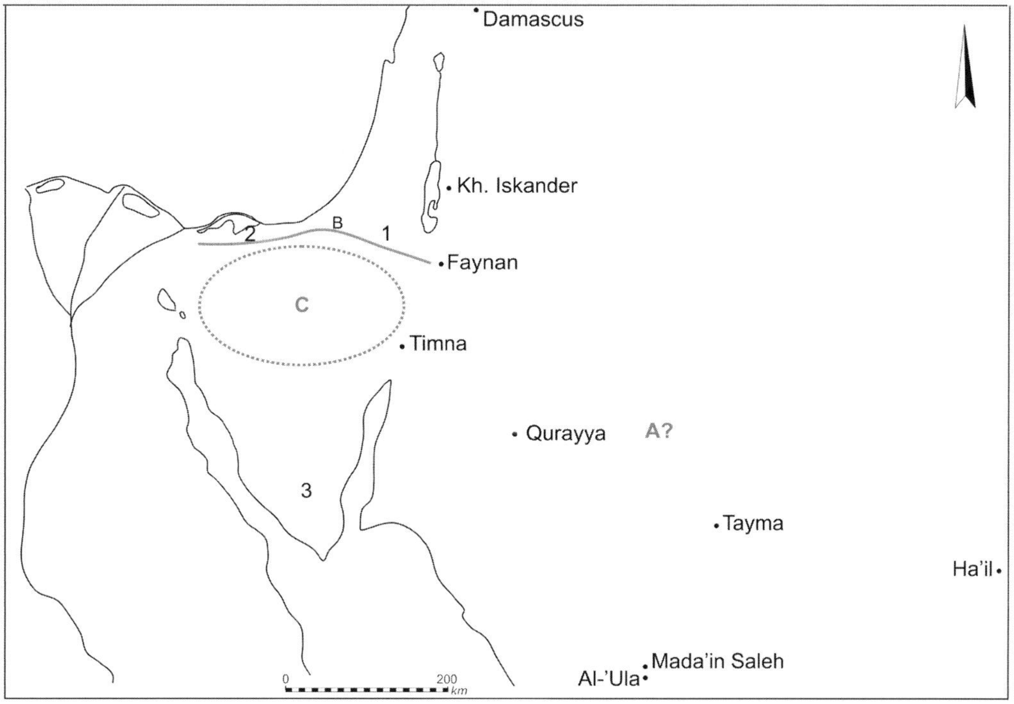

Fig. 18.1: The EB IV settlement pattern in the south Levantine deserts. 1. The Negev Highlands sites, e.g.: Har Yeroham, ʿEn Ziq, Beʾer Mashabim and Beʾer Resisim; 2. northern Sinai; 3. southern Sinai; A. oases of Northern Arabia; B. permanent sites; C. temporary sites.

The goal of this paper is to explain the phenomenon of the emergence of this large-scale settlement in the heart of the desert—despite environmental conditions, and to discuss the possible origin of the population.[1] An analysis of the finds shows two groups of sites with significant differences between them: one group is made up of large sites which are located near water sources and can be considered permanent; and another group which consists of poorly built structures, located far from water sources, and can be considered as temporary nomadic locales.

PERMANENT SITES

Seven large excavated sites (Cohen 1999: 83–298; Dunseth 2013) are situated in the northern part of the Negev Highlands. Several other sites are situated along the northern Sinai coast, some of them—for example, near Bir el-Abed—are quite large (Oren and Yekutieli 1990). All together, the site distribution has a pattern connecting the Negev Highlands toward Egypt (Fig. 18.1: B). The Negev Highlands sites consist of 100–250 high quality structures which were found near natural water sources. While an unusual number of stone implements and copper ingots was found, an indication copper manufacturing and trade, a lack of sickle blades[2] or animal pens indicates that the population living at these sites was not involved in food production.

1 This paper updates two previous articles on the topic (Haiman 1996, 2009).

2 Studies conducted of flint collections from ʿEn Ziq, Har Yeroham and Beʾer Resisim failed to find evidence of a sickle blade industry. The finds at the sites included only a single sickle blade from ʿEn Ziq, found on the surface (by the present author), and four touched blades from Har Yeroham, inconclusively identified as sickle blades (Gilead 1973: 133).

The excavations at the EB IV site of Har Yeroham in the 1960s yielded many stone implements, a copper hoard of 12 ingot bars, copper chips and industrial ashes in all the rooms, especially in Area C, which was identified as a copper industry area (Kochavi 1967). In retrospect, it seems that Kochavi's assumption concerning the multi-source economy of the inhabitants should be revised. In light of the consistent lack of sickle blades and animal pens at the small sites, the only definite economic activity is copper trade and tool processing.

ʿEn Ziq, the largest EB IV site in the Negev, excavated by Cohen (1999: 137–188), further demonstrates the role copper played for that group of sites. Despite its location near a water source, the area does not allow for any agriculture or animal husbandry. About 30 of the 250 structures at the site were excavated; every room yielded an enormous number of stone implements—up to 40 per room: work tables, fine grinding stones and small hammerstones. In addition, dozens of copper ingots and copper chips were found in many of the rooms. The entire site looks like a very busy factory, without, as mentioned, any evidence of other activity. Based on the study of similar finds at Har Yeroham, the stone implements reflect the processing of copper tools from ingots using cold hammering technology (Kochavi 1967: 35–46).

The same is true for the rest of the sites of that group, such as Har Zayyad (Cohen 1999: 94–103) and Beʾer Resisim (Cohen 1999: 200–224) and Mashabe Sade (Dunseth 2013).

The finding of 1000 copper molds in Kh. Hamra Ifdan by Adams,[3] of a size that fits the ingots from the Negev (Adams 2000), leaves no doubt that there was mass production of copper ingots in the Faynan area (Hauptmann *et al.* 2015). Thus, this group of sites in the Negev represents copper specialized groups that penetrated the desert through south Jordan in order to transport the copper, probably to Egypt.

TEMPORARY SITES

Most of the EB IV sites belong to this group (Haiman 2009). The sites are small, consist of a few poorly built structures, and are not in proximity to water sources—all indicating short term habitation. While no evidence of involvement in copper was found, the animal pens, which are lacking at the permanent sites, are prevalent, indicating animal husbandry. Sickle blades found at several sites are indicative of occasional agriculture.

It is difficult to understand the subsistence on animal husbandry and occasional agriculture—the two traditional occupations of the desert nomads —in the specific environment of central Sinai. The floor area of the animal pens in 250 examined sites demonstrates an average of ca. 50 sq m or 25 heads per site—a quarter of the size that allows minimal subsistence. However, this low figure corresponds with herd sizes found in that area in other periods, as well as among present day Bedouins (Baily 1977; Abu Rabia 1994). The same is true for the evidence of agricultural activity found in some sites, since the average annual rainfall of less than 80 mm does not allow for the cultivation of crops.

Another characteristic of this group is the various cult features found near the sites, such as rectangular and round platforms, and one dome-like roofed structure with a "tail" of small stone piles of the kind found near the rectangular platforms.[4]

3 I thank R. B. Adams for the information.

4 In an excavation conducted by Erickson-Gini in one of these rectangular platforms, EB II potsherds were identified. I disagree since the pottery is identical to that of the ʿEn Ziq jars. I thank Erickson-Gini for the information.

THE SETTLEMENT PATTERN AND THE ORIGIN OF THE POPULATION

The pattern suggested here is of a complex system consisting of two desert groups (B and C) and one group (A) which is still enigmatic and is assumed to be located in the sedentary land or in one of the northwest Arabian Desert oases (Fig. 18.1).

1. The existence of A as a theoretical assumption is vital not only to determine the orchestrators of the copper trade system, but also as the source for the food chain as none of the other organs of that system produced food. It is assumed that A is a political entity in charge of copper mining in Faynan, casting standard ingots in commercial quantities with a concentration of copper specialists. The findings at Kh. Iskander include an abundance of stone implements (Richard 1988; Richard and Long 2009), making the site a possible candidate for A. It is also possible to assume that the orchestrator of the system is located in one of the oases of the northwestern Arabian Desert, similar to the later presence of the "Midianites" at Timna (Rothenberg 1988). Petrographic analysis conducted by Goren on the Negev EB IV ceramics has demonstrated that some of the pottery is from the area of Faynan and central Jordan in general (Goren 1996).

   Among the EB IV finds in the permanent sites, there is a group of flat stone figurines or stelae that have been ignored in previous studies. Cohen's publication includes five stelae from four sites (Fig. 18.2; Cohen 1999: 266–267). In addition, more fragments of similar stelae were recorded in the excavation's logs.[5] It is possible to find similarities in their body and neck design (Steimer-Herbert 2010). Based on the Levantine EB IV material culture's "North-South" geographical divide (Prag 2009), it is possible to assume that the entire "South" group originated in the oases of the Arabian Desert.
2. This group represents copper specialized sites distributed along the route from Faynan to Egypt. Since the inhabitants were not involved in food production, the environmental conditions (besides the locations near water sources) are not relevant.
3. The distribution of copper specialized settlements in the desert led to the development of a local nomadic ecosystem on their periphery, attested by the distribution of nomads' sites south of group B, despite its location in the harsh desert away from water sources.

Similar correspondence is prevalent among modern day Bedouin who prefer to locate their settlements near permanent settlements rather than near unoccupied water sources. The harsh desert Bedouin economy is based on close contact with sedentary settlers, working as salaried labor, while the traditional occupations of pastoralism and agriculture are of secondary importance. Therefore, the discovery of animal pens or sickle blades in the EB IV nomad sites do not necessarily demonstrate full subsistence on those occupations.

As mentioned above, one of the characteristics of the temporary sites are common satellite installations which were identified as cultic. A dense accumulation of similar features, as well as others, can be identified east of the sedentary Levant (Kennedy 2011), attesting to the possible origins of that group's population. However, the chronological correlation between these particular installations and the Negev EB IV is not possible (Fig. 18.3).

CONCLUSIONS

The suggested settlement pattern as part of a complex system best explains the rise of the settlements in the harsh desert of central Sinai as a bubble which was not affected by environmental restrictions. The

5 I thank Y. Israel, an area supervisor in Cohen's excavations at ʿEn Ziq and at other EB IV sites in the Negev, for the information.

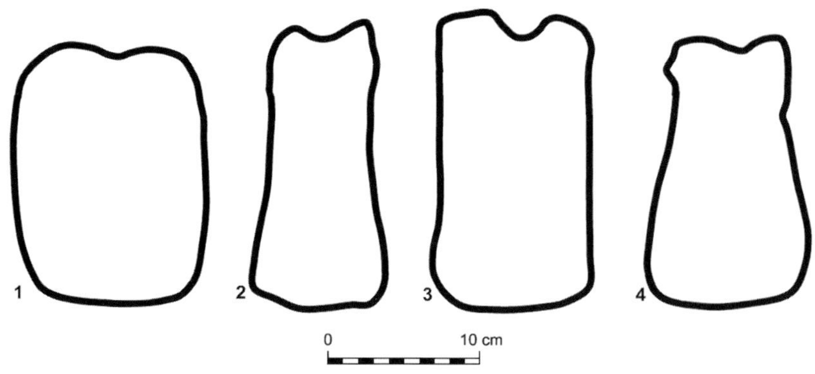

Fig. 18.2: Stelae from excavated EB IV sites in the Negev: 1. Beʾer Resisim (Cohen 1999: 222, Fig. 139A: 10); 2. Har Dimon (Cohen 1999: 93, Fig. 53A: 11); 3. ʿEn Ziq (Cohen 1999: 182, Fig. 112: 24); 4. Mashabei Sade (Cohen 1999: 130, Fig. 80: 1) (courtesy of the Israel Antiquities Authority).

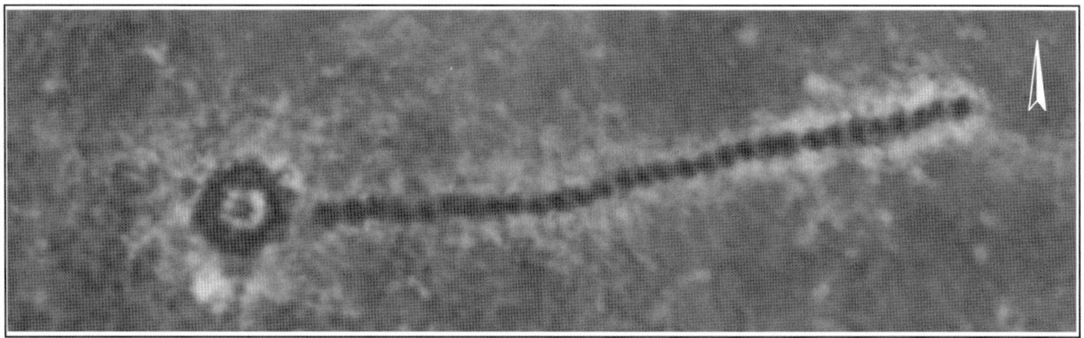

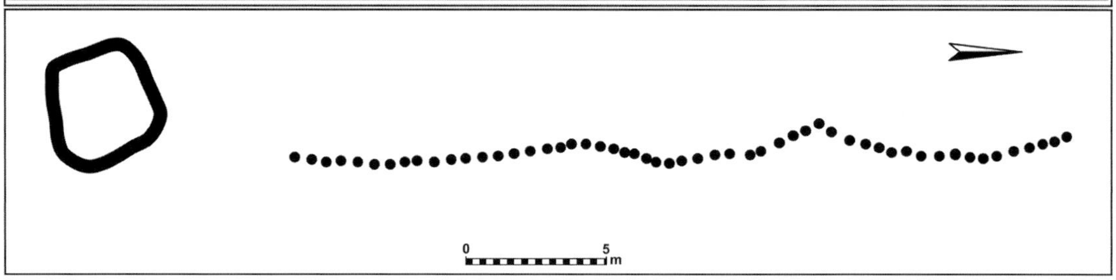

Fig. 18.3: Top: *Google Earth* Satellite image of an installation near Zarqa, Jordan. Bottom: Installation near an EB IV site in the Negev (Haiman 1991: 69–71, Site 156; Courtesy of the Israel Antiquities Authority).

orchestrator of the system, located in the sedentary land, exploited the unique geopolitical circumstances that allowed Asiatics to control the copper transport to Egypt while the desert groups exploited the permanent settlements and relied on the poor environment only to a limited extent. The growth of that system was affected only by the scope of the copper trade. However, it could only have existed as long as geopolitical circumstances allowed.

Throughout its history Egypt was dependent on the importation of copper. During periods of political strength Egypt organized delegations to mine turquoise and copper in the Sinai and the Negev. It seems that the temporary weakness of Egypt during the 1st Intermediate Period (Ward 1971) provided an opportunity for an Asiatic political entity to control the copper transported to Egypt, assisted by Asiatic settlements along the route. The EB IV Asiatic settlements most probably ceased to exist as a result of political changes at the beginning of the Middle Kingdom.[6] The abandonment of the permanent sites left the nomads exposed to the poor environment, where they could not survive and the whole ecosystem collapsed.

REFERENCES

Abu Rabia, A. 1994. *The Negev Bedouin and Livestock Rearing*. Oxford.

Adams, R.B. 2000. The Early Bronze Age III-IV Transition in Southern Jordan: Evidence from Khirbet Hamra Ifdan. In: Philip, G. and Baird, D., eds. *Ceramic and Change in the Early Bronze Age of the South Levant* (Levantine Archaeology 2). Sheffield: 379–401.

Baily, C. 1977. The Bedouins of Sinai. In: Sohar, E., ed. *The Desert*. Tel Aviv: 240–247 (Hebrew).

Cohen, R. 1999. *Ancient Settlement of the Central Negev,* Vol. 1. Jerusalem (Hebrew with English summary).

Dunseth, Z. C. 2013. *Subsistence Practices in the Negev Highlands during the Intermediate Bronze Age: A Microarchaeological Investigation at Mashabe Sade* (M.A. thesis, Tel Aviv University). Tel Aviv.

Gilead, D. 1973. Flint Industry of the Early Bronze Age from Har Yeruham and Tel Nagila. In: Aharoni, Y., ed. *Excavations and Studies* (Publications of the Institute of Archaeology of Tel Aviv University 1). Tel Aviv: 133–141 (Hebrew with English summary).

Goren, Y. 1996. The Southern Levant in the Early Bronze Age IV: The Petrographic Perspective. *Bulletin of the American Schools of Oriental Research* 303: 33–72.

Haiman, M. 1991. *Map of Mizpe Ramon (200)*. Jerusalem.

Haiman, M. 1996. Early Bronze Age IV Settlement Pattern of the Negev and Sinai Deserts: View from Small Marginal Temporary Sites. *Bulletin of The American Schools of Oriental Research* 303: 1–32.

Haiman, M. 2009. Copper Trade and Pastoralism in the Negev and Sinai Deserts in the EBIV. In: Parr, P.J., ed. *The Levant in Transition* (Palestine Exploration Fund Annual 9). London: 38–42.

Hauptmann, A., Schmitt-Strecker, S., Levy, T.E. and Begemann, F. 2015. On Early Bronze Age Copper Bar Ingots from the Southern Levant. *Bulletin of the American Schools of Oriental Research* 373: 1–24.

Kennedy, D. 2011. The "Works of the Old Men" in Arabia: Remote Sensing in Interior Arabia. *Journal of Archaeological Science* 38: 3185–3203.

Kochavi, M. 1967. *The Settlement of the Negev in the Middle Bronze (Canaanite) I Age (*Ph.D. dissertation, The Hebrew University). Jerusalem (Hebrew with English Summary).

Oren, E.D. and Yekutieli, Y. 1990. North Sinai during the MBI Period—Pastoral Nomadism and Sedentary Settlement. *Eretz Israel* 21: 6–22 (Hebrew with English Summary).

Prag, K. 2009. The Late Third Millennium in the Levant: A Reappraisal of the North-South Divide. In: Parr, P.J., ed. *The Levant in Transition* (Palestine Exploration Fund Annual 9). London: 80–89.

Richard, S. 1988. The Early Bronze IV Fortified Site of Khirbet Iskander, Jordan: Third Preliminary Report, 1984 Season. *Bulletin of the American Schools of Oriental Research*, Supplement 25: 107–130.

Richard, S. and Long, J.C. 2009. Khirbet Iskander, Jordan and Early Bronze Age IV Studies: A View from a Tell. In: Parr, P.J., ed. *The Levant in Transition* (Palestine Exploration Fund Annual 9). London: 90–100.

Rothenberg, B. 1988. *The Egyptian Mining Temple at Timna*. London.

Steimer-Herbert, T. 2010. Three Funerary Stelae From the 4th Millennium B.C. In: Al-Ghabban, A.I., Andre-Salvini, B., Demange, F., Juvin C. and Cotty, M., eds. *Roads of Arabia*. Paris: 166–170.

Ward, W.A. 1971. *Egypt and the East Mediterranean World 2200–1900 B.C.* Beirut.

6 It is worth mentioning that the radiocarbon dating from Kh. Iskander and Kh. Hamra Ifdan indicates EB III (Adams 2000), while the pottery from the northern Sinai sites includes "Meidum Pottery" of the Early Middle Kingdom (Oren and Yekutieli 1990).

CHAPTER 19

EXTRACTIVE METALLURGY IN THE CHALCOLITHIC SOUTHERN LEVANT: ASSESSMENT OF COPPER ORES FROM ABU MATAR

Aaron N. Shugar

Copper ores from the Ghassulian Beer-sheba site of Abu Matar were investigated for their mineralogy, micro-chemistry and lead isotope ratios. The results indicate that the majority of the ores excavated on site correspond with the mining site of Faynan. An additional ore type does not correspond with Faynan and offers closer association with ore formations in Anatolia and the Caucasus.

INTRODUCTION

Recent investigations of the elaborate metal artifacts found in the Nahal Mishmar hoard have renewed interest in identifying the potential sites in which they were manufactured. This in turn has revitalized interest in the Ghassulian sites in the Beer-sheba Valley that have demonstrated evidence for metal production. Two main sites which have produced evidence for extractive metallurgical activities are Shiqmim (Levy and Shalev 1989; Golden 2009) and Abu Matar (Golden 1998; Shugar 2001). Several other sites (Arad, Bir es-Safadi, Nevatim, and Neve Noy[1] [Shugar and Gohm 2011]) have produced evidence of crucible use, suggesting that these sites may have been involved in casting metal artifacts. Among all these sites, Abu Matar has provided the largest evidence for extractive metallurgical activities in the region (Fig. 19.1). The extensive range of materials including furnace and crucible fragments, ores and metal artifacts, demonstrates the metallurgical activities that took place on site (Shugar 2001). But with the exotic elemental composition identified with many of the artifacts in the Nahal Mishmar hoard, in conjunction with the recent evidence that the casting clays derived from local soils (Goren 2008), the focus of attention has once again been directed towards the extractive copper production in the region in an attempt to provenance the raw materials used for their construction.

Several other metal artifacts found in the region have shown similar alloy composition to those in the Nahal Mishmar hoard (Namdar *et al.* 2004). There has yet to be evidence of the casting process that reveals slag (dross) or prills that resemble these complex compositions, which may suggest the potential for transport/trade of complex As/Sb metal to the region from afar. Even so, it is likely that the extractive stages of producing complex alloyed bronze were undertaken in the southern Levant, making the transporting/trading of unique polymetallic ore is much more likely as a solid model of similar trade activities already exists with ores traveling from Faynan (Shugar 2001). The best chance to provenance the raw materials used for their production has been to characterize the ores found at local production sites.

---

1 Bir es-Safadi, Nevatim and Neve Noy may be considered extensions of the site of Abu Matar, leaving Arad as the only other Ghassulian sites providing evidence of copper ores.

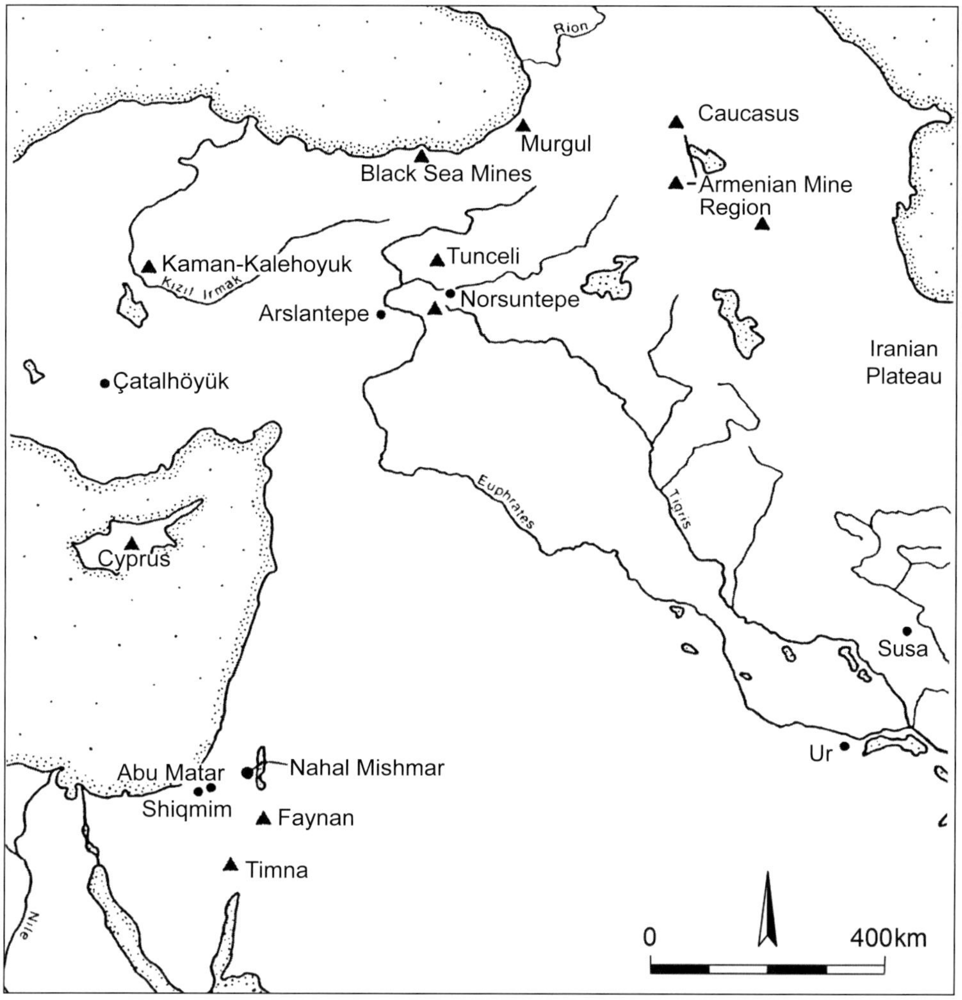

Fig. 19.1: Geographical map of the Near East showing key ore deposits (triangles) and archaeological sites (circles) and archaeological sites mentioned in the text.

Previous analyses of ores found at several sites (Abu Matar–Perrot excavations, Shiqmim and Bir es-Safadi) have shown a strong correlation to the ore formations from Faynan with a minor assemblage possibly coming from Timna (Hauptmann 1989; Rothenberg 1991; Hauptmann *et al.* 1992; Shugar 1998). No evidence for ores from outside the Arabah Valley containing arsenic, nickel or antimony has yet been identified—making a strong case for the transport of ore from the Faynan (and maybe Timna) region alone (Golden, Levy and Hauptmann 2001).

Abu Matar is considered by many scholars to be one of the most important Chalcolithic sites of the Ghassulian-Beer-sheba culture in the southern Levant. The archaeological site of Abu Matar extends over an area of about 19,000 sq m. The excavated portion of the Chalcolithic site of Bir es-Safadi is approximately 8000 sq m (Perrot 1984) and is situated approximately 600 m south-southeast of Abu Matar, directly across Nahal Beer-sheba (Bir es-Safadi encompasses the site of Neve Noy). The sites are not only geographically contiguous, but their occupational sequence and cultural remains are almost identical, especially the fine artistic workmanship seen in ivory carvings, remains from flint knapping centers (Hermon 2008) and evidence of metallurgical activities. With these points in mind it has been suggested that Abu Matar and Bir es-Safadi were actually one large Chalcolithic community that spanned both sides of Nahal Beer-sheba (I. Gilead, personal communication). As the first systematically excavated Chalcolithic site in the region, Abu Matar raised the level of interest in this period and led to the detailed excavations of sites such as Bir es-Safadi (Perrot 1984), Horvat Beter (Dothan 1959), Neve Noy (Eldar and Baumgarten 1985), Gilat (Alon 1977), Grar (Gilead 1989) and Shiqmim (Levy 1987).

Abu Matar has provided the definitive evidence for interpreting the metallurgical activities of the Chalcolithic people in the Beer-sheba Valley. The archaeometallurgical material discovered during Perrot's excavations was extraordinary and has been studied and analyzed numerous times (Tylecote, Rothenberg and Lupu 1974; Merkel 1977; Perrot 1984; Levy 1986; Weisgerber and Hauptmann 1988; Hauptmann 1989; Hauptmann *et al.* 1992; Shalev and Northover 1993; Levy 1995; Golden 1998; Shugar 1998; 2001; Golden, Levy and Hauptmann 2001).

The more recent excavations of Abu Matar conducted by Gilead, Rosen and Fabian (Gilead *et al.* 1992) expanded on Perrot's excavations to establish the areal extent of the Chalcolithic site and to augment the excavations carried out by Perrot 40 years earlier (I. Gilead, personal communication). More recent excavation revealed that Abu Matar must be seen as an extensive metallurgical processing site. Metallurgical processes covered the range from ore processing to smelting and casting (Shugar 2001). In addition, the first evidence for arsenic alloying was discovered among these finds as arsenic rich prills entrapped in slag, suggesting that some level of alloying took place on site. It was suggested that higher concentrations of arsenic alloys could be achieved by separating prills based on color alone (*ibid*). The ores discovered in these later excavations of Abu Matar were investigated in order to determine their mineralogy and to potentially allow us to discover the source of raw materials used to produce the exotic alloys found in the Nahal Mishmar assemblage.

The ores analyzed here comprise 39 samples excavated from three locations at Abu Matar during the 1990/1991 field seasons (Areas A [(10], M [22], and B [7]). In addition, seven samples from the 1950 Perrot excavations of Abu Matar, and 12 samples from the contemporary site of Nevatim were included for comparison. The total assemblage of ore recovered across the site includes a total of 135 ore fragments with a total weight of 1772 g.

METHODOLOGY–INSTRUMENTATION

INTRODUCTION

Investigation started with a description based on visual assessment. Reflected and transmitted optical microscopy was used to identify readily recognizable minerals (as described by Ixer 1990). Samples for reflected light microscopy and electron probe microanalysis (EPMA) were mounted in a two-part epoxy. They were then ground and polished to 1μm. The results established through microscopy and microanalysis were confirmed by X-ray diffraction (XRD). Inductively coupled plasma spectroscopy (ICP-AES) was performed to obtain bulk chemical compositional data. Finally, lead isotope analysis (LIA) was undertaken to provenance the ores.

XRD

A Philips X'Pert-MPD with a PW3040 Console was used for XRD analysis. The system was run with a Mo tube anode with a Beta Filter of Zr. Sollers slit was set at 0.04 radians. The divergence slit was set to 4 degrees and the beam mask and collimator were set to 15 mm. Powdered samples were irradiated from 2° to 52° at a step rate of 0.05° with a count time of 3.0–4.0 sec per step. Peak determination was done using PC-APD software.

EPMA

A JEOL JXA-8600 superprobe was used at 25 kV with a constant current maintained at 5–8 mA. The main crystals used for detection were TAP and LIF. Samples were prepared in the same way as the samples for optical microscopy. In addition, the samples were carbon coated with a standard carbon sputter coater to reduce charging. Elemental quantification was achieved using pure element standards.

ICP

An ICP-AES Jobin-Yvon polychromator (JY-48) was used for bulk chemical analysis.[2] The samples were crushed and powdered to 200 microns using the same method as for the XRD and LIA samples. The samples were then mixed in a zirconium crucible with sodium peroxide in a ratio of 0.5 g of sample per 2 g of Na_2O_2. They were then fused for 20–30 minutes in a furnace at 530 °C. Then they were taken up in aqueous solution with 20 ml of HNO_3 and up to 100 ml of distilled water along with 5 ppm of Sc, which was used as an internal standard. Blanks were also obtained from the Na_2O_2 matrix (Segal, Kloner and Brenner 1994).

LIA

LIA was performed using a thermal ionizing mass spectrometer with a magnetic sector mass spectrometer (TIMS) at the University of Oxford RLAHA Isotrace Laboratories.[3] The precision of the instrumentation is within 0.1% (Gale 1995). Method of sample preparation followed Arden and Gale (1974).

RESULTS

Identification of Mineralization

The ores found at the Chalcolithic sites of Abu Matar and Nevatim are clearly complex copper ores and can be classified into five groups based on gross visual assessment, optical microscopy and mineralization present. They contain relatively high, but varying proportions of copper (oxides/carbonates), copper sulfides and iron (hydr)oxides. Minerals identified definitively through optical microscopy are malachite, quartz, hematite, chalcocite, and occasionally covellite, azurite and chrysocolla.

Some of the ores are, as expected, clearly comparable to the ore formations at Faynan (Hauptmann 1991). Observed similarities are indicated by the tectonic influenced textures, seen in cracking of quartz grains and the shifted and twisted banding found in some of the copper-sulfide mineralization, matching known samples from the Faynan region (Weisgerber and Hauptmann 1988).

Five Ore Types Identified:

The "Type 1" group consists mainly of secondary copper mineral including malachite, atacamite and occasionally azurite in a sandstone matrix (Fig. 19.2a). These ores have high levels of silica and low overall copper content. The grains of quartz are irregular in shape and are densely packed with intergrowth of the copper ore. Occasionally these ores have small inclusions of cuprite or hematite. Chrysocolla was identified in only one ore specimen as a small vein intersecting a quartz grain. Microanalysis confirmed the minerals identified under optical microscopy and XRD. Sample M-O-12 is representative of the assemblage having a very high silica content (78%) and low copper content (14.5%)[4] (Table 19.1). Small

2 The ICP is located at the Israel Geological Service in Jerusalem. The analysis was done by Dr. I. Segal.

3 Samples prepared by the author and run by Prof. N. Gale and Dr. S. Stos-Gale.

4 Scans of large areas can provide a general idea of the bulk composition of a sample but should not be considered a quantitative method.

Zr-rich inclusions were also found in this sample. The results show that this ore type is a good quality ore in its own right (Merkel 1985) but when compared to the other ores in the Abu Matar assemblage, it is considered of relatively poorer quality.

The "Type 2" group is "Tile ore," which takes its name from its characteristic brick red color which is due to the presence of cuprite and iron oxides, most often limonite and hematite (Weisgerber and Hauptmann, 1988) (Fig. 19.2b). On some occasions the ore is brecciated. The ore is usually intermixed with some sulfides and chalcocite is readily apparent. Secondary oxide mineralization often forms on the surfaces of these ores. Isolated islands of barium sulphate have also been identified in some of the ores. The microanalysis of this ore type shows a matrix rich in iron and copper oxides, consisting of intermixed cuprite and hematite with isolated banding layers of chalcocite (Cu_2S) (Table 19.2). Sample M-O-01 (Fig. 19.3) is a good example of the chalcocite inclusions in the tile ore. A line-scan across the banding shows the relationship between copper and sulfur and the increased iron levels in the general matrix.

"Type 3" ores consist of copper-sulfides consisting mainly of chalcocite (Cu_2S) and covellite (CuS) with an iron (hydr)oxide and copper-oxide-rich matrix. These sulfide-rich copper ores are dark, lustrous and relatively dense (Fig. 19.2c). They consist of chalcocite bands which have intergrowths of copper ore and iron ore (cuprite and some hematite) which sometimes contain thin malachite streaks. Covellite is also commonly seen in these samples. Secondary oxide mineralization is not often associated with these ores and the silica content is low.

Microanalysis confirmed the presence of a high ratio of copper sulfides with a proportionally lesser amount of an iron-rich copper ore (Fig. 19.4). This can be seen in sample M-O-15 which contains large amounts of chalcocite and covellite with limited intergrowth of iron-rich copper ore (Fe_2O_3 57% and Cu 16%) (see Table 19.3).

TABLE 19.1: GENERAL SCAN* AND SPOT ANALYSIS RESULTS FROM SEM-EDS (IN WT.%) ANALYSIS OF SAMPLE M-O-12

Sample		*Cu*	*S*	*Fe_2O_3*	*SiO_2*	*MgO*	*Al_2O_3*	*Sn*
M-O-12	General Scan	14.5	0.18	0.186	78.02	0.58	1.74	0.2

* General scans reveal a high silica content and relatively low copper content when compared to the other ore "types" found at Abu Matar.

TABLE 19.2: SEM-EDS SPOT ANALYSIS (IN WT.%) OF CHALCOCITE INCLUSIONS AND THE GENERAL MATRIX OF SAMPLE M-O-01 BASED ON THE LINESCAN TAKEN (FIG. 19.3)

Sample		*Cu*	*S*	*Fe*	*Fe_2O_3*
M-O-01	Chalcocite	73.49	20.29	0.17	0.24
	Matrix	33.77	0.28	17.59	24.63

TABLE 19.3: SEM-EDS SPOT ANALYSIS (IN WT.%) OF CHALCOCITE AND LIMITED INTERGROWTH OF IRON-RICH COPPER ORE TAKEN FROM SAMPLE M-O-15

Sample	*Spot*		*Cu*	*S*	*Fe*	*Fe_2O_3*
M-O-15	1	Chalcocite	72.59	19.88	0.14	0.19
	2	Matrix	15.97	0.08	40.9	57.3

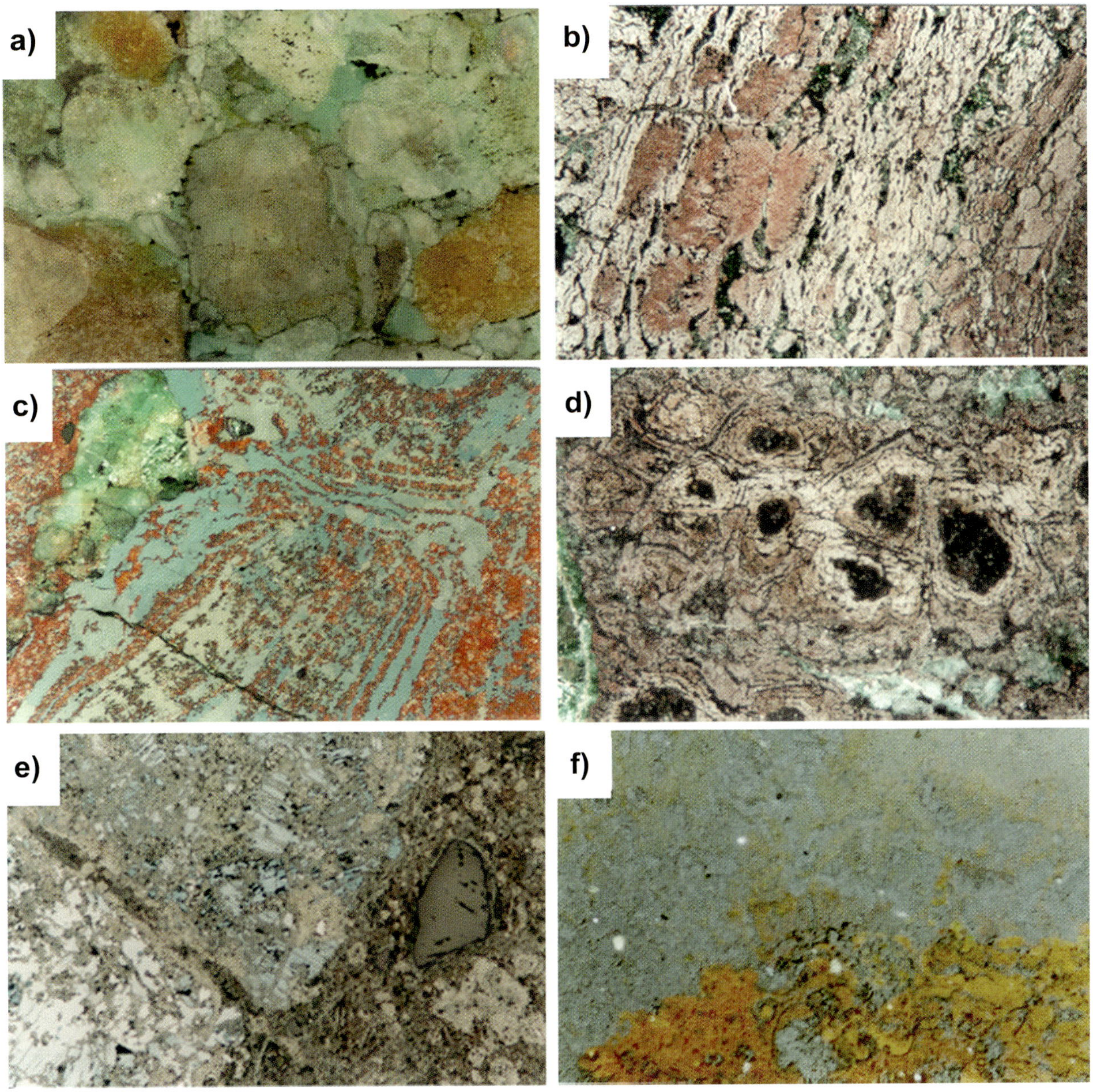

Fig. 19.2: (a) Type 1 ore–Sample M-O-14 (under cross polars width = 2.2 mm). High quartz content with multi-coloration from green malachite to blue azurite. Likely associated with the lower BS layer ores from Faynan where the quartz is being replaced by copper silicates; (b) Type 2 ore–Sample A-O-09 (width = 10 mm). Banded layers of chalcocite (white) and iron oxides (reddish color) with occasional inclusions of malachite (green) are visible under binocular microscopy; (c) Type 3 ore–Sample M-O-15 under cross polars (width = 2.1 mm). Similar in structure to Type 2 ores but only in isolated areas. There are increased proportions of copper sulfides, including chalcocite and covellite (alternate colors of blue). The reddish color is an iron-rich copper ore. Small areas of malachite are present as well (green); (d) Type 4 ore –Sample M-O-3–chalcocite nodules surrounded by heamatite and malachite; (e) Type 4 ore–Sample A-O-03 under plain polarized light (width = 0.36 mm). Intense formation of covellite (indigo blue), chalcocite (white), hematite (light gray), and malachite (dark gray); (f) Type 5 ore– Sample A-O-10 under plain polarized light (width = 2.1 mm). The ore is composed mainly of hematite. The black spots visible in the upper area were confirmed by SEM to be TiO_2.

"Type 4" ores consist of nodular copper-sulfide ores of chalcocite and covellite intermixed with a relatively high level of hematite and silica. Although this ore type is similar to the previous two groups, it exhibits combined characteristics. Banded layers of copper sulfide are apparent but they occur around nodule chalcocite formations (Fig. 19.2d). These nodules are surrounded by a relatively rich matrix of iron oxides with silica grains. There are also areas of highly concentrated covellite surrounded by hematite

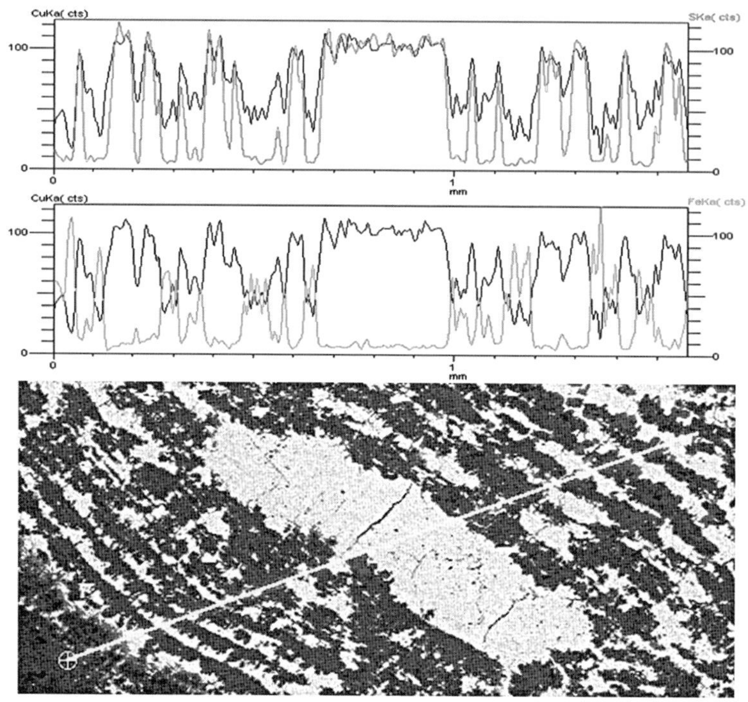

Fig. 19.3: Sample M-O-01. Linescan across the banding of a chalcocite (Cu_2S) inclusion with an iron-rich copper matrix in a Type 2 ore (tile ore). The sulfur positively correlates to the copper while the iron negatively correlates to both copper and sulfur.

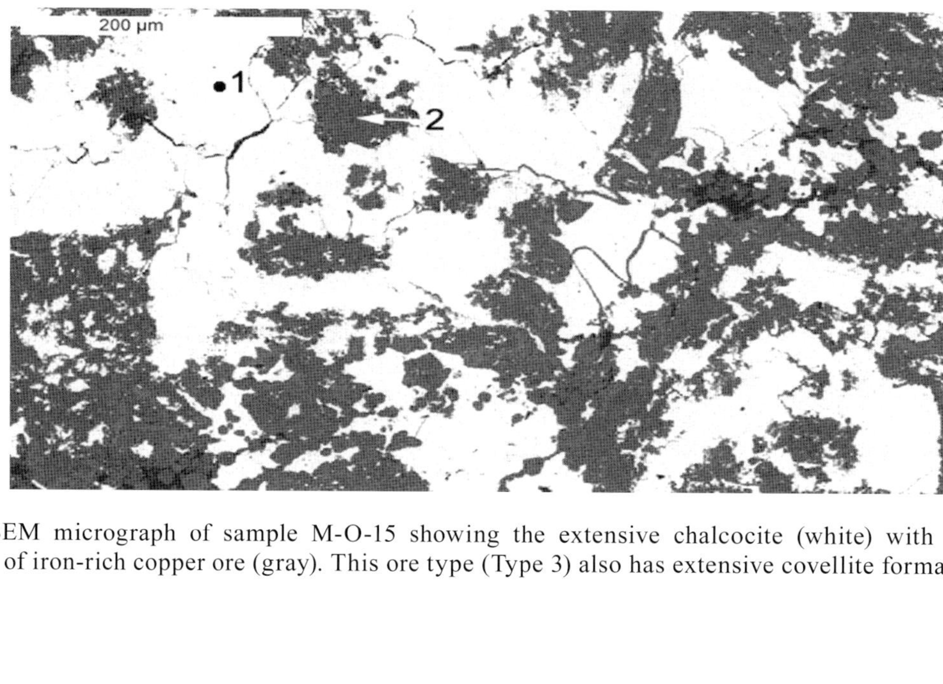

Fig. 19.4: SEM micrograph of sample M-O-15 showing the extensive chalcocite (white) with more limited intergrowth of iron-rich copper ore (gray). This ore type (Type 3) also has extensive covellite formations.

and malachite (Fig. 19.2e). Although Type 4 ores have mineralization similar to Types 2 and 3 ores, they do not resemble the collection of ore types found in the Faynan region. Microanalysis identified various minerals including chalcocite, hematite, malachite, and complex copper, lead and vanadium aggregates. The presence of vanadium in lead is not uncommon (T. Rehren, personal communication), but it is not a common feature of copper ores and it has not been reported in either of the Faynan ore deposits. The nodular formations visible under optical microscopy were confirmed to be chalcocite (Fig. 19.2d). The complexity of these ores may be best seen in sample M-O-06, which has various mineral aggregates within the ore fragment analyzed (Fig. 19.5 and Table 19.4).

"Type 5" ores are fluxes consisting of iron ore (hematite, and ferroan dolomite). Whether fluxes were used by Ghassulian metalsmiths is a controversial subject, with arguments both for and against (Rothenberg, Tylecote and Boydell 1978; Merkel 1985; Hauptmann 1989, 1991; Hauptmann *et al.* 1992; Golden, Levy and Hauptmann 2001; Shugar 2001; Golden 2009). Nevertheless, the samples discovered here are considered to be fluxes based on the assemblage of "Type 1" ores present on the site, which would need a flux to be properly smelted, and represent a new aspect of Chalcolithic copper smelting further suggesting the possible deliberate and controlled use of flux. Two samples were found alongside the copper ore. One was a large hematite nodule (A-O-10) (Fig. 19.2f) and the other appeared to be a gossan type material of ferroan dolomite [Ca (Mg, Fe) $(CO_3)_2$] and was confirmed so through XRD analysis. Optical microscopy of the nodule of hematite revealed small dark inclusions which were confirmed by SEM analysis to be TiO_2.

Table 19.4: SEM-EDS Spot Analysis (in wt.%) of Ore Sample M-O-06. Spots are Indicated in Fig. 19.5 (n/d Indicates Not Detected)

Sample	*Spot*		*Cu*	*S*	*Fe*	*Fe_2O_3*	*Pb*	*V*
M-O-06	1	Complex	14	0.33	6.68	9.35	49.57	9.78
	2	Complex	24.06	1.17	2.92	4.09	43.73	9.11
	3	Hematite	0.53	n/d	67.59	94.63	n/d	n/d
	4	Chalcocite	76.79	19.91	0.29	0.4	n/d	n/d
	5	Matrix	13.8	n/d	46.21	64.69	n/d	n/d

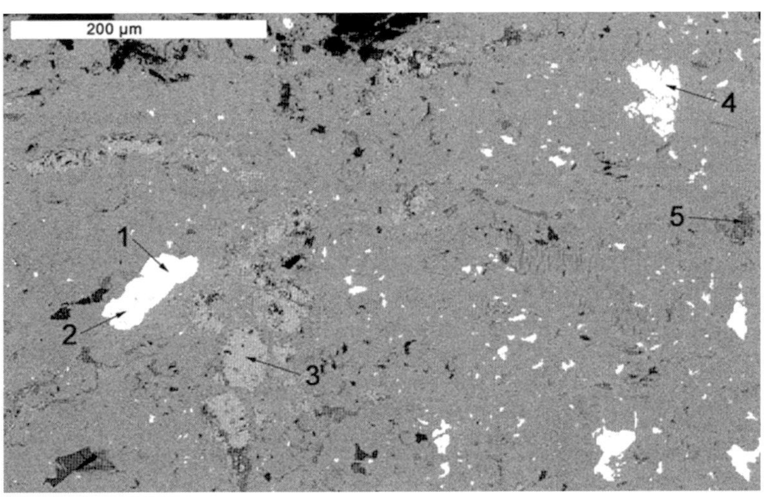

Fig. 19.5: SEM micrograph of ore sample M-O-06 showing multiple mineral formations. Data for spot analysis are presented in Table 19.4.

X-ray Diffraction

X-ray diffraction was used to determine/confirm the ore minerals observed through optical microscopy. A total of 13 samples were analyzed by XRD. All the samples had complex spectra which were difficult to interpret. Only the major peaks where discernible and identified, meaning that minor mineral content was ignored. The main difficulty in interpreting the spectra arises from the overlap of lines for different minerals. For example, djurleite ($Cu_{31}S_{16}$) is basically a form of copper-sulfide and it has almost identical peaks as chalcocite and its presence here may be questioned. The characterizing feature is alteration in peak height, which was difficult to determine here with extensive peak overlap. In addition, some minerals apparently identified, such as aresenopyrite in sample M-O-06, had to be discounted based on their low intensity and their peak overlap with other minerals.

Four main minerals were readily identified throughout most samples, quartz, malachite, chalcolite, and hematite. Additional minerals identified include cuprite, tenorite, chalcopyrite, pyrite, djurleite, iron sulfide, and covellite (see Table 19.5 for details).

Bulk Compositional Analysis

Twenty-two samples were selected for ICP analysis. Selection was based primarily on the amount of sample available to powder for analysis. Six samples come from Area A, 11 samples from Area M, and five samples are from Area B. All samples are copper ore except for sample A-O-10 which is an iron-rich ore flux (hematite). The separation of the Abu Matar ores into five groups based on their optical properties was confirmed through bulk chemical compositional analysis which showed distinct differences between the five "Type" groupings based on optical properties (Table 19.8).

The high silica low quality copper ore (Type 1) was found to have an average of 68.5% SiO_2 with only 17.4% Cu. This group had low iron levels (2.8%) and low sulfur levels (1%), but did have relatively high amounts of Al_2O_3 (2.7%), which is directly related to the silica. This group of copper ores is of good quality in itself (Merkel 1985) but contains the lowest average of copper from all five groups discussed here. The "tile ore" (Type 2) had a lower silica content (11.6%) but very high levels of Fe_2O_3 (27%). The copper content of this ore was much higher than in the silica-rich ores (38.9%) while the sulfur levels were also higher (4.2%). This ore appears to be very good quality copper ore. The sulfide-rich copper ores (Type 3) are clearly distinguished by high amounts of sulfur (12.7%) in addition to the highest copper content of the assemblage (53.4%). The silica levels are extremely low (4.3%) while iron levels are medium (11.3%). Although the high sulfur content might impede smelting without first roasting the ore, this group of ores is a high quality copper ore. The nodular sulfide-rich copper ores (Type 4) have relatively high silica content (23.9%). The iron and sulfur levels are both fairly low at 9.5% and 3.6% respectively. The copper level averages 33.8%. Only one of the fluxes (Type 5) was analyzed and it proved to be a high quality hematite nodule with 55% iron (76.83% Fe_2O_3). All the ores show low concentrations of trace elements including As, Sb, Ni, Zn and Pb.

As expected, certain differentiating features can be seen in the major elemental data from ICP for each "type" group. The levels of copper, iron oxide, silica and sulfur, when combined, display unique characteristics for each group which are specific enough to be diagnostic (Table 19.6). Type 1 ores can roughly be classified as having <20% copper, >50% silica, are low in iron oxide (<5%), and very low in sulfur (<2%). Type 2 ores have 30–50% copper, variable silica (10–20%), relatively high iron oxide (20–30%). Type 3 ores have high levels of copper (> 50%), iron oxide ranging between 10–20%, silica levels below 5%, and high sulfur (>10%). Type 4 ores range between 30–50% in copper and have iron levels below 20%, silica ranging between 15–30%, and sulfur ranging between 1–6%.

TABLE 19.5: XRD DATA LISTING MINERALS FOUND IN EACH SAMPLE IN ORDER OF CONTENT; CHEMICAL FORMULA FOR EACH MINERAL

Sample	*M-O-02*	*M-O-01*	*A-O-06*	*M-O-19*	*B-O-04*	*B-O-03*	*M-O-03*	*Chemical Formulas*
Type	*1*	*2*	*2*	*3*	*4*	*4*	*4*	
Content	Malachite	Malachite	Pyrite	Chalcocite	Hematite	Quartz	Quartz	Digenite $[Cu_9S_5]$
	Quartz	Hematite	Malachite	Iron Sulfide	Malachite	Cuprite	Hematite	Hematite $[Fe_2O_3]$
	Hematite	Chalcocite	Iron Sulfide	Cuprite	Chalcocite	Hematite	Djurleite	Chalcocite $[Cu_2O]$
	Cuprite	Quartz	Quartz	Malachite	Tenorite	Malachite	Chacopyrite	Quartz $[SiO_2]$
	Pyrite	Djurleite		Calcite	Cuprite	Chacopyrite	Lime	Malachite $[Cu_2(CO_3)(OH)_2]$. $[Cu_2CO_3(OH)_2]$
		Chacopyrite			Quartz	Tenorite	Malachite	Dolomite ferroan $[Ca(Mg,Fe)(CO_3)_2]$
		Cuprite			Chacopyrite	Pyrite	Calcite	Covellite $[CuS]$
							Chalcocite	Chacopyrite $[CuFeS_2]$
								Tenorite $[CuO]$
Sample	*M-O-04*	*M-O-06*	*A-O-03*	*M-O-08*	*B-O-06*	*M-O-07*		*Cuprite* $[Cu_2O]$
Type	*4*	*4*	*4*	*4*	*4*	*5*		Pyrite $[FeS_2]$
Content	Hematite	Quartz	Digenite	Quartz	Chalcocite	Dolomite ferroan		Iron Sulfide $[FeSO_3]$
	Quartz	Hematite	Hematite	Hematite	Cuprite			Djurleite $[Cu_{31}S_{16}]$
	Malachite	Digenite	Chalcocite	Digenite	Digenite			Calcite $CaCO_3$
	Cuprite	Malachite	Quartz	Cuprite	Quartz			
	Djurleite		Malachite	Djurleite	Malachite			
	Chacopyrite		Dolomite	Malachite				
	Chalcocite		Covellite	Chacopyrite				
				Chalcocite				

These differences may be best expressed in the ratios of copper to sulfur, copper to iron oxide, and copper to silica (Table 19.7). Type 1 has very equal ratios of silica to copper but a very high ratio of copper to sulfur. Type 2 has relatively equal ratios of iron oxide to copper and of silica to copper, while sulfur and copper have an even closer ratio. Type 3 has similar ratios for iron oxide to copper and sulfur to copper, while silica to copper ratios are very low. Type 4 has an equal ratio of silica to copper, while there is about 4 times as much copper to iron oxide and 10 times as much copper as sulfur. Type 1 ores are highest in silica to copper, Type 2 ores are highest in iron oxides to copper, Type 3 ores are highest in sulfur to copper, while Type 4 ores have variable levels for all ratios but are most equal in silica to copper at 1:1.

The relative levels of trace elements in the ore types are less diagnostic than the major elements. Looking at the averages, Type 1 ores appear to have higher amounts of siderophile elements (i.e. Ni, Mn, Cr, and Ce) than the other types, but this may be somewhat misleading. Only four samples of Type 1 were analyzed and although the average values for the siderophile elements are high, individually they are radically different with usually only one or two outliers accounting for the increased average. For example, nickel has an average of 155 ppm which is twice as much as in any other group (excluding fluxes).

On the whole, the trace element differences between the groups are limited. What may be most interesting is the thorium to uranium ratios since the decay of lead isotopes is directly related to the thorium and uranium levels (see Table 19.7). Differences in this ratio in the five groups may have clear relations to the LIA data obtained, and provide not only a textural and chemical characterization of Type 4 ores, but also a provenance difference. The relatively close ratio of Th:U (1:2) in the Type 4 ores becomes even greater at 1:1 when the outlier ratio of sample M-O-04 is removed.

Table 19.6: Diagnostic Features of the Major Elements and Compounds for the Four Ore Types*

	Cu	*Iron Oxide*	*Silica*	*Sulfur*
Type 1	10–30%	1-– 6%	50-–80%	0–2%
Type 2	30–50%	20–30%	10–20%	1–10%
Type 3	46–60%	10–20%	1–8%	10–15%
Type 4	30–50%	2–20%	15–30%	1–6%

* Type 1 is characterized by high silica, Type 2 is characterized by high iron oxide, Type 3 is characterized by high copper, and Type 4 is characterized by average values.

Table 19.7: Ratio for Each Ore Type Based on ICP Data*

Ratios	*Type 1*	*Type 2*	*Type 3*	*Type 4*
$Cu:Fe_2O_3$	6:1	1:1	5:1	4:1
Cu:S	18:1	9:1	4:1	10:1
$Cu:SiO_2$	1:0.3	1:3	1:12	1:1
Th:U	1:3	1:10	1:6	1:2
	Without sample B-O-04 Th:U			1:1

*The differences in ratios are clear, with Type 1 having the highest silica to copper ratio, Type 2 having the highest iron oxide to copper ratio, Type 3 having the highest sulfur to copper ratio and Type 4 having variable levels.

TABLE 19.8: ICP ANALYSIS OF COPPER AND IRON ORES FOUND AT ABU MATAR DURING THE 1990/1991 EXCAVATIONS*

	Wt. %									*PPM*																				
	SiO_2	*Al_2O_3*	*CaO*	*MgO*	*Fe_2O_3*	*TiO_2*	*Cu*	*P_2O_5*	*S*	*Mo*	*Zn*	*Pb*	*Co*	*Ni*	*Ba*	*Mn*	*Cr*	*V*	*Be*	*La*	*Y*	*Eu*	*Sr*	*Ce*	*As*	*Ag*	*Cd*	*Sn*	*Sb*	*Th*	*U*
Type 1																															
A-O-02	69.8	4.9	<0.5	0.2	1.5	0.1	8.8	0.2	0.1	11	96	254	11	284	1744	49	615	77	1.6	13	5.4	0.7	191	18	2.1	1	0.03	1.2	0.2	6.0	23.9
M-O-02	48.6	0.9	<0.5	n/d	6.5	0.2	29.3	0.4	1.0	8	300	148	13	29	809	114	28	154	1.1	n/d	4.6	0.4	104	7	5.1	5	n/d	2.4	2.6	3.5	14.5
M-O-09	58.3	2.7	<0.5	0.1	2.7	0.2	15.6	0.6	2.5	18	90	71	10	287	149	42	490	66	0.7	n/d	5.5	0.2	129	12	3.8	4	0.03	1.6	1.0	4.0	6.8
M-O-14	77.4	2.2	0.2	n/d	0.5	0.1	16.1	0.3	0.2	2	213	223	18	22	310	546	12	55	4.5	21	19.1	0.7	161	40	12.6	n/d	n/d	4.4	0.2	4.6	8.9
Average	*63.5*	*2.7*	*0.1*	*0.1*	*2.8*	*0.1*	*17.4*	*0.4*	*1.0*	*10*	*175*	*174*	*13*	*155*	*753*	*188*	*286*	*88*	*2*	*8*	*9*	*0*	*146*	*19*	*6*	*2*	*0*	*2*	*1*	*5*	*13*
Type 2																															
A-O-06	2.7	0.3	0.6	n/d	26.0	0.0	46.9	0.8	5.4	30	315	150	5	10	28	n/d	27	282	1.1	n/d	n/d	0.0	47	1	28.0	5	n/d	1.6	11.5	0.4	10.8
A-O-09	9.4	0.9	<0.5	n/d	32.3	0.1	39.3	0.7	1.9	25	291	60	7	37	3550	n/d	26	169	2.3	n/d	n/d	1.0	120	3	37.7	11	1.0	0.6	5.7	1.1	5.4
B-O-05	23.7	1.2	<0.5	n/d	30.1	0.1	33.2	0.9	0.4	22	308	138	6	39	391	n/d	51	211	2.8	n/d	n/d	0.3	68	7	11.5	1	n/d	0.6	10.3	1.5	17.4
M-O-01	10.6	0.4	<0.5	0.2	19.7	0.0	36.2	1.7	9.2	15	191	171	11	60	269	n/d	57	72	0.8	n/d	1.3	0.1	50	2	35.2	17	0.31	0.1	3.8	0.8	5.9
Average	*11.6*	*0.7*	*0.2*	*0.0*	*27.0*	*0.1*	*38.9*	*1.0*	*4.2*	*23*	*276*	*130*	*7*	*36*	*1060*		*40*	*183*	*2*		*0*	*0*	*71*	*3*	*28*	*9*	*0*	*1*	*8*	*1*	*10*
Type 3																															
A-O-07	1.7	0.7	1.2	n/d	9.7	0.0	57.3	1.0	11.5	13	336	127	6	27	133	27	70	147	0.3	n/d	n/d	0.1	240	4	43.6	7	n/d	0.6	2.3	0.9	3.5
M-O-05	8.2	0.4	<0.5	0.3	1.1	0.0	46.2	2.1	11.9	7	424	142	12	65	204	33	59	47	0.4	n/d	1.5	0.0	87	1	1.4	13	0.88	0.1	0.4	0.7	2.5
M-O-15	4.9	0.3	<0.5	n/d	17.9	0.0	52.2	1.1	12.6	25	418	99	6	20	368	n/d	19	192	0.6	n/d	n/d	0.1	65	2	12.0	15	n/d	1.6	6.9	0.5	4.8
M-O-19	2.5	0.4	<0.5	n/d	16.4	0.0	57.9	1.3	14.7	30	470	612	11	46	296	72	116	419	1.0	n/d	n/d	0.1	37	2	13.8	11	n/d	0.5	7.0	0.5	6.2
Average	*4.3*	*0.4*	*0.3*	*0.1*	*11.3*	*0.0*	*53.4*	*1.4*	*12.7*	*19*	*412*	*245*	*8*	*39*	*250*	*33*	*66*	*201*	*1*		*0*	*0*	*107*	*2*	*18*	*11*	*0*	*1*	*4*	*1*	*4*
Type 4																															
A-O-03	9.6	4.8	<0.5	0.2	13.7	0.2	36.2	1.8	5.9	27	333	244	11	86	707	25	247	284	1.1	10	18.1	0.4	209	18	4.8	10	0.03	0.9	6.3	7.8	9.0
B-O-01	22.4	1.0	<0.5	n/d	4.0	0.2	48.4	0.9	3.8	7	453	385	8	27	2306	16	69	133	0.6	n/d	n/d	0.7	67	6	5.4	4	n/d	0.6	2.0	2.0	4.3
B-O-03	23.3	1.0	<0.5	0.3	9.1	0.1	29.0	1.3	0.8	14	521	162	14	179	68	19	285	150	0.9	n/d	2.5	0.0	38	3	4.9	34	2.96	0.9	1.0	2.9	4.2
B-O-04	15.1	1.0	<0.5	0.2	11.8	0.0	33.5	1.5	3.2	23	295	375	18	86	570	172	196	255	1.5	n/d	3.3	0.1	102	1	7.9	12	0.09	0.3	5.7	0.8	10.7
B-O-06	19.4	0.3	<0.5	0.2	3.6	0.1	35.4	1.8	5.6	11	281	183	12	115	307	29	172	66	0.4	n/d	1.9	0.1	93	2	7.3	17	0.08	0.6	0.6	2.8	6.5
M-O-03	17.1	0.8	<0.5	0.2	20.8	0.1	28.7	1.3	2.8	16	298	252	11	57	1200	n/d	68	474	0.8	n/d	4.2	0.3	138	5	6.3	15	0.81	0.4	7.3	2.7	9.4
M-O-04	32.5	4.3	<0.5	n/d	9.3	0.3	30.2	0.8	2.4	8	379	294	10	35	1653	14	54	103	1.0	29	19.3	1.3	585	80	14.7	12	n/d	1.3	1.5	9.7	7.1
M-O-06	44.4	2.0	<0.5	0.2	2.6	0.2	22.5	0.9	2.1	3	201	300	8	175	441	31	321	107	0.6	n/d	5.6	0.2	106	11	2.9	10	0.18	0.8	0.7	6.0	6.9
M-O-08	31.1	1.6	<0.5	n/d	10.7	0.2	40.0	0.7	5.5	23	367	115	6	30	391	51	73	156	0.3	n/d	6.7	0.3	78	9	24.8	13	0.6	3.3	11.3	2.6	4.9
Average	*23.9*	*1.9*	*<0.5*	*0.1*	*9.5*	*0.1*	*33.8*	*1.2*	*3.6*	*15*	*348*	*257*	*11*	*88*	*849*	*40*	*165*	*192*	*1*	*4*	*7*	*0*	*157*	*15*	*9*	*14*	*1*	*1*	*4*	*4*	*7*
Type 5																															
A-O-10	6.3	0.3	<0.5	n/d	76.8	0.0	0.5	0.1	0.2	125	525	57	13	1459	1971	n/d	79	827	1.3	n/d	n/d	0.7	147	2	39.9	n/d	2.3	0.8	3.7	0.4	16.2

* The separation of the ores into five groups based on optical microscopy also seemed valid in the bulk chemical analysis (n/d indicates not detected).

Lead Isotope Analysis

Lead isotope analysis (LIA) was performed on the ores from both the 1950s and the 1990/1991 excavations of Abu Matar. In addition, samples from the contemporary site of Nevatim were analyzed to complement the data from the Abu Matar ore assemblage.[5] In total, 40 samples were analyzed (see table 19.9).

Results

Separation into groups was performed by visual assignment as suggested by Gale (1989). Samples from all three sites (Bir Abu Matar (1950), Abu Matar (1990) and Nevatim) fall into two distinct clusters (Table 19.9, Fig. 19.6). The first cluster closely matches the isotope ratios from Faynan. The second cluster does not and appears to have closer ties to an alternative ore source, possibly in Anatolia. Although no specific ore source can be named, the ores match best the general location of the mines in southeast Anatolia, with possible connections to northern Lebanon, northwest Syria, the Iranian Plateau (S. Stos-Gale, personal communication) or Armenia (Meliksetian *et al.* 2003). Unfortunately, the copper ore sources in northern Lebanon, northwest Syria and the Iranian Plateau have not been as extensively explored as the ones in Anatolia and lead isotope data is not readily available. What is clear is that there are two distinct clusters, or groups of lead isotope ratios, found in the ores from Abu Matar and Nevatim, and only one matched the ores of Faynan.

DISCUSSION

Interregional interaction between the Beer-sheba Valley and Faynan has previously been established through ore assemblages that have been analyzed from the region. Previous reports of chemical and mineralogical analysis of these ores suggest that the copper ores used for metallurgical activities by the Ghassulians represent a fairly good geochemical match to the pattern of mineralization in the ores from Faynan (Weisgerber and Hauptmann 1988; Hauptmann 1989, 1991). Typically, the ores have the characteristic "...tectonically influenced texture of the brecciated ores and the net like replacement of chalcocite which clearly points to the mineralization in Qalb Ratiye and Wadi Abiad in the Faynan area" (Hauptmann, 1991: 405). Type 1, 2, and 3 ores can clearly be associated with Faynan based on mineralization and texture similarities. Type 4 ores have less association with Faynan ores and are more likely from an alternative source. The hematite iron flux is readily found throughout the southern Levant but the ferroan dolomite fragments could have come from Faynan as well. Type 1 ores also have some similarity to the ore formations at Timna, but are more strongly linked with the ores from Faynan.

It has been suggested that the Ghassulain culture had more extensive interregional interaction which occurred to the south, in the Sinai and throughout Egypt (Rowan 2014). Interaction to the north was confirmed to have occurred for the trade of obsidian during the Neolithic period (Dixon, Cann and Renfrew 1968; Renfrew, Dixon and Cann 1968) and confirmed, albeit to a limited amount, in the Chalcolithic period with only a few samples, including one obsidian sample from the Ghassulian site of Gilat fitting into the prescribed context (Yellin, Levy and Rowan 1996). This exchange indicates at least some continued connection to the north. In addition, there are some reports of copper ores (with arsenic and antimony concentrations) from Ugarit and Ras ibn Hanni in Syria, and some potential lead isotope links to those sources such as a mace head discovered in the Judean Desert (Shalev 1996).

5 Samples from Nevatim were provided by Prof. Isaac Gilead.

Table 19.9: Lead Isotope Ratios Acquired from Lead Isotope Analysis*

Site	*Sample Name*	*208/206*	*207/206*	*206/204*
Abu Matar	B-O-04	2.1152	0.866	18.056
Abu Matar	M-O-19	2.1162	0.867	18.062
Abu Matar	BAM-O-06	2.1165	0.866	18.080
Abu Matar	A-O-02	2.1172	0.866	18.087
Abu Matar	M-O-14	2.1146	0.866	18.088
Nevatim	NEV-O-16	2.1164	0.866	18.089
Nevatim	NEV-O-14	2.1169	0.866	18.089
Nevatim	NEV-O-10	2.1120	0.865	18.094
Abu Matar	BAM-O-02	2.1129	0.864	18.095
Nevatim	NEV-O-08	2.1160	0.866	18.099
Nevatim	NEV-O-05	2.1106	0.865	18.105
Abu Matar	BAM-O-01	2.1141	0.865	18.108
Abu Matar	BAM-O-04	2.1146	0.865	18.112
Abu Matar	BAM-O-07	2.1122	0.864	18.112
Abu Matar	M-O-15	2.1114	0.863	18.114
Nevatim	NEV-O-02	2.1132	0.863	18.132
Abu Matar	A-O-07	2.1101	0.863	18.135
Nevatim	NEV-O-04	2.1144	0.864	18.137
Abu Matar	M-O-02	2.1190	0.866	18.142
Abu Matar	M-O-01	2.1110	0.864	18.143
Abu Matar	B-O-01	2.1250	0.868	18.145
Abu Matar	BAM-O-05	2.1103	0.863	18.147
Nevatim	NEV-O-01	2.1102	0.862	18.155
Nevatim	NEV-O-03	2.1108	0.862	18.164
Abu Matar	A-O-09	2.1052	0.861	18.165
Abu Matar	B-O-05	2.1060	0.861	18.182
Abu Matar	A-O-06	2.1136	0.867	18.185
Nevatim	NEV-O-12	2.1096	0.861	18.189
Abu Matar	M-O-09	2.1104	0.863	18.217
Abu Matar	BAM-O-03	2.0966	0.855	18.338
Abu Matar	M-O-08	2.0982	0.854	18.340
Abu Matar	A-O-03	2.1002	0.854	18.397
Abu Matar	B-O-06	2.0921	0.850	18.427
Abu Matar	M-O-06	2.0887	0.849	18.464
Abu Matar	M-O-04	2.0932	0.850	18.466
Abu Matar	B-O-03	2.0883	0.848	18.478
Abu Matar	M-O-03	2.0892	0.849	18.479
Nevatim	NEV-O-17	2.0876	0.848	18.517
Nevatim	NEV-O-09	2.0861	0.844	18.571

*Samples included copper ores from the 1950 excavations of Abu Matar (BAM samples), the 1990/1991 excavations of Abu Matar (A, B, or M samples), and excavations of Nevatim (NEV samples). The data are separated into the two clustered groups visually represented in Fig. 19.6.

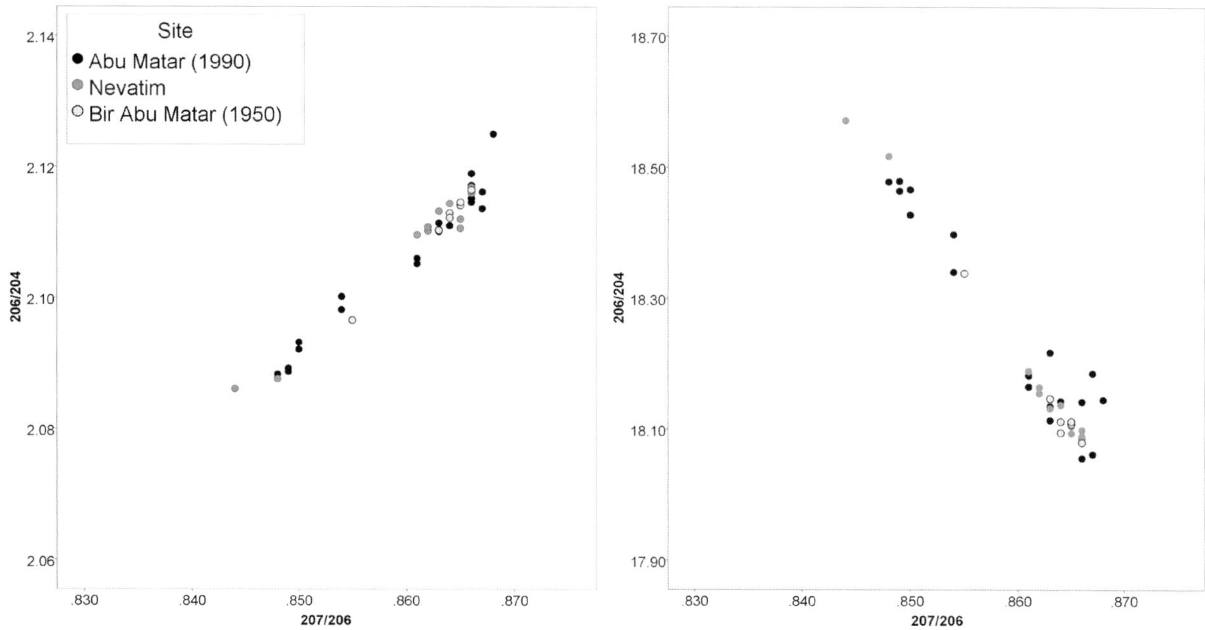

Fig. 19.6: Lead isotope diagram showing two clear clusters from data in Table 19.9. The samples are from the sites of Abu Matar, Bir Abu Matar and the contemporary site of Nevatim. Error is ± 0.1%.

The lead isotope data presented here support the view that an alternative source for copper ore found at Abu Matar should be considered. The data suggests that the source could be in Central Anatolia, Eastern Anatolia or farther afield. When the data from lead isotope analysis of the Abu Matar ores (including Bir Abu Matar and Nevatim) are plotted against various copper ore sources in Anatolia (Wagner *et al.* 1989; Hirao, Enomoto and Tachikawa 1995), Armenia (Meliksetian *et al.* 2003) as well as Faynan (Hauptmann *et al.* 1992) and Timna (Gale *et al.* 1990), associations can be seen. Some of the ores match the lead isotopes of Faynan, as expected, while another cluster best match ores from Kaman-Kalehoyuk, the Eastern Black Sea or several Armenian ore sources (Fig. 19.4). The Anatolian sources have copper-rich ores (Özbal personal communication) as do sites across the Caucasus, including the Armenian region. All these mines could have provided additional ore for the metallurgical activities at Abu Matar.

The overlap of the Abu Matar ores with the Kaman-Kalehoyuk mines needs further investigation since the elliptical range from the lead isotope ratios are widespread, and the few ore samples from Abu Matar and Nevatim analyzed here overlap only the upper range of this ellipse. Sources from the Black Sea mines offer a tighter match while the lead isotope analysis from Armenian ore sources offer a potential closer match (Meliksetian *et al.* 2003) (Fig. 19.7). In addition, the reported composition of the Armenian ores is mostly copper and arsenical copper, with some indication of the potential for nickel in the ores from Talin (*ibid.*). This would provide a potential source for the arsenic rich copper prills discovered in slags from Abu Matar (Shugar 2001), even though no significant levels of arsenic were found in the ores analyzed from Abu Matar.

Moving farther east and plausibly into the Iranian Plarteau, additional sources may be found. The mining sites in the Central Iran region of Anarak contain copper sulfide and arsenical copper ores, and could be considered as a possible source (Nezafati, Momenzadeh and Pernicka 2008). The Zagros Mountains can be considered an extension of the Taurus Mountains in Anatolia and have similarities that cannot be overlooked. They are exceptionally rich in minerals associated with early metallurgy and have

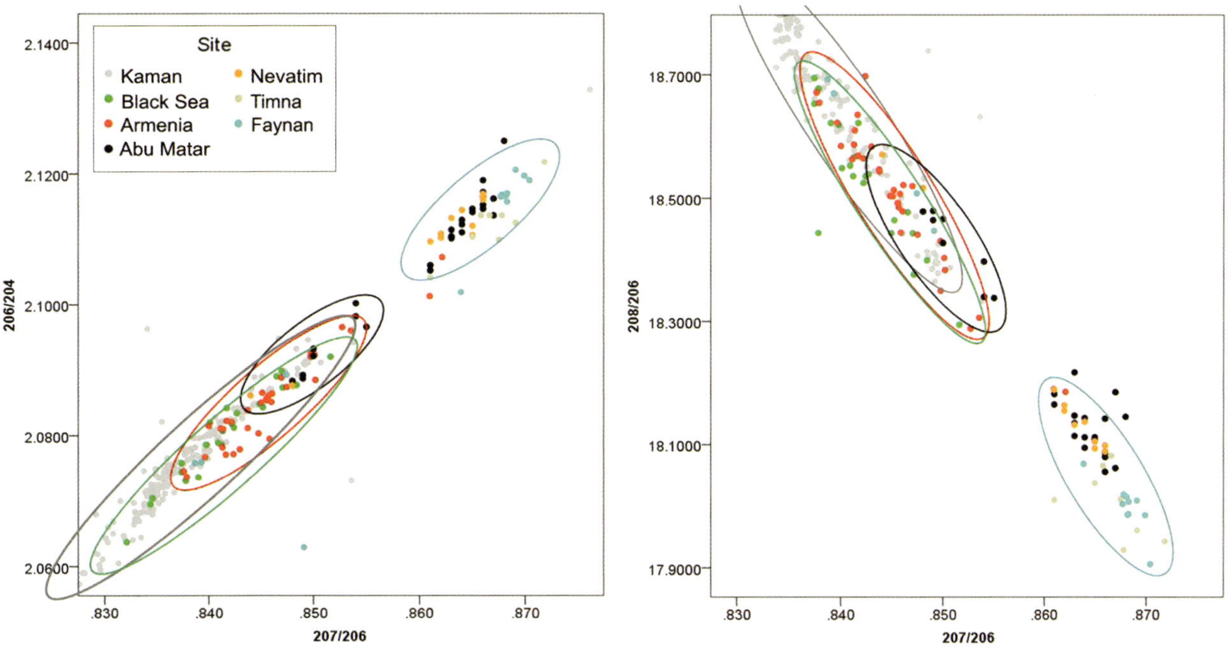

Fig. 19.7: Lead isotope diagram for the isotopic ratios of copper ores found at Abu Matar, Bir Abu Matar (combined as Abu Matar), and Nevatim overlaid on various copper ore sources in the Arabah Valley (Faynan and Timna), Anatolia and Armenia. The ores analyzed in the study fall into the Faynan region, the Kaman–Kalehoyuk region, the Eastern Black Sea region and the Armenian region. Error is ± 0.1%.

been deemed by some to be the birthplace of mining and metallurgy (Pigott 1999). The Anarak region in central Iran has extensive arsenical copper ores including algodonite (Cu_6As) and domeykite (Cu_3As) (Pigott 1999; Nezafati, Momenzadeh and Pernicka 2008), and may provide a more direct link to early copper arsenic alloying found at Abu Matar, although more research needs to be done in that region.

RELATIONSHIP TO THE NAHAL MISHMAR HOARD

The discovery of arsenical copper prills entrapped in the smelting slag at Abu Matar suggests the potential for co-smelting of arsenic-rich ores with copper ores (both carbonate/oxide and sulfide rich). It was considered that Ghassulian interest in the arsenical copper was culturally driven at least with regard to the final cast production of artifacts similar to those found in the Nahal Mishmar hoard (i.e., "prestigious" items, including standards, crowns, and some elaborate mace heads).

With the lack of any complex fahlerz ores found at Ghassulian sites, there is a clear link missing from this story and the connection between the manufacturing of the Nahal Mishmar artifacts and the currently discovered smelting sites in the Beer-sheba Valley. One suggestion that may offer a solution is the potential melting of local copper ores with imported antimony-arsenic-rich metal (Tadmor *et al.* 1995). It could be possible that the copper-arsenic-antimony alloy found in artifacts from the Nahal Mishmar hoard may have been imported as ingots and melted for casting. Ingots have been found in the Faynan region as trade goods in the Early Bronze age (Levy *et al.* 2002; Hauptmann 2007) and the idea that ingots could have been manufactured and traded earlier is not farfetched as it makes more sense than trading for the bulkier/heavier ores. What is clear is that if the evidence supports local manufacturing of complex exotic compositional artifacts, as those found at Givʿat ha-ʾOranim or in the Nahal Mishmar hoard (Goren 2008), the raw materials must have been imported from a distance, whether it be ore or highly alloyed ingots.

As suggested by Tadmor *et al.* (1995), we must look for alternative ore sources to the north and the east. Mining sites such as Norsuntempe and Arslantepe show indication of antimony–arsenic rich ores, and these regions would fall within an acceptable lead isotope signature. It is also likely that the rest of the Caucasus would provide ample complex fahlerz ores. The mines in Azaerbaijan or the ores from Gornaya Racha in Georgia may be valid sources, linking closely to those already explored in Armenia. Of course the sources in central Iran cannot be discounted. But we may not be that far off. Three of the artifacts analyzed by Tadmor *et al.* (1995) have lead isotope values that fall directly in line with the Type 4 ores presented here (61–52, 61–145 and 61–150).

CONCLUSIONS

Based on the combined analyses of the ores, a diagnostic description of the five ore types can be made. These descriptions consider textural observations, bulk and trace chemistry, and lead isotope analysis.

Type 1 ores (six samples) can be described as texturally consisting of a rich silica matrix with intergrowth of secondary copper minerals, mainly malachite, atacamite and chrysocolla, with occasional inclusions of cuprite and hematite. The relative quality of this copper ore was confirmed through bulk chemistry showing a high level of silica (> 50% mean = 63.5%) and a low level of copper (10–30% mean = 17.4%). The inclusions of hematite likely account for the low iron oxide levels found (1–6% mean = 2.8%). Trace elemental composition does not reveal any unique feature to this ore type. Lead isotopes place it within the Faynan cluster.

Type 2 ores (eight samples) can be described as "tile ores" consisting mainly of a mixture of hematite and copper oxides for the general matrix, with some isolated banding of chalcocite. Secondary oxide minerals often form on the surface of these ores. The diagnostically high level of iron oxide seen under optical microscopy was confirmed with bulk chemistry with iron oxide levels consistently between 20–30% (mean = 27%). Copper is also relatively high as the major element in the ore matrix as well as accounting for approximately 78% of the chalcocite. The presence of copper sulfides was confirmed by bulk chemistry, indicating higher levels of sulfur (up to 9%, mean = 4.2%). Variable silica levels of around 10 to 20% (mean = 11.6%) would be expected from the typically silica-rich secondary oxide mineralization that can form on the surfaces of these ores. Trace elemental composition does not reveal much except that the ratio of U:Th is very high at 10:1. The lead isotopes of this ore type place it within the Faynan cluster but three of the ores, samples A-O-06, A-O-09, and B-O-05 all fall at the lower outer edge of the cluster.

Type 3 ores (nine samples) can be described as sulfide-rich copper ores associated with increased amounts of chalcocite and covellite with, to a lesser extent, an iron (hydr)oxide and copper oxide matrix. The ores are dark and relatively dense. Secondary mineralization is not normally witnessed in these samples. The bulk chemistry confirms the high levels of sulfur found in these ores with an average of 12.7%. Moderate iron levels (10–20% mean = 11.3%) are to be expected with the iron (hydr)oxide and copper oxide matrix. With copper deriving from both the matrix and the copper sulfides, its rich copper content of > 50% (mean = 53.4%) is to be expected and is diagnostic. With little to no secondary oxide mineralization and the odd inclusion of malachite in the ore, the low levels of silica (1–8%, mean = 4.3%) are to be expected. The trace elemental composition does not reveal any unique features. All the ores of this type have lead isotope association with the Faynan cluster and they are spread throughout the cluster range.

Type 4 ores (13 samples) can be described as copper sulfide nodular ores with a matrix of a relatively high level of hematite and silica. This ore type combines the features of Type 2 and Type 3 but is still texturally unique within the assemblage. The nodular banded chaococite and covellite are the most diagnostic features of this ore type. Bulk chemistry shows average values for all the major elements. Silica is relatively higher than Type 2 and Type 3 ores, ranging between 10–40% (mean = 23.9%). Trace elemental composition shows few unique variations except in sample B-O-04, which has a Th:U ratio of 1:11 as compared to the other samples which have an average Th:U ratio of 1:1. Based on this it would be expected that the lead isotope ratio for this sample would not fit with the other ores in the group. As expected, while lead isotope analysis did show this group of ores does form a separate cluster from the Faynan cluster, two samples fell into the Faynan cluster.

Type 5 ores, the fluxes (two samples), are iron based ores. They each are distinctive in their composition and mineralization. Sample A-O-10 is a hematite nodule and sample M-O-07 is a fragment of ferroan dolomite. The trace elemental chemistry of sample A-O-10 is relatively usual, but the level of Ni at 1459 ppm is higher than any other ore fragment. This is likely due to the element's siderophile properties. Sample M-O-07 has close ties to the Dolomite Limestone Shale formations at Faynan.

The source of Type 4 ores is still in question but relationships to the ore sources in the Caucasus looks promising and we should not discount the extensive ore formations in Iran. Additional ore samples from the region should be investigated to extend this research and to see if they fit into the same patterns.

The view of metallurgical research in the southern Levant being Levantine-centric as suggested by Thornton (2009b) is not unreasonable. I agree with this view and think a broader scope of investigation, or a more global perspective, needs to be considered when talking about early aspects of metallurgy. For example, a closer look at the relationship between the early copper smelting activities at Tal-i-Iblis (Frame 2004) and those in Faynan and the Beer-sheba Valley should be considered relevant. I look forward to great expansion of investigation to the north and east of the Levant. I anticipate that it may provide new insight into the inter-regional trade networks and offer a larger vision of metallurgical activities in the region as a whole, allowing for metallurgy to be understood on local, regional and global frameworks (Meliksetian *et al.* 2003; Frame 2004; Pigott 2004; Nezafati, Momenzadeh and Pernicka 2008; Thornton 2009b, 2009a).

ACKNOWLEDGMENTS

First, I have to offer my thanks and a debt of gratitude to Beno Rothenberg. Beno introduced me to the material at Abu Matar, Timna and the region, encouraged me to investigate it, mentored me (not only with interpreting the data, but in developing my ideas and sharing his), and supported my research. Thanks also go to Dr. John Merkel for supervising my research and providing clear insight into issues raised, and to Prof. Thilo Rehren for reviewing and commenting on the work. I would also like to thank Rob Ixer for his comments on this chapter, and Professors Isaac Gilead and Steve Rosen for providing access to material and thoughtful discussion.

REFERENCES

Alon, D. 1977. A Chalcolithic Temple at Gilat. *Biblical Archaeologist* 40: 63–65.

Arden, J.W., and Gale, N.H. 1974. A New Electrochemical Technique for the Separation of Lead at Trace Levels from Natural Silicates. *Analytical Chemistry* 46: 2–9.

Dixon, J.E., Cann, J.R. and Renfrew, C. 1968. Obsidian and the Origins of Trade. *Scientific American* 218: 38–46.

Dothan, M. 1959. Excavations at Horvat Beter (Beersheva). *ʿAtiqot* 2: 1–71.

Eldar, I. and Baumgarten, Y. 1985. Neve Noy: A Chalcolithic Site of the Beer-Sheba Culture. *Biblical Archaeologist* 48: 134–139.

Frame, L. 2004. *Investigations at Tal-i Iblis: Evidence for Copper Smelting during the Chalcolithic Period* (B.Sc. thesis, Massachusetts Institute of Technology). Boston.

Gale, N.H. 1989. Lead Isotope Analysis Applied to Provenance Studies—A Brief Review. In: Maniatis, Y., ed. *Archaeometry: Proceedings of the 25th International Symposium*. New York: 469–502.

Gale, N.H. 1995. Some Aspects of Lead Isotope Analysis of Ores and Artefacts. In: Esmael, F.A., ed. *Ancient Egyptian Mining and Metallurgy and Conservation of Metallic Artifacts*. Cairo: 251–272.

Gale, N.H., Bachmann, H.G., Rothenberg, B., Stos-Gale, Z. and Tylecote, R.F. 1990. The Adventitious Production of Iron in the Smelting of Copper. In: Rothenberg, B., ed. *The Ancient Metallurgy of Copper*, Vol. 2. London: 191.

Gilead, I. 1989. Grar: A Chalcolithic Site in the Northern Negev, Israel. *Journal of Field Archaeology* 16: 377–394.

Gilead, I., Rosen, S., Fabian, P. and Rothenberg, B. 1992. New Evidence for the Beginnings of Metallurgy in the Levant Based on Recent Excavations at Abu Matar, Beer-sheba (Israel). *Institute for Archaeo-Metallurgical Studies* 18: 11–14.

Golden, J.M. 1998. *The Dawn of the Metal Age: Social Complexity and the Rise of Copper Metallurgy during the Chalcolithic in the Southern Levant, ca. 4500–3500 BC* (Ph.D. dissertation, University of Pennsylvania). Philadelphia.

Golden, J. 2009. New Light on the Development of Chalcolithic Metal Technology in the Southern Levant. *Journal of World Prehistory* 22(3): 283–300.

Golden, J., Levy, T.E. and Hauptmann, A. 2001. Recent Discoveries Concerning Chalcolithic Metallurgy at Shiqmim, Israel. *Journal of Archaeological Science* 28(9): 951–963.

Goren, Y. 2008. The Location of Specialized Copper Production by the Lost Wax Technique in the Chalcolithic Southern Levant. *Geoarchaeology* 23(3): 374–397.

Hauptmann, A. 1989. The Earliest Periods of Copper Metallurgy in Faynan/Jordan. In: Hauptmann, A., Pernicka, E. and Wagner, G.A., eds. *Old World Archaeometallurgy*. Bochum: 119–136.

Hauptmann, A. 1991. From the Use of Ore to the Production of Metal. In: Eluère, C. and Mohen, J., eds. *La découverte du métal*. Paris: 397–412.

Hauptmann, A. 2007. *The Archaeometallurgy of Copper. Evidence from Faynan, Jordan*. Berlin.

Hauptmann, A., Begemann, F., Heitkemper, E., Pernicka, E. and Schmitt-Strecker, S. 1992. Early Copper Production at Faynan, Wadi Araba, Jordan: The Composition of Ores and Copper. *Archaeomaterials* 6(2): 1–33.

Hermon, S. 2008. *Socio-economic Aspects of Chalcolithic (4500–3500 BC) Societies in the Southern Levant: A Lithic Perspective* (BAR International Series 1744). Oxford.

Hirao, Y., Enomoto, J. and Tachikawa, H. 1995. Lead Isotope Ratios of Copper, Zinc, and Lead Minerals in Turkey in Relation to the Provenance Study of Artefacts. In: Mikasa, H.I.H.P.T., ed. *Essays on Ancient Anatolia and its Surrounding Civilisations*. Wiesbaden: 89–114.

Ixer, R.A. 1990. *Atlas of Opaque and Ore Minerals in Their Associations*. Milton Keynes.

Levy, T.E. 1986. Archaeological Sources for the Study of Palestine: The Chalcolithic Period. *Biblical Archaeologist* 49(2): 82–108.

Levy, T.E., ed. 1987. *Shiqmim I: Studies Concerning Chalcolithic Societies in the Northern Negev Desert, Israel (1982–1984)*, Vol. 1. Oxford.

Levy, T.E. 1995. Cult, Metallurgy and Rank Societies—Chalcolithic Period (Ca. 4500–3500 BCE). In: Levy, T.E., ed. *The Archaeology of Society in the Holy Land*. London: 226–244.

Levy, T.E. *et al.* 2002. Early Bronze Age Metallurgy: A Newly Discovered Copper Manufactory in Southern Jordan. *Antiquity* 76(292): 425–437.

Levy, T.E. and Shalev, S. 1989. Prehistoric Metalworking in the Southern Levant: Archaeometallurgical and Social Perspectives. *World Archaeology* 20(3): 352–372.

Meliksetian, K., Pernicka, E., Badalyan, R. and Avetissyan, P. 2003. Geochemical Characterisation of Armenian Early Bronze Age Metal Artefacts and Their Relation to Copper Ores. *Archaeometallurgy in Europe: International Conference, 24-25-26 September 2003, Milan, Italy: Proceedings*, Vol. 1. Milan: 597–606.

Merkel, J. 1977. *Neutron Activation Analysis of Copper* (M.Sc., University of Minnesota). Minnesota.

Merkel, J. 1985. Ore Beneficiation during the Late Bronze/Early Iron Age at Timna, Israel. *Museum Applied Science Center for Archaeology Journal* 3(5): 164–169.

Namdar, D., Segal, I., Goren, Y. and Shalev, S. 2004. Chalcolithic Copper Artifacts. In: Sass, B. and Roll, I., eds. *Givʿat Ha-Oranim: A Chalcolithic Site* (Salvage Excavation Reports 1). Tel Aviv: 70–83.

Nezafati, N., Momenzadeh, M. and Pernicka, E. 2008. New Insights into the Ancient Mining and Metallurgical Researches in Iran. In: Yalcin, H.z. and Paşamehmetoğlu, A.G., eds. *Ancient Mining in Turkey and the Eastern Mediterranean*. Ankara,: 307–328.

Perrot, J. 1984. Structures d'habitat, mode de la vie et environement: Les villages souterrains des pasteurs de Beer-sheba dans le Sud d'Israel, au IVe millenaire avant l'ere chritienne. *Paleorient* 10(1): 75–92.

Pigott, V.C. 1999. The Development of Metal Production on the Iranian Plateau. *The Archaeometallurgy of the Asian Old World.* Philadelphia: 73–106.

Pigott, V. 2004. On the Importance of Iran in the Study of Prehistoric Copper-Base Metallurgy. In: Stöllner T., Slotta R. and Vatandoust A., eds. *Persia's Ancient Splendour, Mining, Handicraft and Archaeology.* Bochum: 28–43.

Renfrew, C., Dixon, J.E. and Cann, J.R. 1968. Further Analysis of Near Eastern Obsidians. *Proceedings of the Prehistoric Society* 34: 319–331.

Rothenberg, B. 1991. The Ghassulian–Beer-sheba Chalcolithic Enigma. *Institute for Archaeo-Metallurgical Studies* 17: 6–7.

Rothenberg, B., Tylecote, R.F. and Boydell, P.J. 1978. *Chalcolithic Copper Smelting: Excavations and Experiments.* London.

Rowan, Y. 2014. The Southern Levant (Cisjordan) during the Chalcolithic Period. In: Steiner, M. and Killebrew, A., eds. *The Oxford Handbook of the Archaeology of the Levant (c. 8000–332 BCE)*. Oxford: 223–236.

Segal, I., Kloner, A. and Brenner, I.B. 1994. Multi-element Analysis of Archaeological Bronze Objects Using Inductively Coupled Plasma Atomic Emission Spectrometry: Aspects of Sample Preparation and Apectral Line Selection. *Journal of Analytical Atomic Spectrometry* 9: 737–742.

Shalev, S. 1996. Metallurgical and Metallographic Studies. *The Nahal Qanah Cave: Earliest Gold in the Southern Levant* (Monograph Series of the Institute of Archaeology of Tel Aviv University 12). Tel Aviv: 155–174.

Shalev, S. and Northover, J.P. 1993. The Metallurgy of the Nahal Mishmar Hoard Reconsidered. *Archaeometry* 34(1): 35–41.

Shugar, A.N. 1998. Recent Research in Chalcolithic Metallurgy: Investigation of *Abu Matar*, Israel. *International Mining and Minerals* 1(5): 114–116.

Shugar, A.N. 2001. *Archaeometrical Investigation of the Chalcolithic Site of Abu Matar, Israel: A Reasseement of Technologyand Its Implications for the Ghassuluian Culture*. London.

Shugar, A.N. and Gohm, C.J. 2011. Developmental Trends in Chalcolithic Copper Metallurgy: A Radiometric Perspective. In: Rowan, Y.M. and Lovell, J., eds. *Culture, Chronology and the Chalcolithic: Theory and Transition*. Oxford: 133–148.

Tadmor, M., *et al.* 1995. The Nahal Mishmar Hoard from the Judean Desert: Technology, Composition, and Provenance. *'Atiqot* 27: 95–148.

Thornton, C.P. 2009a. *The Chalcolithic and Early Bronze Age Metallurgy of Tepe Hissar, Northeast Iran: A Challenge to the "Levantine Paradigm"*. (Ph.D. dissertation, University of Pennsylvania). Philadelphia.

Thornton, C.P. 2009b. The Emergence of Complex Metallurgy on the Iranian Plateau: Escaping the Levantine Paradigm. *Journal of World Prehistory* 22(3): 301–327.

Tylecote, R.F., Rothenberg, B. and Lupu, A. 1974. The Examination of Metallurgical Material from Abu Matar, Israel. *Historical Metallurgy* 8(2): 32–34.

Wagner, G.A., *et al.* 1989. Archäometallurgische Untersuchungen an Rohstoffquellen des frühen Kupfers Ostanatoliens. *Jahrbuch des Römisch-Germanischen Zentralmuseums Mainz* 36: 637–686 .

Weisgerber, G. and Hauptmann, A. 1988. Early Copper Mining and Smelting in Palestine. In: Maddin, R., ed. *The Beginning of the Use of Metals and Alloys*. London: 52–62.

Yellin, J., Levy, T.E. and Rowan, Y.M. 1996. New Evidence on Prehistoric Trade Routes: The Obsidian Evidence from Gilat, Israel. *Journal of Field Archaeology* 23(3): 361–368.

CHAPTER 20

BRONZE CHISEL AT HORVAT HALUQIM (CENTRAL NEGEV HIGHLANDS) IN A SEQUENCE OF RADIOCARBON DATED LATE BRONZE TO IRON I LAYERS

Hendrik J. Bruins, Irina Segal and Johannes Van der Plicht

The rural settlements in the central Negev highlands and adjacent Sinai area, usually assigned to the Iron IIA, are enigmatic in terms of origin, duration and ethnicity. Very few metal objects have thus far been found at these sites. Excavations at Horvat Haluqim at the edge of an ancient terraced field in the Eastern wadi revealed a continuous series of finely stratified layers in which a bronze chisel was found. Radiocarbon dates for these layers range from the 14th century BCE in the deeper layers to the 12th century BCE in the upper layers. The composition of the bronze chisel, dated to the 12th century BCE, is 89.7% copper, 2.5% tin and 1% lead. The lead isotope ratios show that the origin of the copper ore is from the Faynan Dolomite-Limestone-Shale (DLS) unit. Therefore, trade relations probably existed in the 12th century BCE between Horvat Haluqim and the Faynan area. The Zin Canyon with its springs is a convenient natural east–west route that connects the Faynan area with Horvat Haluqim. Various metallurgic sites in the Faynan area, as well as in the Timna area, have radiometric absolute dates within the Late Bronze and early Iron Ages. The so-called empty second millennium BCE in the region is gradually becoming more visible with the use of geoarchaeology and radiocarbon dating. It seems that local desert inhabitants are related to these settlements, possibly the biblical Amalekites, as proposed by Beno Rothenberg more than 40 years ago.

INTRODUCTION

The rural archaeological site of Horvat Haluqim is situated in the central Negev highlands a few km northeast of Kibbutz Sede Boker. The site is spread along three terraced wadis of the first stream order,[1] sloping southward from the Haluqim Anticline (Fig. 20.1). The average annual rainfall at the nearby Sede Boker meteorological station during the decade of 1990–2000 was 93.3 mm (Bruins 2012: 39). According to Cohen (1976), who excavated a number of structures at the site in 1971–1972, Horvat Haluqim consists of 25 buildings and four cisterns. Besides a Roman watchtower, most of the remains at the site were dated by him to the Iron II (10th century BCE) (*ibid.*; Cohen and Cohen-Amin 2004).

Cohen classified the largest building as an oval fortress, 23 × 21 m in diameter, characterized by casemate rooms surrounding an inner courtyard. The building is situated along the Eastern Wadi, about 100 m downstream (south) from the nearest ancient agricultural terraced fields. Such comparatively

1 The initial part of a wadi in a natural valley system, draining the runoff waters of the related hilly catchment at the Haluqim Anticline.

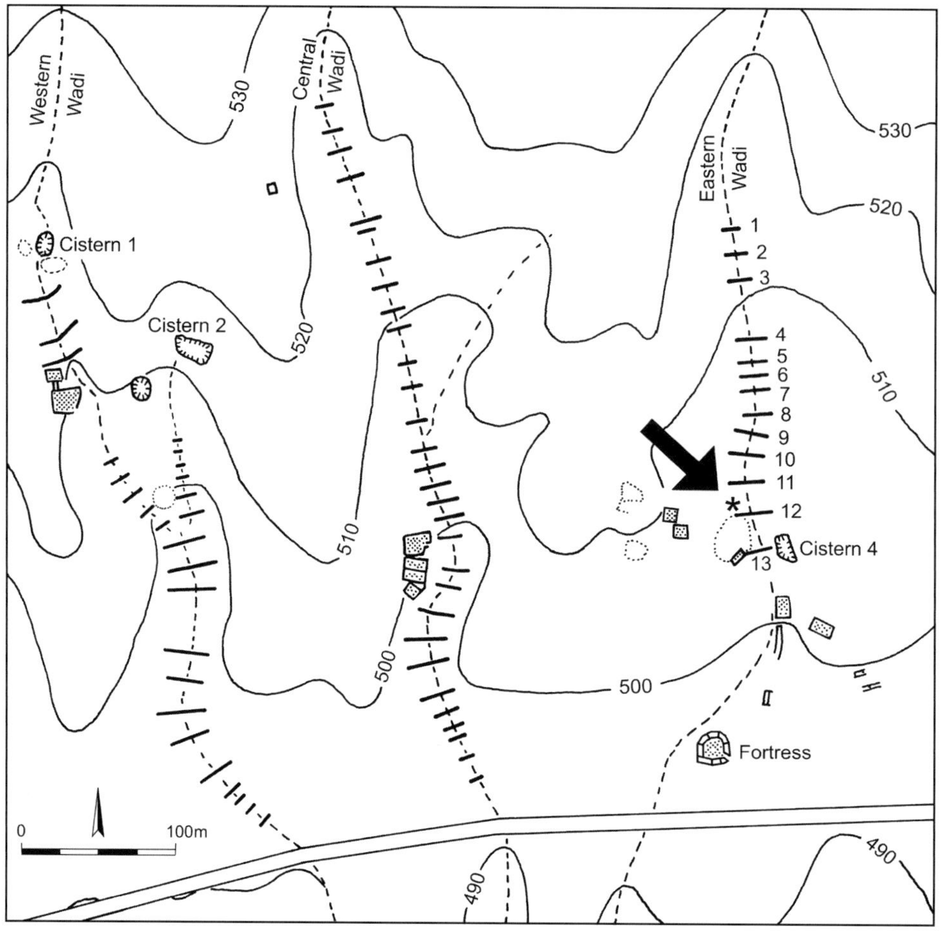

Fig. 20.1: Map of the rural desert site of Horvat Haluqim, showing the three terraced wadis and the adjacent buildings and cisterns. The bronze chisel was uncovered at the westernmost edge of Terraced Field 12, as indicated by an arrow and star *, about 4 m upstream (north) of Terrace Wall 12.

large buildings with oval, irregular or rectangular architectural outlines have been found at dozens of sites in the central Negev highlands and the adjacent northeastern Sinai (*ibid.*). The various scholarly opinions regarding these types of structures is beyond the scope of this article. We only mention these structures here as one of the important features at the site of Horvat Haluqim.

It was Cohen's assessment (1976: 34, 40) that the ancient terraced agricultural fields in the three wadis at Horvat Haluqim probably dated to the Iron Age, in view of the position of some building remains next to such terraced fields. At the time (1970s, 1980s), Cohen's proposition was innovative, since even today many archaeologists maintain that the terraced agricultural fields in the Negev highlands do not predate the Byzantine or late Nabatean period. However, all these opinions were based on circumstantial evidence, as ancient agricultural fields in the Negev highlands had not been excavated at the time.

Bruins renewed excavations at Horvat Haluqim since 1982 (1986, 2007; Bruins, Van der Plicht and Haiman 2012), in the context of the Negev Emergency Survey and following consultations with Rudolf Cohen. His initial investigations dealt with the terraced agricultural fields in the realm of landscape archaeology and geoarchaeology, with special focus on soil science, geomorphology and agriculture based on runoff/floodwater (Bruins, Evenari and Nessler 1986; Bruins 1990, 2007, 2012; Bruins and

Ore 2009; Bruins and Van der Plicht 2017). Often there is no organic material for radiocarbon dating in ancient terraced fields and OSL dating is required to obtain data concerning the age of the terrace soils and underlying sediments (Avni, Avni and Porat 2009; Avni, Avni and Porat 2012; Avni *et al.* 2006).

In this article we present the discovery of a bronze chisel, unearthed *in situ* at the edge of Terraced Field 12 (Area 5), situated in the Eastern Wadi of Horvat Haluqim (Figs. 20.1–2). The bronze chisel was found in a sequence of alternating dark and light-colored layers. Total station measurements of the chisel and five organic samples excavated in its vicinity resulted in a detailed stratigraphic resolution and radiocarbon chronology.

BRONZE OBJECTS IN IRON AGE SITES OF THE NEGEV HIGHLANDS

Following extensive surveys and excavations of many Iron Age rural settlements in the Negev highlands, Cohen (1986: 395) emphasized that only very few metal objects were uncovered, as observed also by Meshel (2000: 68). These objects include a bronze axe found in the Nahal Zena Fortress (Cohen and Cohen-Amin 2004: 65–67), an arrowhead in Atar Ha-Roᶜa (Cohen 1986: 395), an arrowhead in casemate Room 14 of the Har Boqer Fortress (Cohen and Cohen-Amin 2004: 45–47), three copper awls in Building 4 at the Ramat Matred site, and a needle in the Nahal Sirpad Fortress (Cohen 1986: 395).

Therefore, the excavation of a bronze chisel at Horvat Haluqim adds new information to the very limited corpus of metal objects found thus far in these rural central Negev settlements. Our investigation of its stratigraphic position, radiocarbon chronology and chemical composition yielded new data concerning the age of certain rural elements in these settlements, as well as regional copper trade relations.

METHODOLOGY

Excavations were conducted during 2007 at the edge of Terraced Field 12, in the northwestern part of Area 5. An electronic Pentax Total Station PCS/2s theodolite was used to measure various excavation positions and establish absolute elevation levels for the bronze chisel and stratigraphically related organic samples used for radiocarbon dating.

Fig. 20.2: The terraced Eastern Wadi at Horvat Haluqim, looking northward. Terrace Wall 12, oriented east–west, is shown in the foreground. The Terrace Wall once continued where the person on the left is standing. The bronze chisel was excavated in Area 5.

The organic samples were measured at the Radiocarbon Laboratory of the University of Groningen (the Netherlands) with standard procedures for the materials involved. The samples, composed of charred organic material, small black particles or flecks in the terrace soil layers, were pretreated using conventional chemical approaches to remove modern organic material and soil carbonates. The pretreatment of animal bones is somewhat different, as the collagen fraction was separated from the bone sample and subsequently purified. In each case, the purified organic matter was converted into CO_2 and subsequently into solid carbon to prepare the graphite targets for the ion source of the Accelerator Mass Spectrometer (AMS) at Groningen University, according to the methodology described by Van der Plicht *et al.* (2000) and Aerts-Bijma, Van der Plicht and Meijer (2001).

The bronze chisel was put through chemical and isotopic analyses, conducted by Irina Segal in the Geochemistry Department of the Israel Geological Survey in Jerusalem. Drillings were made in the chisel to obtain samples of the internal non-weathered part of the metal. A sample (25 mg) of the metal was dissolved in aqua regia (HNO_3: HCl = 1:3). This solution was analyzed for chemical composition using an Inductively Coupled Plasma Atomic Emission Spectrometer (ICP-AES, Optima 3300, Perkin Elmer). The same solution was used for lead isotope measurements by a Multiple Collector ICP Mass Spectrometer (MC-ICP-MS, NU Plasma). A provenance study was made, based on the lead isotope ratios measurements, in order to investigate the source of the copper in the chisel.

RESULTS

THE BRONZE CHISEL

The chisel was found in a section of Area 5A termed the "copper spot," following its discovery (Fig. 20.3). The precise location of the chisel is indicated by an arrow in Fig. 20.4. The bronze chisel is weathered on the surface, showing various green spots of copper oxidation (Fig. 20.5). It is 4.6 cm long, is (maximum) 0.8 cm wide at its broad end and (maximum) 0.6 cm thick. The narrow end of the chisel is 0.4 cm wide and 0.25 cm thick. Bronze chisels of this period are usually longer and it is quite possible that the chisel broke at its thicker end during use.

Radiocarbon Dating of Layers above and below the Bronze Chisel

The excavated layers in Area 5 of Terraced Field 12 are characterized by thin layers running more or less parallel to each other (Figs. 20.3–4). The colors vary, as some are dark and others lighter. The layers are not inside a building, but are located on the western edge of Terraced Field 12. However, a badly preserved building structure is present to the south of Area 5. This structure, to be excavated at a later date, forms a kind of extension of Terrace Wall 12, subsequently continuing as a low wall surrounding an oval courtyard, situated to the west of Terraced Field 13 (Fig. 20.1). Micromorphology samples were taken of the excavated layers, but these detailed stratigraphic results on the microscopic level will be published elsewhere.

The total station measurements determined the elevation of the bronze chisel to be in the range of 502.64 m to 502.63 m (Table 20.1). A sample of charred organic material (a dark fleck in the finely stratified layers) and a bone fragment were found at a similar level of 502.64 m (0 to 1 cm above the level of the bronze chisel. Both radiocarbon dates (GrA-37496 and GrA-37497) are quite similar, within 1σ of each other (Table 20.1). Their full 1σ and 2σ calibrated age ranges are shown in Table 20.1. The median calibrated values are 1108 BCE and 1146 BCE, respectively. Both dates can be classified in cultural archaeological terminology as belonging, on the basis of time, to the Iron I.

Fig. 20.3: Area 5A, 5B and 5C in the westernmost part of Terraced Field 12. The bronze chisel (BrCh) was uncovered in the "Copper Spot" area, located in Area 5A. The parallel oriented layers (possibly living floors) are visible above the measuring staff. Two stones (Ref#1 and Ref#2) served as measurement points in relation to relative depth and absolute height. An *in situ* grinding stone (GS), typical for crushing cereal grains, was found in the upper part of the finely stratified layers. The Roman watchtower can be seen on the hill spur in the west.

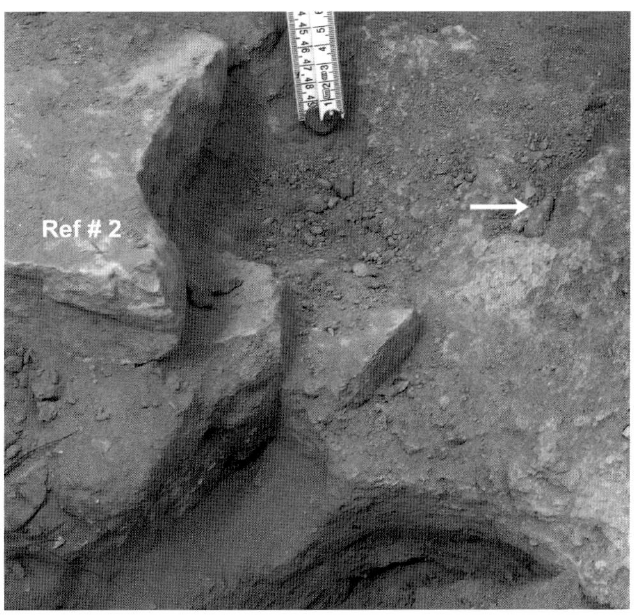

Fig. 20.4: The location of the bronze chisel in Area 5A, upon discovery, is indicated by a white arrow. It was found in a sequence of fine layers, visible as thin parallel layers in the vertical section below the white arrow.

TABLE 20.1: STRATIGRAPHIC POSITION OF THE BRONZE CHISEL IN RELATION TO RADIOCARBON DATED ORGANIC MATERIALS IN ADJACENT LEVELS WITHIN THE FINELY STRATIFIED LAYERS OF AREA 5 (FIGS. 20.1–3), TERRACED FIELD 12, EASTERN WADI, HORVAT HALUQIM

Sample # and Date	*Area and Depth**	*Organic Material Dated*	*Groningen Sample #*	*^{14}C Date (BP)***	*$\delta^{13}C$ (‰) ****	*Calibrated Date Year BCE (1σ) *******	*Calibrated Date Year BCE (2σ)*	*Calibrated Date Year BCE (Median)*
HH-2007-03 11/03/2007	Area 5A, 502.64 m	Charred organic material	GrA-37496	2915 ± 40	-22.95	1192–1143 (21.8%) 1132–1046 (46.4%)	1256–1252 (0.3%) 1231– 997 (95.1%)	1108
HH-2007-05 11/03/2007	Area 5B, 502.64 m	Bone, indeterminable	GrA-37497	2940 ± 40	-18.23	1215–1083 (66.3%) 1064–1059 (1.9%)	1261–1018 (95.4%)	1146
Bronze chisel 11/03/2007	Area 5A, 502.64 m 502.63 m							
HH-2007-01 11/03/2007	Area 5A, 502.61 m	Bone, sheep/goat, upper jaw M3 right, ca 2 years old	GrA-37493	2960 ± 40	-17.12	1257–1251 (2.1%) 1231–1116 (66.1%)	1286–1031 (95.4%)	1171
HH-2007-16 25/01/2007	Area 5B, 502.58 m	Bone, rib, medium-sized mammal	GrA-37515	3020 ± 40	-18.27	1378–1345 (14.2%) 1305–1211 (54.0%)	1397–1156 (91.5%) 1147–1128 (3.9%)	1267
HH-2007-24 13/03/2007	Area 5A, 502.50 m	Charred organic material	GrA-48874	3115 ± 40	-23.14	1433–1376 (40.4%) 1346–1304 (27.8%)	1494–1479 (2.1%) 1456–1269 (93.3%)	1379

* The depth values are absolute elevation levels based on measurements with a Total Station. The soil surface levels of Terraced Field 12 in 2007 next to excavation Area 5 ranged from 503.17 m in the west to 502.94 m in the northeast (Fig. 20.2).

** The radiocarbon dates are reported by convention in BP, i.e., based on the Oxalic Acid standard, using correction for isotopic fractionation with $\delta^{13}C$, and using the so-called Libby half-life value (Mook and Streurman 1983).

*** The $\delta^{13}C$ is the $^{13}C/^{12}C$ ratio of the sample material, expressed in per mille difference from the standard (Mook and Streurman 1983).

**** The radiocarbon dates, measured by AMS at Groningen University, were calibrated with OxCal v. 4.2.2 (Bronk Ramsey 2001, 2013) using the ratified IntCal13 calibration curve (Reimer *et al.* 2013). The calibrated ages are given in three columns: 1σ (68.2% p) represents the age range(s) with highest relative probability; 2σ (95.4% p) gives most of the possible age ranges including those of relatively low probability; the median is a "central" numerical value in the sense that half of the total calibrated age is older than the Median and the other half is younger.

Two other bone fragments and a sample of charred organic material (dark fleck in the finely stratified layers) were found at different elevations below the stratigraphic position of the bronze chisel (Table 20.1). The 1σ and 2σ calibrated age ranges of GrA-37493, GrA-37515 and GrA-48874 are shown in Table 20.1. The radiocarbon results are very consistent with increasing stratigraphic depth. The respective median calibrated values are 1171 BCE, 1267 BCE and 1379 BCE. We have here evidence of a stratified sequence from the Late Bronze Age to the Iron I, suggesting continuous occupation on the western edge of Terraced Field 12 in the Eastern Wadi of Horvat Haluqim during this time period. The bronze chisel, situated stratigraphically between GrA-37497 and GrA-37493, dates most likely to the 12th century BCE.

Chemical Composition of the Bronze Chisel

The geochemistry of the chisel shows a composition of tin bronze, containing 89.7% copper, 2.5% tin and 1% lead. All other elements are present in trace concentrations (Table 20.2).

The lead isotope ratio measurements ($^{204/206}$Pb, $^{208/206}$Pb, $^{207/206}$Pb) of the bronze chisel are shown in Table 20.3. Comparison of these results with lead isotope ratios of relevant copper ores in the region, Faynan (Hauptmann *et al.* 1992), Timna (Gale *et al.* 1990) and Sinai (Hauptmann, Begemann and Schmitt-Strecker 1999), are shown in Figs. 20.6–7. It is clear that the bronze chisel from Horvat Haluqim has lead isotope ratios that are most similar to Faynan DLS, i.e. the copper ore found at Faynan in the Dolomite-Limestone-Shale unit of Cambrian Age (Hauptmann *et al.* 1992). The distance from Horvat Haluqim to Faynan, as the crow flies, is about 73 km, while the distance to Timna is considerably longer, about 125 km.

Fig. 20.5: Close-up of the Horvat Haluqim copper chisel, 4.7 cm long and 0.9 to 0.5 cm wide. Its tapered end to the right was probably hardened and flattened by repeated heating and hammering. The chisel may have been broken during usage in the past at its wider and thicker left end. Here a sample was drilled for our chemical investigation, exposing the bronze metal.

Table 20.2: Chemical Composition, in Weight Percentages, of the Bronze Chisel from Horvat Haluqim

S	*Pb*	*Co*	*Ni*	*Mn*	*Fe*	*Cu*	*Ag*	*Sb*	*As*	*Sn*	*Zn*	*Au*
0.3	1.04	0.026	0.031	<0.001	0.05	89.7	<0.005	<0.01	<0.01	2.51	0.03	<0.01

Table 20.3: Lead Isotope Ratios of the Bronze Chisel from Horvat Haluqim

	$^{204/206}Pb$	σ	$^{208/206}Pb$	σ	$^{207/206}Pb$	σ
Measurement 1	0.05563	0.00002	2.1203	0.0012	0.8691	0.0002
Measurement 2	0.05565	0.00003	2.1200	0.0014	0.8692	0.0003
Measurement 3	0.05563	0.00002	2.1215	0.0010	0.8696	0.0002
Average	0.05563	0.00001	2.1204	0.0005	0.8693	0.0002

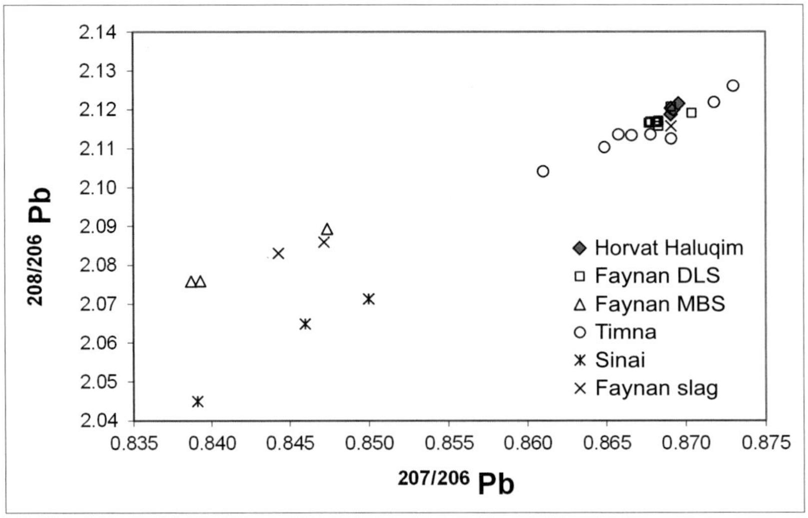

Fig. 20.6: Lead isotope ratios of $^{208/206}$ Pb and $^{207/206}$ Pb of the bronze chisel from Horvat Haluqim, in comparison to Faynan ores and slag (Hauptmann *et al.* 1992), Timna ores (Gale *et al.* 1990) and Sinai ores (Hauptmann, Begemann and Schmitt-Strecker 1999).

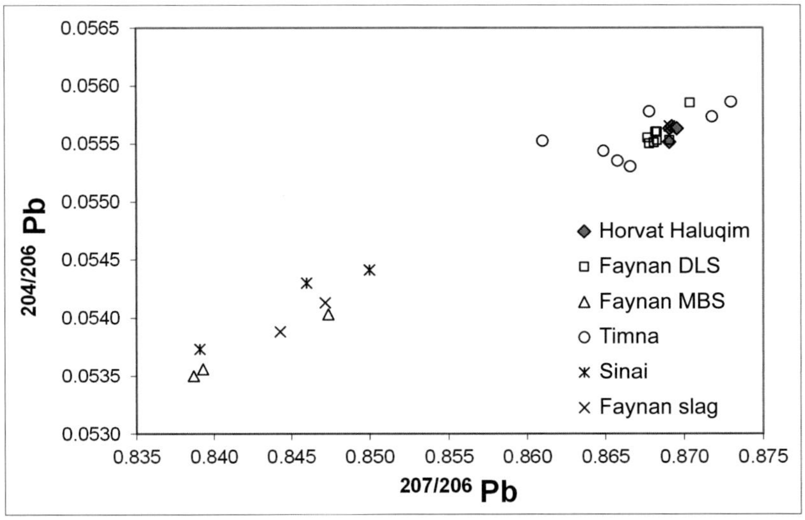

Fig. 20.7: Lead isotope ratios of $^{204/206}$ Pb and $^{207/206}$ Pb of the bronze chisel from Horvat Haluqim, in comparison to Faynan ores and slag (Hauptmann *et al.* 1992), Timna ores (Gale *et al.* 1990) and Sinai ores (Hauptmann, Begemann and Schmitt-Strecker 1999).

DISCUSSION

The bronze chisel from Horvat Haluqim, situated in a sequence of finely stratified layers in an open rural context at the western edge of Terraced Field 12, is dated by radiocarbon to the 12th century BCE. Layers below the bronze chisel become gradually older, reaching the 14th century BCE. A weathered bronze chisel in the Petrie Museum of Egyptian Archaeology UCL (London), dated on archaeological criteria (*not* by radiocarbon) to the 19th Dynasty (Fig. 20.8), looks rather similar to the Horvat Haluqim chisel. They have analogous shapes and dimensions in terms of width and thickness. The Horvat Haluqim chisel appears too short, 4.6 cm, for convenient handling with a hammer. It was probably longer, like the Egyptian 19th Dynasty example (Fig. 20.8), and seems to have broken in the past at its broad end (Fig. 20.5).

Fig. 20.8: A weathered bronze chisel, similar to the one found at Horvat Haluqim. The Egyptian flat chisel, 8.5 cm long and 0.8 cm wide, flattened and flared at the left end, is dated to the 19th Dynasty. Copyright University College London, the Petrie Museum of Egyptian Archaeology UCL, in London, catalogue number UC29303, published here by permission.

The five radiocarbon dates of the sequenced layers are consistent in stratigraphic order. They constitute a problem for the conventional archaeological framework in the area. Extensive surveys and excavations in the central Negev and northeastern Sinai during the 1970s and 1980s convinced Cohen and others that the entire 2nd millennium BCE (Middle Bronze II and III, Late Bronze and Iron I) is absent in this region. However, ^{14}C results of Tell el-Qudeirat, the Qudeirat Valley and Horvat Haluqim yielded dates in the second millennium BCE (Bruins 1986; Bruins and Van der Plicht 2005, 2007, 2017; Carmi and Segal 2007; Van der Plicht, Bruins and Nijboer 2009; Bruins, Nijboer and Van der Plicht 2011).

Concerning the vexing question of origin and ethnicity of these settlements, Cohen (1976: 49) interpreted the site of Horvat Haluqim and similar rural settlements as "the expansion of the Solomonic kingdom into the central Negev highlands." Hence these sites were considered by Cohen as Israelite. The end of these settlements was attributed by him to the campaign by Pharaoh Shishak (Cohen 1976: 50; Cohen and Cohen-Amin 2004). Therefore, the time of human activity in these settlements, from establishment to abandonment, was rather short in the above interpretation, only a few decades from about the middle of the 10th century BCE to roughly around 925 BCE. Although ceramic finds at Horvat Haluqim are comparatively rare, hand-made Negbite pottery dominates, which cannot be used for dating, due to its long time range (Cohen 1976: 44). Only a small fraction of the ceramic finds at Horvat Haluqim, i.e., wheel-made pottery, was included in the above archaeological age assessment.

A comprehensive discussion of the Iron Age sites in the Negev is beyond the scope of this article. Nevertheless, it is appropriate to include here the viewpoints expressed by Beno Rothenberg. His excavations at Timna (Rothenberg 1972: 153–170) showed the presence of hand-made Negbite pottery in contexts with Egyptian material remains of the 19th and 20th Dynasties, ranging from Sethos I (1318–1304 BCE) to Ramesses V (1160–1156 BCE). Therefore, he suggested that the rural settlements in the central Negev, dominated by Negbite pottery, may have a longer history that preceded the Iron IIA, and include the Late Bronze Age and Iron I. These propositions by Rothenberg are indeed supported by the above series of radiocarbon dates from Horvat Haluqim.

Rothenberg (1967: 86–100; 1972: 153–154; 1979: 122–123; 1996–1997: 36) proposed that the biblical Amalekites established the central Negev settlements in the Late Bronze Age. The word Amalek or Amalekites occurs about 50 times in the Hebrew Bible, for example Num 13:29–עֲמָלֵק יוֹשֵׁב בְּאֶרֶץ הַנֶּגֶב– which can be translated as "Amalek resides (dwells) in the land of the Negev." Rothenberg (1972: 181–182) also suggested that the population of the Negev highlands may have been involved in the copper industry at Timna.

The chemical composition of the bronze chisel from Horvat Haluqim (Figs. 20.6–7) points to Faynan, rather than Timna, as the source of the copper ore. Radiocarbon dates from a variety of metallurgic sites in the southern Levant were compiled by Ben-Yosef *et al.* (2010). These dates show quite a number of sites in both the Faynan–Timna areas having radiocarbon ages in the Late Bronze and Iron I. Also archaeomagnetic dating of Site F2 at Timna, as well as ^{14}C and TL dates, show that copper production took place in the region during the Late Bronze and early Iron Ages, 13th to 11th century BCE (Ben-Yosef, Tauxe and Levy 2010). Later in the Iron Age, copper production was prominent at Khirbat en-Nahas, covering the 10th and 9th centuries BCE (Levy *et al.* 2008) and also at Timna Site 30 during the 11th to 9th centuries BCE (Ben-Yosef *et al.* 2012).

CONCLUSIONS

Detailed radiocarbon dating of a sequence of layers on the edge of Terraced Field 12 in the Eastern Wadi of Horvat Haluqim gives evidence of continuous occupation from the Late Bronze Age to the Iron I (14th to 12th centuries BCE). A bronze chisel excavated in these layers was dated by radiocarbon to the 12th century BCE. Lead isotope ratios of the bronze chisel show that the origin of the copper ore is from the Dolomite-Limestone-Shale unit of the Cambrian Age in the Faynan area. Apparently trade relations existed between Horvat Haluqim and the Faynan area in the 12th century BCE. The most convenient east–west route from the Faynan area through the Negev to Horvat Haluqim is via the Zin Canyon, which contains a number of springs. Various metallurgic sites are known in the Faynan area, as well as in the Timna area, having absolute dates within the Late Bronze and early Iron Age. The so-called empty second millennium BCE in the central Negev highlands and northeastern Sinai is gradually showing more evidence of archaeological features that had remained "invisible" with traditional archaeological methodologies, lacking or ignoring both geoarchaeology and radiocarbon dating. Local desert inhabitants probably established these settlements, possibly the biblical Amalekites, as proposed by Beno Rothenberg more than 40 years ago.

ACKNOWLEDGMENTS

We express our thanks to the technicians of the Geochemistry Department at the Israel Geological Survey (Jerusalem) for the chemical measurements of the bronze chisel. Likewise, we thank the technical staff of the Center for Isotope Research at the University of Groningen for the AMS radiocarbon measurements. Thanks to Channa Cohen-Stuart, who participated in part of the excavations during March 2007. Our sincere appreciations to the Petrie Museum of Egyptian and Sudanese Archaeology (University College London) for their permission to publish a photograph of an Egyptian chisel in their collection. We are grateful to Gabriel Ore, who conducted measurements with the electronic total station theodolite. The research was partly funded by the Stichting Midbar Foundation.

REFERENCES

Aerts-Bijma, A.T., Van der Plicht, J. and Meijer, H.A.J. 2001. Automatic AMS Sample Combustion and CO_2 Collection. *Radiocarbon* 43(2A): 293–298.

Avni, G., Avni, Y. and Porat, N. 2009. A New Look at Ancient Agriculture in the Negev. *Cathedra* 133: 13–44 (Hebrew with English abstract).

Avni, Y., Porat, N. and Avni, G. 2012. Pre-Farming Environment and OSL Chronology in the Negev Highlands, Israel. *Journal of Arid Environments* 86: 12–27.

Avni, Y., Porat, N., Plakht, J. and Avni, G. 2006. Geomorphologic Changes Leading to Natural Desertification Processes versus Anthropogenic Land Conservation in an Arid Environment, the Negev Highlands, Israel. *Geomorphology* 82: 177–200.

Ben-Yosef, E., Levy, T.E., Higham, T., Najjar, M. and Tauxe, L. 2010. The Beginning of Iron Age Copper Production in the Southern Levant: New Evidence from Khirbat al-Jariya, Faynan, Jordan. *Antiquity* 84(325): 724–746.

Ben-Yosef, E., Shaar, R., Tauxe, L. and Ron, H. 2012. A New Chronological Framework for Iron Age Copper Production in Timna (Israel). *Bulletin of the American Schools of Oriental Research* 366: 1–41.

Ben-Yosef, E., Tauxe, L. and Levy, T.E. 2010. Archaeomagnetic Dating of Copper Smelting Site F2 in Timna Valley (Israel) and Its Implication on Modeling Ancient Technological Developments. *Archaeometry* 52(6): 1110–1121.

Bronk Ramsey, C. 2001. Development of the Radiocarbon Calibration Program OxCal. *Radiocarbon* 43(2A): 355–363.

Bronk Ramsey, C. 2013. OxCal Program Version 4.2.2. *Radiocarbon Accelerator Unit, University of Oxford.* Available at: http://c14.arch.ox.ac.uk/oxcal.html

Bruins, H.J. 1986. *Desert Environment and Agriculture in the Central Negev and Kadesh-Barnea during Historical Times (*Ph.D. dissertation, University of Wageningen). Nijkerk.

Bruins, H.J. 1990. Ancient Agricultural Terraces at Nahal Mitnan. *ʿAtiqot* 10: 127–128 (Hebrew), 22*–28* (English).

Bruins, H.J. 2007. Runoff Terraces in the Negev Highlands during the Iron Age: Nomads Settling Down or Farmers Living in the Desert? In: Van der Steen, E.J. and Saidel, B.A., eds. *On the Fringe of Society: Archaeological and Ethnoarchaeological Perspectives on Pastoral and Agricultural Societies* (BAR International Series 1657). Oxford: 37–43.

Bruins, H.J. 2012. Ancient Desert Agriculture in the Negev and Climate-Zone Boundary Changes during Average, Wet and Drought Years. *Journal of Arid Environments* 86: 28–42.

Bruins, H.J., Evenari, M. and Nessler, U. 1986. Rainwater-Harvesting Agriculture for Food Production in Arid Zones: The Challenge of the African Famine. *Applied Geography* 6: 13–32.

Bruins, H.J., Nijboer, A.J. and Van der Plicht, J. 2011. Iron Age Mediterranean Chronology: A Reply. *Radiocarbon* 53(1): 199–220.

Bruins, H.J. and Ore, G. 2009. Runoff from Loess or Bedrock? Hillslope Geoarchaeology of Ancient Runoff Farming Systems at Horvat Haluqim and Har Eldad in the Central Negev Desert. *Israel Journal of Earth Sciences* 57: 231–247.

Bruins, H.J. and Van der Plicht, J. 2005. Desert Settlement through the Iron Age: Radiocarbon Dates from Sinai and the Negev Highlands. In: Levy, T.E. and Higham, T., eds. *The Bible and Radiocarbon Dating: Archaeology, Text and Science.* London: 349–366.

Bruins, H.J. and Van der Plicht, J. 2007. Radiocarbon Dating the "Wilderness of Zin." *Radiocarbon* 49(2): 481–497.

Bruins, H.J. and Van der Plicht, J. 2017. Iron Age Agriculture —A Critical Rejoinder to "Settlement Oscillations in the Negev Highlands Revisited: The Impact of Microarchaeological Methods." *Radiocarbon* 59(1): 1–16.

Bruins, H.J., Van der Plicht, J. and Haiman, M. 2012. Desert Habitation History by ^{14}C Dating of Soil Layers in Rural Building Structures (Negev, Israel): Preliminary Results from Horvat Haluqim. *Radiocarbon* 54(3–4): 391–406.

Carmi, I., and Segal, D. 2007. Radiocarbon Dates. In: Cohen, R. and Bernick-Greenberg, H., eds. *Excavations at Kadesh Barnea (Tell el-Qudeirat) 1976–1982,* Part 1 (IAA Reports 34). Jerusalem: 303–306.

Cohen, R. 1976. Excavations at Horvat Haluqim. *ʿAtiqot* 11: 34–50.

Cohen, R. 1986. *The Settlement of the Central Negev in the Light of Archaeology and Literary Sources during the 4th—1st Millennia B.C.E.* (Ph.D. dissertation. The Hebrew University). Jerusalem.

Cohen, R. and Cohen-Amin, R. 2004. *Ancient Settlement of the Negev Highlands, Vol. II: The Iron Age and the Persian Period* (IAA Reports 20). Jerusalem (Hebrew with English summary).

Gale, N.H., Bachmann, H.G., Rothenberg, B., Stos-Gale, Z.A. and Tylecote, R.F. 1990. The Adventitious Production of Iron in the Smelting of Copper. In: Rothenberg, B., ed. *The Ancient Metallurgy of Copper.* London: 182–191.

Hauptmann, A., Begemann, F., Heitkemper, E., Pernicka, E. and Schmitt-Strecker, S. 1992. Early Copper Produced at Feinan, Wadi Araba, Jordan: The Composition of Ores and Copper. *Archeomaterials* 6: 1–33.

Hauptmann, A., Begemann, F. and Schmitt-Strecker S. 1999. Copper Objects from Arad—Their Composition and Provenance. *Bulletin of the American Schools of Oriental Research* 314: 1–17.

Levy, T.E., *et al*. 2008. High-Precision Radiocarbon Dating and Historical Biblical Archaeology in Southern Jordan. *Proceedings of the National Academy of Sciences* 105: 16460–16465.

Meshel, Z. 2000. *Sinai: Excavations and Studies* (BAR International Series 876). Oxford.

Mook, W.G. and Streurman, H.J. 1983. Physical and Chemical Aspects of Radiocarbon Dating. In: Mook, W.G. and Waterbolk, H.T., eds. Proceedings, Groningen Symposium ^{14}C and Archaeology. *PACT Journal* 8: 31–55.

Reimer, P.J., *et al*. 2013. IntCal13 and Marine13 Radiocarbon Age Calibration Curves, 0–50,000 Years Cal BP. *Radiocarbon* 55: 1869–1887.

Rothenberg, B. 1967. *Tsefunot Negev: Archaeology in the Negev and the Arabah*. Tel Aviv (Hebrew).

Rothenberg, B. 1972. *Timna*. London.

Rothenberg, B. 1979. *Sinai: Pharaohs, Miners, Pilgrims and Soldiers*. Berne.

Rothenberg, B. 1996–1997. Researches in the Southern Arabah 1959–1990. Summary of Thirty Years of Archaeo-Metallurgical Field Work in the Timna Valley, the Wadi Amram and the Southern Arabah (Israel). *World Journal of Prehistoric and Ancient Studies* 2–3: 5–42.

Van der Plicht, J., Bruins, H.J. and Nijboer, A.J. 2009. The Iron Age Around the Mediterranean: A High-Chronology Perspective from the Groningen Radiocarbon Database. *Radiocarbon* 51(1): 213–242.

Van der Plicht, J., Wijma, S., Aerts A.T., Pertuisot, M.H. and Meijer, H.A.J. 2000. The Groningen AMS Facility: Status Report. *Nuclear Instruments and Methods in Physics Research Section B* 172(1–4): 58–65.

CHAPTER 21

THE DISCOVERY OF THE SINAITIC SITE KUNTILLET ʿAJRUD[1]

Ze'ev Meshel

Kuntillet ʿAjrud is the only Israelite site (800 BCE) in Sinai. It was built on the eastern edge of the Kingdom of Judah by King Joash of Israel near the Darb Ghazza, the important trade-route that connected the Red Sea with the Mediterranean Sea. Kuntillet ʿAjrud was mainly a religious site. It also served as a way-station but we have no evidence of the kind of goods and commodities that were conveyed.

Two famous explorers of the Near East were the first modern visitors to ʿAjrud: Palmer, on January 1870 and Musil 32 years later. Each described the site and his adventures in arriving there. Their narratives provide vivid impressions of travel conditions in Sinai and the Negev about 150 years ago. Beno Rothenberg was the first Israeli to visit Kuntillet ʿAjrud. He dated it correctly to the Iron II which was his contribution to the site.

INTRODUCTION

Kuntillet ʿAjrud ("the solitary hill of the wells," Ḥorvat Teman in Hebrew), is a single-period Israelite site, unique in its character and finds. It is located in eastern Sinai, 50 km south of Kadesh-Barnea, on the edge of the territory of Judah, near the ancient road leading to Eilat and southern Sinai (Fig. 21.1).

The site was excavated in 1975–1976 by an expedition of the Institute of Archaeology of Tel Aviv University, led by the author.

The site is unique in both its unusual nature and finds. The most important finds are the inscriptions, the murals and the drawings, most of which are religious, mentioning "YHWH of SAMARIA and His ASHERAH" and "YHWH of TEMAN and His ASHERAH." Some of the finds reflect a strong northern influence originating in the Kingdom of Israel.

The site was apparently constructed in about 800 BCE, by order of King Joash of Israel, who had defeated King Amaziah of Judah and taken control of the territory in Judah. The war between them seemingly began with a dispute over renewed trade from the Red Sea and its profits.

The inscriptions and most of the finds attest to the unparalleled uniqueness of the site as a religious center related to the royal journeys to Eilat and Ezion-Geber, and perhaps also to pilgrimages to southern Sinai.

THE FIRST EXPLORERS

Two famous 19th and early 20th century explorers of the Near East, Palmer and Musil, visited Kuntillet ʿAjrud and documented their visits. They are considered the discoverers of the site. Edward Palmer was the first. The following excerpts have been taken from Palmer's description (Palmer 1871: 341–342, 422–423):

---

1 In Memory of Beno Rothenberg, a friend and colleague, and the first Israeli scholar who visited Kuntillet ʿAjrud.

Fig. 21.1: Aerial view of the lonely hill ("*kuntilleh*") on which the site was built.

> We were bound for some ruins, called Contellet Garaiyeh, of which we had heard; the weather was frightfully hot, and as a sandstorm had been blowing with great violence for two days, it was not by any means a comfortable journey....
>
> ...Contellet Garaiyeh we found to be a white hill, with a slight depression on the summit surrounded by what at first appeared to be a natural parapet or rampart. On digging into the latter, however, we discovered that it was composed entirely of debris and concealed the foundations of an old wall composed of sun-dried bricks and containing beams of wood with signs of mortices, bolts, &c. The most curious feature about it was that at regular intervals we came upon fragments of large amphorae or jars; these had been built into the wall in regular sets of four, carefully packed with straw and protected by a framework of wood. The place appeared to have been an ancient fort, and the jars were probably arranged to serve as reservoirs for water in lieu of cisterns. One jar which we extracted was marked upon the shoulder with a Phoenician aleph, but this circumstance is not to be relied upon for fixing the date, as even at the present day the potters of Palestine and Syria make use of the most ancient signs for their trade-marks, having probably had them handed down from generation to generation. Yet the wood used in the frame-work seems to indicate a period when the country was better cultivated than it is at present, for there is not now a single tree to be found, save one small sidr-tree in the fort at Nakhl and an acacia in Wády Fahdí dedicated to the sheikh of that name.

According to his itinerary, Palmer remained at the site on January 28–29, 1870, in the depth of winter. It is thus peculiar that he would note that "the weather was frightfully hot." More appropriate to the season are sandstorms, such as the one he described, and another we experienced during the second excavation season. He describes the site as the ruins of an ancient fort enclosed by a wall which was found upon "digging into it." He erred in his assessment that it was built of sun-dried bricks. However, he correctly pointed out that beams of wood were integrated into the construction. We identified the beams we found as tamarisk, a bushy tree found even in the wadi below the site. Upon comparing his findings with ours, it became clear that he had excavated in the western storeroom, uncovering a few of the pithoi lying there, and discerning the letter *alef* on the shoulder of one of them, the likes of which we would find often in our excavations. Palmer thought it was the Greek *alpha*, and erred in dating the site, proposing to identify it with the Roman way-station Gypsaria on the Elusa-Eilat road that appears in the *Tabula Peutingeriana*. It seems that the "frame-work of wood" Palmer described, which "protected the vessels," consisted of fallen branches from the ceiling.

During our excavations, we got the impression that robber's pits were dug in more than one place. It is difficult to know which were Palmer's investigations and which the result of subsequent rummaging by local Bedouin who hoped to uncover valuables. One can suppose that the single incised letter Palmer found, as well as other inscriptions perhaps exposed by the Bedouin, helped spread the rumor that there were inscriptions at the site. Fortunately for us, the Czech investigator Alois Musil, who visited the site on September 7, 1902, some 32 years after Palmer, failed to discover any of the inscriptions, leaving them all to be unearthed during our expedition.

Musil's (1907: 170–175) account is no less intriguing than Palmer's. The dangers, struggles and adventures involved in such a journey to the region—encounters with Bedouin and the swindling by the guides, a result of the explorers' utter dependency on them—are reflected in the account. The following are a few excerpts, translated from the German (*ibid.*):

> [The Bedouin guide] wanted to show me another inscription, located above a cave next to Kuntille Ajrud, where there are also structural remains. Since I had already heard a lot about this site, I agreed to go.... We came upon an isolated hill with a flat summit, rising some 20 m above the surroundings and appearing to be nearly insurmountable. This is the Kuntille Ajrud where the inscriptions are said to be.... We dismounted from the camels and rushed to find them. The guide did not agree to join us, preferring to stay with the camels. It was an arduous climb up the hill, which is composed of limestone blocks joined by thick layers of salt, its summit measuring some 60 by 10 m. It was once home to a small fort, the foundations of which are still visible, and is now used for burials.
>
> Despite our hopes, we encountered no cave, cranny, ancient tomb, or inscription.
>
> Dejected, we made the descent to Wadi el-Karaje, which runs along the base of the western slope of the hill. My trusty escort Muhammad was ready to clash with the guide once he announced that he no longer wanted to guide us and requested an exorbitant fee for his services. We would have freely struck him, had we not been in such a delicate and precarious situation, given that he was of the Tihi ibn Breik tribe, in whose territory we were presently situated. He could have turned his tribal brethren against us. Furthermore, we feared he would prevent us from receiving an alternate guide, and, should we pay the fee he requested, we would be asked to pay even more to the next guide, only increasing the risk of falling prey to further opportunistic brigands. Upon feigning destitution and pleading with him, he ultimately agreed to get the going rate....
>
> We rode with Muhammad to Birkat Ajrud, located on the right bank of W. el-Karaje, some 400 m northwest of the Kuntille hill. The watering hole is fed by an underground stream. As we joined together, our guide took notice of a few Bedouin, immediately sounding a war cry and yelling that we had climbed onto the holy spot of Kuntille. At once we were inundated by battle cries. Muhammad and his camel were taken captive.
>
> What ensued is difficult to retell....The enemy split into two groups: those of the smaller group took our camels, those of the larger surrounded us while hiding behind bushes. The chief approached us on a white mare and sought our capitulation. Since it was clear to me that we had no chance should we resist, I called to my companions to lower their rifles and planned to negotiate with the Sheikh. However Barakhat did not heed my order and struck the Sheikh with the butt of his rifle, attempting to fell him from his mare. There were immediate shouts of rage, the Bedouin

> converged upon us and a brawl ensued, at the end of which we lay on the ground, without our weapons....Surrounding us stood the enemy, enraged that we would dare climb up to their holy place. They cursed us and the Sheikh swore that he was going to shoot that dog Barakhat. Once they were calmed and began discussing how to divide the loot and what to do with us...the Sheikh finally announced that, although our rifles and possessions are quite tempting, we could continue on our way should we pay a fine for damaging their holy place, though Barakhat he wished to kill with his own hands. The *dijje*, or blood ransom, for the murder he would pay with money from selling the Kuntille salt. After further negotiations we gave them more gifts, received the camels in return, and saw the release of Barakhat. Amidst reciprocal curses, we loaded our belongings onto the camels and quit the place without a guide.

The account includes a few interesting details on Kuntillet ʿAjrud, including the fact that rumors of the existence of inscriptions at the site had already become fairly widespread; wall remains were visible at the site; the place was used for burial; and, most interesting, it was considered a holy place by the local Bedouin. It should be stressed in this regard that none of the Bedouin who visited us during our excavation at the site made such a claim. Concerning the graves, we only encountered a single burial, situated in the rubble of one of the walls. Perhaps Musil was misled by the vestiges of Palmer's activity at the site. Another inaccuracy by Musil is geological: the hill is not of limestone, but chalk, with thin layers composed primarily of gypsum.

Prof. Beno Rothenberg visited the site in 1967 and identified it as: "an Iron Age fort (8th century BCE). We have not found another Iron Age site anywhere else in the Sinai" (1972: 91).

REFERENCES

Musil, A. 1907. *Arabia Petraea II: Edom*. Vienna.

Palmer, E.H. 1871. *The Desert of the Exodus II*. Cambridge.

Rothenberg, B. 1972. *An Archaeological Survey in Sinai, 1967–1972* (Annual of the Land of Israel Museum 14: 89–100) (Hebrew).

CHAPTER 22

THE ORIGIN OF THE COPPER USED IN CANAAN DURING THE LATE BRONZE/IRON AGE TRANSITION

Naama Yahalom-Mack and Irina Segal

The accepted interpretation of the archaeological and metallurgical data from the site of Timna in the Arabah in Israel as having been entirely under Egyptian auspices/control during the time of the 19th and 20th Dynasties has been recently and successfully challenged by renewed excavations at the site. Both the nature of the Egyptian influence as well as the chronology of the site, have undergone revision. The present study examines this issue from the perspective of the finished products. Securely dated bronze objects from four settlement sites in Canaan dating to the second half of the second millennium BCE were subjected to chemical and lead isotope analyses, showing that the use of copper from Timna was prevalent mainly during the 11th century BCE, following the withdrawal of the Egyptians from Canaan and not beforehand. This result is examined in light of the geo-political situation following the demise of the Mediterranean-basin trade at the end of the Late Bronze Age, particularly vis-à-vis Cyprus and its lively copper trade, as well as the floruit of metalworking in Iron Age I Canaan.

INTRODUCTION

POTENTIAL COPPER SOURCES

Archaeological and textual evidence has shown that there were a number of possible sources of copper available to Canaan during the second half of the second millennium BCE. The main sources to consider during this period are Cyprus and the Arabah Valley in Israel, namely the Faynan and Timna copper ore deposits.

Cyprus was the most important source of copper in the Eastern Mediterranean in antiquity (e.g., Knapp 1986; Zwicker 1986; Muhly 1989; Stos-Gale *et al.* 1997). The circulation of Cypriot copper in the form of oxhide ingots through maritime trade is well known in the Late Bronze Age (e.g., Gale and Stos-Gale 1986; Gale 1991; Knapp and Cherry 1994; Stos-Gale *et al.* 1997; Jones 2007; Kassianidou 2009; Lo Schiavo *et al.* 2009). This is evidenced in Canaan by a number of copper oxhide ingots, probably dating to the Late Bronze Age, found in underwater excavations and surveys off the northern coast of Israel (Galili, Shmueli and Artzy 1986; Galili, Gale and Rosen 2011; Galili, Gale and Rosen 2013). Based on the occurrence of oxhide ingots in contexts dated to the 12th and 11th century BCE in Sardinia, it had been suggested that Cypriot copper was still traded in the form of oxhide ingots until the 11th century (e.g., Begemann *et al.* 2001; Kassianidou 2012). Since Cyprus seems to have survived the crisis that occurred at the end of the Late Bronze Age, ca. 1200 BCE in the Mediterranean basin, and to have maintained varying trade relations with the Levant during parts of the Iron Age I (Muhly 1989; Karageorghis 1992; Muhly 1992; Knapp 1994; Mazar 1994; Sherratt 1998; Gilboa 2005), the importation of copper from Cyprus after 1200 BCE remains a possibility.

Local copper ore deposits are known from Faynan on the eastern side and Timna on the western side of the Arabah Valley (see Fig. 22.1). At Timna, evidence for exploitation of copper has been identified as early as the fifth millennium BCE (Hauptmann and Wagner 2007; Ben-Yosef *et al.* 2008). Sporadic traces of mining and smelting activities were found to date throughout the second millennium BCE, with the main period of exploitation attributed to the Egyptians of the 19th and 20th Dynasties between the late 14th and 12th centuries BCE, along with desert groups (Rothenberg 1988, 1990). Recently, based on renewed excavations at the site and radiocarbon dating, an alternative scenario has been proposed. Avner (2014) showed that metallurgical activities in Timna were already taking place when the Egyptians set foot on the site, based mainly on his re-evaluation of the stratigraphy and concept of the Hathor shrine at Site 200. Ben Yosef *et al.* (2012), based on renewed excavations at Site 30, which yielded a series of radiocarbon dates, proposed that the main period of exploitation in Timna was in fact between the 11th and 9th centuries BCE, parallel to the activities in Faynan (see below).

Evidence for copper mining in Faynan has been recovered from as early as the fourth millennium BCE, with intense and well-organized smelting activity conducted throughout the second half of the third millennium BCE (Levy *et al.* 2002; Weisgerber 2006; Hauptmann 2007). Activity in Faynan was

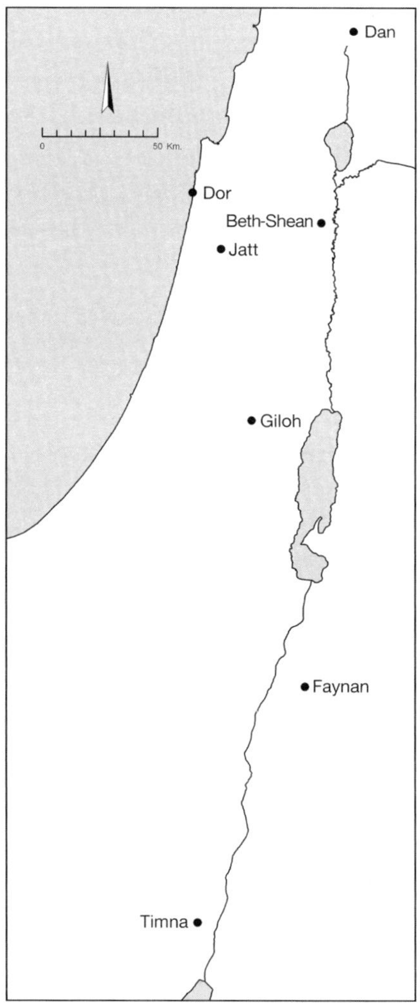

Fig. 22.1: Map of sites mentioned in this chapter.

minimal during most of the second millennium BCE, while by the 11th century BCE, and mainly during the 10th and 9th centuries BCE (Iron IIA), the site became the largest Iron Age copper-smelting site in the Levant (e.g., Bachmann and Hauptmann 1984; Levy *et al.* 2004; Hauptmann 2007; Levy *et al.* 2008; Ben-Yosef 2010; Ben-Yosef *et al.* 2010).

Copper production in the Sinai Peninsula related to the New Kingdom Egyptians must have taken place on a considerable scale (Rothenberg 1987: 7). The largest smelting site on the Sinai Peninsula was Bir Nasib I, with an estimated 100,000 tons of slag (Abdel-Motelib *et al.* 2012). It is described as a supra-regional center for copper production at least during the Late Bronze and Iron I, but to date remains unexcavated.

The possibility that recycled metal was used during this time must also be considered. Although this practice is as old as the beginning of the use of metal, there is evidence indicating that it had increased considerably during the transition from the Late Bronze to the Iron Age. For example, in the Cape Gelidonya shipwreck, dated to ca. 1200 BCE, large amounts of scrap metal intended for recycling were found alongside copper ingots (Bass 1967). Recycling is also assumed from the growing evidence for metal hoarding during this time (Knapp, Muhly and Muhly 1988; Knapp 2000).

The goal of the present study is to establish the source of copper during the second half of the second millennium BCE in Canaan through the analysis of bronze objects from secure stratigraphic settlement contexts related to this time. Identifying the source of the copper is an important component of our understanding of the trade patterns between Cyprus and Canaan, and particularly changes in these patterns around 1200 BCE, with the demise of international trade, and around the third quarter of the 12th century BCE, with the withdrawal of the Egyptian empire from Canaan.

Lead Isotope Analysis (LIA)

Beginning in the 1980s, Lead Isotope Analysis (LIA) has helped to establish the sources of copper in use around the Mediterranean (e.g., Gale and Stos-Gale 1982; Pernicka *et al.* 1984; Gale 1991; Hauptmann, Begemann and Schmitt-Strecker 1999; Begemann *et al.* 2001; Stos-Gale and Gale 2009). This method is based on the assumption that no significant isotope fractionation occurs during the smelting and the re-melting processes. The lead isotope ratios thus serve as a "fingerprint" of the mineral ore deposits, which can be compared with the end-product (Gale and Stos-Gale 1982; Pernicka *et al.* 1984). Together with chemical analysis, LIA is useful for provenance studies of metallurgical remains.

Alongside the advantages of this method, several caveats discussed in various studies in the past should be noted (e.g., Budd *et al.* 1995; Sayre, Yener and Joel 1995; Knapp 2000; Sayre *et al.* 2001; Hauptmann 2007: 31–38; Stos-Gale and Gale 2009), particularly when dealing with analyses of finished products. One problem exists when copper is alloyed with tin, which may contain some lead, or in the case of lead added to the copper, affecting the lead isotopic ratios. Another potential problem is that of recycling, wherein the lead can reflect more than one source of copper. Finally, the overlap between the lead isotope ratio of different copper ore sources can make the provenience data indecisive. In the latter case, archaeological and historical considerations can occasionally resolve the uncertainty, as does the combination of analytical methods (Hauptmann 2007: 38). One way to overcome these difficulties is to increase the amount of sampled items as much as possible, in order to achieve maximum probability of the copper ore sources.

In Israel, LIA was first used in the mid-1990s by the Israel Geological Survey, to establish the origin of copper in objects from the fourth and third millennia BCE, the Chalcolithic through the Intermediate Bronze Ages (Segal, Halicz and Cohen 1999; Segal, Halicz and Kamenski 2002; Segal, Kamenski and Merkel 2002; Namdar *et al.* 2004; Segal 2009; Segal and Rosen 2011). However, this method was largely avoided for periods later than the third millennium BCE mainly because of the above-mentioned recycling hypothesis.

LIA was employed in studies of the mining districts of Timna and Faynan. Over 40 ores from Timna sandstones (Amir-Avrona Formation), as well as metal objects from the Hathor shrine, were analyzed for LI by Gale *et al.* (1990). It was found that the objects from the shrine are consistent with the sandstone ores. A much smaller group of samples was taken from Faynan. These included a few ore samples from the Massive-Brown-Sandstone unit (MBS), used mainly during the Chalcolithic and Roman periods (Weisgerber 2006; Hauptmann 2007), and eight ore samples from the underlying Dolomite-Limestone-Shale unit (DLS), which was the main formation used during the Bronze and Iron Ages (Hauptmann *et al.* 1992; Hauptmann 2007). Several metallurgical remains (slag and metal fragments) from the site of Khirbet en-Nahas at Faynan were also subjected to LIA. It was shown that the LI ratios of the Faynan DLS ores cluster together with most of these metallurgical remains, overlapping part of the Timna ores (see Fig. 22.4).

It is important to mention that Sardinian ores overlap with both Timna and Faynan (see, e.g., Hauptmann 2007: 82–83). Copper deposits in Sardinia were apparently exploited during the Nuragic period, from the 15th century BCE and onwards (Balmuth and Tylecote 1976; Begemann *et al.* 2001). However the occurrence of Cypriot oxhide ingots on the island and the decentralized sociopolitical organization does not support the possibility that Sardinian copper was exported. About 100 Nuragic hoards do testify to intensification of metallurgical practices in Sardinia at the end of the second millennium BCE and to extensive long-distance connections with the Aegean in the east and the Iberian Peninsula in the west (Begemann, Schmitt-Strecker, Pernicka and Lo Schiavo 2001), but there is still no evidence for trade relations between Sardinia and the Levant at this time. Based on these considerations, Sardinia at this point in research will not be considered as a possible ore source of copper in Canaan throughout the periods in question, even though the remote possibility of the use of Sardinian copper cannot be completely dismissed.

MATERIALS AND METHODS

ANALYTICAL PROCEDURE AND INSTRUMENTATION

After surface drillings were discarded, the 20–25 mg drillings were dissolved in aqua regia (HNO_3:HCl = 1:3) and diluted with double distilled water up to 25–50 ml. This solution was then subjected to chemical analysis using ICP-AES (Optima3300, Perkin Elmer) and after additional dilution to LIA using MC-ICP-MS (NU plasma, Wrexham, UK), coupled with an Aridus desolvation system (CETAC technologies). The methods were previously described (Segal and Halicz 2005). In analysis of chemical composition the standard deviation for major elements was about 1%, for traces—up to 5%. The error of lead isotope ratio analysis (1σ) is less than 0.01% for $^{204/206}$Pb, and less than 0.005% for $^{208/206}$Pb and $^{207/206}$Pb.

ARTIFACTS ANALYZED

In the present study, 47 artifacts from four different settlement sites were sampled for chemical and LI analyses (for description, see Table 22.1); three of the sites are located in different geographic areas; Tel Beth-Shean in the Jordan Valley, Tel Dan in the Upper Galilee and Tel Dor in the coastal strip (Fig. 22.1). A single object from Giloh, in southern Jerusalem, was also sampled.

The artifacts sampled for this study are dated to the 15th–10th centuries BCE. Five artifacts are dated to the LB I (15th century BCE) and 11 artifacts are dated to the LB II and mainly to the later part of the 13th century BCE (LB IIB). Of the latter, six especially well-preserved specimens originated in two adjacent contemporary buildings (Buildings NB and NC) at Tel Beth-Shean that are assumed to have served the Egyptian administration at the site (Panitz-Cohen and Mazar 2009). These include a complete Mycenaean-type knife (BS-14, Fig. 22.2: 1) and a complete unique fork-like object (BS-15, Fig. 22.2: 3), both of which are considered to be imported based on typological considerations (Yahalom-Mack 2009b), as well as two stunning bolts (BS-16, 17), a complete Egyptian 19th Dynasty-type razor (BS-18, Fig. 22.2: 2) and a needle (BS-19). Another Egyptian-type razor (BS-6) originated from a different context at Tel Beth-Shean and is dated to the same period.

The largest group of samples in this study (31) dates to the Iron I (12th–11th centuries BCE). This group includes seven artifacts dating to the early part of this period (Iron IA) from Tel Beth-Shean, occupied at the time by Egyptians of the 20th Dynasty, and three artifacts from Tel Dan, Stratum VI, which also dates to this time period. Fourteen artifacts are dated to the later part of the Iron I (Iron IB, late 12th–11th centuries BCE), five generally to the Iron I, and two to the IRI–II transition.

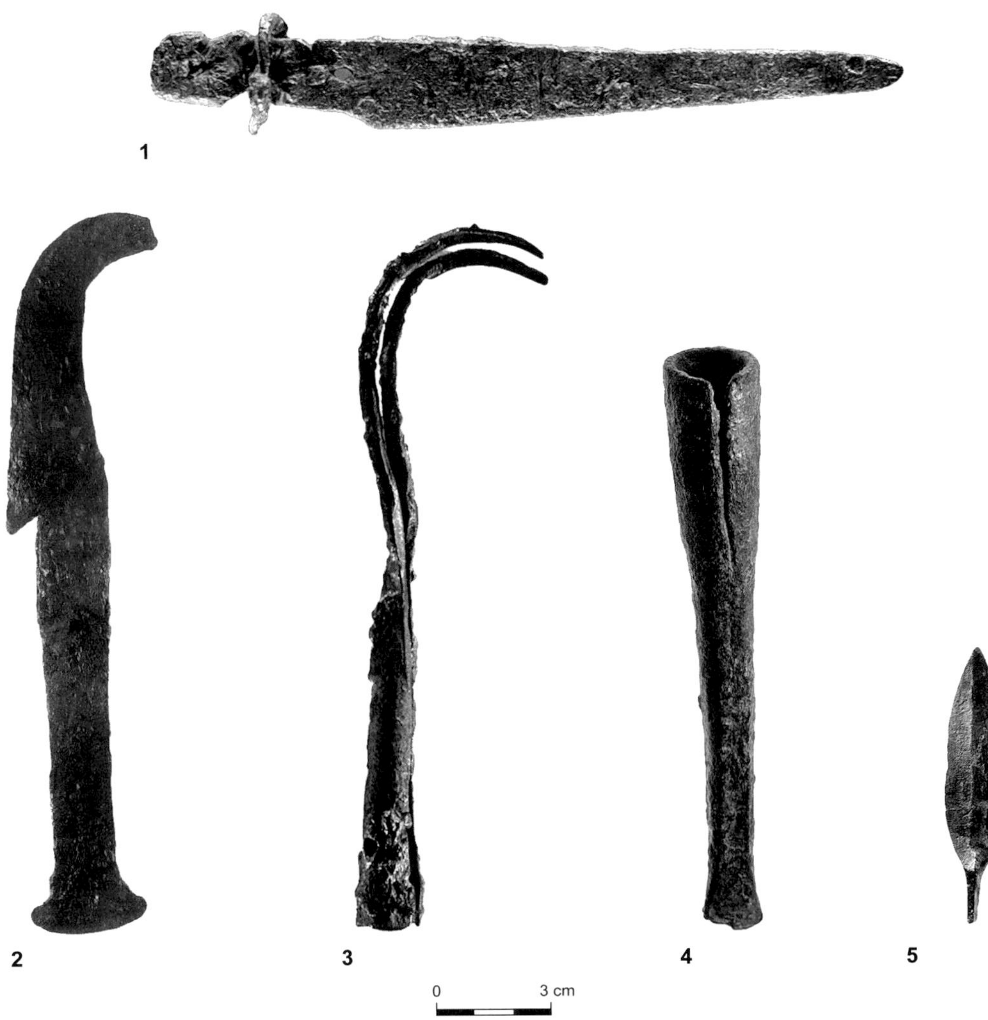

Fig. 22.2: Selected artifacts from Tel Beth-Shean: (1) Knife (BS-14); (2) Razor (BS-18); (3) Fork (BS-15); (4) Socket chisel (BS-4); (5) Arrowhead (BS-8) (courtesy of the Tel Beth-Shean Expedition).

TABLE 22.1: OBJECTS SUBJECTED TO LIA FROM TEL BETH-SHEAN, TEL DAN, TEL DOR AND GILOH

No.	*Site*	*Sample*	*Reg. No.*	*Locus*	*Stratum*	*Century [BCE]*	*Type*	*Reference*	*Fig.*
1	Beth-Shean	BS-2	683139	68316	R-2	15th	Cauldron handle	Yahalom-Mack 2007	
2	Dan	Dan-2	16794	6156	VIII	15th	Dagger handle	Unpublished	
3	Dan	Dan-5	24928	7246	VIII	15th	Perforated disc	Unpublished	
4	Dan	Dan-1	24700	7225	VIII	15th	Rod	Unpublished	
5	Dan	Dan-19	23132	4643	VIII	15th	Slag	Unpublished	
6	Dan	Dan-9	20857	8224	VII	14th–13th	Arrowhead	Unpublished	
7	Dor	Dor-5	186050	18414	G-12b	13th	Ring	Unpublished	
8	Dor	Dor-6	186026	18414	G-12b	13th	Waste product	Unpublished	
9	Beth-Shean	BS-15	184158	18411	N-4	End of 13th	Fork	Yahalom-Mack 2009b	22.2: 3
10	Beth-Shean	BS-14	184143	98417	N-4	End of 13th	Knife	Yahalom-Mack 2009b	22.2: 1
11	Beth-Shean	BS-19	184250	18411	N-4	End of 13th	Needle	Yahalom-Mack 2009b	
12	Beth-Shean	BS-18	387624	38703	N-4	End of 13th	Razor	Yahalom-Mack 2009b	22.2: 2
13	Beth-Shean	BS-16	104359	10429	N-4	End of 13th	Stunning bolt	Yahalom-Mack 2009b	
14	Beth-Shean	BS-17	387514	38703	N-4	End of 13th	Stunning bolt	Yahalom-Mack 2009b	
15	Beth-Shean	BS-6	108228	10852	S-5	End of 13th	Razor	Yahalom-Mack 2009b	
16	Dor	Dor-8	187208	18511	G-11	End of 13th	Hoe	Unpublished	
17	Beth-Shean	BS-8	288051	28814	S-3a	12th	Arrowhead	Yahalom-Mack 2009b	22.2: 5
18	Beth-Shean	BS-12	107208	10728	S-4	12th	Arrowhead	Yahalom-Mack 2009b	
19	Beth-Shean	BS-20	188096	18804	S-3a	12th	Chisel	Yahalom-Mack 2009b	
20	Beth-Shean	BS-11	108163	10831	S-4	12th	Figurine fragment	Yahalom-Mack 2009b	22.3
21	Beth-Shean	BS-5	288283	28831	S-3a	12th	Stick pin	Yahalom-Mack 2009b	
22	Beth-Shean	BS-9	288060	28814	S-3a	12th	Stunning bolt	Yahalom-Mack 2009b	
23	Beth-Shean	BS-1	288039	28807	S-3a	12th	Arrowhead	Yahalom-Mack 2009b	
24	Dan	Dan-3	23957	7169	VI	12th	Chisel	Shalev 1993	
25	Dan	Dan-14	23477/8	7079	VI	12th	Crucible slag	Shalev 1993	
26	Dan	Dan-12	24114	7020	VI	12th	Needle	Shalev 1993	
27	Giloh	BS-3	-	-	-	12th–11th	Dagger	Mazar 1981	
28	Dan	Dan-10	23450/9	7065	V	12th–11th	Arrowhead	Shalev 1993	
29	Dan	Dan-13	9812	622	IRI	12th–11th	Crucible slag	Unpublished	
30	Dan	Dan-18	23377	7061	V	12th–11th	Crucible slag?	Unpublished	
31	Dan	Dan-20	18536	4323	IRI	12th –11th	Crucible slag?	Unpublished	
32	Dor	Dor-4	182131	18285	G-10	Late 12th –11th	Chisel	Unpublished	

No.	*Site*	*Sample*	*Reg. No.*	*Locus*	*Stratum*	*Century [BCE]*	*Type*	*Reference*	*Fig.*
33	Dor	Dor-3	181194	18078	G-10a	Late 12th –11th	Rod	Unpublished	
34	Dor	Dor-1	184080	18330	G-10c	Late 12th –11th	Unidentified	Unpublished	
35	Dor	Dor-2	183983	18316	G-10b	Late 12th –11th	Waste product	Unpublished	
36	Dor	Dor-7	183742	18312	G-10c	Late 12th –11th	Waste product	Unpublished	
37	Beth-Shean	BS-7	887086	68709	S-2	11th	Socket chisel	Yahalom-Mack 2009b	
38	Dan	Dan-11	18059/2	4202	IVB	11th	Arrowhead	Shalev 1993	
39	Dan	Dan-8	23667/1	7114	IV	11th	Chisel	Shalev 1993	
40	Dan	Dan-17	18059/17	4202	IVB	11th	Crucible slag	Unpublished	
41	Dan	Dan-16	24106	7015	IV	11th	Crucible slag?	Shalev 1993	
42	Dan	Dan-7	18101/10	4202	IVB	11th	Handle	Shalev 1993	
43	Dan	Dan-4	24083/20	7015	IV	11th	Knife / (ingot)	Shalev 1993	
44	Dan	Dan-6	23392	7060	IV	11th	Strip	Shalev 1993	
45	Dan	Dan-15	23864	7152	IV	11th	Worked?	Unpublished	
46	Beth-Shean	BS-4	388240	38841	S-2/ S-1b	11th –10th	Socket chisel	Yahalom-Mack 2009b	22.2: 4
47	Beth-Shean	BS-10	388227	38841	S-2/ S-1b	11th –10th	Socket chisel	Yahalom-Mack 2009b	

Note: *TBS III* = Yahalom-Mack 2009b

RESULTS

CHEMICAL COMPOSITION

The chemical composition of the items is presented in Table 22.2. The results show that the majority of the objects are made of tin bronze, with an average tin content of 4.4±3% and 5.9±3.2% for objects dated to the LB II and Iron I, respectively. Lead content varies from 0.05–3%, reaching ca. 21% in the figurine fragment from Beth-Shean (BS-11B). Trace elements are usually low. Five artifacts—a cauldron handle (BS-2b) and fork (BS-15b) from Tel Beth-Shean, a rod from Tel Dan (Dan-1), an unidentified object (Dor-1) and a waste product (Dor-6) from Tel Dor—are made of relatively pure copper. In addition, an arrowhead from Tel Dan (Dan-10) is made of arsenical bronze (1.5% As).

The unusually high lead content of the figurine fragment from Beth-Shean (BS-11, Fig. 22.3) suggests that it had been added intentionally, most likely in order to facilitate casting. Nickel content varies in the range of 0.005–0.36%. In the slag samples, iron content reaches up to ca. 16%.

Lead Isotope Analyses (LIA)

The results of the LIA are compiled in Table 22.3 and are plotted in Figs. 22.4–6 against LI ratios of the ingots and ores from Cyprus (http://oxalid.arch.ox.ac.uk), Timna (sandstones) and Faynan Dolomite Limestone Shale units (DLS) in the Arabah (Gale *et al.* 1990; Hauptmann *et al.* 1992; Hauptmann 2007) and Anatolia (Seeliger *et al.* 1985; Wagner *et al.* 1986; Yener *et al.* 1991; Sayre *et al.* 1992; Hirao, Enomoto and Tachikawa 1995). Because of the small number of measurements available from Faynan, metallurgical remains were also plotted. Note that LI ratios of copper ores from the MBS unit in Faynan were not included as they were exploited mainly in Chalcolithic and Roman times (see above). LI ratios of two slag samples from Bir Nasib, Sinai, were included (Pfeiffer 2009; Abdel-Motelib *et al.* 2012 after), as well as selected copper ingots, mostly oxhide in shape (Gale and Stos-Gale 1987; Gale 1991; Stos-Gale *et al.* 1997; Stos-Gale *et al.* 1998; Galili, Gale and Rosen 2012).

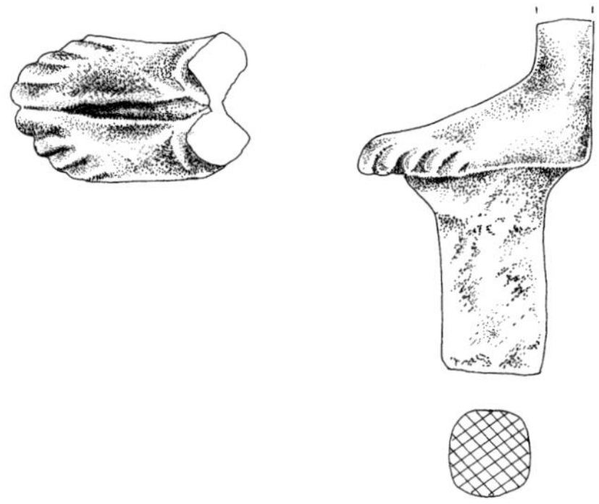

Fig. 22.3: Figurine fragment from Tel Beth-Shean (BS-11). Scale 1:1. Courtesy of the Tel Beth-Shean Expedition.

Table 22.2. Chemical Composition of Copper-based Oobjects, in wt. %

Century [BCE]	*Sample*	*S*	*Pb*	*Co*	*Ni*	*Mn*	*Fe*	*Cu*	*Ag*	*Sb*	*As*	*Sn*	*Zn*	*Au*
15th	BS-2	0.1	0.07	0.02	0.03	<0.001	0.11	91.9	<0.01	<0.01	0.26	<0.1	<0.1	<0.01
	Dan-1	<0.1	0.11	<0.005	0.01	<0.001	0.41	89.5	<0.01	0.05	0.58	<0.1	<0.1	<0.01
	Dan-2	<0.1	0.14	<0.005	0.02	<0.001	0.14	85.9	0.03	0.04	0.31	8.6	<0.1	<0.01
	Dan-5	0.1	0.31	0.01	0.03	<0.001	0.48	82.8	0.01	0.04	0.38	7.6	<0.1	<0.01
	Dan-19	0.1	0.27	0.04	0.32	0.051	4.59	21.8	<0.01	0.02	0.08	5.3	<0.1	<0.01
14th–13th	Dan-9	0.1	0.29	0.02	0.04	0.005	0.22	90.0	0.01	0.04	0.48	6.6	<0.1	0.01
13th	Dor-5*	0.5	0.15	0.03	0.01	0.002	0.95	75.5	<0.01	0.03	0.39	1.6	0.6	<0.01
	Dor-6*	<0.1	0.05	<0.005	0.01	<0.001	0.17	82.3	<0.01	<0.01	0.06	<0.1	<0.1	<0.01
End of 13th	BS-6	0.1	3.39	0.02	0.04	<0.001	0.02	82.4	0.03	0.09	0.48	6.1	<0.1	<0.01
	BS-14	0.1	0.15	0.03	0.05	<0.001	0.08	83.9	0.04	0.04	0.40	5.6	<0.1	<0.01
	BS-15	0.1	0.09	0.04	0.04	<0.001	0.23	88.8	0.03	0.02	0.69	0.3	<0.1	<0.01
	BS-16	0.1	1.07	0.09	0.05	<0.001	0.21	85.2	0.02	0.05	0.47	3.0	<0.1	<0.01
	BS-17	0.2	0.26	0.05	0.04	<0.001	0.47	80.9	0.02	0.03	0.31	7.8	<0.1	<0.01
	BS-18	0.2	0.44	0.03	0.03	<0.001	0.41	82.9	<0.01	0.02	0.29	6.2	<0.1	<0.01
	BS-19	0.1	0.12	0.02	0.02	<0.001	0.12	86.9	<0.01	0.04	0.32	2.1	<0.1	<0.01
	Dor-8	0.1	0.25	0.05	0.09	<0.001	0.06	85.6	<0.01	0.05	0.50	8.6	<0.1	<0.01
12th	BS-1	0.1	0.11	0.04	0.03	<0.001	0.05	79.3	<0.01	0.02	0.40	9.9	<0.1	<0.01
	BS-5	0.2	3.09	0.03	0.04	<0.001	0.17	82.3	<0.01	0.03	0.40	7.5	<0.1	<0.01
	BS-8	0.2	0.21	0.02	0.03	<0.001	0.15	78.6	0.02	0.02	0.38	12	<0.1	0.02
	BS-9	0.2	0.49	0.01	0.05	<0.001	0.08	82.1	0.03	0.03	0.35	7.8	<0.1	<0.01
	BS-11	0.2	20.9	0.02	0.03	<0.001	0.02	62.5	0.02	0.02	0.23	6.3	<0.1	<0.01
	BS-12	0.3	0.56	0.07	0.03	<0.001	0.44	81.9	0.03	0.02	0.59	6.2	<0.1	<0.01
	BS-20	0.2	0.15	0.09	0.05	<0.001	0.12	82.8	0.02	0.02	0.51	5.9	<0.1	<0.01
	Dan-3*	0.1	0.22	0.04	0.07	<0.001	0.10	80.5	<0.01	0.03	0.15	7.0	<0.1	<0.01
	Dan12	0.1	0.13	0.01	0.04	<0.001	0.32	96.7	<0.01	0.01	0.53	1.7	<0.1	<0.01
	Dan-14	0.1	0.12	0.02	0.02	0.038	7.52	31.3	<0.01	0.10	0.21	3.0	0.1	<0.01

Century [BCE]	*Sample*	*S*	*Pb*	*Co*	*Ni*	*Mn*	*Fe*	*Cu*	*Ag*	*Sb*	*As*	*Sn*	*Zn*	*Au*
Late 12th–11th	Dor-1	0.5	0.39	0.05	0.03	0.006	0.26	94.2	<0.01	<0.01	0.04	0.1	0.1	<0.01
	Dor-2	0.1	1.24	<0.005	0.02	<0.001	0.05	69.4	0.23	0.10	0.19	12	<0.1	0.03
	Dor-3	0.1	0.28	<0.005	0.02	<0.001	0.09	90.9	0.01	0.05	0.41	4.4	<0.1	0.01
	Dor-4	0.1	0.45	0.02	0.04	<0.001	0.24	80.3	0.04	0.03	0.25	9.2	<0.1	0.01
	Dor-7*	0.2	0.17	<0.005	0.01	0.002	0.10	74.1	<0.01	0.01	0.05	4.6	0.4	<0.01
12th –11th	BS-3	0.2	3.03	0.02	0.04	<0.001	0.31	82.2	<0.01	<0.01	0.06	4.8	<0.1	<0.01
	Dan-10	<0.1	0.15	0.02	0.06	<0.001	0.69	97.2	0.05	0.08	1.50	0.4	<0.1	<0.01
	Dan-13**	<0.1	0.14	0.10	0.04	1.85	4.75	19.9	<0.01	0.02	0.06	9.8	0.1	<0.01
	Dan-18	0.1	0.09	0.07	0.06	0.213	16.1	22.7	0.02	0.02	0.11	3.2	<0.1	<0.01
	Dan-20	0.1	0.30	0.16	0.03	0.398	13.8	30.8	<0.01	0.01	0.09	1.6	0.1	<0.01
11th	BS-7	0.2	0.51	0.02	0.03	<0.001	0.1	81.0	0.05	<0.01	0.04	8.9	<0.1	<0.01
	Dan-4	<0.1	0.27	0.01	0.03	<0.001	0.08	84.6	0.02	0.07	0.33	6.6	<0.1	<0.01
	Dan-6	<0.1	0.30	0.01	0.04	<0.001	0.11	92.5	<0.01	0.05	0.27	3.2	<0.1	0.01
	Dan-7*	0.1	0.25	0.03	0.02	<0.001	0.25	79.3	0.02	0.03	0.33	9.6	<0.1	0.01
	Dan-8	0.1	0.27	<0.005	0.02	<0.001	0.41	92.1	<0.01	0.05	0.38	2.5	<0.1	<0.01
	Dan-11	<0.1	0.60	<0.005	0.03	<0.001	0.04	91.1	<0.01	0.01	0.12	3.4	<0.1	<0.01
	Dan-15	0.1	0.19	0.01	0.03	<0.001	0.07	87.5	0.01	0.06	0.37	5.2	<0.1	0.01
	Dan-16	0.1	0.11	0.04	0.03	0.078	6.87	29.4	<0.01	0.02	0.12	4.2	<0.1	<0.01
	Dan-17	<0.1	0.32	0.05	0.12	0.055	13.8	12.3	<0.01	0.08	0.10	5.6	<0.1	<0.01
11th –10th	BS-4	0.2	0.31	0.01	0.03	<0.001	0.26	82.6	0.02	0.07	0.31	6.3	<0.1	<0.01
	BS-10	0.1	0.29	0.01	0.03	<0.001	0.09	81.9	<0.01	0.02	0.12	10.0	<0.1	<0.01

* Corroded samples

** Slag samples were dissolved in aqua regia as the rest of the samples, and analyses were done on soluble parts only.

TABLE 22.3. LI RATIOS OF THE STUDIED ARTIFACTS

Century BCE	*Sample*	$^{208/206}Pb$	σ	$^{207/206}Pb$	σ	$^{204/206}Pb$	σ
15th	BS-2	2.09415	0.00008	0.85486	0.00003	0.054607	0.000017
	Dan-1	2.05656	0.00011	0.82747	0.00004	0.052739	0.000005
	Dan-2	2.07256	0.00006	0.83594	0.00003	0.053383	0.000003
	Dan-5	2.10381	0.00008	0.85911	0.00004	0.054898	0.000002
	Dan-19	2.11348	0.00009	0.86674	0.00005	0.055383	0.000004
14th–13th	Dan-9	2.09783	0.00008	0.85598	0.00004	0.05464	0.000004
13th	Dor -5	2.0748	0.00008	0.84099	0.00003	0.05368	0.000003
	Dor -6	2.08171	0.00011	0.84739	0.00004	0.054333	0.000024
End of 13th	BS-6	2.10298	0.00006	0.85998	0.00002	0.054902	0.000004
	BS-14	2.06219	0.0001	0.83079	0.00005	0.052956	0.000006
	BS-15	2.08127	0.00007	0.84618	0.00003	0.054234	0.000009
	BS-16	2.07025	0.00007	0.83702	0.00003	0.05336	0.000005
	BS-17	2.10084	0.00007	0.85829	0.00003	0.054791	0.000004
	BS-18	2.08083	0.00008	0.84527	0.00004	0.053936	0.000007

Century BCE	*Sample*	$^{208/206}Pb$	σ	$^{207/206}Pb$	σ	$^{204/206}Pb$	σ
	BS-19	2.06555	0.00011	0.83293	0.00005	0.053074	0.000009
	Dor -8	2.10546	0.00005	0.86124	0.00002	0.055006	0.000003
12th	BS-1	2.07978	0.00007	0.84271	0.00003	0.053789	0.000004
	BS-5	2.12153	0.00005	0.87314	0.00002	0.055793	0.000002
	BS-8	2.10223	0.00005	0.85925	0.00002	0.054855	0.000002
	BS-9	2.10042	0.00005	0.8585	0.00002	0.054816	0.000004
	BS-11	2.12014	0.00005	0.87216	0.00002	0.055727	0.000003
	BS-12	2.09288	0.00008	0.85219	0.00003	0.054343	0.000005
	BS-20	2.08233	0.00009	0.84476	0.00004	0.05389	0.000009
	Dan-3	2.10378	0.00009	0.85898	0.00005	0.054882	0.000004
	Dan-12	2.06158	0.00006	0.83151	0.00002	0.053025	0.000003
	Dan-14	2.09614	0.00005	0.85363	0.00002	0.054519	0.000002
Late 12th–11th	Dor-1	2.11615	0.00005	0.86667	0.00002	0.055445	0.000003
	Dor-2	2.09063	0.00006	0.84745	0.00002	0.054074	0.000002
	Dor- 3	2.10135	0.00005	0.85754	0.00003	0.054757	0.000003
	Dor--4	2.11298	0.00006	0.86724	0.00004	0.055407	0.000003
	Dor -7	2.11486	0.00006	0.86782	0.00002	0.055501	0.000004
12th–11th	BS-3	2.12104	0.00006	0.87096	0.00003	0.055749	0.000003
	Dan-10	2.07477	0.00007	0.83741	0.00004	0.053543	0.000003
	Dan-13	2.112	0.00005	0.86569	0.00002	0.055367	0.000002
	Dan-18	2.10429	0.00007	0.85883	0.00003	0.054854	0.000003
	Dan-20	2.11906	0.00006	0.86988	0.00002	0.055675	0.000002
11th	BS-7	2.1183	0.00006	0.86913	0.00003	0.055608	0.000005
	Dan-4	2.10485	0.00009	0.85993	0.00004	0.054934	0.000005
	Dan-6	2.10611	0.00011	0.86062	0.00004	0.054948	0.000003
	Dan-7	2.08923	0.00005	0.84985	0.00002	0.054235	0.000003
	Dan-8	2.08491	0.00005	0.84393	0.00002	0.053862	0.000003
	Dan-11	2.11941	0.00004	0.86935	0.00002	0.055611	0.000002
	Dan-15	2.09847	0.00004	0.85551	0.00002	0.054622	0.000003
	Dan-16	2.11352	0.00007	0.86588	0.00003	0.055398	0.000005
	Dan-17	2.0845	0.00008	0.84352	0.00003	0.053842	0.000003
11th–10th	BS-4	2.10341	0.00006	0.85859	0.00004	0.05483	0.000007
	BS-10	2.11701	0.00011	0.86809	0.00004	0.055522	0.000008

The results show that the LI for most of the artifacts sampled for this study are inconsistent with Cypriot ores and alternatively have LI which are consistent with ores from Timna and Faynan DLS. Of the five artifacts dated to the LB I, two (Dan-5 and Dan-19) are consistent with Timna sandstones and another (BS-2) falls in the range between two slag fragments from Bir Nasib. The remaining two have LI consistent with ores from Anatolia. One of them (Dan-1), is also consistent with Cypriot ores.

Of the 11 artifacts dated to the LB II and mainly LB IIB, four are consistent with Timna ores (BS-6, BS-17, Dan-9 and Dor-8), while the remaining seven are inconsistent with Timna or with Faynan DLS copper ores. One of the seven (Dor-5) and possibly another, that is, possibly imported based on

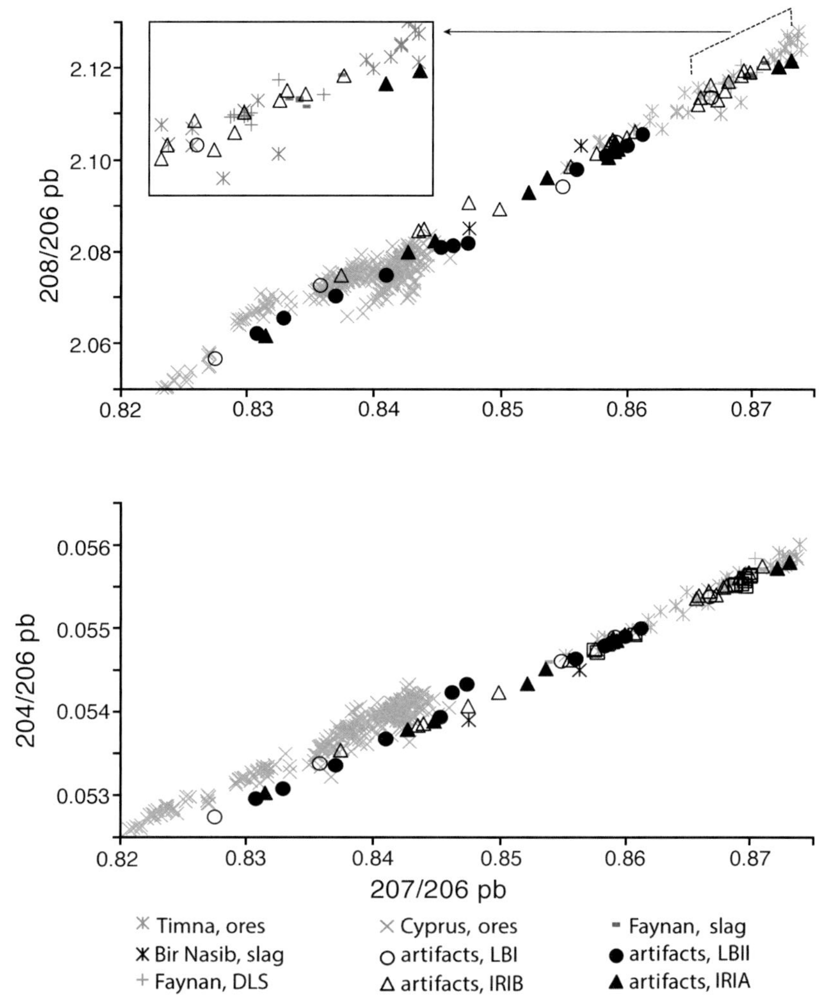

Fig. 22.4: LI ratios for studied artifacts arranged by periods, compared with LI ratios of sandstone ores from Timna (Amir and Avrona Formations), Faynan DLS ores and Cyprus as well as slag from Faynan and Bir Nasib, Sinai (for references see text).

its unique shape (BS-15) have LI which are consistent with LI of Cypriot ores and oxhide ingots from the Cape Gelidonya shipwreck (Fig. 22.6). Notably, three other artifacts (Dor-6, BS-14 and BS-19) are consistent with slab ingots from the Cape Gelidonya shipwreck (Fig. 22.6), while one other artifact (BS-16) has LI ratios which fall in their range.

Most of the Iron I artifacts analyzed in the present study (N=21/31) are consistent with copper ores from Timna (Fig. 22.4). Of these, six artifacts (BS-3, BS-7, BS-10, Dan-11, Dan-20 and Dor-7) are also consistent with DLS ores and metallurgical remains from Faynan (Fig. 22.4). This indicates that copper from the Arabah was used for the production of artifacts during the Iron I.

For the ten items which are inconsistent with copper ores from Timna and the Faynan DLS unit, Cyprus does not appear to be a possible copper ore source during this period, nor do the Late Bronze–Iron I oxhide ingots (Figs. 22.5–6). One artifact dated to the 12th century BCE has LI ratios consistent with the above-mentioned slab ingots from the Cape Gelidonya shipwreck (Fig. 22.6).

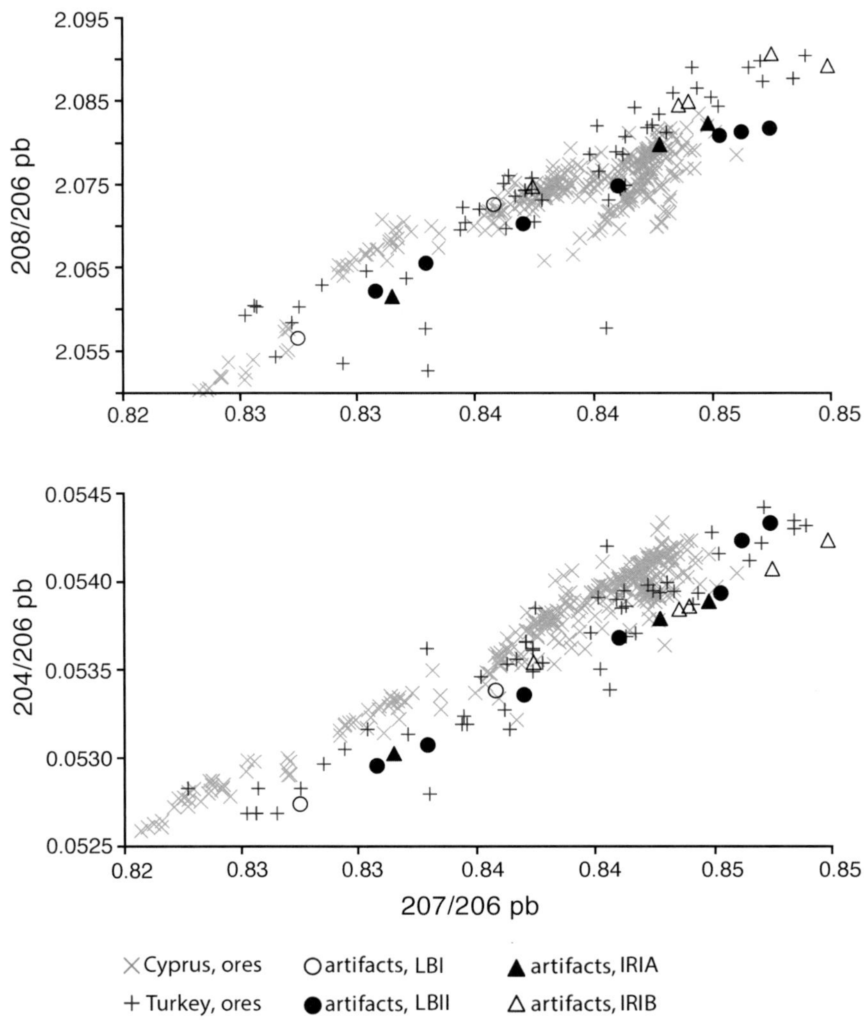

Fig. 22.5: LI ratios for studied artifacts, arranged by periods, compared with LI ratios of ores from Cyprus and Anatolia (for references see text).

DISCUSSION

The results of the LIA presented above show that in all three consecutive periods (LB I, LB II and Iron I) there appears to have been exploitation of the mines in the Arabah, and especially in the Iron I. Eleven objects from Jatt in the Sharon Plain (see location in Fig. 22.1) dated to the Iron IB (11th century BCE) that were analyzed for LI by Stos-Gale (2006), support this conclusion. Although this picture may be the result of the larger Iron I sample available for the present study, it also reflects a reality that is supported by other evidence, as discussed below.

Although based on a small sample, the LIA of objects dated to the LB I raises the possibility that objects made of copper ores from the Arabah were in use at settlement sites during this period. This appears to be in agreement with the reevaluation of the Hathor shrine at Timna, which suggests that the Egyptians of the 19th Dynasty did not initiate metallurgical activities there, but rather took control of existing local activity (Avner 2014). In light of this reevaluation, the previous understanding that

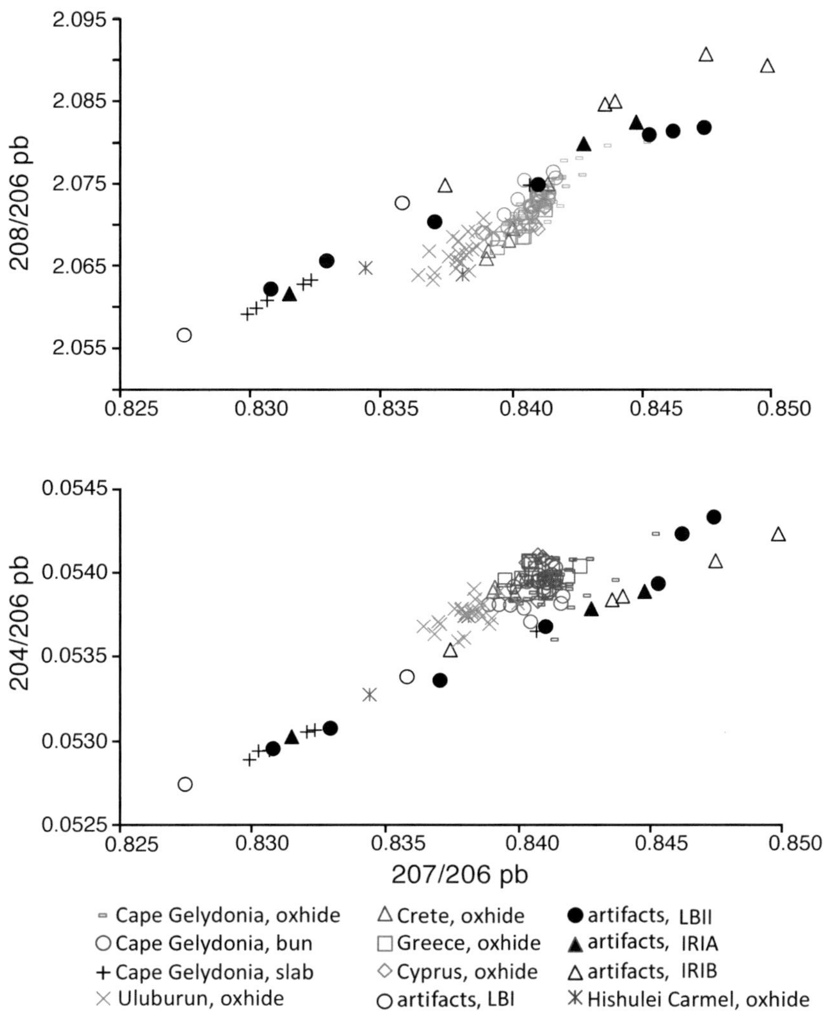

Fig. 22.6: LI ratios for studied artifacts, arranged by periods, compared with LI ratios of selected ingots (for references see text).

the Egyptians were responsible for the technological innovations at Timna during this time needs to be reconsidered. The nature of these continuous metallurgical activities at Timna perhaps included the distribution of the copper to settlement sites, and were more than ad-hoc, sporadic and small-scale.

The consistency of the four objects dated to the LB II with copper ores from Timna, although also comprising a small sample, is possibly a reflection of the above-mentioned Egyptian exploitation of Timna. Note that while only two of the seven items from Beth-Shean—an Egyptian stronghold—were consistent with Timna ores, at least two of the remaining five items, a Mycenaean-type knife (BS-14) and a fork (BS-15), were suspected of being imported, based on their typology. Since these two were unlikely to have been produced in Canaan they cannot attest to the origin of copper used at the site. Interestingly, one of the items from Beth-Shean that was consistent with Timna ores was an Egyptian-type razor (BS-6), indicating that copper from Timna was used for the production of Egyptian-style objects.

The Egyptian interest in controlling the Timna mines might have been prompted by the economic advantage of exporting the copper to Egypt; however, we lack evidence for such a process. Alternatively, and most likely, this control was geared at supplying the needs of the Egyptians in their occupation of Canaan (which intensified in the period of the 19th Dynasty: see, e.g., Weinstein 1981: 17–22), as well as to prevent others from obtaining this important raw material, particularly for the manufacture of weapons. The lack of Egyptian presence at contemporary Faynan is notable and should be examined against the background of the geo-political situation of the time.

Interestingly, LI measurements of the only leaded tin-bronze object included in this study (BS-11), with over 20% Pb, correspond with LI measurements of copper ores from Timna. This, along with the few lead objects which were found in the Hathor shrine at Timna and analyzed by Gale *et al.* (1990), may indicate the addition of lead from the Arabah or Sinai for the production of selected bronze objects.

Following the 19th and 20th Dynasties' Egyptian episode at Timna, which was considered its main period of exploitation, a hiatus was postulated (albeit following a short "Midianite" phase) (Rothenberg 1990). The results of the present study, showing that the majority of the objects dated to the Iron I, and particularly to the Iron IB (11th century BCE), were most likely made of copper from Timna and possibly Faynan, suggest that the main period of exploitation was in fact subsequent to the Egyptian occupation. This conclusion is supported by the excavations at Faynan and the renewed excavations at Timna, which produced a sequence of radiocarbon dates which attribute the bulk of activity at both sites to the 11th–9th centuries BCE (Levy *et al.* 2008; Ben-Yosef *et al.* 2012).

Thus, rather than marking a demise in metallurgical production activity, the withdrawal of the Egyptians from Canaan during the latter part of the 12th century BCE appears to have done just the opposite. Remains of bronze-working have been found at 12 sites dispersed throughout Canaan, doubling that of such sites dating to the 19th and 20th Dynasties (six and five respectively, see Yahalom-Mack 2009a). Iron was used for utilitarian purposes for the first time (*ibid.*) and copper production in the Arabah Valley had been resumed. The latter was due to the disruption of the vast maritime trade at the end of the 13th century BCE in addition to the political vacuum that was created by the Egyptian withdrawal (e.g., Levy and Najjar 2006; Finkelstein and Piasetzky 2008; Levy *et al.* 2008; Ben-Yosef 2010).

We have shown that although Faynan DLS ores are completely overlapped by Timna, the consistency of objects both with Timna and Faynan DLS ores is indicated only during the Iron I. This could be the result of the small sample, or could indicate that copper from Faynan was not distributed to settlement sites in Canaan before the Iron I. The possibility that copper from Faynan was in use in Canaan during the Iron IB further demonstrates the continuity in the exploitation of copper ores in the Arabah and forms a link between the operations at Timna during the Late Bronze and Iron IIA periods and the large-scale operations in Faynan during the Iron I–Iron IIA. This is in agreement with Ben-Yosef's (2010) conclusions regarding the socio-economic circumstances of copper exploitation in both districts. His work noted similarities between the material culture of Faynan and Timna, related to the copper-production process, from the 12th to the end of the 9th centuries BCE, which indicated, accordingly, a regional social cohesiveness.

It can be established that Cypriot copper ores were not used for the production of local artifacts during the Iron I. There is some indication, however, for the use of Cypriot copper during the LB II and possibly during the LB I. A single item dated to the LB I (Dan-1) has LI ratios that are consistent with those of Cypriot copper ores, while two artifacts dated to the LB II (Dor-5 and BS-15), have LI ratios that are consistent with Cypriot copper ores and oxhide ingots. In addition, three and possibly four artifacts have LI ratios which are consistent with slab ingots from the Cape Gelidonya shipwreck. Impurities found in slab ingots showed that they were made directly from smelted sulfide ores (Maddin and Stech-Wheeler 1976: 171–173), but certainly not from Cypriot ores.

The relations with Cyprus in regard to metallurgy during the LB II–Iron I transition may have been more in the transfer of luxury items, such as bronze cult stands and bi-metallic knives, rather than of the raw material proper. This corresponds to the shift that occurred in Cyprus itself during the 13th–12th centuries BCE, from the production of copper ingots to the production of metal end-products (e.g., Sherratt 1994).

The results of this study shed new light on trade in metals and their circulation in the Levant during the Late Bronze–Iron I sequence and is of particular significance, as it is based on the sampling of artifacts recovered from secure stratigraphic contexts. In this study we had determined mainly the consistency or better yet, the inconsistency of items with ores from Timna, Faynan DLS and Cyprus. Items which were inconsistent with any of these sources can sometimes be positively attributed to certain ore sources in Anatolia or in the Aegean but could have LI ratios which are the result of mixing and recycling. Increasing the sample size in future studies will undoubtedly decrease the possibility that we are dealing with recycled metals and increase our ability to trace destinations for the copper from the Arabah and other regional sources.

ACKNOWLEDGMENTS

We thank the excavators, A. Mazar, D. Ilan, R. Ben-Dov, A. Gilboa and I. Sharon for permission to sample the objects and for funding the analyses. This study was conducted as part of a doctoral research under the supervision of Amihai Mazar, Hebrew University, and Sariel Shalev, Haifa University. We would like to thank U. Avner and N. Panitz-Cohen for their input and valuable comments.

REFERENCES

Abdel-Motelib, A., *et al.* 2012. Archaeometallurgical Expeditions to the Sinai Peninsula and the Eastern Desert of Egypt (2006, 2008). *Metalla* 19: 3–59.

Avner, U. 2014. Egyptian Timna—Reconsidered. In: Tebes, J., ed. *Unearthing the Wilderness: Studies on the History and Archaeology of the Negev and Edom in the Iron Age*. Leuven: 103–162.

Bachmann, H.G. and Hauptmann, A. 1984. Zur alten Kupfergewinnung in Fenan und Khirbet en-Nahas im Wadi Arabah in Südjordanien. *Der Anschnitt* 35(4): 110–123.

Balmuth, M.S. and Tylecote, R.F. 1976. Ancient Copper and Bronze in Sardinia: Excavation and Analysis. *Journal of Field Archaeology* 3(2): 195–201.

Bass, G.F. 1967. *Cape Gelidonya: A Bronze Age Shipwreck* (Transactions of the American Philosophical Society, New Series, Vol. 57). Philadelphia.

Begemann, F., Schmitt-Strecker, S., Pernicka, E. and Lo Schiavo, F. 2001. Chemical Composition and Lead Isotopy of Copper and Bronze from Nuragic Sardinia. *European Journal of Archaeology* 4(1): 43–85.

Ben-Yosef, E. 2010. *Technology and Social Process: Oscillations in Iron Age Copper Production and Power in Southern Jordan* (Ph.D. dissertation, University of California). San Diego.

Ben-Yosef, E., Levy, T.E., Higham, T., Najjar, M. and Tauxe, L. 2010. The Beginning of Iron Age Copper Production in the Southern Levant: New Evidence from Khirbat al-Jariya, Faynan, Jordan. *Antiquity* 84: 724–746.

Ben-Yosef, E., Shaar, R., Tauxe, L. and Ron, H. 2012. A New Chronological Framework for Iron Age Copper Production at Timna (Israel). *Bulletin of the American Schools of Oriental Research* 367: 31–71.

Ben-Yosef, E., *et al.* 2008. A New Approach for Geomagnetic Archaeointensity Research: Insights on Ancient Metallurgy in the Southern Levant. *Journal of Archaeological Science.* 35: 2863–2879.

Budd, P., Pollard, A.M., Scaife, B. and Thomas, R.G. 1995. Oxhide Ingots, Recycling and the Mediterranean Metals Trade. *Journal of Mediterranean Archaeology* 8(1): 1–32.

Finkelstein, I. and Piasetzky, E. 2008. Radiocarbon and the History of Copper Production at Khirbet en-Nahas. *Tel-Aviv* 35: 82–95.

Gale, N.H. 1991. Copper Oxhide Ingots: Their Origins and Their Place in the Bronze Age Metals Trade in the Mediterranean. In: Gale, N.H., ed. *Bronze Age Trade in the Mediterranean* (Studies in Mediterranean Archaeology 100 Vol. 90). Jonsered: 197–239.

Gale, N.H., Bachmann, H.G., Rothenberg, B., Stos-Gale, Z.A. and Tylecote, R.F. 1990. The Adventitious Production of Iron in the Smelting of Copper. In: Rothenberg, B., ed. *The Ancient Metallurgy of Copper* Vol. 2. London: 182–190.

Gale, N.H. and Stos-Gale, Z.A. 1982. Bronze Age Copper Sources in the Mediterranean: A New Approach. *Science* 216: 11–18.

Gale, N.H. and Stos-Gale, Z.A. 1986. Oxhide Copper Ingots in Crete and Cyprus and the Bronze Age Metals Trade. *The Annual of The British School at Athens* 81: 81–100.

Gale, N.H. and Stos-Gale, Z.A. 1987. Oxhide Ingots from Sardinia, Crete and Cyprus and the Bronze Age Copper Trade: New Scientific Evidence. In: Balmuth, M.S., ed. *Studies in Sardinian Archaeology III: Nuragic Sardinia and the Mycenaean World.* Oxford: 159–198.

Galili, E., Gale, N. and Rosen, B. 2011. Bronze Age Metal Cargoes off the Israeli Coast. *Skyllis: Zeitschrift för Unterwasser Archäologie* 11: 64–73.

Galili, E., Gale, N. and Rosen, B. 2013. A Late Bronze Age Shipwreck with a Metal Cargo from Hishuley Carmel, Israel. *International Journal of Nautical Archaeology* 42: 2–23.

Galili, E., Shmueli, N. and Artzy, M. 1986. Bronze Age Ship's Cargo of Copper and Tin. *International Journal of Nautical Archaeology* 15(1): 25–37.

Gilboa, A. 2005. Sea Peoples and Phoenicians along the Southern Phoenician Coast—A Reconciliation: An Interpretation of Šikila (SKL) Material Culture. *Bulletin of the American Schools of Oriental Research* 337: 1–37.

Hauptmann, A. 2007. *The Archaeometallurgy of Copper: Evidence from Faynan, Jordan*. Berlin.

Hauptmann, A., Begemann, F., Heitkemper, E., Pernicka, E. and Schmitt-Strecker, S. 1992. Early Copper Produced at Feinan, Wadi Araba, Jordan: The Composition of Ores and Copper. *Archeomaterials* 6: 1–33.

Hauptmann, A., Begemann, F. and Schmitt-Strecker, S. 1999. Copper Objects from Arad—Their Composition and Provenance. *Bulletin of the American Schools of Oriental Research* 314: 1–17.

Hauptmann, A. and Wagner, I. 2007. Prehistoric Copper Production at Timna: TL-Dating and Evidence from the East. In: La Niece, D.H. and Craddock, P., eds. *Metals and Mines: Studies in Archaeometallurgy*. London: 67–75.

Hirao, Y., Enomoto, J. and Tachikawa, H. 1995. Lead Isotopes Ratios of Copper, Zinc and Lead Minerals in Turkey—In Relation to the Provenance Study of Artifacts. In: Mikasa, H.I.H.P.T., ed. *Essays on Ancient Anatolia and Its Surrounding Civilizations.* Wiesbaden.

Jones, M.R. 2007. *Oxhide Ingots, Copper Production, and the Mediterranean Trade in Copper and Other Metals in the Bronze Age* (M.A. thesis, Texas A&M University). College Station, Texas.

Karageorghis, V. 1992. The Crisis Years: Cyprus. In: Ward, W.A. and Joukowsky, M., eds. *The Crisis Years: The 12th Century B.C. —From Beyond the Danube to the Tigris*. Dubuque, IA: 79–86.

Kassianidou, V. 2009. Oxhide Ingots in Cyprus. In: Lo Schiavo, F., Muhly, J.D., Maddin, R. and Giumlia Mair, A., eds. *Oxhide Ingots in the Central Mediterranean*. Rome: 41–81.

Kassianidou, V. 2012. The Origin and Use of Metals in Iron Age Cyprus. In: Iacovou, M., ed. *Cyprus and the Aegean in the Early Iron Age: The Legacy of Nicolas Coldstream*. Nicosia: 229–260.

Knapp, A.B. 1986. *Copper Production and Divine Protection: Archaeology, Ideology and Social Complexity on Bronze Age Cyprus*. Goteborg.

Knapp, A.B. 1994. Emergence, Development and Decline on Bronze Age Cyprus. In: Mathers, C. and Stoddart, S., eds. *Development and Decline in the Mediterranean Bronze Age*, Vol. 8. Sheffield: 271–304.

Knapp, A.B. 2000. Archaeology, Science-based Archaeology and the Mediterranean Bronze Age Metal Trade *European Journal of Archaeology* 3(1): 31–56.

Knapp, A.B. and Cherry, J.F. 1994. *Provenience Studies and Bronze Age Cyprus: Production, Exchange and Politico-Economic Change*. Madison.

Knapp, A.B., Muhly, J.D. and Muhly, P.M. 1988. To Hoard is Human: Late Bronze Age Metal Deposits in Cyprus and the Aegean. *Report of the Department of Antiquities, Cyprus*: 233–262.

Levy, T.E., *et al*. 2004. Reassessing the Chronology of Biblical Edom: New Excavations and 14C Dates from Khirbet en-Nahas (Jordan). *Antiquity* 78: 865–879.

Levy, T.E., Adams, R.B., Prange, M., Schmitt-Strecker, S. and Najjar, M. 2002. Early Bronze Age Metallurgy: A Newly Discovered Copper Manufactory in Southern Jordan. *Antiquity* 76: 425–437.

Levy, T.E., *et al*. 2008. High-Precision Radiocarbon Dating and Historical Biblical Archaeology in Southern Jordan. *Proceedings of the National Academy of Sciences* 105(43): 16460–16465.

Levy, T.E. and Najjar, M. 2006. Some Thoughts on Khirbat en-Nahas, Edom, Biblical History and Anthropology—A Response to Israel Finkelstein. *Tel Aviv* 33: 107–122.

Lo Schiavo, F., Muhly, J.D., Maddin, R. and Giumlia Mair, A. 2009. *Oxhide Ingots in the Central Mediterranean*. Rome.

Maddin, R. and Stech-Wheeler, T. 1976. Metallurgical Study of Seven Bar Ingots. *Israel Exploration Journal* 26: 170–173.

Mazar, A. 1981. Giloh: An Early Israelite Settlement Site near Jerusalem. *Israel Exploration Journal* 31(1–2): 1–36.

Mazar, A. 1994. The 11th Century B.C. in the Land of Israel. In: Karageorghis, V., ed. *Proceedings of an International Symposium: Cyprus in the 11th Century B.C.* Nicosia: 39–57.

Muhly, J.D. 1989. The Organization of the Copper Industry in Late Bronze Age Cyprus. In: Peltenburg, E., ed. *Early Society in Cyprus*. Edinburgh: 298–314.

Muhly, J.D. 1992. The Crisis Years in the Mediterranean World: Transition or Cultural Disinitegration? In: Ward, W.A. and Joukowsky, M., eds. *The Crisis Years: The 12th Century B.C. — From Beyond the Danube to the Tigris*. Dubuque, Iowa: 10–26.

Namdar, D., Segal, I., Goren, Y. and Shalev, S. 2004. Chalcolithic Copper Artifacts. In: Scheftelowitz, N. and Oren, R., eds. *Giv'at Ha-Oranim, A Chalcolithic Site* (Salvage Excavation Reports No. 1). Tel Aviv: 70–83.

Panitz-Cohen, N. and Mazar, A. 2009. *Excavations at Tel Beth-Shean 1989–1996. Vol. III: The Early Iron Age in Areas S and N*. Jerusalem.

Pernicka, E., *et al*. 1984. Archäometallurgische Untersuchungen in Nordwestanatolien. *Jahrbuch des Römisch-Germanischen Zentralmuseums Mainz* 31: 533–599.

Pfeiffer, K. 2009. *Neue Untersuchungen zur Archäometallurgie des Sinai. Die Entwicklungsgeschichte der Innovation „Kupfermetallurgie“* (Ph.D. dissertation, Freie Universitat Berlin). Berlin.

Rothenberg, B. 1987. Pharaonic Copper Mines in Southern Sinai. *Institute for Archaeo-Metallurgical Studies (IAMS) Newsletter* 10/11: 1–7.

Rothenberg, B. 1988. *The Egyptian Mining Temple at Timna,* Vol. I. London.

Rothenberg, B. 1990. *The Ancient Metallurgy of Copper,* Vol. II. London.

Sayre, E.V., Joel, E.C., Blackman, M.J., Yener, K.A. and Özbal, H. 2001. Stable Lead Isotope Studies of Black Sea Anatolian Ore Sources and Related Bronze Age and Phrygian Artefacts from Nearby Archaeological Sites. Appendix: New Central Taurus Ore Data. *Archaeometry* 43(1): 77–115.

Sayre, E.V., Yener, K.A. and Joel, E.C. 1995. Comments on `Oxhide Ingots, Recycling and the Mediterranean Metals Trade. *Journal of Mediterranean Archaeology* 8: 45–53.

Sayre, E.V., Yener, K.A., Joel, E.C. and Barnes, I.L. 1992. Statistical Evaluation of the Presently Accumulated Lead Isotope Data from Anatolia and Surrounding Regions. *Archaeometry* 34(1): 73–105.

Seeliger, T.C., *et al.* 1985. Archäometallurgische Untersuchungen in Nord- und Ostanatolien. *Jahrbuch des Römisch-Germanischen Zentralmuseums Mainz* 32: 597–659.

Segal, I. 2009. Middle Bronze Age I Tools from the Central Negev (Israel): Technology and Provenance. In: *Archaeometallurgy in Europe 2007*. Aquileia: 135–144.

Segal, I. and Halicz, L. 2005. Provenance Studies in Archaeometallurgy Using Lead Isotope Analysis Ratios Determination by Q-ICP-MS and MC-ICP-MS. *Israel Journal of Earth Sciences* 54(2): 87–96.

Segal, I., Halicz, L. and Cohen, R. 1999. A Study of Ingots and Metallurgical Remains from ʻEin Ziq and Be'er Resisim, Central Negev, Israel. In: Young, S.M.M., Pollard, A.M., Budd, P. and Ixer, R.A., eds. *Metals in Antiquity* (BAR International Series 792). Oxford: 179–186.

Segal, I., Halicz, L. and Kamenski, A. 2002. The Metallurgical Remains from Ashqelon-Afridar, Israel, and Their Source. In: Jerem, E. and Biro, K., eds. *Archaeometry 98, Proceedings of the 31st Symposium, Budapest, 1998*. Oxford: 461–473.

Segal, I., Kamenski, A. and Merkel, J. 2002. The Cave of the Sandal, Ketef Jericho: New Evidence from Recent Chalcolithic Copper Finds. *Institute for Archaeo-Metallurgical Studies (IAMS) Newsletter* 22: 7–10.

Segal, I. and Rosen, S.A. 2011. Early Desert Pastoralism: Copper Objects from the Camel Site. In: Rosen, S.A. *An Investigation into Early Desert Pastoralism: Excavations at the Camel Site, Negev*. Los Angeles: 123–132.

Shalev, S. 1993. Metal Production and Society at Tel Dan. In: Biran, A. and Aviram, J., eds. *Biblical Archaeology Today, 1990: Proceedings of the Second International Congress on Biblical Archaeology (Supplement)*. Jerusalem: 57–65.

Sherratt, S. 1994. Commernce, Iron, and Ideology: Metallurgical Innovation in 12th-11th Century Cyprus. In: Karageorghis, V., ed. *Proceedings of the International Symposium: Cyprus in the 11th Century B.C.* Nicosia: 59–108.

Sherratt, S. 1998. 'Sea Peoples' and the Economic Structure of the Late Second Millennium in the Eastern Mediterranean. In: Gitin, S., Mazar, A. and Stern, E., eds. *Mediterranean Peoples in Transition, Thirteenth to Early Tenth Centuries BCE*. Jerusalem: 292–313.

Stos-Gale, Z.A. 2006. Provenance of Metals from Tel Jatt Based on Their Lead Isotope Analyses. In: Artzy, M., ed. *The Jatt Metal Hoard in Northern Canaanite/Phoenician and Cypriote Context*. Barcelona: 115–120.

Stos-Gale, Z.A. and Gale, N.H. 2009. Metal Provenancing Using Isotopes and the Oxford Archaeological Lead Isotope Database (OXALID). *Archaeol Anthropol Sci* 1: 195–213.

Stos-Gale, Z.A., *et al.* 1998. The Copper and Tin Ingots of the Late Bronze Age Mediterranean: New Scientific Evidence. Paper presented at the The Fourth International Conference on the Begining of the Use of Metals and Alloys (BUMA IV). Matsue, Shimane, Japan.

Stos-Gale, Z.A., Maliotis, G., Gale, N.H. and Annetts, N. 1997. Lead Isotope Characteristics of the Cyprus Copper Ore Deposits Applied to Provenance Studies of Copper Oxhide Ingots. *Archaeometry* 39(1): 83–123.

Wagner, G.A., *et al.* 1986. Geochemische und isotopische Charakteristika früher Rohstoffquellen für Kupfer, Blei, Silber und Gold in der Türkei. *Jahrbuch des Römisch-Germanischen Zentralmuseums Mainz* 33: 723–752.

Weinstein, J.M. 1981. The Egyptian Empire in Palestine: A Reassessment. *Bulletin of the American Schools of Oriental Research* 241: 1–28.

Weisgerber, G. 2006. The Mineral Wealth of Ancient Arabia and Its Use I: Copper Mining and Smelting at Feinan and Timna – Comparison and Evaluation of Techniques, Production, and Strategies. *Arabian Archaeology and Epigraphy* 17(1): 1–30.

Yahalom-Mack, N. 2007. The Metal Objects: Catalogue and Discussion. In: Mazar, A. and Mullins, R.A., eds. *Excavations at Tel Beth-Shean 1989–1996, Vol. II: The Middle and Late Bronze Age Strata in Area R*. Jerusalem: 606–620.

Yahalom-Mack, N. 2009a. *Bronze in the Beginning of the Iron Age in the Land of Israel: Production and Utilization in a Diverse Ethno-political Setting* (Ph.D dissertation, The Hebrew University). Jerusalem.

Yahalom-Mack, N. 2009b. The Typology of the Metal Objects. In: Panitz-Cohen, N. and Mazar, A., eds. *Excavations at Tel Beth-Shean 1989–1996, Volume III: The Early Iron Age in Areas S and N*. Jerusalem: 564–584.

Yener, K.A., *et al.* 1991. Stable Lead Isotope Studies of Central Taurus Ore Sources and Related Artifacts from Eastern Mediterranean Chalcolithic and Bronze Age Sites. *Journal of Archaeological Science* 18: 541–577.

Zwicker, U. 1986. Ancient Metallurgical Methods for Copper Production in Cyprus: Part II, Sulphide Ore and Copper-Arsenic-Alloy Production. *Bulletin of the Cyprus Association of Geologists and Mining Engineers* 3: 92–111.

CHAPTER 23

THE ARABAH COPPER INDUSTRY IN THE ISLAMIC PERIOD: VIEWS FROM FAYNAN AND TIMNA

Ian W. N. Jones, Mohammad Najjar and Thomas E. Levy

The Faynan District, located at the northeastern end of the Arabah Valley in Jordan, and Timna, 105 km to its southwest in Israel, contain the two primary copper ore bodies of the southern Levant. Though the sites were intensively mined in antiquity, copper production continued in each location during the Islamic period, with both showing distinctive patterns of exploitation. Excavation of Islamic period (600–1800 CE) copper production sites in Timna began with the work of the late Professor Beno Rothenberg and the Aravah Expedition, but similar research in Faynan has been limited primarily to surveys. With this in mind, the Edom Lowlands Regional Archaeology Project (ELRAP) recently expanded its research program to investigate this period, which represents the last phase of copper production in the region. This paper reviews the archaeological evidence from Timna and summarizes Early (600–1000 CE) and Middle (1000–1400 CE) Islamic period material from ELRAP excavations in Faynan in 2000, 2011 and 2012. It then situates the changing patterns of Islamic period copper production in the Arabah Valley within shifts in the geographic and productive foci of the economy of the southern Levant.

INTRODUCTION

Faynan and Timna, located 105 km apart at opposite ends of the Arabah Valley (Fig. 23.1), contain the two primary copper ore bodies of the southern Levant. While excavation of Islamic period (ca. 600–1900 CE) copper production sites in Timna—notably Site 28, also called Be'er Orah (Rothenberg 1988; 1999: 166–168)—began with the work of the late Beno Rothenberg and the Aravah Expedition, research on the Islamic periods in Faynan, it was until recently limited primarily to survey (e.g., Levy *et al.* 2003; Barker, Gilberston and Mattingly 2007; Hauptmann 2007; Jones, Levy and Najjar 2012). With this in mind, the Edom Lowlands Regional Archaeology Project (ELRAP) recently expanded its research program to investigate the Middle Islamic period (1000–1400 CE),[1] the last phase of copper production in the region. In this paper, we summarize the evidence for Islamic period copper resource exploitation in Timna and present preliminary results from our excavations in Faynan. Finally, we present an argument for why Timna and Faynan show such different patterns of Islamic period settlement and copper exploitation, rooted in major shifts in the economy of the southern Levant in the 11th and 12th centuries CE.

1 ELRAP has adopted Whitcomb's (1992a) archaeological chronology, which divides the Islamic period into Early (600–1000 CE), Middle (1000–1400 CE) and Late (1400–1800 CE) phases.

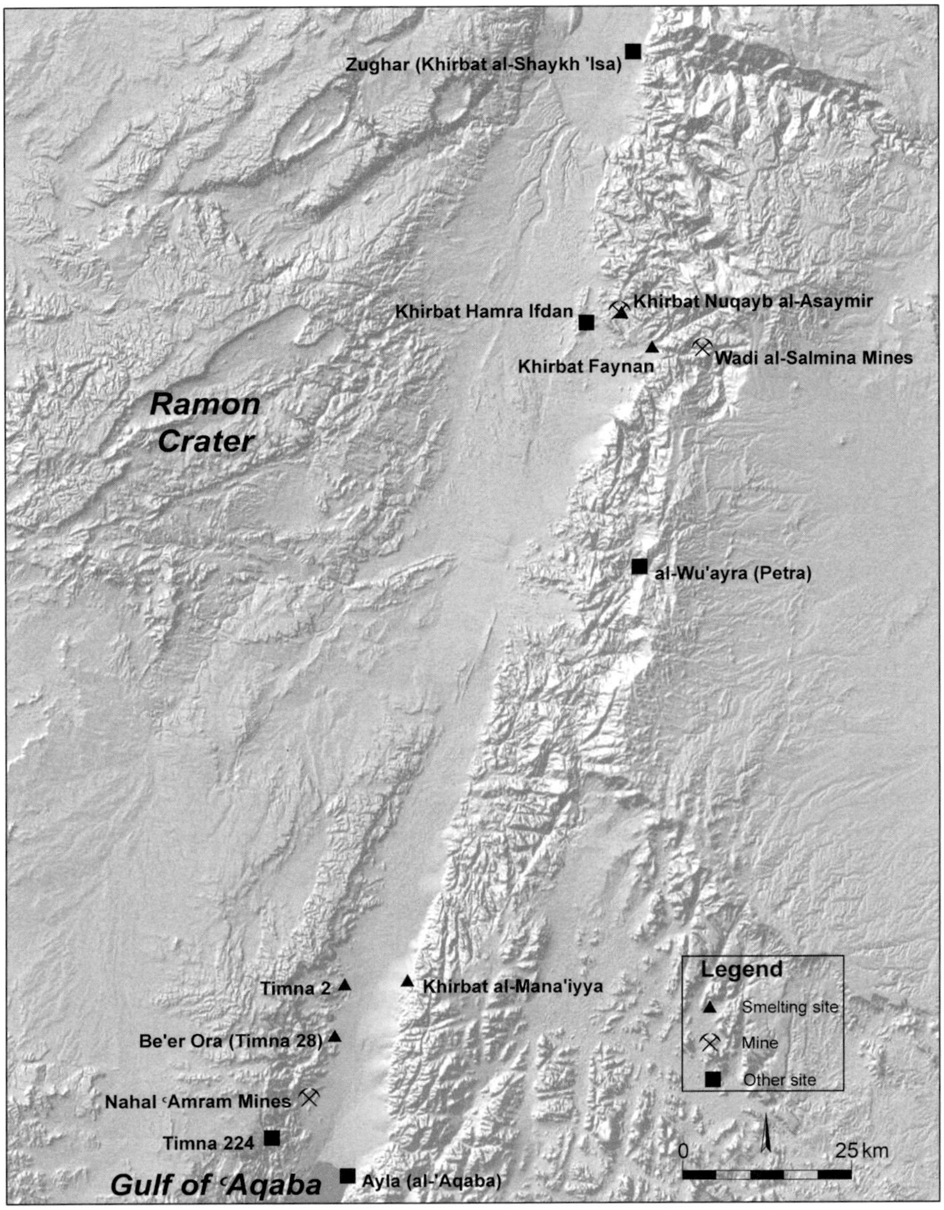

Fig. 23.1: Map of selected sites mentioned in the text. Background image: © 2013 Esri.

THE EARLY ISLAMIC PERIOD

Settlement in the southern Arabah Valley during the Early Islamic period is now fairly well understood, and has been summarized by Avner and Magness (1998), Whitcomb (2006) and Damgaard (2009) among others. The key sites for copper production are the Beʾer Orah smelting camp—initially dated to Roman and Byzantine periods (Rothenberg 1962: 62; 1972: 222–223; 1985: 131–132), but re-dated in the mid-1980s to the Early Islamic period on the basis of radiocarbon evidence (Rothenberg 1988)—and the mines in Nahal ᶜAmram (Willies 1990, 1991). Some smelting also took place in Nahal ᶜAmram at Timna 33 (Willies 1991: 113), although archaeomagnetic data suggests that this was primarily Early Roman smelting, with limited reuse in the Early Islamic period (Ben-Yosef *et al.* 2008: 2876). Some

Early Islamic reuse of furnaces is also evident at Timna 2 (Rothenberg 1988: 2). Finally, recent ELRAP excavations at Khirbat al-Manaîyya—previously misidentified as Iron Age (Ben-Yosef 2012), in an interesting parallel to Beʾer Orah—have provided evidence for significant Early Islamic copper smelting in the Jordanian southern Arabah Valley (Jones *et al.* Forthcoming). This is not an exhaustive overview, but demonstrates that in the southern Arabah Valley copper was produced at a number of sites and on different scales, ranging from fairly intense production to opportunistic reuse.

In Faynan there is, thus far, no evidence of copper production in the Early Islamic period. ELRAP excavations in 2000, 2011 and 2012 have shown that Byzantine settlement at Khirbat Hamra Ifdan and Khirbat Faynan continued into the Early Islamic period, perhaps as late as the 9th century CE; however, there is no evidence for copper smelting. Weisgerber (2006: 25) mentions a site north of Faynan where he identified Early Islamic ring slag, but this is an error in the text, and the site is actually in the southern Arabah (Jones *et al.* forthcoming). In this period, then, we see continuity of Roman and Byzantine patterns of trade at Khirbat Hamra Ifdan, and Byzantine rural or monastic settlement at Khirbat Faynan, but little evidence for even a small copper industry.

This absence is both surprising and revealing. Whitcomb (2006: 241–242) raised the possibility of an Early Islamic settlement system in the northern Arabah Valley, supporting the city of Zughar, a large town in Ghawr al-Safi, represented archaeologically by Khirbat al-Shaykh ʿIsa (see Politis, O'Hea and Papaioannou 2007). If copper production was part of this system, as it was in Ayla's settlement system, one would expect to find it in Faynan, yet this is not the case. It is possible that, as Whitcomb suggested, we should instead look for this activity across the Arabah Valley. Copper was produced in the northwestern Arabah Valley as late as the Roman period at ʿEn Yahav (Yekutieli, Shilstein and Shalev 2005), and Early Islamic settlement in this region is known from inscriptions and ceramics found at agricultural sites (Porath 1987, 2016; Avner and Magness 1998: 50). To the north, some pieces of copper ore found on the road leading east from Zughar (Politis, O'Hea and Papaioannou 2007: 203) may indicate some copper production in Ghawr al-Safi, but given the evidence it is most likely that Zughar obtained copper from Ayla.

THE MIDDLE ISLAMIC PERIOD

Compared to the Early Islamic period, evidence for Middle Islamic period copper production in the southern Arabah Valley is very limited. Willies (1991: 136) collected a small number of Middle Islamic sherds associated with mining activities in Nahal ʿAmram, and mines at Timna 37 were probably also reused (Rothenberg 1972: 224), but copper production seems to have taken place only on a small scale during this period. The best evidence for any sort of metal production is a blacksmith's workshop at Site 224, associated with repairs to the Egyptian Hajj route in the mid-14th century (Rothenberg 1972: 224–228). This ironworking site is unique, and deserves more attention than it has received, but it is of limited relevance to the present topic as there is no evidence of copper production.

As with the Early Islamic period, activities at Faynan during the Middle Islamic period do not match those at Timna. While Middle Islamic production at Timna was rather limited, this period represents the peak of Islamic period activity in Faynan.

KHIRBAT NUQAYB AL-ASAYMIR

The key site for Middle Islamic period copper production in Faynan is Khirbat Nuqayb al-Asaymir (Fig. 23.2)—called al-Furn by some scholars (e.g., Hauptmann 2007)—a small village of the late 12th and early 13th centuries CE, first identified by Nelson Glueck (1935, 1940). Previous surveys dated Khirbat

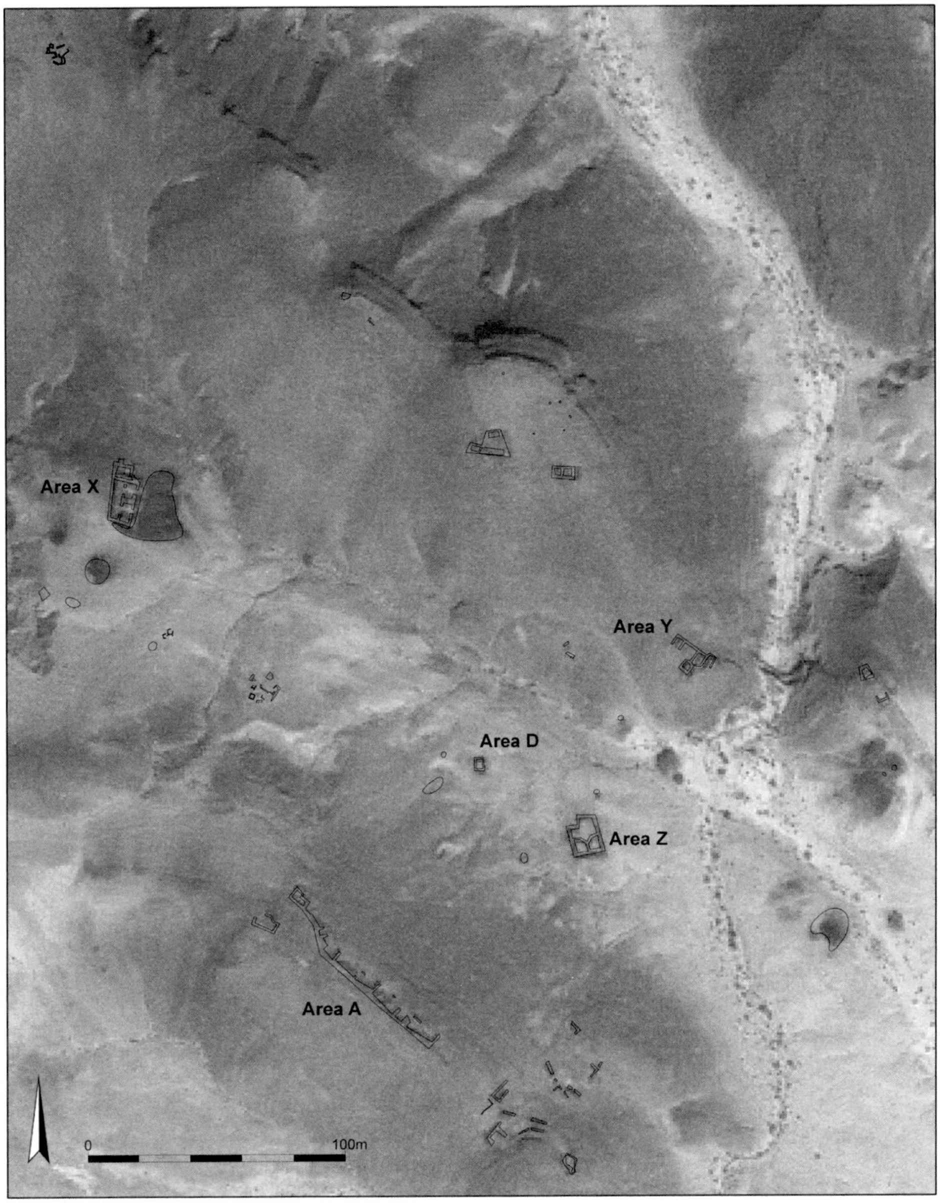

Fig. 23.2: Map of Khirbat Nuqayb al-Asaymir, with excavation areas labelled. Background image: IKONOS satellite imagery courtesy of GeoEye. GeoEye data is owned by GeoEye, Inc. All rights reserved by GeoEye, Inc.

Nuqayb al-Asaymir as either early Mamluk (Hauptmann, Weisgerber and Knauf 1985: 171) or "Ayyubid/ Mamluk" (Hauptmann 2007: 126).[2] Our analysis of ceramics collected during a 2002 survey, however, led us to suggest that the site was founded as early as the late 12th century CE and primarily occupied during the first half of the 13th century (Jones, Levy and Najjar 2012: 88–89). This places the occupation of the site in the Ayyubid period (1187–1260 CE), with no solid evidence of continuity into the Mamluk

2 Hauptmann has consistently specified a 13th century date for the site, in addition to using the broader term "Mamluk." In his monograph, in fact, he specifically suggests 1206–1236 CE (Hauptmann 2007: 126), based on coins published by Kind *et al.* (2005). Newson *et al.* (2007: 364), however, bring up the "Mamluk" designation as one of their reasons for suggesting that copper production at Khirbat Nuqayb al-Asaymir might have continued into the 14th century.

period (1260–1516 CE). Our 2011 and 2012 excavations support this dating. In Area Z, finds from the earliest stratum (Z2b) included both late 12th century Syrian glazed stonepaste ceramics (Fig. 23.3) and an early 13th century coin, and charcoal from the smelting workshop in Area X was radiocarbon dated to 1049–1210 CE (AA–99135, cal. 1-sigma, Fig. 23.4).

Although preliminary, we can offer the following reconstruction. Khirbat Nuqayb al-Asaymir seems to have been founded not long after the Crusaders were ousted from Transjordan in the late 1180s, and was active into the early 13th century. Numismatic data collected by Kind *et al.* (2005: 188) indicate that the site was occupied through at least 1234 CE. Continuity of copper production into the Mamluk period is possible, but presently there is no evidence that this was the case. While some ceramics found at Khirbat Nuqayb al-Asaymir are difficult to date with certainty to the late 12th or early 13th century, nothing is definitively diagnostic of the late 13th or 14th centuries. This makes Khirbat Nuqayb al-Asaymir an even more intriguing site. In addition to being one of the best-preserved copper smelting sites of the Middle Islamic period in the southern Levant, it also provides a unique opportunity to fill in the so-called "Ayyubid gap," a perceived lack of Ayyubid material at multi-period sites "created by over-categorizing ceramic types" (Walker 1999: 211), or the less optimistic "imagined Byzantine-Mamluk settlement 'gap'" in southern Jordan (Walmsley and Grey 2001: 163).

KHIRBAT FAYNAN

Copper was also produced during this period at Khirbat Faynan, at two slag mounds designated Faynan 2 and Faynan 6 by the DBM (Hauptmann 2007: 97, 103). The exact timing of this production in relation to Khirbat Nuqayb al-Asaymir is not entirely clear. A DBM team found 14th century coins in the area surrounding Faynan 6 (Kind *et al.* 2005: 188), suggesting that this area's activity dates to the century after Khirbat Nuqayb al-Asaymir went out of use. In 2012, an ELRAP team conducted a small sounding at the Faynan 6 mound, known as Area 15 on the ELRAP site grid. Two charcoal samples have produced radiocarbon dates similar to those from Khirbat Nuqayb al-Asaymir (AA–102544, 1042–1155 CE, cal.

Fig. 23.3: Late 12th century Syrian underglaze painted stonepaste bowl sherd found at Khirbat Nuqayb al-Asaymir Area Z. Photo: Leah Trujillo.

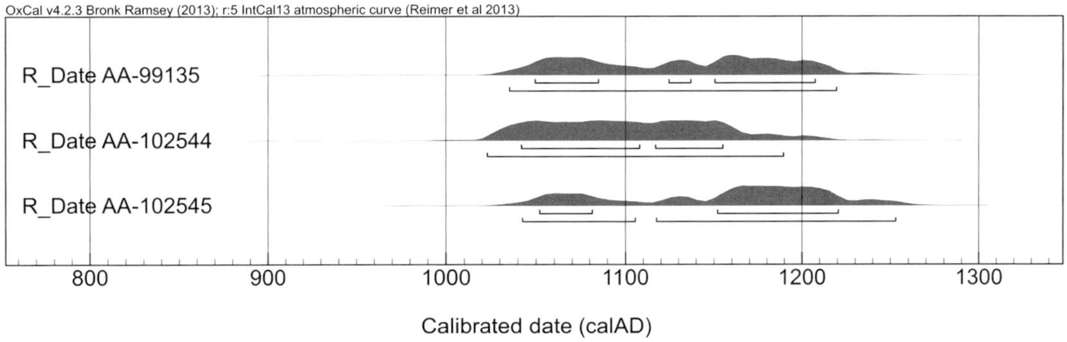

Fig. 23.4: Calibrated radiocarbon dates from KNA Area X (top) and Khirbat Faynan Area 15 (bottom).

1-sigma and AA–102545, 1052–1220 CE, cal. 1-sigma, Fig. 23.4), suggesting that production at Khirbat Faynan was contemporary with, and possibly began earlier than, that at Khirbat Nuqayb al-Asaymir.

While much about Middle Islamic smelting at Khirbat Faynan remains uncertain, this activity should not be dismissed as simply opportunistic re-smelting of old slag, as Weisgerber (2006: 26) did. In 1989, a DBM team analyzed charcoal from Faynan 2 and 6, and determined that the species represented were primarily juniper and Palestine oak, with smaller quantities of olive and tamarisk at Faynan 6 (Baierle *et al.* 1989). All of these, with the exception of tamarisk, are trees that grow on the plateau, a minimum of 7 km to the east, and would not have been found in Faynan in the Islamic period. This indicates a significant effort for charcoal provisioning. Likewise, a 2009 ELRAP survey led by Erez Ben-Yosef found evidence for Middle Islamic period copper mining in Wadi al-Salmina, a small drainage near the plateau (Ben-Yosef 2010: 101; summarized in Jones, Levy and Najjar 2012). The motivation for transporting ore and charcoal this distance is still an open question that is currently under investigation in collaboration with Dr. Brita Lorentzen, who is analyzing the large charcoal assemblages from ELRAP excavations at Khirbat Nuqayb al-Asaymir and Khirbat Faynan (see Jones 2017).

ECONOMIC SHIFTS

As Avner and Magness (1998), Damgaard (2009) and Whitcomb (2006) have argued, the Early Islamic settlements of the southern Arabah Valley served primarily to support the growing city of Ayla. Some of the copper produced in the southern Arabah Valley would have been consumed in the city, but certainly much was destined for other southern Levantine cities and Red Sea ports.[3] During this period Ayla was the northern terminus of the Red Sea and Indian Ocean trade—"the great port of Palestine and the emporium of the Hijjâz," as al-Muqaddasi (1896: 64) described it in the late 10th century—and its hinterland would have been linked to this trade system.

We can ask, like Whitcomb (2006), whether the city of Zughar would have required a similar hinterland settlement system, but it is important to consider that copper production may not have been part of this system. Some copper certainly would have been required in Zughar, but its economy was

3 Oman, with an estimated output of 48,000–60,000 tons of copper in the Early Islamic period (Hauptmann 1985: 115), was a much more important producer for the Indian Ocean trade. Nonetheless, 300–500 tons of copper were produced at Be'er Orah (Avner and Magness 1998: 42), "dozens of tons" at Khirbat al-Manaʿyya (Ben-Yosef 2012: 66) and unknown amounts from other southern Arabah Valley smelting sites, including a fairly large one that was destroyed in the 1950s (Avner and Magness 1998: 42). This, together with the copper produced alongside gold and silver in the Hijaz (see Heck 1999, 2003), would have been significant on a local level.

focused on agriculture rather than trade. In the Early Islamic period, although also famous for dates, the primary cash crop of Zughar was indigo (al-Muqaddasi 1896: 69; Whitcomb 1992b: 117). Given the differences in economy, it may not be entirely surprising that there is little evidence for Early Islamic copper production in the northern Arabah Valley.

Returning to Ayla, the 11th century was a rather difficult period. The Jarrahids captured the city in 1024 and a major earthquake struck in 1068 (Whitcomb 1987: 248; Power 2012: 137). When the Crusaders took the city in 1116 CE its connections to the Red Sea trade were essentially severed (Whitcomb 1988: 222). Whitcomb's (*ibid.*: 2010) excavations at Ayla showed little activity post-dating Phase E (1050–1100 CE), suggesting that the settlement significantly contracted, or was simply abandoned, in the aftermath of the Crusader conquest. Likewise, the majority of the hinterland settlements that had survived through the 11th century, including any remaining copper producing settlements, were abandoned by the early 12th century (Power 2012: 137).

The resumption of copper production in Faynan, on the other hand, was linked to the increasing economic importance of sugar production in the late 12th century (Jones, Levy and Najjar 2012). While the earliest dates for sugar production in the Jordan Valley and Dead Sea region are not entirely clear,[4] indigo remained the primary agricultural product of this region into the mid-12th century (al-Idrisi 1836: 339; LaGro 2002: 26). By the 13th century, however, sugar production had become one of the most important aspects of the economy of Transjordan (*ibid.*). This is reflected archaeologically by a large increase in the number of sugar mills compared to the 12th century (Stern 1999; Burke 2004).

Sugar production also requires copper, in the form of large boiling vessels known in Arabic as *dusut*. Each of these vessels contained roughly 150 kilograms of copper, and eight were required for each millstone at a sugar factory. Each sugar factory would, therefore, have required at least one ton of copper to operate (for a full summary of these calculations, see Jones, Levy and Najjar 2012: 94–95). Our estimate of the copper requirements of the medieval sugar industry (46–78 tons) and Hauptmann's (2007: 126) estimate of the amount of copper that was produced at Khirbat Nuqayb al-Asaymir (65–100 tons) are surprisingly close. Khirbat Nuqayb al-Asaymir could have provisioned this industry and produced a small surplus for use in other copper goods.

This does leave open the question of why copper was mined in Faynan, rather than imported from Europe or elsewhere. We suggest, following Coughenour's (1976) arguments about iron production in the 'Ajlun region, that this may be related to 12th century Papal bans on the trade of metal and timber with the Muslim world. While these were enforced only rarely, the threat may have been sufficient to make a local source of copper very attractive. This explains both the relatively small amount of copper produced, and the relatively short duration of production. Even early on, the Mamluks seem to have relied primarily on European copper, and by the 14th century copper production in Faynan seems to have ceased. By the 15th century, sugar production in the southern Levant had diminished significantly (Ashtor 1977: 257–262; Walker 2007).

CONCLUSION

As we have demonstrated, the views from Timna and Faynan in the Islamic periods are quite different. In Timna, copper was produced throughout the Early Islamic period, and primarily destined for the city of Ayla and other Red Sea ports. Following the removal of Ayla from this trade system, copper production in Timna essentially ceased, with very limited production evident in the Middle Islamic period. In Faynan,

4 Sato (2009: 10; 2015: 23) argues that sugar cultivation was introduced into the southern Levant before the 10th century, and Galloway (1977: 180) suggests the early 7th century.

on the other hand, there is no evidence for copper exploitation in the Early Islamic period. Instead, copper production began at Khirbat Nuqayb al-Asaymir and Khirbat Faynan only in the mid- to late 12th century, continuing into the 13th. Smaller-scale activity may have continued into the 14th century at Khirbat Faynan, but our excavation produced no evidence of this. Copper production in Faynan was primarily undertaken to support the expanding sugar industry in the Jordan Valley and Dead Sea region. Continuing analysis of the material from Khirbat Nuqayb al-Asaymir and Khirbat Faynan will allow us to further expand on the relationship between the copper and sugar industries in the future.

ACKNOWLEDGMENTS

We thank the Department of Antiquities of Jordan and especially Director of Excavations and Surveys, Jehad Haroun, for support of the excavations described here. The excavations at Khirbat Nuqayb al-Asaymir were made possible by grants from the Kershaw and Katzin Family Foundations to TEL. Portions of this work were supported by the National Science Foundation under IGERT Award #DGE-0966375, "Training, Research and Education in Engineering for Cultural Heritage Diagnostics" at the UC San Diego's Center for Interdisciplinary Science in Art, Architecture and Archaeology (CISA3) at the Qualcomm Institute, to IWNJ. We are grateful for the logistical and moral support provided by the American Center of Oriental Research in Amman, Jordan, especially Director Dr. Barbara Porter and former Associate Director Dr. Christopher Tuttle. We thank our ʿAzazmeh, ʿAmarin and al-Manaja Bedouin workers, ELRAP field staff, UC San Diego undergraduates and volunteers, and especially Khirbat Nuqayb al-Asaymir assistant supervisors Kathleen Bennallack, Aaron Gidding and Claire McConnell. We are grateful to Kathleen Bennallack, Alicia Boswell, Kyle Knabb, Jessica Novak and Elizabeth Plunger for insightful comments on earlier drafts of this chapter, and Dr. Erez Ben-Yosef for many valuable discussions, and for inviting us to contribute to both the Mining for Copper conference and this volume. Any errors in this work are, of course, our own. We would like to acknowledge GeoEye for generously providing the IKONOS satellite imagery used in this chapter. Finally, we thank the Royal Society for the Conservation of Nature (RSCN) in Jordan for their help and cooperation. Khirbat Nuqayb al-Asaymir is located in the Dana UNESCO Biosphere protected area.

REFERENCES

al-Idrisi. 1836. Géographie d'Édrisi Traduite de l'Arabe en Français d'Après Deux Manuscrits de la Bibliothèque du Roi et Accompagnée de Notes. In: Jaubert, P.A., trans. *Recueil de Voyage et de Mémoires publié par la Société de Géographie* 5. Paris.

al-Muqaddasi. 1896. Description of Syria, Including Palestine. In: Le Strange, G., trans. *The Library of the Palestine Pilgrims' Text Society*, Vol. 3. New York: 1–103.

Ashtor, E. 1977. Levantine Sugar Industry in the Later Middle Ages—An Example of Technological Decline. *Israel Oriental Studies* 7: 226–280.

Avner, U. and Magness, J. 1998. Early Islamic Settlement in the Southern Negev. *Bulletin of the American Schools of Oriental Research* 310: 39–57.

Baierle, H.U., Frey, W., Jagiella, C. and Kürschner, H. 1989. Die Brennstoffressourcen im Raum Fenan (Wadi Araba, Jordanien) und die bei der Kupferzverhüttung verwendeten Brennstoffe. In: Hauptmann, A., Pernicka, E. and Wagner, G.A., eds. *Archäometallurgie der Alten Welt: Beiträge zum Internationalen Symposium "Old World Archaeometallurgy," Heidelberg 1987* (Montanhistorische Zeitschrift Der Anschnitt, Vol. 7). Bochum: 213–222.

Barker, G., Gilberston, D. and Mattingly, D., eds. 2007. *Archaeology and Desertification: The Wadi Faynan Landscape Survey, Southern Jordan* (Wadi Faynan Series 2). Oxford.

Ben-Yosef, E. 2010. *Technology and Social Process: Oscillations in Iron Age Copper Production and Power in Southern Jordan* (Ph.D. dissertation, University of California). San Diego.

Ben-Yosef, E. 2012. Environmental Constraints on Ancient Copper Production in the Aravah Valley: Implications of the Newly Discovered Site of Khirbet Mana'iyah in Southern Jordan. *Tel Aviv* 39: 58–74.

Ben-Yosef, E., *et al*. 2008. A New Approach for Geomagnetic Archaeointensity Research: Insights on Ancient Metallurgy in the Southern Levant. *Journal of Archaeological Science* 35: 2863–2879.

Burke, K.S. 2004. A Note on Archaeological Evidence for Sugar Production in the Middle Islamic Periods in Bilād al-Shām. *Mamluk Studies Review* 8: 109–118.

Coughenour, R.A. 1976. Preliminary Report on the Exploration and Excavation of Mugharat el-Wardeh and Abu Thawab. *Annual of the Department of Antiquities of Jordan* 21: 71–78, 186–189.

Damgaard, K. 2009. A Palestinian Red Sea Port on the Egyptian Road to Arabia: Early Islamic Aqaba and Its Many Hinterlands. In: Blue, L., Cooper, J., Thomas, R. and Whitewright, J., eds. *Connected Hinterlands: Proceedings of Red Sea Project IV* (Society for Arabian Studies Monographs 8). Oxford: 85–97.

Galloway, J.H. 1977. The Mediterranean Sugar Industry. *Geographical Review* 67: 177–194.

Glueck, N. 1935. *Explorations in Eastern Palestine, II* (The Annual of the American Schools of Oriental Research 15). New Haven.

Glueck, N. 1940. *The Other Side of the Jordan.* New Haven.

Hauptmann, A. 1985. *5000 Jahre Kupfer in Oman, Band 1: Die Entwicklung der Kupfermetallurgie vom 3. Jahrtausend bis zur Neuzeit* (Der Anschnitt, Beiheft 4). Bochum.

Hauptmann, A. 2007. *The Archaeometallurgy of Copper: Evidence from Faynan, Jordan* (Natural Science in Archaeology). New York.

Hauptmann, A., Weisgerber, G. and Knauf, E.A. 1985. Archäometallurgische und bergbauarchäologische Untersuchungen im Gebiet von Fenan, Wadi Arabah (Jordanien). *Der Anschnitt* 37: 163–195.

Heck, G.W. 1999. Gold Mining in Arabia and the Rise of the Islamic State. *Journal of the Economic and Social History of the Orient* 42: 364–395.

Heck, G.W. 2003. *The Precious Metals of West Arabia and Their Role in Forging the Economic Dynamic of the Early Islamic State.* Riyadh.

Jones, I.W.N. 2017. Fuel Provisioning for the Copper Industry in Islamic Period Southern Jordan. *Palestine Exploration Quarterly* 149: 87–90.

Jones, I.W.N., Ben-Yosef, E., Lorentzen, B., Najjar, M. and Levy, T.E. Forthcoming. Khirbat al-Mana'iyya: An Early Islamic Period Copper Smelting Site in the Southeastern Wadi ʿAraba, Jordan. Submitted to *Arabian Archaeology and Epigraphy.*

Jones, I.W.N., Levy, T.E. and Najjar, M. 2012. Khirbat Nuqayb al-Asaymir and Middle Islamic Metallurgy in Faynan: Surveys of Wadi al-Ghuwayb and Wadi al-Jariya in Faynan, Southern Jordan. *Bulletin of the American Schools of Oriental Research* 368: 67–102.

Kind, H.D., Gilles, K.J., Hauptmann, A. and Weisgerber, G. 2005. Coins from Faynan, Jordan. *Levant* 37: 169–195.

LaGro, H.E. 2002. *An Insight into Ayyubid-Mamluk Pottery: Description and Analysis of a Corpus of Mediaeval Pottery from the Cane Sugar Production and Village Occupation at Tell Abu Sarbut in Jordan* (Ph.D. dissertation, Universiteit Leiden). Leiden.

Levy, T.E., *et al*. 2003. An Iron Age Landscape in the Edomite Lowlands: Archaeological Surveys Along Wādī al-Ghuwayb and Wādī al-Jāriya, Jabal Ḥamrat Fīdān, Jordan, 2002. *Annual of the Department of Antiquities of Jordan* 47: 247–277.

Newson, P., *et al*. 2007. The Islamic and Ottoman periods. In: Barker, G., Gilbertson, D. and Mattingly, D., eds. *Archaeology and Desertification: The Wadi Faynan Landscape Survey, Southern Jordan* (Wadi Faynan Series 2). Oxford: 349–368.

Politis, K.D., O'Hea, M. and Papaioannou, G.A. 2007. Ghawr as-Safi Survey and Excavations 2006–2007. *Annual of the Department of Antiquities of Jordan* 51: 199–210.

Porath, Y. 1987. Qanats in the ᶜAravah. *Qadmoniot* 20: 106–114 (Hebrew).

Porath, Y., Liphschitz, N., Wiesel, Y. and Sharon, M. 2016. Tunnel-Well (Qanat) Systems and Settlements from the Early Islamic Period in the ᶜArava. *ᶜAtiqot* 86: 1–81, 113–116. (Hebrew with English summary).

Power, T. 2012. Trade Cycles and Settlement Patterns in the Red Sea Region (ca. AD 1050–1250). In: Agius, D.A., Cooper, J.P., Trakadas, A. and Zazzaro, C., eds. *Navigated Spaces, Connected Places: Proceedings of Red Sea Project V, Held at University of Exeter, 16–19 September 2010* (British Foundation for the Study of Arabia Monographs 12). Oxford: 137–145.

Rothenberg, B. 1962. Ancient Copper Industries in the Western Arabah: An Archaeological Survey of the Arabah, Part I. *Palestine Exploration Quarterly* 94: 5–71.

Rothenberg, B. 1972. *Were These King Solomon's Mines? Excavations in the Timna Valley* (New Aspects of Archaeology). New York.

Rothenberg, B. 1985. Copper Smelting Furnaces in the Arabah, Israel: The Archaeological Evidence. In: Craddock, P.T. and Hughes, M.J., eds. *Furnaces and Smelting Technology in Antiquity* (British Museum Occasional Papers 48). London: 123–150.

Rothenberg, B. 1988. Early Islamic Copper Smelting—and Worship—at Beer Ora, Southern Arabah (Israel). *Institute for Archaeo-Metallurgical Studies Newsletter* 12: 1–4.

Rothenberg, B. 1999. Archaeo-Metallurgical Researches in the Southern Arabah 1959–1990, Part 2: Egyptian New Kingdom (Ramesside) to Early Islam. *Palestine Exploration Quarterly* 131: 149–175.

Sato, T. 2009. The Origin and Expansion of Sugar Production in the Islamic World. *Bulletin of the Graduate Division of Literature of Waseda University* 54: 5–20.

Sato, T. 2015. *Sugar in the Social Life of Medieval Islam* (Islamic Area Studies, Vol. 1). Leiden.

Stern, E.J. 1999. *The Sugar Industry in Palestine during the Crusader, Ayyubid and Mamluk Periods in Light of the Archeological Finds* (M.A. thesis, The Hebrew University). Jerusalem (Hebrew with English summary).

Walker, B.J. 1999. Militarization to Nomadization: The Middle and Late Islamic Periods. *Near Eastern Archaeology* 62: 202–232.

Walker, B.J. 2007. Sowing the Seeds of Rural Decline? Agriculture as an Economic Barometer for Late Mamluk Jordan. *Mamluk Studies Review* 11: 173–199.

Walmsley, A.G. and Grey, A.D. 2001. An Interim Report on the Pottery from Gharandal (Arindela), Jordan. *Levant* 33: 139–164.

Weisgerber, G. 2006. The Mineral Wealth of Ancient Arabia and Its Use I: Copper Mining and Smelting at Feinan and Timna—Comparison and Evaluation of Techniques, Production, and Strategies. *Arabian Archaeology and Epigraphy* 17: 1–30.

Whitcomb, D.S. 1987. Excavations in ʿAqaba: First Preliminary Report. *Annual of the Department of Antiquities of Jordan* 31: 247–266.

Whitcomb, D.S. 1988. A Fatimid Residence at Aqaba, Jordan. *Annual of the Department of Antiquities of Jordan* 32: 207–224.

Whitcomb, D.S. 1992a. Reassessing the Archaeology of Jordan of the Abbasid Period. In: Tell, S., ed. *Studies in the History and Archaeology of Jordan IV.* Amman: 385–390.

Whitcomb, D.S. 1992b. The Islamic Period as Seen from Selected Sites. In: MacDonald, B., ed. *The Southern Ghors and Northeast 'Arabah Archaeological Survey* (Sheffield Archaeological Monographs 5). Dorchester: 113–118.

Whitcomb, D.S. 2006. Land behind Aqaba: The Wadi Arabah during the Early Islamic Period. In: Bienkowski, P. and Galor, K., eds. *Crossing the Rift: Resources, Routes, Settlement Patterns and Interactions in the Wadi Arabah* (Levant Supplementary Series 3). Oxford: 239–242.

Whitcomb, D.S. 2010. Ayla at the Millennium: Archaeology and History. *Annual of the Department of Antiquities of Jordan* 54: 167–176.

Willies, L. 1990. Exploring the Ancient Copper Mines of the Wadi Amram (South Arabah). *Institute for Archaeo-Metallurgical Studies Newsletter* 15–16: 12–15.

Willies, L. 1991. Ancient Copper Mining at Wadi Amram, Israel: An Archaeological Survey. *Bulletin of the Peak District Mines Historical Society* 11: 109–138.

Yekutieli, Y., Shilstein, S. and Shalev, S. 2005. ʿEn Yahav—A Copper Smelting Site in the ʿArava. *Bulletin of the American Schools of Oriental Research* 340: 1–21.

SECTION IV

BEYOND THE SOUTHERN LEVANT: CYPRUS, OMAN, GREECE AND BRITAIN

CHAPTER 24

APLIKI *KARAMALLOS* ON CYPRUS: THE 13TH CENTURY BCE MINERS' SETTLEMENT IN CONTEXT

Vasiliki Kassianidou[1]

Thanks to rich copper ore deposits in the geological formation known as the Pillow Lavas in the Troodos Foothills, Cyprus is one of the richest countries in copper per surface area in the world. Because of this mineral wealth the island played a leading role in the metals trade of the Late Bronze Age. Production and export of copper developed significantly in the Late Cypriot (1600–1050 BCE), reaching a peak in the 13th century BCE. From 1450 BCE onwards, Cypriot copper, in the shape of oxhide ingots, was exported far and wide, and according to Lead Isotope Analysis the copper used to produce the ingots was extracted from the ore deposit of Apliki.

The Apliki mine lies within Cyprus' richest mining district, which in modern times produced more than 80% of the total copper ore concentrate that was exported from the island. In 1938 the Cyprus Mining Corporation began to exploit the Apliki deposit and the remains of a Late Bronze Age miners' settlement came to light. The purpose of this paper is to present some of this extraordinary material from one of the most infamous, and yet little known, metallurgical sites of 13th century BCE Cyprus.

INTRODUCTION

In the northwestern foothills of the Troodos Mountains lies the Solea Valley. It takes its name from the ancient kingdom of Soloi, one of the wealthiest Iron Age kingdoms of ancient Cyprus (Stylianou 1992: 512). The valley is defined by two rivers, the Atsas and the Karyiotis, which run from the highest peak of the Troodos Mountains to sea level. As the distance covered is short, the Karyotis River has the ability to carry the greatest volume of water of any river on the island, and as a result it is the last river to dry up in times of drought and the first to flood (Boutin and Given 2013: 54). Consequently the Solea Valley is extremely fertile, as Dr. J. Sibthorp, the editor of *Flora Graeca*, who visited the area in 1787 (Cobham 1895: 247), eloquently described:

> We now entered the vale of Soulea, the most beautiful we had yet seen in the island; well watered and richly cultivated. Green meadows contrasted with the corn now ripe, hamlets shaded with mulberry-trees, and healthy peasantry busily employed with their harvest, and the care of their silk-worms, enlivened the scenery. (Cobham 1895: 277)

---

1 I would like to thank the organizers and especially Erez Ben-Yosef for inviting me to take part in this tribute to Beno Rothenberg, whom I am proud to call my mentor. Many of the speakers in the conference have, like myself, benefited from the IAMS scholarships set up by Beno and were able to push ahead with their research and receive their Masters' or Doctoral degrees as a result of his efforts. Thus, although Beno made history through his work at Timna and in southwest Spain, I believe that in many ways an even more important legacy he has left behind is a whole new generation of archaeometallurgists.

The Valley of Solea, however, is best known for its copper mines, as it includes the richest ore deposits on the island of Cyprus (Fig. 24.1). The most important are found at Mavrovouni, Skouriotissa and Apliki, which are part of the so-called Solea axis copper ore deposits (Ingham 1959: 140–55). The Apliki Mine is famous, or infamous, among the archaeological and archaeometallurgical community because, according to Lead Isotope Analysis, it is most probably the mine that generated the copper used to produce the vast majority of copper oxhide ingots found in Cyprus and abroad (Gale 2011: 218). In modern times the three mines produced more than 85% of the total copper ore concentrate that was exported from Cyprus (Constantinou 1982: 15). Furthermore, based on the size of the ancient slag heaps that dominate the landscape it is estimated that in antiquity this area probably produced more than 50% of the total amount of copper produced on the island (*ibid.* 2007: 339). The average copper content in these three ore deposits varies from 2–4.5%, but in the secondary enrichment zone—which is the one that would have been exploited by the ancients—it reached as much as 20% (*ibid.*: 342). The ore deposits were not only rich in copper, they were also easy to mine. According to Constantinou (*ibid.*: 341):

> The human energy necessary for the mining of one ton of conglomeratic ore from these three ore bodies [Apliki, Mavrovouni and Skouriotissa] of this area is orders of magnitude less than that required for the mining of the same amounts of massive sulphide ore from Ergani Maden [Turkey], Ermioni [Greece], Jabel Said [Saudi Arabia], Rio Tinto [Spain] or the copper ores from Timna and Feinan.

In 1912, Charles Godfrey Gunther, a prospector hired to investigate the possibility of copper mining in Cyprus, visited Skouriotissa (Lavender 1962: 66–77). He was the first to identify and interpret the remains of the ancient copper industry, from both the mining and the smelting stages of production that

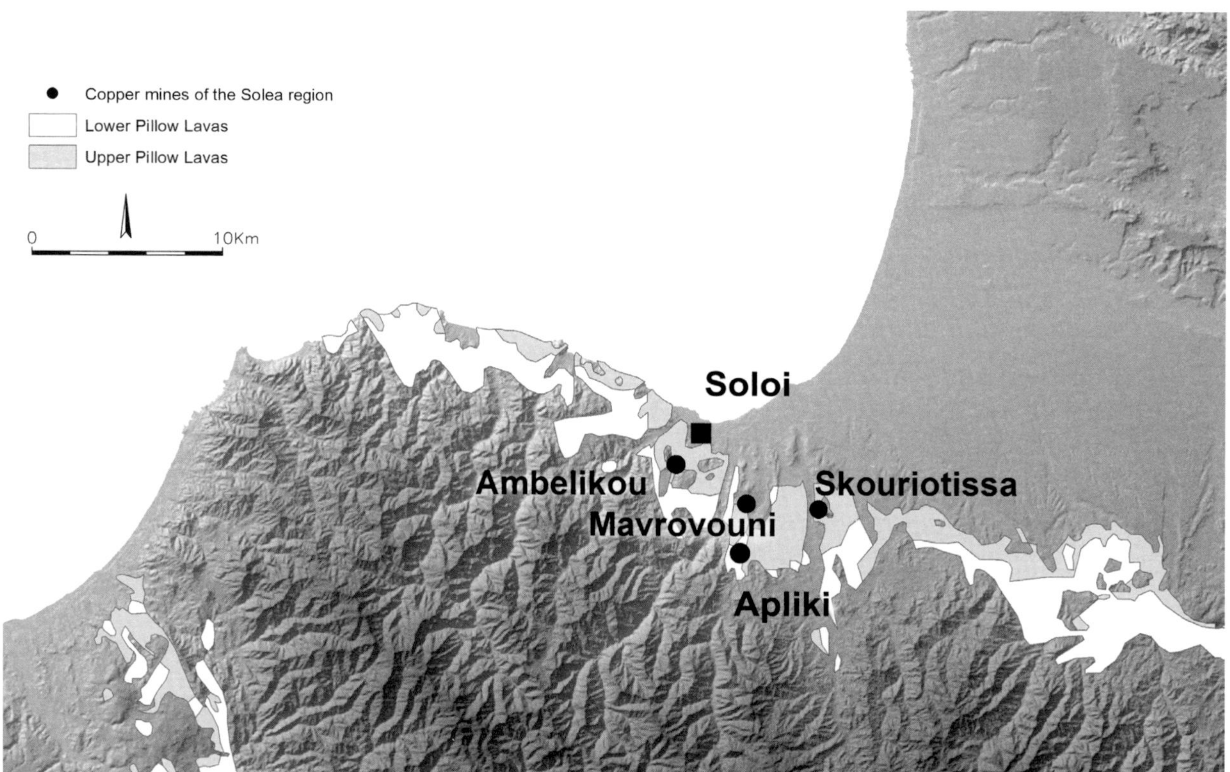

Fig. 24.1: Map of the mining district of Skouriotissa in the Solea Valley (produced by V. Kassianidou with digital geological data provided by the Cyprus Geological Survey). For location see Fig. 24.2.

he observed there (Kassianidou 2013a: 115). His efforts eventually led to the formation of the Cyprus Copper Mines Corporation (CMC), the American company that had the lease to exploit Skouriotissa, Apliki and Mavrovouni from 1921 until the Turkish invasion of 1974, after which their operations ceased.

In 1938 CMC began to exploit the Apliki deposit and the remains of a Late Bronze Age settlement came to light. Joan Du Plat Taylor, a young woman working for the department of Antiquities, was asked to carry out the rescue excavations (Kling and Muhly 2007: xi). She did a remarkably good job but she only published a few preliminary reports (Du Plat Taylor 1940; 1952). In the sixties, unfortunately the site was destroyed by the open cast mine. The finds and the excavation diaries, however, are stored in the Cyprus Museum and they were used by Kling and Muhly to produce a monograph on the excavation, which includes several specialist studies (Kling and Muhly 2007). Muhly presented some of the preliminary results of his study on the archaeometallurgical material from the site in a paper published in 1989 (Muhly 1989) and I have now taken over the task of completing this study.

CYPRUS AND COPPER IN THE LATE BRONZE AGE

The Late Bronze Age brought many changes to the island of Cyprus, including economic growth, social complexity and writing (for a recent review of the evidence, see Knapp 2013: 348–447). Urban centers were established along the coast (Fig. 24.2). Their prosperity is believed to have depended on trade with the Near East and the Aegean. Important settlements located inland possessed monumental architecture and large storage facilities for olive oil. Furthermore, in all excavated sites a great number of objects

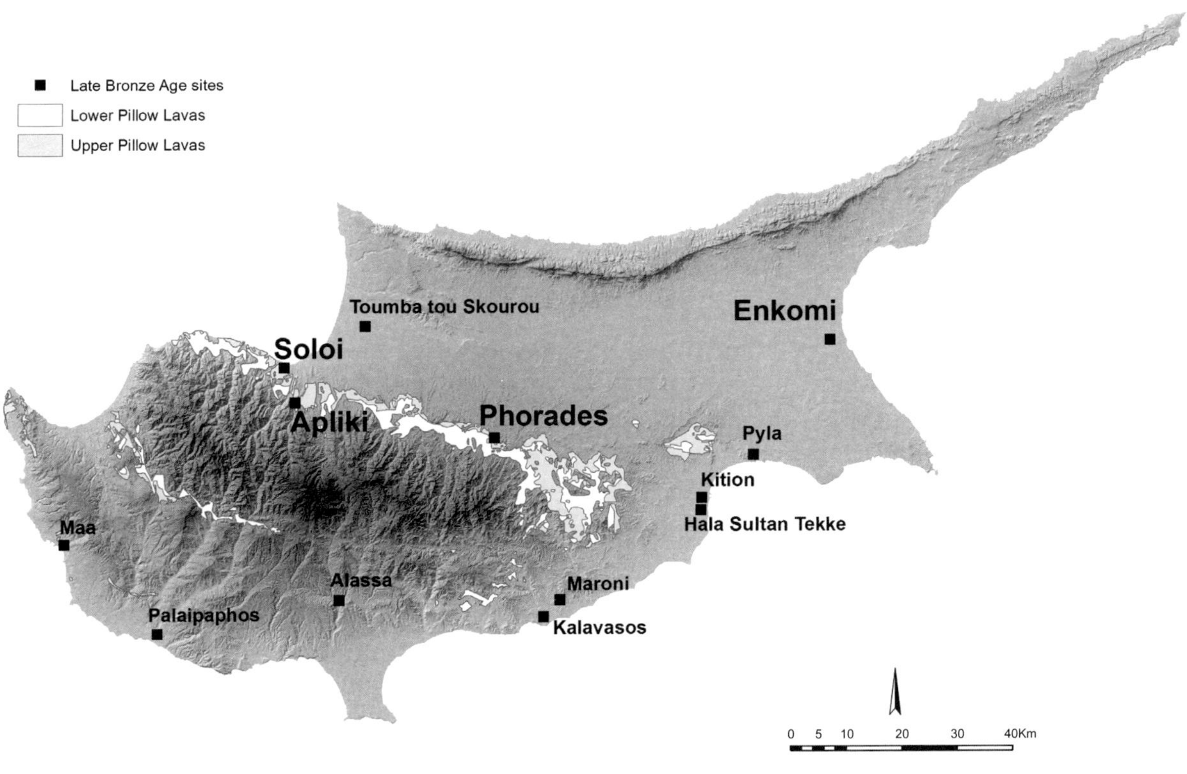

Fig. 24.2: Map of Cyprus with sites mentioned in the text (produced by V. Kassianidou with digital geological data provided by the Cyprus Geological Survey).

related to the production of copper have been recovered (for recent papers on the subject, see Kassianidou and Papasavvas 2012).

Written sources from sites in Syria dating from the 19th to the 17th centuries BCE suggest that Cyprus, which is called Alashiya in its neighbors' texts, exported some of the copper produced on the island to the East (Knapp 2011: 250). It was, however, in the Late Bronze Age that the production and export of copper truly developed, reaching a peak in the 13th century BCE. This marked increase in production (and subsequently export), is due to significant technological developments in the smelting installations: the furnaces were equipped with a system of bellows and tuyères, the use of which led to a significant increase in the operating temperature and thus to the melting of their contents which could easily and efficiently be separated without a significant loss of metal in the slag (Kassianidou 2011: 45). From the 16th century BCE onwards Cypriot copper began being traded mainly in the form of oxhide ingots (Kassianidou 2013b: 136). The importance of the copper industry for Late Bronze Age Cypriots is clearly seen in the depictions of oxhide ingots in various media (Papasavvas 2009). According to Lead Isotope Analysis (Stos 2009: 172–173), in the 14th century BCE Cyprus was able to load a ship, which sank at Uluburun, with ten tons of copper in the form of 354 oxhide and 121 plano-convex ingots (Pulak 2005: 59–63). Furthermore, according to the eight letters sent from Alashiya to Egypt and found in the Amarna archive, approximately 29 tons of copper were exported from Cyprus to Egypt in a period of 15 to 30 years (Knapp and Kassianidou 2008: 135).

From the 13th century and until the 11th century BCE copper in the form of oxhide ingots was exported far and wide (see papers in Lo Schiavo *et al.* 2009). According to Lead Isotope Analysis the copper that was used to produce all these ingots is consistent with a Cypriot provenance and more specifically it is consistent with the ore body of Apliki (Gale 2011: 218).

THE SETTLEMENT OF APLIKI *KARAMALLOS*

Du Plat Taylor opened up a series of trenches on the hill of Apliki in order to investigate the extent of the archaeological site. The most important finds come from Trench A, where the remains of an L-shaped building destroyed by fire were uncovered (Du Plat Taylor and Kling 2007: 10). Trench B, which was very long and narrow, also contained architectural remains (*ibid.*: 41). Unlike the building in Trench A, the houses in this area were not destroyed by fire and seem to have been abandoned at the end of the occupation of the site. In Trench C, which is called the opencast, additional architectural remains were uncovered but these were not documented (*ibid.*: 64). In Trenches D–H no additional architectural remains were recovered but all the pottery collected was dated to the Late Bronze Age (*ibid.*: 65). Furthermore, in Trench F an irregular pit was cut into the rock and was identified as a possible old shaft entrance (*ibid.*:66).

The most important finds came from the building in Trench A (Fig. 24.3). Due to the fire that destroyed this structure, the floor and the objects that were lying on it were sealed by collapsing roof timbers and other debris. Radiocarbon dating of samples taken from the wooden beams and carburized cereals demonstrated that the buildings were constructed just after 1300 BCE and were destroyed between 1200 and 1180 BCE (Manning and Kuniholm 2007: 328). In other words, radiocarbon analysis confirmed the Late Cypriote IIC date provided by the pottery assemblage (Kling 2007a: 168).

The excavation revealed the remains of an L-shaped building with several rooms arranged around a courtyard (Du Plat Taylor and Kling 2007: 10). They were full of large storage vessels which perhaps held olive oil, while baskets contained legumes and cereals. Helbaek (1962: 185–186) identified hulled six-row barley, bread wheat, horsebean, lentil, grape, olive and coriander. None of these plants could

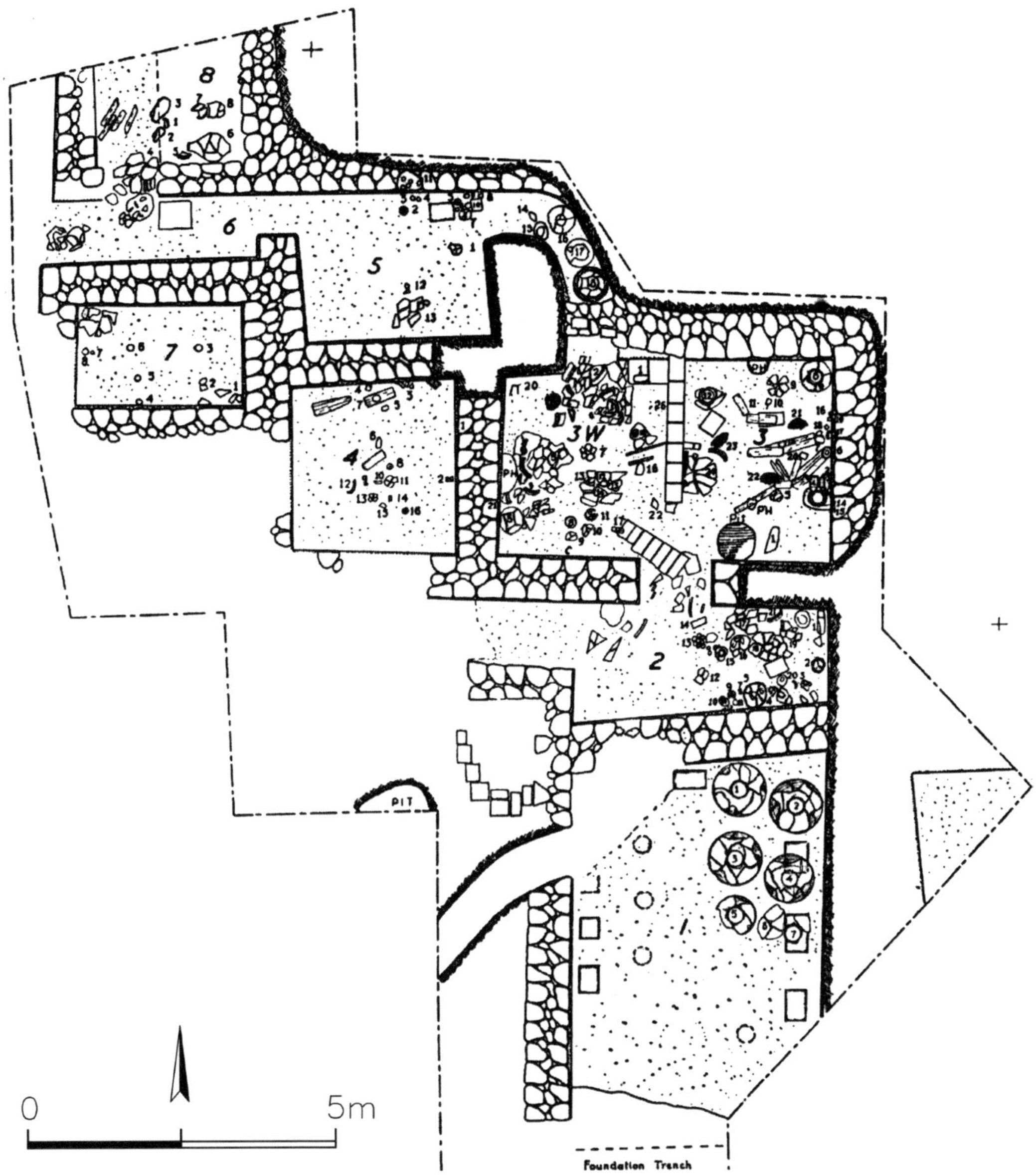

Fig. 24.3: Apliki *Karamallos* Plan of Area A (after Du Plat Taylor 1952: Fig. 3). Slag was collected in Rooms 2, 5 and 7 while the tuyères were found in Rooms 2, 3, 5 and 6.

have been cultivated on the steep, rocky slopes of Apliki and they would have been brought in. Some of the numerous stone tools that were recovered were clearly used to process cereals, while others, such as perforated stone hammers and large and heavy pounders (some of which are stained with iron oxides) must have been used for ore beneficiation (Kassianidou 2007: 283).

Large amounts of slag and a handful of broken tuyères denoted metallurgical activity, but they were both found in storage rooms in which no metallurgical installations were identified. More importantly none of the tuyères seems to have ever been used, as they bear no evidence of slagging (Fig. 24.4). With perhaps one exception (the one from House B III Pit 3), they do not even seem to have been fired at a high temperature. Furthermore, compared to tuyères from other contemporary sites such as Enkomi (Kassianidou 2012: 99) the ones from Apliki are truly massive.

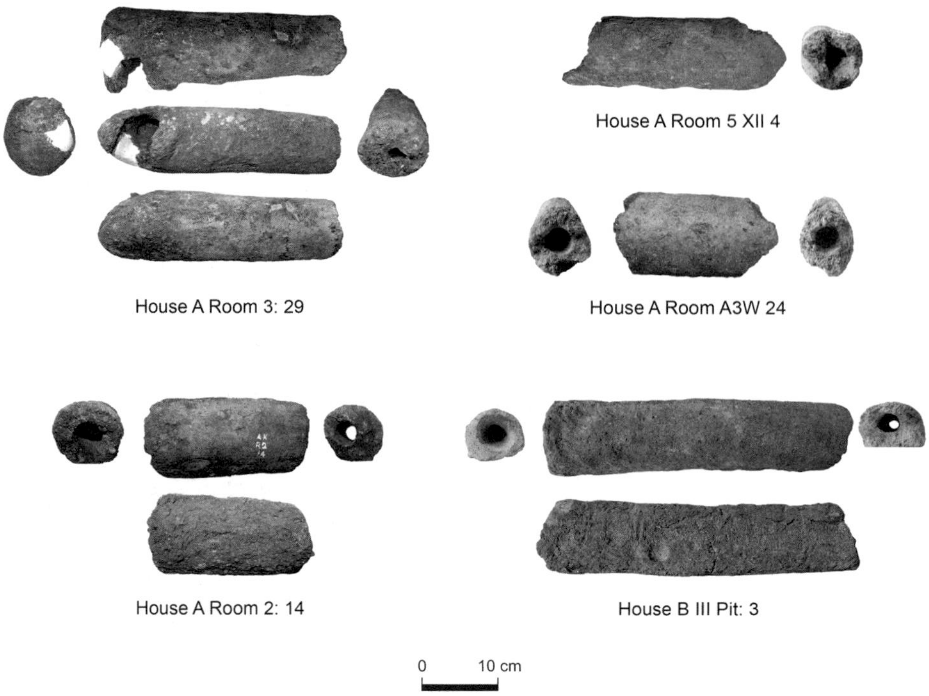

Fig. 24.4: The tuyères from Area A and Area B of Apliki *Karamallos* (photographed by V. Kassianidou).

The first to study these artifacts was Tylecote, who identified three types: the straight cylindrical, the straight "D" shaped and the elbow tuyère (Tylecote 1982: 92–93). Thanks to Tylecote, one of the tuyères from Apliki (House B III Pit: 3) is one of the best known archaeometallurgical finds from Cyprus, as it was featured in many of his publications. However, a few years ago I realized that Tylecote actually made a mistake that has become a factoid of our field (Kassianidou 2008: 252). Although when he initially published a drawing of tuyères from Cyprus he clearly stated in the legend that they came from Enkomi and Apliki (Tylecote 1981: 111); in later publications (e.g., Tylecote 1982: 93) he used the same drawing but in the legend he mistakenly stated that one of the D shaped tuyères came from the Middle Cypriot site of Ambelikou *Aletri*. He thus suggested a reconstruction of how the Ambelikou crucible would have been used with the D-shaped tuyère (*ibid.*). In fact the two are separated by several centuries and this reconstruction is wrong. Tuyères and bellows were not used in the Middle Bronze Age; they are a technological innovation of the Late Bronze Age (Kassianidou 2011).

The other so called D-shaped tuyère (House A Room 6: 5) is completely different and even more massive than the rest (Fig. 24.5). I would like to suggest that this in fact is not a tuyère but rather the nozzle of a pot bellows. I am basing this suggestion on the similarity with an unpublished example from Enkomi and the almost complete example in the collection of Pancyprian Gymnasium Museum (*ibid.*: 43). Like the tuyères from Apliki this artifact has not been fired to a high temperature nor does it seem to have been used.

What about the slag, and how can its presence in several rooms of House A be explained? Muhly contemplated this question in 1989:

> But why copper smelting slag in such a domestic context, in the rooms of what must have been a private house? Smelting activity certainly was not being carried out in the rooms of House A, rooms that had been cut out of the rocky hillside and surrounding an open courtyard. We have to assume that, for whatever reason, the slag was being stored in the houses of the miners' village. (Muhly 1989: 309)

Fig. 24.5: The so-called D shaped tuyère from House A (House A Room 6: 5) (photographed by V. Kassianidou).

Perhaps the slag was brought to the house to be crushed in order to collect even the smallest amount of metal trapped in its matrix. Apart from the pottery the largest assemblage of artifacts from House A are the stone tools, some of which were used to crush the slag, perhaps even by women. The smelting furnaces were clearly located somewhere else. Du Plat Taylor (1952: 150) notes that there was a slag heap at the bottom of the hill but she believed this to have been Roman in date. The Late Bronze Age smelting workshop must also have been somewhere on the hill but unfortunately it too has now been obliterated by the open cast mine.

In comparison with the metallurgical remains from other Cypriot copper workshops such as Phorades (Knapp and Kassianidou 2008) or Enkomi (Kassianidou 2012) it must be admitted that the ones from Apliki *Karamallos* are not that impressive. The metallic artifacts from the site seem to be even less remarkable. The total assemblage from the *Karamallos* includes 11 objects made of copper alloys weighing less than 100 g, two made of lead and one gold earring (Kling 2007b: 311–313). Note that not a single piece of oxhide ingot was found in the settlement which is believed to have produced all the oxhide ingots dating post-1450 BCE, including the 354 carried on the ship that sank at Uluburun. Equally striking is the fact that more than half of the 11 copper alloy artifacts, which are described as drills, awls or chisels in the monograph, are in fact styli. Papasavvas was the first to identify bronze styli from sites in Cyprus (namely Enkomi and Kalavasos *Agios Dhimitrios*) dating to the 13th century BCE (Papasavvas 2003). It is astonishing that more styli have been found in the few excavated rooms of House A at Apliki *Karamallos* than in the administrative building at Kalavasos *Agios Dhimitrios*, which covers an area of several hundred square meters. Someone was clearly busy recording something in that house. The evidence, in other words, supports Keswani's (1993: 77) idea that House A was an official residence and workplace from which the production of copper was controlled and the subsistence foodstuffs were stored for distribution to the labor force.

But Apliki *Karamallos* was not an affluent settlement. Apart from the single gold earring and the few fragments of ivory (Kling 2007b) there are no luxury goods such as the ones found in Enkomi, for example.[2] Furthermore, there are only a few examples of imported Mycenaean and Canaanite pottery

2 For an overview, see the British Museum online catalogue of finds from Enkomi, http://www.britishmuseum.org/research/publications/online_research_catalogues/ancient_cyprus_british_museum/enkomi/introduction.aspx

and even local fine wares such as White slip and Base ring are not common (Kling 2007a: 168). This led Webb to state that:

> Notably there is little in House A to suggest that elite goods acquired from the coastal emporia were reaching Apliki in substantial numbers. Rather the complement of finds gives the impression of a relatively small-scale venture operating within a local exchange system in which relations with agricultural villages which supplied basic foodstuffs and other essential commodities were of more immediate import than those with distant coastal centers. (Webb 2007: 271)

APLIKI *KARAMALLOS* IN CONTEXT

Clearly there must have been an administrative center, an emporium, on the coast that would have been in charge of the export of the huge amounts of copper, which according to Lead Isotope Analysis were produced in Apliki. Since 1974 this section of the coast, together with 38% of the island, has been under Turkish occupation and thus no archaeological work has been conducted there for almost 40 years. In the 1970s an important site was excavated at Morphou *Toumba tou Skourou* (Vermeule 1974; Vermeule and Wolsky 1990) and this is often considered to be the main settlement in this part of the island in this period, and perhaps the port from which the copper from Apliki was exported (Keswani 1993: 77). However, *Toumba tou Skourou* is more than 6 km from the coast and 20 km from Apliki. Furthermore, the study of the pottery found at *Karamallos* detected no special relation between the two sites (Kling 2007a: 168).

I strongly believe that we should look for an emporium on the coast below Apliki where the modern village of Karavostasi is located. Karavostasi means ship-stop in Greek, and this is where in the 20th century, the Cyprus Mines Corporation had built a pier to ship the ore extracted from the mines of the area. This is also where the Iron Age kingdom of Soloi, one of the richest and most important on the island throughout the first millennium BCE, was established (Stylianou 1992: 512–513). Undoubtedly there must have been a Late Bronze Age predecessor to Soloi from which copper was shipped.

It is unclear whether the copper was cast into oxhide ingots in this area and then shipped abroad, or whether black copper or even matte was transported either overland or by sea to Enkomi, where it was turned into copper oxhide ingots and then exported. Enkomi was the most cosmopolitan port on the island throughout the Late Bronze Age. Muhly (1989: 299) even suggested that at least for the first half of the Late Bronze Age, Enkomi may have been the capital of Cyprus and the seat of the king of Alashiya.

Although it may seem irrational today that matte or black copper would have been carried from the ore bearing and forested regions of the Troodos foothills all the way to Enkomi (which is about 95 km east—as the crow flies—from Apliki), it is not impossible. Dikaios (1969: 10), who attempted to understand the presence of extensive metallurgical workshops in Enkomi, wondered why copper ore could not have been carried to Enkomi from the Troodos foothills when at the beginning of the 20th century ice was carried by donkeys from the mountain top to Nicosia overnight during the summer.

During his excavations in the northern part of the site, Dikaios uncovered a series of metallurgical workshops, the earliest phase of which dates to the 16th century BCE (Kassianidou 2012). In the Late Cypriot IIC, a new building consisting of different sectors was erected in this part of the town (Dikaios 1969: 46). The western sector constitutes a vast group of copper workshops organized around a central court (*ibid.*: 56).

The cylindrical crucible (Inventory number 1640), one of the best known archaeometallurgical finds from Enkomi, was found in one of these rooms (*ibid.*: 58; *ibid.* 1971: 644). Tylecote (1982: 92) suggested that this was a smelting furnace. Recent analysis of slag samples from Enkomi, however, led Hauptmann (2011: 200) to suggest that it is a product of secondary metal or slag processing rather than of primary smelting. Elbow tuyères were found in the same rooms (Inventory number 1617) (*ibid.*: 644). The Late Cypriot IIC tuyères from Enkomi, which continue to be of the elbow type, are very different from the massive contemporary examples from Apliki (Fig. 24.6). They are probably different since the Enkomi installation, which consisted of a set of pot bellows, elbow tuyères and a cylindrical crucible, was most likely used in the secondary processing of the metal or perhaps even for melting the metal (Kassianidou 2016: 83–84). It is, I believe, very likely that some of the main products of the extensive Enkomi metallurgical workshops were copper oxhide ingots (Kassianidou 2013b: 143). A slag heap, located to the northwest of the building, supports the idea of extensive metallurgical activities in this area (*ibid.* 1969: 61). We have already started sampling the slag from Enkomi and hopefully the subsequent analytical study will finally reveal the metallurgical processes that took place in these workshops.

Apliki was not the only copper ore deposit to be exploited in this period. Lead Isotope Analysis demonstrates that several bronze artifacts from a number of Late Cypriot IIC sites are consistent with other Cypriot ore deposits (Gale and Stos-Gale 2012: 71). Unfortunately no Late Bronze Age mines or primary smelting sites have yet been discovered in any of the other mining regions because of the modern exploitation of the copper ore deposits. Nevertheless, the vibrant Cypriot copper industry of the 13th century is clearly illustrated by the abundance of archaeometallurgical finds such as slag, tuyères and crucible fragments that have been recovered in other contemporary Late Cypriot IIC settlements, such as Kalavasos *Ayios Dhimitrios* (South 2012) and Alassa *Pano Mandilaris* (Hadjisavvas 2011). Both are in close proximity to important copper ore deposits and presumably both would have been involved in the exploitation of these deposits. Unfortunately we still know very little about the political organization of the island in this period and how the copper production and trade was organized. However, a text from Ugarit mentions the king of Alashiya whose name is *Kusmesusa* (Malbrat-Labat 1999).

Fig. 24.6: An elbow tuyère (inventory number 1617) from Enkomi (dating to the Late Cypriot IIC) compared to a contemporary one from Apliki *Karamallos*.

Peltenburg has recently combined the above information with the absence of any clearly identifiable palaces on the island (although there were large administrative buildings in several settlements), and proposed a model for how the copper production and trade were organized:

> The incentive for accepting the institution of a *primus inter pares,* referred to in the texts as a "king," was to ensure and expand the successful export of high value copper and other insular resources by joining the existing framework of the international gift exchange system. That system required a "king," and, as far as Alašiya was concerned, it enabled merchants to operate effectively under the aegis of the Alašiyan leader. No doubt merchants worked independently too, but the trade in metals was especially the concern of elites of the highest rank. (Peltenburg 2012: 18)

CONCLUSIONS

According to Lead Isotope Analysis the copper used to produce all the oxhide ingots that date later than 1450 BCE are consistent with the ore deposit of Apliki. On the hill of Apliki at the site of *Karamallos,* Du Plat Taylor excavated the remains of a miners' settlement dating to the 13th century BCE. An analysis of the finds from Apliki *Karamallos* shows that the people who mined the copper ore and extracted the metal had little to do with the outside world. Clearly someone else was in charge of exporting the metal and reaped the benefits of this economic venture. Whether this someone was a king seated in Enkomi or a local ruler based in the Bay of Morphou remains for the time being an open question.

REFERENCES

Boutin, A. and Given, M. 2013. The Karkortis Valley. In: Given, M., Knapp, A.B., Noller, J., Sollars, L. and Kassianidou, V., eds. *Landscape and Interaction: The Troodos Archaeological and Environmental Survey Project, Cyprus, Vol. 2: The TAESP Landscape* (Levant Supplementary Series 15). Oxford and Oakville: 51–151.

Cobham, C.D. 1895. *Excerpta Cypria.* Nicosia.

Constantinou, G. 1982. Geological Features and Ancient Exploitation of the Cupriferous Sulphide Orebodies of Cyprus. In: Muhly, J.D., Maddin, R. and Karageorghis, V., eds. *Early Metallurgy in Cyprus, 4000–500 B.C.* Nicosia: 13–23.

Constantinou, G. 2007. Contribution of the Geology to the Early Exploitation of the Cupriferous Sulphide Deposits of the Skouriotissa Mining District. In: Kling, B. and Muhly, J.D., eds. *Joan du Plat Taylor's Excavations at the Late Bronze Age Mining Settlement at Apliki Karamallos, Cyprus* (Studies in Mediterranean Archaeology 94: 1). Sävedalen: 337–345.

Dikaios, P. 1969. *Enkomi: Excavations 1948–1958*, Vol. 1. Mainz.

Dikaios, P. 1971. *Enkomi: Excavations 1948–1958*, Vol. 2. Mainz.

Du Plat Taylor, J. 1940. Mines where the Mycenaeans Got Their Copper Discovered in Cyprus. *Illustrated London News* 196: 251.

Du Plat Taylor, J. 1952. A Late Bronze Age Settlement at Apliki, Cyprus. *The Antiquaries Journal* 32: 133–167.

Du Plat Taylor, J. and Kling, B. 2007. Overview of the Excavated Areas. In: Kling, B. and Muhly, J.D., eds. *Joan du Plat Taylor's Excavations at the Late Bronze Age Mining Settlement at Apliki Karamallos, Cyprus* (Studies in Mediterranean Archaeology 94:1). Sävedalen: 7–91.

Gale, N.H. 2011. Copper Oxhide Ingots and Lead Isotope Provenancing. In: Betancourt, P.P. and Ferrence, S.C., eds. *Metallurgy: Understanding How, Learning Why: Studies in Honor of James D. Muhly* (Prehistory Monographs 29). Philadelphia: 213–220.

Gale, N.H. and Stos-Gale, S. 2012. The Role of the Apliki Mine Region in the Post *c.* 1400 BC Copper Production and Trade Networks in Cyprus and in the Wider Mediterranean. In: Kassianidou, V. and Papasavvas, G., eds. *Eastern Mediterranean Metallurgy and Metalwork in the Second Millennium BC.* Oxford: 70–82.

Hadjisavvas, S. 2011. Broken Symbols: Aspects of Metallurgy at Alassa. In: Betancourt, P.P. and Ferrence, S.C., eds. *Metallurgy: Understanding How, Learning Why: Studies in Honor of James D. Muhly* (Prehistory Monographs 29). Philadelphia: 21–27.

Hauptmann, A. 2011. Slags from the Late Bronze Age Metal Workshops at Kition and Enkomi, Cyprus. In: Betancourt, P.P. and Ferrence, S.C., eds. *Metallurgy: Understanding How, Learning Why: Studies in Honor of James D. Muhly* (Prehistory Monographs 29). Philadelphia: 189–202.

Helbaek, H. 1962. Late Cypriot Vegetable Diet at Apliki. *Opuscula Atheniensia* 4: 171–186.

Ingham, F.T. 1959. Economic Geology. In: Wilson, R.A.M., ed. *The Geology of the Xeros-Troodos Area* (Geological Survey Department Cyprus Memoir 1). Nicosia: 137–177.

Kassianidou, V. 2007. Ground Stone Tools from Apliki *Karamallos*. In: Kling, B. and Muhly, J.D., eds. *Joan du Plat Taylor's Excavations at the Late Bronze Age Mining Settlement at Apliki Karamallos, Cyprus* (Studies in Mediterranean Archaeology 94:1). Sävedalen: 277–306.

Kassianidou, V. 2008. The Formative Years of the Cypriot Copper Industry. In: Tzachilli, I., ed. *Aegean Metallurgy in the Bronze Age. Proceedings of an International Symposium held at the University of Crete, Rethymnon, Greece, on November 19–21, 2004*. Athens: 249–267.

Kassianidou, V. 2011. Blowing the Wind of Change: The Introduction of Bellows in Late Bronze Age Cyprus. In: Betancourt, P.P. and Ferrence, S.C., eds. *Metallurgy: Understanding How, Learning Why. Studies in Honor of James D. Muhly* (Prehistory Monographs 29). Philadelphia: 41–48.

Kassianidou, V. 2012. Metallurgy and Metalwork in Enkomi: The Early Phases. In: Kassianidou, V. and Papasavvas, G., eds. *Eastern Mediterranean Metallurgy and Metalwork in the Second Millennium BC.* Oxford: 94–106.

Kassianidou, V. 2013a. Ancient and Modern Copper Production at Skouriotissa *Vouppes*. In: *Given, M., Knapp, A.B., Noller, J., Sollars, L. and Kassianidou, V., eds. Landscape and Interaction: The Troodos Archaeological and Environmental Survey Project, Cyprus*, *Vol. 2: The TAESP Landscape* (Levant Supplementary Series 15). Oxford and Oakville: 114–138.

Kassianidou, V. 2013b. The Production and Trade of Cypriot Copper in the Late Bronze Age. An Analysis of the Evidence. *Pasiphae* 7: 133–146.

Kassianidou, V. 2016. Metallurgy and Metalwork in Enkomi—Revisiting Porphyrios Dikaios' Excavations. In: Bourogiannis, G. and Muhlenbock, C., eds. *Ancient Cyprus Today: Museum Collections and New Research* (Studies in Mediterranean Archaeology Pocket-book 184). Uppsala: 79–90.

Kassianidou, V. and Papasavvas, G., eds. 2012. *Eastern Mediterranean Metallurgy and Metalwork in the Second Millennium BC.* Oxford and Oakville.

Keswani, P. 1993. Models of Local Exchange in Late Bronze Age Cyprus. *Bulletin of the American Schools of Oriental Research* 292: 73–83.

Kling, B. 2007a. Pottery from Apliki *Karamallos*. In: Kling, B. and Muhly, J.D., eds. *Joan du Plat Taylor's Excavations at the Late Bronze Age Mining Settlement at Apliki Karamallos, Cyprus* (Studies in Mediterranean Archaeology 94:1). Sävedalen: 95–227.

Kling, B. 2007b. Other Objects from Apliki *Karamallos*. In: Kling, B. and Muhly, J.D., eds. *Joan du Plat Taylor's Excavations at the Late Bronze Age Mining Settlement at Apliki Karamallos, Cyprus* (Studies in Mediterranean Archaeology 94:1). Sävedalen: 307–320.

Kling, B. and Muhly, J.D., eds. 2007. *Joan du Plat Taylor's Excavations at the Late Bronze Age Mining Settlement at Apliki Karamallos, Cyprus* (Studies in Mediterranean Archaeology 94:1). Sävedalen.

Knapp, A.B. 2011. Cyprus, Copper and Alashiya. In: Betancourt, P.P. and Ferrence, S.C., eds. *Metallurgy: Understanding How, Learning Why: Studies in Honor of James D. Muhly* (Prehistory Monographs 29). Philadelphia: 249–254.

Knapp, A.B. 2013. *The Archaeology of Cyprus: From Earliest Prehistory through the Bronze Age* (Cambridge World Archaeology). Cambridge.

Knapp, A.B. and Kassianidou, V. 2008. The Archaeology of Late Bronze Age Copper Production: Politiko *Phorades* on Cyprus. In: Yalçin, Ü., ed. *Anatolian Metal IV* (Der Anschnitt, Beiheft 21). Bochum: 135–147.

Lavender, D. 1962. *The History of the Cyprus Mines Corporation*. San Marino.

Lo Schiavo, F., Muhly, J.D., Maddin, R. and Giumlia Mair, A., eds. 2009. *Oxhide Ingots in the Central Mediterranean*. Rome.

Malbran-Labat, F. 1999. Nouvelles données épigraphiques sur Chypre et Ugarit. *Report of the Department of Antiquities, Cyprus,* 121–123.

Manning, S.W. and Kuniholm, P.I. 2007. Absolute Dating at Apliki *Karamallos*. In: Kling, B. and Muhly, J.D., eds. *Joan du Plat Taylor's Excavations at the Late Bronze Age Mining Settlement at Apliki Karamallos, Cyprus* (Studies in Mediterranean Archaeology 94: 1). Sävedalen: 325–335.

Muhly, J.D. 1989. The Organisation of the Copper Industry in Late Bronze Age Cyprus. In: Peltenburg, E., ed. *Early Society in Cyprus*. Edinburgh: 298–314.

Papasavvas, G. 2003. Writing on Cyprus: Some Silent Witnesses. *Report of the Department of Antiquities, Cyprus*, 79–94.

Papasavvas, G. 2009. The Iconography of the Oxhide Ingots. In: Lo Schiavo, F., Muhly, J.D., Maddin, R., and Giumlia Mair, A., eds. *Oxhide Ingots in the Central Mediterranean*. Rome: 83–132.

Peltenburg, E. 2012. Text Meets Material in Late Bronze Age Cyprus. In: Georgiou, A., ed. *Cyprus an Island Culture*. Oxford: 1–23.

Pulak, C. 2005. Das Schiffswrack von Uluburun. In: Yalçin, Ü., Pulak, C. and Slotta, T., eds. *Das Schiff von Uluburun. Welthandel vor 3000 Jahren.* Bochum: 55–102.

South, A. 2012. Tinker, Tailor, Farmer, Miner: Metals in the Late Bronze Age Economy at Kalavasos. In: Kassianidou, V. and Papasavvas, G., eds. *Eastern Mediterranean Metallurgy and Metalwork in the Second Millennium BC*. Oxford and Oakville: 35–47.

Stos, Z.A. 2009. Across the Wine Dark Seas…Sailor Tinkers and Royal Cargoes in the Late Bronze Age Eastern Mediterranean. In: Shortland, A.J., Freestone, I.C. and Rehren, Th., eds. *From Mine to Microscope. Advances in the Study of Ancient Technology.* Oxford: 163–180.

Stylianou, P.J. 1992. The Age of the Kingdoms: A Political History of Cyprus in the Archaic and Classical Periods. In: Papadopoulos, T., ed. *Travaux et Mémoires* 2. Nicosia: 375–530.

Tylecote, R.F. 1981. From Pot Bellows to Tuyères. *Levant* 13: 107–118.

Tylecote, R.F. 1982. The Late Bronze Age: Copper and Bronze Metallurgy at Enkomi and Kition. In: Muhly, J.D., Maddin, R. and Karageorghis, V., eds. *Early Metallurgy in Cyprus, 4000–500 B.C.* Nicosia: 81–100.

Vermeule, E.D.T. 1974. *Toumba tou Skourou: The Mound of Darkness: A Bronze Age Town on Morphou Bay in Cyprus*. Boston.

Vermeule, E.D.T. and Wolsky, F.Z. 1990. *Toumba tou Skourou: A Bronze Age Potters' Quarter on Morphou Bay in Cyprus*. Boston.

Webb, J.M. 2007. Stamp Seal of Serpentine from Apliki *Karamallos*. In: Kling, B. and Muhly, J.D., eds. *Joan du Plat Taylor's Excavations at the Late Bronze Age Mining Settlement at Apliki Karamallos, Cyprus* (Studies in Mediterranean Archaeology 94:1). Sävedalen: 269–275.

CHAPTER 25

KING HEROD AND THE CYPRUS COPPER MINES

Shimon Dar

Where did Herod the Great acquire the large sums of money he needed for his construction enterprises in Judea and other countries around the Mediterranean? According to Josephus Flavius, annual income from taxes in Judea was 1,050 talents, a great deal of money, but hardly enough to cover Herod's expenditures. Two examples demonstrate the high cost of the king's construction enterprises: The outlay for the Caesarea harbor was 1,750 talents; the expenditure for the Temple Mount was 1,200 talents. But it seems that the direct taxes Herod collected in Judea were but a fraction of the kingdom's budget: He engaged in commercial activities with the Nabateans and received royalties from the lucrative spice trade. He also received taxes from Augustus in Syria and part of Asia Minor. However, it seems that the main source of his income came from the right to exploit the rich copper mines of Cyprus, which, according to Josephus (Ant. 16.128), was granted him by his patron, Augustus. We estimate that it was these funds that enabled Herod to achieve his grandiose construction enterprises.

HISTORICAL SOURCES

Regarding the copper mines of Cyprus (Fig. 25.1), Josephus Flavius recounts:

> In the following days Herod made a present of 300 talents to Caesar, who was providing spectacles and doles for the people of Rome, while Caesar gave him half the revenue from the copper mines of Cyprus, and entrusted him with the management of the other half. In addition he honored him with hospitality and lodging. And he allowed him to have authority to dispose of his kingdom and to appoint as his successor whichever of his sons he chose or even to apportion it among them so that the share of the honor would go to each. (*Antiquities* 16: 128–129; trans R. Marcus and A. Wikgren1969)

According to Nikos Kokkinos (1998: 160), this event occurred in 12/13 BCE. Generations of scholars have sought to explain Augustus's moves and the social and economic implications of this act upon the rule of Herod, the Judean king who reigned by the grace of Rome. Scores of monumental construction activities that were carried out during Herod's reign in Israel proper (Fig. 25.2) and abroad were, at the time, unprecedented in the Mediterranean Basin, and even surpassed those of the imperial capital in Rome (Netzer 1980; Broshi 1985).

Josephus further states that Herod was obsessed with honor and never missed an opportunity to secure glory during his lifetime and afterwards, and spent money beyond his means (*Antiquities* 16: 153–154).

As his many building ventures required enormous sums of money, scholars have attempted to reconstruct the sources of wealth at the king's disposal. In addition to direct taxation that the royal treasury levied and which is estimated at approximately 1,050 talents per annum (Gabba 1990: 175), Herod had additional sources of income, such as the export of agricultural produce from his estates,

Fig. 25.1: Cyprus and its major settlements, with places mentioned in text (from Geva and Dar 2009: Fig.5).

Fig. 25.2: Caesarea—The Sunken Herodian Harbor (from Raban and Holum, 1996: Figure 1).

the cultivation of aromatic plants, involvement in the Arabian luxury trade, the half-shekel tax that the Judean diaspora contributed to the Temple, revenues from pilgrimages to Jerusalem and the Temple, loans to his neighbors (e.g., the Nabataean kings) and more (Broshi 1985; Gabba 1990). We would not be surprised if King Herod and the Nabatean kings were involved in copper mining and trade from the mines in the southern Arabah Valley. Apparently, an additional and central source of his capital was income from the operation of the Cyprus copper mines, rights to which he was granted by his patron Augustus, ruler of Rome.

The importance of copper and bronze for human culture in antiquity cannot be exaggerated and there was no cultural center in which they were not used. Even after the introduction of ironworking, the importance of copper did not decline. Cyprus was blessed with mines abundant in copper that had been exploited from early in human history (Fig. 25.3).

Copper was used to create vessels for everyday use, decorative vessels and even for imperial coins. Thus, it is easy to understand the unending demand for copper, and for its high price.

The importance of copper for the Cypriot economy in the Roman period is well attested by archaeology, as well as by written sources like Pliny and Galen (Michaelides 1996: 144–146; Kassianidou 2011: 540–541).

Pliny frequently mentions Cyprus copper and extols the abundance of medicines that were produced from copper and its compounds (Pliny, Book 34: 32–33, 100–127; trans. H. Rackman, 1968). These medicines were used for various ulcers, the clotting and drying of wounds, treating boils, leprosy, external and internal tumors, for healing diseases of the eyes, ears, nose and throat, treatment of intestinal worms, hemorrhoids, and diseases of the genitals and uterus. These pharmaceuticals were produced in well-known centers and consumed with honey, wine, vinegar, linseed, flour, etc. According to modern pharmacists, even today cuprous sulfate is used for the treatment of fungi and intestinal parasites.

Scholars specializing in copper of the Roman period hypothesize that, at its height, copper production in the Empire was 15,000–17,000 tons per annum, a quantity only exceeded during the

Fig. 25.3: Troodos Mountains—ancient mines in Amiandos (photograph by S. Dar).

Industrial Revolution (Dartmouth Toxic Metals 2013). It may be presumed that Cyprus was one of the more important areas from which copper was produced during the Roman period and that between a third and a fourth of Roman production originated on Cyprus. According to the metallurgist Bachman, there is more than four million tons of copper slag around the Troodos Mountains, most of which is from the Roman period (Wertime 1982: 357; Kassianidou 2000: 747). Modern scholars estimate that such a quantity of slag is evidence of 400,000 tons of copper that was produced in the area.

Exploitation of copper on Cyprus and other mining areas resulted in the destruction of forests and brush to a point where wood and charcoal had to be imported from Cilicia in Asia Minor, as attested in detail by Strabo the Cappadocian (Strabo: 14.6.1–6). Presumably, the reasons that led Augustus and his economic advisors to grant Herod the mining rights to Cyprus were political and economic. Herod, the loyal vassal of Augustus Caesar, acted as his representative in the Roman East and not only as the ruler of Judea.

Josephus recounts the following story, which can teach us about Augustus and his court's viewpoint regarding Herod's loyalty to Rome:

> "And they say that Caesar himself and Agrippa often remarked that the extent of Herod's realm was not equal to magnanimity, for he deserved to be king of all Syria and Egypt" (*Antiquities* 16, 141). Some scholars doubt the veracity of the account and claim that the story originated in Herod's court. (Kasher and Witztum 2007; 242–243)

In addition to his loyalty to Rome, Herod was known as an energetic and capable ruler, and his involvement in economic activities was known by all. Herod's operation of the Cypriot copper mines was based on another factor—the existence of Jewish communities all over the island that could have been a loyal and trustworthy administrative foundation (see below).

The annexation of the island to Rome and the expropriation of its abundant and valuable resources took place one generation before Octavian (Augustus) took control of Rome, after the Battle of Actium. In 58 BCE, Marcus Porcius Cato was sent to Cyprus as a representative of Rome in order to organize the administration and expropriate the final capital of the Egyptian house of Ptolemy (Badian 1965). Cato the Younger, great-grandson of Cato the Elder, political leader and noted agronomist, was one of Rome's leading statesmen, a supporter of Pompey who also served as governor of Sicily, North Africa and Asia.

On Cyprus, Cato confiscated the royal treasury of 7,000 talents, an act which led the final rulers of the House of Ptolemy to commit suicide. This large sum of money indicates the sizeable wealth that had accumulated on Cyprus as a result of the production and sale of copper.

Even after its annexation to Rome, Cyprus remained a source of great revenue. The island's inhabitants were required to pay 200 talents as a special tax in return for not having Roman soldiers posted on the island. This money was collected by brutal force (Jones 1971: 371).

ANNUAL INCOME FROM THE CYPRIOT COPPER MINES

The information regarding this issue is indirect, but large sums of money were definitely involved. Some scholars claim that the 300 talents given by Herod to Augustus, for the "bread and circuses" of the people of Rome, were the amount produced by the Cypriot mines (Gabba 1990: 176). This estimate is apparently too low. If those scholars who claim that 4,000–5,000 tons of copper were produced on Cyprus each year are correct, then the revenues were even higher, as hinted by the lofty sums expropriated by Cato from the Cypriot treasury in 59 BCE.

According to Roman Imperial law, the minerals and mines were the property of the conqueror, and a special Procurator was in charge of them (Kassianidou 2000: 751–752) (Fig. 25.4). From the time of

Hadrian (117–138 CE), we have written evidence from Portugal concerning the management of mines. The inscriptions affirm the full ownership of the mines by the Caesar and the state fiscus. A free citizen was appointed as a Procurator of the mines.

Contractors who leased the mines were responsible for paying an agreed sum of money, and in the meanwhile half of the copper was owned by the Procurator. It was forbidden to remove the copper from the mines until the leasing funds were paid. It seems that King Herod was entrusted with the same conditions in the time of Augustus (Fig. 25.5).

As noted above, the Jewish community on Cyprus was most likely a factor that enabled Herod to operate the mines. Jews had reached Cyprus during the First Temple period, but organized communities are known to have spread out over all parts of the island only from the Hellenistic/Hasmonean period (Roth-Gerson 1983: 94–97) (Fig. 25.6), as attested by epigraphic and archaeological evidence.

Fig. 25.4: Spanish miners in a relief from Ancient Linares (from Rostovtzeff 1966: Plates 35, 62).

Fig. 25.5: Roman ships that took part in the metal trade.

Fig. 25.6: Cyprus—Jewish settlements in antiquity.

The mines were operated in antiquity by contractors. Some of the miners were convicts, prisoners of war or rebels, all of whom were sent to work with shackles on their feet; their lifespan in the mines was short. Abraham Shalit claims that among those Herod sent to work in the mines were convicts and rebels who were sentenced to life in prison (Schalit, 1964: 131–132). This was a policy embraced by Augustus in Italy. Another way in which Herod strengthened his relationship with the Jews of Cyprus was by marrying Alexandra, the daughter of his brother, Phasael II, to Timius, who belonged to a local aristocratic Jewish family. According to Kokkinos and others, Timius was nominated as Procurator of the copper mines on Cyprus; Alexandra lived with him on the island (Kokkinos, 1998: 160). Similarly, it is presumed that the city Soloi, which had a long-standing Jewish community, served as the center for the mining activity (Kassianidou 2011: 540–541) (Figs. 25.7, 25.8).

Hasmonean and Herodian coins have been found on Cyprus; Kokkinos claims that an inscription from Paphos is also from Herod's times. (Kokkinos, 1998, 160). It was part of Herod's policy to cultivate relations with Jews in the diaspora and he was quick to protect them from their persecutors in many centers in the Mediterranean Basin. His actions on Cyprus were part of this policy, which brought him prestige among the Jews of the diaspora and among their Gentile neighbors. These relations had economic aspects, which Herod exploited in his own interest (Kasher 1996: 22).

In nearby Asia Minor, Jews were active in mining and quarrying activities during the Roman period, as can be attested by funerary stones discovered at Central Asia Minor sites (Levick and Mitchell 1993: 9) (Fig. 25.9). We may assume that some kind of connections existed between the Jewish communities in both provinces.

Fig. 25.7: The Bay of Soloi (photograph by S. Dar).

Fig. 25.8: Soloi—The loading plant for copper in modern times (photograph by S. Dar).

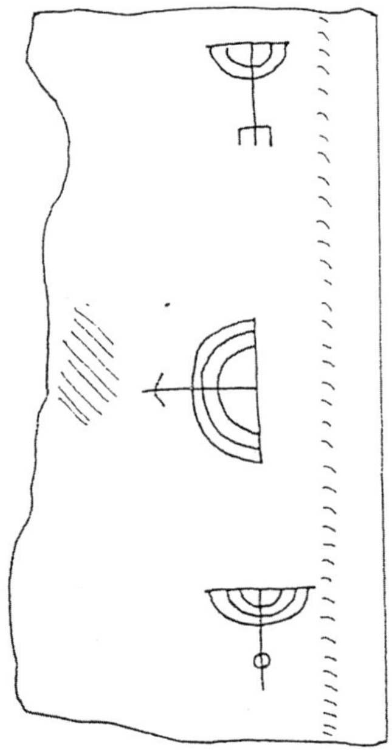

Fig. 25.9: Three incised menorahs in Roman period quarries—Upper Tembris Valley, Central Asia Minor (from Levick and Mitchell 1993: 9).

CONCLUSION

The expenses Herod incurred from his construction activities in the Land of Israel and abroad reached thousands of talents, judging by the construction costs of the Temple Mount and the Temple in Jerusalem, and of the construction of the Caesarea port (Warshavski and Peretz 1992: 44–46; Kost 2011: 203). There is little doubt that the lease on the Cypriot mines and their operation contributed much revenue that financed these endeavors. In Joseph Geiger's opinion (Geiger 2009), these construction ventures had domestic considerations, such as the employment of thousands of urban laborers and the prevention of social disorder among the unemployed. Augustus's Rome utilized a similar policy.

REFERENCES

Geva, H. and Dar, S., eds. 2009. *Guide to the Trip to North Cyprus*. Jerusalem (Hebrew).

Badian, E. 1965. M. Porcius Cato and the Annexation and Early Administration of Cyprus. *Journal of Roman Studies* 55: 110–121.

Broshi, M. 1985. Basic Lines in the Economy of Eretz Israel in the Period of Herod. In: Gross, N., ed. *Jews in the Economy*. Jerusalem: 27–33 (Hebrew).

Dartmouth Toxic Metals.. 2013. http://www.dartmouth.edu/~toxmetal/toxic-metals/more-metals/copper-history.html

Gabba, E. 1990. The Finances of King Herod. In: Kasher, A., Rappaport, U. and Fuks, G., eds. *Greece and Rome in Eretz Israel: Collected Essays*. Jerusalem: 160–168.

Geiger, J. 2009. Rome and Jerusalem: Public Building and the Economy. In: Jacobson, D.M. and Kokkinos, N., eds. *Herod and Augustus*. Leiden and Boston: 157–169.

Jones, A.H.M. 1971. *The Cities of the Eastern Roman Provinces,* 2nd ed. Oxford.

Kasher, A. 1996. Herod and the Jewish Diaspora. In: Isaac, B. and Oppenheimer, A., eds. *Studies on the Jewish Diaspora in the Hellenistic and Roman Periods*. Tel Aviv: 11–22 (Hebrew).

Kasher, A. and Witztum, E. 2007. *King Herod: A Persecuted Persecutor*. Haifa (Hebrew).

Kassianidou, V. 2000. Hellenistic and Roman Mining in Cyprus. In: Ioannides, G.K. and Hadjistyllis, S.A., eds. *Acts of the Third International Congress of Cypriot Studies*. Nicosia: 745–756.

Kassianidou, V. 2011. The Production of Copper in Cyprus during the Roman Period. In: Demetriou, A., ed. *Society of Cypriot Studies. Acts of the Fourth International Conference for Cypriot Studies*, Vol. 1:2 (Antiquity Section). Nicosia: 539–547.

Kokkinos, N. 1998. *The Herodian Dynasty: Origins, Role in Society and Eclipse*. Sheffield.

Kost, Y. 2011. The Inputs in Building Sebastos, Herod's Port of Caesarea. In: Ayalon, E. and Izdarechet, A., eds. *Caesarea Treasures*, Vol. 1. Jerusalem: 275–282 (Hebrew).

Levick, B. and Mitchell, S., eds. 1993. *Monuments from Appia and the Upper Tembris Valley, Cotiaeum, Cadi, Synaus, Ancyra Sidera and Tiberiopolis*. London.

Michaelides, D. 1996. The Economy of Cyprus during the Hellenistic and Roman Periods. In: Karageorghis, V. and Michaelides, D., eds. *The Development of the Cypriot Economy*. Nicosia: 139–152.

Netzer, E. 1980. Herod's Building Projects: State Necessity or Personal Necessity. *Cathedra* 15: 38–51 (Hebrew).

Raban, A. and Holum, G.H., eds. 1996. *Caesarea Maritima. A Retrospective after Two Millennia*. Leiden.

Rostovtzeff, M. 1966. *The Social and Economic History of the Roman Empire*, Vol. 1. Oxford.

Roth-Gerson, L. 1983. The Jews in Asia Minor. In: Stern, M., ed. *The Jewish Diaspora in the Hellenistic-Roman World*. Jerusalem and Tel Aviv: 77–100 (Hebrew).

Schalit, A. 1964. *King Herod: Portrait of a Ruler*. Jerusalem (Hebrew).

Strabo. 2000. *Geography*. Translated by H.L. Jones (Loeb Classical Library). Cambridge.

Warshavski, A. and Peretz, A. 1992. Building the Temple Mount: Organization and Execution. *Cathedra* 66: 3–46 (Hebrew).

Wertime, T.A. 1982. Cypriot Metallurgy Against the Backdrop of Mediterranean Pyrotechnology: Energy Reconsidered. In: Muhly, J.D., Maddin, R. and Karageorghis, V., eds. *Early Metallurgy in Cyprus 4000–500 BC: Acts of the International Archaeological Symposium*. Nicosia: 351–361.

CHAPTER 26

ARCHAEOMETALLURGICAL RESEARCH ON IRON AGE (1250-300 BCE) COPPER PRODUCTION IN THE NORTHERN AL-HAJJAR MOUNTAINS (OMAN PENINSULA)

Julie Goy, Michele Degli Esposti, Cécile Le Carlier de Veslud and Anne Benoist

Recent archaeological investigations by the French Archaeological Mission in the Emirate of Fujairah (U.A.E.), directed by Anne Benoist, revealed evidence of intensive extraction of copper ores and extensive use of copper and copper alloys during the Iron Age.

This substantial metallurgical production stemmed from the favorable geological environment of the Oman Peninsula and specifically from the ophiolitic al-Hajjar Mountains, which fostered the development of metallurgy as early as the Bronze Age. While much work has been done on third millennium BCE metallurgy, little is known about the processes carried out more than 1,000 years later, as well as about the economic and social context of this production.

This situation prompted the undertaking of new investigations, comprising stratigraphic excavations and surveys, which will allow a consideration of the complete chaîne opératoire, from the ore to the final object. The preliminary results of this project, coming from three recent surveys and a first test excavation, are summarized here, along with the earliest results of chemical analysis carried out on ingots, slags and objects of Iron Age sites located in the Emirate of Fujairah.

INTRODUCTION

GEOGRAPHICAL AND GEOLOGICAL CONTEXT

The Oman Peninsula, located in the southeastern corner of Arabia, comprises the entire territory of the United Arab Emirates and the northern regions of the Sultanate of Oman.

The imposing al-Hajjar Mountains, with an altitude reaching in places to ca. 3000 m, wind their way across the whole peninsula for almost 600 km; they constitute the southeastern margin of the Arabian continental plate and represent a late Cretaceous continent-ocean collision boundary (Batchelor 1992: B108).

The geological process that took place along this boundary caused the obduction of what is today the world's best-exposed ophiolite nappe complex: the Semail Ophiolite, which hosts all the metalliferous deposits currently worked (Ag, Au, Cr, Cu and Zn). Occurrences of copper mineralization are mainly found as vein fillings in faults (*ibid.*: B114) or massive sulphide deposits (Fig. 26.1) (Hauptmann, Weisgerber and Bachmann 1988: 35). Conversely, the Cyprus-type massive sulphide deposits which mostly occur at or near the top of the lower lava series, are extremely rare in the Semail Ophiolite but well attested in the Masirah Ophiolite, which lies off the southeast coast of Oman (Batchelor 1992: B117).

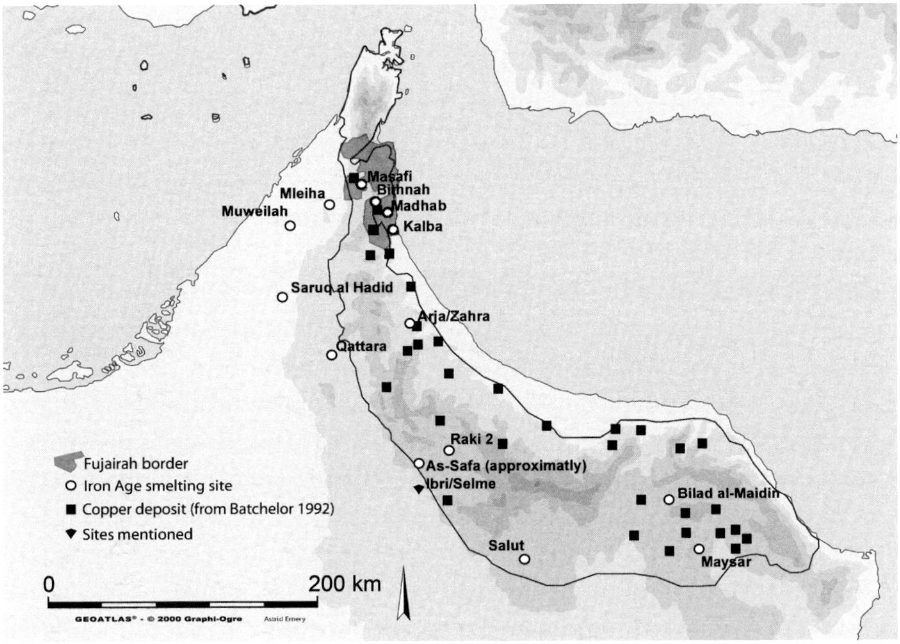

Fig. 26.1: Map of the Oman Peninsula with copper mineralizations and Iron Age smelting sites.

A BRIEF OUTLINE OF PREHISTORIC COPPER PRODUCTION IN THE OMAN PENINSULA

This favorable geological setting led to an initial exploitation of copper deposits as early as the end of the fourth millennium BCE, during the local Hafit period (3100–2700 BCE). While this initial production has to be considered quite limited, evidenced only by the objects found in collective graves (Weeks 2003),[1] the following Umm an-Nar period (2700–2000 BCE) saw an intensification in copper working that resulted in a flourishing exchange, especially with Mesopotamia, as recorded in Sumerian texts from the time of the great Sargon of Akkad onwards.[2] Archaeologically, this evolution is reflected in the discovery of several mining and smelting sites, among which a paramount role in the development of studies has been played by Maysar (today al-Moyassar) 1, thoroughly investigated by the mission of the Deutsche Bergbau Museum (Weisgerber 1980; 1981; 2007; Hauptmann 1985; Hauptmann, Weisgerber and Bachmann 1988; Prange *et al.* 1999).

The following periods of the local Bronze Age, i.e., Wadi Suq and Late Bronze Age (2000–1600 and 1600–1250 BCE) are still largely unknown to archaeologists. Few human settlements have been found and most of the material comes again from burial structures. Copper production seems to decrease, and textual evidence suggests that export towards Mesopotamia decreased as well, due to the competition of Cypriot copper (Oppenheim 1954; Millard 1973; Weisgerber 2007).

A new rise in copper production characterizes the Iron Age (1250–300 BCE) (See Weeks 2003: Fig. 2.12), alongside the emergence of a widespread settlement pattern that is beginning to be well known by scholars. Despite the large numbers of copper items found thus far, mining and metallurgical sites are almost unknown, probably due to some extent to the exploitation of the same sources in Islamic times (Weisgerber 2007). Exchanges with Mesopotamia are again attested (Lombard 1984), although the form in which copper was traded is unknown.

1 But see recently discovered evidence for a Hafit period smelting site in Kutterer, Neureiter and Jasim 2013.

2 Among others, Glassner 1989.

THE IRON AGE IN SOUTHEASTERN ARABIA

In the second half of the second millennium BCE, a rather distinct change in the material culture[3] defines the beginning of the local Iron Age, displaying a remarkable degree of homogeneity throughout the Oman Peninsula, not unlike the previous periods. The development of an irrigation based on the partially underground *aflaj* (sing. *falaj*) system towards the turn of the first millennium, stands as the basis for a substantial demographic growth, mirrored in the number of newly founded settlements. These can now comprise differentiated, specific areas for meeting, dwelling, cultic practices, as well as agricultural land and fortifications.[4] The communities living in these settlements were not isolated, and many archaeological indicators reflect permanent contacts with Mesopotamia, Iran and Bahrein, i.e., the neighboring Assyrian, Median and Achaemenid empires (Potts 1990: 84).

Abundant evidence speaks of a definite increase in copper production; outstanding are the production sites of Rākī II (Weisgerber and Yule 1999: 108–115) and Sārūq al-Ḥadīd (Nashef 2010), as well as the metal hoard discovered in 1979 at ʾIbrī/Selme (Yule and Weisgerber 2001),[5] now paralleled by the recent discovery of a large metal production site at asʾSafa (Fig. 26.1) (Al-Bakri, Genchi and Tosi 2013). Excavation by the French Archaeological Mission in Masāfī (Emirate of Fujairah) unearthed a different kind of evidence, recovering 217 ingots, "furnace bottoms"[6] and scraps from two large storage jars buried beneath the floor of a collective building at Masāfī 1 (Benoist *et al.* 2012).

The absence of iron metallurgy is a striking feature of Iron Age southeastern Arabia, clearly contrasting with the period's definition (Lombard 1989); the discovery of iron production at Sārūq al-Ḥadīd, initially reported as dating to the Iron Age and thus neatly redefining this picture, should most likely be placed several centuries later (Herman, Cassan and Qandil 2012)[7].

The role of metallurgy within contemporary society went beyond the utilitarian function, and was endowed with a cultic aspect which is difficult to define precisely but that had among its main symbols the snake, as widely attested at many Iron Age sites of the region, where it is also independently found both in the form of applied and painted decoration on pottery and on bronze/copper figurines (Fig. 26.2).[8]

BACKGROUND AND METHODOLOGY

Since the late twenties of the last century, and actually well before Gulf archaeology really started with the Danish expeditions in Bahrein (Bibby 1969), the study of ancient, namely third millennium metallurgy, has attracted scholars towards southeastern Arabia, in search of the geographical location of the land of Magan mentioned in Mesopotamian texts.[9] This has led to an in-depth investigation of the many aspects of early Bronze Age metallurgy, with the above-mentioned German mission working in

3 The situation is not equally clear all over the peninsula, and some shapes and fabrics of the Late Bronze Age are quite similar to the subsequent ones, but it is beyond the scope of this paper to discuss the issue in more detail.

4 It is the case of the settlement networks at Bithnah (Benoist 2013a) and Masāfī (Benoist 2013b).

5 The hoard, which probably also resulted from tomb robbing, contained earlier items as well.

6 This term will be used here instead of the more common "slag cake" in order to underline the fact that these masses of solidified metal were left to cool down inside the furnace.

7 The renewal and intensification of archaeological research at this site after this paper was presented is actually confirming an Iron Age iron production.

8 This issue has been addressed by the authors elsewhere (*Benoist et al. 2015).*

9 Peake 1928, and see n. 6.

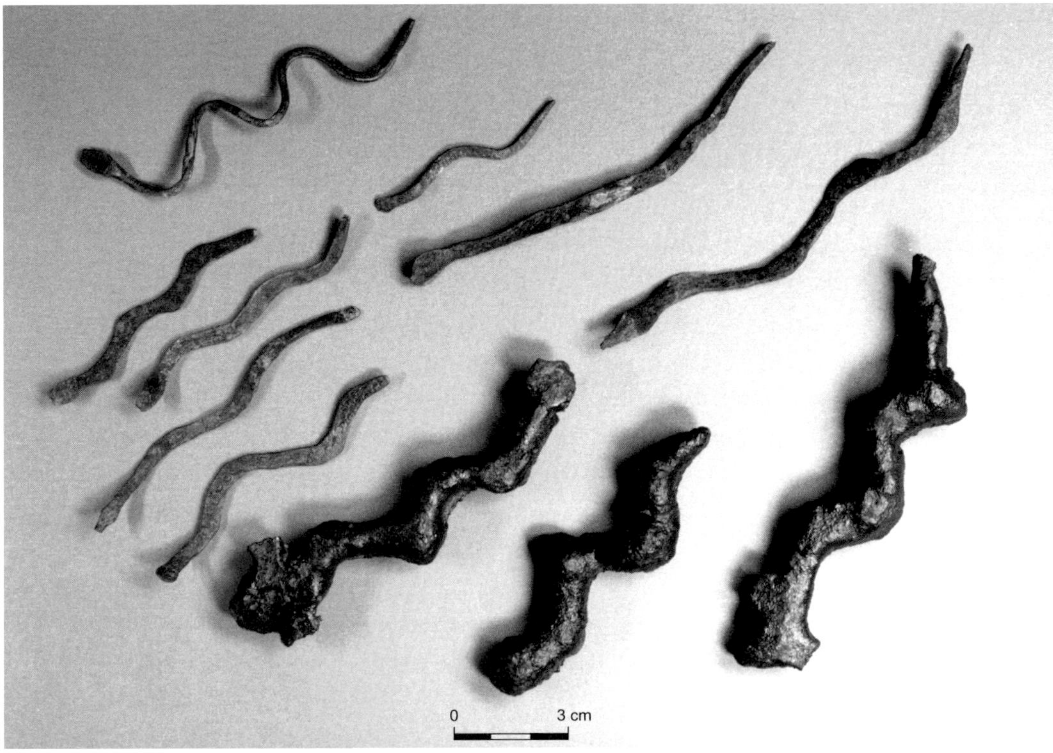

Fig. 26.2: Some examples of snake decorations and bronze snakes.

the context of the southern al-Hajjar Mountains, while L. Weeks (1997; 2003) carried out an extensive overview of the known data and further analytical work in their northern part.

Iron Age metallurgy in this region has conversely received much less attention, and the extensive chemical characterizations of the ʾIbrī/Selme hoard objects (Prange and Hauptmann 2001) stand almost isolated in the literature.

Nor has archaeology supplied comparable evidence concerning safely dated production sites; at this time, the key site of asʾSafa and Sārūq al-Ḥadīd promise definite advancements in the field, but we must wait for the initial results.

Given this state of the research, the exceptional discovery of the buried copper hoard of Masāfī 1 prompted the undertaking of a comprehensive study of the contemporary metallurgy in the northern part of the Oman Peninsula, starting from the territory of the Emirate of Fujairah, with whose government the French archaeological mission has an active agreement for excavations and survey.

The first steps in this direction had already been taken in 2001, when A. Ploquin and M. Ohnenstetter carried out a first survey of the area (Benoist *et al.* 2003). The project was restarted in 2009 with the first survey, followed by new campaigns in 2011 and 2013[10]; a test excavation was conducted in January, 2013 at one of the sites which had been identified in 2011.

Slags were collected from several surveyed sites in order to be analyzed along with similar finds coming from excavated contexts such as the jars of Masāfī 1. The latter contained different types of items, which are represented in Fig. 26.3.[11] Moreover, a few finished objects from Masāfī 1 and Masāfī 3 were also studied (Table 26.1).

10 For details on the survey strategy and chronology, see Goy *et al.* 2013: 127–128.

11 The typology was proposed in Goy 2012.

Table 26.1: General List of Samples and Provenance

Sample Code	*Object Type*	*Location*	*Analyses*
436/E345-03	Slag	Bithnah 44	SEM
436/E345-07	Slag	Bithnah 44	SEM
447/E368-03	Slag	Bithnah 44	SEM
452/E370	Slag	Bithnah 44	SEM
125/E63-01	Slag	Bithnah 44	SEM
154/E 192-01	Slag	Bithnah 44	SEM
B54	Slag	Bithnah 54	SEM
S05	Slag	BIT-90.054	SEM
S07	Slag	BIT-90.054	SEM
S09	Slag	BIT-90.054	SEM
S10	Slag	BIT-90.054	SEM
S13	Slag	BIT-90.054	SEM
S48	Slag	MAD-09.005	SEM
S50	Slag	MAD-09.005	SEM
S51	Slag	MAD-09.005	SEM
S52	Slag	MAD-09.005	SEM
S30	Slag	Aswani A	SEM
S41	Slag	MAD-09.001	SEM
S42	Slag	MAD-09.001	SEM
S45	Slag	MAD-09.001	SEM
S55	Slag	MAY-11.001	SEM
S59	Slag	MAY-11.001	SEM
S63	Slag	ANO-11.001	SEM
S67	Slag	ANO-11.001	SEM
S02	Copper prill	AWH-09.001	ICP
S35	Copper prill	ASW-11.001	ICP
S37	Copper prill	ASW-11.001	ICP
S67 (2)	Copper prill	ANO-11.001	SEM
046	F. bottom	MSF1	ICP-Xray
051	F. bottom	MSF1	ICP
053	F. bottom	MSF1	ICP
056	F. bottom	MSF1	ICP
110	F. bottom	MSF1	ICP-Xray
111	F. bottom	MSF1	ICP-Xray
140	F. bottom	MSF1	ICP-Xray
147	F. bottom	MSF1	ICP-Xray
148	F. bottom	MSF1	ICP-Xray
165	F. bottom	MSF1	ICP-Xray
001	F. bottom	MSF1	ICP
010	F. bottom	MSF1	ICP
011	F. bottom	MSF1	ICP
013	F. bottom	MSF1	ICP

Sample Code	*Object Type*	*Location*	*Analyses*
015	F. bottom	MSF1	ICP
002	F. bottom	MSF1	ICP
003	F. bottom	MSF1	ICP
047	F. bottom	MSF1	ICP
006	F. bottom	MSF1	ICP
009	F. bottom	MSF1	ICP
170	Ingot	MSF1	ICP
101	Ingot	MSF1	ICP-Xray
029	Ingot	MSF1	ICP-Xray
030	Ingot	MSF1	ICP-Xray
031	Ingot	MSF1	ICP
033	Ingot	MSF1	ICP
034	Ingot	MSF1	ICP
035	Ingot	MSF1	ICP
039	Ingot	MSF1	ICP
040	Ingot	MSF1	ICP-Xray
043	Ingot	MSF1	ICP
045	Ingot	MSF1	ICP-Xray
005	Ingot	MSF1	ICP
050	Ingot	MSF1	ICP
057	Ingot	MSF1	ICP-Xray
067	Ingot	MSF1	ICP-Xray
070 down	Ingot	MSF1	ICP-Xray
070 up	Ingot	MSF1	ICP-Xray
071	Ingot	MSF1	ICP-Xray
076	Ingot	MSF1	ICP-Xray
082	Ingot	MSF1	ICP-Xray
084	Ingot	MSF1	ICP-Xray
012	Ingot	MSF1	ICP
014	Ingot	MSF1	ICP
028	Ingot	MSF1	ICP
205-16	Snake	MSF3	ICP
205-35	Arrowhead	MSF3	ICP
205-56	Snake	MSF3	ICP
213-44	Snake	MSF3	ICP
213-60	Snake	MSF3	ICP
051-009	Knife	MSF1	ICP
094-2	Bowl	MSF1	ICP

Twenty-four pieces of slag were sampled and analyzed using SEM-EDS, in order to document the compositional data in oxide percentages, which will allow a future comparison with the rocks of the region (Goy *et al.* 2013: 130). The data-set was broadened with the inclusion of seven previously analyzed samples (Table 26.2) (Pilault 2005).

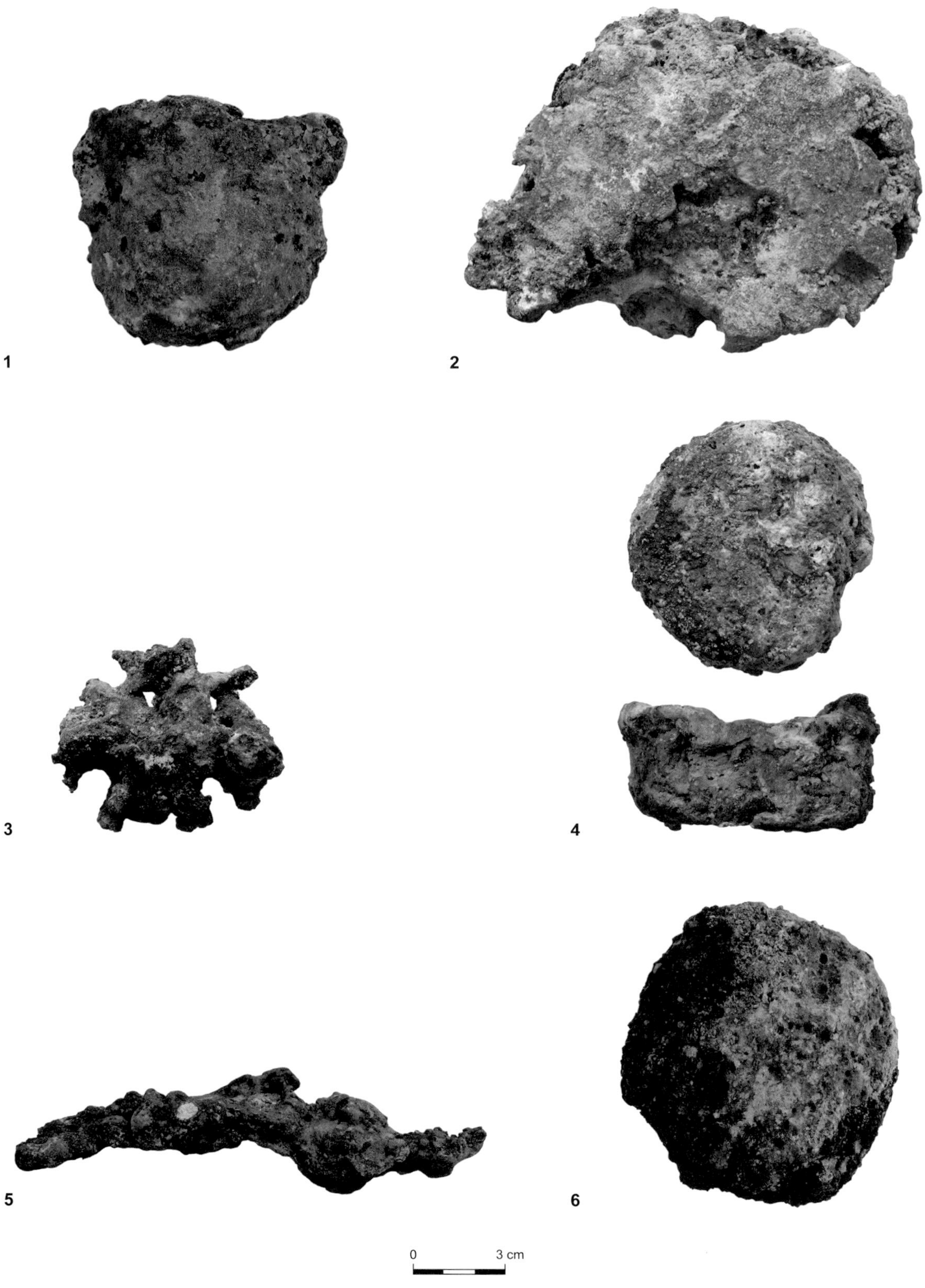

Fig. 26.3: Different types of ingots recorded at Masafi.

Table 26.2: SEM Results for Slag

Sample Code	*Location*	*MgO*	*Al2O3*	*SiO2*	*K2O*	*CaO*	*TiO2*	*FeO*	*CuO*
S5	BIT-90.054	8.65	10.47	37.05	0.56	5.73	0.33	32.03	0.96
S7	BIT-90.054	12.05	11.93	45.78	0.69	7.22	0.38	35.40	1.09
S9	BIT-90.054	8.77	14.13	38.17	0.54	6.41	0.24	24.42	0.85
S10	BIT-90.054	9.09	11.45	38.29	0.64	6.84	0.31	34.46	1.51
S13	BIT-90.054	9.58	11.38	38.77	0.44	6.23	0.04	29.95	0.60
B54	Bithnah 54	7.06	10.14	33.96	0.59	6.23	-	40.60	1.39
452/E370	Bithnah 44	6.84	14.46	42.48	0.37	6.96	-	27.90	0.99
436/E345-03	Bithnah 44	8.38	11.96	35.43	0.34	5.93	-	36.88	1.08
436/E345-07	Bithnah 44	8.58	10.38	38.53	0.73	3.95	-	32.79	5.03
154/E 192-01	Bithnah 44	4.37	7.51	42.93	0.38	6.66	-	36.92	1.01
125/E63-01	Bithnah 44	7.86	10.84	34.40	0.46	6.44	-	37.79	1.21
447/E368-03	Bithnah 44	9.83	11.62	33.93	0.31	6.03	-	36.93	1.36
S48	MAD-09.005	7.78	17.57	42.68	0.35	8.28	0.74	18.38	2.44
S50	MAD-09.005	6.00	22.67	44.01	0.48	6.05	2.43	16.69	0.77
S51	MAD-09.005	7.05	17.99	46.73	0.52	9.67	2.43	22.86	1.02
S52	MAD-09.005	6.71	17.88	41.11	0.39	8.66	2.18	17.65	1.26
S30	Aswani A	4.95	11.06	34.37	0.97	3.70	0.32	36.44	4.52
S41	MAD-09.001	7.06	17.09	40.98	0.35	8.71	1.83	19.67	1.21
S42	MAD-09.001	6.92	15.96	40.78	0.84	7.40	1.39	22.75	0.59
S45	MAD-09.001	7.69	20.19	45.30	0.68	8.43	2.83	32.83	2.24
S55	MAY-11.001	7.38	15.18	38.60	0.18	7.93	0.47	28.58	0.85
S59	MAY-11.001	8.86	15.02	39.97	0.43	8.24	0.23	23.28	0.28
S63	ANO-11.001	4.56	15.66	39.91	0.29	7.35	0.42	31.09	0.74
S67	ANO-11.001	4.51	16.15	41.65	0.28	8.69	0.36	25.78	0.69

ICP-AES was instead used for 20 "furnace bottoms" plus 25 ingots, for metallic copper prills spotted in the cut sections of four pieces of slag (Table 26.3) and for the seven finished objects (Table 26.4).[12]

PRELIMINARY RESULTS

SURVEY

During the surveys several smelting sites were located, as well as three open-cast mines (Fig. 26.4). The latter, all close to each other in Wādī al Hayl, provided no dating evidence, but should probably be correlated with the adjacent Islamic fort and settlement. The majority of the smelting sites also date to the Islamic period, or lack any dating evidence.[13]

12 SEM-EDS analyses were conducted in the laboratory of CMEBA (Centre de Microscopie Electronique a Balayage et microAnalyse) in Rennes, while ICP-AES analyses were conducted in the laboratory of Rennes 1 University.

13 See Goy *et al.* 2013 for more details on the main Islamic sites located during the survey.

Table 26.3: ICP Results for Ingots and Prills*

Object	*Ag ppm*	*As %*	*Bi ppm*	*Co ppm*	*Cu %*	*Fe %*	*Ni %*	*Pb ppm*	*Sb ppm*	*Sn %*	*Zn ppm*
Ingots (L1: Oval; L2: Plano-convex; L3: Flat)											
L1	63.32	0.16	644.38	631.14	97.38	2.08	0.13	118.74	45.10	0.00	906.59
L1	0.00	19.02	241.52	11,108.49	55.34	16.69	0.00	73,973.09	249.52	0.30	875.00
L2	34.44	0.00	313.14	158.15	94.73	5.04	0.09	128.80	36.98	0.00	648.27
L2	0.00	0.08	285.28	798.65	92.48	7.19	0.00	374.75	71.47	0.01	919.76
L2	39.38	0.00	509.76	243.06	97.12	2.59	0.11	0.00	3.70	0.00	1,027.70
L2	7.20	0.00	311.12	108.52	97.78	2.07	0.03	0.00	3.78	0.00	693.95
L3a	69.45	2.06	291.19	2,062.32	90.91	4.40	2.30	156.74	84.98	0.00	676.13
L3a	66.30	0.02	319.82	379.20	97.98	1.58	0.27	127.49	35.49	0.00	622.97
L3a	0.00	0.36	299.35	1,975.76	96.34	0.42	0.22	108.63	140.55	2.32	923.45
L3a	8.67	0.42	324.80	1,628.43	55.12	43.86	0.27	13.88	14.36	0.00	1,318.88
L3a	10.23	0.86	354.17	2,716.93	88.06	8.07	2.50	197.66	321.98	0.00	1,436.07
L3a	20.45	0.01	394.05	432.44	89.64	10.12	0.06	0.00	3.52	0.00	857.17
L3b	42.64	0.01	650.93	840.10	94.08	5.35	0.31	123.61	37.67	0.00	725.34
L3b	58.46	0.00	762.87	219.87	99.17	0.57	0.07	127.46	35.73	0.00	695.93
L3b	106.43	1.22	608.14	2,633.04	88.73	7.03	2.58	206.73	65.99	0.00	804.11
L3b	0.00	0.43	253.84	1,414.86	74.40	24.55	0.00	3,498.26	109.32	0.01	768.63
L3b	55.70	9.18	185.22	16,570.23	22.32	5.04	60.31	2,133.41	8,880.10	0.00	3,584.25
L3b	28.91	0.21	421.44	479.03	93.66	5.80	0.12	67.17	10.98	0.00	1,043.42
L3b	25.05	0.14	342.12	1,050.72	91.27	7.84	0.49	0.00	6.83	0.00	1,040.13
L3b	46.74	0.64	346.69	1,973.25	84.34	14.09	0.59	0.00	3.84	0.00	981.50
L3b	44.90	0.58	322.35	1,413.64	90.35	8.31	0.49	0.00	4.55	0.00	937.07
L3b	41.72	0.16	307.96	3,320.64	80.25	18.65	0.47	0.00	47.17	0.00	941.21
L3b	29.23	0.19	503.33	790.52	98.45	0.51	0.60	0.00	16.97	0.00	1,240.41
L3b	32.32	7.30	384.46	3,448.41	81.92	2.96	7.10	690.89	444.77	0.00	2,197.86
L3a	30.64	0.00	321.05	823.10	91.37	7.79	0.62	0.00	2.83	0.00	1,010.64
Copper prills											
S67	24.78	0.03	223.32	947.39	47.56	51.93	0.04	0.00	0.00	0.19	1,456.45
S37	22.37	0.00	331.60	2,394.96	63.26	36.19	0.23	0.00	0.00	0.00	443.16
S35	9.81	0.00	258.60	4,166.03	54.27	44.52	0.70	0.00	0.00	0.00	584.18
S02	17.84	0.00	158.86	1,109.47	31.77	67.98	0.07	0.00	0.00	0.00	493.44

* The Results have been updated since the Timna Conference and the last publication: Benoist *et al.* 2015.

The only possible Iron Age slag dump, likely hinting at smelting operations carried out in its proximity, was located in Wādī Madhab (MAD-09.005), immediately west of the city area of Fujairah (Fig. 26.4).

It is worth noting that all the smelting sites, MAD-09.005 included, are situated on flat terraces flanking the numerous *wydian* (sing. *wādī*) that cut through the mountains; although no mine was spotted close to any of them, such a location clearly assures the proximity to mineral resources on the one hand and to fuel, namely trees, on the other.

TABLE 26.4: ICP RESULTS FOR OBJECTS*

Cu %	*Fe %*	*Ni %*	*Pb ppm*	*Sb ppm*	*Sn %*	*Zn ppm*
99.01	0.13	0.06	361.20	992.96	0.00	723.32
99.54	0.13	0.13	136.67	68.92	0.00	712.53
99.37	0.24	0.17	151.90	64.69	0.00	739.99
99.36	0.28	0.16	147.86	66.18	0.00	660.17
99.13	0.38	0.19	159.98	125.88	0.00	809.43
95.55	2.50	1.37	163.70	88.20	0.00	923.73
86.37	0.77	0.04	520.48	365.15	12.28	656.85

* The Results have been updated since the Timna Conference and the last publication: Benoist *et al.* 2015.

The same site of MAD-09.005 suggests that a spatial connection could have existed during the Iron Age between smelting sites and fortifications: the valley where it is located is in fact controlled from the opposite side by the coeval fortress of Husn Madhab (Benoist and Corboud 1998). To some extent, this is also the situation at Bithnah, where the now destroyed smelting site Bithnah 54 stands on the opposite side of the valley to Bithnah 24 Fortress (Benoist 2013a: Fig. 1). A similar situation is reported by Weisgerber for Wadi Miadin (Weisgerber 2007: 303) and can be hypothesized for the smelting site near the fort of Awhala (AWH-09.001). Here, in fact, the smelting site can be dated to the Islamic period, while a nearby fort was erected above the remains of an Iron Age fort (Potts *et al.* 1996; Petrie 1998).

A similar connection, in this case between the fortified settlement of Masāfī 2 and a possible mine exploiting the copper vein which crosses Jebel al-Haryan (on which the settlement was erected) was initially postulated (Goy *et al.* 2013: 131 and Fig. 3 left). Test excavations carried out in early 2013 showed that there was no mine but just a small natural cave.

TEST EXCAVATION AT WADI MADHAB (MAD-09.005)

In January, 2013, a test excavation was begun on the possibly Iron Age smelting site located in Wādī Madhab, in order to test its contents for consistence and collect some dating evidence. A large amount of slag[14] was collected, both from the surface and from excavated contexts. This slag has been sampled for future analyses, and will be compared with securely dated Iron Age slag previously recovered from the excavations at Bithnah 44, Bithnah 54, Masāfī 1 and Masāfī 5 (Pillault, Le Carlier and Ploquin 2013; Benoist 2013b: 59).

The whole area surrounding the site had been deeply impacted by water erosion and by construction work for a modern dam. The site itself, which occupies a triangular area of ca. 220 sq m, is shaded by four acacia trees standing on an elevated area, and associated with the main stone accumulations visible on the surface.

Three possible structures, comprising stone alignments, which may indicate the presence of dry stone walls, were easily identified, while larger boulders were clustered around at least two of the acacia trees. This suggested that the tree growth could be preferentially associated with the presence of buried layers rich in organic remains, possibly connected with anthropic activity.

14 The final estimate is around 2,215 kg.

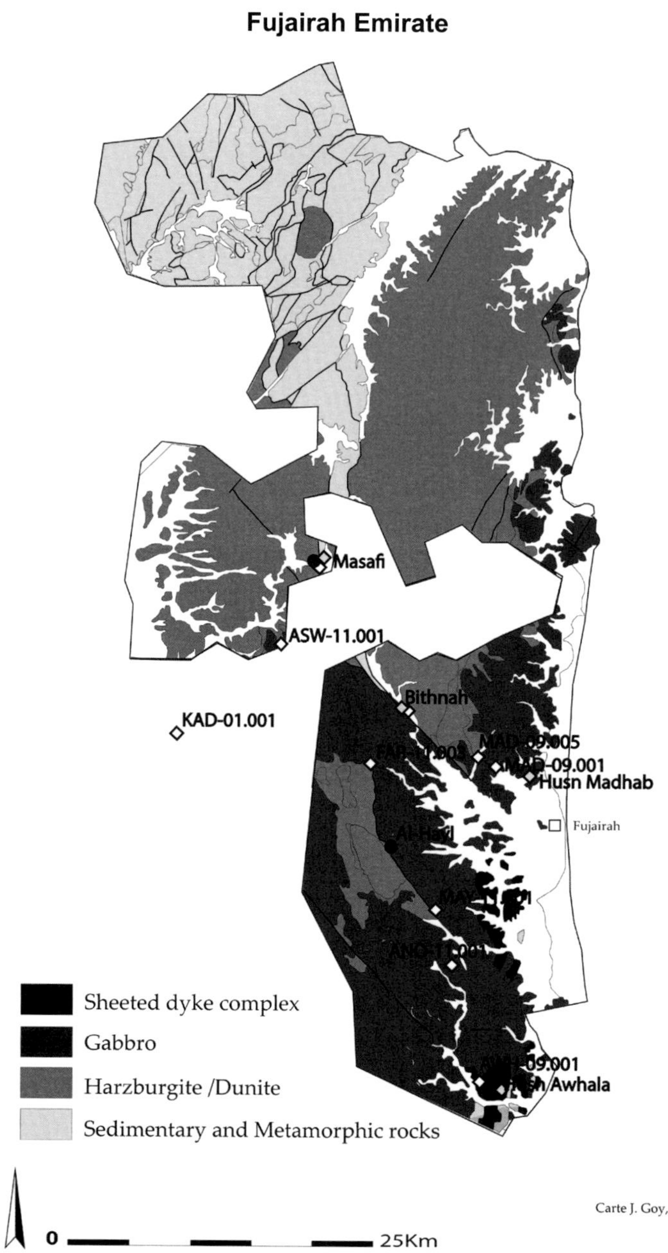

Site	x	y	type	datation
HAY-11.001	56.1328	25.045233	mine	Islamic
HAY-11.002	56.1322	25.045855	mine	Islamic
HAY-11.003	56.1317	25.045509	mine	Islamic
FAR-11.003	56.12032	25.08464	slag	Islamic
MAY-11.001	56.15207	25.02176	slag	Islamic
ASW-11.001	56.08043	25.1359	slag	Islamic
ANO-11.001	56.16041	24.59369	slag	Islamic
MAD-09.001	56.173975	25.090374	slag	Islamic
MAD-09.005	56.183256	25.083183	slag	Iron Age
MSF-10.005	56.09274	25.1803	slag	Islamic
BIT-90.054	56.143164	25.1111	slag	Islamic
AWH-09.001	56.17306	24.54324	slag	Islamic

Carte J. Goy, d'après Thomas et al. 2006

Fig. 26.4: Location map and list of surveyed sites. and list of surveyed sites.

A benchmark was established in the middle of the site, with an elevation of +15 m ASL. Two test trenches were laid down, of which only T1, shaped like an L so as to include the visible alignments, comprised structural remains that will be briefly described here, leaving aside the full stratigraphic description.[15]

Two of the mentioned alignments proved to correspond to two poor, dry walls, almost parallel with each other, running in a roughly north–south direction (Fig. 26.5).

15 A comprehensive excavation report has been submitted to the Fujairah Authority for Tourism and Archaeology.

The western wall (Wall 1) comprises two rows of medium sized, unhewn stones, with just a single course surviving above ground, most of which was actually visible on the surface. It is ca. 0.50 m wide and 3.70 m long. At its southern end the wall shows an eastward bend, unfortunately incomplete. Several slags were visible among its stones.

Wall 2, roughly parallel with Wall 1 though with slightly larger stones, comprises only a single row of stones, of which only one course has survived. It has an average width of 0.30 m and is preserved

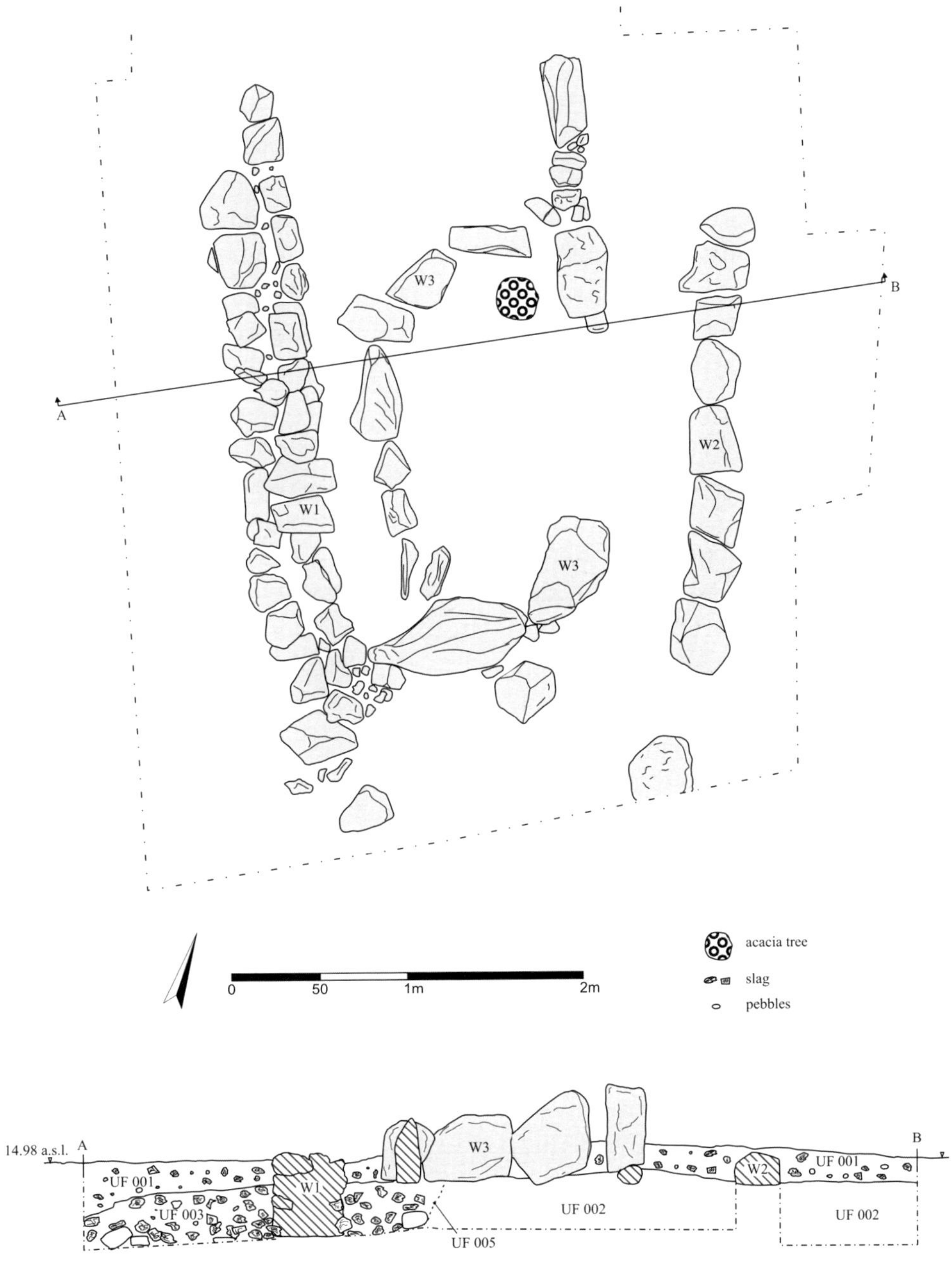

Fig. 26.5: Madhab 5 plan and section.

over a length of 2.70 m. In contrast to Wall 1, the stones that had been identified on the surface probably belonged to the second course of Wall 2, and had been displaced mainly by root action.

Between Wall 1 and Wall 2, large unshaped boulders defined the oval feature Wall 3. Roughly shaped, Wall 3 was also partially destroyed by root action. Its main axes measure 2 and 1.40 m respectively, the main one running parallel to Wall 1 and Wall 2.

Both Wall 2 and Wall 3 were erected above ground, resting on a soil labelled as UF 002. No evidence of anthropic activity was recovered from this layer. Slags and shells collected during its excavation were situated only in the upper 5 cm, and should thus be considered intrusions from the overlying deposit.

Abundant slag pieces were stuck much deeper in the underlying soil west of Wall 3, leading to the distinction of another layer, UF 003. In order to investigate the nature of UF 003, a deep sounding was cut transverse to the mentioned structures (Fig. 26.5), reaching a depth corresponding to the bottom of UF 002. This demonstrated that Wall 1 had an underground foundation comprised of several courses of small and medium sized stones and embedding a large number of smelting slags, suggesting that it could belong to a phase earlier than Wall 2 and Wall 3.

The wall may have cut through UF 003 when it was erected, though another possibility is that this slag-rich layer actually represents the backfilling of Wall 1's foundation trench. The possible trench cut, visible only in the eastern part of the section, has been labelled as UF 005. The real nature of UF 003 must remain a conjecture, as further excavation at the site is not currently scheduled.

It is relevant that the unearthed structures were surely erected after a substantial mass of slags had been produced. Besides, they were certainly not massive structures and the overall elevation must be calculated as very limited, likely comprising light and perishable materials. Considering the depression characterizing the center of the investigated area, one hypothesis could be that the depression itself coincides with the original furnace, if one envisages it as comprising a partially excavated chamber.

Another possibility could be that the actual furnaces stood a few meters to the west, on the actual foothills, and in this case they would be lost below the dam works debris.

Several stone tools (mainly crushers but also larger anvils and a fragment of a small saddle quern) were recovered from the excavation; some copper ore was collected as well. The quantity of shells, mainly *terebralia palustris*, a well-known species that thrives in mangroves, was remarkable.

Potsherds were extremely rare, and no diagnostic fragments were recovered. Even so, the fabrics compare well with the Iron Age pottery known from the nearby site of Husn Madhab, as well as with the assemblages of the coeval settlements of Bitnah and Masafi.

ANALYSIS

Overall, the elemental composition and phase content of the slags shows good homogeneity, with no clustering of the data which could be related to chronological differences. This means that compositional data are of no help in dating sites for which other indicators are missing—a possibility that we wanted to test. This is peculiarly evident when Madhab samples are considered. The only clusterings that emerge further underline how the slag composition depends on the ore and not on a hypothetically different smelting process. Higher Ti content in fact distinguishes the slag from the two sites in Wādī Madhab (Iron Age MAD-09.005 and Islamic MAD-00.001) from the others, but not between themselves (Fig. 26.6a). The slag from Bithnah 44, on the other hand, clusters due to the absence of the same element.

A similar distinction is shown by the Al and Fe contents: Wādī Madhab samples have higher Al and lower Fe (Fig. 26.6b) than the rest of the sample. Madhab samples were actually unique

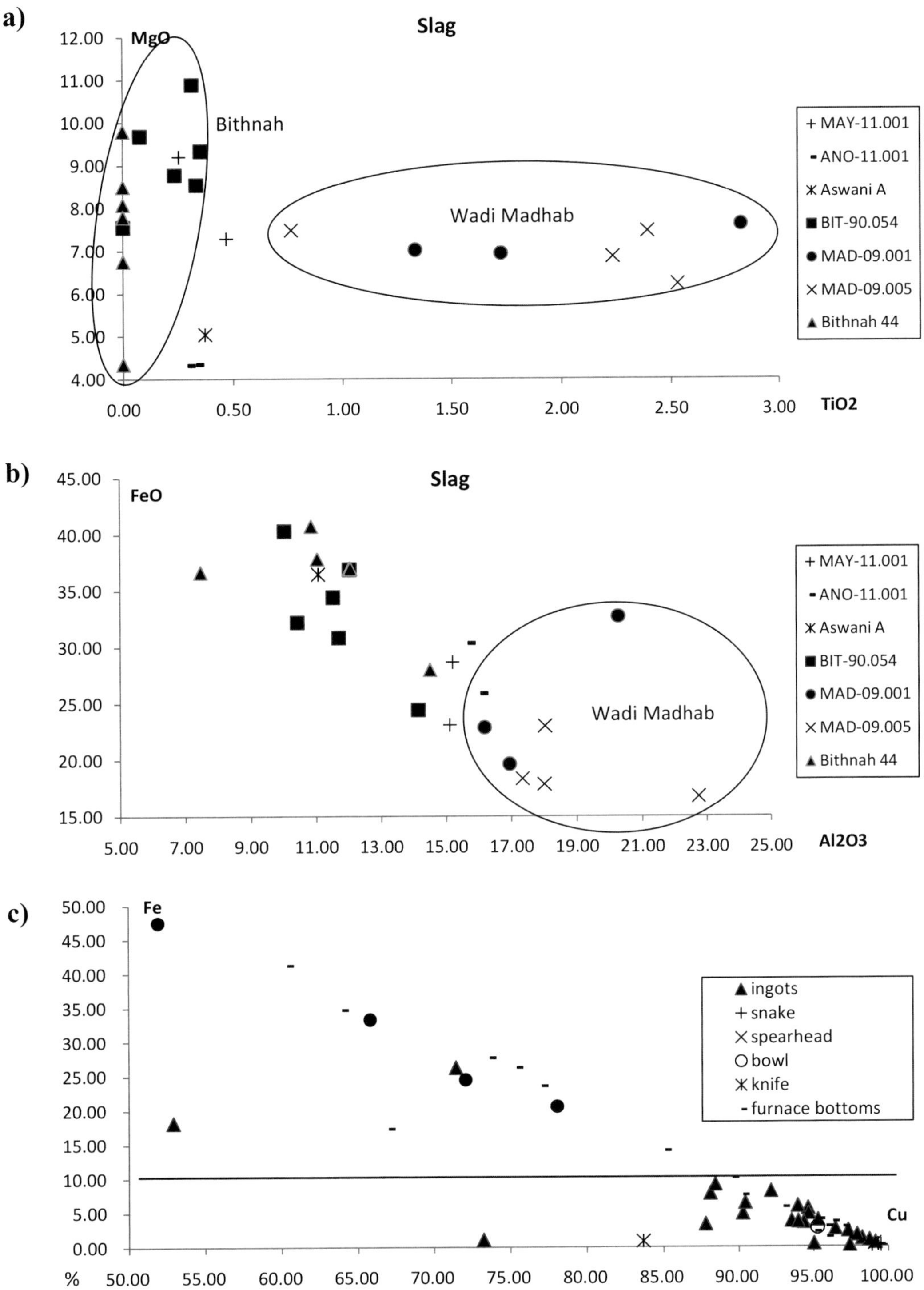

Fig. 26.6: (a, b) data in oxide%: binary graphs showing the clustering of the slags from the two sites in Wadi Madhab (MAD-09.001 and MAD-09.005), on the basis of Al, Ti and Fe contents; (c) data in weight%: the distribution of the analyzed items (slag excluded) on a Fe-Cu binary diagram clearly shows the increasing refinement of the metal from ingots to finished objects, and the fact that Fe is the main impurity. Outliers are linked with occasional high As, Ni or Sn content.

already at macroscopic levels, showing a distinctive rusty-red color, although this may derive from the burial environment.

The main mineral components of the slags were determined using SEM-EDS and studying back-scattered electron images.

The most widely represented phases are olivine (fayalite and forsterite), spinel (magnetite and hercynite) and wüstite. The presence of such siliceous and ferrous minerals could indicate the use of a flux consisting predominantly of iron ores to remove silica gangue, possibly also mirrored in the oxide contents (w%) of the slags, where SiO_2 and FeO are the two main components (Table 26.2). This range of siliceous and ferrous minerals is known in slag from a large number of sites over a wide area of the Near East, covering a wide chronological span as well.[16]

Results are not yet complete, and data on the sulphidic component are awaited, which will help determine the different stages of metal production; besides, the incidental inclusion of matte in the metal/slag would be highly useful for the identification of the ore.

Further work will address other aspects of the smelting process, such as the slag's liquidus temperature, its viscosity and the physicochemical conditions in the smelting furnace.[17]

The analyzed ingots consist of unalloyed Cu, with contents mainly above 90% (16 samples), and a few particularly low exceptions (Table 26.3).[18] Only ingot 33 shows a Sn content of 2.32%, while the copper prill found in slag S67 comprises 0.19% of this element. Such percentages could not derive from ores available in the area (Hauptmann, Weisgerber and Bachmann 1988: 46), and should thus be regarded as an addition. When this percentage is compared with that of the only analyzed true bronze object (see below), it seems likely that the ingots derived from the recycling of scrap metal possibly added to fresh Cu.

The main trace elements are Fe, Ni and As. Fe could be tentatively correlated both to the use of mixed Cu-Fe sulphidic ores[19] which are abundant in the whole Samail Ophiolite (Weeks 2003: 12), and to the already-discussed use of Fe-rich flux. The Iron Age increase in copper production has already been linked to the initial (for southeastern Arabia) intensive exploitation of extensive primary Cu ores incorporating unaltered sulphides (Weisgerber 1988: 256). At the same time, more easily reachable and exposed ore bodies were likely exploited, and the use of gossan minerals in the charge can also have contributed to raise the Fe content.

High Fe percentages can be linked to the use of particularly Fe-rich minerals and/or flux, rather than to a generically poorly controlled process. Its general efficiency is in fact reflected in the minimal quantity of Cu lost in the slag. This hypothesis can find support in the possible distinction between a high- and a low-Fe type of ingot, which could reflect different degrees of purification (see Fig. 26.6c).[20] Most of the ingots in fact have Fe content which would make their direct use for casting rather difficult, if not impossible, but it is known that a simple re-melting operation is generally sufficient to lower such percentages to as low as 0.05% (Craddock 2000: 154).

A similar grouping is consistent for the furnace bottoms as well (Fig. 26.6c), although this could also indicate their connection with different stages of copper extraction and purification—high-Fe examples deriving from primary smelting and the remaining ones from a refining operation.

16 E.g., Hauptmann (1985) on Medieval Omani slags and Ploquin *et al.* (1994) on the early Medieval slag from Mleiha (Emirate of Sharjah). See also Moorey 1994: 244.

17 This will allow comparison with data from Bithnah (Pillault, Le Carlier and Ploquin 2013: 281) and also with existing data on Bronze Age metallurgical procedures (see Hauptmann, Weisgerber and Bachmann 1988).

18 The results have been updated since the Timna conference and the last publication: Benoist *et al.* 2015.

19 A correlation between Fe content and such mixed sulphidic ores was already assumed by Tylecote (1962).

20 E.g., Klemenc *et al.* 1999: 147–148. The two broad groups could as well indicate two different refining stages, the work of two distinct workshops, or more likely the use of ores with different iron concentrations.

Refining—prior to the manufacture of the various objects—also seems to be indicated by the decrease in levels of Ni and As from the ingots to the objects (Tables 26.3 and 26.4).

However, two unexpected results from ingots recorded Ni content of 60.31% and 7.10% respectively (Table 26.3). This alloy has a completely different whitish color that makes these ingots easily distinguishable from the others.

It is for the moment impossible to formulate convincing hypotheses to explain such high values but, awaiting further analyses, it is worth mentioning here that a deliberate production of high Ni copper alloys seems unlikely, as Cu-Ni alloys have no recognized advantages over unalloyed copper apart from the silvery color and a better resistance to corrosion. In any case, it has to be stressed that the analyses of slag from Bithnah indicated a furnace temperature between 1250 and 1500 °C (Pillault, Le Carlier and Ploquin 2013: 281), thus capable of also melting a Ni-rich alloy like the ones possibly witnessed in our two samples, whose melting point would be well above 1300 °C (ASM 2007).

The high As values of some ingots are slightly easier to explain. First, As-rich mixed mineralizations are known within deposits from the lower part of the Ophiolitic sequence in the territory of the U.A.E. Second, in equilibrium condition, As can form pure intermetallic phases with Cu, with As ratios around 30% (Thornton, Rehren and Pigott 2009: 309), and the formation of arsenates also avoids much of the As loss which would be due to the element volatility (Pernicka 1999). Since the objects from Masafi resulted almost As-free (Table 26.4), it may be assumed that the final casting eliminated these impurities, or that ingots with relevant As content were intentionally not used to produce objects.

Finally, below are some comments on the (admittedly small quantity of) analyzed objects. All are made of unalloyed Cu, with the exception of the knife blade MSF1-010-56 containing almost 12.28% Sn. It may be of significance that this is possibly the only sampled utilitarian object; it would be tempting to connect the rarity of Sn exclusively with tools which required a greater hardness. The remaining objects are in fact a copper bowl, a pointed fragment whose original shape and function cannot be asserted and four copper snakes. The latter in particular belongs to a category widely attested at southeastern Arabian Iron Age sites, surely bearing some cultic value, with growing evidence of a link between such animals and metallurgy (Benoist 2007; Potts 2007 ; Benoist *et al.* 2015).

CONCLUSIONS

One of the main points that can be drawn from the data presented here is that the analyzed ingots from Masāfī 1, not to mention the metal chunks of furnace bottoms, were probably not intended to be cast into objects, given their rather random composition and inaccurate refining. Their archaeological context suggests as well that the hoard may have had a non-technological value, albeit intrinsically still bearing an economic one. Therefore, purely technological explanations for their characteristics could be misguided. These items may have been deposited inside the jars as a votive offering just by virtue of a general connection with metallurgy. This somehow endowed them with some sort of "importance," regardless of clumsy production. In addition, the absence of further refinement could be the consequence of a will for heavier (by not removing iron impurities) or nice looking (silvery appearance of the high-nickel ingots) votive offerings. This consideration calls into question the wider-ranging issue of the social role of metallurgy within Iron Age communities, which is beyond our scope here, but that has been addressed by the authors in Benoist *et al.* 2015.

The preliminary results presented here are meant to be integrated into a more comprehensive, interconnected archaeological and analytical program, which was initiated with the aim of overcoming the recognized "dearth of archaeometallurgical research on Iron Age copper smelting" (Weeks 2003: 42). Further surveying is planned along the *wydian*, cutting through the mountains of the U.A.E. in order to build a comprehensive database of smelting sites (from different periods) as well as to enlarge the analytical sample. The chemical and mineralogical composition of slag will then be compared with that of the possible mother rock,[21] while similar finds from contemporary sites in the Oman Peninsula will also be analyzed, and available data from other teams will be integrated into a general discussion.

For the time being, we can confirm that there seems to be a general homogeneity in the composition of the different items from these sites, but we will have to wait for the final analytical results in order to draw detailed parallels. In this way we hope to be able to define the general use of copper and its alloys in this area during the Iron Age, and possibly identify modalities of metal exchange, whether in the form of finished objects or as intermediate products.

REFERENCES

ASM. 2007. *ASM Handbook online, Vol. 3: Alloy Phase Diagrams*. Materials Park, OH.

Bachmann, H.G. 1982. *The Identification of Slags from Archaeological Sites* (Institute of Archaeology Occasional Publication 6). London.

al-Bakri, S.S., Genchi, F. and Tosi, M. 2013. Early Iron Age Tribal Alliances in Eastern Arabia: The Archaeological Evidence from the Daba Corridor Grave and the Metal Manufacturing Site of as'Safah. *Paper Presented at the 46th Seminar for Arabian Studies, London, 26th–28th July, 2013.*

Batchelor, D.A.F. 1992. *The Identification of Slags from Archaeological Sites* (Institute of Archaeology, Occasional Publications 6). London.

Benoist, A. 2007. An Iron Age II Snake Cult in the Oman Peninsula: Evidence from Bithnah (Emirate of Fujairah). *Arabian Archaeology and Epigraphy* 18(1): 34–54.

Benoist, A., ed. 2013a. *La Vallée de Bithnah au Cours de l'Age du Fer* (BAR International Series 2510). Oxford.

Benoist, A. 2013b. A Green Paradise: Economic Strategies, Collective Practices, and Local Ancestors of the Iron Age Community of Masāfī (Emirate of Fujairah, UAE). *Proceedings of the Seminar for Arabian Studies* 43: 47–62.

Benoist, A. and Corboud, P. 1998. La céramique du site fortifié de Husn Madhab dans l'émirat de Fujairah. In: Mouton, M., ed. *Assemblages céramiques des sites de l'âge du fer de la péninsule d'Oman* (Documents d'archéologie de l'Arabie 1). Lyon (CD-Rom).

Benoist, A., Bernard, V., Charbonnier, J., Goy, J., Hamel, A. and Sagory, T. 2012. Une occupation de l'âge du fer à Masafi. Travaux récents de la Mission archéologique française aux E.A.U. dans l'émirat de Fujairah (Émirats Arabes Unis). *Chroniques Yéménites* 17 (Available online at: <http: //cy.revues.org/1803>).

Benoist, A., Bernard, V., Ohnenstetter, M., Ploquin, A., Saint-Genez, F. and Schiettecatte, J. 2003. *French Archaeological Mission in Fujairah 2001: First Campaign at Bithnah.* Lyon.

Benoist, A. and Corboud, P. 1998. La céramique du site fortifié de Husn Madhab dans l'émirat de Fujairah. In: Mouton, M., ed. *Assemblages céramiques des sites de l'âge du fer de la péninsule d'Oman* (Documents d'archéologie de l'Arabie 1). Lyon (CD-Rom).

Benoist, A., Le Carlier, C., Goy, J., Degli Esposti, M., Armbruster, B. and Attaelmanan, G. 2015. Snake, Copper and Water in the North-Eastern Arabian Religion during the Iron Age: The Bithnah and Masafi Evidence. *Proceedings of the 17th Rencontres Sabéennes, Paris, 6th-8th July 2013*. Oxford: 21–36.

Bibby, T.G. 1969. *Looking for Dilmun*. New York.

21 The work has already begun on the first analyzed samples from Bithnah (Pillault, Le Carlier and Ploquin 2013).

Craddock, P.T. 2000. From Earth to Furnace: Evidences for the Earliest Metal Smelting Technologies in the Eastern Mediterranean. *Paléorient* 26: 151–165.

Glassner, J.J. 1989. Mesopotamian Textual Evidence on Magan/Makan in the Late 3rd Millennium B.C. In: Costa, P.M. and Tosi, M., eds. *Oman Studies. Papers on the Archaeology and History of Oman*. Rome: 181–191.

Goy, J. 2012. *Typologie et analyses chimiques des déchets de production (scories) et des productions métalliques dans l'émirat de Fudjairah* (M.A. thesis, Université Paris 1). Paris.

Goy, J., Le Carlier de Veslud, C., Degli Esposti, M. and Attaelmanan, A.G. 2013. Archaeometallurgical Survey in the Area of Masafi (Fujairah, UAE): Preliminary Data from an Integrated Programme of Survey, Excavation, and Physicochemical Analyses. *Proceedings of the Seminar for Arabian Studies* 43: 127–144.

Hauptmann, A. 1985. *5000 Jahre Kupfer in Oman. Band 1: Die Entwicklung der Kupfermetallurgie vom 3. Jahrtausend bis zur Neuzeit* (Der Anschnitt, Beiheft 4). Bochum.

Hauptmann, A., Weisgerber, G. and Bachmann, H.G. 1988. Early Copper Metallurgy in Oman. In: Maddin, R., ed. *The Beginnings of the Use of Metals and Alloys. Papers from the 2nd International Conference, Zhangzhou, China, 21–26 October 1986*. Cambridge: 34–51.

Herman, J.T., Cassan, J. and Qandil, H.S. 2012. A Sequence of Inland Desert Settlement in the Oman Peninsula: 2008–2009 Excavations at Saruq al-Hadid, Dubai, UAE. *Arabian Archaeology and Epigraphy* 23(1): 50–69.

Kutterer, J., Neureiter, C. and Jasim, S. 2013. Second Report on the Copper Smelting Site HLO1 in Wādī al-Hilo (Sharjah, UAE). *Proceedings of the Seminar for Arabian Studies* 43: 185–196.

Lombard, P. 1984. Quelques Eléments sur la Métallurgie de l'Age du Fer aux E.A.U. In: Boucharlat, R. and Salles, J.-F., eds. *Arabie orientale, Mésopotamie et Iran méridional de l'âge du fer au début de la période islamique: réunion de travail, Lyon, Maison de l'Orient, (28 Juin-2 Juillet) 1982* (Recherche Sur Les Civilisations 38). Paris: 225–235.

Lombard, P. 1989. Âge du fer sans fer: le cas de la péninsule d'Oman au Ier millénaire avant J.-C. In: Fahd, T., ed. *L'Arabie préislamique et son environnement historique et culturel* (Travaux du Centre de recherche sur le Proche-Orient et la Grèce antique de l'Université des sciences humaines de Strasbourg 10). Leiden: 25–27.

Millard, A.R. 1973. Cypriot Copper in Babylonia, c. 1785 B.C. *Journal of Cuneiform Studies* 25(4): 211–214.

Moorey, P.R.S. 1994. *Ancient Mesopotamian Materials and Industries: The Archaeological Evidence*. Oxford.

Nashef, K. 2010. Saruq al Hadid. An Industrial Complex of the Iron Age II Period. In: Avanzini, A., ed. *Eastern Arabia in the First Millennium BC*. Rome: 213–226.

Oppenheim, A.L. 1954. Seafaring Merchants of Ur. *Journal of the American Oriental Society* 74(1): 6–17.

Peake, H. 1928. The Copper Mountain of Magan. *Antiquity* 2: 452–457.

Pernicka, E. 1999. Trace Element Fingerprinting of Ancient Copper: A Guide to Technology or Provenance? In: Young, S., Pollard, P., Budd, P. and Ixer, R. eds. *Metals in Antiquity* (BAR International Series 792). Oxford: 163–171.

Petrie, C. 1998. The Iron Age Fortification of Husn Awhala (Fujairah, U.A.E.). *Arabian Archaeology and Epigraphy* 9(2): 246–260.

Pillault, S., Le Carlier, C. and Ploquin, A. 2013. Quelques interprétations paléométallurgiques à partir de l'analyse morphologique, microscopique et chimique de spécimens métalliques et scorifiés de Bithnah. In: Benoist, A., ed. *La vallée de Bithnah au cours de l'âge du fer* (BAR International Series 2510). Oxford: 276–290.

Ploquin, A., Orzechowski, S., Mouton, M. and Briand, B. 1994. Paléométallurgie à Mleiha. Une première approche. In: Mouton, M., ed. *Mleiha I. Environnement, stratégies de subsistance et artisanats* (Travaux de la Maison de l'Orient méditerranéen 29). Lyon: 171–190.

Potts, D.T. 1990. *The Arabian Gulf in Antiquity, Vol. 1: From Prehistory to the Fall of the Achaemenid Empire*. Oxford.

Potts, D.T. 2007. Revisiting the Snake Burials in the Late Dilmun Building Complex on Bahrain. *Arabian Archaeology and Epigraphy* 18(1): 55–74.

Potts, D.T., Weeks, L., Magee, P., Thompson, E. and Smart, P. 1996. Husn Awhala: A Late Prehistoric Settlement in Southern Fujairah. *Arabian Archaeology and Epigraphy* 7(2): 214–239.

Prange, M. and Hauptmann, A. 2001. The Chemical Composition of Bronze Objects from 'Ibrī/Selme. In: Yule, P. and Weisgerber, G., eds. *The Metal Hoard from 'Ibrī/Selme, Sultanate of Oman* (Prähistorische Bronzefunde XX,7). Stuttgart: 75–84.

Prange, M., Götze, H.J., Hauptmann, A. and Weisgerber, G. 1999. Is Oman Ancient Magan? Analytical Studies of Copper from Oman. In: Young, S.M.M., Pollard, A.M., Budd, P. and Ixer, R.A., eds. *Metals in Antiquity* (BAR International Series 792). Oxford: 187–192.

Thornton, C.P., Rehren, Th. and Pigott, V. 2009. The Production of Speiss (Iron Arsenide) during the Early Bronze Age in Iran. *Journal of Archaeological Science* 36: 308–316.

Tylecote, R.F. 1962. *Metallurgy in Archaeology*. London.

Weeks, L.R. 1997. Prehistoric Metallurgy at Tell Abraq. *Arabian Archaeology and Epigraphy* 8(1): 11–85.

Weeks, L.R. 2003. *Early Metallurgy in the Persian Gulf. Technology, Trade, and the Bronze Age World.* Boston and Leiden.

Weisgerber, G. 1980. "...und Kupfer in Oman" —Das Oman-Projekt des Deutschen Bergbau-Museum. *Der Anschnitt* 32: 62–110.

Weisgerber, G. 1981. Mehr als Kupfer in Oman—Ergebnisse der Expedition 1981. *Der Anschnitt* 33: 174–263.

Weisgerber, G. 1988. Oman: A Bronze-producing Centre during the 1st Half of the 1st Millennium BC. In: Curtis J. ed. *Bronze Working Centres of Western Asia c. 1000-539 BC.* London and New York: 285–295.

Weisgerber, G. 2007. Iron Age Mining and Smelting (Lizq Period). In: Cleuziou, S. and Tosi, M., eds. *In the Shadow of the Ancestors: The Prehistoric Foundations of the Early Arabian Civilization in Oman*. Muscat: 302–303.

Weisgerber, G. and Yule, P. 1999. Preliminary Report of the 1996 Season of Excavation in the Sultanate of Oman. In: Yule, P., ed. *Studies in the Archaeology of the Sultanate of Oman* (Orient-Archälogie Band 2). Leidorf: 97–117.

Yule, P. and Weisgerber, G. 2001. *The Metal Hoard from 'Ibrī/Selme, Sultanate of Oman* (Prähistorische Bronzefunde 20,7). Stuttgart.

CHAPTER 27

ANCIENT MINING AND METALLURGICAL ACTIVITY AT THE GOLD-SILVER-COPPER ORE DEPOSITS IN MAVROKORFI AREA, MOUNT PANGAEON (NORTHEAST GREECE)

Markos Vaxevanopoulos, Michail Vavelidis, Vasilios Melfos, Dimitra Malamidou, Spyros Pavlides

Mount Pangaeon in northeastern Greece was mentioned by Herodotus (5th century BCE), among other ancient writers, as one of the richest gold prospecting sites in antiquity. This is evidenced by the vein and shear-hosted precious and base metal deposits in the rocks of Pangaeon. The ore deposits are related to the hydrothermal activity associated with the intrusion of a granodioritic magma in the Pangaeon lithological units.

The present research focuses on the "Mavrokorfi-1" mine in one of the mountain's highest peaks. Ancient mining techniques dated from Roman to Ottoman times were studied, and samples from the gold-silver-copper ore deposit were microscopically observed and chemically analyzed. Slags from the adjacent smelting site, Valtouda, were studied and compared with the Mavrokorfi ore deposit. The samples from the mine and the metallurgical site were isotopically analyzed. Archaeological and geological information resulting from this documentation imply a possible gold and silver exploitation with secondary extraction of copper in Roman times.

INTRODUCTION

Mount Pangaeon is an isolated mountain in eastern Macedonia, northeastern Greece, with Mount Symvolon to the south and the plain of Philippoi on its northeast edge (Fig. 27.1). It is located 15 km west of the city of Kavala, along one of the main routes in antiquity that connected the Balkans with Asia, best known in Roman times as Via Egnatia. In eastern Macedonia, there are indications of indigenous metal production in the Late Neolithic (fifth millennium BCE) in the prehistoric settlements of Sitagroi (Renfrew and Slater 2003) and Dikili Tash (Séfériadès 1992). According to Herodotus (*The Histories*, VII, 112), Pangaeon was well-known in antiquity for precious metals such as gold and silver, but also as a sacred mountain associated with the myth of Dionysus. It is believed that the mythical King Cadmus and the Phoenicians were the first to extract gold from Pangaeon (Pliny the Elder, *Natural History*, 7.57). Thracians exploited these metal ores in archaic and classical times, though they were an item of dispute between the Thasians and the Athenians, until Philip II's conquest (4th century BCE). The Tyrant of Athens, Peisistratos, who was exiled in the area ca. 550 BCE, acquired enough wealth and experience to exploit the Lavrion silver mines when he returned in power to Athens. The adjacent Mounts Lekani and Symvolon (Fig. 27.1) have also been recorded as ancient prospecting areas with considerable mining works and metallurgical sites (Photos *et al.* 1989; Koukouli-Chrysanthaki 1990; Vavelidis, Christofides and Melfos 1997a, b).

A detailed survey, including a geological and archaeological study of the ancient mining and smelting sites in the broader area of Mount Pangaeon, was carried out during the years 2009–2011. An

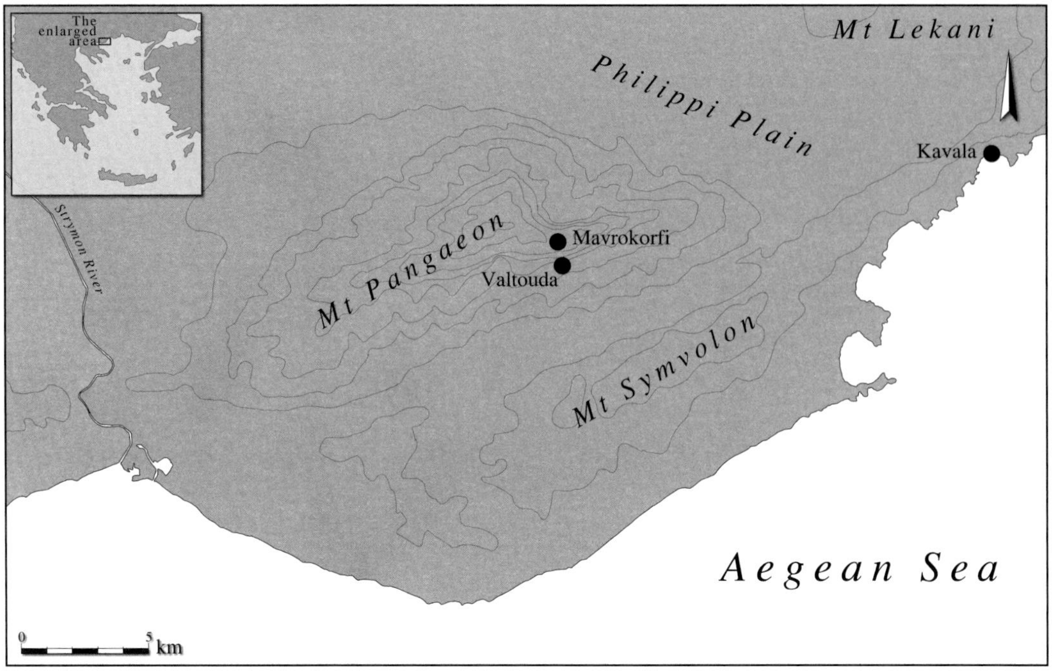

Fig. 27.1: Geographic map of the broader Pangaeon area.

area of about 65 sq km was archaeologically and geologically explored for new mining and metallurgical localities, apart from those which had been previously identified. The underground adits located during this survey were investigated and mapped. A total of 24 underground and surface mines were studied in relation to the 11 metallurgical sites found in the adjacent areas. Metallurgical slags from these sites were obtained for comparative analysis with the samples from the ore deposits.

Previous combined archaeological and geological surveys have demonstrated the importance of the area as an ancient metallurgical center that extends from Strymon River to Mount Lekani (Koukouli-Chrysanthaki 1990, Vavelidis, Christofides and Melfos 1997a, b). Six well-fortified establishments all around the mountain, dated from the Hellenistic to Byzantine times, also imply the significant geopolitical and strategic role of Mount Pangaeon due to its rich ore deposits (Fig. 27.2).

This paper focuses on the Mavrokorfi ancient mining area, with the most remarkable underground mine referred to as "Mavrokorfi-1," and the adjacent Valtouda metallurgical site. Considerable information was gained by the archaeometallurgical excavation, which took place in November, 2012 under the auspices of the Department of Geology of the Aristotle University of Thessaloniki in collaboration with the 18th Ephorate of Prehistoric and Classical Antiquities at the metallurgical site of Valtouda. The aim of this research was to investigate the extent of the ancient mining and metallurgical works at Pangaeon and the level of expertise that the ancient Greeks used for extracting the precious metals from the ore.

GEOLOGICAL SETTING

Mount Pangaeon belongs geotectonically to the homonymous Lower Tectonic Unit of the Southern Rhodope Core Complex (SRCC) and forms a highly uplifted, dome-like, horst structure (Brun and Sokoutis 2007; Mountrakis 2010). It comprises mainly metamorphic rocks of Palaeozoic-Mesozoic Age, Oligocene-Miocene magmatic rocks and a sequence of sediments with an Upper Miocene to recent age (Fig. 27.3).

The core of the mountain dome is dominated by the presence of granitoids with a Lower Miocene Age which can be described as tonalites, granodiorites and granites (Eleftheriadis, Frank and Petrakakis 1999, 2001; Eleftheriadis and Koroneos 2003). The granitoids intruded in the metamorphic rocks of the Pangaeon Unit of the SRCC, which comprises massive coarse grained calcitic marbles with dolomitic lenses as intercalations, mica gneisses, mica schists, calc-silicate schists and amphibolites, and was metamorphosed to the upper greenschist-lower amphibolite facies (Baker *et al.* 1993; Mposkos and Liati 1993).

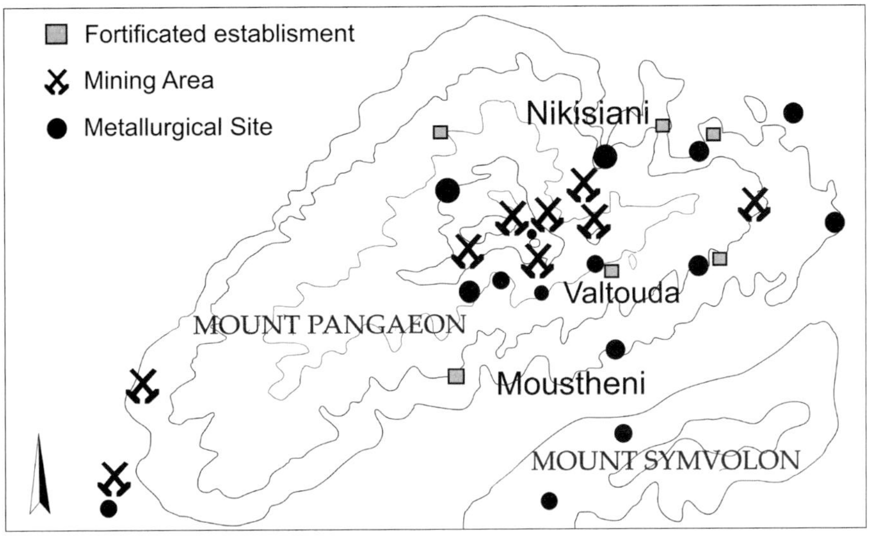

Fig. 27.2: Pangaeon metallurgical and mining sites, with six fortified establishments depicted by gray squares.

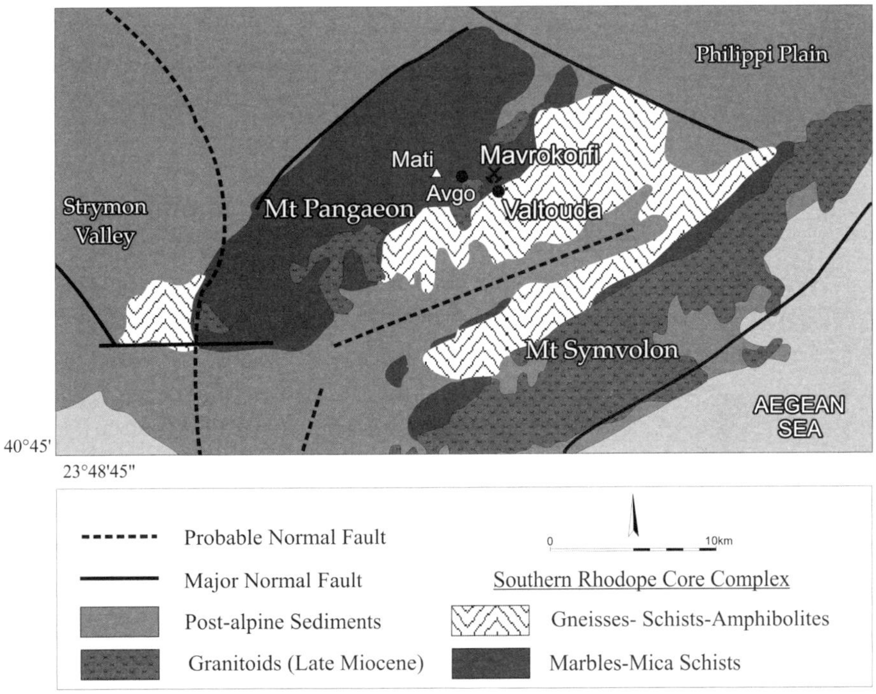

Fig. 27.3: Geologic map of Pangaeon and adjacent areas.

Several ore deposits rich in gold, silver and base metals are hosted at shear zones and the vertical and sub-vertical tectonical discontinuities which intersect all the lithologies of Mount Pangaeon. Replacement-type deposits in the marble and ore mineralization in the gneiss and the amphibolite are also found. At the central-eastern part of the mountain the ores contain Fe, Pb, Zn, As, Cu, Ag, Au and substantial quantities of W, Bi and Sb. These elements are related to various mineral assemblages including pyrite, arsenopyrite, galena, chalcopyrite and sphalerite, as well as minor bismuthinite, tetrahedrite, malachite, azurite, limonite and rutile.

The ores in the western part of the mountain contain Fe, Mn, Pb, Zn, As, Cu, Ag, Au and minor Cd, Ba and Sb. The minerals related to these elements are mostly oxidized and the most common are pyrite, baryte and smithsonite.

The deposits are related to the hydrothermal activity associated with the intrusion of the granodioritic magma in the marbles, schists, gneisses and amphibolites (Eliopoulos and Kilias 2011). The granitoids are exposed due to the erosion process in several parts of the mountain and in some places a thin aureole is developed in the contacts with the adjacent marbles (Eleftheriadis and Koroneos 2003).

The Pangaeon horst structure is considered a result of the North Aegean extensional process which started during the Middle Eocene, 40 million years ago (Brun and Sokoutis 2007). From mid-Eocene to mid-Oligocene the metamorphic dome was exhumated along the detachment fault zone at the western part of Pangaeon in a north northwest–south southeast direction. The exhumation continued during the late Miocene along steep vertical and sub-vertical normal faults in a southwest–northeast direction. They belong to the final extensional period when the plutonic bodies of Pangaeon intruded the metamorphic core complex and several basins were developed simultaneously (Brun and Sokoutis 2007).

ANALYTICAL METHODS

The ancient mine "Mavrokorfi-1" was topographically and geologically mapped. A study of mining traces was carried out in order to investigate the exploration tools.

Ten samples of the ore deposits were collected at the "Mavrokorfi 1" mine (PMM) and the surrounding mining area (Aghia Triada-PAT, Asimotrypes-PAM and Avgo-PAV). Five metallurgical slags were sampled from the Valtouda excavation (PVA) and three from the Avgo mining district.

During the 2012 archaeometallurgical excavation two trenches (VAL_1 and VAL_2) were excavated on the western hill of Valtouda. The dimensions of the VAL_1 trench were 4 × 4 m and 1.05 m deep. The dimensions of the VAL_2 trench were 2 × 1 m with a depth of 1.95 m. Slag samples from Valtouda were sampled from each stratigraphical unit of the VAL_2 excavation trench. The sampling depths varied from 0 to 1.95 m. PVA_1 and PVA_2 were obtained from a depth of 10 cm, PVA_3 from a depth of 50 cm, PVA_4 from a depth of 100 cm, and PVA_5 from the bottom of the stratigraphy at 195 cm. Using optical microscopy and a JEOL 840A scanning electron microscope, equipped with back-scattered imaging capabilities, the mineralogical composition, textures and other microscopic features and structures of 15 polished and thin sections from the ore deposits and the slag samples were studied at the Faculty of Sciences, Aristotle University of Thessaloniki. Operating conditions were 20 kV and 20 nA, with a beam diameter of 1 μm. The following X-ray lines were used: AuMα, BiMα, FeKα, PbMα, SKα, SbLα, and TeLα. Pure elements (for Au, Ag and Sb), pyrite (for Fe and S), galena (for Pb), and synthetic BiTe (for Bi and Te) were used as standards.

Chemical compositions of the selected samples were determined using ICP Mass Spec analysis of each sample after Aqua Regia digestion at the ACME analytical laboratories in Vancouver, Canada. Analysis of Pb isotopes (Pb^{204}, Pb^{206}, Pb^{207}, Pb^{208}) was carried out in order to determine the isotopic signature of each sample for provenance issues.

RESULTS AND DISCUSSION

MAVROKORFI MINING AREA

Mavrokorfi constitutes one of the highest peaks of Pangaeon with an altitude of 1479 m ASL. It is located almost 5 km east of the mountain's highest peak, Mati (1956 m), and 2.5 km east of the Avgo peak (1836 m) (Fig. 27.2). The Mavrokorfi district covers an area of about 6 sq km with surface and underground mining works (Krikelas 1973). The ore deposit is hosted in the discontinuities of the marbles such as the karstic voids and the tectonic structures and, at the lower part of the gneisses and amphibolites, along its schistosity voids and at the intersections between tectonic faults and the marbles' schistosity. The mineralization is extensively oxidized at the contacts between the marbles and the gneiss-amphibolite intercalations. The Avgo peak is also surrounded by several surface mining works and slag remains which should probably be dated to Roman times. This site is characterized by abundant mining debris covering an area of 1.5 sq km at an altitude of 1725 m. The ore deposits are located in the marbles, just below or in the gneiss and amphibolitic layers.

The most representative and significant ancient mine, "Mavrokorfi-1," is located in the northern part of the Mavrokorfi area at an altitude of 1299 m. Its entrance is 5 × 1.6 m opened in the marbles with gneiss and amphibolite intercalations (Fig. 27.5a, b and g). The mine consists of galleries with a total length of 120 m and a vertical distance of 18 m between the upper and the lower level. The area outside the entrance is characterized by the presence of a 25 m long steep slope with mining waste material. Oxidized ore occurrences are visible mainly at a fault scarp with a northwest–southeast direction, 40 m northwest of the mine's entrance. The mine was opened along the tectonic discontinuities, mainly faults, following the ore deposit (Fig. 27.4, 5c and d).

The "Mavrokorfi-1" mine is characterized by the presence of two distinct phases of exploitation. The older is characterized by narrow and square cross-sectioned galleries with an average height of about 1 m (Fig. 27.5).

The second phase covers mainly the area of the mine's entrance and enlarges the former galleries, thus destroying their characteristics. The mining process of this second phase is connected with the refilling of marble debris of the former galleries (Fig. 27.5e). The ore in the inner part of the mine is oxidized and semi-oxidized with malachite and limonite being the main minerals.

Based on the pottery finds during the geological and archaeological mapping in the galleries of the "Mavrokorfi-1" mine, the older phase with small and square cross-sectioned galleries is dated to the Roman period, whereas the later one is dated to Ottoman times. Study of the mining traces has shown that the main mining tools in both exploitation phases were a combination of the classical pick and hammer-chisel. The estimated total extracted volume of ore is approximately 85 cu m. The exploitation of the older phase focused only on the totally oxidized ore, while the latter phase of exploitation concentrated on the partly oxidized ore consisting of pyrite, arsenopyrite and chalcopyrite in the gneiss and amphibolitic intercalations. It is very difficult to distinguish other exploitation periods.

VALTOUDA METALLURGICAL SITE

The Valtouda metallurgical site is closely related to the Mavrokorfi ancient mines, and lies 1200 m to the south, covering an area of 0.045 sq km with a large amount of surface slags. Very close to this site (only several meters from the southern edge) there is a natural spring and a lake covering an area of 0.002 sq km (Fig. 27.6a). A fortified, undated establishment 200 m from the site implies the importance of this

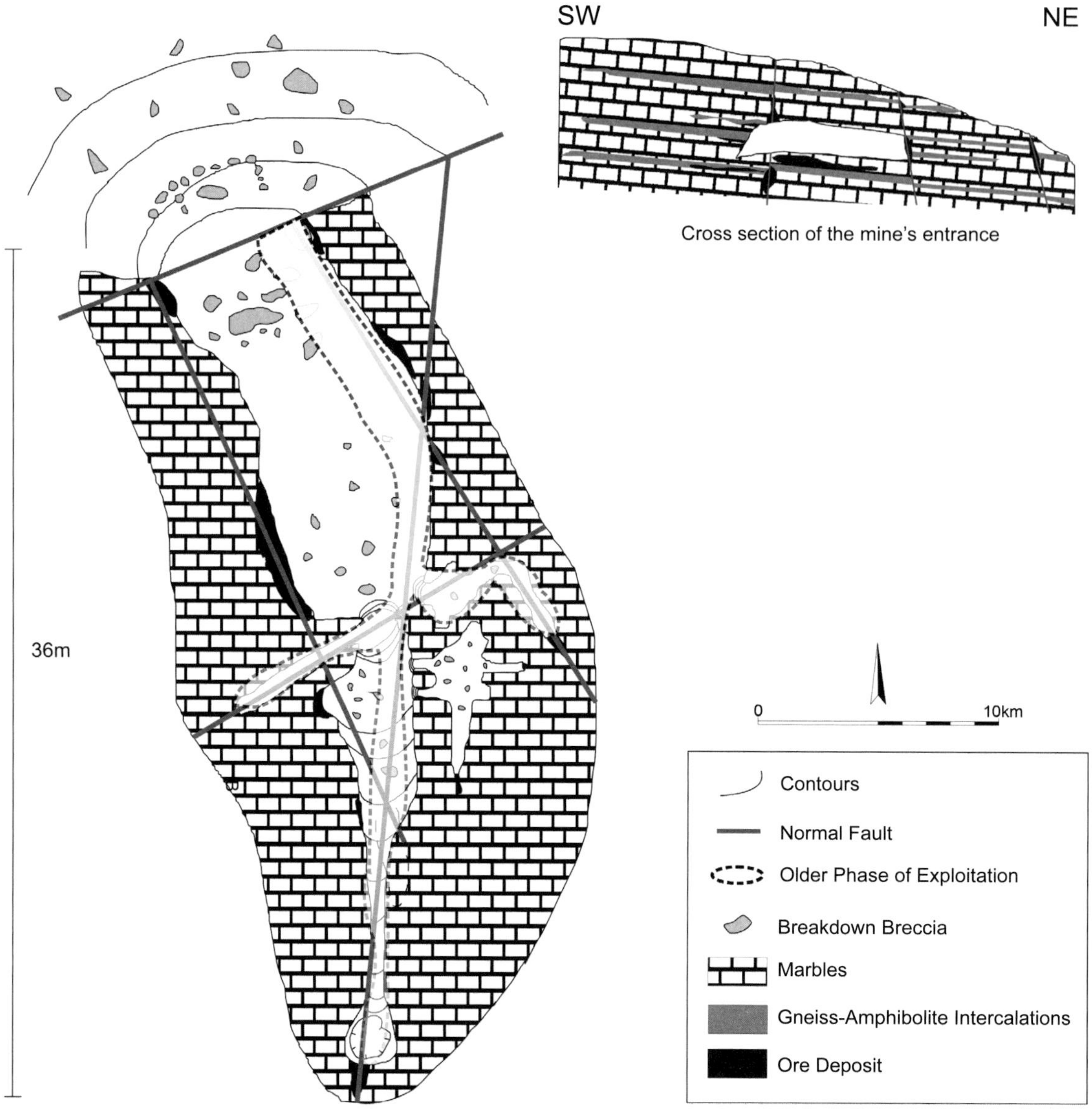

Fig. 27.4: Map of the "Mavrokorfi-1" mine with the tectonic faults and the ore deposit distribution. At the upper-right corner a southwest-northeast cross-section is shown.

location and suggests that the site was exploited over long periods of time (Fig. 27.6b). The fortification has a trapezoidal shape with a diameter of almost 330 m, covers an area of 0.6 hectares, and is defined by a surrounding dry-stone wall preserved to a height of 1 m.

At the trench VAL_1 a circular construction with a total diameter of 150 cm, preserved to a height of 30 cm, was uncovered (Fig. 27.7). It was based at a depth of 77 cm below the surface. The construction is characterized by a concentric placement of cobbles with an average dimension at their main axis of 27 cm. The cobbles consist of gneiss and marble probably of a local provenance.

Based on the slag remains in the inner wall and the clay lining found at the surface of several cobbles, it may be assumed that this construction was related to the smelting procedure. The inner sides

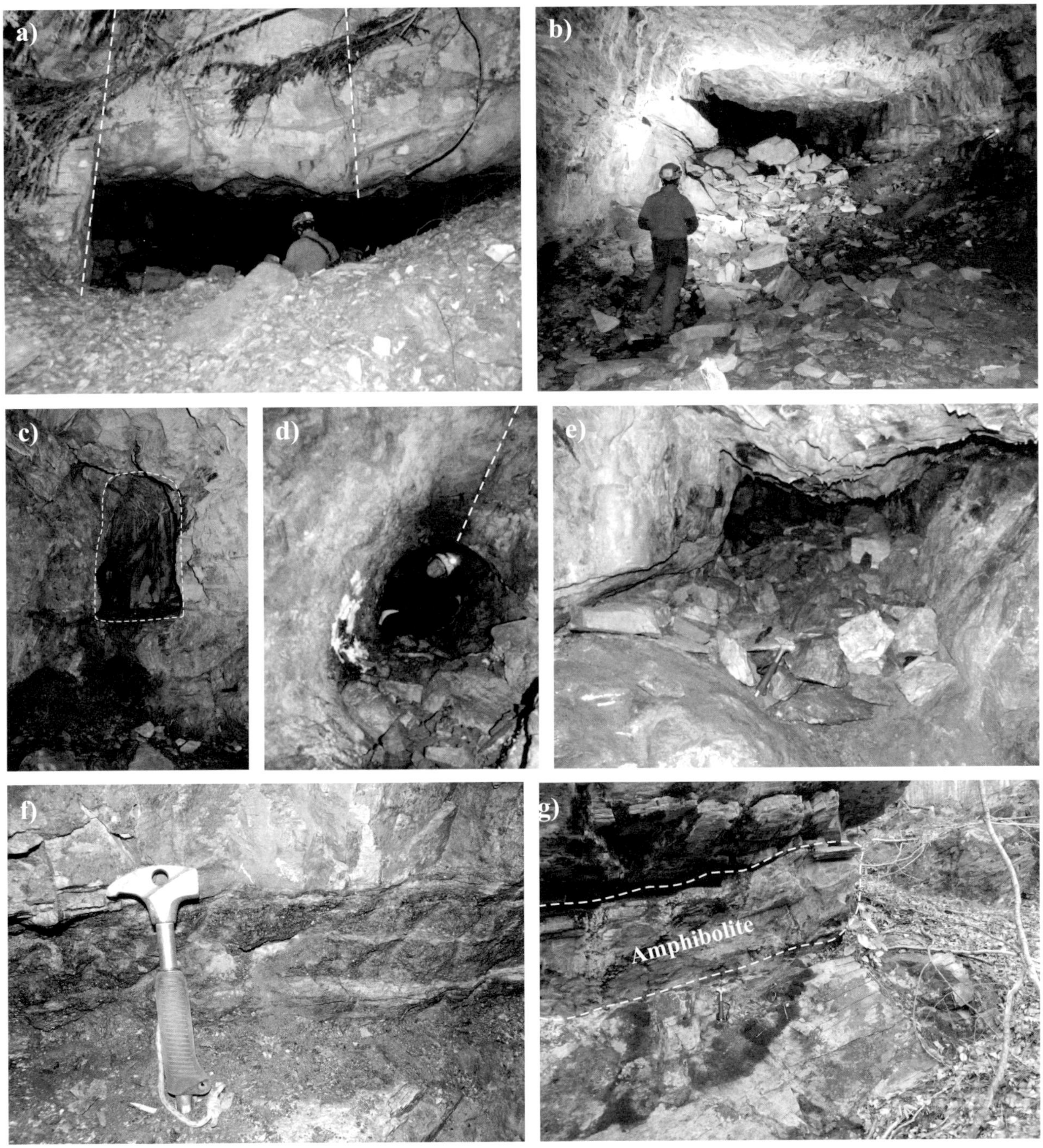

Fig. 27.5: Photographs form the ore deposit and the "Mavrokorfi-1" mine: (a) view of the mine's entrance. Tectonic faults are depicted with dotted line; (b) view of the entrance from the inner part of the mine; (c) cross-section of the mining phases. The Roman gallery (with dotted line) has an average height of 1 m and is intersected by the younger and largest Ottoman chamber; (d) narrow Roman gallery with an average height of about 1 m along a tectonic fault (depicted with dotted line); (e) older phase gallery refilled with mining debris; (f) partly oxidized ore deposit at the first and largest chamber of the mine; (g) amphibolite intercalation in marbles outside the mine.

Fig. 27.6: The Valtouda metallurgical site: (a) view to the north showing the locations of the Mavrokorfi peak (black arrow) and the excavation trenches VAL-1 and VAL-2 (white arrow); (b) view to the south showing the locations of the fortificated establishment (black arrow) and the excavation trenches (white arrow).

Fig. 27.7: The circular construction found at the VAL_1 trench during the 2012 archaeometallurgical excavation.

of some of the clay fragments were burnt, and the outer sides were also affected. Only a few fragments were not affected, according to macroscopical observation. They were probably lining material from a smelting furnace (Fig. 27.8). Most of the cobble gneiss surfaces have been vitrified due to the high furnace temperatures which affected the silicic minerals (e.g., quartz) accompanying the ore mineralization.

The second excavation trench, VAL_2, was opened with the aim of documenting the stratigraphy of the western hill of the site. The trench reaches a depth of 1.95 m, and is comprised of several inclined layers of small to medium gravel, sand and clay. Slags and furnace linings are abundant in each stratum. Small pieces of gravel of the ore mineralization containing malachite have been found at the bottom layers of the VAL_2 trench.

Fig. 27.8: Sherds of Roman oinochoae found in trench VAL_2 (1–3); lining fragments from VAL_1 (4, 5).

The archaeological finds from both trenches are mainly pottery sherds of unpainted wheel-made vessels (Fig. 27.8). They belong to small amphoras, oinochoae and pots dated to Roman times (from the 1st to 4th centuries CE). Their surface bears stains from smelted metals.

MINERALOGICAL OBSERVATIONS

The study under the microscope showed that the gneiss in "Mavrokorfi-1" mine is characterized by the presence of K-feldspar, green hornblende, quartz and chlorite. Green hornblende in the gneiss and the amphibolite is extensively altered to chlorite, which was determined mainly as prochlore $(Mg,Fe,Al)_6(Al,Si)_4O_{10}(OH)_8$, based on the microprobe analyses. Small quartz crystals are embedded in pyrite of the ore mineralization.

The ore deposit at the "Mavrokorfi-1" consists of pyrite, chalcopyrite, chalcocite as well as bismuthinite and rutile hosted in the gneiss and the amphibolite. The pyrite forms euhedral grains which are partly and peripherally altered to limonite and other iron oxides. Chalcopyrite grains also show an incipient decomposition reaction rim of chalcocite. Total altered chalcopyrites and chalcocites to malachite are widespread in the mine's ore deposit pockets. The bismuthinite is found either in the host rocks as disseminated grains with a length up to 200 μm or as inclusions in pyrite with an average length of 40 μm. Small rutile grains are also observed in the cloritized gneisses. Well formed pyrite crystals include a considerable number of tiny rutile and chalcopyrite grains (Fig. 27.9). At the non oxidized ore, gold is probably "invisible" and hosted in the pyrites' lattice.

The slags of Valtouda present three main different textures and mineralogical compositions. The first group of slags contains siliceous matrix with wustite dendrites and unreacted chalcopyrite and pyrite crystals (Fig. 27.10a). Prills of metallic Pb and Cu are also present as well as a copper phase resembling composition of CuO originated from a possible roasting procedure of the sulfidic ore. The glassy matrix consists mainly of monticellite (Fe,Mg,Mn)$CaSiO_4$. Fayalite is rare and its scarcity can be explained by the

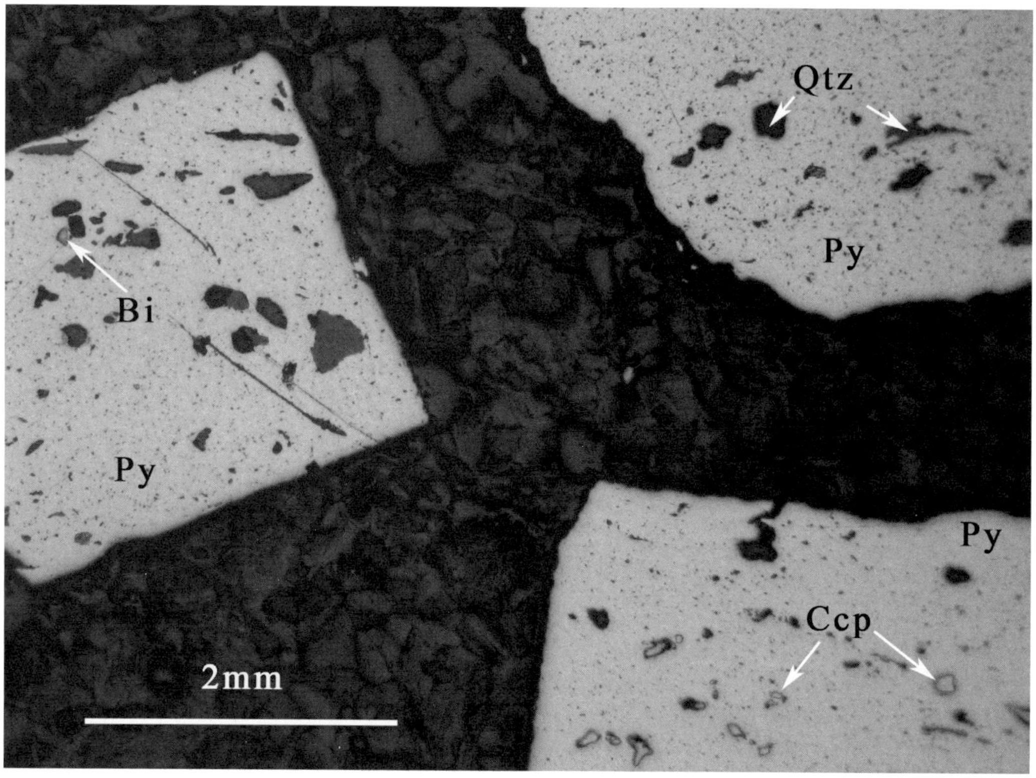

Fig. 27.9: Reflected light microphotograph of the "Mavrokorfi-1" non-oxidized ore deposit with three pyrte crystals including quartz (Qtz), chalcopyrite (Ccp) and bismuthinite (Bi).

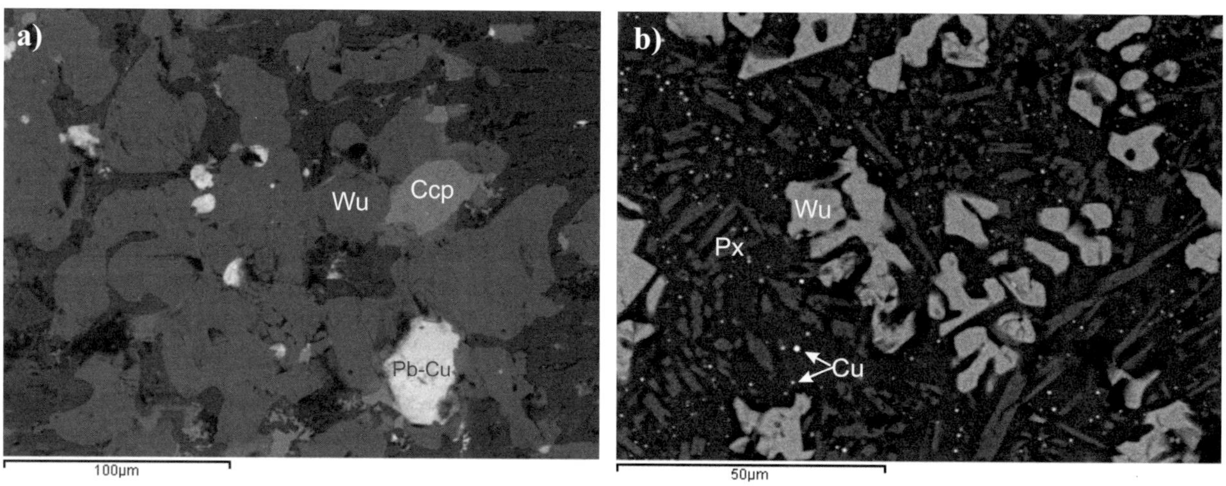

Fig. 27.10: Backscattered electron images of Valtouda slags: (a) unaffected chalcopyrite (Ccp) with Pb-Cu prills and wustite crystals (Wu) in the glassy matrix; (b) copper prills in glassy matrix of pyroxenes.

high amounts of MgO and CaO (Bachmann 1982) in the ore. The second group of slags contains copper prills in a glassy matrix of pyroxenes with a (Fe,Mn,Mg)$CaSiO_6$ composition and wustite dendrites (Fig. 27.10b). Spinels are present in this group of slags mainly in the form of magnetite as angular small grains in a glassy matrix or crystal skeletons. Wustite is more abundant in relation to the magnetite and other iron oxides. The third slag group contains only PbO prills and probably resulted from a cupellation process.

Chemical-Isotopic Analyses

According to ICP-MS chemical analyses of the mineralization at the Mavrokorfi mine, Au ranges from 7.3ppm in the partly oxidized ore to 37.4 ppm in the totally oxidized ore and exceeds 100ppm in the ore deposits of the surrounding mines. Silver exceeds 100ppm and copper in the oxidized ore ranges from 0.1% to 1% (Table 27.1).

The Au concentrations at the Valtouda metallurgical slags vary from 0.2 to 0.9ppm, whereas silver ranges between 20 and 65ppm. The first group of slags (the result of the probable roasting of waste material) demonstrates the highest values of gold, silver and copper, 0.0009%wt, 0.0021%wt and 0.26%wt respectively. It is also noteworthy that Au, Ag and Cu concentrations present a remarkable reduction in the slags compared to the adjacent ores.

The Pb-isotopic results show that the ore samples from the "Mavrokorfi-1" mine and the slag from Valtouda plot in the same area in the diagrams of Pb^{207}/Pb^{204} versus Pb^{207}/Pb^{204} and Pb^{208}/Pb^{204} versus Pb^{206}/Pb^{204} (Fig. 27.11). Samples from Avgo and Korifi ore deposits also imply a strong relation with Valtouda and Mavrokorfi.

Lead isotopic composition reveals a connection between the Mavrokorfi mine's ore deposit and the Valtouda metallurgical slags. They slightly differ from other Pb isotopic signatures of Pb ore deposits from northeastern Greece (e.g., Palea Kavala, Thasos, Chalkidiki) that have been previously studied (Wagner *et al.* 1981, 1986; Chalkias and Vavelidis 1989). It is very difficult to distinguish between the ore provenance of the other sites and that from the eastern part of Pangaeon (Avgo and Korifi). However, the strong relation between the Mavrokorfi and Avgo ore deposits and the Valtouda slags is clearly depicted in the Pb isotopic diagrams comparing to west Pangaeon ore deposit sample and the slag from Domatia metallurgical area.

Table 27.1. Compositional Data from Chemical Analyses of the Ore and the Slags at the Mavrokorfi and Adjacent Areas*

Sample	*Au*	*Ag*	*Cu*	*As*	*Pb*	*Zn*	*Mn*	*Fe (%)*	*Ba*	*Bi*	*Sb*	*Cd*	*W*
Ore													
PMM_1 Ore	7.3	12.5	9872.0	710.1	24.0	39.0	500.0	27.7	43.0	169.3	0.4	0.3	8.8
PMM_2 Ore	37.4	16.2	>10000	583.7	59.0	43.3	101.0	17.2	18.1	32.4	6.1	0.0	40.2
PAV_5 Ore	9.4	10.4	>10000	>10000	17827.0	>10000	>10000	>40.0	11.3	5.6	437.0	797.3	0.3
PAT_4 Ore	51.2	>100	>10000	422.8	2684.9	335.0	5417.0	24.0	35.9	387.5	8.83	1.6	63.6
PAM_4 Ore	>100	6.3	1588.2	>10000	2613.5	1985.9	1299.0	21.5	148.1	35.63	67.28	47.9	42.9
Slags													
PVA_2 slag	0.4	65.2	1944.6	5661.4	16981	>10000	1940.0	>40.0	100.7	0.7	311.0	1.0	0.9
PVA_3 slag	0.2	23.0	1420.8	6562.9	16303	>10000	>10000	36.5	190.4	0.3	121.0	3.6	1.8
PVA_4 slag	0.9	20.8	2619.4	6099.6	17332	8023.1	>10000	>40.0	144.8	2.7	309.0	2.4	2.9
PAV_6 slag	0.1	2.6	252.7	1211.4	10661.9	2617.0	7680.0	23.3	179.8	0.22	432.2	0.16	8.3

*Metal compositions are given in ppm. Fe is given in weight percent.

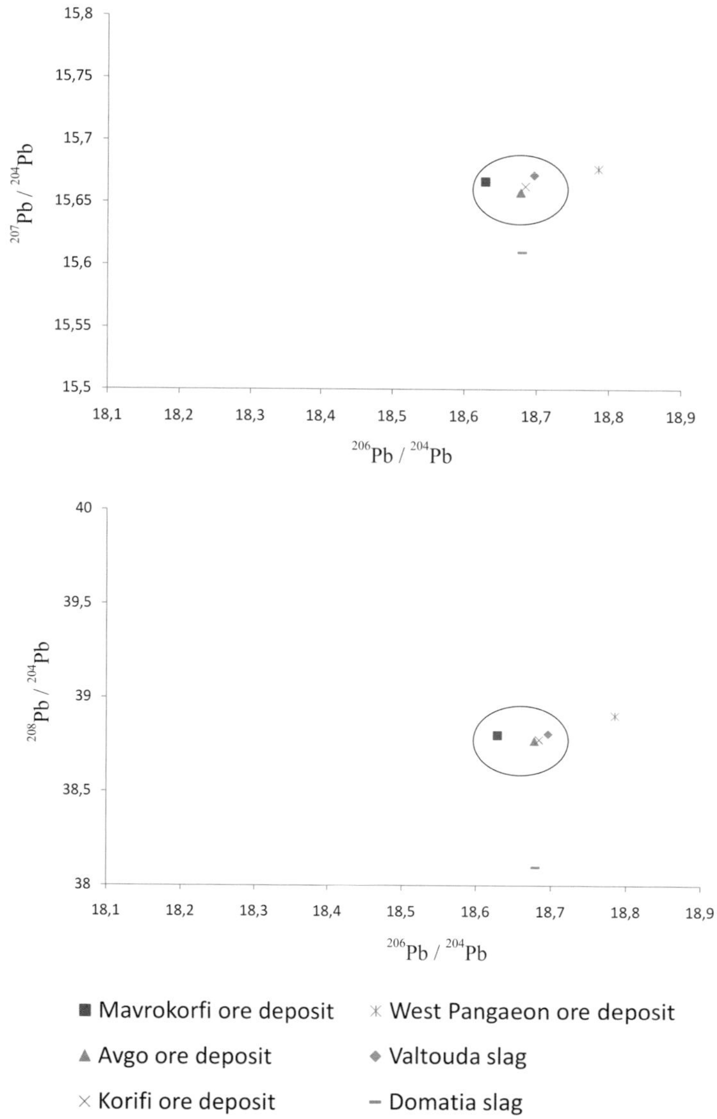

Fig. 27.11: Lead isotope abudance ratios of Mavrokorfi and adjacent ancient mining area compared with the Valtouda and Domatia slags.

CONCLUSIONS

The Mavrokorfi ore deposits of Mount Pangaeon are rich in Au, Ag and Cu, among other metals. The results of this study show that in the Mavrokorfi-1 mine two discrete phases of the mining process are evidenced. The older phase is dated to the Roman period and the later phase to Ottoman times. The Roman exploitation is characterized by narrow galleries which follow only the oxidized ore mineralization. The Ottoman period extraction focused on the oxidized mineralization and the primary ores. It is assumed that the Au-, Ag- and Cu-bearing ore mineralization extracted from the Mavrokorfi-1 mine was carried to the Valtouda metallurgical site.

Valtouda constitutes one of the most important metallurgical sites of Mount Pangaeon where for the first time the copper extraction process is documented in the broader area of Pangaeon, along with gold and silver production. According to our research, smelting activities at Valtouda are entirely related to the Mavrokorfi and the Avgo ancient mining works. Ore deposits from these areas were smelted at Valtouda, probably for the production of copper and precious metals (Au and Ag). Copper smelting is documented by malachite gravels and the copper slags found in the stratigraphy of VAL_2 trench. Lead and smelted copper material might have been cupellated for precious metals extraction.

This process is dated to the Roman era based on ceramic evidence. Byzantine or Ottoman period pottery has not been found at the Valtouda metallurgical site, demonstrating that the ore from the adjacent mines was carried to other smelting sites. There is no evidence for earlier periods either. During the Roman period the area was part of the territory of the city of Philippoi, which had the status of a Roman colony (Colonia Augusta Philippensis). The city was a powerful economic and political center and most probably controlled the mines of Mount Pangaeon. Further study on the pyro-metallurgical remains will clarify the extraction process and the metallic products, thus illuminating the whole *chaîne opératoire*.

ACKNOWLEDGMENTS

This paper is based on research done as part of the first author's Ph.D. dissertation in the Department of Mineralogy, Petrology, Economic Geology, at the Faculty of Geology of Aristotle University of Thessaloniki. We acknowledge financial support of the first author by the Alexander S. Onassis Public Benefit Foundation of Greece, during 2009 and 2011. We would also like to thank the Institute for the Archaeo-Metallurgical Studies of the University College London (U.C.L.) for the financial support of the first author's travel expenses to attend the Timna Park International Conference, Israel, 2013.

REFERENCES

Bachmann, H.G. 1982. *The Identification of Slags from Archaeological Sites*. London.

Baker, J.H., Eliopoulos, D.G., Arvanitides, N.D. and Chatzipanagis, J. 1993. *Carbonate-Hosted Precious and Base Metal Mineralizations in Northern Greece: Development of New Exploration Strategies. Fertile Lineaments and Regional Variations of Metallic Elements: 3.4. The Pangeon-Symvolon Regions* (Institute of Geology and Mineral Exploration, Final Report, EEC Project MA2M-CT-90-0015). Athens: 42–48.

Brun, J.P. and Sokoutis, D. 2007. Kinematics of the Southern Rhodope Core Complex (North Greece). *International Journal of Earth Sciences (Geol Rundsch)* 96: 1079–1099.

Chalkias, S. and Vavelidis, M. 1989. Interpretation of Lead-Isotope Data from Greek Pb-Zn Deposits, Based on an Empirical Two Stage Model. *Bulletin of the Geological Society of Greece* 23/2: 177–193.

Eleftheriadis, G., Frank, W. and Petrakakis, K. 1999. 40Ar/39Ar Geochronology of the Pangeon Granitoids, Rhodope Unit (Northern Greece). *European Journal of Mineralogy* 11: 62.

Eleftheriadis, G., Frank, W. and Petrakakis, K. 2001. 40Ar/39Ar Dating and Cooling History of the Pangeon Granitoids, Rhodope Massif (Eastern Macedonia, Greece). *Bulletin of the Geological Society of Greece* 34: 911–916.

Eleftheriadis, G. and Koroneos, A. 2003. Geochemistry and Petrogenesis of Post-Collisonal Pangeon Granitoids in Central Macedonia, Northern Greece. *Chemie der Erde* 63: 364–389 (Greek).

Eliopoulos, D. and Kilias, S.P. 2011. Marble-Hosted Submicroscopic Gold Mineralization at Asimotrypes Area, Mount Pangeon, Southern Rhodope Core Complex, Greece. *Economic Geology* 106: 751–780.

Koukouli-Chrysanthaki, Ch. 1990. The Mines of the Thasians' Coast. In:. *Mélanges D. Lazaridis: Cité et territoire en Macédoine et Thrace antiques* (Recherches franco-helléniques 1). Thessaloniki: 493–514 (Greek).

Krikelas, O. 1973. *Report on the Study of Primary and Alluvial Gold Deposits in E. Macedonia*. Athens.

Mountrakis, D. 2010. *Geology and Geotectonic Evolution of Greece*. Thessaloniki.

Mposkos, E. and Liati, A. 1993. Metamorphic Evolution of Metapelites in the High Pressure Terrane of the Rhodope Zone, Northern Greece. *Canadian Mineralogist* 31: 401–424.

Photos, E., Koukouli-Chrysanthaki, C., Tylecote R.F. and Gialoglou G. 1989. *Precious Metals Extraction in Palaia Kavala, NE Greece, An Archaeometallurgical Attempt to Locate Skapte Hyle* (International Old World Archaeometallurgy Conference). Heidelberg: 179–190.

Renfrew, C. and Slater, E.A. 2003. Metal Artifacts and Metallurgy. In: Elster, E.S. and Renfrew, C., eds. *Prehistoric Sitagroi: Excavations in Northern Greece, 1968–1970, Vol. 2: The Final Report* (Monumenta Arcaeologica 20). Los Angeles: 301–318.

Sériadès, M. 1992. Le métal. In: Treuil, R., ed. *Dikili Tash, village préhistorique de Macédoine orientale I. Fouilles de Jean Deshayes (1961–1975)*, Vol. 1 (Bulletin de correspondance hellénique supplément 24). Athens: 113–119.

Vavelidis, M., Christofides, G. and Melfos, V. 1997a. Ancient Mining Activity in the Broader Area of Kavala-Filippoi. In: Palyvou, C., ed. *Ancient Greek Technology*. Thessaloniki: 95–101.

Vavelidis, M., Christofides, G. and Melfos, V. 1997b. Ancient Gold Mines Near Krinides in Philippi Area (Macedonia, N. Greece). Dans Archäometrie und Denkmalpflege: Kurz-berichte. Gemeinsame Jahrestagung des Arbeitskreises Archäometrie de GDCh und des Arbeitskreises Archä ometrie und Denkmalpfleg der Deutschen Mineralogischen Gesellschaft in Wien, 24–26.3.1997: 84–86.

Wagner, G.A., Pernicka, E., Gialoglou, G. and Vavelidis, M. 1981. Ancient Gold Mines on Thasos. *Naturwissenschaften* 68: 263–264.

Wagner, G.A., Pernicka, E., Vavelidis, M., Baranyl, I. and Bassiakos, I. 1986. Archäometallurgische Untersuchungen auf Chalkidiki. *Der Anschnitt* 38: 5–6.

CHAPTER 28

THE GREAT ORME BRONZE AGE COPPER MINE IN NORTH WALES: OPPORTUNITIES TO LINK ORE TO METAL

Robert Alan Williams

The Great Orme Copper Mine with over 6 km of accessible workings is one of the most extensive surviving Bronze Age copper mines in Europe. In combination with the nearby Pentrwyn Bronze Age smelting site, the only one known in Britain, they offer unique archaeological opportunities. In particular, there is the potential to geochemically define a mine-based Bronze Age metal group based on extensive sampling and analysis of ores from the entire Bronze Age workings, slags with copper prills and bronze particles found in the mine. This mine-based approach provides a potentially radical alternative to defining Bronze Age metal groups based only on artifact analyses and will assist in the assessment of recycling. The combination of geochemical and lead isotope analyses will define a Great Orme mine-based metal group that will allow the identification of artifacts in existing analytical databases with which it is consistent. Combining this work with spatial and temporal evidence will reveal the contribution of the Great Orme metal supply to the various phases of the British Bronze Age in a clearer light. This paper was written in 2013 during the initial stages of the project and sets out the methodology being used and some very preliminarily findings. These are starting to resolve past mineralogical issues and are enabling a major reassessment of the importance of the mine.

GREAT ORME: GEOLOGICAL AND ARCHAEOLOGICAL SETTING

The Great Orme is a prominent peninsula approximately 3 km by 2 km, rising to 207 m ASL, and overlooking the seaside resort of Llandudno in north Wales in western Britain (Figs. 28.1, 28.2). This headland is an isolated outcrop of Lower Carboniferous Limestone rocks (Late Dinantian–Asbian and Brigantian stages) around 330 million years old, and have been deformed into a gentle faulted syncline. They consist of gray massive and rubbly limestones, with some dolomitization, interbedded with mudstones and rare sandstones (Jenkins and Lewis 1991). Primary and secondary copper mineralization occurs mainly along the Pyllau Valley and is associated with dolomites and a series of north–south faults. Ixer and Davies (1996) classified the deposit as a "copper-dolomite" association possibly late Mesozoic to early Tertiary in age (100–50 Ma) and post-dating the associated lead vein which is considered to be no earlier than early to middle Trias (around 240 Ma).

The Great Orme's archaeological landscape includes burials from the Late Upper Palaeolithic (Lynch, Aldhouse-Green and Davies 2000), a Mesolithic stone shelter and adjacent to the main copper mine, a Neolithic stone cromlech (James 2011). Elsewhere there is a hillfort at Pen-y-Dinas and small Roman finds but currently no evidence of Roman mining. Copper mining had resumed by the late 17th century and continued with interruptions until about 1880, mainly exploiting areas beneath or adjacent to the Bronze Age workings. The 19th century metal mining landscape of old shafts and

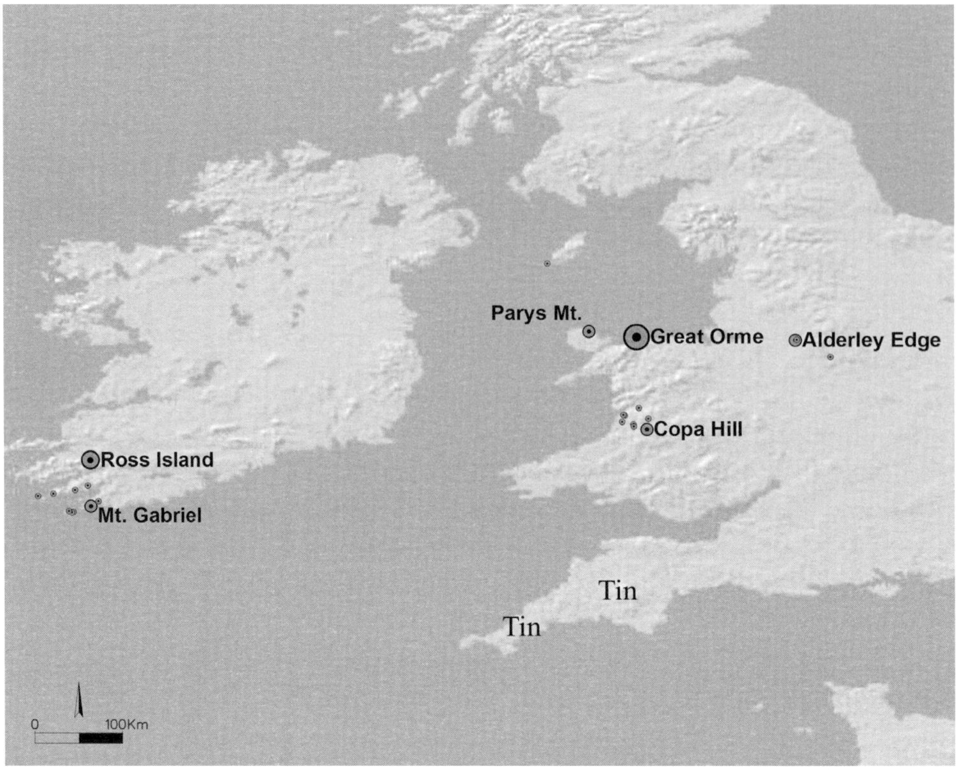

Fig. 28.1: Location map for the Great Orme mine and other dated and probable Bronze Age copper mines and trials in Britain and Ireland. Prominent mine sites are highlighted with larger circles (basemap credit: Copyright:© 2014 Esri).

Fig. 28.2: Oblique satellite image of the Great Orme headland showing the mine and smelting site locations (basemap image credits: Getmapping plc, NASA, Landsat/Copernicus).

large waste tips covered the Bronze Age working until they were uncovered between 1987 and 1991. Several springs on the Great Orme were used for ore washing and have shown some evidence of Bronze Age usage (Lewis 1996).

The Bronze Age copper mine (Fig. 28.3) is one of the most extensive surviving complexes in Europe, with over 6 km of accessible tunnels within an area about 245 × 140 m and up to 70 m below the surface (Lewis 1996). They are a complex of often tight, irregular passages with several large, mined-out chambers, including a large, "opencast" area at surface which may have resulted from a past collapse. Artifacts include over 2,400 stone mauls, around 30,000 bone fragments and over 50 small bronze fragments. There is evidence of the veins in softer areas of dolomite being worked by bone tools and some evidence for firesetting. The workings have been dated using charcoal and bone with 17 Bronze Age ^{14}C dates whose two sigma limits range from 1884cal BC to 933cal BC (James 2011) and with one Iron Age date (763-406cal BC). Past studies includes a major surface excavation report (Dutton and Fasham 1994), detailed underground surveys establishing criteria for Bronze Age workings (Lewis 1996), studies of the bone fragments (James 2011) and a reassessment of earlier studies (Wager 2001).

The Pentrwyn copper smelting site on a cliff ledge, 1.2 km east of the mines (Fig. 28.2), is the only one thus far discovered in Britain dating from the Bronze Age. The site has been mainly lost by the construction of a road in the 19th century but a few hundred grams of smelting residues (mostly <10 mm particles) have been recovered from a series of excavations in 1998 and 2011 (Hopewell and Jones 1999; Smith 2012).

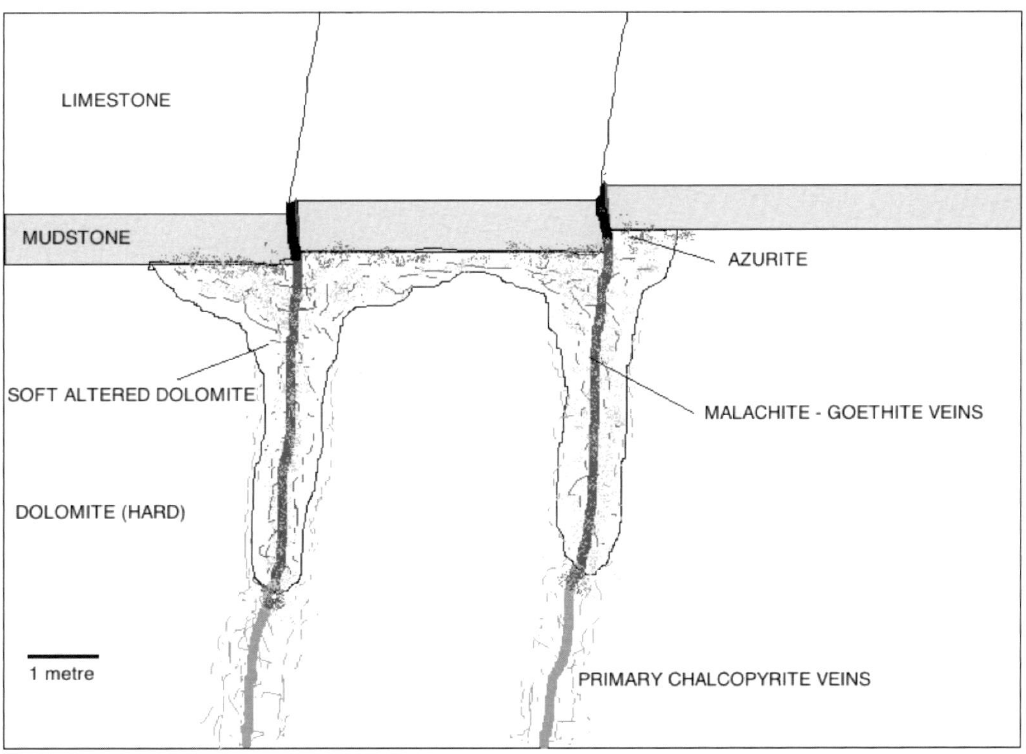

Fig. 28.3: Simplified section of the mineralized veins at the Great Orme and the formation of the secondary malachite-goethite veins with soft wall rocks (based on Lewis 1996).

THE CHALLENGE OF LINKING ORE TO METAL

Tracing Bronze Age metal artifacts back to specific mines or mining regions has long been a desirable archaeological objective in order to establish important centers of production, to understand the metal supply and to deduce exchange networks. The two principle methods that are used to link ore or slag to artifacts are chemical analysis (minor and trace elements) to establish chemical groups, and in more recent decades the use of lead isotopes to define isotopic groups. While these two independent techniques are sometimes applied separately, ideally they should be used in combination to provide stronger provenance evidence (Pernicka 2004).

The challenges of linking metal artifacts to ore sources include the variability of trace elements in natural ore deposits, the lack of samples of the actual ores worked in the Bronze Age, the addition of trace elements from additives used during smelting and from tin addition, the split of trace elements between ore and slag (Hauptmann 2007), the effects of different smelting techniques (Pollard *et al.* 1991), the redox conditions and the effects of recycling (Bray and Pollard 2012).

Various scholars have proposed metal artifact groups for Britain with Northover (1980, 1991) listing 16 groups (A to T) for various stages of the Bronze Age based mainly on levels of arsenic, antimony, silver and nickel (Table 28.1). Group A has been linked with Chalcolithic/Early Bronze Age southwest Irish copper mines, particularly the Ross Island mine (O'Brien 2004). Modified groups have been proposed by Rohl and Needham (1998), Needham (2002) and Bray (2009) particularly for the Early Bronze Age.

Despite the Great Orme mine being much larger than any other known Bronze Age mine in Britain (Fig. 28.1), with estimated copper metal outputs ranging from 25 to 1769t (Timberlake 2009), detecting this copper in British and Irish Bronze Age copper alloy artifact analyses has proved to be a puzzle for researchers. Influential mineralogical studies by Ixer and Budd (1998) and Ixer (2001) concluded that "the mineralogy of these [Great Orme] ores shows that they could only produce trace element poor copper metal." These views were reflected in a number of other papers, "copper from the Great Orme cannot be identified in the contemporary bronzes...the ore is characterized by very low nickel contents" (Craddock 1994) and "...at a time when most artifacts contained some degree of arsenic, the Great Orme

Table 28.1: A Summary Table of the Metal Groups Proposed by Northover (1980 and 1991)*

	Group	*Principle Impurities*
EBA	A	Arsenic, Antimony & Silver (Ross Island, Ireland)
	B	Arsenic & Nickel combinations (As <0.25% to >1.25%; Ni <0.06% to >0.06%)
	C	Traces only (<0.1%)
	D	Arsenic, Antimony, Silver & Nickel
	E	Arsenic, Antimony & Nickel
	F	Arsenic & Silver
	G	Nickel
MBA	M,N,O,P,R	Mainly Arsenic & Nickel combinations (As >0.1% to >1.05%; Ni<0.2% to >0.5%) some with traces of Co, Sb, Ag
LBA	MBA groups plus S and T	MBA types plus some high Antimony and Silver types

* Alternative metal (and isotopic) groups have been proposed by Rohl and Needham (1998), Needham (2002) and Bray (2009).

ores contained no arsenic whatsoever" (Budd 2000). All this contrasts with the fact that the largest artifact groups do contain one or more of these impurities and especially during the Middle Bronze Age when arsenic-nickel groups dominated British metalwork. Metal with low impurities (all less than 0.1%) is only a minor group (Northover 1980 and 1991).

Ixer's conclusions were based mainly on the ore minerals observed by reflected light microscopy, using the implicit assumption that without the presence of significant levels of arsenic or nickel minerals there could not be enough of those elements to give significant levels in the copper metal. The small number of chemical analyses that had been done on ore minerals gave conflicting results and even those with potentially significant arsenic and nickel were dismissed by Ixer as "...unlikely to be reflected in any copper metal produced from these ores" (Ixer 2001). A small bronze artifact and particle found underground with high levels of trace elements were interpreted as being "...unlikely to be the product of local copper ores and that the metal came from elsewhere" (Jenkins and Lewis 1991). A study by Northover and Salter (1999) for Gwynedd Archaeological Trust analyzed three very small copper prills from the Pentrwyn smelting sites using EMPA and showed significant levels of arsenic (in two of them) and nickel (Table 28.4 bottom). They struggled to resolve this with Ixer's work but concluded that minerals containing these elements could be present at Great Orme but "for the present, the connection cannot be regarded as completely certain."

Other scholars (Rohl and Needham 1998; Northover in Lynch, Aldhouse-Green and Davies 2000; Timberlake 2009; Bray 2012) have expressed the view that the Great Orme mines could have been an important source of Middle Bronze Age metal, implying significant arsenic and nickel levels (e.g., Acton Park type) but without resolving the conflicting complex mineralogical, chemical, smelting, isotopic and artifact analysis evidence. A review of this evidence by the author concluded that a major sampling and analysis program of relevant Great Orme materials was required to provide clearer evidence to resolve the situation and define a Great Orme metal group. In conjunction with the OXSAM and OXALID databases of British Bronze Age metalwork, the aim is to allow for a major re-evaluation of the Great Orme mine's contribution to the metal supply during the various phases of the Bronze Age in Britain and beyond.

The seven year research project began in 2011, bringing in specialist mineralogical, geological and geochemical expertise and this interim report was written in 2013. The research objective is to establish the key characteristics and variations in the Bronze Age Great Orme ores, smelting slags/prills and bronze particles in the mine in order to define a Great Orme metal group. Thus, artifacts with which it is consistent could be identified in the British Bronze Age metalwork databases and the archaeological implications re-assessed.

MATERIALS AND METHODS

Three types of materials were used for the research:

1. Copper ores systematically collected from as many parts as possible of the 6 km of Bronze Age workings, including all surface and underground veins. Over 50 samples have been collected thus far and are ready for analysis.
2. Slag particles and copper prills from the 2011 excavations at the Pentrwyn smelting site, which, with the original 1997 excavation, only amounts to about 600 g (most are less than 10 mm in size). Planned smelting experiments with Great Orme ore will also provide further samples for analysis.
3. Bronze particles found in the underground mine at location 21 in 1991/2. Over 50 particles were found but only two were chemically analyzed.

In addition, past work on these three types of materials was compiled from unpublished sources.

A variety of analytical techniques are being used to allow comparison of results between techniques. Reflected light microscopy (RLM) yields valuable information on mineralogy, smelting techniques and metallography. Overall, chemical composition is being revealed from XRF (PTM Ltd) and SEM-EDS (University of Liverpool) with key minor and trace element analyses being obtained from AAS (University of Liverpool) and LA-ICP-MS (University of Cardiff). Useful additional semi-quantitative information has been obtained on site with a high specification pXRF. The use of LA-ICP-MS (laser ablation inductively coupled plasma mass spectroscopy) is novel in this type of provenance study of copper ores/prills and results will be compared with those from more traditional methods. Some new lead isotope analyses have been determined by the NERC Isotope Laboratory at Keyworth on ores and prills using MC-ICP-MS.

GEOCHEMICAL ANALYSES

ORES

An extensive program of ore sampling is nearing completion and an analysis program has begun. The systematic sampling has provided strong evidence that the principal Bronze Age ore was a hard intimate mixture of malachite (hydrated copper carbonate) and goethite (hydrated iron oxide) that was easily worked from a local zone of very soft altered dolomite (Figs. 28.3, 28.4). These secondary minerals were formed from primary chalcopyrite (copper iron sulfide) by interaction with ground waters, and the residual sulfide present in certain samples. While in some classic supergene mineral deposits the copper moves away leaving a goethite rich gossan, in this deposit the copper has generally not moved far and has been mainly re-deposited within and around the goethite. Less altered chalcopyrite is present in the hard rock on each side of the altered zone and this material may have been intentionally left behind by Bronze Age

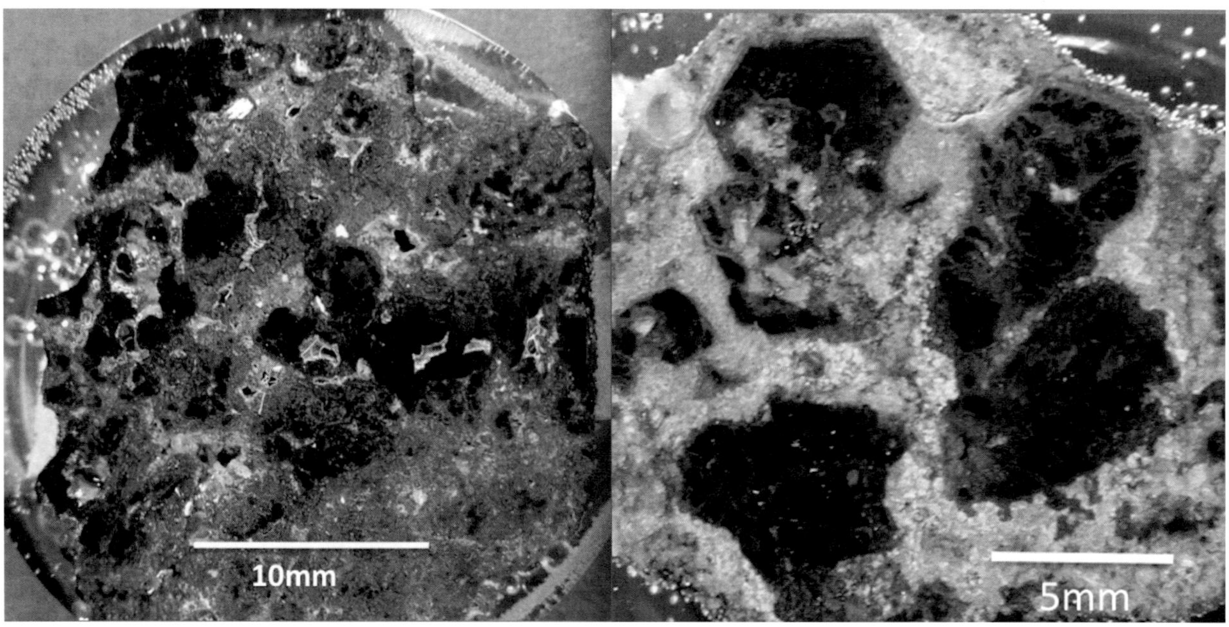

Fig. 28.4: Polished sections of the typical brown-green ores of intimately mixed-goethite with some calcite and dolomite. The goethite on the right has taken the form of the chalcopyrite crystals it replaced (pseudomorphs).

miners. Small amounts of secondary copper mineralization deposited away from the goethite in small malachite veinlets and along bedding planes in the mudstones, often as azurite pisoliths, appear to be a much less important ore as the copper content is much lower and apparently occurs in smaller volumes. Gangue (waste) minerals in the ore include variable amounts of calcite and dolomite and low levels of silica. Microscopic amounts of other ore and gangue minerals are present (Ixer 2001).

XRF analyses (semi-quantitative) on three bulk ore samples from different locations immediately demonstrated substantial levels of arsenic (up to 1%) and nickel (up to 0.5%) before calcination and very low levels of antimony and silver (Table 28.2). LA-ICP-MS allows measurement of minor and trace elements in individual mineral crystals and so far 17 analyses on various minerals in the ten ores have been completed and examples of provisional results are given in Table 28.3. Chalcopyrites contained only trace amounts (<0.05%) of arsenic and nickel. However, the malachites had much more variable levels of arsenic (0.01% to 0.85%) and nickel (<0.05% to 0.59%). The most surprising finding was that all the goethites contain much higher levels of arsenic (mostly >1%) than any of the malachites and in some cases significant levels of nickel (up to 0.69%) as well. Goethite is revealed for the first time as a key potential source of several trace elements in the copper metal as it would have inevitably formed part of the charge used in the smelt. Therefore, only analyzing individual mineral particles of malachite or chalcopyrite from the ore, rather than the ore in bulk, would give rise to the misleading and confusing

TABLE 28.2: GREAT ORME BULK ORE ANALYSES ON UNCALCINED ORE. ANALYZED ON PHILLIPS XL30 COURTESY PTM LTD., SEMI-QUANTITATIVE (+/-15%)

Wt%	*Ore G33*	*Ore G15*	*Ore G30*
CuO	25.3	28.8	27.6
Fe2O3	19.0	22.9	20.4
As2O3	1.0	0.8	0.2
NiO	0.5	0.2	0.1
Sb2O3	<0.05	<0.05	<0.05
Ag2O	<0.05	<0.05	<0.05
Co3O4	0.3	<0.05	0.1
CaO	10.7	6.9	9.7
MgO	4.9	3.6	4.9
SiO2	5.6	3.8	3.6
Al2O3	0.7	1.2	0.7
TiO2	<0.05	0.1	<0.05
P2O5	0.1	<0.05	0.1
K2O	0.1	0.3	0.1
Pb0	<0.05	<0.05	0.1
ZnO	<0.05	<0.05	<0.05
MoO3	0.1	<0.05	<0.05
Mn3O4	0.3	0.2	0.2
SO3	0.1	1.1	0.5
LoI 1200°C	31.4	30.1	31.8
Total	*100.1*	*100.0*	*100.1*

TABLE 28.3: EXAMPLES OF GREAT ORME ORE MINERAL ANALYSES BY LA-ICP-MS AND SEM-EDS*

Mineral	*Method*	*Cu* %	*Fe* %	*As* ppm or %	*Ni* %	*Sb* ppm	*Ag* ppm	*Pb* ppm	*Co* ppm or %	*U* ppm	*S* %	*Mg* ppm or %	*O* %	*Ca* %	*Si* %	% *Totals****	*C+H* Theor
Malachite G35	LA-ICP-MS**	47.5	0.86	640	0.14	6	<3	12	331	15	0.20	2887		<0.05	2.05		
	SEM-EDS	47.5	2.3	0.1	0.0							0.3	38.1	0.0	5.2	93.7	6.3
Malachite G33	LA-ICP-MS**	55.8	5.02	10801	0.75	18	<3	22	2259	105	<0.2	29119		4.01	<0.06		
	SEM-EDS	55.8	2.2	0.7	0.7				0.1			0.0	34.0	0.0	0.3	93.7	6.3
Chalcopyrite G15	LA-ICP-MS**	35.4	27.13	93	<0.05	44	8	<8	<4	<0.04	31.92	<85			0.05		
	SEM-EDS	35.4	29.5								35.1					100.0	
Chalcopyrite 109	LA-ICP-MS**	36.8	29.45	<15	<0.05	<5	<3	<8	<4	<0.04	34.45	<85			0.10		
	SEM-EDS	36.8	28.8								34.4					100.0	
Goethite OC37	LA-ICP-MS**	3.43	57.8	19820	0.58	77	5	162	1775	138	<0.2	702		0.13	0.98		
	SEM-EDS	6.6	57.8	1.2	0.6				1.3				29.2	0.8	1.3	98.9	1.1
Goethite E1	LA-ICP-MS**	7.66	42.3	3857	0.18	249	10	19694	615	196	0.19	298		0.26	2.28		
	SEM-EDS	12.6	42.3	0.0	0.0			5.0				0.0	34.0	0.6	4.4	98.9	1.1

* LA-ICP-MS anlayses were done on individual crystals along a 300 micron line. SEM-EDS measurements made adjacent to each line.

** Used SEM-EDS to provide major element internal standard (Cu or Fe) as per Resano *et al.* (2010)

*** Original totals with SEM-EDS mineral standards before normalizing were +/-2% of theoretical total allowing for C & H in the mineral results. This appears to explain the mainly low levels of arsenic and nickel in the handful of past ore mineral analyses (Jenkins and Lewis 1991; Saville 2005). Another important observation is that nearly all the examined goethites contain several weight percent of copper (confirmed by SEM-EDS analysis), which enhances the copper content of the ore and makes it a low grade copper ore mineral in its own right. The special crystal structure of goethite allows it to have exceptional levels of atoms of other elements within its structure (Chang, Howie and Zussman 1998). Antimony and silver levels in the ores are all mostly less than 50ppm except for a few variable goethites. Lead concentration is generally low in the malachite, except near the lead vein, but higher and more variable in the goethite.

The overall picture of minor/trace levels revealed so far by LA-ICP-MS microanalyses and the initial XRF bulk analyses is supported also by semi-quantitative pXRF (hand-held) work. These minor/trace elements results will now be examined against atomic absorption spectroscopy analyses along with a much larger set of ore samples in order to establish variations and averages from different mine areas.

How the levels of minor and trace elements in copper ores split between metal and slag during smelting depends very much on the element and, in some cases, the variations of the smelting conditions (Pollard *et al.* 1991). Studies have shown that both nickel and arsenic have a strong tendency to concentrate in the metal (Hauptmann 2007; Tylecote, Ghaznavi and Boydell 1977). The losses that might be expected from volatilization of arsenic are minimized or eliminated in reducing conditions. Hence, while sulfide ores might lose arsenic (and antimony if present) from the roasting stage in oxidizing conditions to oxidize sulphur before smelting, the oxide ores such as malachite (which do not need roasting) could retain all its arsenic and transfer most of it to the metal. The significance is that oxide ores can have a lower level of arsenic than sulfide ores and still produce the same level in the metal. So if all the measured arsenic and nickel in Great Orme ores partitions into the metal, then normalizing to 100% copper (see Pernicka 2004) suggests high levels of arsenic and nickel might be possible, subject to losses from refining and recycling.

SMELTING SLAGS AND COPPER PRILLS

The Pentrwyn excavations in 1997 and 2011 have yielded around 600 g of smelting residues, mostly as particles 1 to 10 mm in size (Fig. 28.5). As mentioned earlier, Northover and Salter (1999) analyzed three copper prills from the 1997 material (Table 28.4) and found significant arsenic (in 2 of the 3) and nickel levels. They noted the presence of sulfides in the prills and slags and concluded that the material was probably from the smelting of sulfide rich ores "possibly disseminated polymetallic sulfides not yet discovered in the main Great Orme mine complex," referencing Ixer's work.

The material from the 2011 excavations consists of around 500 mostly very weathered small particles (1 to 10 mm). Only the coarsest slag particles and prills still showed the original microstructures (Fig. 28.5). A detailed study of this material is providing strong evidence for the simple smelting of high-grade secondary copper ores without added fluxes, which were never fully liquefied. Residual textures are consistent with the secondary malachite-goethite ores described earlier with only traces of unconverted sulfides. The less weathered samples suggested the goethite converted to magnetite or wustite, depending on the redox conditions, and the malachite was reduced to copper prills which weathered to cuprite (copper oxide) and atacamite (hydrated copper chloride). As the ores are generally low in silica, only a small amount of glassy phase was formed but apparently enough to allow some aggregation of the copper prills to larger prills. The copper prills were recovered by crushing the smelt, possibly with some sort of panning, and then re-melting and refining of the prills. Minimal smelting waste was produced from the rich ores and there is no evidence for additives being used. It was not a fully slagging process that produced a mainly liquid slag in which the copper droplets sink to form an ingot and the slag can be

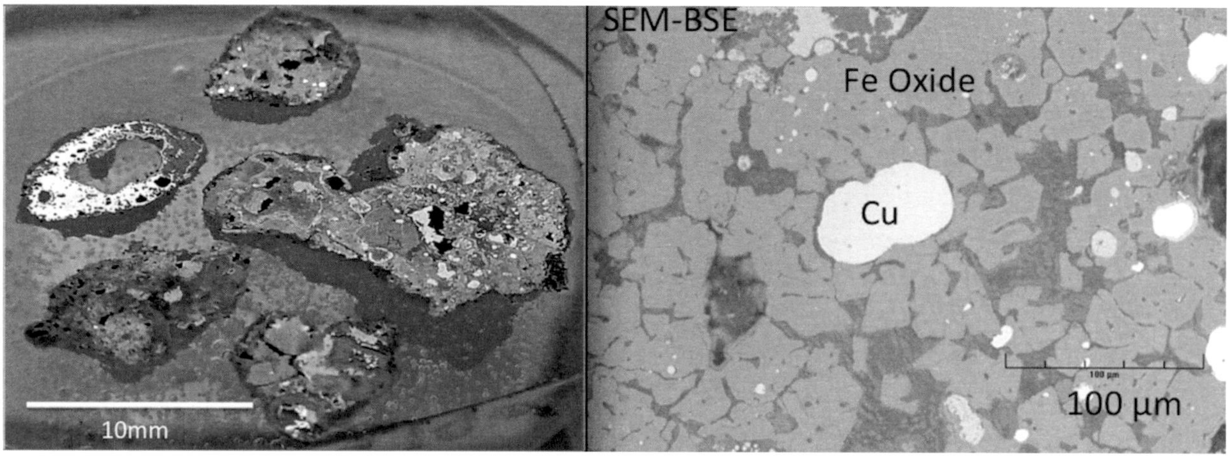

Fig. 28.5: Smelting residues from the Pentrwyn smelting site with the polished section (left) showing a copper prill (white) and inhomogeneous slags which have never been fully molten. SEM-EDS image (right) of the products of smelting a malachite-goethite ore, namely copper prills, magnetite and a small amount of glassy phase.

tapped off. Whether a fully slagging process would have offered any decisive advantages when smelting the plentiful high-grade, low silica copper ores of the Great Orme is debatable. The simple smelting process without additives would have produced only small amounts of slag and crushing makes the slag remains less visible archaeologically.

Analysis of copper prills from the 2011 excavations using LA-ICP-MS has shown arsenic levels up to 0.28% but with nickel levels from <0.05% up to 0.13%. These results are noticeably lower than those found by Northover and Salter in the prills from the 1997 excavation (arsenic 0 to 0.62% and nickel 0.20 to 0.82%). Also the latest study found more variable iron levels (3% to <0.03%); the latter probably reflecting refined metal. All these results serve to demonstrate the likely variation in Great Orme metal. Further analyses using AAS are planned.

BRONZE FRAGMENTS FOUND IN THE MINE

Besides ores and slags/prills, another potential source of information about the metal compositions produced at the mine are the bronze fragments found in the mine (Fig. 28.6). This is based on the reasonable assumption that if they are contemporary with the Bronze Age workings they are very likely to be made of the metal the mine was producing. In 1831 a small bronze tip, possibly broken off a bar or pick (Rohl and Needham 1998), was found underground, and in 1991 excavations in a narrow underground vein location yielded over 50 irregular bronze particles (typically 5 to 15 mm). While they were associated with stone and bone tools and charcoal, dated to 1410 to 1053 BCE, the materials were probably redeposited in a later period.

To date only the bronze tip and two of the bronze particles have been fully analyzed (Table 28.2) and they show metal with significant levels of arsenic and nickel and low levels of antimony and silver (Lewis 1996; Budd 2000; Rohl and Needham 1998). One of the particles analyzed by SEM-EDS gave much higher values than with AAS but this probably reflects the limitation of the former method at low concentrations. Recently the location of the remaining 40 particles of bronze has been tracked down and a more comprehensive set of analyses is now planned. The Smithsonian Institute did some qualitative XRF spectrums on several particles in 1993 and detected some nickel and arsenic as well as other elements, but at unquantified levels.

Table 28.4: Examples of Analyses of Copper Prills from Pentrwyn Smelting Site and Bronze Fragments from Underground at the Great Orme Mine by LA-ICP-MS and SEM-EDS*

Sample Ref.	*Method*	*Cu***	*Fe*	*As*	*Ni*	*Sb*	*Ag*	*Pb*	*Zn*	*Co*	*Sn*	*S*	*Cl*	*Si*	*O*	Au	Bi	*Total****
Copper prills from Pentrwyn smelting site: Samples from 2011 Excavation (SEM-EDS University of Liverpool & LA at University of Cardiff)																		
P5	SEM-ED	93.0	0.2	0.3	0.2	0.0	0.0	0.2	1.3	0.0	0.0	0.2	0.6	0.2	3.7			100.0
	LA-ICP-MS	93.0	<0.06	500ppm	0.13	19ppm	10ppm	106ppm	<20ppm	121ppm	<18ppm	<0.2		<0.06				
P12B	SEM-ED	88.4	2.1	1.1	0.1	0.0	0.0	1.5	0.8	0.0	0.2	0.0	0.6	0.4	4.8			100.0
	LA-ICP-MS	88.4	0.71	616ppm	0.08	132ppm	173ppm	12417ppm	346ppm	345ppm	<18ppm	0.29		<0.06				
P10B	SEM-ED	86.3	3.5	0.4	0.3	0.0	0.0	0.5	1.1	0.0	0.1	0.9	0.7	0.3	5.8			100.0
	LA-ICP-MS	86.3	0.97	2848ppm	0.09	32ppm	244ppm	68.00	268.00	204ppm	<18ppm	0.47		<0.06				
Copper prills from Pentrwyn smelting site: Samples from 1998 Excavation. University of Oxford: Northover and Salter (1999) in GAT report G1497 app. II.																		
2222 (av. of 3)	EMPA	96.95	2.83	0.62	0.29	0.00	0.03	0.08	0.00	0.09	0.00	0.02				0.00	0.00	100.91
2227	EMPA	98.24	2.44	0.00	0.82	0.01	0.16	0.02	0.04	0.17	0.01	0.08				0.02	0.00	102.01
2237 (av. of 13)	EMPA	94.39	2.74	0.52	0.20	0.01	0.07	0.04	0.02	0.04	0.01	0.04				0.01	0.01	98.10
Bronze particles from underground in Great Orme mines - Analyzed by Craddock and Hook (1991) at British Museum																		
1246 (Tip)	AAS	84.8	0.1697	1.1574	0.5542	<0.0377	0.0509	0.51	0.0036	0.0509	14.89					<0.005	<0.018	99.6
1089 U/G (Loc. 21)	AAS	82.8	0.1970	1.0600	0.4160	0.0500	0.0510	0.09	0.0150	0.0530	10.70					<0.003	0.0170	95.5

* Wt % unless stated

** For samples from 2011: SEM-EDS used to provide major element internal standard (Cu) as per Resano *et al* (2010).

*** For samples from 2011: The SEM-EDS data is normalized from data using certified reference materials. Original totals between 98 and 102%

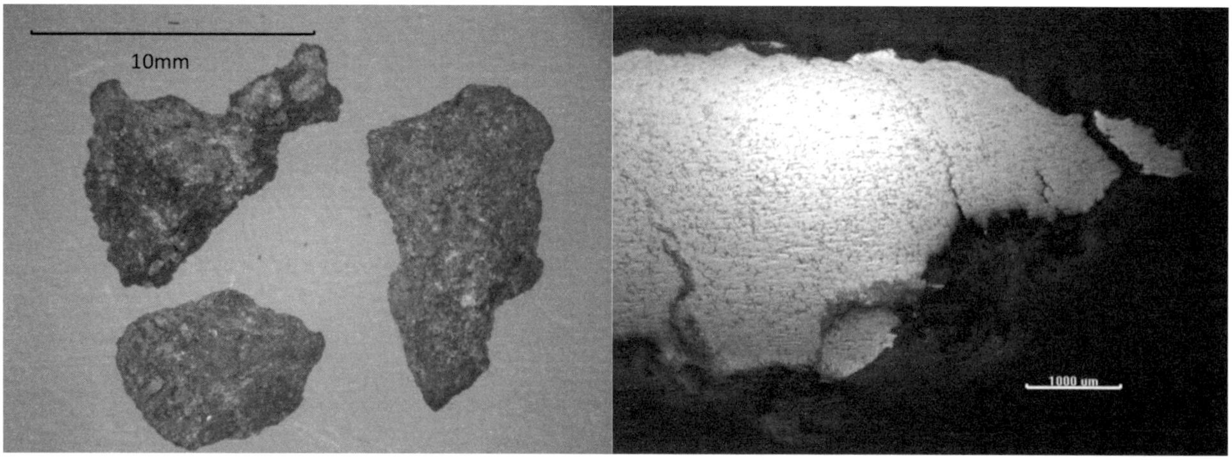

Fig. 28.6: Weathered bronze particles from Location 21 underground and a polished section showing weathering and irregular fractures.

GEOCHEMICAL SUMMARY

Some of the new geochemical data produced thus far in the project and the compilation of old data is shown together in Fig. 28.7 in a plot of arsenic versus nickel for all the materials discussed, plus a range of British Bronze Age metalwork analyses (OXALID database). Plotting the analyses of just the malachite in the ore (adjusted to 100% copper) shows that, given minimal losses in a smelt with reducing conditions, the ore is capable of producing an arsenic-nickel rich metal common in many artifacts, and covers the whole composition range of both the prills and mine bronze fragments. The arsenic and nickel in the goethite might be partially available to boost levels further. All this refutes the view that the Great Orme mines could only produce copper with low levels of impurities. Figure 28.8 shows that all Great Orme metal would have had very low levels of antimony and mostly low silver levels. All this supports the view that the Great Orme may have been a major source of copper, especially in the Middle Bronze Age. Defining a distinct geochemical Great Orme metal group will allow new insights from existing databases. The probable limits for the key trace/minor elements of the mine group await analyses of the much larger sample set of ores and also further slags/prills and mine bronze particles.

LEAD ISOTOPE ANALYSIS

The mine-based metal group can be further refined by the use of the independent technique of lead isotope analysis; this is particularly important if there is any geochemical overlap with metal from other British or European mines. Data on Great Orme materials were produced by Rohl and Needham (1998), Joel *et al.* (1997), and new measurements for this project on ores and copper prills.

Lead isotope ^{204}Pb is constant since the formation of the Earth while other lead isotopes (^{206}Pb, ^{207}Pb and ^{208}Pb) are formed by the decay of uranium and thorium. In ideal circumstances the graphs of the ratios of the various lead isotopes define a small discreet area (an isotopic "signature") for an ore deposit and this is retained in the produced metal. Providing the metal artifact is wholly from that source without recycling or lead additions from another source the two "signatures" should generally match. However, the Great Orme ores do not define a small discreet area on the ratio graphs; instead they define a long zone because the ores are radiogenic and this presents a challenge and an opportunity (Fig. 28.9). An understanding of the cause of this phenomenon has been provided by the LA-ICP-MS results which

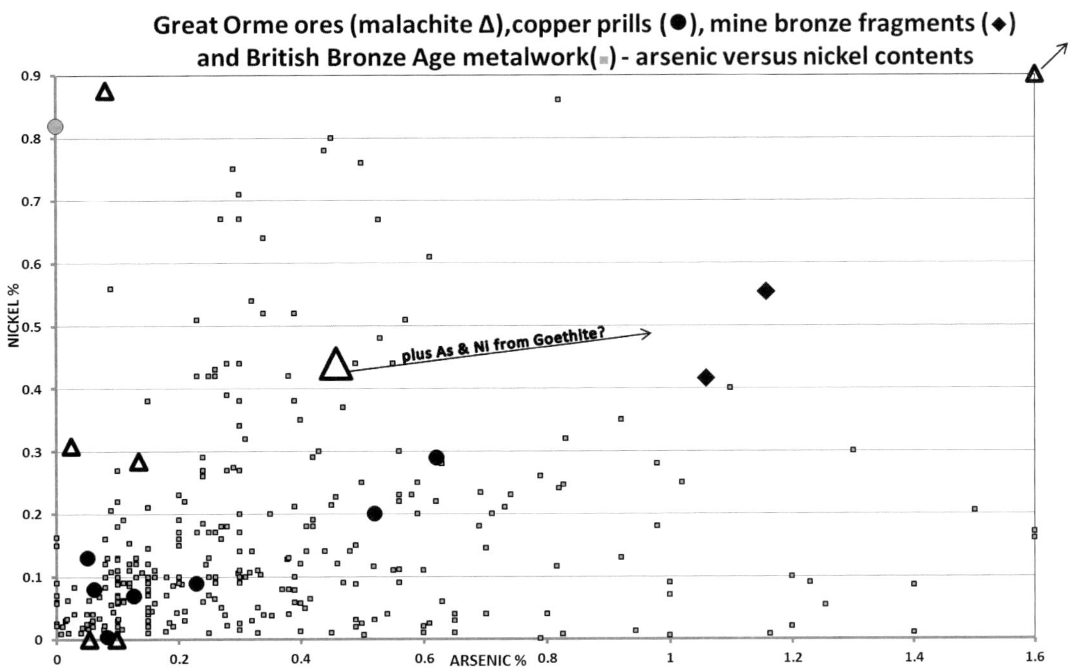

Fig. 28.7: Plot of arsenic versus nickel showing only the malachite portion of ore (normalized to 100% copper) could produce arsenic-nickel rich metal which matches the copper prills and mine bronze particles and a wide range Bronze Age metalwork analyses (OXALID). The large triangle is an average of six malachite analyses without additions from goethite.

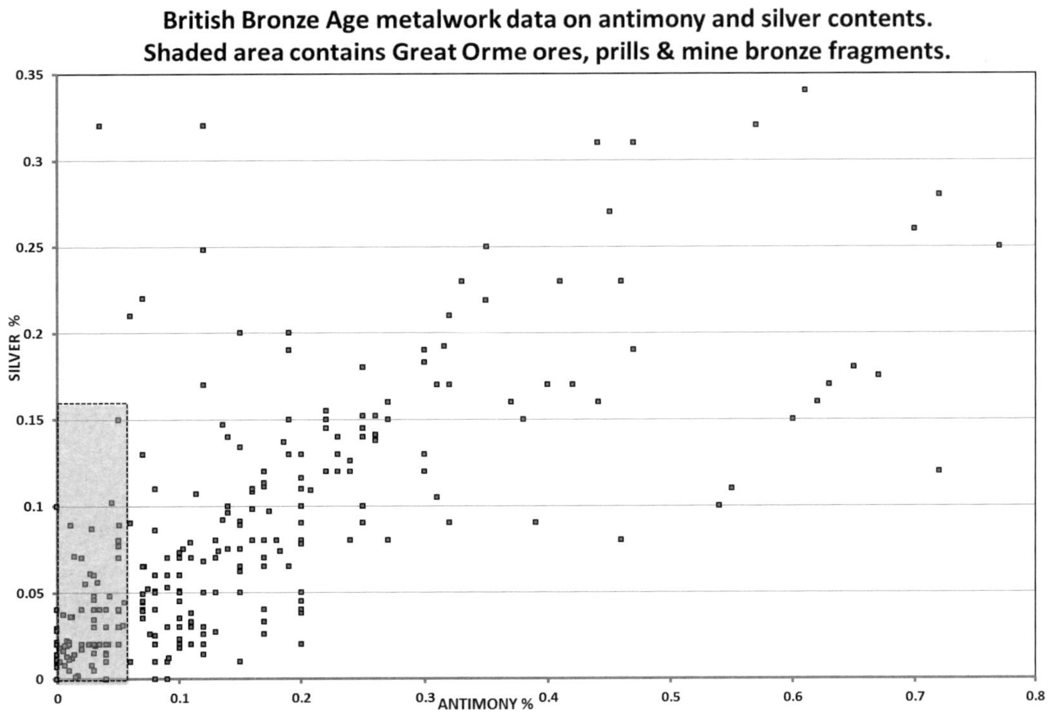

Fig. 28.8: Plot of antimony versus silver, which are generally very low for all Great Orme materials (ores, prills and mine bronze fragments). Bronze Age metalwork analyses from OXALID.

indicate significant levels of uranium (up to around 195 ppm) and low levels of lead in some of the copper minerals (<8 ppm), but also locally high lead levels (>1000 ppm), probably related to areas around the lead (galena) vein(s) at the Great Orme which result from a separate mineralizing event. For the copper ore with very low levels of lead it takes very little lead from the lead-rich areas (or absorbed by the lead-rich goethite; Table 28.3) to shift the ratios towards the area (top right area in Fig. 28.9) where nearly all British Bronze Age metalwork artifacts lie.

Plotting the available data from other dated British Bronze Age mines shows that they plot separately to Great Orme with the exception of the Alderley Edge mine, whose ore genesis may be linked to the Great Orme (Fig. 28.10). Plotting all the analyzed artifacts from each Bronze Age period defines many objects potentially consistent with the Great Orme from all periods (see Figs. 28.11–14 with caption comments) excluding the Chalcolithic. One of several complications to consider is that some artifacts contained substantial amounts of added lead, and so the lead source will potentially swamp the copper source "signature" unless the lead used was from the Great Orme. A further complication, especially in the Late Bronze Age, is that some imported continental copper sources could fall in the Great Orme "signature" zone; this possibility is being investigated. In addition, the challenges of recycling have to be considered in detail but having another fixed point source like Ross Island may be helpful. Overall, there are many complications still to sort out but it seems reasonable that this will allow us to narrow down the metalwork consistent with Great Orme. For instance, Great Orme ores show a range of isotopic values (Rohl and Needham 1998) and initial trace element results that are both consistent with most Acton Park metalwork and potentially with some other assemblages as well.

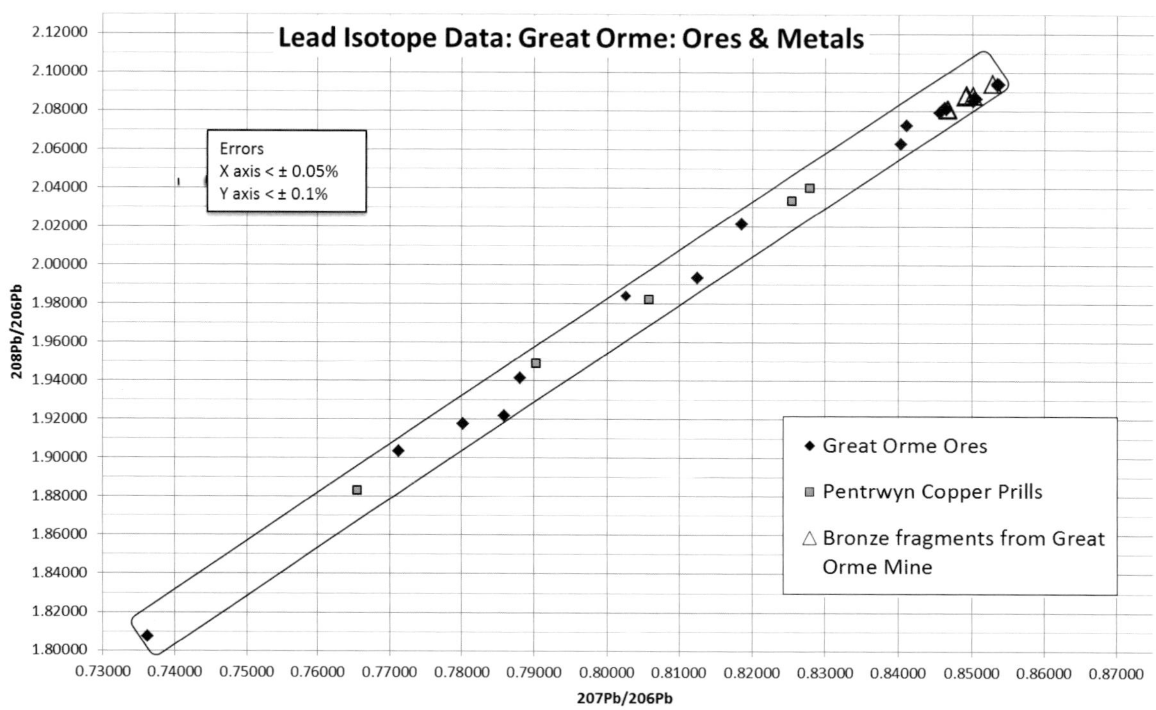

Fig. 28.9: The Great Orme copper ores are radiogenic (low lead–high uranium) but need very little lead from the separate lead vein to cluster in the top right area where most Bronze Age metalwork lies. All ores, prills and bronze fragments match up. Data from OXALID, Rohl and Needham (1998), Joel *et al.* (1997) and six new measurements.

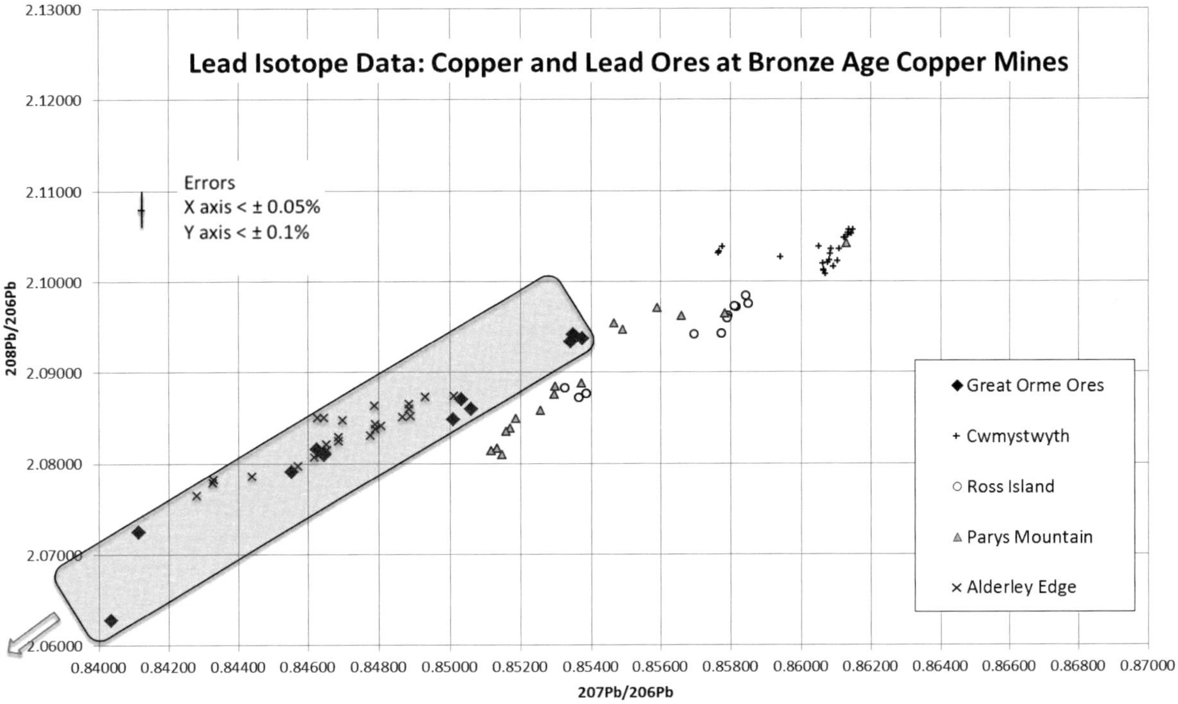

Fig. 28.10: Lead isotope measurements of copper and lead ores from dated British and Irish Bronze Age copper mines. The zone of radiogenic Great Orme ores overlapping only with Alderley Edge ores. Data from OXALID, Rohl and Needham (1998), Joel *et al.* (1997) and six new measurements.

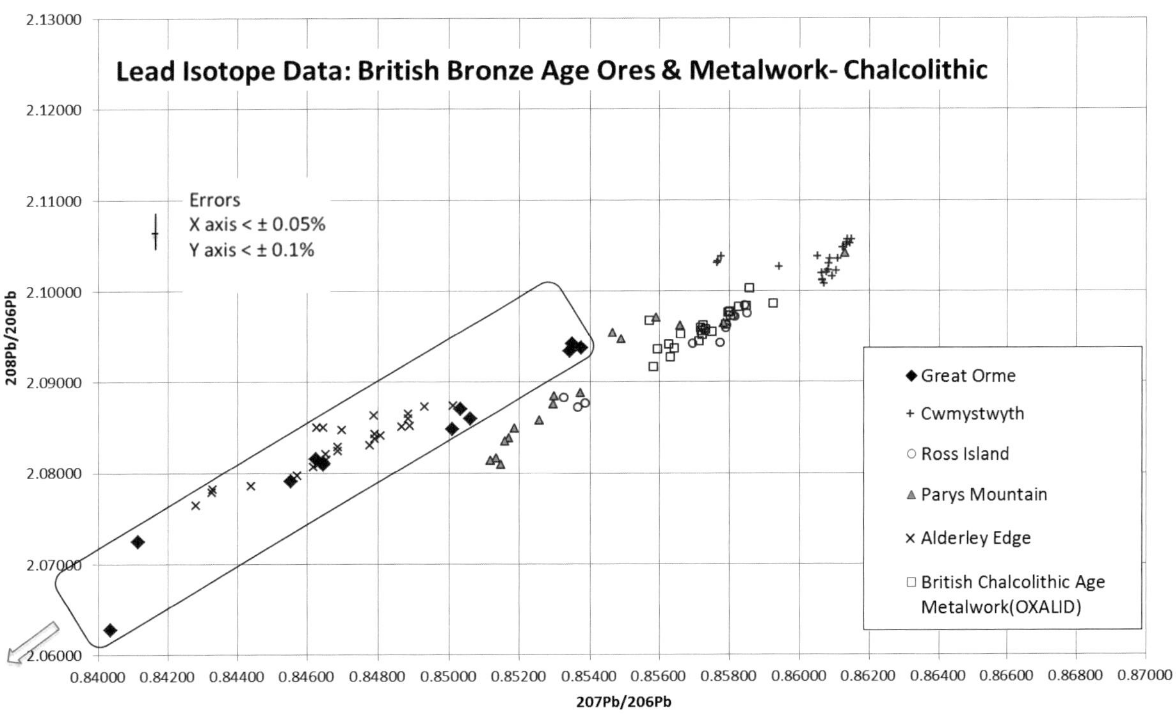

Fig. 28.11: Mainly Chalcolithic copper artifacts cluster around the Ross Island mine in S.W. Ireland and possibly Parys Mountain. Data sources–see Figure 28.10.

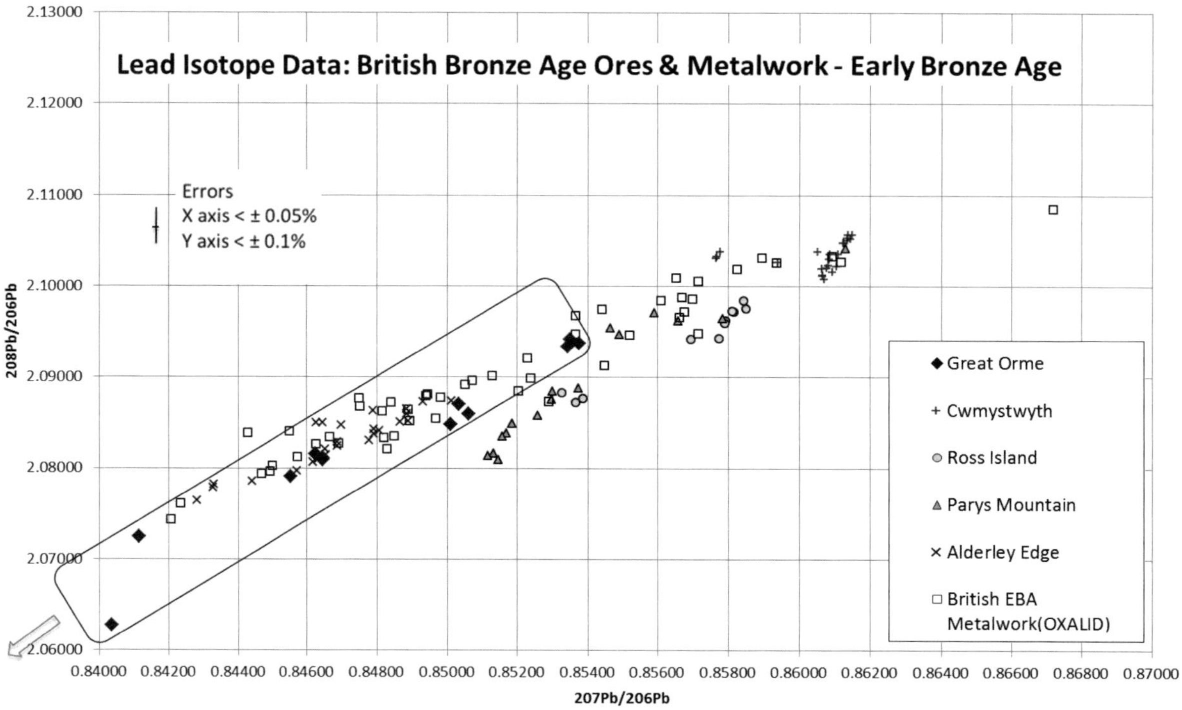

Fig. 28.12: Early Bronze Age artifacts match various mines including Cwmystwyth, Parys Mountain, Alderley Edge and Great Orme, which matches ^{14}C dating. Data sources–see Figure 28.10.

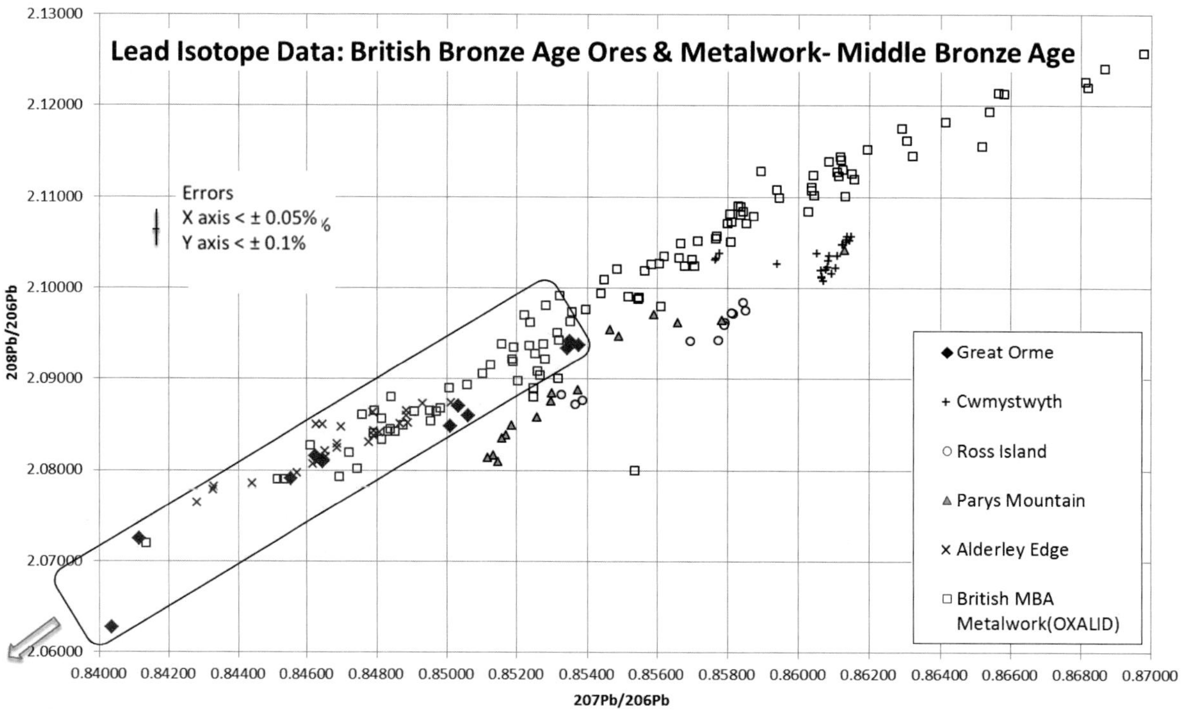

Fig. 28.13: Many Middle Bronze Age artifacts match Great Orme but the data indicates a currently unknown source in Britain/Ireland or an imported source. Lead additions, when present, would reflect the only lead ore source. Data source–see Figure 28.10.

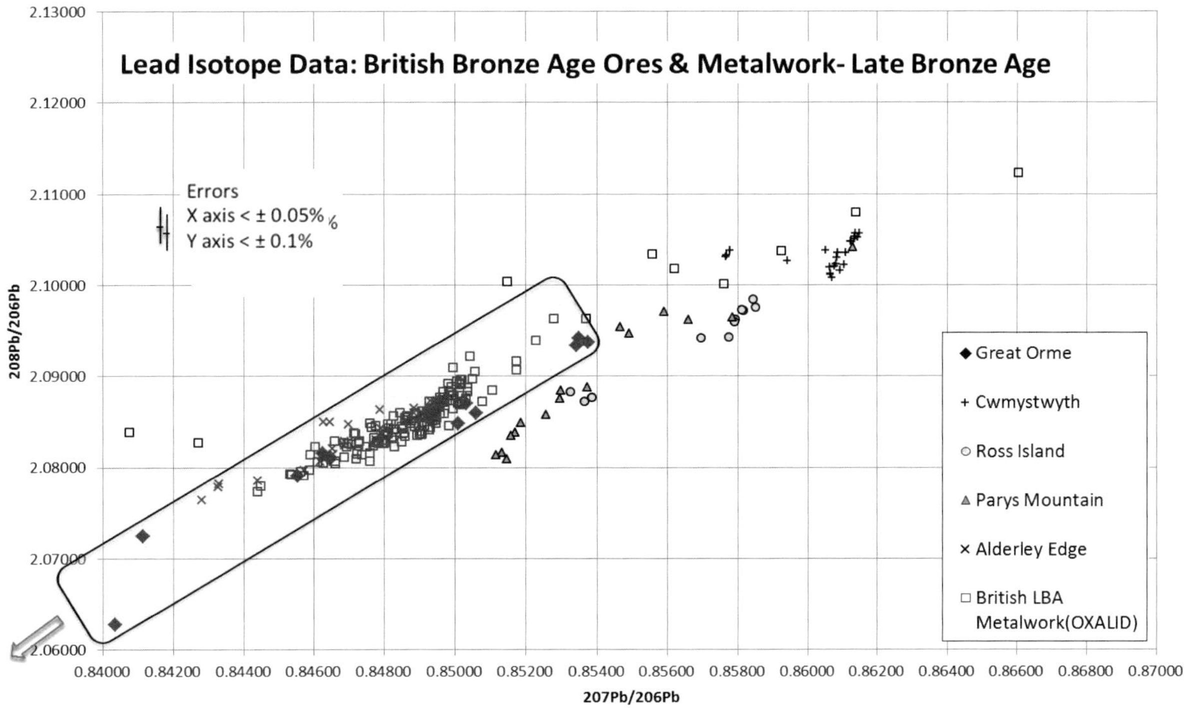

Fig. 28.14: Late Bronze Age metalwork clusters in the Great Orme area but complications like overlap with some continental sources, lead additions and incompatible trace element chemistry reduces possible matches. Data sources–see Figure 28.10.

SUMMARY

The ongoing systematic sampling and analysis program of Great Orme ores, slags/prills and bronze particles from the mine, in combination with existing published and unpublished data, is starting to define a mine-based metal group founded on a combination of geochemical and lead isotope information. Databases of Bronze Age artifacts (OXSAM & OXALID) are being used to define artifacts consistent with Great Orme metal. In combination with other spatial and temporal evidence, the foundation is being laid for detailed studies of the Great Orme metal supply and the associated social and technological implications. At the time of writing in 2013 the project is still at an early stage but preliminary results are suggesting the mine produced copper with arsenic-nickel impurities but with low antimony and silver levels. This composition type was particularly common in the early Middle Bronze Age and may result in a major reassessment of the importance of these mines.

ACKNOWLEDGMENTS

Great Orme Mines Ltd.* (Tony Hammond, Andy Lewis, Nick Jowett, Edric Roberts, Sian James), Great Orme Exploration Society (Dave Wrenall and members), Gwynedd Archaeological Trust (George Smith),* CADW,* Historical Metallurgy Society,* NERC Isotope Geosciences Laboratory (Jane Evans), University of Cardiff (Iain McDonald), University of Leicester (Nick Marsh), University of Nottingham (Chris Sommerfield), Ancient Arts (Dave and Sue Chapman), University of Oxford (Peter Bray, Mark Pollard, Peter Northover, OXALID and OXSAM databases), Simon Timberlake, Robert Ixer,

Christopher Williams, University of Rennes (Cecile Le Carlier de Veslud), Johan Ling, Lene Melheim), Helen Thomas and many others up to the time of writing (2013) this preliminary report. University of Liverpool PhD Supervisors, Matthew Ponting, Duncan Garrow and Rachel Pope. Analytical funding was provided by organizations marked with an asterisk.

REFERENCES

Hopewell, D. and Jones, S. 1999. *Pentrwyn Metal Working Site, Great Orme. Project G1497* (GAT Report no. 321. Unpublished report, Gwynedd Archaeological Trust, March 1999).

Bray, P.J. 2009. *Exploring the Social Basis of Technology: Re-Analysing Regional Archaeometric Studies of the First Copper and Tin-Bronze Use in Britain and Ireland* (Ph.D. dissertation, Oxford University). Oxford.

Bray, P.J. 2012. Before ^{29}Cu Became Copper: Tracing the Recognition and Invention of Metalleity in Britain and Ireland during the Third Millennium BC. In: Allen, M., Gardiner, J. and Sheridan, A., eds. *Is There a British Chalcolithic? People, Place and Polity in the Later Third Millennium* (Prehistoric Society Research Paper 4). Oxford.

Bray, P.J. and Pollard, A.M. 2012. A New Interpretive Approach to the Chemistry of Copper-Alloy Objects: Source, Recycling and Technology. *Antiquity* 86: 853–867.

Budd, P. 2000. Meet the Metal Makers. *British Archaeology* 56: 12–17.

Chang, L.L., Howie, R.A. and Zussman, J. 1998. *Rock Forming Minerals, Vol. 5B: Non-silicates* (2nd edition). Bath.

Craddock, P.T. 1994. Recent Progress in the Study of Early Mining and Metallurgy in the British Isles. *Journal of the Historical Metallurgy Society* 28(2): 69–83.

Craddock, P.T. and Hook, D.R. 1991. *Report of the Compositions of Three Copper Alloy Artifacts from Within the Mine on the Great Orme's Head, Llandudno* (British Museum Research Laboratory Report). London.

Dutton, L. and Fasham, P. 1994. Prehistoric Copper Mining on the Great Orme, Llandudno, Gwynedd. *Proceedings of the Prehistoric Society* 60: 245–286.

Hauptmann, A. 2007. *The Archaeometallurgy of Copper: Evidence from Faynan, Jordan.* London.

Ixer, R.A. 2001. An Assessment of Copper Mineralisation from the Great Orme Mine, Llandudno, North Wales, as Ore in the Bronze Age. *Proceedings of the Yorkshire Geological Society* 53(3): 213–219.

Ixer, R.A. and Budd, P. 1998. The Mineralogy of Bronze Age Copper Ores from the British Isles: Implications for the Composition of Early Metalwork. *Oxford Journal of Archaeology* 17(1): 15–41.

Ixer, R.A. and Davies, J. 1996. Mineralisation at the Great Orme Copper Mines, Llandudno, North Wales. *UK Journal of Mines and Minerals* 17: 7–14.

James, S.E. 2011. *The Economic, Social and Environmental Implications of Faunal Remains from the Bronze Age Copper Mines at Great Orme, North Wales* (Ph.D. dissertation, University of Liverpool). Liverpool.

Jenkins, D.A. and Lewis, C.A. 1991. Prehistoric Mining for Copper in the Great Orme, Llandudno. In: Budd, P., Jackson, C. and Chapman, B., eds. *Proceedings of a Conference on the Application of Scientific Techniques to Archaeology, Bradford, September 1989* (Oxbow Archaeological Monograph 9). Oxford: 151–161.

Joel, E.C., Taylor, J.J., Ixer, R.A. and Goodway, M. 1997. Lead Isotope Analysis and the Great Orme Mine. In: Sinclair, A., Slater, E. and Gowlett, J., eds. *Archaeological Sciences 1995: Proceedings of a Conference on the Application of Scientific Techniques to the Study of Archaeology; Liverpool, July 1995* (Oxbow Monograph 64). Oxford: 123–131.

Lewis, C.A. 1996. *Prehistoric Mining at the Great Orme: Criteria for the Identification of Early Mining* (M.A. thesis, University of Wales). Bangor.

Lynch, F., Aldhouse-Green, S. and Davies, J.L. 2000. *Prehistoric Wales*. Sutton.

Needham, S.P. 2002. Analytical Implications for Beaker Metallurgy in North-West Europe. In: Bartelheim, M., Pernicka, E. and Krause, R., eds. *Die Anfänge der Metallurgie in der Alten Welt [The Beginnings of Metallurgy in the Old World]* (Forschungen zur Archäometrie und Altertumswissenschaft 1). Rahden: 99–133.

Northover, J.P. 1980. The Analysis of Welsh Bronze Age Metalwork. In: Savory, H., ed. *Guide Catalogue of the Bronze Age Collections, National Museum of Wales*. Cardiff: 229–243.

Northover, J.P. 1991. Bronze in the British Bronze Age. In: Oddy, W.A., ed. *Aspects of Early Metallurgy* (British Museum Occasional Paper 17). London: 63–70.

Northover, J.P. and Salter, C.J. 1999. Interim Report on the Metal-Working Debris from Pentrwyn, Great Orme, Gwynedd, Site G1497 (Appendix II). In: Hopewell, D. and Jones, S. 1999, eds., *Pentrwyn Metal Working Site, Great Orme, Project G1497* (GAT Report no. 321). Gwynedd.

O'Brien, W. 2004. *Ross Island: Mining, Metal and Society in Early Ireland* (Bronze Age Studies 6). Galway.

Pernicka, E. 2004. Archaeometallurgy: Examples of the Application of Scientific Methods to the Provenance of Archeological Metal Objects. In: Martini, M. and Piacentini, M., eds. *Physics Methods in Archaeometry* (Proceedings of the International School of Physics, Enrico Fermi. Course CLIV). Bologna: 309–329.

Pollard, A.M., Thomas, R.G., Ware, D.P. and Williams, P.A. 1991. Experimental Smelting of Secondary Copper Minerals: Implications for Early Bronze Age Metallurgy in Britain. In: Pernicka, E. and Wagner, G.A., eds. *Archaeometry '90, Proceedings of the 27th Symposium on Archaeometry held in Heidelberg April 2–6, 1990*. Heidelberg: 127–136.

Resano, M., Garcia-Ruiz, E. and Vanhaecke, F. 2010. Laser Ablation-Inductively Couples Plasma Mass Spectrometry in Archaeometric Research. *Mass Spectrometry Review* 29: 55–78.

Rohl, B. and Needham, S. 1998. *The Circulation of Metal in the British Bronze Age: The Application of Lead Isotope Analysis* (British Museum Occasional Paper 102). London.

Saville, R.J. 2005. *Identifying the Copper Source of a Selection of Bronze Age Artifacts through Trace Elements Analysis* (M.Sc. thesis, University of Wales). Aberystwyth.

Smith, G. 2012. *Pentrwyn Copper Smelting Sites Excavation, Great Orme, Llandudno, Conwy (GAT Project G2178): Preliminary Report* (Report No. 1029: Gwynedd Archaeological Trust). Gwynedd.

Timberlake, S. 2009. Copper Mining and Production at the Beginning of the British Bronze Age. In: Clark, P., ed. *Bronze Age Connections: Cultural Contact in Prehistoric Europe*. Oxford: 9.

Tylecote, R.F, Ghaznavi H.A. and Boydell, P.J. 1977. Partitioning of Trace Elements between Ores, Fluxes, Slags and Metal during the Smelting of Copper. *Journal of Archaeological Science* 4: 305–333.

Wager, E. 2001. *The Character and Context of Bronze Age Mining on the Great Orme, North Wales, UK* (Ph.D. dissertation, University of Sheffield). Sheffield.

CHAPTER 29

COPPER MINING AND SMELTING IN THE BRITISH BRONZE AGE: NEW EVIDENCE OF MINE SITES INCLUDING SOME RE-ANALYSES OF DATES AND ORE SOURCES

Simon Timberlake and Peter Marshall

Twelve Early Middle Bronze copper mining sites have now been identified within the UK as a result of archaeological excavation and the radiocarbon dating of fire setting remains, namely bone and antler mining tools. This evidence suggests a fairly widespread phase of small-scale mining and prospecting for local ore sources within the coastal and upland areas of western Britain, beginning around 2000 BCE and terminating around 1500 BCE—this termination perhaps coinciding with the opening-up of new deposits of chalcopyrite ore in the Alps and the import of metal from Europe. An improved comprehension of the sequence of metal exploitation in Britain has recently been acquired through Bayesian modelling of more than 91 extant radiocarbon dates from these mines. Current re-examination of the totality of this excavation evidence has since provided us with a more realistic model of the mining process, the types of ores extracted, and with the strategies adopted for prospection during a significant period of metal exploitation beginning with the Beaker phenomenon in Britain and Ireland.

INTRODUCTION

Over the last 25 years the discovery and excavation of some 12 Bronze Age metal mines or prospection sites, and the probable identification of at least eight others within England and Wales, have provided a suitable time-frame for the first exploitation of metal within the UK (Timberlake 2009, 2010). This exploitation consisted of a widespread phase of early prospection, most of which took place between 2000 and 1650 BCE, followed by production at a very limited number of sites (such as the Great Orme in North Wales) which continued right up until the Late Bronze Age (Lewis 1996; Timberlake 2009). All of these investigations, except for that on the Great Orme, form part of a long term program of study by the Early Mines Research Group (EMRG, Crew and Crew 1990).

The majority of these mining sites lie on or close to the western seaboard of Britain, with a concentration in North and mid-Wales, but particularly in mid-Wales, where these sites seem to be associated with both lead and copper veins, where the copper mineralization lies close to the surface (Fig. 29.1). Most of these mines were originally identified on the basis of finds of cobblestone tools, the majority of which were collected as beach pebbles along the coast. Only at distances greater than 20 km from the coast do we find the regular use of river pebbles and glacial erratics. Both morphometric analysis and archaeological experiment have suggested there could have been distinct advantages in the choice of beach cobbles over riverine cobbles, in terms of shape, smoothness and competency (lack of flaws), wherever the former were available (Jenkins and Timberlake 1997; Timberlake and Craddock 2013: 42).

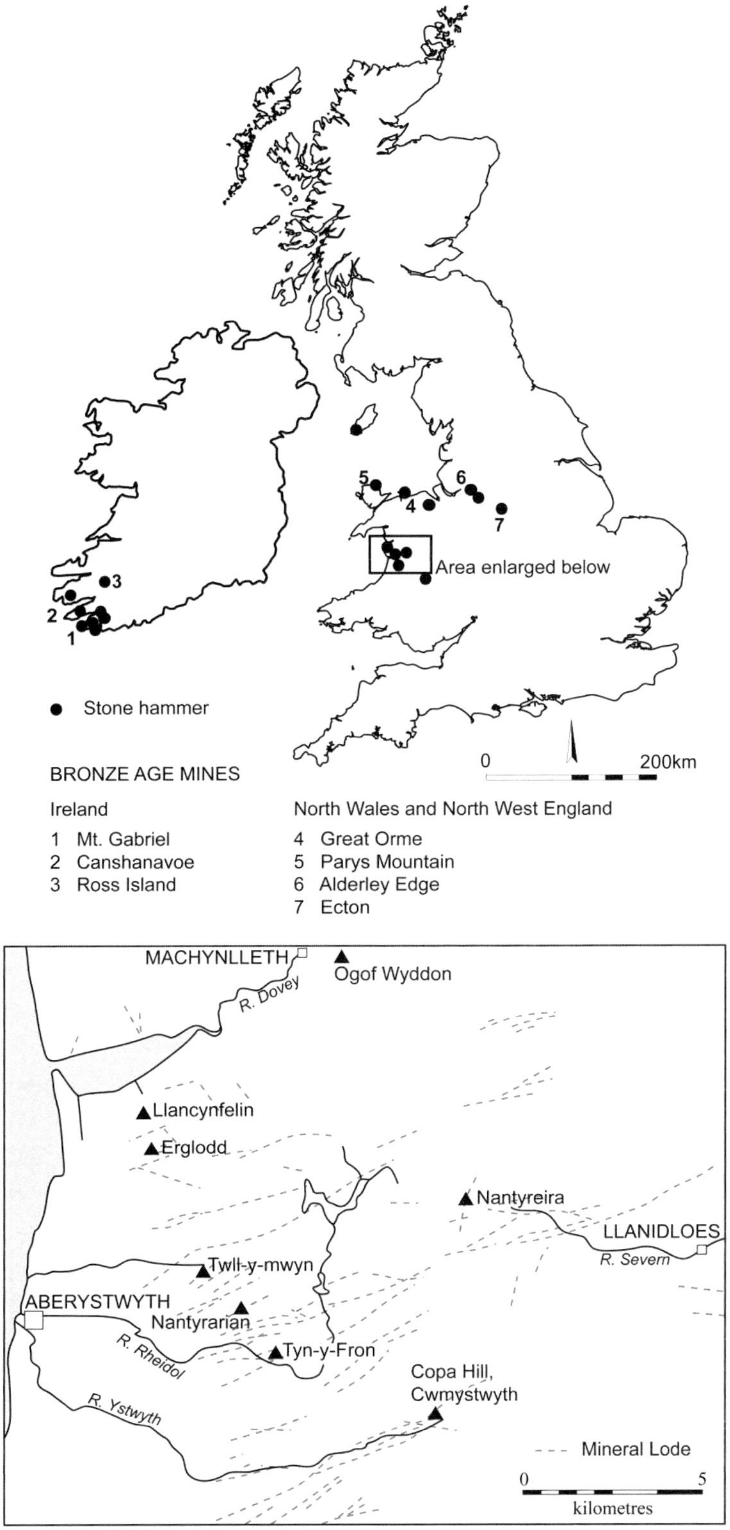

Fig. 29.1: Map of British Isles showing occurrence of stone mining hammers and locations of Bronze Age copper mines. Inset map shows the group of eight Bronze Age mines within Central Wales (drawing B. Craddock).

Given that we know that mining in southwest Ireland began around 2400 BCE at the Ross Island Mine, Killarney (O'Brien 2004), it would seem reasonable to view the current pattern of sites as simply an eastward migration of copper prospecting and mining into mainland Britain following the decline of production at the former site, yet in reality this pattern is far from simple.

BRONZE AGE COPPER MINES

ISLE OF MAN

We find evidence for copper prospection on the Isle of Man, at a location close to one of the shortest sea crossings between Ireland and Britain. Traces of this exploitation are found on the stunning cliffs of Bradda Head, where stone hammers were first found in 1989 (Pickin and Worthington 1989; Doonan and Hunt 1999). The mineralization here can be seen far out to sea, much as it would have been when Bronze Age peoples crossed in their plank boats or perhaps skin *curraghs* from Ireland. The presence of this "metal route" seems all the clearer when we examine the Copper Age/Early Bronze Age metalwork from the island, which shows affinities with both Irish and early Welsh and Scottish (Migdale) axes, along with the mixing (or recycling) of Ross Island metal with copper from other sources (Rohl and Needham 1998; Davey *et al.* 1999).

NORTH WALES

On the island of Anglesey just off the North Welsh coast, evidence of Early Bronze Age copper mining has been found in at least four to five locations underground on Parys Mountain (Fig. 29.2), where David Jenkins has been working since the early 1990s (Jenkins 1995, 2002). The first miners here may have worked in opencast drifts dug from the surface to depths of over 20 m with the help of fire setting and stone tools, mining small shoots of oxidized minerals along faults close to the junction with the sulphide zone (Timberlake 2010). In the 18th century CE when the deposit was rediscovered, lumps of native copper weighing up to 15 kilos were found close to the surface (Lentin 1800), most probably near the contact of the gossan/supergene zone.

Some 20 km to the southeast of Parys Mountain, along the North Welsh coast, lie the important Bronze Age mines located on the Carboniferous Limestone headland of the Great Orme. These form the largest Bronze Age copper mines in the British Isles, and are arguably among the largest in Europe. Within the area of the main opencast in the Pyllau Valley close to the summit of this headland a stock work of intersecting copper veins carrying oxidized ores (mostly malachite together with goethite in dolomite and calcite) within a weathered and partly-rotted dolomitic limestone were worked by Bronze Age miners by means of an opencast linked to several kilometers of narrow galleries and larger chambers underground extending to depths of over 30 m (Lewis 1990; Dutton and Fasham 1994). Estimates of Bronze Age copper production here range from 30 to nearly 1800 tons of metal in total (Lewis 1996; Timberlake 2009: 97).

CENTRAL WALES

Copa Hill, Cwmystwyth

Approximately 60 km due south of Parys Mountain and the Great Orme is the mid-Wales group of mines, chief among which is the Comet Lode opencast on Copa Hill, Cwmystwyth (Fig. 29.3). This is a good example of a small to medium sized Bronze Age mine that is also unique in terms of its more or

Fig. 29.2: Bronze Age mining site underground on Parys Mountain, Anglesey, north Wales. David Jenkins excavating (photo ST).

Fig. 29.3: View of Copa Hill, Cwmystwyth in the Cambrian Mountains of Central Wales. The Bronze Age opencast shows up in sunlight on the distant hill (photo ST).

less complete survival, and also lack of later disturbance (Timberlake 2002). Its pristine state made it an ideal choice for archaeological investigation; 5000 tons of prehistoric mine spoil cover the hillside below the top of the 10 m+ deep opencast (Timberlake and Switsur 1988).

Some 2.5 m below the surface of the now infilled working, a short fireset gallery with the very well-preserved marks of stone tools in its roof (Fig. 29.4) was uncovered when this site was first excavated in 1989 (Timberlake 1990).

Since then almost a third of the mine has been excavated, and many thousands of stone tools have been recovered and recorded, with a significant percentage of these tools showing a range of utilitarian uses. With the exception of just one or two examples, none were grooved, and only 9% showed some evidence of modification for the purposes of hafting (Timberlake and Craddock 2013). However, more than 41% appear to have been re-used, some up to three or four times, and all for different purposes (*ibid.*: 91). The toolset includes mining hammers, chisels, wedges, crushing stones and ore anvils—and very rarely grindingstones (*ibid.*: 49, Fig. 29.13).

Fig. 29.4: Roof of mine gallery within the Bronze Age mine on Copa Hill. The imprint of the stone tools used in mining can be clearly seen (photo ST).

The waterlogged condition of the mine sediments overlying some of the rock floors within the Comet Lode Opencast on Copa Hill have enabled the survival of many of the objects and tools of daily working life. This includes some of the twisted handles made of hazel that once held hammerstones used as mining tools, fragments of kreel-type baskets used to carry or to wash ore, ropes for lifting bags or for carrying wood, and the remains of antler picks also used as mallets with wedges and stone chisels. When these were discarded, some were also thrown as fuel onto the fires used to break up the rocks (Timberlake 2003).

Clearly, however, the most sensational find from these excavations was the discovery of the hollowed-out 5 m-long log launder (Fig. 29.5), perhaps one of the very earliest examples of mine drainage known. One of three such launders excavated was found *in situ* in its last functional position within the entrance cutting to the mine. The sides reveal the marks of the metal axes used to carve it (*ibid.*: 70, Fig. 75a).

The original intention of these launders may have been to tap and channel away the water coming into the working at a spring located on the southern edge of the opencast. For instance, we know that flooding was a serious problem in the deeper workings, and that this contributed to the eventual demise of the mine in or about 1500 BCE.

Nevertheless, this same water may also have been used for washing and separating the crushed ore within the entrance cutting (working floor) located at the front of the opencast; it was here that some of the ore was separated from the gangue, and perhaps lead minerals from copper, and possibly also the malachite from the chalcopyrite.

Micro-excavation of some of the lenses of crushed mineral present within the Bronze Age mine spoil has revealed considerable amounts of relict chalcopyrite (part of this oxidized to goethite), most of this apparently discarded during the hand-picking and concentration of ore (Jenkins and Timberlake 1997). This evidence supports the idea that secondary copper minerals such as malachite, but perhaps also native copper and copper oxide, could have been the minerals preferentially sought. If this was the case they would have been very thoroughly collected, with the result that little or no trace of these survive today. However, a recent investigation by Alan Williams (see Chapter 28, this volume) has identified small amounts of all of these minerals surviving as a relict ore within the prehistoric tips on

Fig. 29.5: Views of the 4000 year old wooden drainage launder in the entrance cutting to the Bronze Age mine, Copa Hill (photos ST).

Copa Hill. More significantly, wherever we find weathered vein outcrops that were not discovered (or at least worked) in the Bronze Age (such as at Guefron Mine to the east of Plynlimon) malachite and the other secondary copper minerals are more abundant.

On Copa Hill drill cores were taken in 1996 from the sides of the emptied Bronze Age workings. These cores show that the prehistoric miners worked through the relatively oxidized carbonate-quartz vein right up to the solid, unaltered sulphides and galena, then stopped (Jenkins and Timberlake 1997: 33–35).

CENTRAL–NORTHWEST ENGLAND

Alderley Edge

Early in the second millennium BCE people were also exploiting the malachite and azurite impregnated sandstones and shale bands present within the Triassic (Bunter) sandstones of Alderley Edge in Cheshire, a site close to present day Manchester (Timberlake and Prag 2005). This mine lies more than 100 km to the east of the Great Orme and the Central Wales Orefield.

A Bronze Age wooden shovel was recovered from here in the 19th century (Sainter 1878), while the sometimes complex-grooved stone mining hammers from this locality are unique within the British Isles (Timberlake 2005a). Pit workings containing grooved stone tools were found at Brynlow in 1874 (Dawkins

1875) and at Engine Vein in 1901 (Roeder 1901), while another undisturbed example was discovered and archaeologically excavated close to the southern edge of Engine Vein in 1997 (Fig. 29.6) (Timberlake and King 2005). This has provided us with an interesting record for the prospection for copper dating from around 1900 BCE, and the subsequent use of this pit for ore processing, and then finally as a hearth around 1700 BCE.

Within this and some of the neighboring emptied-out pit workings on Engine Vein it is still possible to see the imprint of the use of these stone tools within the baritized sandstone, while the base of the excavated pit preserves the soot stains from fire setting. In its base a layer of placed cobbles was found, which included some neatly broken and deposited fragments of worn stone tools (Timberlake 2010). The mines of Alderley Edge may have produced as little as one to two tons of copper metal during the Early Bronze Age.

Ecton Copper Mine

The most recent archaeological work carried out by the EMRG has focused on the survey and excavation of several small prehistoric workings which form part of the Ecton Copper mines in northwest Staffordshire, the latter located on the edge of the limestone Peak District of Central England. Here three possible sites of Bronze Age extraction for copper carbonate ores were identified on the mineralized outcrops close to the summit of Ecton Hill, two of which, Stone Quarry Mine and The Lumb, were investigated in 2008 and 2009. On the better preserved Lumb, one of at least four short bedding plane opencuts was partially

Fig. 29.6: View of Engine Vein, Alderley Edge, Cheshire in England. On the left hand side of the opencut the remains of cut-away Bronze Age pit workings can be seen (photo ST).

excavated to reveal a series of small working hollows mined using hand-held hammerstones, bone picks, scrapers and antler points (Timberlake 2010). The discovery of an Early Bronze Age flint scraper nearby was a rare find for this site.

The bone tools from here have been dated and used to model the periodicity and sequence of mining within this series of hilltop sites. Some of these were recovered from the spoil of a 17th century re-working of the prehistoric Stone Quarry Mine, while the remainder come from re-deposited as well as original contexts that have been excavated on The Lumb. The dating of these mining tools suggest a main period of working between 1800–1700 BCE.

RADIOCARBON DATING

Following the fieldwork we have established a research framework to help guide future investigations and link the results of this with future ore and metal provenancing studies (see Alan Williams, Chapter 28, this volume). An equally important role has been to provide a more meaningful analysis of the radiocarbon chronologies. The latter study has proved extremely important on account of the absence of cultural and chronologically meaningful finds (such as pottery) at the mines themselves, alongside the negative evidence for associated smelting and metalworking sites. Because of the lack of any chronological phase markers, the date ranges for these sites have been both broad and overlapping, the latter covering many hundreds of years, with quite uncertain start and end dates for the mining activity.

The current dating project by Peter Marshall has to some extent helped to resolve these difficulties, and while still probabilistic, the trends presented here are generally more meaningful. The resulting model charts a phenomenon of indigenous exploitation which appears to follow the adoption of metallurgy spreading from mainland Europe through Ireland into Britain.

BAYESIAN MODELLING

In order to identify the accurate measurements (i.e., where the radiocarbon concentration in the sample has been accurately measured) and the accurate dates (i.e., those accurate measurements from samples with good taphonomic provenance) an assessment of the existing 91 radiocarbon determinations from early mining activity in Wales and England was undertaken in 2012.

Following this assessment a series of site-based Bayesian models were constructed for interpreting the chronology from the following sites: Alderley Edge, Copa Hill, Ecton, Parys Mountain and the Great Orme.

This approach (which is encapsulated by Bayes' theorem) is fundamentally probabilistic and contextual, and incorporates both our existing knowledge as well as our data (or "posterior beliefs") (Bronk Ramsey 2009).

The Bayesian model for Ecton incorporates radiocarbon dates (shown in outline) from two areas of the mine working. The samples are all from fragments of tools used for mining (and therefore have excellent taphonomic integrity), although there is no stratigraphic relationship between them, the "prior belief" incorporated into our model is that they are all related. This model therefore provides a series of *estimates* for the dates of the tools used for mining (shown as solid distributions), and for archaeologists more importantly, an estimate of the beginning and ending of activity – i.e., *start, The Lumb.*

If one assumes that mining might have shifted from one of these locations to the other, then there is a good correlation here with the model which suggests that Stone Quarry tools are older than those on The Lumb, therefore the duration of mining within the deeper workings at Stone Quarry may have been greater (between 30 and 120 years) than on The Lumb (between 30 and 75 years). The probability of course is that mining was much shorter than this at both locations. Interestingly, there seems to have been a short recurrence of mining or prospecting activity on The Lumb sometime during the Middle Bronze Age.

A model for the currency of early mining in Wales and England is presented here which has been constructed by taking estimates for the start and end of mining activity at individual sites and then combining these with some of the dates from these sites. This model shows good overall agreement between the radiocarbon dates and the prior information, suggesting that this activity most probably started in the last quarter of the third millennium cal. BCE.

British mines fall into three spatially distinct groups—mid-Wales (the Plynlimon area), the north Wales coast and northwest central England. By estimating the first dated mining activity within each of these areas (i.e., *First mid-Wales*) an evaluation of the temporal spread of mining activity can be determined. It is clear from this that the earliest dated mining activity took place in the mid-Wales group of mines (80% probability).

The spread of mining activity across Wales and England might best be depicted in Fig. 29.7. It shows the most likely order for the start of mining activity in these three areas as follows: mid-Wales>north Wales coast> northwest central England> =43.3% probability. The statistical trend shows an obvious movement eastwards from mid-Wales to England over a relatively short period of time. When compared with the similarities in mining methods and tools used, this suggests that a mutual knowledge and experience once linked the miners of mid-Wales and central England.

Currently all of the Irish mine dates are in the process of being amalgamated into a chronological model for the British Isles. Although it is early to draw a finite conclusion, the evidence would seem to suggest that mining and the search for metals migrated to mainland Britain during the most active period of exploitation of the Ross Island mine, rather than just being a consequence of its demise; i.e., it would appear that the Mount Gabriel-type mines appeared somewhat later than the transmission of mining expertise and prospection interests to west-central Wales.

Recent re-analysis of the dates for the Great Orme Mines suggests that mining began there around 1700 BCE but then had its most active period around 1500–1300 BCE, when most of the copper was being produced. Some of the latest Middle Bronze Age–Late Bronze Age dates appear to come from the deepest levels in the mine.

DISCUSSION AND CONCLUSIONS

The adoption of this approach of pan-regional fieldwork to the investigation of prehistoric mining in Britain has met with some success in helping to provide sufficient dating evidence and archaeological data to understand the ore deposits, the ores mined, the technology adopted, and just as important, the spread of indigenous mining and extractive metallurgy eastwards across Britain from the end of the third millennium to the middle of the second millennium BCE.

Bronze Age copper mines have not yet been identified within the Lake District (Cumbria), in Cornwall, or in Scotland (although a grooved stone "mining tool" was found at the Wanlockhead Mine in 1929 (Pickin 2008) and others are alluded to in earlier 19th century descriptions (see Hunter 1884). Furthermore, prehistoric tin extraction in southwest England (judging from the finds of antler picks and other objects within later "stream works" (Penhallurick 1986) appears almost exclusively to have been alluvial, and on account of this it seems likely that many of the sites would have been destroyed by later re-working. It is possible therefore that we might be dealing with the totality of the surviving evidence. A summary of what we currently know might thus provide a useful and moderately reliable account of the "story" of early mining and the beginnings of metallurgy in the British Isles.

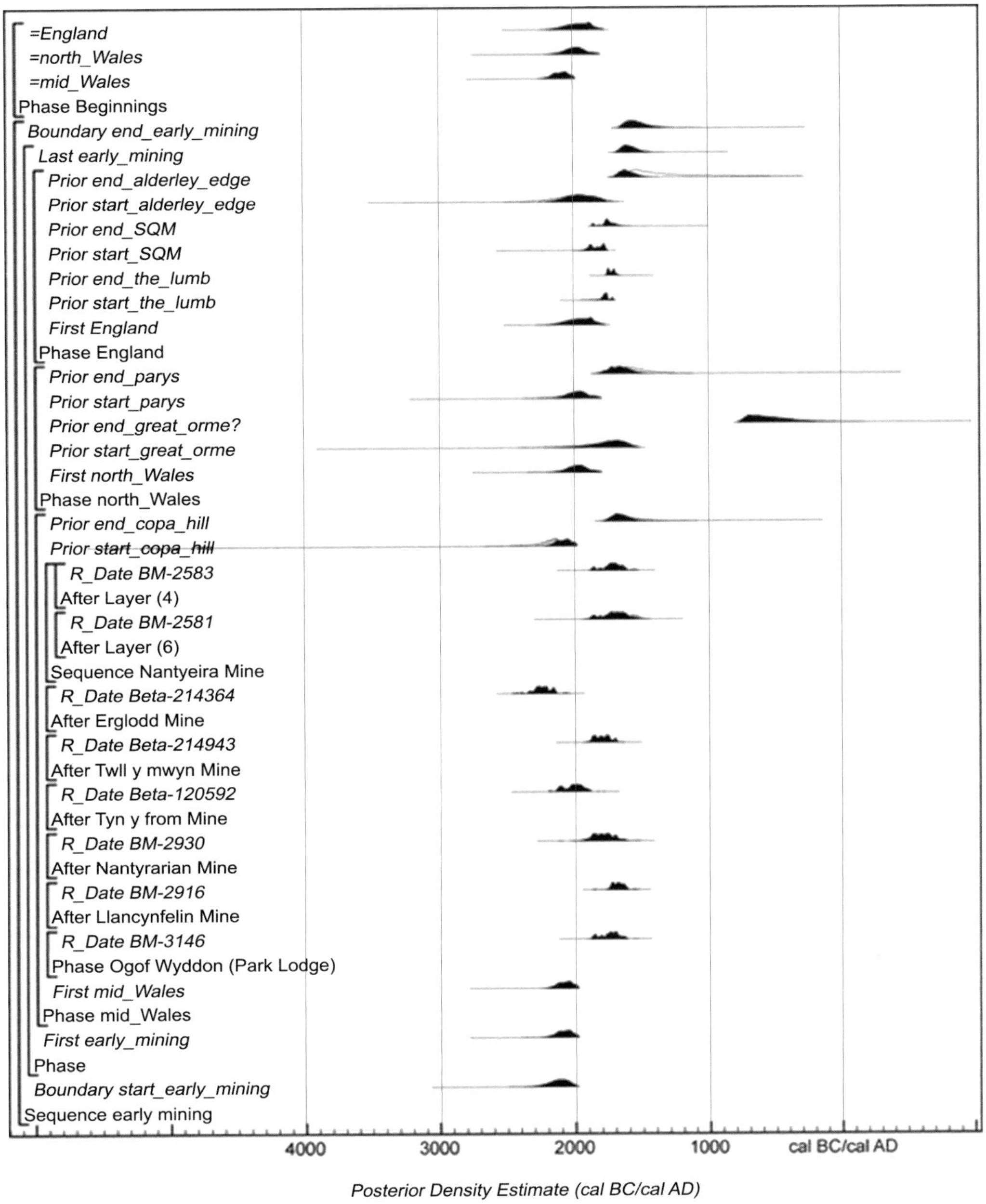

Fig. 29.7: Bayesian modelled dates (posterior density estimates shown in black) for the main British Bronze Age mines. The changes in start and end dates indicate the probable timewise shift in prospection and mining: Central Wales >North Wales> Central England (P. Marshall, English Heritage).

THE SEARCH FOR COPPER

Our re-analysis of the radiocarbon chronologies of the mines suggests that early prospecting for new copper sources in Britain begins before the decline of the Ross Island Mine.

Central Wales appears to have been one of the earliest areas of mining interest in Britain, suggesting trans-Irish Sea routes across to southwest Wales or north Wales via the Isle of Man, and along the coast to the area of the Dyfi Estuary (and the hinterland of Plynlimon).

Prospectors or perhaps just the ideas then spread rapidly northwards and also eastwards into north Wales and central-northwest England.

Possibly we are looking at two phases or waves of prospection; a movement from southwest Ireland to central Wales, north Wales then northwest England, or simultaneously from Ireland to central Wales and Ireland to north Wales, then to northwest England.

The Great Orme was probably discovered early on, at the very beginning of the second millennium BCE, but it was not worked on any significant scale until later.

The Earliest Mines

Both at Alderley Edge (campsites and flint tools) and at Copa Hill (some early radiocarbon dates associated with surface working) there is some minor evidence for a Mesolithic-Neolithic interest in minerals—possibly for pigments. At Alderley Edge some of the mineralized soft sandstones are both varied and strongly colored (Timberlake and Prag 2005; Timberlake 2009).

There may well have been a Beaker "association" with the initiation of metal mining at a number of sites (e.g., Ross Island, Copa Hill and Alderley Edge). The Banc Tynddol Bell-Beaker-type gold foil disc was found at the foot of the mineral vein on Copa Hill (Timberlake 2009: 102–104). The date of the latter find may be consistent with the very earliest metal mining taking place at this mine (i.e., 2200–2100 BCE).

The earliest Bronze Age miners were perhaps pastoralists engaged in transhumance agriculture. They appear to have been working on Copa Hill mostly in the summer–autumn months, and at the beginning of each season they may have brought animals and fresh stone tools with them from the coast (Timberlake 2002). Wood for fuel came from the clearance of new pastures.

The copper ore may never have been smelted by the miners, which might explain the lack of smelting sites. In fact, rich, hand-picked ore (concentrate) may even have been traded for metal, simply as a means to enter the exchange economy.

By 1600–1500 BCE most of these upland mines (including Parys Mountain) were probably flooded and abandoned.

Copper Mining and Production in the Middle–Late Bronze Age

The Great Orme Mine appears to have been the main source of British copper between 1600–1400 BCE. Here miners may have lived in or close to the mine; they were probably supplied with meat and tool material, and working may have been continuous (Lewis 1996; James 2011)

The deepest workings on the Great Orme suggest that some mining took place in the Late Bronze Age, perhaps reaching the water table and the sulphide zone. By now metal was coming from other (Continental) sources.

Metal Production

From 2300 to 1800 BCE the arsenical signature of Ross Island-sourced metal within British copper and bronze artifacts becomes weaker as this metal is recycled and local metal is added (Bray 2012).

Lead isotope analysis can sometimes be problematic with regard to ore/metal provenancing studies in the British Bronze Age. There would appear to be an issue of overlap in lead isotope signatures for potential British ore deposits, while one also needs to consider the problem of "pooling of metal" from a number of different mined sources (Timberlake 2009; Bray 2012).

It now seems likely that much of the smelting of ore probably took place away from the mines themselves (Timberlake 2005b).

Increasingly we are realizing that it is important to understand metalworking processes and also original compositions of ore and metals through experiment. Recent work is helping to frame models for smelting and simple metal artifact production as it might have been done in the Early Bronze Age (Timberlake 2005b, 2007).

Great Orme metal would appear to be the main source of copper used in the important Middle Bronze Age Acton Park metalwork industry (1500–1400 BCE) which has a North Wales origin (Rohl and Needham 1998 [see also Williams this volume]).

New metal from Alpine sources (perhaps with its origin in the deeper-mined chalcopyrite being exploited on the Mitterberg and at other sites) reaches the UK as recycled scrap from 1400 BCE onwards.

ACKNOWLEDGMENTS

We would like to thank David Jenkins and Andy Lewis for providing us with the ^{14}C data for Parys Mountain and the Great Orme, and John Meadows (formerly English Heritage) for the original analyses of the Ecton dates. Alan Williams provided us with useful discussion on the subject of ores mined, while Brenda Craddock helped with the analysis of hammerstones and also the production of some of the graphics used in this paper. I am grateful to the entire team of the Early Mines Research Group and the many people who made these excavations and research projects possible.

REFERENCES

Bray, P. 2012. Before 29Cu Became Copper: Tracing the Recognition and Invention of Metalleity in Britain and Ireland during the Third Millennium BC. In: Allen, M.J., Gardiner, J. and Sheridan, A., eds. *Is There a British Chalcolithic? People, Place and Polity in the Late Third Millennium* (Prehistoric Society Research Paper 4). Oxford: 56–70.

Bronk Ramsey, C. 2009. Bayesian Analysis of Radiocarbon Dates. *Radiocarbon* 51: 337–360.

Crew, P. and Crew, S., eds. 1990. *Early Mining in the British Isles: Proceedings of the Early Mining Workshop at Plas Tan y Bwlch, 17–19 November, 1989* (Plas Tan y Bwlch Occasional Paper 1). Plas Tan y Bwlch, Snowdonia National Park, Wales.

Davey, J., Northover, P., O'Connor, B. and Woodcock, J. 1999. Bronze Age Metallurgy on the Isle of Man: A Symposium. In: Davey, P., ed. *Recent Archaeological Research on the Isle of Man* (British Archaeological Reports [British Series] 278). Oxford: 39–60.

Dawkins, W.B. 1875. On the Stone Mining Tools from Alderley Edge. *Proceedings of the Manchester Literary and Philosophical Society* 14: 74–79.

Doonan, R. and Hunt, A. 1999. Assessing the Research Potential for Prehistoric Mining and Metallurgy on the Isle of Man. In: Darvill, T., ed. *Billown Neolithic Landscape Project, Isle of Man: 5th Report 1999* (Bournemouth University School of Conservation Sciences Research Report 7). Bournemouth and Douglas: 64–69.

Dutton, L. and Fasham, P. 1994. Prehistoric Copper Mining on the Great Orme, Llandudno, Gwynnedd. *Proceedings of the Prehistoric Society* 60: 245–286.

Hunter, J.R.S. 1884. The Silurian Districts of Leadhills and Wanlockhead and Their Early and Recent Mining History. *Transactions of the Geological Society of Glasgow* 7: 372–392.

James, S.E. 2011. *The Economic, Social and Environmental Implications of Faunal Remains from the Bronze Age Copper Mines at Great Orme, North Wales* (Ph.D. dissertation, University of Liverpool). Liverpool.

Jenkins, D.A. 1995. Mynydd Parys Copper Mines. *Archaeology in Wales* 35: 35–37.

Jenkins, D.A. 2002. *Potential Investigations on Mynydd Parys, October 2002: ^{14}C Dates* (Unpublished Report). Bangor, Wales.

Jenkins, D.A. and Timberlake, S. 1997. *Geo-Archaeological Research into Prehistoric Mining for Copper in Wales* (Unpublished Report). Bangor, Wales.

Lentin, A.G.L. 1800. *Briefe über die Insel Anglesea*. Leipzig.

Lewis, A. 1990. Underground Exploration of the Great Orme Copper Mines. In: Crew, P. and Crew, S., eds. *Early Mining in the British Isles: Proceedings of the Early Mining Workshop at Plas Tan y Bwlch, 17–19 November, 1989* (Plas Tan y Bwlch Occasional Paper 1). Plas Tan y Bwlch, Snowdonia National Park, Wales: 5–10.

Lewis, A. 1996. *Prehistoric Mining at the Great Orme: Criteria for the Identification of Early Mining* (M.A. thesis, University of Wales). Bangor, Wales.

O'Brien, W. 2004. *Ross Island—Mining, Metal and Society in Early Ireland* (Bronze Age Studies 6). Galway.

Penhallurick, R. 1986. *Tin in Antiquity*. London.

Pickin, J. 2008. A Grooved Stone Tool from Wanlockhead—Loom Weight or Prehistoric Mining Hammer? *Addenda Antiquaria. Transactions of the Dumfriesshire and Galloway Natural History and Antiquarian Society* 82: 141–143.

Pickin, J. and Worthington, T. 1989. Prehistoric Mining Hammers from Bradda Head, Isle of Man. *Bulletin of the Peak District Mines Historical Society* 10(5): 274–275.

Roeder, C. 1901. *Prehistoric and Subsequent Mining at Alderley Edge, with a Sketch of the Archaeological Features of the Neighbourhood.* Transactions of the Lancashire and Cheshire Antiquarian Society 19: 77–118.

Rohl, B. and Needham, S. 1998. *The Circulation of Metal in the British Bronze Age: The Application of Lead Isotope Analysis* (British Museum Occasional Paper 102). London.

Timberlake, S. 1990. Excavations and Fieldwork on Copa Hill, Cwmystwyth, Dyfed, 1989. In: Crew, P. and Crew, S., eds. *Early Mining in the British Isles: Proceedings of the Early Mining Workshop at Plas Tan y Bwlch, 17-19 November, 1989* (Plas Tan y Bwlch Occasional Paper 1). Plas Tan y Bwlch, Snowdonia National Park, Wales: 22–29.

Timberlake, S. 2002. Mining and Prospection for Metals in Early Bronze Age Britain—Making Claims within the Archaeological Landscape. In: Brück, J., ed. *Bronze Age Landscapes: Tradition and Transformation*. Oxford: 179–192.

Timberlake, S. 2003. *Excavations on Copa Hill, Cwmystwyth (1986–1999): An Early Bronze Age Copper Mine within the Uplands of Central Wales* (British Archaeological Reports [British Series] 348). Oxford.

Timberlake, S. 2005a. Stone Mining Tools from Alderley Edge: A Re-examination of Hammer-stones in the Collections of The Manchester Museum and from Archaeological Excavations on Engine Vein (1997). In: Timberlake, S. and Prag, J., eds. *The Archaeology of Alderley Edge—Survey, Excavation and Experiment in an Ancient Mining Landscape*. London: 58–78.

Timberlake, S. 2005b. In Search of the First Melting Pot. *British Archaeology* 82: 32–33.

Timberlake, S. 2007. The Use of Experimental Archaeology/Archaeometallurgy for the Understanding and Reconstruction of Early Bronze Age Mining and Smelting Technology. In: La Niece, S., Hook, D. and Craddock, P., eds. *Metals and Mines: Studies in Archaeometallurgy.* London: 27–36.

Timberlake, S. 2009. Copper Mining and Metal Production at the Beginning of the British Bronze Age. In: Clark, P., ed. *Bronze Age Connections*. Oxford: 956–122.

Timberlake, S. 2010. Geological, Mineralogical and Environmental Controls on the Extraction of Copper Ores in the British Bronze Age. In: Anreiter, P., ed. *Mining in European History and its Impact on Environment and Human Societies: Proceedings for the 1st Mining in European History—Conference of the SFB-HIMAT, 12–15 November, 2009, Innsbruck*. Innsbruck: 289–296.

Timberlake, S. and Craddock, B. 2013. Prehistoric Metal Mining in Britain: The Study of Cobble Stone Mining Tools based on Artefact Study, Ethnography and Experimentation. *Chungara: Revista de Antropología Chilena* 45(1): 33–59.

Timberlake, S. and King, C. 2005. Archaeological Excavations at Engine Vein, Alderley Edge, 1997. In: Timberlake, S. and Prag, J., eds. *The Archaeology of Alderley Edge—Survey, Excavation and Experiment in an Ancient Mining Landscape* (British Archaeological Reports [British Series] 396). Oxford: 33–57.

Timberlake, S. and Prag, A., eds. 2005. *The Archaeology of Alderley Edge—Survey, Excavation and Experiment in an Ancient Mining Landscape* (British Archaeological Reports [British Series] 396). Oxford.

Timberlake, S. and Switsur, R. 1988. An Archaeological Investigation of Early Mineworkings on Copa Hill, Cwmystwyth: New Evidence of Prehistoric Mining. *Proceedings of the Prehistoric Society* 54: 329–333.

SECTION V

METALWORKING

CHAPTER 30

JUDAH OF IRON VS. ISRAEL OF COPPER: THE METALWORKING DEVELOPMENT IN THE LAND OF ISRAEL AND ITS HISTORICAL IMPLICATIONS

Yulia Gottlieb

The article examines the discrepancies in evidence of metalworking in the northern and southern Land of Israel in the Iron I–IIB. In the south, iron-working was adopted immediately upon the foundation of the first rural Iron Age settlements; throughout the Iron Age the metallurgy in the Beer-sheba Valley and Shephelah remained iron-dominated. In the north, on the other hand, iron was not appreciated for its utilitarian potential for a longer period, and copper-working remained a principal metal industry well into the Iron IIB. The article suggests that the discrepancy in metalworking stems from and reflects the different processes of state formation in the Kingdoms of Israel and Judah. It further discusses the implications of this evidence on comprehending the broader regional phenomenon—the copper production enterprise in the Arabah Valley (in particular at Faynan) and Timna. Issues explored include the relation of the Northern and Southern Kingdoms to the Faynan copper; the role of the Beer-sheba Valley in the copper industry; reconstruction of the copper trade network; and control over the Faynan copper production system.

INTRODUCTION

In a recent article I presented the results of the renewed analysis of the process of introduction and development of iron-working in the Land of Israel in the Iron I–Iron IIB (Gottlieb 2010). This analysis was based on a reexamination of the available data, namely its stratigraphic/chronological contexts, as well as on new, unpublished material, taking into consideration the revision of the Iron I to Iron IIA chronology.[1] The analysis revealed a striking discrepancy in the progress of technological (iron- and copper-working) development in the country as a whole.

Since this publication appeared, new evidence has accumulated, which now allows a better understanding of this phenomenon and its setting in a context of the historical and socio-economical situation in the southern Levant. The present article integrates the new data and offers further insight into this issue.[2]

---

1 The Iron I–IIA chronology, terminology and strata correlation on which the present study is based is that suggested by Herzog and Singer-Avitz (2004, 2006, 2011).

2 New material published since the time of my previous article does not alter the quantitative and qualitative correlation of the data; hence the previously archived conclusions remain valid. Nonetheless, the comparative tables were updated with the new data: for Megiddo: Blockman and Sass 2013; Eliyahu-Behar *et al.* 2013; for Beth-Shean: Yahalom-Mack 2009a, 2009b; for Hazor: Ben-Tor 2012; Yahalom-Mack and Gadot *et al.* 2014; for Tel ʿAroer: Thareani 2011; for Beer-sheba: Paz 2016; for Beth-Shemesh: Bunimovitz and Lederman 2016. Sites which provided a single metal object as well as sites with unclear stratigraphic sequence are not included in tables and maps (see Gottlieb 2010). Briefly mentioned and partly published data on metallurgical evidence from Tel Rehov Strata V–IV, Megiddo Levels Q-5–Q-4 and Khirbat Qeiyafa, which still awaits investigation and full publication (Yahalom-Mack and Eliyahu-Behar 2015) is excluded as well.

THE METALWORK EVIDENCE: A SURVEY[3]

THE NORTH (FIGS. 30.1, 30.2)

The earliest iron objects—jewelry—appeared in the north in the Late Bronze III.

In the late Iron I and early Iron IIA, the iron industry at major northern sites was still at an early stage of development, producing mainly jewelry, knives and small implements. It thus appears that from the time of the first occurrence of iron in the 12th century until the end of the 10th century BCE, the iron industry remained largely ignored.[4] Copper-working remained the principal metal industry in the north during the Iron I and early Iron IIA. This is evidenced both by the amount and diversity of objects and by the extensive remains of metallurgical activity (mainly in the Iron I) (Yahalom-Mack 2009a).

It was not until the late Iron IIA (9th century BCE)—with the emergence of the Omride Kingdom—that iron was actually adopted in the north. Iron agricultural tools appear at most of the sites. The metallographic evidence from the Tell Taanach smithy—the deliberate carburization of iron—demonstrated signs of technological knowledge.

However, copper-working was not interrupted and production continued side by side with the new iron-working industry. At some of the sites (such as Megiddo) copper-based objects still outnumbered iron by a ratio of 2:1. The metallurgical remains from Tel Taanach testify to both iron and copper production at the site. The recently discovered remains of a metal industry from Hazor include products indicative of both copper and iron production in Strata X–IX (Gadot 2012: 541; Yahalom-Mack and Gadot *et al.* 2014)[5]. Similar evidence was recently reported from Megiddo and Tel Rehov (Yahalom-Mack and Eliyahu-Behar 2015, and see n. 2).

In the subsequent Iron IIB period as well, copper/bronze continued to be used extensively for the manufacture of small tools, weapons and jewelry; copper-working was conducted concurrently with iron-working at sites such as Hazor Strata VIII–V (Yahalom-Mack and Gadot *et al.* 2014; see also Yahalom-Mack and Eliyahu-Behar 2015).

THE BEER-SHEBA VALLEY AND NEGEV HIGHLANDS (FIGS. 30.1, 30.3)

The evidence for metalworking revealed at the Beer-sheba Valley sites was not homogeneous.

Tel Masos

In terms of metalwork both Iron Age Strata III and II at Tel Masos reflect the same picture that was revealed at the northern sites, mainly in the late Iron I—a flourishing bronze industry and a disregard for iron. This tendency intensifies in the early Iron IIA Stratum II, with more than a twofold increase in the number of copper-based objects.

The Negev Highland Settlements

The system of early Iron IIA sites in the Negev Highlands provided only copper-based objects (these, however, are very few, n=7).

---

3 The present discussion focuses on the evidence from the northern (Israel) and the southern (Judah) regions. For a detailed discussion of the metalworking evidence from the sites in other regions, see Gottlieb 2010.

4 This disregard is all the more intriguing in view of three isolated examples of iron agricultural tools, which might imply the advent of iron for practical applications already in the Iron I (Yoqne'am XVII, XVI; Megiddo H-7). Moreover, evidence for iron production alongside bronze working, yet to be examined, was recently reported for Megiddo related to the Iron II Levels Q-5-2 (Yahalom-Mack and Eliyahu-Behar 2015: 291–292).

5 Ilan (1999: 230) pointed to a similarity of the copper-working industry from Stratum IX at Hazor to the one found in the Iron I levels at Tel Dan. Yahalom-Mack (2009a: 39) assigns this metallurgical activity to the MB II and the LB II.

The picture that emerged at Arad and Beer-sheba was completely different. Iron first appears there in the late Iron I; Beer-sheba Stratum VIII was characterized by the appearance of the first permanent dwellings! The first Iron Age settlement at Arad (Stratum XII) yielded exclusively iron implements, weapons and tools, including those used for agriculture. Only a few copper-based artifacts—jewelry items—were found in the early Iron IIA strata of Beer-sheba and Arad. The number of iron objects at both sites intensifies in the subsequent strata (late Iron IIA–Iron IIB) with dozens and hundreds of

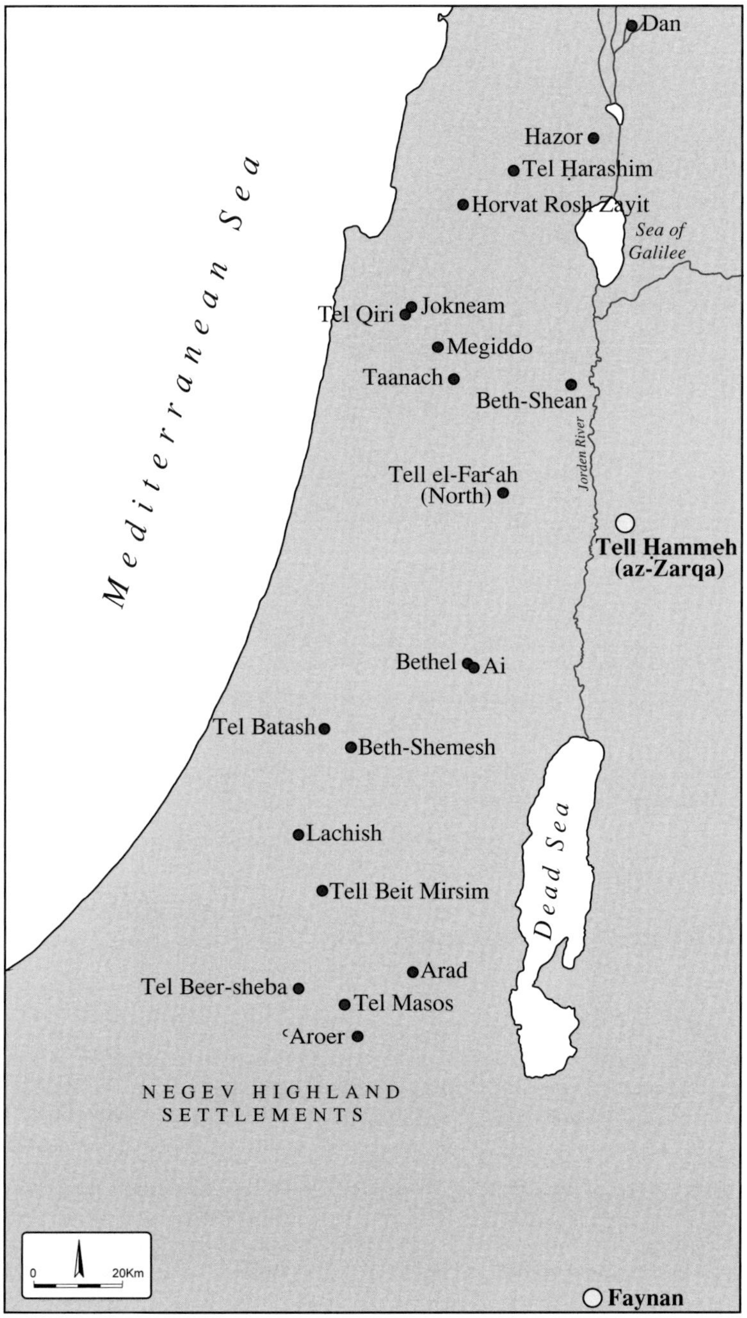

Fig. 30.1: Map with main sites investigated in this study.

utilitarian tools representing a wide repertoire of types, while the use of copper is restricted chiefly to decorative and nonfunctional objects (see Paz 2016). By-products of iron-working activities—slags, pieces of ore and wrought iron—appear at these two sites from the late Iron I (Beer-sheba Stratum VIII) on. Clear evidence for iron production at the site was recently obtained through the metallographic analysis of the residual materials from Strata III–II (Eliyahu-Behar 2016: 1425–1429). The analysis

Site	Copper Total	Copper Agric. tools	Iron Total	Iron Agric. tools	Metallurgical Remains
LATE IRON I					
Megiddo VIA (K-4; H-9; M-4)	ca. 197	+	13	-	Copper-working[1]
Beth Shean VI (S-2)	ca. 64	+	5	-	Copper-working
Hazor XII-XI	ca. 28		1	-	
Yoqneᶜam XVII	6	-	3	+	Copper-working
Taanach IB	12	-	1	-	
Early Iron IIA					
Megiddo VB (K-3; H-7)	7	-	4	+	
Beth-Shean Lower V	1	-	2	-	
	13[2]		9[2]		
Late Iron IIA					
Megiddo VA–IVB (K-2; H-5)	61[3]	-	25[3]	+	
Taanach IIB	2	-	9	+	**Iron**- & Copper-working
Tell el-Farᶜah (N) VIIb	43	-	37	+	
Ḥorvat Rosh Zayit II[4]	3	-	Over 60	+	
Hazor X–IX	11	-	15	+	**Iron**- & Copper-working
Iron IIB					
Megiddo IVA (H-3; L-2)	37[5]	-	24[5]	+	**Iron**-working[6]
Tell el-Farᶜah (N) VIIc–VIId	17	-	17	-	
Hazor VIII–V	23	-	25	+	**Iron**- & Copper-working
Beth-Shean P-8, P-7, P-6	3	-	12	-	

[1]Copper-working activity in Level K-4 was recently questioned, but not ruled out (Eliyahu-Behar *et al* 2013:1281)

[2]From the Double Temples Complex (see Gottlieb 2010: 92-93, n. 9).

[3]Does not include eight copper and four iron objects from Schumacher's excavations.

[4]Note, however, that Ḥorvat Rosh Zayit stands apart from the other northern sites presented above, in view of its distinct cultural affiliation. The fort and its material culture feature strong Phoenician impact and, according to Gal and Alexandre (2000, 2003) might represent a Phoenician royal estate.

[5]Does not include four copper-based and 34 iron tools from Schumacher's excavation from the smithy, in view of unclear stratigraphic affiliation: they may belong either to Stratum IVA or III (Gottlieb 2010: 92, n. 7).

[6]Remains belong either to Stratum IVA or III (*ibid.*).

Fig. 30.2: Metalwork evidence for the north, late Iron I–Iron IIB.

Site	Copper		Iron		Metallurgical Remains
	Total	Agric. tools	Total	Agric. tools	
LATE IRON I					
Tel Masos III	15	+	1	-	Copper-working
Beer-sheba VIII	2	-	1	-	**Iron-working**
Early Iron IIA					
Tel Masos II	ca. 34	-	3	+	Copper-working
Negev Highlands	7	-	-	-	
Arad XII	-	-	8	+	**Iron-working**
Beer-sheba VII	2	-	16	+	**Iron-working**
Late Iron IIA					
Arad XI	1	-	11	+	**Iron-working**
Beer-sheba VI–V–IV	14	-	ca. 64	+	**Iron-working**
Iron IIB					
Arad X–IX–VIII	13	-	58	+	**Iron-working**
Beer-sheba III–II	ca. 77	-	ca. 400	+	**Iron-working**
Aroer IV-III	2	-	35	+	**Iron-working**

Fig. 30.3: Metalwork evidence for the Beer-sheba Valley, late Iron I–Iron IIB.

revealed that both stages of iron-working—primary smithing and forging were conducted at the site in the 9th and 8th centuries BCE! According to Behar (*ibid.*: 1428), iron could be smelted directly at the site. Finally, no traces of copper-working activity, characteristic of the Iron I and IIA sites both in the north and at Tell Masos, have been found at these two sites.

Thus, from the time of the first small rural Iron Age settlements, the metal industry at Arad and Beer-sheba may be characterized as essentially iron-dominated (the same picture was recently observed in the Iron IIB–C strata at Tel ʿAroer: only five copper-based jewelry items and a decorative object were recorded, compared to 64 various iron objects and by-products of iron-working activity (Thareani 2011: 240–241).

THE SHEPHELAH[6] (FIGS. 30.1, 30.4)

Very few iron objects appear in the well-stratified pre-late Iron IIA contexts in the Shephelah (Gottlieb 2010: 97–98).[7] Although scarce, the evidence from Tel Batash (Stratum IV) revealed the

6 Does not refer to the Philistine city of Tell es-Safi/Gath, where remains associated with copper and iron production were recently detected in Stratum A–4 (radiocarbon dated between 935–800 BCE (Eliyahu-Behar *et al.* 2012). First, the site belongs to a different cultural sphere (for the evidence from the other Philistine sites, see Gottlieb 2010: 98–99) and second, no metal objects—either of iron or of copper—were associated with these remains (Eliyahu-Behar *et al.* 2012: 263).

7 Recently, iron and copper-based objects and evidence of copper-working related either to the early Iron IIA or the late Iron I were reported for Khirbat Qeiyafa (Garfinkel, Ganor and Hasel 2011: 8). The full data is not yet published.

predominance of iron over copper; it also includes iron slag, possibly indicating some iron-working activity (Mazar 2001: 217).

The late Iron IIA is marked by a great leap forward in iron-working, with a threefold predominance of iron- over copper-based objects, the appearance of the first iron agricultural tools and the foundation of a blacksmith's workshop at Beth-Shemesh (Level 3),[8] which testifies to a firmly-established iron technology on a significant scale (Bunimovitz and Lederman 2012).

This tendency persists into the following Iron IIB.

NORTH VS SOUTH: SUMMARY AND DISCUSSION

In the late Iron I the metallurgical evidence at both the northern and southern sites is similar and characterized by extensive copper-working, a low ratio of iron to copper and by a meager repertoire of iron objects compared to copper-based ones (Figs. 30.5, 30.6).

The picture revealed in the northern sites in the early Iron IIA remained virtually unchanged, while at the southern sites of Arad and Beer-sheba a completely different picture emerges. By the early Iron IIA the Beer-sheba Valley had already entered full-scale production of iron, which remained the metalworking industry of choice throughout the Iron Age (Figs. 30.7, 30.8).

It would be more reasonable to assume that the well established centers of the north, rather than the newly-founded poor villages of the Negev, would have been the first to adopt the new technology. Indeed, the Canaanite tradition of copper metallurgy that characterizes the northern sites is considered in the factors for the development of iron technology there. However, copper-working was not relinquished in the north even by the late Iron IIA (Figs. 30.9, 30.10) and Iron IIB (Fig. 30.11), when the necessary technological information and experience for effective iron-working had already been acquired.

The technological situation in the north may be interpreted as a "strategy of local optimization" (Renfrew 1984: 404), which is expressed in a clear preference for the tried and tested practice of bronze-working over the new iron-working one. It was influenced by two main factors: The first is *cultural background*, namely cultural conservatism: the local, deeply-rooted Canaanite tradition with its developed bronze industry would be expected to persist longer at the northern sites (in particular Megiddo and Beth-Shean) than it would in other parts of the country. The second is *economic well being*: the environmental settings of the northern sites—favorable climate and fertile soils provided the firm base for their ability to maintain and intensify trade connections with Cyprus, the Phoenician coast and, possibly, the Arabah Valley for copper (see below). This could allow them to continue utilizing the more expensive and highly effective bronze for a longer period.

As for the southern sites, the cultural background of the early settlers of Arad XII and Beer-sheba VII is unknown. What is clear is that immediately upon the foundation of the first Iron Age settlements they adopted the new iron-working technology.

In this respect, the copper-production system at Faynan seems to provide some relevant insights. Levy, Najjar and Ben-Yosef demonstrated that the resumption of copper production at Faynan which was later developed into an industrial-scale international enterprise, was initiated by a local, semi-nomadic tribal society (Levy and Najjar 2006; Levy 2009; Ben-Yosef 2010: 371–372; Ben-Yosef *et al.* 2010: 743–744; 2012: 64–65).

8 For the debated ^{14}C dating of Level 3, see Finkelstein and Piasetzky 2009: 270; Bunimovitz and Lederman 2011: 45-46, n. 7–8.

Site	Copper Total	Copper Agric. tools		Iron Total	Iron Agric. tools	Metallurgical Remains
			Late Iron I			
Beth-Shemesh III (Levels 7-4)	ca. 22	+		ca. 4	-	Copper-working
			Early Iron IIA			
Tel Batash IV	3	-		6	-	**Iron-working?**
			Late Iron IIA			
Lachish IV	11	-		31	+	
Beth-Shemesh IIa-IIb* (Level 3)	10	-		13 350**	+	**Iron-working**
			Iron IIB			
Lachish III ***	ca. 20	-		ca. 170	+	
Tell Beit Mirsim St. A	ca. 10	-		ca. 60	+	
Tel Batash III	-	-		5	-	

* Does not include objects assigned generally to Stratum II (14 copper/bronze and seven iron).

** The renewed excavations revealed "more than 350 iron artefacts and small slags".

***Does not include about 1000 iron arrowheads and armour scales from the destruction level.

Fig: 30.4: Metalwork evidence for the Shephelah, late Iron I–Iron IIB.

Site	Copper Total	Copper Agric. tools		Iron Total	Iron Agric. tools	Metallurgical Remains
			NORTH			
Megiddo VIA (K-4; H-9; M-4)	ca. 197	+		13	-	Copper-working*
Beth-Shean VI (S-2)	ca. 64	+		5	-	Copper-working
Hazor XII-XI	ca. 28	-		1	-	
Yoqneᶜam XVII	6	-		3	+	Copper-working
Taanach IB	12	-		1	-	
			SOUTH			
NEGEV						
Tel Masos III	15	+		1	-	Copper-working
SHEPHELAH						
Beth-Shemesh III (Levels 7-4)	ca. 22	+		ca. 4	-	Copper-working

* Copper-working activity in Level K-4 was recently questioned (Eliyahu-Behar et al 2013:1281)

Fig. 30.5: Comparative table: North vs south in the late Iron I.

It is interesting that the beginning of the iron smelting activities revealed at Tel Hammeh in Jordan, dated to the late 10th century BCE at the latest, coincides with the emergence of the first settlements of Arad and Beer-sheba. Just as with the copper smelting in the Arabah (Ben-Yosef 2010: 964–965; Ben-Yosef *et al.* 2012: 64, n. 16), the evidence for iron smelting at Hammeh reflects a "well-established and substantial operation" rather than an experimental stage, suggesting that iron production was known and practiced in the region even earlier (Veldhuijzen and Rehren 2007: 199).

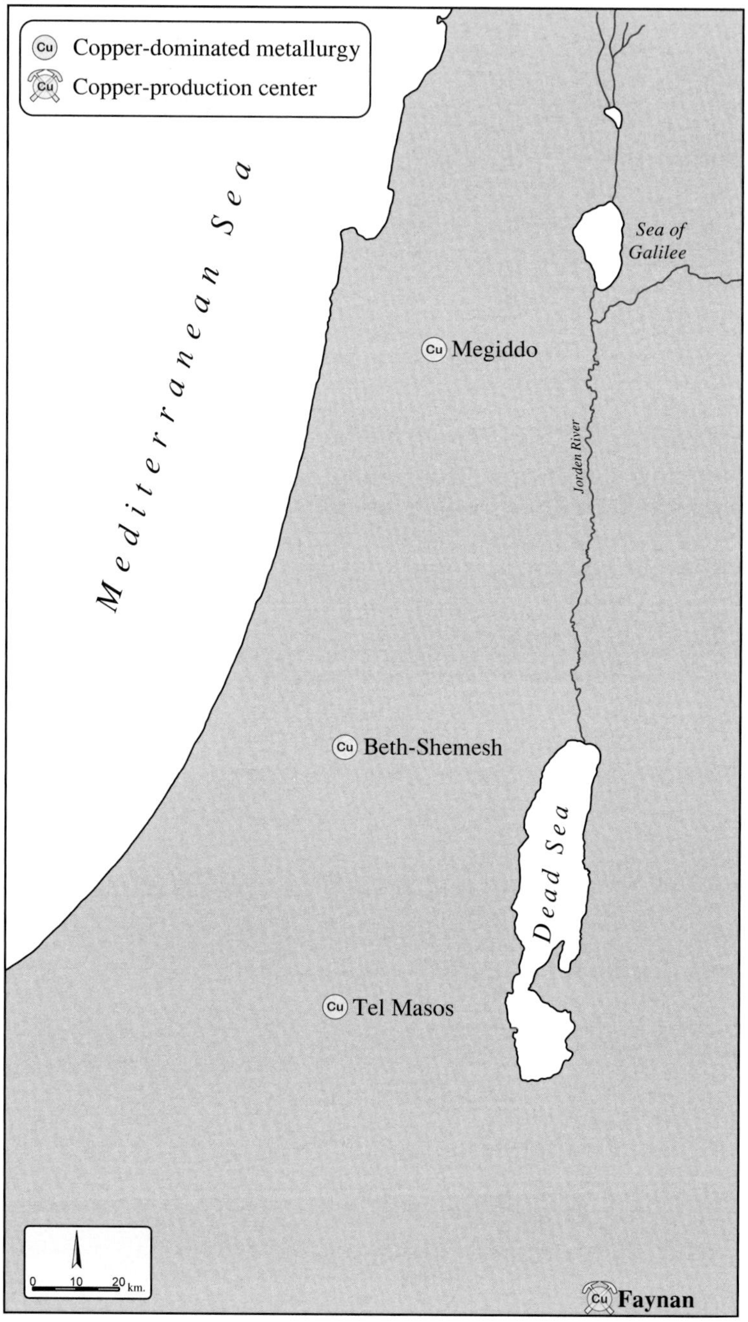

Fig. 30.6: North vs south in the late Iron I.

Site	Copper		Iron		Metallurgical Remains
	Total	Agric. tools	Total	Agric. tools	
NORTH					
Megiddo VB (K-3; H-7)	7	-	4	+	
Beth-Shean Lower V	1	-	2	-	
	13*		9*		
SOUTH					
NEGEV					
Tel Masos II	ca. 34	-	3	+	Copper-working
Arad XII	-	-	8	+	**Iron-working**
Beer-sheba VII	2	-	16	+	**Iron-working**
SHEPHELAH					
Tel Batash IV	3	-	6	-	**Iron-working?**

* From the Double Temples Complex (see Gottlieb 2010: 92-93, n. 9).

Fig. 30.7: Comparative table: North vs south in the early Iron IIA.

It cannot be ruled out that Negev nomads were involved in iron production long before the appearance of their first permanent settlements, in particular at Arad and Beer-sheba.[9]

It may be further argued that the immediate adoption of iron by the early settlers of the Beer-sheba Valley was conditioned by a socio-economic factor. Contrary to the northern sites, which could allow themselves to make a choice in favor of one or another metalworking industries, the primary motive behind the adoption of metal (iron) in the early settlements of Arad and Beer-sheba was a necessity, dictated by their economic insufficiency. It seems highly likely that such a necessity was conditioned by a monopoly on the copper production at Faynan. One would speculate that the emerging new power (later to become the Kingdom of Judah) was deprived of access to the copper supply, which was alternately controlled by the local Edomite polity and, possibly, by Egypt (for foreign vs. local control of copper production, see Levy, Najjar and Ben-Yosef 2014: 977–994). Consequently, it was forced to search for an alternative and developed its own independent metal technology.

ISRAEL OF COPPER VS JUDAH OF IRON: METALLURGY AND STATE

Recently, the study of ceramic assemblages, stratigraphy and architecture of the key northern and southern sites by Herzog and Singer-Avitz (2006) revealed clear differences in the social processes that

9 It is worth noting that a "pocket" of effective iron-working, which could be dated as early as the Iron I, was observed at the sites of Bethel and et-Tell of the Central Hill Country (Gottlieb 2010: 99–100). The early adoption of iron in these small rural settlements might be linked to their close proximity to the Transjordan iron ore (Yahalom-Mack 2009a: 274) and might imply that their knowledge of iron-working was acquired during engagement in (Hammah) iron production activities.

led to state formation in Judah and Israel. The discrepancy in metalworking evidence between the two regions perfectly reflects these variegated processes (Fig. 30.12).

The Northern Kingdom of Israel, which benefited from favorable environmental conditions and economic rewards of western and northern trade (including, in particular, the flow of Cypriot copper), is characterized by an emphasis on ceremonial apparatus, monumental architecture and a "tendency to flaunt" (Herzog and Singer-Avitz 2006: 187). Its continued use of copper/bronze—probably also

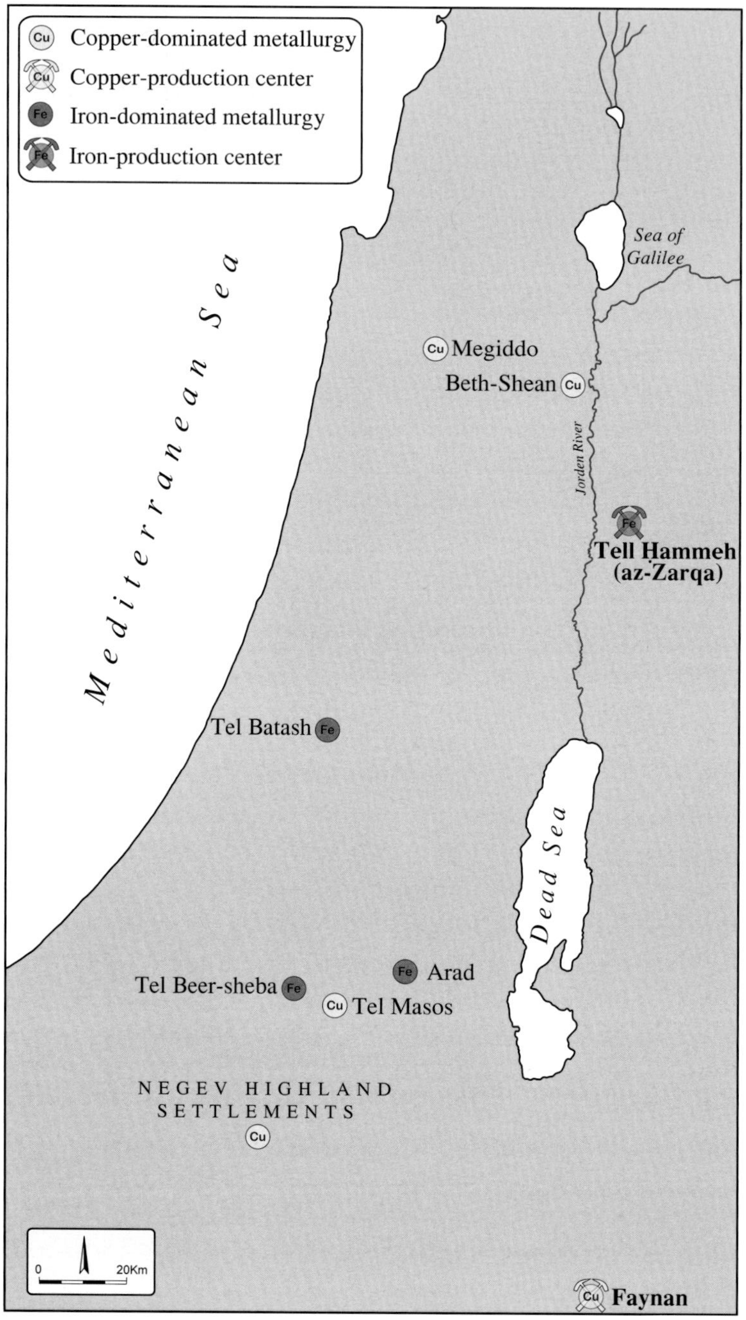

Fig. 30.8: North vs south in the early Iron IIA.

Site	Copper Total	Copper Agric. tools	Iron Total	Iron Agric. tools	Metallurgical Remains
NORTH					
Megiddo VA–IVB (K-2; H-5)	61*	-	25*	+	
Taanach IIB	2	-	9	+	**Iron- &** Copper-working
Tell el-Farah (N) VIIb	43	-	37	+	
Ḥorvat Rosh Zayit II	3	-	Over 60	+	
Hazor X–IX	11	-	15	+	**Iron- & Copper-**working
SOUTH					
NEGEV					
Arad XI	1	-	11	+	**Iron-working**
Beer-sheba VI–V–IV	14	-	ca. 64	+	**Iron-working**
SHEPHELAH					
Lachish IV	11	-	31	+	
Beth-Shemesh IIa-IIb** (Level 3)	10	-	13 350***	+	**Iron-working**

* Does not include eight copper and four iron objects from Schumacher's excavations.

** Does not include objects assigned generally to Stratum II (14 copper/bronze and seven iron).

*** The renewed excavations revealed "more than 350 iron artefacts and small slags".

Fig. 30.9: Comparative table: North vs south in the late Iron IIA.

considered more prestigious than iron—should be viewed as a reflection of prosperity and of high social status.

The Kingdom of Judah, on the other hand, developed as a reaction to harsh conditions resulting from environmental pressure and economic burden following political conflicts and Egyptian campaigns to the region. Judah's concerns were characterized by a military emphasis. The predominance of this concern is indicated starting from the earliest Iron Age settlements (the "enclosed settlements" of Arad XII and Beer-sheba VII) till the erection of fortified cities and fortresses (Beer-sheba V and Arad XI, Lachish IV). Erecting solid city walls—stronger, more labor intensive and more expensive than the casemate walls in the north—also testifies to a deep concern for protection (Herzog and Singer-Avitz 2004: 228–229, 232–233; 2006).

Starting from the early Iron IIA onward, iron-working was the principal metal industry in Judah. The choice of iron—a practical, cheaper metal with easily accessible and controlled resources, was closely connected with, and responded to, all the main goals of the Southern Kingdom. As the state took form, we witness development from the first small-scale iron-working at the Beer-sheba Valley settlements to the well-established industry in the Shephelah with the foundation of the administrative and military centers of Lachish Level IV and Beth-Shemesh Level 3. It is worth noting that the majority of the Beth-Shemesh blacksmiths' products were arrowheads (Bunimovitz and Lederman 2012: 109), which further implies the interconnection between the military concern and the role of iron in Judah.

To conclude this point, the choice and preference of one metalworking industry over the other proved to be tightly connected to diverse economic and cultural factors, as well as to diverse state priorities of the Kingdoms of Israel and Judah.

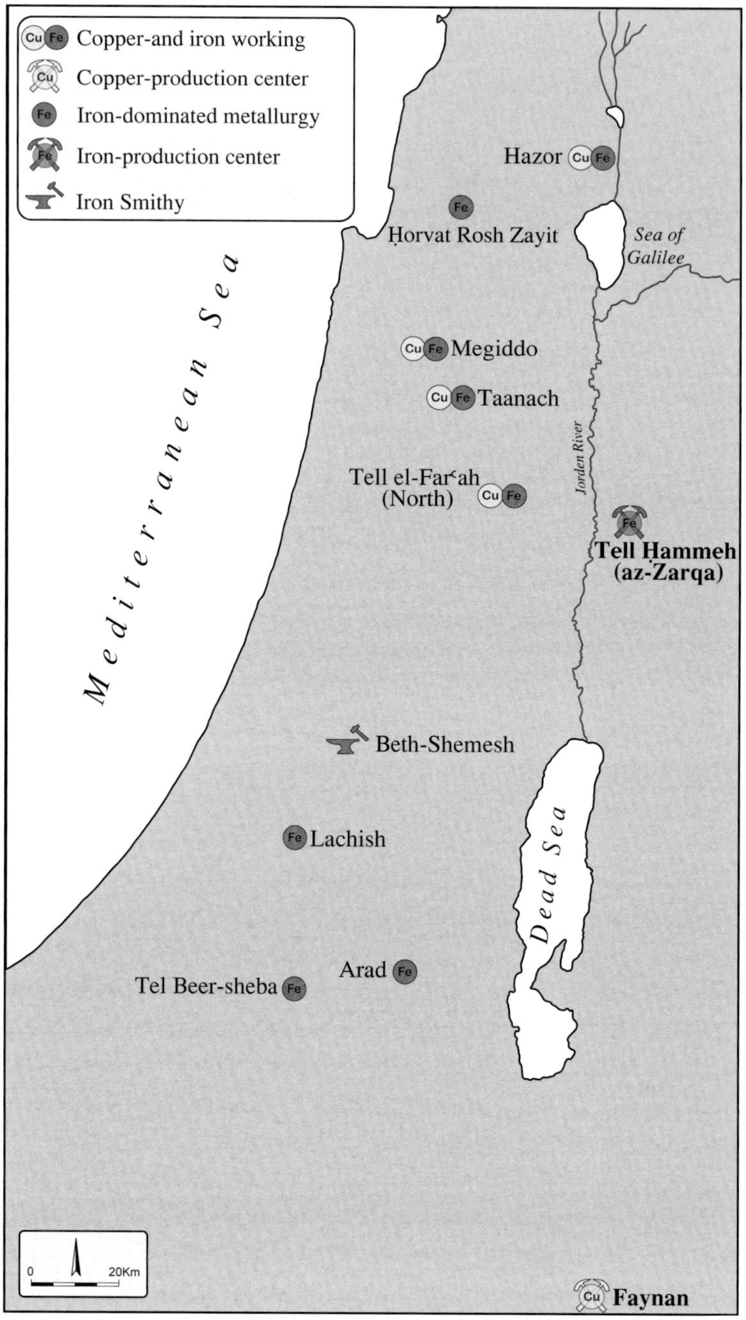

Fig. 30.10: North vs south in the late Iron IIA.

Site	Copper Total	Copper Agric. tools	Iron Total	Iron Agric. tools	Metallurgical Remains
NORTH					
Megiddo IVA (H-3; L-2)	37*	-	24*	+	**Iron-working****
Tell el-Farʿah (N) VIIc–VIId	17	-	17	-	
Hazor VIII–V	23	-	25	+	**Iron**- & Copper-working
Beth-Shean P-8, P-7, P-6	3	-	12	-	
SOUTH					
NEGEV					
Arad X–IX–VIII	13	-	58	+	**Iron-working**
Beer-sheba III–II	ca. 77	-	ca. 400	+	**Iron-working**
ʿAroer IV-III	2	-	35	+	**Iron-working**
SHEPHELAH					
Lachish III ***	ca. 20	-	ca. 170	+	
Beth-Shemesh (Level 2)	3	-	3	-	
Tell Beit Mirsim St. A	ca. 10	-	ca. 60	+	
Tel Batash III			5	-	

* Does not include four copper-based and 34 iron tools from Schumacher's excavation.

** Schumacher's excavation, either Stratum IVA or III (Gottlieb 2010: 92, n. 7).

*** Does not include about 1000 iron arrowheads and armour scales from the destruction level.

Fig. 30.11: Comparative table: North vs south in the Iron IIB.

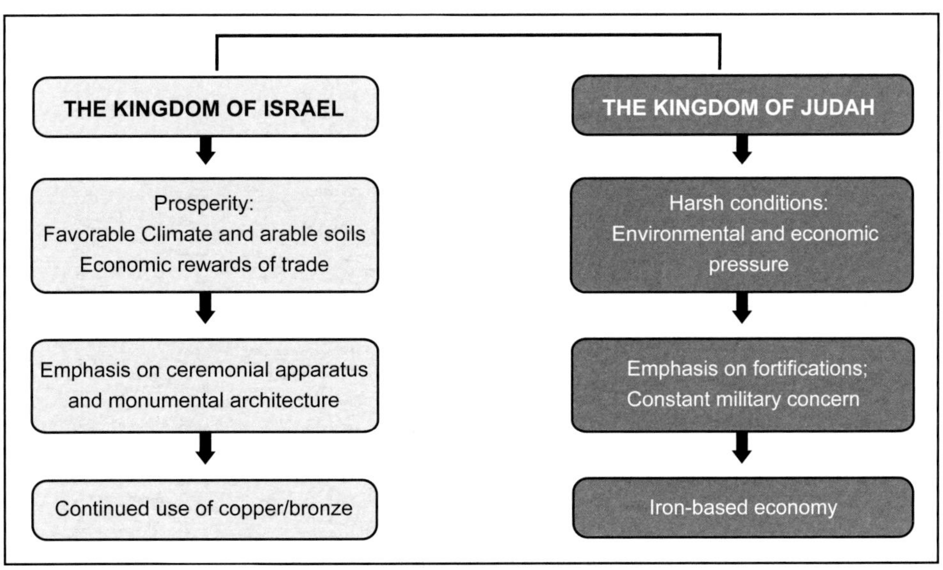

Fig. 30.12: Metalworking and state formation.

ISRAEL, JUDAH AND ARABAH COPPER—A METALWORKING PERSPECTIVE

The above evidence is crucial for understanding the relation of the Northern and Southern Kingdoms to copper production at Faynan. Recently, several scenarios were suggested concerning the possible involvement of these two states with the Arabah copper-production system (Fantalkin and Finkelstein 2006; Finkelstein and Lipschits 2011; Ben-Yosef *et al.* 2012; Na'aman 2013).

According to Na'aman (2013: 261) following the Egyptian withdrawal from the region, the Kingdom of Judah made an attempt to take over the copper trade in the Arabah and the Beer-sheba Valleys in order to take part in its gains. Ben-Yosef and others (2012: 64–65) pointed to a rather abrupt cessation of the copper-production activity at Fayan and raised the possibility of a military conflict with Judah as a possible scenario (the other candidate being Hazael, king of Aram). The obtained metalworking evidence clearly indicates that, contrary to its northern neighbor (as well as to Egypt), Judah was not a consumer of copper. This implies that any interaction with the Edomite polity did not bring Judah access to the copper supply.[10] Judah, which had its own iron-working metal industry would not lose much with the elimination of the Arabah copper-production system.

It should also be emphasized that either an attempt by Judah to seize control over the copper trade or the demolition of the Edomite copper-production system would probably have run counter to the interests of the Northern Kingdom of Israel. The interest of the latter in the supply of copper is well corroborated by the metalworking evidence from the northern sites. The Kingdom of Israel remained a copper consumer well into the Iron IIB and provided a potential market both for Cypriot and Arabah[11] copper.

It has recently been suggested that the supply of copper was a primary motive behind the Omride expansion in Transjordan (Fantalkin and Finkelstein 2006: 29–30; Finkelstein and Lipschits 2011: 148–149). The metalworking evidence supports this hypothesis: like Egypt in the early Iron IIA, gaining access to the Arabah Valley copper supply and control over the trade route could certainly have become a reason for the Omride's involvement in Transjordan.

THE BEER-SHEBA VALLEY AND COPPER PRODUCTION AT FAYNAN

THE "TEL MASOS CHIEFDOM"

Another related aspect concerns a rather paradoxical discrepancy in the metalworking evidence between the neighboring, contemporary settlements of the Beer-sheba Valley themselves; while Tel Masos II showed extensive evidence of copper-working, Arad XII and Beer-sheba VII demonstrated an iron-dominated metal industry. This is especially striking in light of the fact that all three sites are located in close proximity to the Arabah copper-production center (Fig. 30.1).

Finkelstein and others have recently suggested that the beginning of sedentarization in the Iron I and the strong settlement activity in the Beer-sheba Valley (including Tel Masos II, Arad XII and Beer-sheba VII) and in the Negev Highlands in the early Iron IIA were connected to the copper production at Khirbat en-Naḥas. It was further suggested that this system—called the "Tel Masos chiefdom"—administered the Faynan copper production system (Martin and Finkelstein 2013: 9–10, 29–39, with references)

10 It may be assumed that the only benefit to Judah would have been the imposition of taxation on the transportation of copper throughout its territory. Otherwise, one would have to suppose that by succeeding in gaining access to copper, Judah turned this metal into an export commodity while using iron for its local demands.

11 For the Arabah (and in particular Faynan) ores as a source of copper for the northern Land of Israel in the Iron I, see Yahalom-Mack and Segal 2009; Yahalom-Mack 2009a: 188–253; Yahalom-Mack and Eliyahu-Behar 2015:300).

A clear connection between the Negev Highlands' settlements and the Arabah (either Fayan or Timna) copper production system was recently established though the petrographic and typological pottery analysis (Martin and Finkelstein 2013). It clearly demonstrated that their population was "engaged in mining and smelting activities in the Nahal Arabah copper districts." This was further confirmed by a study of camel remains: camels were common in the early Iron Age IIA in the Negev Highlands sites, reflecting their trade connections with the Arabah (Sapir-Hen and Ben-Yosef 2013). This is also corroborated by the fact that this settlement system provided solely copper-based objects (although extremely few [n=7]).

At Tel Masos, both Strata III and II provided many copper-based objects and remains of a copper-working industry; the number of objects even doubles itself in Stratum II, which favors its involvement in the copper trade.[12] However, the supposed role of this site as a trading post for Faynan copper was recently questioned (Yahalom-Mack 2009a: 272–273; Ben-Yosef 2010: 946–947). The examination of the site's metallurgical evidence demonstrated that it "represents at most a small bronze workshop, similar to many others in the Iron Age southern Levant" (*ibid.*).[13] Besides, camel bones, which could reflect the involvement of the site in trade, were not recorded in either of its strata (Sapir-Hen and Ben-Yosef 2013: 281, Table 2). Finally, the Stratum II settlement came to an end when the industrial copper production at Faynan was still at its peak (Levy *et al.* 2004; 2005; Levy, Ben-Yosef and Najjar 2012: 210).

Although the relation of the Negev Highlands settlements to the settlement of Tel Masos II, as well as the relation of the latter to the copper-production at Faynan, are still unclear, they share a common feature—copper-based objects.

As to the settlements of Arad and Beer-sheba, according to the "Tel Masos chiefdom" theory, they must have belonged to the same socio-economic entity as Tel Masos (and the Negev Highlands sites) (Tebes 2003: 68–69); as such they must have been involved in (or affected by) the complex Faynan copper production enterprise. Yet, although located just off one of the supposed main copper routes (see Ben-Yosef 2010: 78, Fig. 2.4; Ben-Yosef, Najjar and Levy 2014: 540–547), they hardly yielded any copper objects at all, and developed an entirely different metal industry instead. Besides, no camel bones—which would imply engagement in trade, and which are evident at Faynan, Timna and in the Negev Highlands settlements—were found at these sites (camels first appeared at Beer-sheba in the Iron IIB Stratum II–when the city became involved in the southern Arabian long distance trade (Singer-Avitz 1999).

One way to explain the discrepancy in metalwork between Tel Masos II and Arad XII and Beer-sheba VII, is that they do not completely match chronologically. Both in terms of its character and its bronze industry, the prosperous settlement of Stratum II of Tel Masos parallels the late Iron I Strata of Megiddo VIA and Beth-Shean Upper VI (Gottlieb 2010: 102–103). It could be assumed that Tel Masos II was founded in the later years of the late Iron I, and that Arad XII and Beer-sheba VII were established somewhat later (and see Mazar 2011: 111, Table 2)[14] (Fig. 30.13).

---

12 A few locally made Egyptian pottery sherds were recently taken as an implication for Egyptian presence in the settlement of Stratum II, which in turn was linked to the involvement of Egypt in the Arabah copper production enterprise (Ben-Dor Evian 2011: 111, 113). For the problematic stratigraphic affiliation of these sherds, see Herzog and Singer-Avitz 2004: 222–223.
An Egyptian pottery vessel was also reported from the contemporaneous Stratum III of Tel Esdar (Ben-Dor Evian 2011: 103–104). It is worth noting that the only metal object found in this stratum was a copper-based item (Kochavi 1969: 33).

13 Yahalom-Mack (2009a) dated the copper-working activity at Tel Masos II to the Iron I.

14 According to his Modified Conventional Chronology, Mazar places both Stratum II of Tel Masos and the later part of Megiddo VIA in the Iron IB (Mazar 2011: 111, Table 2; however, see Na'aman 2013: 261). With this dating, the peak of the copper production system at Faynan occurs only *after* the Stratum II settlement came to an end.

PERIOD	SITE				
Late Iron I	Masos III Masos II ↓		Beer-sheba IX-VIII		COPPER PRODUCTION AT FAYNAN
Early Iron IIA	↓	Arad XII	Beer-sheba VII	Negev Highland settlements	COPPER PRODUCTION AT FAYNAN
Late Iron IIA		Arad XI	Beer-sheba V		COPPER PRODUCTION AT FAYNAN

Fig. 30.13: Chronological scheme: suggested correlation of strata in settlements of Beer-sheba Valley, the Negev Highland and copper production at Faynan.

But even so, the copper production at Khirbat en-Nahas flourished during most of the 9th century BCE, yet the settlements of Arad (Stratum XI) and Beer-sheba (Stratum V)—by now Judean strongholds—continued to maintain the iron-based economy.

Be that as it may, the cumulative evidence of metalworking, pottery and faunal remains leads to the following conclusions: First, throughout the Iron IIA (both early and late) we witness in the south the co-existence of two different and independent systems: the copper-working industry represented by Faynan, Tel Masos II (but see above) and the Negev Highlands sites; and the iron-working industry at the settlements of Arad XII–XI and Beer-sheba (VII–IV).

Second, from the very beginning of their settlement in the Beer-sheba Valley, the sites of Arad and Beer-sheba represented an isolated independent socio-economic entity. They were neither connected to the copper production system at Faynan, nor did they constitute part of the "Tel Masos chiedom."

THE BEER-SHEBA VALLEY AND THE ARABAH COPPER NETWORK

The above arguments have in turn important implications on clarifying two interrelated issues:

1. Reconstruction of the trade routes for Faynan copper.
2. Establishing the source of administration and control over the copper-production industry.

According to the commonly-held assumption, Faynan copper was transported to the southern Coastal Plain, the south Levantine centers and Egypt via the Beer-sheba Valley (Martin and Finkelstein 2013: 9; Yahalom-Mack and Galili *et al.* 2014: 173–174). Indeed, the Beer-sheba Valley functioned as an important artery in the Arabian trade (Ben-Yosef 2010: 942; 946; 78, Fig. 2.4), which was especially prominent in the Iron IIB (Singer-Avitz 1999; Thareani 2011: 305–307). Yet, the evidence from the Iron IIA settlements of Arad and especially from Beer-sheba—the next station along the route to the coast after Tel Masos—contradicts the claim for transportation of the Arabah copper via the Beer-sheba Valley. Besides, it has recently been demonstrated that the southern Coastal Plain could hardly be involved with the "affluent international trade" (Herzog and Singer-Avitz 2011: 172). Hence, it may be deduced that the main bulk of the Faynan copper was transported to Egypt and the north by alternative networks (for trade routes, see Ben-Yosef 2010: 76–79; Ben-Yosef, Najjar and Levy 2014: 540–547). Thus, for example, as Finkelstein and Lipschits recently suggested (2011), the transportation of the Faynan copper to the north—and in particular to the Northern Israelite Kingdom under Omride rule—was conducted via the King's Highway on the Transjordanian plateau.

Furthermore, two principal theories concerning the administration and control of the copper-production system at Faynan were recently examined (Martin and Finkelstein 2013, with references). These were defined as "Copper Networks A"—the Edomite-related control—and "B"— control from Cisjordan operated by the "Tel Masos chiefdom" (*ibid.*).[15] The obtained evidence leaves us with only two regional entities that took part in the Faynan copper processing and transportation: the local Edomite entity in Transjordan and the population of the Negev Highlands settlements in Cisjordan. Considering Na'aman's suggestion that the latter was inherently Edomite (Na'aman 2013: 257), the evidence clearly favors the first theory. It indicates that both copper production and networks were controlled by the chiefdom-based tribal society of Edom. This is not to say that there were no attempts (at times possibly successful) by neighboring kingdoms to gain control over this important economic resource (Levy, Ben-Yosef and Najjar 2012: 212; Na'aman 2013; Levy, Najjar and Ben-Yosef 2014: 977–994). In the Iron I–early Iron IIA, possible candidates for such actions could be the Egyptians whose interest in the Arabah copper supply and involvement (either military or economic/commercial) in the Faynan copper enterprise, is supported by the archaeological evidence (Levy *et al.* 2008; Levy 2009: 161; Ben-Yosef 2010: 371–373, 971–977; Ben-Yosef *et al.* 2012: 47–48, 52, 59, 63–65; Levy, Ben-Yosef and Najjar 2012: 210; Levy, Najjar and Ben-Yosef 2014: 982–990; Munger and Levy 2014: 741–759). In the early and late Iron IIA a possible candidate could be the Northern Kingdom of Israel.

CONCLUSIONS

In this article I explored the evidence for metalworking in the northern and southern regions of the Land of Israel, against the background of historical and socio-economic processes in the region in the Iron I–IIA. I reached several interrelated conclusions:

1. I established a connection between the disparity in metalworking evidence between the north and the south, and the peculiarities of state formation in the Kingdoms of Israel and Judah. The Kindgom of Israel comes across as a constant consumer of copper throughout the Iron Age, while the formation and consolidation of the Kingdom of Judah was closely tied to the development of iron-working metallurgy.
2. Aside from Egypt, I isolated an additional power in the region that was probably a major player in consumption and trade of the Arabah copper—the Northern Kingdom of Israel.
3. The connection of the Beer-sheba Valley to the Faynan copper-production enterprise was proved problematic, as was the character of the supposed "Tel Masos chiefdom" entity itself. My suggestions are: One, this region played only a minor role in transportation of the Arabah copper, and two, the theory of control from this region over the Faynan copper industry cannot be sustained.

15 Recently Levy, Najjar and Ben-Yosef pointed out the problems of recognition of the controlling group in the archaeological record, especially if the later is also a tribal confederation (2014: 982–983).

REFERENCES

Ben Dor Evian, S. 2011. Egypt and the Levant in the Iron Age I–IIA: The Ceramic Evidence. *Tel Aviv* 38: 94–119.

Ben-Tor, A. 2012. Tools and Weapons. In: Ben-Tor, A., Ben-Ami, D. and Sandhaus, D. 2012. *Hazor 6: The 1990–2009 Excavations—The Iron Age*. Jerusalem: 530–541.

Ben-Yosef, E. 2010. *Technology and Social Process: Oscillations in Iron Age Copper Production and Power in Southern Jordan* (Ph.D. dissertation, UC San Diego). San Diego.

Ben-Yosef, E., Levy, T.E., Higham, T., Najjar, M. and Tauxe, L. 2010. The Beginning of Iron Age Copper Production in the Southern Levant: New Evidence from Khirbat al-Jariya, Faynan, Jordan. *Antiquity* 84: 724–746.

Ben-Yosef, E., Najjar, M. and Levy, T.E. 2014. Local Iron Age Trade Routes in Northern Edom. From the Faynan Copper Ore District to the Highlands. In: Levy, T.E., Najjar, M. and Ben-Yosef, E. 2014. *New Insights into the Iron Age Archaeology of Edom, Southern Jordan*, Vol. 2. Los Angeles: 493–575.

Ben-Yosef, E., Shaar, R., Tauxe, L. and Ron, H. 2012. A New Chronological Framework for Iron Age Copper Production at Timna (Israel). *Bulletin of the American School of Oriental Reasearch* 367: 31–71.

Blockman, N. and Sass, B. 2013. The Small Finds. In: Finkelstein, I., Ussishkin, D. and Cline, E.H., eds. *Megiddo V: The 2004–2008 Seasons* (Monograph Series of the Institute of Archaeology of Tel Aviv University 31). Tel Aviv: 866–930.

Bunimovitz, S. and Lederman, Z. 2011. Close Yet Apart: Diverse Cultural Dynamics at Iron Age Beth-Shemesh and Lachish. In: Finkelstein, I. and Na'aman, N., eds. *The Fire Signals of Lachish: Studies in the Archaeology and History of Israel in the Late Bronze Age, Iron Age and Persian Period in Honor of David Ussishkin*. Winona Lake: 33–53.

Bunimovitz, S. and Lederman, Z. 2012. Iron Age Iron: From Invention to Innovation. In: Webb, J.M. and Frankel, D., eds. *Studies in Mediterranean Archaeology: Fifty Years On* (*SIMA* 137). Uppsala: 103–112.

Bunimovitz, S. and Lederman, Z. 2016. *The Tel Beth-Shemesh. A Border Community in Judah. Renewed Excavations 1990–2000: The Iron Age.* (Monograph Series of the Institute of Archaeology of Tel Aviv University 34). Tel Aviv.

Eliyahu-Behar, A. 2016. Evidence for Iron Working. In: Herzog, Z. and Singer-Avitz, L., eds. *Beer-Sheba III: The Early Iron IIA Enclosed Settlement and the Late Iron IIA–Iron IIB Cities*, Vol. 3 (Monograph Series of the Institute of Archaeology of Tel Aviv University 33). Tel Aviv: 1425–1429.

Eliyahu-Behar, A., Yahalom-Mack, N., Gadot, Y. and Finkelstein, I. 2013. Metalworking in Area K: A Reevaluation. In: Finkelstein, I., Ussishkin, D. and Cline, E.H., eds. *Megiddo V: The 2004–2008 Seasons* (Monograph Series of the Institute of Archaeology of Tel Aviv University 31). Tel Aviv: 1271–1284.

Eliyahu-Behar, A., *et al*. 2012. Iron and Bronze Production in Iron Age IIA Philistia: New Evidence from Tell es-Safi/Gath, Israel. *Journal of Archaeological Science* 39: 255–257.

Fantalkin, A. and Finkelstein, I. 2006. The Sheshonq I Campaign and the 8th-Century-BCE Earthquake—More on the Archaeology and History of the South in the Iron I–IIA. *Tel Aviv* 33: 18–42.

Finkelstein, I. and Lipschits, O. 2011. The Genesis of Moab: A Proposal. *Levant* 43: 139–152.

Finkelstein, I. and Piasetzky, E. 2009. Radiocarbon-Dated Destruction Layers: A Skeleton for Iron Age Chronology in the Levant. *Oxford Journal of Archaeology* 28: 255–274.

Gadot, Y. 2012. The Metal Industry at Hazor. In: Ben-Tor, A., Ben-Ami, D. and Sandhaus, D. 2012. *Hazor VI: The 1990–2009 Excavations—The Iron Age*. Jerusalem: 541.

Gal, Z. and Alexandre, Y. 2000. *Horbat Rosh Zayit* (IAA Reports 8). Jerusalem.

Gal, Z. and Alexandre, Y. 2003. The Iron Age "Low Chronology" in Light of the Excavations at Horvat Rosh Zayit. *Israel Exploration Journal* 53: 147–150.

Garfinkel, Y., Ganor, S., Hasel, M. 2011. The First Four Years of Excavation at Khirbet Qeiyafa, A Fortified City of the Judean Shephelah from the Beginning of the 10th Century BCE. *Qadmoniot* 44(141): 2–12 (Hebrew).

Gottlieb, Y. 2010. The Advent of the Age of Iron in the Land of Israel: A Review and Reassessment. *Tel Aviv* 37: 89–110.

Herzog, Z. and Singer-Avitz, L. 2004. Redefining the Centre: the Emergence of State in Judah. *Tel Aviv* 31: 209–244.

Herzog, Z. and Singer-Avitz, L. 2006. Sub-Dividing the Iron Age IIA in Northern Israel: A Suggested Solution to the Chronological Debate. *Tel Aviv* 33: 163–195.

Herzog, Z. and Singer-Avitz, L. 2011. Iron Age IIA Occupational Phases in the Coastal Plain of Israel. In: Finkelstein, I. and Na'aman, N., eds. *The Fire Signals of Lachish: Studies in the Archaeology and History of Israel in the Late Bronze Age, Iron Age and Persian Period in Honor of David Ussishkin*. Winona Lake: 159–174.

Ilan, D. 1999. *Northeastern Israel in the Iron Age I: Cultural, Socioeconomic and Political Perspectives* (Ph.D. dissertation, Tel Aviv University). Tel Aviv.

Kochavi, M. 1969. Excavations at Tel Esdar. *ʿAtiqot* 5: 14–48 (Hebrew).

Levy, T.E. 2009. Pastoral Nomads and Iron Age Metal Production in Ancient Edom. In: Szuchman, J., ed. *Nomads, Tribes, and the State in the Ancient Near East. Cross-Disciplinary Perspectives* (The Oriental Institute of the University of Chicago, Oriental Institute Seminars 5). Chicago: 147–177.

Levy, T.E., *et al.* 2004. Reassessing the Chronology of Biblical Edom: New Excavations and 14C Dates from Khirbet en-Naḥas (Jordan). *Antiquity* 78: 865–879.

Levy, T.E., Ben-Yosef, E. and Najjar, M. 2012. New Perspectives on Iron Age Copper Production and Society in the Faynan Region, Jordan. In: Kassianidou, V., ed, *Eastern Mediterranean Metallurgy and Metalwork in the Second Millennium BC*. Oxford: 197–214.

Levy, T.E., *et al.* 2008. High Precision Radiocarbon Dating and Historical Biblical Archaeology in Southern Jordan. *Proceedings of the National Academy of Science* 105: 16460–16465.

Levy, T.E. and Najjar, M. 2006. Some Thoughts on Khirbet en-Nahas, Edom, Biblical History and Anthropology—a Response to Israel Finkelstein. *Tel Aviv* 33: 3–17.

Levy, T.E., Najjar, M. and Ben-Yosef, E. 2014. Conclusion. In: *New Insights into the Iron Age Archaeology of Edom, Southern Jordan*, Vol. 2. Los Angeles: 977–1001.

Levy, T.E., Najjar, M., van der Plicht, J., Higham, T. and Bruins, H.J. 2005. Lowland Edom and the High and Low Chronologies: Edomite State Formation, the Bible and Recent Archaeological Research in Southern Jordan. In: Levy, T.E. and Higham, T., eds. *The Bible and Radiocarbon Dating: Archaeology, Text and Science*. London: 129–163.

Martin, M.A.S. and Finkelstein, I. 2013. Iron IIA Pottery from the Negev Highlands: Petrographic Investigation and Historical Implications. *Tel Aviv* 40: 6–45.

Mazar, A. 2001. *Timna (Tel Batash) II* (Qedem 42). Jerusalem.

Mazar, A. 2011. The Iron Age Chronology Debate: Is the Gap Narrowing? Another Viewpoint. *Near Eastern Archaeology* 74: 105–111.

Munger, S. and Levy, T.E. 2014. The Iron Age Egyptian Amulet Assemblage. In: Levy, T.E., Najjar, M. and Ben-Yosef, E. 2014. *New Insights into the Iron Age Archaeology of Edom, Southern Jordan*, Vol. 2. Los Angeles: 741–765.

Na'aman, N. 2013. The Kingdom of Judah in the 9th Century BCE: Text Analysis versus Archaeological Research. *Tel Aviv* 40: 246–298.

Paz, I. 2016. Metal Objects. In: Herzog, Z. and Singer-Avitz, L., eds. *Beer-Sheba III: The Early Iron IIA Enclosed Settlement and the Late Iron IIA–Iron IIB Cities*, Vol. 3. Tel Aviv: 1156–1188.

Renfrew, C. 1984. *Approaches to Social Archaeology*. Edinburgh.

Sapir-Hen, L. and Ben-Yosef, E. 2013. The Introduction of Domestic Camels to the Southern Levant: Evidence from the Aravah Valley. *Tel Aviv* 40: 277–285.

Singer-Avitz, L. 1999. Beersheba—A Gateway Community in Southern Arabian Long-Distance Trade in the Eighth Century B.C.E. *Tel Aviv* 26: 3–74.

Tebes, J.M. 2003. A New Analysis of the Iron Age I 'Chiefdom' of Tel Masos (Beersheba Valley). *Aula Orientalis* 21: 63–78.

Thareani, Y. 2011. *Tel ʿAroer: The Iron Age II Caravan Town and the Hellenistic–Early Roman Settlement* (Annual of the Nelson Glueck School of Biblical Archaeology, Hebrew Union College—Jewish Institute of Religion 8). Jerusalem.

Veldhuijzen, H.A. and Rehren, T. 2007. Slags and the City: Early Iron Production at Tell Hammeh, Jordan, and Tel Beth-Shemesh, Israel. In: La Niece, S., Hook, D. and Craddock, P., eds. *Metals and Mines—Studies in Archaeometallurgy*. London: 189–201.

Yahalom-Mack, N. 2009a. *Bronze in the Beginning of the Iron Age in the Land of Israel: Production and Utilization in a Diverse Ethno-Political Setting* (Ph.D. dissertation, the Hebrew University of Jerusalem). Jerusalem.

Yahalom-Mack, N. 2009b. Typology of the Metal Objects. In: Panitz-Cohen, N. and Mazar, A., eds. *Excavations at Tel Beth-Shean 1989–1996, Vol. 3: The 13th–11th Centuries BCE (Areas S and N)*. Jerusalem: 564–584.

Yahalom-Mack, N. and Eliyahu-Behar, A. 2015. The Transition from Bronze to Iron in Canaan: Chronology, Technology and Context. *Radiocarbon* 57(2): 285–305.

Yahalom-Mack, N., Gadot, Y., Eliyahu-Behar, A., Bechar, S., Shilstein, S. and Finkelstein, I. 2014. Metalworking at Hazor: A Long-Term Perspective. *Oxford Journal of Archaeology* 33(1): 19–45.

Yahalom-Mack, N., *et al.* 2014. New Insights into Levantine Copper Trade: Analysis of Ingots from the Bronze and Iron Ages in Israel. *Journal of Archaeological Science* 45: 159–177.

Yahalom-Mack, N. and Segal, I. 2009. Provenancing Copper-Based Objects Using Lead Isotope Analisys. In: Panitz-Cohen, N. and Mazar, A., eds. *Excavations at Tel Beth-Shean 1989–1996, Vol. 3: The 13th–11th Centuries BCE (Areas S and N)*. Jerusalem: 589–596.

CHAPTER 31

TRADITION AND CONTINUITY IN CRAFT WORKSHOPS AT TEL DAN

Rachel Ben-Dov

The southern side of Tel Dan served in various periods as a public area allocated for storage and crafts. Structures surrounding courtyards were revealed in the strata dating to the beginning of the Iron Age. This area yielded numerous installations and artifacts attesting to metallurgical activity. The aim of this paper is to present the evidence for this activity and to examine the function of the installations and objects involved in the process.

The metallurgical activities conducted in the courtyards utilized a range of installations; some were stone-built and others were dug into the ground. Inside and around the installations were pounders and large stones, crucibles with slag adhering to their bottoms, slag, tuyères, fragments of metal and bone objects, pot bellows, burnt wood, pottery vessels, an abundance of animal bones, a large concentration of flint and brittle clay—worked and smoothed on the interior—perhaps attesting to crumbled molds.

Broken metal implements and scraps were melted down in the crucibles and recycled. Analysis of the bronze artifacts demonstrates that the source of the copper which was used to manufacture them was Timna or Faynan.

INTRODUCTION

The Hebrew term *ḫarash* in the Old Testament designates a craftsman in stone, wood or metal (Abramsky 1958: 304–305). Coppersmiths were considered rich in wisdom, understanding and the skill "for making any work in bronze." (1 Kgs 7:14). Great importance was ascribed to them. In peacetime they were sent abroad to exercise their expertise, and in wartime, if their lands were conquered, they were exiled with their national leaders: "He carried away all Jerusalem...all the mighty men of valor...and all the craftsmen and the smiths" (2 Kgs 24:14). In excavations in Areas AB and B1 on the south margin of Tel Dan (Fig. 31.1; Biran 2008: 1686),[1] an early Iron Age quarter was exposed that yielded numerous artifacts associated with metallurgy.

The quarter is of considerable size—35 × 20 m—and extends under the remains of the Iron Age II upper gate and the paved road that follows the slope northward into the city (Biran 1994: 249). The relevant architectural remains include a variety of installations; some were stone-built and others were

1 The excavations at Tel Dan were directed by the late Prof. Avraham Biran under the auspices of the Israel Department of Antiquities and Museums (1966–1973) and the Nelson Glueck School of Biblical Archaeology, Hebrew Union College–Jerusalem (1974–1999). The staff included: G. Cook (surveyor); D. Packman (processing and typological classification of the pottery); N. Cohen (pottery restoration); M. Ben-Gal, N. Z'evi and A. Ruban (drawing of finds); A. Biran (field photography); and Z. Radovan (artifact photography). The area supervisors included: Y. Porath (1974–1975) and N. Heidebrecht (1984–1985) in Area B1; R. Hestrin (1979) and the author (1984–1988) in Area AB.

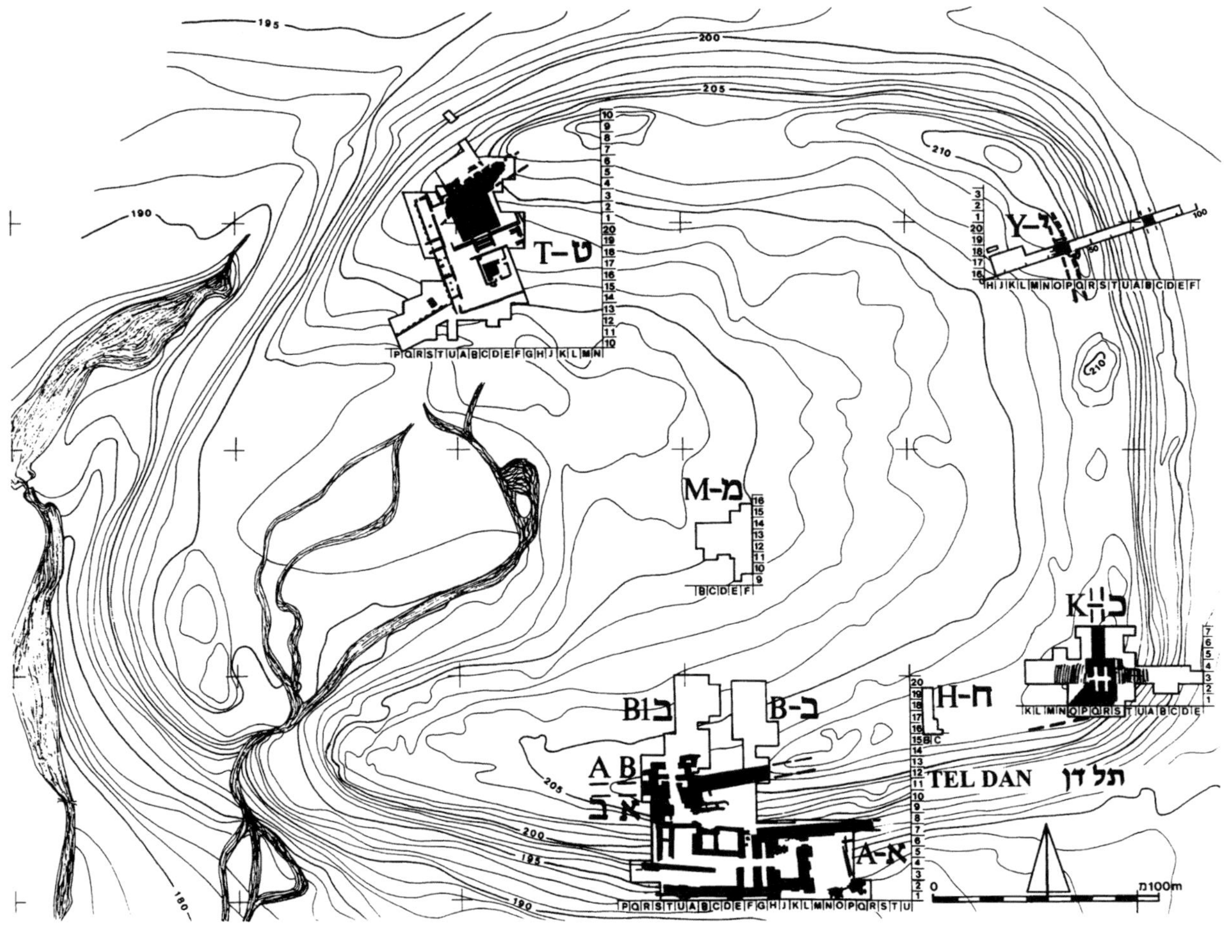

Fig. 31.1: Topographic map of Tel Dan and excavation areas.

dug into the ground. Near these installations were ceramic crucibles, tuyères, stone tools, fragments of metal and bone objects, as well as burnt animal bones (Biran 1994: color plates 21–23). Area AB, comprised of four stages evincing metallurgical activity, which Biran ascribed to Strata VI–IV, dated from the 12th to the beginning of the 10th centuries BCE (e.g., Biran 1989a; 1989b; 1994: 147–157; 1996: 26–28).[2] The aim of this paper is to present the nature of the activities performed here exclusively on the basis of the metallurgical evidence and to examine the function of every installation, tool and other object in the process.

The earliest evidence of metallurgical activity at Tel Dan was uncovered in Area B1, Stratum VIII, ascribed to the LB I (Ben-Dov 2011a: 349–366). Continuous occupation was observed in Areas AB–B1 from the LB I (Stratum VIII) until the end of the Iron Age I (Stratum IV), during which the buildings of each successive stage made use of the walls from its predecessor, along with new walls being added and floors raised (Ben-Dov 2011a: 15–31).

2 Area AB is of singular importance for the Iron Age I stratigraphy at the site. In addition to the architectural changes from one stratum to another, differences were observed in the typology of the pottery, in the composition of the metal slag and in the faunal remains.

THE ARCHITECTURAL REMAINS OF STRATUM V IN AREA AB

Although evidence for metalworking was found in all Iron Age I strata, it is best exemplified in the finds of a Stratum V building in Area AB (Fig. 31.2; cf. Biran 1996: 27, Phase B; Ben-Dov 2011b: 77–79). This stratum is clearly defined by a ca. 50 cm thick layer of burnt red mudbrick debris that covered and preserved a rich assemblage of pottery and other finds *in situ* on the floor. The building was partially cleared over an area of ca. 13 × 13 m and comprises rooms and courtyards (Fig. 31.2). Three rooms of different sizes are defined by walls built of fieldstones. The finds on the floors varied from room to room, indicating a varied function for each. A completely exposed room in the middle of the area (Locus 7063) had benches along its south and west walls, and a *bamah* (platform, Locus 7082) in the southwest corner on which an assemblage of vessels bearing a cultic character was found: bowls, a chalice and a vessel in the shape of a temple (Biran 1989a: Figs. 20–22: 5; 1994, Ills. 110–112). Room 7067, southeast of Room 7063, contained an assortment of vessels, including cooking-pots, chalices, and a complete crucible (Fig. 31.9: 10 below), and was apparently used as a storeroom. Part of another room (Locus 7065), east of Room 7063, had a *tabun* containing cooking-pot fragments and stone tools.

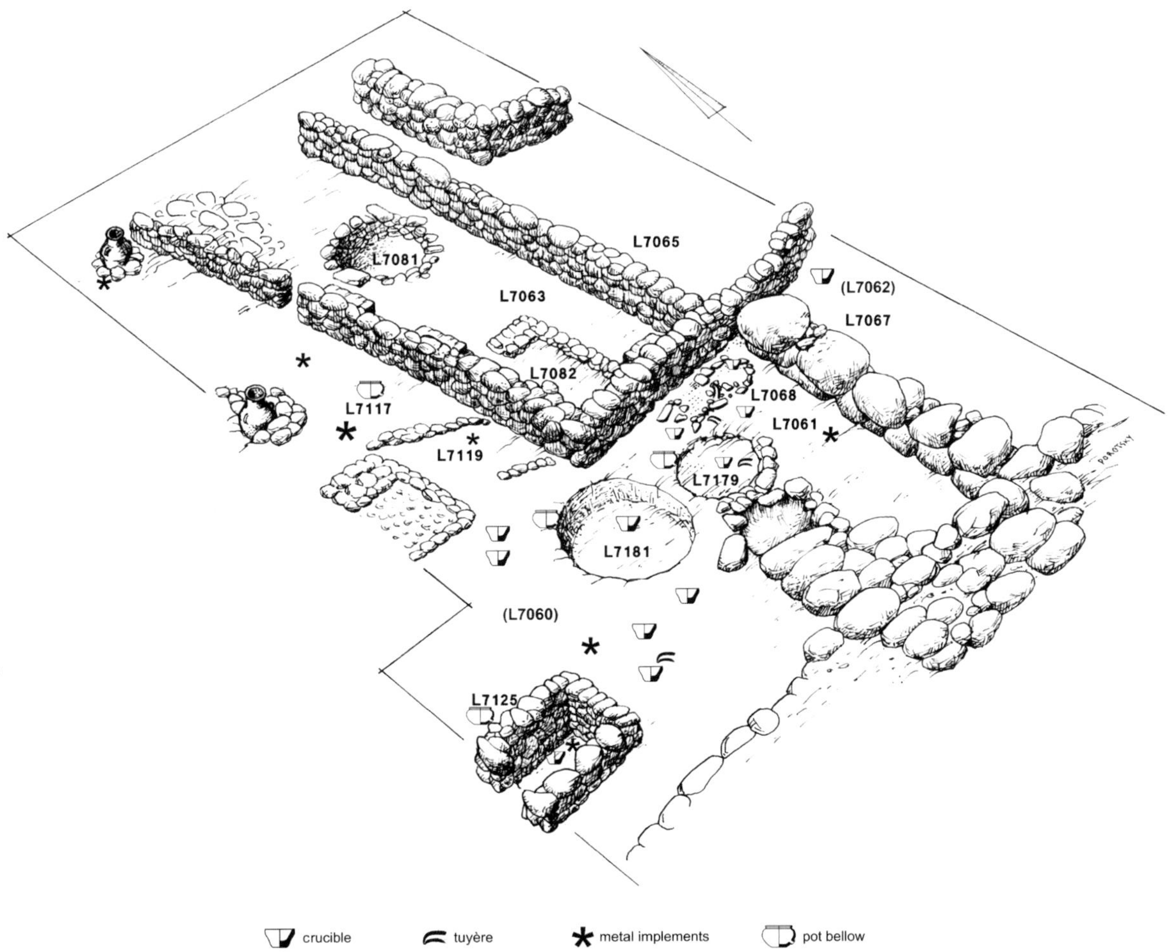

Fig. 31.2: The proposed restoration of the Area AB Stratum V building, the location of the installations, and the distribution of the finds.

THE COURTYARDS

South Courtyard (Locus 7061; Figs. 31.2, 31.3)

This 5 m wide space (north–south) was exposed only along 9 m of its length. This open space contained a range of installations and furnaces. Pounders and large stones of vesicular basalt inside and around these installations suggest they served as working surfaces. There were also pottery vessels, crucibles (Figs. 31.2, 31.3a), tuyères (Fig. 31.3: b1–3) and fragments of pot bellows (Fig. 31.2), as well as burnt wood, olive pits and an abundance of animal bones.

West Courtyard (Locus 7117, Locus 7119; Figs. 31.2, 31.4)

This courtyard contained charcoal piles found over small ash pits, metal tools and fragments, slag, a large concentration of flint (Fig. 31.4), pieces of bone implements, animal bones, fragments of pot bellows and brittle clay, worked and smoothed on the interior. An opening in the structure's west wall led to Room 7063.

THE INSTALLATIONS

These consisted both of round pits dug into the ground, which vary in diameter and depth, and those whose walls were built from stone.

Unlined Pits

Small Charcoal Piles

Charcoal and ash piles, 0.10–0.30 m in diameter and ca. 0.10 m in depth, some containing crucible fragments, were found in all strata.

Ash Pits

This pit type was found in the South Courtyard near the furnaces and, in one case (Locus 7181; Fig. 31.2), reached 2 m in diameter and ca. 0.50 m in depth. They were filled with ash and contained fragments of crucibles, tuyères and slag.

Storage Pits/Silos

These pits are exemplified by Locus 7081 in the north portion of Room 7063, which was 1.2 m in diameter and 2.5 m in depth and was dug into a gravel layer topped by three stone courses.[3] The original function of Pit 7081 is unclear; conceivably, it could have served for the storage of raw materials. However, toward the end of its use, earth, sherds and other artifacts were cast into it. Storage pits have been found at Tel Dan in all the Iron I strata, many in Areas AB–B1 (Biran 1994: 126, Ills. 88–90).[4]

Pits with a similar diameter and depth to those at Tel Dan are known from the workshops at Kition (Cyprus) (Karageorghis and Demas 1985: 82).

---

3 The area's gravel surfaces were laid during the LB I (Ben-Dov 2011a:15, 16).

4 For an extensive discussion of storage pits and silos at the beginning of the Iron Age, see Greenhut 2006: 22–115, 292–294.

Fig. 31.3: The Southern Courtyard (Locus 7061). Installation 7068: crucible (a); tuyères (b1, b2, b3.); and adjoining an ash pit (c); looking north.

Fig. 31.4: The Western Courtyard (Locus 7117). in the foreground (near arrow), a concentration of flint implements adjacent to a wall, looking east.

CONSTRUCTED INSTALLATIONS/FURNACES

Stone Circles

Furnace 7068 is one of a series of similar circles in close proximity to one another in the South Courtyard, and is the most complete and clearest of all (Figs. 31.2, 31.3, 31.5). It is located in the northeast corner of the courtyard and consists of a circle of stones; it has an internal diameter of 0.45 m and is 0.40 m high. The outer wall was lined with cooking-pot sherds (Biran 1989a: Fig. 31.17: 1, 2). The stones were coated on the inside with a thick layer of coarse clay (*tabun* material) and the bottom was formed using a fragment of a pithos coated with plaster. Inside the furnace was the base of a crucible with slag adhering to its bottom (Fig. 31.3: a). On the other side, between the stones, was the end of a tuyère (Fig. 31.3: b1 and Fig. 31.9: 14 below). The furnace was filled to the brim with ash mixed with slag, quartz stones, shells and animal bones (Fig. 31.6). They were covered with pieces of coarse clay from a kind of dome which apparently served as the furnace's roof (Figs. 31.5: d; 31.10: 9 below).

West of this furnace was a second, poorly preserved stone circle; a tuyère was found between its stones (Fig. 31.3: b2). Fragments of another tuyère were found on the other side (Figs. 31.3: b3; 31.9: 11, 13). Incorporated into and near the furnaces were basalt and flint pounders, as well as large flat stones—"working tables"—of vesicular basalt. Crucible fragments, slag, bronze objects and engraved bones were strewn about the area. Such furnaces were also found in Strata VI and IV (Biran 1989b: Pl. 2).

SEMICIRCLES (FIGS. 31.2, 31.7)

Two semicircular installations, built from a single course of stones, were located in the South Courtyard. The stones were set around the rim of a pit (Locus 7179), 1.20 m in diameter, which contained ash, crucible fragments, a tuyère and part of a pot bellows; another part of the pot bellows was found on the floor of the South Courtyard (Fig. 31.10: 1 below). South of this installation was a second semicircle (Fig. 31.7), built over a gravel floor. Both installations were ascribed to Stratum V, while another pair was observed in Stratum VI. Semicircular installations were also reported in Room 428 in the industrial area at Beth-Shemesh (Grant and Wright 1939: 56).

Fig. 31.5: The Southern Courtyard (Locus 7061), top layer. Installation 7068: a. crucible, b1, b2. tuyères, d. the collapsed dome that covered Locus 7068 (Fig. 31.10: 9).

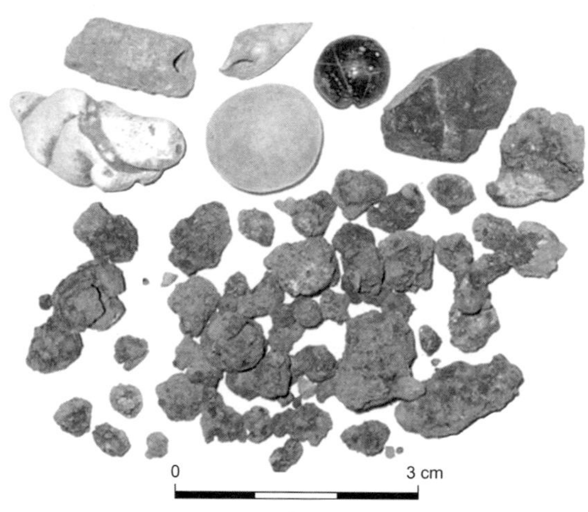

Fig. 31.6: The contents of the ash fill inside Furnace 7068.

Rectangular Installation (Locus 7125; Fig. 31.2)

Only the eastern portion of this installation was exposed for a length of 1.5 m; it was 1.2 m in width and 1.45 m in depth, and its eastern wall was preserved to a height of four courses. The large stones at its base were covered with a layer of small stones, including fragments of crucibles, a pot bellows and a bronze implement (Fig. 31.11: 11 below). The installation's mode of operation is unclear.

Another type of installation was found in Stratum IV above Locus 7125 (Locus 7060; Fig. 31.8). Four of these installations were revealed, each consisting of a flat-based stone, 0.30 × 0.40 m, surrounded by a ring of small stones, ca. 0.30 m in height. Their openings face south and in front of the openings is a slanted stone,

Fig. 31.7: Two stone semicircles, Stratum V, looking northeast.

Fig. 31.8: Installation 7060, Stratum IV, looking northwest.

also facing south. These installations were found filled to the brim with ash and contained crucible fragments. Nearby were fragments of bronze implements and slag. Their function and mode of operation are uncertain.

FINDS RELATED TO METALLURGY FROM AREAS AB AND B1

CRUCIBLES (FIG. 31.9: 1–10)

The typical crucible at Tel Dan is a deep bowl handmade from coarse clay; it has a thick, flat base, high, thick walls, and a flat or rounded rim. The bases of 23 crucibles as well as numerous rim fragments were recorded of a type characteristic of all strata from the Late Bronze Age (Fig. 31.9: 1–2) to the end of the Iron I (Fig. 31.9: 3–10).

The crucibles vary in size, wall thickness and volume. This diversity characterizes all of the strata at Tel Dan. Most exhibit a maximum diameter (just under the pinched mouth) of 14 cm; the maximum height, including the base, is 8–11 cm; the depth is 3–8 cm. Their average volume is 30 cu cm, but one exceptionally large crucible could hold 84 cu cm (Fig. 31.9: 6).

The crucibles have two types of spouts: the more common Type A is pinched (Fig. 31.9: 6–8); examples have been recovered from all the strata at the site. Type B has a U-shaped spout that was cut from the vessel's rim (Fig. 31.9: 10). The reason for this difference is unclear, as the vessels are otherwise identical in size and shape. Can it be that they were used in separate processes or that the distinction is chronological? In the excavated area, Type B is found only in the earlier Strata VI and V.

Crucibles with a pinched spout occur at various sites and in different periods where metalworking is attested; on the temple's courtyard and floor, or in graves (Table 31.1). The closest parallels to the Tel Dan crucibles, in shape and in date, are those from Tel Megiddo (Yahalom-Mack and Shalev 2006: Fig. 18.31, No. 663) and Area K at Qatna (Luciani 2002: 160, Fig. 129). Parallels for Type B have yet to be found at any other site.

Fig. 31.9: A Selection of Crucibles and Tuyères

No.	*Object*	*Locus*	*Registration No.*	*Stratum*	*Context*	*Type*
1	Crucible	4643	23166	VIII	Courtyard	
2	Crucible	4643	23132	VIII	Courtyard	
3	Crucible	1204	10529/2	VI	Pit	A
4	Crucible	7015	24090/16	IV	South Courtyard	
5	Crucible	7015	24090/15	IV	South Courtyard	
6	Crucible	591	9508	IV	Inside installation	A
7	Crucible	7062	23404/8	IV	Room	A
8	Crucible	7062	23372/2	IV	Room	A
9	Crucible	7015	24102/6	V	South Courtyard	A
10	Crucible	7067	23443/1	V	Room	B
11	Tuyère	7061	23451/3	V	South Courtyard, part of installation	End A
12	Tuyère	4202	18059	IV	South Courtyard	End A
13	Tuyère	7061	23451/4	V	South Courtyard, part of installation	End A
14	Tuyère	7068	23453/4	V	Part of installation	End A
15	Tuyère	7174	23981	VI	West Courtyard, inside installation	End A
16	Tuyère	1204	10480	VI	Pit	End B
17	Tuyère	7135	23776/1	V	West Courtyard	End B

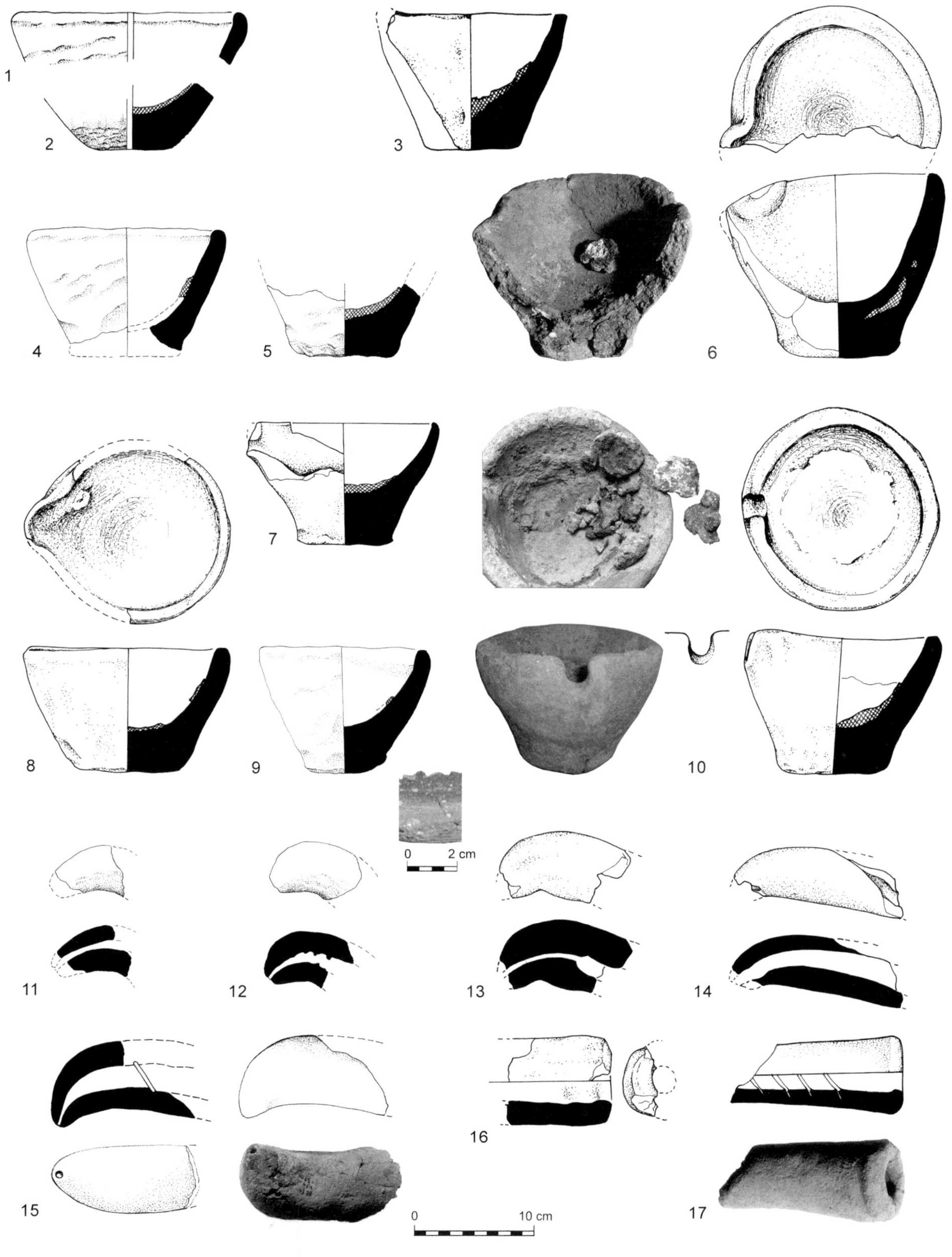

Fig. 31.9: A selection of crucibles (1–10) and tuyères (11–17).

Table 31.1: Sites with Evidence of Metallurgical Activity

Site	*Context*	*Ob*	*Sl*	*In*	*Mo*	*PB*	*Ty*	*Cu*	*Installations*	*Period*	*References*
Abu-Hawam, Tell	Next to Room 45				√					IR I	Hamilton 1935: 58, No. 359
Akko	Area AB	√	√				√	√	Furnace	End of LB	Dothan 1981a: 111
Aphek	Pit 5027 (Philistine pit)	√						√		IR I	Yahalom-Mack and Shalev 2009: 416
Azor	Cemetery							2		IR I	Dothan 1961: 228, Pl. 4a
Be'er Resisim	Central Negev–Ramat Matred	√		√						MB I	Segal, Halicz and Cohen 1999: 179
Beit Mirsim, Tell	Stratum D	√		√	√	2		√		MB IIC	Albright 1938: 53–54, Pls. 31: 7, 40: 7–8, 41: 13
Beth-Shean	Courtyard north of Seti I's temple; L1101		√		√		√	√	√	LB II	Rowe 1940: 9, Pls. LIIA: 8, LXXIA:5; Yahalom-Mack 2009b: Pl 10.17
Beth-Shemesh	St. III: L440, 441	√	√				√		Furnaces	12th-11th cent.	Grant and Wright 1939: 56-57
Beth-Shemesh	St. IIa Area X 27						√		√	10th cent.	Grant and Wright 1939: 75
Dan, Tel	On the city's south margin	√	√		√		√	√	Furnaces	LB I	Ben-Dov 2011a: Figs. 31.208–210
Dan, Tel	Temple, courtyards and workshops	√	√			√	√	√	Installations	IR I	Biran 1989a: 124-125, 1994: 147–157
Deir 'Alla, Tell	Inside a room on slope	√	√				√		Rectangular furnaces; pits	IR I	Franken 1969: 36–38
Dor, Tel	Area G (Phase 10) in the centre of the tel		√			√	√	√	Furnace	IR I	Stern *et al.* 2000: 29*; Yahalom-Mack 2009a: Fig. 31.II.37: 10
'En Ziq	Ore, winter campsite of nomads	√		√						MB I	Segal, Halicz and Cohen 1999: 179
Gerisa, Tel	Rubble covering courtyard						√	√		LB II-III	Herzog 1993: 483
Har Yeroḥam	Area A on the floor of the rooms			√						MB I	Kochavi 1963: 292, Pl. 15: 5
Ḥarashim, Tel	Northwest edge of tell	√	√				√	√	Furnace	IR I	Aharoni 1957: 20
Hayyat, Tell el-	Temple courtyard (figurine)	√			√			√		MB IIA	Falconer and Magness-Gardiner 1989: 259; Falconer and Fall 2006: Figs. 6.7, 6.8, 6.24–6.28
Hazor	Area D				√		√	√		MB II	Yadin *et al.* 1958: Pl. XCVIII: 6–14; XCII: 30–31
Hazor	Areas A, D, F	√	√	√	√		√	√		LB II	Yadin *et al.* 1960: Pl. CXLVII: 10–11; 1961: CLVI: 30, CLVIII: 31
Mount Hebron (north of Dahiriya)	Hoard	√		√						MB I	Dever and Tadmor 1976: 164
Jerusalem	City of David		√				√	√		IR I	Mazar 2009: 54
Kāmid el Lōz	Crucibles on temple floor	√						√		LB	Metzger 1993: Pl. 13: 3, 4

Site	*Context*	*Ob*	*Sl*	*In*	*Mo*	*PB*	*Ty*	*Cu*	*Installations*	*Period*	*References*
Kinneret	On lake shore			3						IR I	Stepansky 2000: 15, Fig. 31.22
Lachish	North of Fosse Temple II		√				√			LB II	Tufnell, Inge and Harding 1940: Pls. LXXIII; XXIX: 16
Lachish	Room 5097 St.P3		√	√		√	√	√	Installations	IR II	Rothenberg 1975: Pl. 38: 6–9; Ussishkin 1983: Pl. 5
Masos, Tel	Area A, room in building 96		√					3	Furnace	11th century	Kempinski and Fritz 1977: 158, Fig. 31.18: 3
Megiddo	Tombs				√	√				LB I	Guy 1938: Pls. 37: 7; 49: 22; 60: 3; 156: 17–18
Megiddo	Southeastern flanks of city	√					√	√		IR I	Yahalom-Mack and Shalev 2006: Fig. 18: 30, 31, 33
Mevorakh, Tel	Figurine of snake from temple courtyard	√						√		LB IIA	Stern 1984: 6–7; Fig. 31.2: 16
Miqne-Ekron, Tel	Acropolis							√		LB II	Killebrew 1996: 94
Mor, Tel	Inside houses next to Egyptian fortress		√				√	√	4 furnaces	End of LB	Barako 2007: Fig. 31.8.1: 2–4
Nahariya	Figurine on temple platform (*Bama*)	√			2					MB IIA	Dothan 1956: 43–44; 1981b, 77, Pl. 5: 4
Nami, Tel	In temple courtyard atop rampart	√						√	Remains of furnace	End of 13th century	Artzy 1993: 1096; 1995: 22
Qasile, Tell	Adjoining temple	√	√					√	Furnaces		Maisler 1950–51: 75
Raddana, Khirbat	Fortified village, Area S	√	√				√	√		IR I	Callaway 1993: 1253
Rehov, Tel	Area D, St. D-9	√					√	√	Installations	LB IIB	Yahalom-Mack 2009a: 51, Figs. 31.II.19, 20
Serabit el-Khadim	Mine L	√			40	√	√	√	√	LB II	Beit-Arieh 1985: Figs. 31.13, 14, Pl. VIII: 1, 2
Timna	Temple courtyard	√	√				√	√	Furnace, Stone installation	LB II	Rothenberg 1988: 192, 195-196, Fig. 31.31–34
Timna	Area S, Site 30		√		√		√	√		IR II	Rothenberg 1990: Figs. 31.54, 73-77, 81-82; Ben-Yosef *et al.* 2012: Figs. 31.16, 20; Ben-Yosef 2012: Fig. 31.20.4-20.5
Yoqne'am	Area B2, Floor 2516, near south slope	√	√				√	√	Installation surrounded by ash	IR I- 11th cent.	Yahalom-Mack and Shalev 2005: 373, Fig. 31.I.35: 3–8; 2009a: 53–55
Zafit/ Tell es-Safi	Area A, St. A4		√				√	√		Early IA IIA	Eliyahu-Behar *et al.* 2012: 261, Figs. 31.3, 6
Zeror, Tel	Craftsmen's quarter on south slope	√	√				√	√	Furnaces	LB IIB	Tylecote 1981: Pl. 13; Kochavi 1965: 254

Ob. = object; Sl. = slag; In. = ingot, Mo. = mold; PB. = pot bellows; Ty. = tuyère; Cu. = crucible

TUYÈRES (FIG. 31.9: 11–17)

Tel Dan's typical tuyère is a ceramic pipe made from coarse, brittle clay mixed with straw and gravel, including small stones. At one end (A; Fig. 31.9: 11–15) the diameter of the bore is only a few millimeters and terminates with a bend. At the other end (B) its bore reaches 2 cm in diameter and terminates in a straight, slightly thickened edge (Fig. 31.9: 16–17).

At least ten large fragments were found, as well as an abundance of smaller non-restorable fragments. The section in Fig. 31.9: 12, 17 shows a spiral groove running lengthwise along the inside of the tuyère. Some of the tuyères were coated with a white slip. End A comprises the majority of finds, this being the part that was inserted into the furnace and baked by the heat. These exhibit clear traces of combustion and drops of metal still adhere to the tip of some bores.

There is no difference between the tuyère types from the LB I and from the Iron I levels (Biran 1989a: Figs. 31.2, 31.8). Such tuyères have also been found at other sites (Table 31.1).

Tuyères are employed in various metallurgical processes, such as smithing, smelting and refining. Tylecote associates the shape of the tuyère with the kind of work it performed and claims that several persons blowing through a set of tuyères could furnish a gust of air sufficient for several local processes (Tylecote 1981: 108, 118).

POT BELLOWS (FIG. 31.10)

Tel Dan yielded only fragments of clay pot bellows; their shape is reconstructed based on examples from other sites (Davey 1979: 102, Figs. 31.3–4). They measure ca. 0.30 m in diameter and their estimated height is ca. 0.20 m. The pots have a wide mouth, an everted rim, thick walls and a flat, thick base. The fragments included part of a spout of one of the pots (Fig. 31.10: 7). The pot's wide mouth was covered with a leather hide held in place by a cord that was fastened to the pot in the groove beneath the everted rim.

Two types have been found together; both are wheel-made and manufactured of coarse clay mixed with white grits and straw. The first (Fig. 31.10: 1, 2) consists of red "sandy" clay mixed with quartz grits, and coated with a greenish slip; it has a carinated shoulder (Fig. 31.10: 1). Its closest parallel comes from Ugarit and was found in a Late Bronze Age stratum (*Récent* 3) (Davey 1979: 108, Pl. 4: 14).

FIG. 31.10: SELECTION OF POT BELLOWS

No.	*Description*	*Locus*	*Registration No.*	*Stratum*	*Context*
1	Sandy paste (2.5YR6/6), white grits and quartz, bright yellow slip (2.5Y8/4)	7061 + 7179	23403 + 24610/1	V	South Courtyard + pit
2	Sandy clay (see No. 1)	7099	23568/5	V	Courtyard
3	Pink paste (5YR7/3), black core	7129	23713/3	V	
4	Red paste (2.5YR5/6), thick black core	7142	23810/1		
5	Red paste (2.5YR5/6), thick black core	7249	24935/1	VI	Pit 7240
6	Red paste (2.5YR5/6), thick black core	7240	24416/2	VI	Pit
7	Spout; coarse red Paste (2.5YR5/6), thick black core	7140	23820/1	V/VI	West Courtyard
8	Reddish-brown paste (5YR6/4), thick black core, white and black grits	7060	23448/1	IV	West Courtyard
9	Top of Installation 7068	7068	23453/5	V	South Courtyard

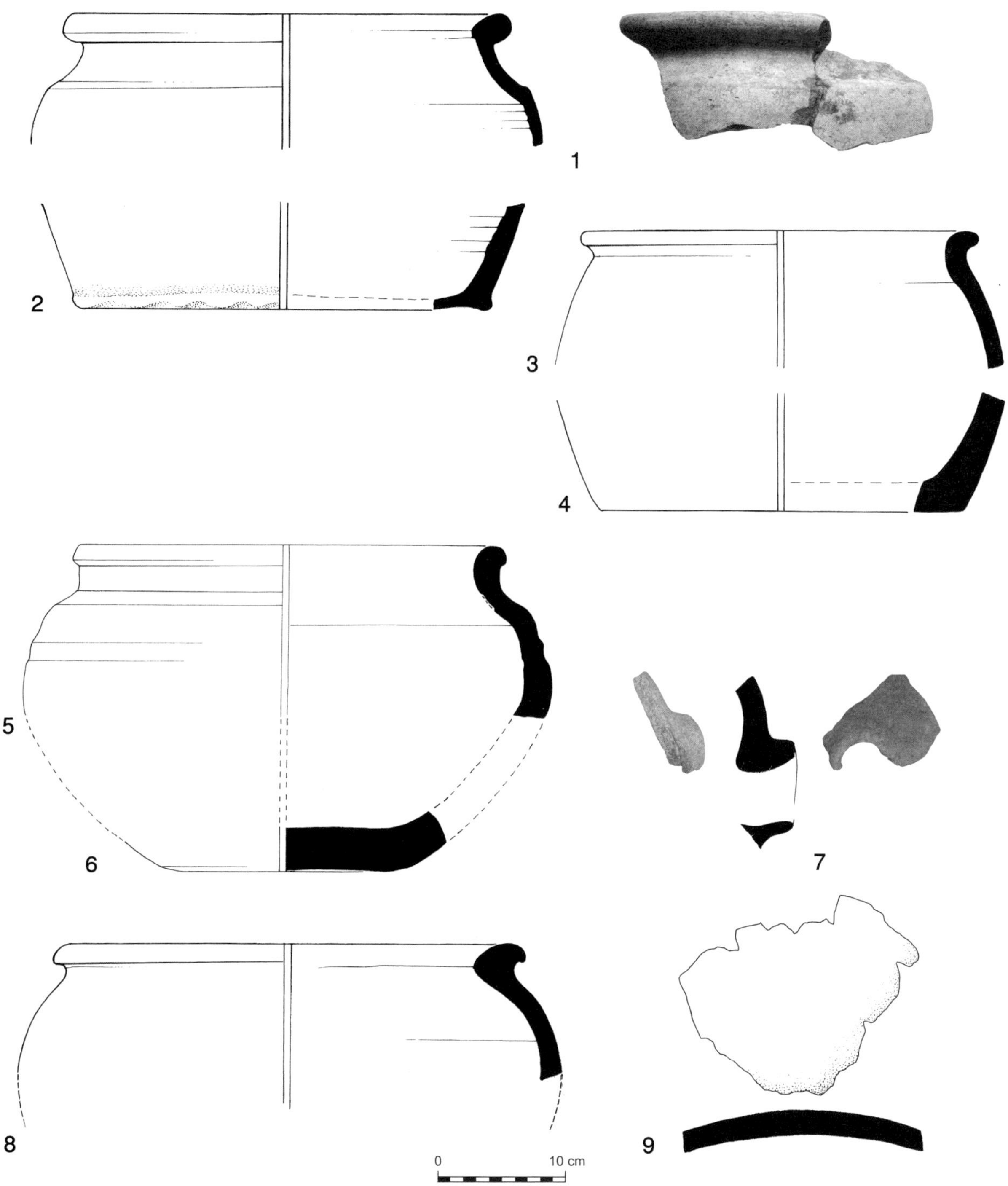

Fig. 31.10: Selection of pot bellows fragments from the courtyards adjoining the installations (1–8); dome that covered Furnace 7068 (9; see Fig. 31.5: d).

The second type (Fig. 10: 3–8) is made of reddish-brown clay and its walls are rounded. Judging by the clay, the spout fragment, mentioned above, belongs to this type. A similar pot bellows was also found in LB I Megiddo Tomb 1145A (Guy 1938: Pl. 49: 22; and Table 31.1).

The pot was activated by hand or foot. This process is depicted in Egyptian wall paintings, such as in the Tomb of *Rekh-mi-re'* from Thebes and in the Tomb of the Vizier H̲epu (Davies 1943, Pl. LII; 1963: 10, Pl. VIII). Davey (1979) distinguishes between the foot-operated pot bellows, and the hand-operated type. In his opinion, the taller pots were hand-operated and a single pot was employed (Davey 1979: 110), whereas the foot-operated pots were generally made of stone, and a pair was employed. If so, the ceramic pots found at Tel Dan belong to the former type.

MOLDS

Areas AB and B1 did not yield items that can, with any confidence, be identified as molds. A similar observation was also made for other sites, such as Tell Deir ʿAlla (Franken 1969: 38), and the workshop at Kition (Cyprus; Stech, Maddin and Muhly 1985: 395). However, pieces of coarse, unbaked clay mixed with small stones and straw, smoothed and plastered on the inside, were found in the West Courtyard alongside metal artifacts. It is conceivable that these represent molds that have crumbled. Additionally, a limestone mold was found in a LB I context in Area K in a building which may have been a temple (Minoff 1992: 87*; Ben-Dov 2011a: Figs. 105, 212).

BRONZE ARTIFACTS (FIG. 31.11)

A range of bronze artifacts, both broken and complete, characterized the area's assemblage. Most derive from Stratum V in the West Courtyard, in or near the installations (especially near the ash pits). They include fragments of needles, awls, thin plates, spear butts (or plough points), as well as a faultily cast axe head (Fig. 31.11: 10). A thin blade wrapped in a linen cloth was found in a pit in Area B1 (Ben-Dov and Gorski 2009: 81, Fig. 1). According to Shalev (1993: 64), these were scraps of metal utensils meant for recycling.

Fig. 31.11: A Selection of Bronze Objects

No.	*Object*	*Locus*	*Register No.*	*Stratum*	*Context*
1	Needle	7117	23682	V	West Courtyard
2	Needle	7020	241114/1	VI	South Courtyard
3	Awl	7114	23667	V	West Courtyard
4	Rod	7060	23383 + 23392	IV	South Courtyard
5	Knife	7015	24023/20	IV	South Courtyard
6	Arrowhead	7065	23450/9	V	Dismantling of installation
7	Awl (?)	7169	23957	IV	West Courtyard
8	Handle	4202	18101/10	IV	South Courtyard
9	Hook	7126	23712	VI	South Courtyard
10	Axe	7131	23755	V	West Courtyard
11	Plough point	7125	23786	V	On bottom of installation
12	Plough point or spear butt	7133	23775	V	West Courtyard

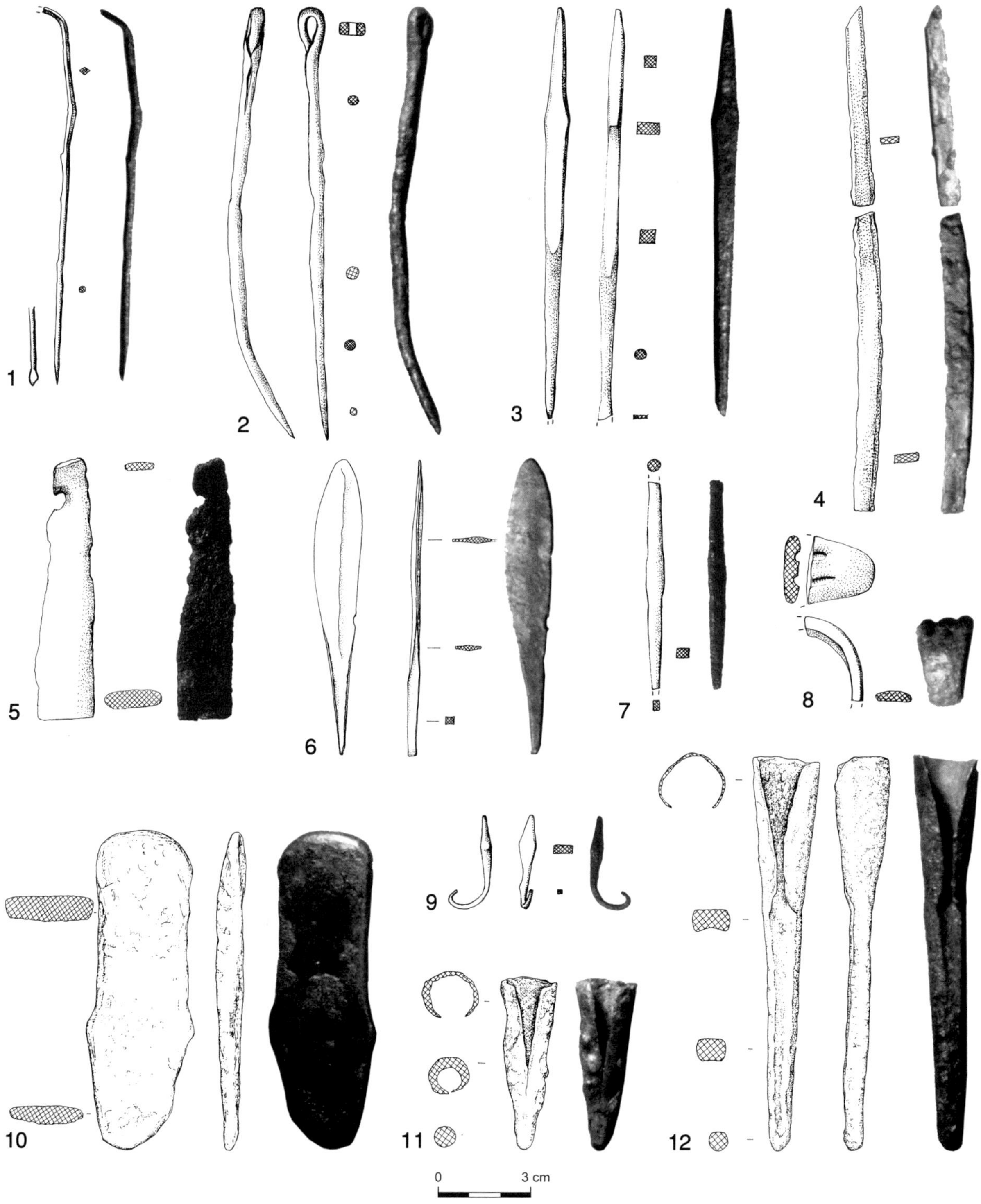

Fig. 31.11: Fragments of metal artifacts near the installations in the courtyards.

STONE TOOLS

These include working surfaces, grindstones and pounders of basalt or flint (Fig. 31.12). They were associated with the installations and served in the industrial activities for crushing slag and bones. The vesicles in the basalt surfaces contained bits of metal.

FLINT TOOLS

An abundance of flint was found in the West Courtyard of Stratum V (Fig. 31.4). The assemblage was examined by Steve Rosen, who concluded that it includes a variety of finds: flint cores, waste, flakes and even blades, attesting to the local production of the latter. Some still retained traces of the glue used to affix them to some implements. At Tell Qasile, flints and animal bones were also among the finds in the artisan quarter (Maisler 1950–1951: 75).

BONE ARTIFACTS

An assemblage of bone artifacts (Fig. 31.13) was found in the West Courtyard together with the bronze objects. They include fragments of pierced artifacts, indicating that they were fastened to some object. Presumably, they were detached from broken implements of metal meant to be recycled. One of the bone handles (Fig. 31.13: 9) still retained the remains of the metal tang whose blade had split off. A conical fragment found on a Stratum VI floor (Fig. 31.13: 6) was made of hippopotamus ivory, an identification made by the zoologist François Poplin, who claims that this fragment is remnant of a raw material for making ivory artifacts.

ANIMAL BONES

A plethora of sheep, goat and cattle bones was found alongside the courtyard installations in all strata (Wapnish 1993: 431, Table 31.2). They included numerous mandibles and horns. Some were burnt and others were green-tinged, the latter indicating that they had been close to metal implements. Such an abundance of bones is also characteristic of other sites yielding evidence of metal workshops: e.g., Tel Dor (Stern *et al.* 2000: 29*); the City of David (Mazar 2009: 53); and Kition (Cyprus), where pits with bones burnt to a fine ash were found in the west workshops (Stech, Maddin and Muhly 1985: 393–394). These bones may have been used as flammable material. The bone ash could also have been used as flux to lower the melting temperature of the slag, and as a temper in the manufacture of crucibles and molds (for other suggestions, see Karageorghis and Kassianidou 1999: 180–183).

THE CERAMIC ASSEMBLAGE FROM THE TEL DAN WORKSHOPS

The following discussion of the pottery is meant to consider why certain vessels were found in the proximity of the workshops. Accordingly, consideration will be given only to the most common types.

CHALICES

Bowls on a high base comprise the dominant vessel in Area AB (cf. Biran 1989a: Figs. 5: 9, 10; 22: 4, 6; 26: 1–5, 7; 27: 5–10). Ten complete vessels and some 18 fragments were found, most in Stratum IV, but they characterize all the strata under discussion. Many were found in the courtyards adjoining the

Fig. 31.12: An assemblage of stones found in the vicinity of the installations.

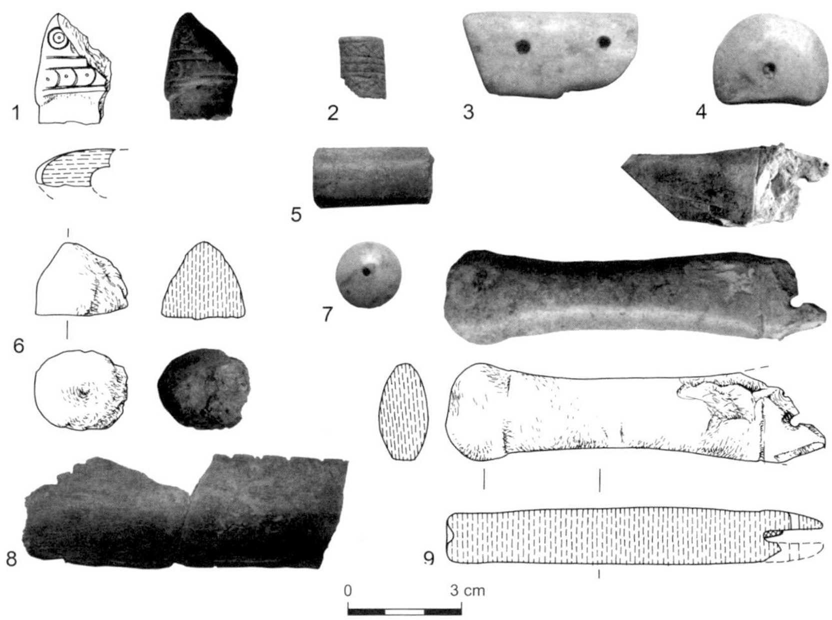

Fig. 31.13: Assortment of fragments of bone and ivory artifacts near the installations in the West Courtyard: handles (1, 2, 3, 5 and 9); spindle whorls (4, 7); industrial raw material (6); and curved bone (8).

installations, while the more complete examples derive from the "storerooms" (Biran 1994: Ills. 116–117: 1, 3). Their bowls were greatly scorched on the inside.

Chalices are usually considered cultic vessels, and, indeed, they have been found in every room identified as a temple. Area AB yielded a complete chalice on Cultic Platform 7082 (Biran 1989a: Fig. 22: 6). When examining the distribution of the chalices in Area AB and their position in relation to the installations, one has to consider other possible functions. For example, a wall painting in the New Kingdom "Tomb of the Two Sculptors" at Thebes in Egypt (Fig. 31.14; Davies 1925: Pl. X) shows a jeweler at work, which is conducted on a bowl with a high foot. One can assume that the chalices from Area AB served the smith in various processes.

COOKING-POTS

The remains of some 20 cooking-pots, mostly just the upper part, were found mainly in the courtyards. Their sherds were used to line the walls of installations in all the strata, including Stratum VIII of LB I (Ben-Dov 2011a: Fig. 34: 1–6, 8) and Strata VI–IV of the Iron I (Biran 1989a, Figs. 5: 1–8; 12: 1–5; 17: 1–2; 18: 2–3; 24: 2–3; 27: 1–4).

PITHOI

The remains of at least five pithoi were recovered (Biran 1989a: Figs. 12: 6–8; 17: 5, 6). Most were sunk into the ground in the corners of the rooms (Fig. 31.2), while a few were buttressed by brick or stone installations which had been built around them (Fig. 31.15). They evidently contained substances used in the production process, perhaps water. Fragments of numerous pithoi were found in Strata VI–IV (Biran 1994: Ill. 91–92). Pithoi were also found in the workshop context at Tel Harashim (Aharoni 1956: 61–62, Fig. 4).

Various scholars conducted laboratory analyses on the Tel Dan metal remains, the slag and the crucibles; the conclusions presented here are based on the results of their studies and their inferences.[5]

Goren conducted petrographic analyses on the fragments of three crucibles. He found that their material consisted of silty clay rich in iron oxide. The matrix contains oblong vesicles and considerable portions have vegetal remains. There were also granules of quartz sand, basalt, limestone and small amounts of flint. All the crucibles were produced on the site or in its vicinity.

Twenty bronze artifacts and slag from Tel Dan that were taken for chemical analysis have demonstrated that, in most cases, the source of the copper used in their manufacture was Timna or Faynan (Yahalom-Mack 2009a: 241; Table V.6). Excavations underway at Khirbat en-Nahas (Transjordan) have demonstrated the existence of a wide-ranging metallurgical center at the beginning of the Iron Age, between 1200 to 900 BCE (Ben-Yosef, Levy and Najjar 2009: 98–99).

SUMMARY

The southern margin of Tel Dan evidently served, during various periods, as a public area allocated to storage and crafts—activities that were carried out in the courtyards. In the building described above, every room was assigned a distinct purpose. The assemblage discovered in Locus 7082 indicates that this room served for cultic activities; other rooms were used as storerooms. Numerous scholars have studied the relationship between metal and commerce, and have also drawn a connection between metal and religious structures (e.g., Artzy 2000: 29). Metallurgical accessories (crucibles, tuyères and molds) have been found in the Levant on the floors and in the courtyards of temples, and as funerary offerings (Table 31.1). Metallurgical activity was carried out inside the temples, e.g., the Middle Bronze Age temple at Nahariya (Dothan 1981b: 77). The remains at Kition (Cyprus) include metallurgical workshops

5 Chemical analyses were conducted by S. Shalev and N. Halperin (Shalev 1993). Tests were conducted by I. Segal of the Geological Survey of Israel in Jerusalem by means of Lead Isotope Analysis with the aim of establishing the source of the copper (Yahalom-Mack and Segal 2011: 367–373).

Fig. 31.14: A depiction of a jeweler blowing into a tuyère inserted into a pile of charcoal in a high-based bowl, as shown in a wall painting from an Egyptian tomb (Davies 1925: Pl. X).

Fig. 31.15: An *in situ* pithos within a stone-built construction.

adjoining and connected with the temple precincts from Late Cypriot IIC onward (Karageorghis and Demas 1985: 253–254, 272). A similar association exists in the Aegean realm (Blakely 1999: 17).

This paper is concerned with the work of the blacksmith in the recycling of metal, a process that was practiced in the courtyards at Tel Dan. The broken metal implements were placed in a crucible that was set into a pit dug into the ground or into a constructed installation that would then be covered with a heap of flammable material (Shalev 1993: 64–65). Tylecote (1979: 16–17) claims that "for the purpose of melting scraps of metal in crucibles a ring of stones, a pile of hot charcoal and a clay tuyère connected to bellows is all that would have been required...." This description matches the remains uncovered at Tel Dan.

A metallurgical tradition existed at Tel Dan from the beginning of the Late Bronze Age. There is considerable evidence for workshops and for the processing and recycling of metal from various sites in the Land of Israel and Transjordan during many periods, but especially at the beginning of the Iron Age (Fig. 31.16 and Table 31.1; see also Waldbaum 1978: 74, Table A–1; Artzy 2006: 9–71).

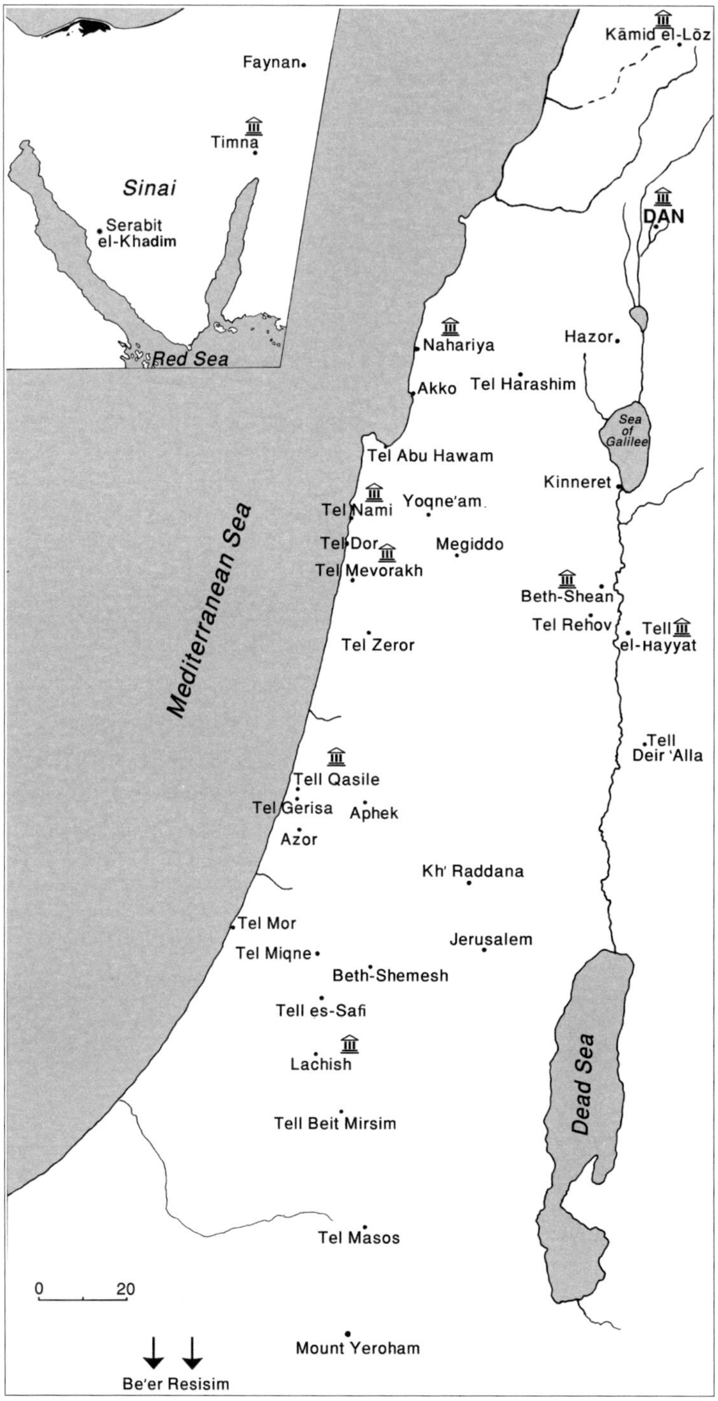

Fig. 31.16: Map of the sites yielding evidence of metalwork; 🏛 sites where the finds were discovered in a cultic context (see Table 31.1).

Apart from metal objects and slag, the courtyards have yielded deposits of flint tools and animal bones, in which mandibles and horns are prominent. One can assume that this area was used for various crafts connected to the production and shaping of tools. The concentration of flint blades alongside flint cores and flakes provides evidence of flint production. It is possible that flint blades were produced and attached to the handles manufactured from wood, bone or mandibles. The animal bones were also used to create various objects. The hippopotamus ivory (mentioned above) and the horns served as raw material for the carving of implements and other items (Ben-Dov 2002: 153). The concentration of various crafts in the area of metallurgical activities is not unique to Tel Dan, but is known from other sites as well. For example, an oil maker's house was found in the vicinity of the metal production area at Tel Yoqne'am (Yahalom-Mack and Shalev 2005: 373). At Tell Qasile, a granary and loom weights attesting to textile production were found in proximity to the metalworking area (Maisler 1950–1951: 75–76). Large storerooms for the olive oil industry were found in the vicinity of the "coppersmith's house" at Kalvasos-Ayios Dhimitrios, Cyprus (South 2012: 45). The excavators of Kition also identified various other crafts which were carried out near the metal workshop (Karageorghis and Demas 1985: 77). The pottery assemblage from the workshop district should be examined with the aim of determining the function of each vessel, as well as other items, in the industrial process.

Finally, the main source of copper during the Late Bronze Age was Cyprus (Knapp, Kassianidou and Donnelly 2001: 204). Recent studies have identified that the source at the beginning of the Iron Age was in Timna and Faynan. As demonstrated above, the provenance of the objects whose remains made their way to the workshops at Tel Dan was indeed Timna or Faynan.

ACKNOWLEDGMENTS

I would like to thank Naamah Kelman-Ezrachi, Dean of the Hebrew Union College, Jerusalem, for her support and encouragement, Eli Minof for his valuable remarks, and all other researchers mentioned above. I am extremely thankful for all their contributions.

REFERENCES

Abramsky, S. 1958. *Haraš.* In: Mazar, B., Tur-Sini, N.H., Yeivin, S., eds. *Encyclopedia Biblica 3.* Jerusalem: 304–308 (Hebrew).

Aharoni, Y. 1956. Galilean Survey: Israelite Settlements and their Pottery. *Eretz-Israel* 4: 56–64 (Hebrew with English summary).

Aharoni, Y. 1957. *The Settlement of the Israelite Tribes in Upper Galilee*. Jerusalem (Hebrew).

Albright, W.F. 1938. *The Excavations of Tell Beit Mirsim II: The Bronze Age* (Annual of the American Schools of Oriental Research 17). New Haven.

Artzy, M. 1993. Nami. In: Stern, E., Lewison-Gilboa, A. and Aviram, J., eds. *The New Encyclopedia of Archaeological Excavations in the Holy Land,* Vol. 3. Jerusalem: 1095–1098.

Artzy, M. 1995. Nami: A Second Millennium International Maritime Trading Center in the Mediterranean. In: Gitin, S., ed. *Recent Discoveries in Israel: A View to the West.* New York: 17–41.

Artzy, M. 2000. Cult and Recycling of Metal at the End of the Late Bronze Age. In: Astrom, P. and Süzenhagen, D., eds. *Periplus, Festschrift für Hans-Günter Buchholz zu seinem achtzigsten Geburtstag am 24. Dezember 1999* (Studies in Mediterranean Archaeology 127). Uppsala, Sweden: 27–32.

Artzy, M. 2006. *The Jatt Metal Hoard in Northern Canaanite/Phoenician and Cypriote Context* (Cuadernos ed Arqueología Mediterránea 14). Saragossa.

Barako, T.J. 2007. *Tel Mor: The Moshe Dothan Excavations, 1959–1960* (Israel Archeology Authority Reports 32). Jerusalem.

Beit-Arieh, I. 1985. Serabit el-Khadim: New Metallurgical and Chronological Aspects. *Levant* 7: 89–116.

Ben-Dov, R. 2002. The Late Bronze Age "Mycenaean" Tomb. In: Biran, A. and Ben-Dov, R., eds. *Dan II: A Chronicle of the Excavations and the Late Bronze Age "Mycenaean" Tomb* (Annual of the Nelson Glueck School of Biblical Archaeology). Jerusalem: 35–234.

Ben-Dov, R. 2011a. *Dan III: Avraham Biran Excavations 1966–1999. The Late Bronze Age* (Annual of the Nelson Glueck School of Biblical Archaeology 9). Jerusalem.

Ben-Dov, R. 2011b. Craft Workshops at Tel Dan. *Eretz-Israel* 30: 77–94 (Hebrew with English summary).

Ben-Dov, R. and Gorski, A. 2009. Metal Implement Wrapped in Linen from Tel Dan. *Israel Exploration Journal* 59: 80–85.

Ben-Yosef, E. 2012. A Unique Casting Mold from the New Excavations at Timna Site 30 (Israel): Evidence of Western Influence? In: Kassianidou, V. and Papasavvas, G., eds. *Eastern Mediterranean Metallurgy and Metalwork in the Second Millennium BC.* Oxford: 188–196.

Ben-Yosef, E., Levy, T.E. and Najjar, M. 2009. New Iron Age Copper-Mine Fields Discovered in Southern Jordan. *Near Eastern Archaeology* 72(2): 98–101.

Ben-Yosef, E., Shaar, R., Tauxe, L. and Ron, H. 2012. A New Chronological Framework for Iron Age Copper Production at Timna (Israel). *Bulletin of the American Schools of Oriental Research* 367: 31–70.

Biran, A. 1989a. The Evidence for Metal Industry at Dan. *Eretz-Israel* 20: 120–134 (Hebrew with English summary).

Biran, A. 1989b. Tel Dan 1987, 1988. *Israel Exploration Journal* 39: 93–96.

Biran, A. 1994. *Biblical Dan*. Jerusalem.

Biran, A. 1996. A Chronicle of the Excavations, 1966–1992. In: Biran, A., Ilan, D. and Greenberg, R., eds. *Dan I: A Chronicle of the Excavations, the Pottery Neolithic, the Early Bronze Age and the Middle Bronze Age Tombs* (Annual of the Nelson Glueck School of Biblical Archaeology). Jerusalem: 1–63.

Biran, A. 2008. Dan. In: Stern, E., ed. *The New Encyclopedia of Archaeological Excavations in the Holy Land,* Vol. 5. Jerusalem: 1686–1689.

Blakely, S. 1999. Metallurgy and Cult in Greece and the Near-East. *American School of Oriental Research Newsletter* 49(1): 1719.

Callaway, J.A. 1993. Raddana, Khirbet. In: Stern, E., Lewison-Gilboa, A. and Aviram, J., eds. *The New Encyclopedia of Archaeological Excavations in the Holy Land*, Vol. 4. Jerusalem: 1253–1254.

Davey, C.J. 1979. Some Ancient Near Eastern Pot Bellows. *Levant* 11: 101–111.

Davies, N.D.G. 1925. *The Tomb of Two Sculptors at Thebes*. New York.

Davies, N.D.G. 1943. *The Tomb Rekh-mi-Rē at Thebes*. New York.

Davies, N.D.G. 1963. *Scenes from Some Theban Tombs*. Oxford.

Dever, W.G. and Tadmor, M. 1976. A Copper Hoard of the Middle Bronze Age I. *Israel Exporation Journal* 26: 163–169.

Dothan, M. 1956. The Sacrificial Mound at Nahariya. *Eretz-Israel* 4: 41–46 (Hebrew with English summary).

Dothan, M. 1961. The Excavations at Azor, 1960. *Yediot, Bulletin of the Israel Exploration Society* 25(4): 224–230 (Hebrew).

Dothan, M. 1981a. Akko 1980. *Israel Exploration Journal* 31: 110–112.

Dothan, M. 1981b. Sanctuaries along the Coast of Canaan in the MB Period: Nahariyah. In: Biran, A., ed. *Temples and High Places in Biblical Times.* Jerusalem: 74–81.

Eliyahu-Behar, A., *et al.* 2012. Iron and Bronze Production in Iron Age IIA Philistia: New Evidence from Tell es-Safi/Gath, Israel. *Journal of Archaeological Science* 39: 255–267.

Falconer, S.E. and Fall, P.L. 2006. *Bronze Age Rural Ecology and Village life at Tell el-Hayyat, Jordan* (BAR International Series 1586). Oxford.

Falconer, S.E. and Magness-Gardiner, B. 1989. Hayyat (Tell El). In: Homès-Fredericq, D. and Hennessy, J.B., eds. *Archaeology of Jordan. Surveys and Sites A-K* (Akkadica Supplementum 7). Leuven: 254–261.

Franken, H.J. 1969. *Excavations at Tell Deir ʿAlla I.* Leiden.

Grant, E. and Wright, G.E. 1939. *Ain Shems Excavations (Palestine) V: Text.* Haverford.

Greenhut, Z. 2006. *Production, Storage and Distribution of Grain during the Iron Age and their Linkage to the Socio-Economic Organization of the Settlement in Israel* (Ph.D. dissertation, Tel Aviv University). Tel Aviv (Hebrew).

Guy, P.L.O. 1938. *Megiddo Tombs*. Chicago.

Hamilton, R.W. 1935. Excavations at Tell Abu Hawam. *Quarterly of the Department of Antiquities in Palestine* 4: 11–60.

Herzog, Z. 1993. Gerisa, Tel. In: Stern, E., Lewison-Gilboa, A. and Aviram, J., eds. *The New Encyclopedia of Archaeological Excavations in the Holy Land*, Vol. 2. Jerusalem: 480–484.

Karageorghis, V. and Demas, M. 1985. *Excavations at Kition V*. Nicosia.

Karageorghis, V. and Kassianidou, V. 1999. Metalworking and Recycling in Late Bronze Age Cyprus—The Evidence from Kition. *Oxford Journal of Archaeology* 18(2): 171–188.

Kempinski, A. and Fritz, V. 1977. Excavation at Tel Masos (Khirbet El-Meshash). *Tel Aviv* 4: 136–159.

Killebrew, A.E. 1996. Report of the 1985–1987 Excavations in Field INE: Areas 5,6,7, The Bronze and Iron Ages, Text and Data Base (Plates, Sections, Plans). In: Gitin, S., ed. *Tel Miqne-Ekron*. Jerusalem.

Knapp, A.B., Kassianidou, V. and Donnelly, M. 2001. Copper Smelting in Late Bronze Age Cyprus: The Excavations at Politiko Phorades. *Near Eastern Archaeology* 64(4): 204–209.

Kochavi, M. 1963. The Excavations at Har Yeruḥam. *Yediot, Bulletin of the Israel Exploration Society* 27(4): 284–292 (Hebrew).

Kochavi, M. 1965. Tel Zeror. *Israel Exploration Journal* 15(1–2): 253–255.

Luciani, M. 2002. Operation K. In: Al-Maqdissi, M., Luciani, M., Morandi Bonacossi, D., Novák, M. and Pfälzner, P., eds. *Excavating Qatna I.* Damascus: 145–168.

Maisler, B. 1950–1951. The Excavations at Tell Qasile: Preliminary Report. *Israel Exploration Journal* 1(2): 61–76.

Mazar, E. 2009. *The Palace of King David, Excavations at the Summit of the City of David. Preliminary Report of Seasons 2005–2007.* Jerusalem.

Metzger, M. 1993. *Kāmid el-Lōz I: Die spätbronzezeitlichen Tempelanlagen, Die Kleinfund.* Bonn.

Minoff, E. 1992. A Mace Head Mold from Tel Dan. *Eretz Israel* 23: 87*–89*.

Rothenberg, B. 1975. Metals and Metallurgy. In: Aharoni, Y., ed. *Investigations at Lachish: The Sanctuary and the Residency (Lachish V,* No. 4*)*. Tel Aviv: 72–83.

Rothenberg, B. 1988. *The Egyptian Mining Temple at Timna*. London.

Rothenberg, B. 1990. Copper Smelting Furnaces, Tuyeres, Slag, Ingot-Molds and Ingots in the Arabah: The Archaeological Data. In: Rothenberg, B., ed. *The Ancient Metallurgy of Copper.* London: 1–77.

Rowe, A. 1940. *The Four Canaanite Temples of Beth-Shan.* Philadelphia.

Segal, I., Halicz, L. and Cohen, R. 1999. A Study of Ingots and Metallurgical Remains from ʻEin Ziq and Be'er Resisim, Central Negev, Israel. In: Young, S.M.M., Pollard, A.M., Budd, P. and Ixer, R.A., eds. *Metals in Antiquity* (BAR International Series 792). Oxford: 179–186.

Shalev, S. 1993. Metal Production and Society at Tel Dan. In: Biran, A. and Aviram, J., eds. *Biblical Archaeology Today, Proceedings of the Second International Congress on Biblical Archaeology, Pre-Congress Symposium: Population, Production and Power; June 1990, Supplement.* Jerusalem: 57–65.

South, A. 2012. Tinker, Tailor, Farmer, Miner: Metals in the Late Bronze Age Economy at Kalavasos. In: Kassianidou, V. and Papasavvas, G., eds. *Eastern Mediterranean Metallurgy and Metalwork in the Second Millennium BC, A Conference in Honour of James D. Muhly, Nicosia, 10th–11th October 2009.* London: 35–47.

Stech, T., Maddin, R. and Muhly, J.D. 1985. Copper Production at Kition in the Late Bronze Age. In: Karageorghis, V. and Demas, M., eds. *Excavations at Kition V: The Pre-Phoenician Levels, Areas I and II*. Nicosia: 388–402.

Stepansky, Y. 2000. Tel Kinneret, Map of Kefar Naḥum. *Hadashot Arkheologiyot* (Excavations and Surveys in Israel) 112: 13–16 (Hebrew).

Stern, E. 1984. *Excavations at Tel Mevorakh (1973–1976) II: The Bronze Age* (Qedem 18). Jerusalem.

Stern, E., *et al.* 2000. Tel Dor 1997–1998. *Hadashot Arkheologiyot* (Excavations and Surveys in Israel) 111: 23*–30*.

Tufnell, O., Inge, C.H. and Harding, L. 1940. *Lachish II: The Fosse Temple*. London.

Tylecote, F.R. 1979. *History of Metallurgy*. London.

Tylecote, F.R. 1981. From Pot Bellows to Tuyères. *Levant* 13: 107–118.

Ussishkin, D. 1983. Excavations at Tel Lachish 1978–1983: Second Preliminary Report. *Tel Aviv* 10(2): 97–185.

Waldbaum, J.C. 1978. *From Bronze to Iron*. Göteborg.

Wapnish, P. 1993. Archaeozoology: The Integration of Faunal Data with Biblical Archaeology. In: Biran, A. and Aviram, J., eds. *Biblical Archaeology Today, Proceedings of the Second International Congress on Biblical Archaeology. June 1990.* Jerusalem: 426–442.

Yadin, Y., *et al.* 1961. *Hazor III–IV: An Account of the Third and Fourth Seasons of Excavations, 1957–1958, Text.* Jerusalem.

Yadin, Y., *et al.* 1958. *Hazor I: An Account of the First Season of Excavations, 1955.* Jerusalem.

Yadin, Y., *et al.* 1960. *Hazor II: An Account of the Second Season of Excavations, 1956*. Jerusalem.

Yahalom-Mack, N. 2009a. *Bronze in the Beginning of the Iron Age in the Land of Israel: Production and Utilization in a Diverse Ethno-Political Setting* (Ph.D. dissertation, The Hebrew University). Jerusalem.

Yahalom-Mack, N. 2009b. The Metal Objects. In: Panitz-Cohen, N. and Mazar, A., eds. *Excavations at Tel Beth-Shean 1989–1996, III*. Jerusalem: 564–584.

Yahalom-Mack, N. and Segal, I. 2011. Provenancing Five Copper-Based Artifacts. In: Ben-Dov, R., ed. *Dan III: Avraham Biran Excavations 1966–1999: The Late Bronze Age* (Annual of the Nelson Glueck School of Biblical Archaeology 9). Jerusalem: 367–373.

Yahalom-Mack, N. and Shalev, S. 2005. Metal Production and Utilization. In: Ben-Tor, A. and Zarzecki-Peleg, A., eds. *Yoqneʿam II: The Iron Age and the Persian Period* (Qedem Reports 6). Jerusalem.

Yahalom-Mack, N. and Shalev, S. 2006. Metallurgical Activity in Levels K-5 and K-4. In: Finkelstein, I., Ussishkin, D. and Halpern, B., eds. *Megiddo IV: The 1998–2002 Seasons.* Tel Aviv: 542–549.

Yahalom-Mack, N. and Shalev, S. 2009. The Metal Objects. In: Gadot, Y. and Yadin, E., eds. *Aphek-Antipatris II: The Remains on the Acropolis: The Moshe Kochavi and Pirhiya Beck Excavations*. Tel Aviv: 416–443.

CHAPTER 32

A FIRST CENTURY JEWISH RECYCLING ECONOMY[1]

Matthew Ponting and Dan Levene

"Recycling Economies, When Efficient, Are By Their Nature Invisible"[2]

Alloying copper with other metals to produce a more workable and durable material is known to have occurred in antiquity. Archaeometallurgy has traditionally concentrated on the application of analytical methods to determine the composition and manufacturing processes of the alloys rather than the process through which such alloys were curated. Control of the proportions of metals added to the crucible, assuming the components were freshly smelted prior to the process of alloying, is easier to determine, and would clearly have been a technique that was used in antiquity. This manner of alloying affords control and is easily reproduced. Controlling the composition and quality of such alloys through the recycling process, however, is much more difficult and in general has been considered a less viable method to produced alloys of consistent quality. In this article a combination of archaeological finds and Jewish literary sources of late antiquity clearly show that the management of scrap metal was actually a very developed and sophisticated industry that provided products which metalworkers could rely on in their curation of these alloys.

INTRODUCTION

Samuel Krauss's introduction to the Hebrew edition of his original German *Talmudische Archaeologie* (Krause 1910–1912) is a testament to the scale of his vision. Though one should never underestimate the magnitude of his scholarly achievement we might acknowledge how naïve his prediction of what his interdisciplinary approach was soon to yield. Krauss's approach is not only still valid but one which needs to be embraced with greater enthusiasm by scholars. If there is any methodological criticism, it would surely be the fact that his interdisciplinary approach is weighted more on an interaction with secondary literature rather than the direct discourse between living practitioners of different disciplines. It is the sad truth that historians, philologists and archaeologists still do not work together often enough; this paper is the product of such collaboration.[3]

We start by discussing why Ponting's analyses of 1st century finds from Second Temple period sites show a high level of control of the types of copper alloy used. We then look at a section from the Mishna, a Jewish text that was redacted soon after these finds were made, that presents a list of

1 The English translations of the Jewish sources are based on Epstein, I. 1935. *The Babylonian Talmud.* London, Danby, H. 1933. *The Mishnah.* Oxford, and Neusner, J. and Sarason, R. S. 1977. *The Tosefta.* 6 vols. New York. In many cases, however, these have been amended by Levene.

2 Needham, S. P., Leese, M. N., Hook, D. R. and Hughes, M. J. 1989. Developments in the Early Bronze Age Metallurgy of Southern Britain. *Archaeometallurgy* 20: 384.

3 We dedicate this article to Beno Rothenberg, the father of modern archaeometallurgy, an inspirational teacher and always a friend.

the raw materials that were available to the metalworker at that time (Stemberger 1996: 108–148). This text provides unique information about metal recycling that we argue suggests the existence of a sophisticated industry of collection and sorting, as well as trade in, disposed-of metal objects that we refer to at present as the "scrap metal" industry, an industry that is, to a great extent, invisible to archaeology. We then argue that what we can learn from the Jewish texts about the way in which metal was recycled—in this period and beyond—offers an explanation for the compositions observed in the archaeological record. Finally, we use our ideas about the existence in late antiquity of a highly organized industry of what we call "scrap metal trade" to explain the make-up of an assemblage of metalwork found in an early medieval workshop in Tiberias (Ponting 2008). We argue that the best way to explain the different groupings of artifactual elements in the Tiberias workshop is that they were sorted according to composition, illustrating a fine ability to distinguish alloys in "scrap" so as to control the products that were then to be re-made from it.

COPPER ALLOYS

Throughout Classical antiquity copper was alloyed with three metals: tin, lead and zinc. Of these metals, tin was the most important alloying component prior to the 1st century BCE, when zinc began to be alloyed with copper on a large scale to make a golden-colored alloy today called brass. In antiquity this alloy was called *oreichalkos* by the Greeks and *aurichalcum* by the Romans. The occurrence of objects made from this alloy in excavated assemblages is of interest because it seems initially to have been reserved for specific uses such as coinage and, especially in the Roman world, military use (Bayley 1998: 19). Indeed, there is an increasing body of evidence to suggest that, from the early 1st century CE, brass was strongly associated with Roman Imperialism and became increasingly the alloy of preference for decorative metal objects of many types across the Roman world and among groups that sought to emulate Roman styles and fashions (Ponting 2002). Tin continued to be alloyed with copper in small amounts (<10%) to make bronze and was used for many common artifact types. Larger amounts of tin (~20%) were added to copper to make special alloys reserved for special items, such as mirrors (Craddock 1985: 64). The addition of lead to copper-alloys can also be of interest, first appearing early in the first millennium BCE, initially as an aid to casting, but increasingly as a cheap bulking agent during the Classical period, reaching the height of its use in Roman Imperial times (Fig. 32.1) (*ibid.*: 61).

Outside of the Mediterranean world of Classical antiquity the picture of copper-alloy choice and use is less clear. East of the Roman Empire the metallurgy of ancient Parthia is little studied, indeed, we have only managed to find a handful of analyses of Parthian copper-alloy coins from 1950 and nothing on other artifact types (Caley 1950). The coins analyzed are all made of bronze that is free of any zinc. However, it is known that alloys containing appreciable levels of zinc were being produced in northern India (Uttar Pradesh) as early as the first millennium BCE but these are rare. According to Craddock (1998: 27) the combination of textual and archaeological evidence suggests that brass was being made by cementation quite regularly in India by the later part of the first millennium BCE and that there is some (textual) evidence that may suggest that the technology to produce brass was introduced by Hellenistic invaders.

The raw materials for copper-alloy production are relatively scarce in the Levant and in antiquity must have involved trade over great distances. Clearly the curation of the metal that was brought in and transformed into artifacts was of considerable importance and the fact that copper-alloys can be melted down and re-cycled into new artifacts must have been heavily exploited as a result. However, evidence for recycling is difficult to find. Such evidence that exists is generally from earlier periods, but strongly

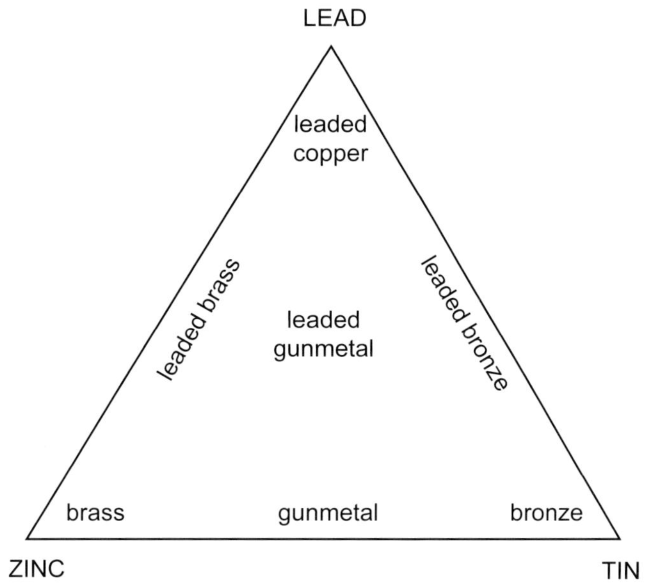

Fig. 32.1: Copper based alloys: bronze, brass, leaded copper and leaded gunmetal (after Bayley 1998: 8).

suggests that recycling of copper alloys was both necessary and commonplace. In particular, recycling appears to increase during periods of economic and social pressure and is often associated with temples and other structures of power and control (Karageorghis and Kassianidou 1999).

JEWISH VS. NON-JEWISH COPPER ALLOYS?

Analysis of copper alloy metalwork from regions on the periphery of Roman influence generally show a gradual increase in the proportion of brass as contacts increase and strengthen (often ending with conquest, as in the case of Britain) (Hamilton 1996; Dungworth 1997). Because of the increased availability of brass, the trace levels of zinc found in all copper-alloys also rise significantly due to the ease with which zinc vapor will contaminate other alloys through the re-use of crucibles and other tools (Dungworth 1995: 134). It is therefore worthy of comment that copper alloy objects found at Jewish sites dated to the period immediately before the Great Revolt are free from even trace levels of zinc, whereas those from earlier Jewish sites and from later non-Jewish sites include brasses, and the other copper alloys are contaminated with significant traces of zinc. The metal objects excavated at Gamla, Yodfat and in the "Burnt house" in Jerusalem have all been found to consist exclusively of bronze and copper with no brasses or other alloys containing significant levels of zinc (Fig. 32.2). This is strange as not only would brass objects be expected in assemblages of this period but the presence of brass in metal destined for re-cycling would, in usual working circumstances,[4] cause zinc to find its way into all copper alloys. This is clearly demonstrated in the analyses of copper-alloy objects from Roman Britain (Fig. 32.3). Dungworth's data from sites in Northern Britain shows how zinc contamination of bronzes increases during the 1st century CE, reflecting increased contacts with the Roman army and other groups regularly trafficking with the Roman world (Dungworth 1997). Indeed, brass objects form

4 One would expect that through recycling at least small quantities of brass, and consequently zinc, would find their way into most produced alloys, even if only through contaminated crucibles, as is shown in Dungworth's analyses of British material and Beck *et al.*'s analyses of French material.

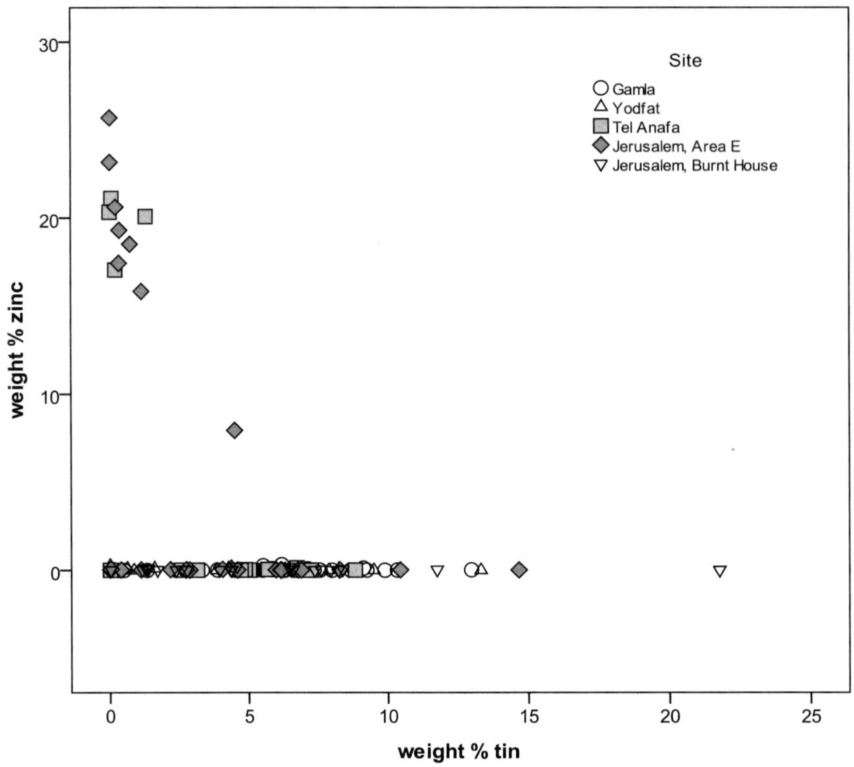

Fig. 32.2: Zinc and tin percentages in copper alloys in Roman Palestine.

a significant part of an assemblage of metalwork from a 1st century BCE house in Jerusalem (Ponting 2006). How it comes to be absent a century later from the copper alloys in Gamla, Yodfat and in the "Burnt house" in Jerusalem is something of a mystery.

What are the possibilities that might have resulted in zinc free copper-alloys in the Jewish assemblages in Palestine during the late Second Temple period?

1. These metal artifacts were a collection of very old objects that predate the arrival of Roman brass.
2. These metal artifacts were made in metal workshops that only used metal from ingots that had never been recycled and therefore could not have been contaminated with zinc.
3. These metal artifacts were made in workshops that used recyclable materials, the alloy types of which they could distinguish to such an extent as to exclude brass and sources of other minor zinc contamination.

The first and second possibilities are unlikely. The archaeological contexts of the objects and their typology clearly indicate that they are contemporary and commonplace. Given the level of recycling generally acknowledged as necessary to maintain metal supplies in the region, the second possibility is highly unlikely. The third possibility seems the most plausible but requires qualification. In order for us to consider it feasible we must try and understand how a workshop could retain both its metal supplies and workshop environment free from even the smallest traces of zinc. This would be especially difficult where the manufacturing cycle involves recycling scrap metal because the metalworker would have to rely on exceptional scrap sorting skills that would exclude not only the obvious brasses, identified by their golden color, but copper alloy objects that themselves were made of recycled copper alloy objects that contained zinc. How could such rigorous scrap-sorting be achieved and maintained?

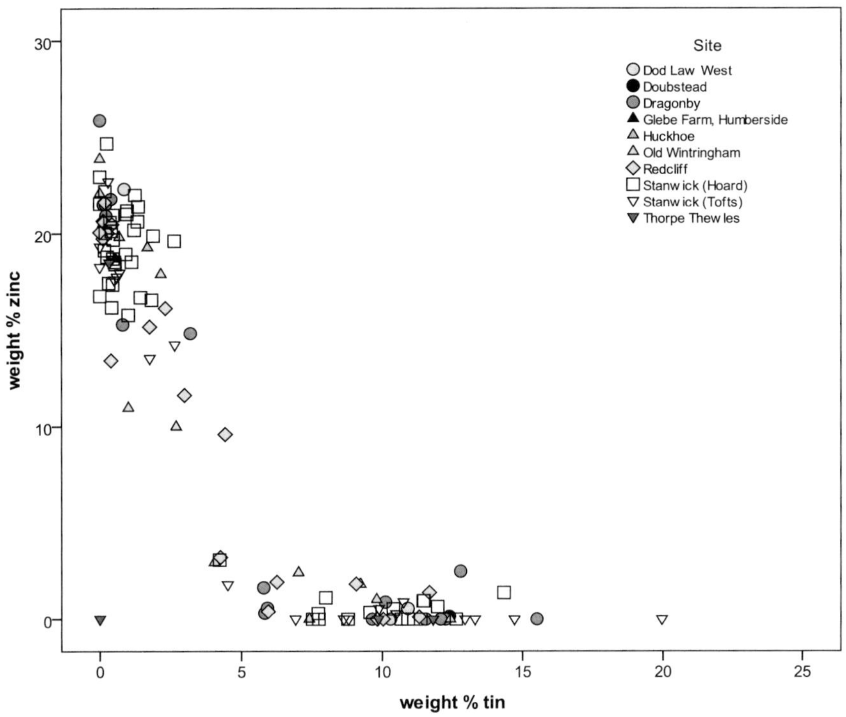

Fig. 32.3: Zinc and tin percentages in copper alloys in Roman Britain.

THE MISHNA—LITERARY EVIDENCE FOR RECYCLING

It has already been noted that the classical corpus of Jewish literature is not concerned with technical metallurgical information, at least not as its primary objective (Levene and Rothenberg 2007: 7–27). If, however, one were to choose a section of text that comes closest to doing so it would undoubtedly be what is found in the order of Tohorot (Purities) of the Mishna (mKel 11–14), Chapters 11–14 of the tractate *Kelim,* which might best be translated as "utensils" or "vessels," and which deals with the rules intended to guarantee that utensils entering the inner parts of the Temple did not contravene its state of ritual purity (Neusner 1974; Douglas 1993).[5] This mishanic text has an equivalent in the Tosefta (Stemberger 1996: 149–163), though it is absent from the Talmud.[6] The study of *Kelim* was largely neglected after the destruction of the Temple because it no longer had practical application, and many of the institutions and networks around it would have likewise disappeared. Thus it is lacking the later layers of interpretation and commentary that accompany other tractates, including commentaries which have often preserved older traditions.[7]

5 The tractate of *Kelim* discusses a variety of categories of materials into which all utensils belong: earthenware (Chapters 2–10), metals (Chapters 11–14), wood, leather and bone (Chapters 15–19), leather and clothing (Chapters 26–28) and glass (Chapter 30).

6 There are two versions of the Talmud, which literally means "study"—the Jerusalem and the Babylonian. The first of these was redacted ca. 450 CE and the latter ca. 600 CE (see Stemberger 1996: 164–223).

7 We do, however, possess a small number of commentaries to aid us in our attempts to understand this text, most significant of which is the medieval gaonic commentary attributed to Hai Gaon (Epstein 1982) and the commentary of Maimonides (Kafih 1963).

Of particular interest are mKel 11:1 and 3 in which we are presented with a confirmation of the part recycling played in the culture of that period as exemplified by the following rule:

כלי מתכות פשוטיהן ומקבליהן טמאין נשברו טהרו חזר ועשה מהן כלים חזרו לטומאתן הישנה

> Utensils of metal, [whether] flat [in shape] or receptacles are impure (i.e., susceptible to impurity). If they broke they became pure [i.e., insusceptible to impurity]. If again one made of them utensils they revert to their impurity [i.e., susceptibility to impurity] (mKel 11:1) (Levene and Rothenberg 2007: 36–41)

This section reiterates the common-sense realization that one of the most obvious properties of metal is its recyclability. This affirms the fact that the metalworker was not dependant on the use of ingots fresh from the smelting sites to ply his trade. Rather, what we see here is that his was an industrial culture of recycling whereby any metal object that outlived its usefulness ceased to be that object per-se and reverted to being raw material; it was transformed, so to speak, from an object with a name and a specific function to raw material, i.e., "scrap."

The second rule regarding metal utensils is as follows:

כל כלי מתכות שיש לו שם בפני עצמו טמא

> Any metal utensil that has a name of its own is impure [i.e., susceptible to impurity] ... (mKel 11:2)

We provide this section to illustrate the sensibility behind the mishnaic taxonomy. As referred to in the comment to 11:1, the recyclability of metal means that an object can be categorized in different ways according to whether it is still deemed to fulfil the function for which it was made. In a moment it can be transformed semantically from "utensil" to "scrap." This semantic change is for the ancient Jew extended to a change also in the morphology of the material. Metal formed into a shape whereby it can then be used to perform a useful function was considered by the ancient Jew to acquire a state of susceptibility to impurity, or just referred to as "impure"—טמא. Once such a metal object/utensil ceased to be considered functional it then became unsusceptible to impurity, referred to succinctly as "pure"—טהור. This could happen when an object broke, decayed, stopped being useful as a result of age, etc. (*ibid.*: 136–144).

The following section, mKel 11:3, is the most significant for our discussion:

העושה כלים מן העשת ומן החררה ומן הסובב של גלגל ומן הטסין ומן הצפויין מכני כלים ומאגני כלים מאזני כלים מן השחולת ומן הגרודות טהורין רבי יוחנן בן נורי אומר אף מן הקצוצות משברי כלים מן הגרוטים ומן המסמרות שידוע שנעשו מכלים טמאין מן המסמרות בית שמאי מטמאין ובית הלל מטהרין.

> One who makes utensils from [either]: the lump of metal, from the ingot, from the hoop of the wheel, from sheet metal, from the metal plating, from bases of utensils, from rims of utensils, from handles of utensils, from what falls in the process of metalwork, from the filings and shavings, they are pure [i.e., unsusceptible to impurity]. R. Johanan ben Nori[8] says: "Also from the cut bits of metal." [Utensils that are made from] broken bits of [metal] utensils, from *scrap metal* and from nails that are known to have been made from [other] utensils are impure (i.e., susceptible to impurity). [Utensils that are made] from nails, the School of Shammai declare them impure, but the School of Hillel declare them pure.[9]

The fact that the raison d'être for this text was to provide metal–workers with suitable guidelines to assure the goods produced in their workshops adhered to the requirements of Temple ritual purity does not detract from what we perceive it to be: a list of raw materials that were available to the metalworker. This list is, to the best of our knowledge, unique for this period and region and must therefore be pivotal

8 First to 2nd century CE.

9 For a discussion of the halakhic significance a metallurgical and philological analyses of this text yields, and some notes on the various Hebrew terms in this text and their meaning, see Levene and Rothenberg 2007: 148–162.

in helping us understand the metal working industry of the late Second Temple period. This is especially pertinent if we wish to consider that the most plausible way of interpreting the archaeological evidence is to take it as an indication that metalworkers were able to ensure that the scrap metal they recycled did not produce coppers or bronzes that were contaminated by zinc. We shall start with some more general observations about this list and then consider if this passage and what we can learn from it can shed light on how the Jews of Roman Palestine managed to keep their copper alloys free of zinc in a way that others, such as in Roman Britain and Gaul, were unable or were uninterested in doing.

One of the most obvious things about this list is that it does not delineate the types of metal that each of its constituents might relate to. This is an issue that has been partly discussed elsewhere (*ibid.*: 141–144); indeed, such an enquiry into the extensive lists of metal objects that exists in the tractate of *Kelim* is a desideratum and one the present authors hope to address at some length in the future.

The most noteworthy aspect of this list for our purposes is the fact that apart from the first two items, the "lump of metal" and the "ingot," the rest of the items are all clearly recyclable. Of the recyclables there are several categories we might distinguish: The first group, "the hoop of the wheel," "sheet metal," "metal plating," "bases of utensils," "rims of utensils" and "handles of utensils" seem to all have in common the fact that they are not broken up but are identified by form and function. The second group, "what falls in the process of metalwork," and "the filings and shavings," are the by-product of manufacturing processes. And the third group, the "cut bits of metal,"[10] "broken bits of [metal] utensils," "*scrap metal*" and "nails" seem to be what we would more commonly refer to as scrap; those objects and parts of them that have ended their life-cycle of usefulness, have become damaged, bits that have broken off, or such objects or parts of them that have been smashed or chopped up.

THERE IS SCRAP AND THERE IS SCRAP – *grwṭym* גרוטים

As can be seen above we have translated the Hebrew term גרוטים (*grwṭym*) as "*scrap metal*" yet have rendered it in italics, as we shall argue that the meaning of the Hebrew term differs in nuance from the meaning of its English rendition. Before looking at the etymology of this word or considering what it might mean from looking at the way it is used in context elsewhere in the literature of the period we should consider its meaning in mKel 11:3. Note that the term *grwṭym* appears within a list alongside the items "what falls in the process of metalwork," "filings," "shavings," "cut bits of metal" and "broken bits of [metal] utensils," which suggests that it is considered as distinct from them. Whereas we consider these other items as sub-categories of what we call "scrap metal" the text presents the term *grwṭym* as an equivalent item. So in the context of our list of raw materials *grwṭym* can be distinguished as scrap metal that is not the by-product of a manufacturing process, it does not consist of parts of metal objects, and it is not processed scrap (cut or broken up). What is left is the disused object, being a very specific type of scrap. We might, therefore, suggest that the contemporary English term "scrap metal" covers a wider variety of things than what the mishnaic Hebrew *grwṭym* encompasses. This is a semantic subtlety but one that indicates that the variety of items that we include as being defined by what we call "scrap[11] metal" were not the same as those included in the term *grwṭym*.

The word גרוטים (*grwṭym*) is a plural noun and is loaned into Hebrew from Greek and attested in Hebrew only from its mishnaic form of the language onwards. It is derived from the Greek *γρύτη* which can mean either a "woman's dressing-case or vanity-bag," a "workman's tool-bag" or "frippery." (See

10 Admittedly, this category could encompass material that is the by-product of a manufacturing process.

11 "Scrap" being a word that in contemporary English is "often confused with waste," <http://en.wikipedia.org/wiki/Scrap>.

Jastrow 1996: 266b; Ben-Jehiel, Kohut and Mussafia 1878: 363; Krauss 1898: 183b–184a; Levy and Leberecht Fleischer 1876, Vol. 1: 358b–359a) It is also used to denote "smallness" in relation to fish (Liddell and Scott 1940). One can possibly see some connection between the Greek and the Hebrew through the term "frippery," maybe even its association of "smallness," but this is a loose connection and it is clear that in Hebrew it acquired a different meaning.

The process of recycling starts at home when a metal object ceases to be of use. We cite two text references that tell us that people had a dedicated pile of *grwṭa'wt* (scrap metal) to which they added other such metal objects. In the Babylonian Talmud Shabbat 123a there is mention of a needle that has become defunct either through the loss of its eye or point that is cast among the owners *grwṭa'wt*,[12] and in the Babylonian Talmud *Bava Metziᶜa* 52b it conversely suggests that the owner of a sela coin (a silver tetradrachm) that had been damaged beyond a certain point may not be cast to his *grwṭa''wt*.[13] Both these references suggest that much like we have recycling bins for certain types of materials the same was true in antiquity. The disused metal object lost the value that was intrinsically connected with its form and design. Yet it was recognized that its intrinsic value as a raw material had not been lost.

What information then do we have about how scrap metal entered the trade cycle and how did this aspect of trade and industry look? Here we present two text references where there is mention of *grwṭa'wt* (metal scrap) as a trade commodity. In the Babylonian Talmud *Avodah Zarah* 53a there is a section that is concerned with what a merchant may do if he finds that he has in his possession a metal idol. What is interesting is the description of the way in which it was acquired; the text presents the circumstances as follows: "if one bought scrap metal from heathens and found an idol amongst it."[14] The implication here is that the merchant bought a job-lot of *grwṭa'wt* among which there was a metal idol. We are not told who the heathen was but it is not impossible that he was, like the man with the needle mentioned in the Babylonian Talmud *Shabbat* 123a (above), a person who amassed a *grwṭa'wt* through buying from large households, various institutions or workshops and then resold this material. The second reference is also from the Babylonian Talmud *Bava Metziᶜa* 73a in which rules regarding cash investments in and loans to merchants who take advantage of price differences in different localities. This aspect of the discussion is of less interest to us; what is of interest is the fact that one of the varieties of commodities that is discussed is *grwṭa'wt*.[15] So it seems that *grwṭa'wt* was a commodity that was traded and, furthermore, as the material presented so far suggests, consists of a jumble of materials that have not been sorted out. One might find in such a

12 אדם זורקה לבין גרוטאות, MS Oxford OPP .ADD FOL. 23 (366), Vatican 108 גרוטאותיו, "his scrap".

13 ולא יזרקנה בין גרוטותיו.

14 הלוקח גרוטאות מן העובדי כוכבים ומצא בהן עבודת כוכבים, the Soncino translates לוקח as "brought" which is amended here to "bought" which is both a possible meaning of the root לקח (Jastrow[1996]: 717) and is supported by the fact that the exchange of money is mentioned in this section later on.

15 As it happens the Talmud advises against investment in *grwṭa'wt* (scrap metal) as opposed to other agricultural produce. The reason is not given; however, one of the medieval commentators, the RABAD (Rabbi Abraham son of David, France, 13th century), suggested that this was due to the fact that *grwṭa'wt* were not a usual commodity in the sense that they were gathered only sporadically. This is not an entirely satisfactory explanation. We suggest that it is more likely to be connected to either the control of the prices of metal or the fact that there is somehow not such a predictable or reliable price to such material before it has been sorted out properly and therefore one might not be able to predict, as with other produce, to the same extent one would the fluctuations of value as with agricultural produce —i.e., the price of apples in different regions, etc. In any case, there might have been a whole variety of reasons that are not apparent to us from the text. What it does tell us, however, is that it was a traded commodity and one upon which the lay person was not advised to speculate.

job lot of *grwṭa'wt* a needle or an idol. What also seems to be indicated by the examples presented so far is that *grwṭa'wt* are discernible objects, ones that are not in use but are still recognizable as what they once were.

A text reference from the Tosefta *Kelim*: *Bava Metziʿa* 1:3 adds another aspect of the scrap metal trade in which a distinction is made between *grwṭa'wt* that came from abroad and those that have come from local sources.[16] This suggests a knowledge and/or perception of important differences between scrap metal from different sources and is discussed later in this paper.

So far we have ascertained from the texts that disused metal objects were collected at source, in the home or any place where they were owned and used (Babylonian Talmud, *Shabbat* 123a and Babylonian Talmud *Bava Metziʿa* 52b). Then at some point these assemblages were sold (Babylonian Talmud *Avodah Zarah* 53a). The word used to describe these assemblages of disused and unsorted metal objects is *grwṭa'wt*. We have seen that such assemblages were a traded commodity. We have also seen that they were traded between towns domestically (Babylonian Talmud *Bava Metziʿa* 73a). The reference in Tosefta *Kelim Bava Metziʿa* 1:3 shows that they were also traded across borders and over long distances.

A SOPHISTICATED CULTURE OF RECYCLING

The evidence that we have thus far shows that at least in the case of the Jewish sources from the later part of antiquity there is compelling evidence for the existence of an organized culture and industry of metal recycling.

In juxtaposition to this evidence we have the copper alloy assemblages from Gamla, Yodfat and the "Burnt house" in Jerusalem that show an ability of metalworkers to control the composition of their alloys to such a degree as to be able to produce copper alloys free of zinc traces all together.

We would argue that the literary evidence that we have presented above provides an explanation for how these Jewish metalworkers might control the composition of their copper alloys by having a long tradition of a highly sophisticated culture of exceedingly selective metal recycling.

PURITY AND SUSCEPTIBILITY TO IMPURITY AS AN INDICATOR OF CAREFUL RECYCLING

"It is hard to imagine that the detailed and technical descriptions of utensils and objects described in the Mishna and Tosefta *Kelim*, both whole and broken, represent technological fantasies rather than material reality" (Schwartz 2006: 149). This statement might seem commonsensical but the transmission of these texts persisted despite the fact that after the destruction of the Jerusalem Temple in 70 CE they had no practical use, and the fact that some of their meaning was lost.[17] Our attempt to recover a better understanding of these texts depends on the chance of coming across meaningful materials that we can cross match or compare, either in the form of other texts and/or archaeological finds (Levene and Rothenberg 2007: 142–43).

There are a number of points that we can deduce from the *Kelim* texts about the use of metals and their related industries in late Second Temple Palestine. The most obvious is that the late Second Temple Jews were concerned with issues of purity and impurity, and furthermore, they had very specific and

16 הגרוטאות בין משל ארץ בין משל חוצה לארץ.

17 See the discussion regarding mKel 11:3 in Levene and Rothenberg 2007: 136 ff.

particular concepts/beliefs that relate to metals' susceptibility to impurity.[18] We may also state with confidence that these concepts/beliefs were directly related to the material properties that this class of substance, namely metals (מתכות), displays.

The tractate *Kelim* provides a list of the types of raw materials that the artisan, in this case the metalworker, would use in his workshop (mKel 11:3). This pericope provides a list of supplies that presumably would have been available to the late antique metalworker in Palestine. This differs to a great extent from most objects made from other materials that are discussed in the other chapters in *Kelim* whose processes of recyclability (where such exist) might well be referred to as "reused garbage," the study of which has been referred to elsewhere as "garbology."(Schwartz 2006) In contrast, the list of metal objects mentioned in mKel 11:3 represent well defined commodities for which the description "garbage" does not apply; metal scrap has always had, as today, a significant commercial value. Considering both current, recent historical, and what we have shown is implied from the Mishna, Tosefta and Talmud, practices of the collection, processing, supply and trade of metal scrap, we would argue that the list in mKel 11:3 represents a list of recognized commercial wares that were the domain of established networks of industrial suppliers.

Early rabbinic texts such as the Mishna and Tosefta indicate that the rules governing the ritual purity of utensils would have been adhered to. Furthermore, we must acknowledge that for this to take place there would have had to have been an industry of manufacture with its suppliers of raw materials that would have catered to the people who adhered to these rules. Indeed, we know that such rules were in evidence in some form already during the First Temple and that these were the kernel from which the rather complex and detailed rules that we have in the Mishna, in the tractate of *Kelim* in the order of *Tohorot*, and other related materials in the Tosefta and Talmud eventually evolved. It is clear that for such rules to have been adhered to there would have had to have been an industry that could produce utensils that were ritually pure, could be maintained in such a state during their journey to the Temple, and could also deal with the recycling of impure utensils in a way that would assure that the impurity could be satisfactorily quarantined and eventually removed. As the mishnaic manual of *Kelim* clearly states, this involved very clear rulings regarding the raw materials that were used (mostly recyclables—mKel 11:3). For this aspect to be maintained very reliable networks of supply would have had to be established and maintained. An example of such an industry is, of course, the stone vessel industry in the Second Temple period. Here, an entire industry developed to service the need for ritually pure vessels at a time when the Tosefta tells us that "purity broke out in Israel" (Tosefta Shabbat 1:14; JT Shabbat 1:3b; Magen 2002),

The lack of zinc in the assemblages of objects from Gamla, Yodfat and the "Burnt house" in Jerusalem may well have been the result of such strictly controlled networks of scrap metal supply that were long established and still existed at the end of the Second Temple period.

18 The tractate of *Kelim* discusses a variety of categories of materials into which all utensils belong: earthenware (Chapters 2–10), metals (Chapters 11–14), wood, leather and bone (Chapters 15–19), leather and clothing (Chapters 26–28) and glass (Chapter 30). These material categories are derived from the following biblical verses, though the Mishna elaborates much beyond the biblical text. Lev. 11:32 "And anything on which one of them falls when dead shall be unclean: be it any article of wood, or a cloth, or a skin, or a sack—any such article that can be put to use shall be dipped in water, and it shall remain unclean until evening; then it shall be clean"; Numbers 31:22–23 "Gold and silver, copper, iron, tin, and lead—any article that [habitually] comes in fire—these you shall pass through fire and they shall be pure, except that they must be purified with water of lustration; and anything that does not [habitually] come in fire you must pass through water"; and Leviticus 11:33 "And if any of those falls into an earthen vessel, everything inside it shall be unclean and the vessel itself you shall break."

Goodman (2008: 288) points out that in the Old Testament it is clearly implied that purity was connected to the Temple rituals and that it was recognized that being in a state of impurity was unavoidable and that there were rituals to remove it when needed. He contrasts this by stating that in "post-biblical texts pollution was sometimes treated as intrinsically undesirable" (*ibid.*: 289).

We know that on the one hand the rules pertaining to purity and impurity are ancient and, though they evolved over a long period of time, are nevertheless sure to reflect earlier practices. A metalworking industry concerned with supply and manufacture to support the community that adhered to such rules is sure to have evolved and is more than likely to have been significant by the late Second Temple period when pilgrimage, as various Jewish sources of the period attest, was very big business (Goodman 1999). Evidence for a significant and long lived demand, are in themselves enough to imply the continued existence of an industrial network of supply and manufacture (Karageorghis and Kassianidou 1999).

MKel 11:3 reveals the product list that the supply network would have traded. The supply and manufacture network would have had to be both a conservative one and a very skilled one in terms of the ability to sort out types of recyclable metal scrap. So much so as to be able to maintain supplies that were completely separate from that of the Roman metal industry and so remain completely free of zinc. In contrast to this is the example from the British Isles shown above, in which contact with the Roman world made its mark in the form of zinc traces across all forms of copper alloy. The ability to retain a completely separate industrial network in the east of the Roman Empire would have relied on a conservative and well-established metal trade network that had evolved to accommodate Temple and related needs. An additional factor that might have helped keep at least part of this conservative supply network both separate from the Roman one and contamination by brass is the fact that, as far as we know, the Persian world had yet to embrace brass technology.[19] In fact, it has been suggested that Iranian production of brass starts only in the 6th century CE (Forbes 1950: 284–285; Allan 1979: 45). If this is indeed the case, and some of the conservative supply networks that Jews who required "purity"/"impurity" controlled bronze ware (as implied in mKel 11:3) were in the Iranian sphere, then any such metalwork they acquired via these channels would have been zinc free. Such circumstances offer a scenario that would explain the zinc free alloy patterns found in Jewish metalwork on the eve of the destruction of the Temple.

The Tosefta *Kelim Bava Metziʿa* 1:3, which we mentioned above, suggests the probability of exactly such a scenario:

> העושה כלים מן הגרוטאות בין משל ארץ בין משל חוצה לארץ טמאין ר' יהודה אומר העושה כלים מן הגרוטאות של חוצה לארץ טהורין
>
> One who makes utensils from the *grwṭa'wt*, whether they are from this country or whether from abroad they are impure (susceptible to impurity). Rabbi Yehuda says: One who makes utensils from the *grwṭa'wt* from abroad they are pure (unsusceptible to impurity).

The "*grwṭa'wt* from abroad" are pure. What is interesting in this statement is the fact that Parthian supplies that would have been included in this as sanctioned for their purity would have also provided the metalworkers with zinc free alloys for recycling. What we see here is literary evidence of the fact that the Jewish metalworking industry of the late Second Temple period was concerned and particular about its metal supplies; which predominantly consisted of recyclable metal. The archaeological finds that are zinc free suggest that such concern and particularity was equally present in regard to the composition of the alloys of the material that came in these supplies.

19 There is so far no evidence for the Parthian zinc, contra Triester 1990.

SCRAP AWARENESS

The evidence that we have presented suggests a very careful and aware culture of recycling, both in terms of sensitivity to the physical properties of the metals as well as their rules of ritual purity. They distinguish the varieties of types of what we call "scrap metal" with a greater lexical range than the English term that we use suggests. To them *grwṭa'wt* has the much more limited meaning of "metal object that has stopped being of use," i.e., it is of no use in its current form and despite being recognizable in its shape for what it used to be, it reverts for them to raw material ready to be refashioned—recycled. Most of the rest of the list in mKel 11:3 which would come under the term "scrap metal" in current English is not referred to in mishanaic Hebrew by such an all-encompassing term. Once the metal is sorted it is not scrap to them; it is what metal utensils are made from.

A MEDIEVAL ASSEMBLAGE OF "SCRAP METAL" FROM TIBERIAS

In 1998 the excavation of Fatimid (11th century CE) buildings at the foot of Mount Berenice on the outskirts of the modern town of Tiberias revealed what appears to be a coppersmith's workshop (Hirschfeld and Gutfeld 2008). Three large ceramic pithoi were found containing over 1,000 copper-alloy objects together with over 200 kg of scrap metal (Fig. 32.4). A sample of 103 objects was analyzed to establish the range of alloys used and it was found that over half of the objects were made of brass, 16% were made of bronze, 20% made of gunmetal and 10% of copper. Clearly brass would appear to be the alloy of preference. Most of the objects were incomplete or fragmentary; all the lampstands had been disassembled and there were groups of separate handles, legs and other parts of objects that did not match. The scrap metal comprised predominantly of cut-off handle-lugs and rims of large copper buckets; very few pieces of gunmetal or brass scrap were identified. There was also a large amount of copper off-cuts from manufacturing processes. This was curious given the low proportion of un-alloyed copper among the objects and suggests that a fairly rigorous process of separation and selection was being applied. Additionally, there were two brass jugs filled with metal turnings from finishing vessels on a simple lathe. Analysis of a sample of the contents indicated that these, too, were un-alloyed copper and did not appear to relate to any of the vessels in the hoard that were almost exclusively made of brass or gunmetal.

Possibly the most surprising element in the assemblage was the inclusion of 85 copper-alloy coins, of which 75 were Byzantine copper folles dating to between 1030 and 1078 CE. All these coins were made of pure copper and were not part of the official Fatimid currency system of the time, which was composed of gold and silver denominations. Therefore, the value of the coins to the hoarders would have been purely intrinsic and would explain their presence alongside other forms of un-alloyed copper scrap.[20]

Because virtually all the identifiable objects in the hoard were essentially disassembled parts of complete pieces of copper-alloy furniture, equipment or table-ware and were accompanied by a large volume of cut-up scraps of metal, it is conceivable that the entire assemblage should be seen as scrap. This scrap, however, was carefully divided-up and selected and there were few, if any, links between individual pieces. It might have been possible to match some of the lampstand elements and some of the legs from coherent groups, but generally speaking the hoard was comprised of separate elements selected by a set of criteria that defies a modern rationale.

20 The interpretation of other coin hoards as "bullion" hoards is well documented; indeed, a larger hoard including anonymous folles from Turkey was similarly interpreted as a "bullion" hoard with it being noted that "this kind of collection could be found in coppersmiths' shops until quite recently" (Lowick, Bendall and Whitting 1977: 16–17).

Fig. 32.4: Examples of some of the categories of raw materials listed in mKel 11:3 that are also found in the Tiberias hoard. From top left: sheet metal/metal plating/cut bits of metal, bases of utensils (examples of two types), rims of utensils (examples of rim and folded rims), handles of utensils (examples of ewer handle, spoon handle and table leg), and broken bits of metal utensils.

The Tiberias hoard becomes much more understandable if viewed alongside the descriptions of the types of metal found in the documents discussed above (especially mKel 11:3). We find that in the Tiberias hoard we have four of the five categories of raw materials that were listed as being necessary for the production of utensils that would be regarded as halakhically pure. There is a lump or ingot of leaded-bronze, cut pieces of sheet-metal, bases, rims and handles of utensils as well as the two jugs full of shavings and a large amount of "cut bits of metal." Indeed, the only category not represented is metal "from the hoop of the wheel" which is explained by the entire assemblage being restricted to copper-alloys while wheel hoops would have been made of iron; a metal with which this particular workshop was not involved. The fact that the make-up of the 11th century CE hoard of copper-alloy objects so closely matches the categories listed in the Mishna from the Second Temple period is remarkable, more so given that many of the objects in the Tiberias assemblage clearly originated in both Christian (the coins) and Moslem (style of and inscriptions on vessels) traditions. However, it seems likely that what we are seeing in the archaeological reality of the Tiberias hoard is the continuation of well-established traditions of the recycling industry of the region, technological traditions of which we have an earlier example dating to the Second Temple period in the form of the list from mKel 11:3—a list that was a product of a preoccupation with ritual purity that reflects, and inadvertently reveals, the very real concerns about the quality and composition of copper-alloys in a region where recycling was the mainstay of the metalworking industry.

CONCLUSIONS

The recycling of copper-alloy metalwork is a human activity that is invisible in the archaeological record and it has usually been assumed that the majority of copper-alloy objects in the past were made from freshly smelted metal. This is implicit in most published works on ancient copper-alloy working; opening as they usually do with a discussion of ore sources, smelting technologies and the archaeological evidence for these. The recycling of copper-alloys is sometimes mentioned in passing as a likely explanation for the increased use of mixed ternary alloys (gunmetals) from the later 1st century CE onwards, but then fades from the discussion (for example Tylecote 1976, Healy 1978 and, more recently, Craddock, in Oleson 2000). However, apart from the fact that the recycling of a precious and scarce resource such as metal is likely to have been "de rigueur" from the inception of metalworking and is clearly attested in the ethnographic record, there is a growing body of analytical and archaeological evidence that points to a widespread reliance on copper alloy recycling. One of the earliest pieces of archaeological evidence is the group of ingots from the Cape Gelidonya wreck (Bronze Age) that includes two stylistically different ingots made of bronze rather than un-alloyed copper and which have been interpreted as being a class of ingot made from scrap (Muhly, Stech-Wheeler and Maddin 1977: 358). The excavations at Kition in Cyprus have revealed Late Bronze Age workshops in the temple precinct where scrap bronze from votive offerings appears to have been recycled on a large scale (Karageorghis and Kassianidou 1999). The scientific analysis of copper-alloy artifacts suggests that low levels of arsenic in early tin-bronze objects indicates the recycling of arsenical coppers to make bronze (Northover 2000) and, similarly, that the gradual reduction of tin contents observed in some quarters is indicative of recycling due to the loss of tin through volatilization every time the alloy is melted (Gowland 1930: 511). However, the topic has yet to be comprehensively studied, although some encouraging research is in progress. This study of textual data alongside archaeological data adds a further dimension to this discussion and provides the important human dimension often lacking in purely scientific investigation. Here we have the voices of the people who regulated the day-to-day activities of the producers and consumers of metal artifacts and which allow us a hazy glimpse into their minds and their concerns and priorities.

REFERENCES

Allan, J. 1979. *Persian Metal Technology, A.D. 700–1300*. London.

Bayley, J. 1998. The Production of Brass in Antiquity with Particular Reference to Roman Britain. In: Craddock, P.T., ed. *2000 Years of Zinc and Brass* (British Museum Occasional Paper 50). London: 7–26.

Beck, F., Menu, M., Berthould, T. and Hurtel, L. 1985. Métallurgie des bronzes. In: Hours, J., Le Gal, S. and Fleury, F., eds. *Recherches gallo-romaines I*. Paris: 68–139.

Ben-Jehiel, N., Kohut, A. and Mussafia, B. 1878. *Sefer Arukh Ha-Shalem*. New York.

Caley, E.R. 1950. Notes on the Chemical Composition of Parthian Coins with Special Reference to the Drachms of Orodes I. *Ohio Journal of Science*: 107–120.

Craddock, P.T. 1985. Three Thousand Years of Copper Alloys: From the Bronze Age to the Industrial Revolution. In: England, P. and van Zelst, L., eds. *The Application of Science in the Examination of Works of Art*. Boston: 59–67.

Craddock, P.T. 1998. Zinc in India. In: Craddock, P.T., ed. *2000 Years of Zinc and Brass*. London: 27–71.

Craddock, P.T. 2008. Mining and Metallurgy. In: Oleson, J.P., ed. *The Oxford Handbook of Engineering and Technology in the Classical World*. Oxford: 93–120.

Danby, H. 1933. *The Mishnah*. Oxford.

Douglas, M. 1993. *In the Wilderness: The Doctrine of Defilement*. Sheffield.

Dungworth, D. 1995. *Iron Age and Roman Copper-Alloys from Northern Britain* (Ph.D. dissertation, University of Durham). Durham.

Dungworth, D. 1997. Iron Age and Roman Copper-Alloys from Northern Britain (Internet Archaeology 2). Sheffield: http://intarch.ac.uk/journal/issue2/dungworth_index.html

Epstein, I. 1935. *The Babylonian Talmud*. London.

Epstein, J.N. 1982. *The Gaonic Commentary on the Order of Tohorot Attributed to Rav Hay Gaon*. Jerusalem.

Forbes, R.J. 1950. *Metallurgy in Antiquity*. Leiden.

Goodman, M. 1999. The Pilgrimage Economy of Jerusalem in the Second Temple Period. In: Levine, L.I., ed. *Jerusalem: Its Sanctity and Centrality to Judaism, Christianity and Islam*. New York: 69–76.

Goodman, M. 2008. *Rome and Jerusalem: The Clash of Ancient Civilizations*. London.

Gowland, W. 1930. *The Metallurgy of the Non-Ferrous Metals*. London.

Hamilton, E.G. 1996. *Technology and Social Change in Belgic Gaul: Copper Working at the Tittelberg, Luxenbourg, 125 BC–AD 300* (Masca Research Papers 13). Pennsylvania.

Healy, J.F. 1978. *Mining and Metallurgy in the Greek and Roman World*. London.

Hirschfeld, Y. and Gutfeld, O. 2008. Stratigraphy and Architecture. In: Hirschfeld, Y. and Gutfeld, O., eds. *Tiberias: Excavations in the House of the Bronzes, Final Report, Vol. I: Architecture, Stratigraphy and Small Finds* (Qedem 48). Jerusalem: 1–34.

Jastrow, M. 1996. *A Dictionary of the Targumim, the Talmud Babli and Yerushalmi, and the Midrashic Literature*. New York.

Kafih, J. 1963. *Mishnah'im Perush Rabbenu Moshe ben Maimon*. Jerusalem.

Karageorghis, V. and Kassianidou, V. 1999. Metalworking and Recycling in Late Bronze Age Cyprus—the Evidence from Kition. *Oxford Journal of Archaeology* 18(2): 171–188.

Krauss, S. 1898. *Griechische und Lateinische Lehnwörter im Talmud, Midrasch und Targum*. Berlin.

Krauss, S. 1910–1912. *Talmudische Archäologie*. Leipzig.

Levene, D. and Rothenberg, B. 2007. *A Metallurgical Gemara: Metals in the Jewish Sources*. London.

Levy, J. and Leberecht Fleischer, H. 1876. *Neuhebräisches und Chaldäisches Wörterbuch über die Talmudim und Midraschim*. Leipzig.

Liddell, H.G. and Scott, R. 1940. *A Greek-English Lexicon*. Oxford.

Lowick, N.M., Bendall, S. and Whitting, P.D. 1977. *The "Mardin" Hoard: Islamic Countermarks on Byzantine Folles*. Hampshire.

Magen, Y. 2002. *The Stone Vessel Industry in the Second Temple Period*. Jerusalem.

Muhly, J.D., Stech-Wheeler, T. and Maddin, R. 1977. The Cape Gelidonya Shipwreck and the Bronze Age Metals Trade in the Eastern Mediterranean. *Journal of Field Archaeology* 4: 353–362.

Needham, S.P., Leese, M.N., Hook, D.R. and Hughes, M.J. 1989. Developments in the Early Bronze Age Metallurgy of Southern Britain. *Archaeometallurgy* 20(3): 383–402.

Neusner, J. 1974. A History of the Mishanaic Laws of Purities, Part 1. In: *Kelim*. Leiden.

Neusner, J. and Sarason, R.S. 1977. *The Tosefta*. New York.

Northover, J.P. 2000. Analysis of Bronze Age Metalwork from Tel Beydar. In: van Lerberghe, K. and Voet, G., eds. *Subartu VI: Tell Beydar: Environmental and Technical Studies.* Turnhout: 117–123.

Ponting, M. 2002. Roman Military Copper-Alloy Artefacts from Israel: Questions of Organisation and Ethnicity. *Archaeometry* 44(4): 555–571.

Ponting, M. 2006. Scientific Analysis and Interpretation of the Copper-Alloy Metalwork. In: Geva, H., ed. *Jewish Quarter Excavations in the Old City of Jerusalem, Vol. 3: Area E and Other Studies.* Jerusalem: 283–300.

Ponting, M.J. 2008. The Scientific Analysis and Investigation of a Selection of the Copper-Alloy Metalwork from Tiberias. In: Hirschfeld, Y. and Gutfeld, O., eds. *Tiberias: Excavations in the House of the Bronzes, Final Report, Vol. I: Architecture, Stratigraphy and Small Finds* (Qedem 48). Jerusalem: 35–62.

Schwartz, J. 2006. "Reduce, Reuse and Recycle," Prolegomena on Breakage and Repair in Ancient Jewish Society: Broken Beds and Chairs in Mishnah Kelim. *Jewish Studies an Internet Journal* 5: 147–180.

Stemberger, G. 1996. *Introduction to the Talmud and the Midrash.* Edinburgh.

Triester, M.Y. 1990. About the Early Production of Metallic Zinc in Parthia. *Bulletin of the Metals Museum* 15: 33–40.

Tylecote, R.F. 1976. *The Early History of Metallurgy in Europe*. London.

CHAPTER 33

EARLY BRONZE AGE REFINING OF COPPER

Christopher John Davey

This paper describes the role and process of copper refining. It then argues that the crucibles depicted on the walls of some Egyptian Old Kingdom tombs and others found at the Isin-Larsa period site of Tell edh-Dhiba'i, were originally devised and used as refining vessels. The reasons why they have not often been discovered at smelting sites are discussed and the terminology used by the ancient Egyptians for crucibles is considered.

INTRODUCTION

The subject of copper refining during the Early Bronze Age has not prompted a great deal of scholarly discussion. Early production of copper from ore-bearing rock has long been recognized to be a two stage process of smelting and refining (Merkel 1990: 78). Nearly all discussion and experimentation has concentrated on the first or smelting stage, while refining has been considered uncomplicated. This may have been the case once smelting had developed to the point where slag could be tapped, but prior to this the actual production of useful copper metal was dependent on the refining process.

Until at least the Late Bronze Age the smelting of copper ores produced prills of copper embedded in slag. The copper produced in this way often contained significant iron (2–5%) and is sometimes called "black copper." It was partially liberated by comminution, after which the copper bearing material was separated by visual or gravity based processes and then melted to separate the heavier copper from the lighter gangue. Early descriptions of copper refining, including fire-refining and poling, are found in Theophilus (Hawthorne and Smith 1963: 144f) and John Percy (1861). Tylecote (1976: 95) describes a Japanese post-medieval method where pieces of black copper would gravitate down through ignited charcoal, pass by a tuyère where impurities were oxidized, and would then be collected in a crucible at the base of the furnace.

The refining procedure is potentially complex and may be ore-specific (Schlesinger and Biswas 2011: 393f). At 1200 ^{o}C copper may contain up to 5% iron and its removal is important if the properties of pure copper are required. Iron is more likely to oxidize than copper so refining processes involve the application of oxygen to the surface of the molten copper to convert the iron to an oxide. The copper will also oxidize if too much oxygen is applied, so an encircling reducing environment may limit this from occurring. Once a satisfactory refining technique was developed in antiquity for copper from a particular ore-body, it might be expected that it would have been followed rigidly thereafter.

Merkel's refining experiments were carried out in shallow pits rather than crucibles in a manner not dissimilar to smelting (Merkel 1990: 107–109). The charge was the material he had previously smelted. The experiment produced plano-convex ingots and after four melts, the iron content was reduced to 0.014%. The last melt involved fire-refining and poling; the former was found to be particularly effective with respect to the removal of Fe, Co and Hg, the latter less so, although that may have been a question of experimental proficiency (*ibid.*: 116–118).

THE MERERUKA–TELL EDH-DHIBAꜤI CRUCIBLES AND THEIR OPERATION

The 3rd Dynasty in Egypt saw the beginning of monumental stone masonry that was to some extent facilitated by the development of a large-scale copper tool producing industry in the Nile Valley (Romer 2007). This industry was depicted on some Old Kingdom tomb-walls showing the use of a distinctive style of crucible, the profile of which became the hieroglyphic ideogram for "metalworker" (Davey 1985; Drenkhahn 1976) (Fig. 33.1). Such crucibles have been found at Tell edh-Dhibaꜥi (Davey 1983; 2007; 2009b; 2012) (Fig. 33.2). However in an upright position, these crucibles will not retain a liquid, raising the question of how they were used for melting copper.

In the preliminary report of the Tell edh-Dhibaꜥi excavation, Winston proposed that the crucibles were used in a "tipped back" position, but the many Old Kingdom tomb images are unequivocal that they were used in an upright orientation (Al-Gailani 1965; Davey 2012). Other later representations of the crucibles' operation reveal that tipping the crucible back may have become common later in the Old Kingdom and then, in the First Intermediate period, the crucibles themselves changed in shape so that an upright crucible with a hole in the side would retain the liquid (Davey 1985; 2009b).

One of the most instructive scenes found in the Tomb of Mereruka, Saqqara, shows the front of the crucible being poked and molten copper pouring out. This image suggests that the opening in the front of

Fig. 33.1: The melting and casting scenes from the Old Kingdom Tomb of Mereruka, Saqqara. The processes depicted are part of the production of the vessels shown in the upper register (from Duell 1938: Pl. 40).

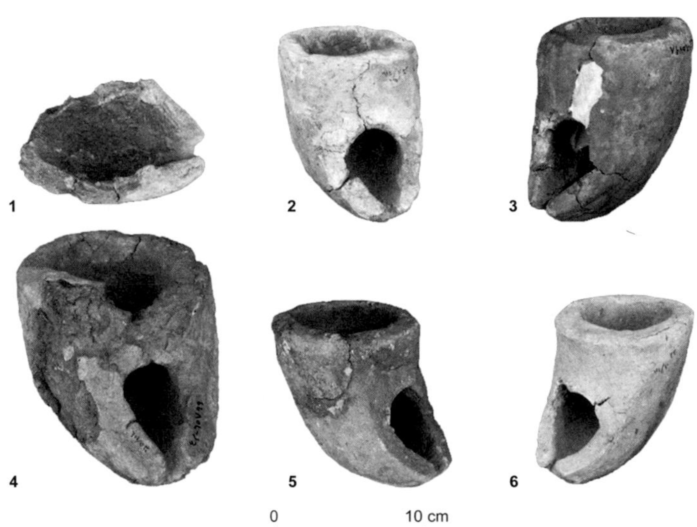

Fig. 33.2: The five complete crucibles and one fragment discovered at Tell edh-Dhibaꜥi from the Isin-Larsa period, ca 1850 BCE. 1. 614/8; 2. 614/7; 3. 614/4; 4. 614/6; 5. 614/5; 6. 614/3 (photos: Christopher J Davey).

the crucible was partially blocked with a ceramic barrier and when it was dislodged, the molten charge flowed out underneath it (Figs. 33.3, 33.4). Being heavier, the metal would discharge first and anything floating above the metal would either be retained or would flow out subsequently.

The result of this process is that lighter charcoal and dross material would be separated and skimmed off from the heavier metal, which would pour from the crucible in a refined form. The volume of copper involved in each pour from the Tell edh-Dhibaᶜi crucibles was of the order of 50 ml. This volume was limited because the barrier only partially blocked the hole in the side of the crucible. The space above the barrier allowed the charcoal inside the crucible to be ventilated with blowpipes and it enabled observation of and access to the molten copper charge.

Since the volume of copper in each melt was small, this technology was manageable. A blowpipe could be used to ventilate the charcoal to achieve the necessary temperature and melt the small volume of copper without exhausting the operators. The refractory and insulating ceramics from which the

Fig. 33.3: A close-up of the metal discharge from the crucible as depicted in the Tomb of Mereruka. The scene is shown in profile for greater clarity and the hands holding the crucible have been rotated to reveal the crucible shape (photo: Christopher J Davey).

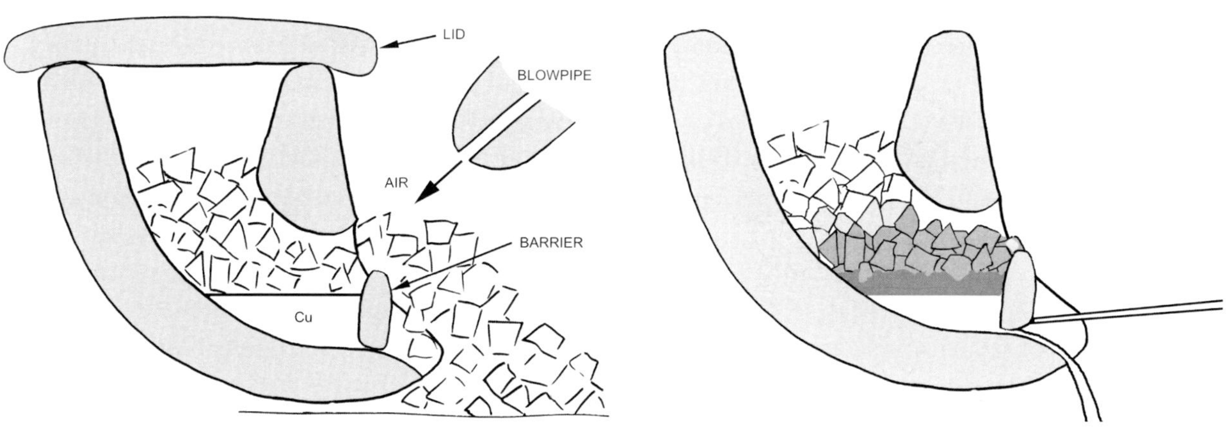

Fig. 33.4: Suggested sections of the crucible during melting and pouring.

crucibles were made were comparatively weak and not capable of carrying a significant weight of molten metal. Large volumes of copper could be melted when bellows were used in later periods and large hot crucibles could be manipulated when iron tools became available.

Early Bronze Age copper refining and melting technology depicted in the Egyptian Old Kingdom tombs depended on the rapid heating of relatively small amounts of copper and employing accurately directed narrow jets of air from blowpipes. The same crucibles would have been used repeatedly, barely cooling down between pours. After each pour, the slag would be removed and the barrier secured with something like a paste of dung. The charge of unrefined copper and charcoal make-up would be added and sometimes a lid was placed on top of the crucible before the fire in the crucible was rekindled by ventilating the hot charcoal in the crucible left from the previous pour.

Some of the Old Kingdom scenes depict a lid on the crucibles during heating (Fig. 33.1). This would have accelerated the rise in temperature, but it would also have promoted reducing conditions in the crucible as the flow of air through the crucible was inhibited. The Mereruka–Tell edh-Dhibaʿi crucibles were therefore reaction vessels in which oxidizing and reducing conditions could be regulated as the lid was opened and closed and the air from the blowpipe was applied to the molten surface of the copper.

The ultimate reason why these apparently unsuitable melting receptacles were devised and used for unrefined copper was most probably their versatility as reaction vessels for the refining process. The surface of the molten copper was directly accessible through the opening above the barrier so that a flow of oxygen from blowpipes could be directed onto it. Charcoal could also be arranged on or around the molten surface and together with the movement of the crucible lid to vary air-flow through the crucible, reducing conditions were controlled within the crucible.

The containment of heat within the crucible enabled operators to work next to it without protection and to even lift it up by hand with the assistance of lumps of wet clay or something similar. The crucible's high level of fuel efficiency was a secondary reason for their utilization. The crucibles and lids were made from an insulating ceramic so that when operated with accurately guided blowpipes the fire was contained within the crucible and heat losses were minimized; charcoal usage was concomitantly minimized. The re-use of the crucibles while still hot would also have increased their efficiency. More importantly, the heat energy needed to raise the temperature of the crucible itself for the first time with the associated endothermic ceramic reactions was significant at this scale and so repeated use of the same crucible was a logical approach (Davey and Edwards 2007).

The application of this refining type process depicted in the Old Kingdom tomb scenes may have been necessary to produce the quality of copper required for making sheet copper which could be beaten and welded into prestige utensils. Bulkier tools or ingots may not have needed copper of such quality.

One consequence of the operation of the Tell edh-Dhibaʿi crucibles with removable ceramic barriers should be commented upon. If the crucibles or the barriers were to vitrify, the barrier would weld to the crucible and become immovable, thus preventing the discharge of the molten metal. One of the Tell edh-Dhibaʿi crucibles (No. 614/7) had been used many times to the point where it was about to disintegrate, but it had no vitrification, demonstrating the refractory nature of the ceramic. This attribute, as well as their shape, means that the Mereruka–Tell edh-Dhibaʿi crucibles are incompatible with the crucible typology proposed by Bayley and Rehren (2007). They are an addition to Tylecote's earlier typology (Tylecote 1976: 13).

A second consequence also needs comment. Hauptmann has referred to the tomb of Mereruka scene as a possible instance of smelting (Hauptmann 2007: 220, Fig. 7.1). His own discussion acknowledges that crucibles for smelting have been found to be larger than melting crucibles (*ibid.*: 218), which do not get much smaller than those found at Tell edh-Dhibaʿi and depicted in the Mereruka scene. The almost complete absence of slag in the Tell edh-Dhibaᶜi crucibles also militates against their use for smelting, as do the attendant Old Kingdom tomb casting scenes.

THE HYPOTHESIS

It is hypothesized that the crucibles depicted in the Old Kingdom tomb scenes were in fact originally used for refining copper at remote arid smelting sites, and were taken to Egypt by metalworkers to carry out refining in association with the production of sheet copper for the fabrication of prestige copper artifacts. The copper brought to Egypt may have been the unrefined material of black copper and slag. There could be many reasons for this change of industry structure. It may have been political, reflecting a change in control of the industry, or at least the refining and fabricating phases of it. Alternatively, the refiners may have been attracted by the availability of charcoal in Egypt—allowing production levels to increase—or they may have just preferred the comforts of Egypt to the privations of the desert.

Elsewhere it has been suggested that the technology associated with the Tell edh-Dhibaᶜi crucibles further developed in Egypt, as did their shape (Davey 2009b). This implies that the refining of copper within Egypt became less important after the Old Kingdom.

APPARENT ABSENCE FROM THE ARCHAEOLOGICAL RECORD

The fuel efficiency of these crucibles suggests that their technology was appropriate for regions where charcoal was difficult to obtain because fuel sources were limited or trade routes were long. Their presence may therefore be expected in at least Mesopotamia, the Levant, the Sinai and the deserts of Egypt. Our knowledge of these crucibles indeed comes from these areas, but the occurrence of such crucibles is limited.

There are a number of reasons why these crucibles have not been identified at smelting and metalworking sites. The crucibles' ceramic fabric was refractory and therefore may have been carefully managed and not discarded. Crucibles were used repeatedly until they disintegrated and their remnants would have been ground up as grog for new crucibles.

The asymmetric crucible shape makes fragments difficult to identify. Even archaeometallurgists have shown little interest in these objects and few are likely to recognize their fragments.

WADI SERABIT EL-KHADIM

Experiments conducted at the Royal Melbourne Institute of Technology with replicas of the Tell edh-Dhibaᶜi crucible revealed that their failure mode was to break horizontally across the lower portion—thus creating a sherd in the shape of crucible fragment No. 614/8 from Tell edh-Dhibaᶜi (Fig. 33.2). Fragments of this shape can be described as having lamp-like spouts and could be reconstructed to form shallow bowls similar to the crucibles discovered at Tell el-Dab'a (Philip 2006: 199–204), rather than the taller Tell edh-Dhibaᶜi crucibles.

Of the 40 crucible fragments found by the Tel Aviv University expedition to the caves of Wadi Serabit el-Khadim, Beit-Arieh published two similar shaped fragments (1985: 111, Fig. 13: 6–7) (Fig. 33.5). Most archaeologists analyzing such objects would assume that they were originally bowl crucibles, but Beit-Arieh correctly compared them to the Tell edh-Dhibaᶜi vessels.

Both the Tell edh-Dhibaᶜi and Wadi Serabit el-Khadim finds date to the early second millennium while the Old Kingdom tomb images date to the third millennium. They therefore do not directly support the chronology of the hypothesis. However, metallurgical traditions were long running. The Tell edh-

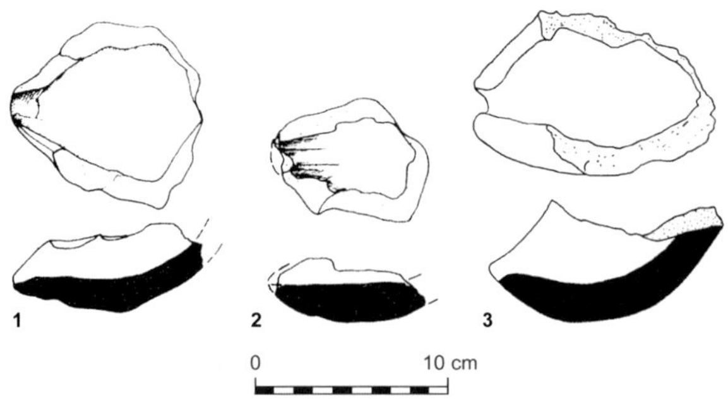

Fig. 33.5: Wadi Serabit el-Khadim crucibles, examples Nos. 1 and 2 (Beit-Arieh 1985: Fig. 13: 6–7) compared to example No. 3, the fragment from Tell edh-Dhiba'i (Davey 1983: No. 614/8).

Dhibaᶜi technology does seem to have been present in Mesopotamia from the early third millennium (Davey 1983; 2009a), and there is no reason to assume that the metalworkers of Wadi Serabit el-Khadim could not also have been the bearers of a long-standing metallurgical tradition. The finds certainly point to a wide distribution of the technology in the arid regions of the ancient Near East. In this context it is worth noting that the equipment found at Wadi Serabit el-Khadim, including foot operated pot-bellows, clay molds, etc., closely parallels that found at Tell edh-Dhibaᶜi (Davey 1988).

EGYPTIAN WORDS FOR CRUCIBLE

There are two words for crucibles found in the Old Kingdom tomb inscriptions, *ds* and *bd (bḏ.t),* and they are often used in parallel. Miller (1990) has discussed *ḏs (ds)* vessels in the context of beer consumption at Deir el Medina. He argues that vessels of the same name were used as separation vessels to refine copper and he envisages a "metaphorical extension" of the term to much larger vessels "used to skim clarified beer off the dregs and sediments." While there is a possibility of a circular argument, it can be concluded that there is a complimentary need for separation and refinement for both brewing and metallurgy, and that utensils called *ds* were used in both instances.

The other term used to describe crucibles is variously spelled *bd, bḏ,* or *bḏ.t*. There is a similar word for bread molds, but where the meaning is crucible, the hieroglyphs normally include a crucible determinative (Fig. 33.6). Scholars have differed when translating these terms. Weeks (1994: 35, Fig. 30) ignores the crucible determinative and translates *bḏ* as "mold," (tomb of Iymery, 5th Dynasty, Giza, G 6020, LG 16, PM III,1 171 [3]), while in the tomb of Senezemib: Mehi (5th Dynasty) (Giza, G 2378, LG 26, II/E PM, III. 1.2 88 [6]) Brovarski (2001: 145, Fig. 116b) translates the same hieroglyphs as "crucible". The tomb of Mehu (6th Dynasty) (Saqqara, II/N, PM III. 2.2 620 [11]) has different hieroglyphs, *bḏ3* and no determinative, and so may actually refer to a mold, though Altenmüller emended the hieroglyphs and translated it as "crucible" (Altenmüller 1998: 146f, Pl. 42). Weeks' rendering of *bḏ* as "mold" has been accepted by Hannig, who also defines *bḏ* and *bḏ.t* as "schmelztiegel" or "melting pot" (2003: 430), as does Scheel (1985: 157).

Receptacles in ancient Egypt are commonly referred to as *ds* and so the term may logically also signify a crucible. The other word, *bḏ,* is less straightforward. There is no reason to link bread molds with crucibles and there are no other parallel Egyptian terms; the range of spellings used in the inscriptions is also interesting and may point to a loan word. The Semitic word *bdd* (Heb בדד בד; Arabic بد) means "separate," "separation" or "scatter." This raises the possibility that the vessel's name may have originated from a Semitic environment where such vessels were used for refining, that is separating

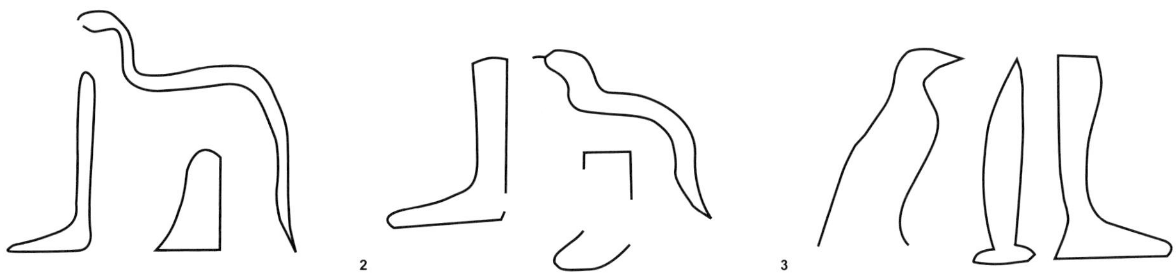

Fig. 33.6: Hieroglyphics referring to crucibles (1 and 2) and mold (3). 1. Senezemib: Mehi (Brovarski 2001: 145, Fig. 116b); 2. Iymery (Weeks 1994: 35, Fig. 30); 3. Mehu (Altenmüller 1998: 146f, Pl. 42).

copper metal from slag material. If the word was not Egyptian, it would explain the range of spellings used to try and represent the correct sound. The range of later Semitic words translated as "refine" (צרף זקק;) are often used in the context of gold and silver and so they may not be relevant to the Early Bronze Age copper refining terminology. In fact Notebaart argues that זקק primarily refers to washing, i.e., sluicing of alluvial gold (2012: 141).

The possible association of the terms used for "crucible" in Old Kingdom tomb inscriptions with the idea of separation may add weight to the hypothesis that the Mereruka—Tell edh-Dhibaʿi crucibles were originally used for separating copper from slag. It may also point to an origin outside Egypt in a Semitic language environment.

CONCLUDING COMMENTS

The Mereruka–Tell edh-Dhibaʿi crucibles and their complex operation have raised questions about their function and origin since they were discovered. Why create melting pots that could not hold a liquid? The proposal that these crucibles were originally devised as reaction vessels to undertake the refining of copper offers a solution to this problem.

The hole in the side of the crucible—that would otherwise be quite awkward—was partially blocked by a removable barrier above which there was an opening into the reaction zone and through which the operation's progress could be monitored. It was also the means whereby the center of the fire inside the crucible could be ventilated and oxygen could be applied to the surface of the molten metal.

The Old Kingdom tomb images depict the occasional use of a cover on the crucible. Experiments have demonstrated that the temperature within the crucible rises rapidly when a lid is used, but more importantly, it may also promote and regulate reducing conditions within the crucible to limit copper losses during the refining process.

The depiction of this process in Old Kingdom tombs indicates that the refining process common at smelting sites was also carried out in the Nile Valley. There is a range of plausible reasons why the metalworkers may have moved to Egypt. The fact that the shape of the crucible they brought with them became the hieroglyph ideogram for metalworkers demonstrates their significance in the development of the Egyptian metalworking industry.

Furthermore, it provides an enduring view of a process that was used at arid zone smelting sites for refining copper. The rareness of these vessels may partly be explained by the re-cycling of the crucible fabric and the difficulty to identify them. With greater recognition of the crucible shape by archaeologists, it is probable that Wadi Serabit el-Khadim will not be the only site to yield fragments of these refining crucibles. Analyses of Old Kingdom copper objects themselves may also shed light on the processes to which they have been subject.

REFERENCES

al-Gailani, L. 1965. Tell edh-Dhibaʿi. *Sumer* 21: 33–40.

Altenmüller, v.H. 1998. *Die Wanddarstellungen im Grab des Mehu in Saqqara.* Mainz-am-Rhein.

Bayley, J. and Rehren, T. 2007. Towards a Functional and Typological Classification of Crucibles. In: La Niece, S., Hook, D. and Craddock, P., eds. *Metals and Mines.* London: 46–55.

Beit-Arieh, I. 1985. Serâbît el-Khâdim: New Metallurgical and Chronological Aspects. *Levant* 17: 89–116.

Brovarski, E. 2001. *The Senedjemib Complex, Part I: The Mastabas of Senedjemib Inti (G 2370), Khnumenti (G 2374) and Senedjemib Mehi (G 2378).* Boston.

Davey, C.J. 1983. The Metalworkers' Tools from Tell edh-Dhibaʿi. *The Bulletin of the Institute of Archaeology University of London* 20: 169–185.

Davey, C.J. 1985. Crucibles in the Petrie Collection and Hieroglyphic Ideograms for Metals. *Journal of Egyptian Archaeology* 71: 142–148.

Davey, C.J. 1988. Tell edh-Dhibaʿi and the Southern Near Eastern Metalworking Tradition. In: Maddin, R., ed. *The Beginning of the Use of Metals and Alloys*. Cambridge, MA: 63–68.

Davey, C.J. 2009a. The Early History of Lost-Wax Casting. In: Mei, J. and Rehren, T., eds. *Metallurgy and Civilisation: Eurasia and Beyond.* London: 147–154.

Davey, C.J., 2009b. A Metalworking Servant Statue from The Oriental Institute, University of Chicago. *The Bulletin of the Australian Centre for Egyptology* 20: 37–46.

Davey, C.J. 2012. Old Kingdom Metallurgy in Memphite Tomb Images. In: Evans, L., ed. *Ancient Memphis: "Enduring Is the Perfection," Proceedings of the International Conference Held at Macquarie University, Sydney on August 14–15, 2008.* Leuven: 85–108.

Davey, C.J. and Edwards, W.I. 2007. Crucibles from the Bronze Age of Egypt and Mesopotamia. *Proceedings of the Royal Society of Victoria* 120(1): 146–154.

Drenkhahn, R. 1976. *Die Handwerker und ihre Tätigkeiten im alten Ägypten.* Wiesbaden.

Duell, P. 1938. *The Mastaba of Mereruka, Part I (Chambers A 1–10, Plates 1–103).* Chicago.

Hannig, R. 2003. *Ägyptisches Wörterbuch.* Mainz-am-Rhein.

Hauptmann, A. 2007. *The Archaeometallurgy of Copper: Evidence from Faynan.* Berlin.

Hawthorne, J.G. and Smith, C.S. 1963. *On Divers Arts. The Foremost Medieval Treatise on Painting, Glassmaking, and Metalwork/Theophilu; Translated from the Latin with Introduction and Notes by John G. Hawthorne and Cyril Stanley Smith.* Chicago.

Merkel, J.F. 1990. Experimental Reconstruction of Bronze Age Copper Smelting Based on Archaeological Evidence. In: Rothenberg, B., ed. *The Ancient Metallurgy of Copper.* London: 78–157.

Miller, R.L. 1990. Ds-Vessels, Beer mugs, Cirrhosis and Casting Slag. *Göttinger Miszellen* 115: 63–83.

Notebaart, C. 2012. *Metallurgical Metaphors in the Hebrew Bible.* Bergambacht.

Percy, J. 1861. *Metallurgy: Fuel; Fire-Clays; Copper; Zinc; Brass; etc.* London.

Philip, G. 2006. *Tell el-Dabʿa, Metalwork and Metalworking Evidence of the Late Middle Kingdom and the Second Intermediate Period.* Vienna.

Romer, J. 2007. *The Great Pyramid: Ancient Egypt Revisited.* Cambridge.

Scheel, B. 1985. Studien zum Metallhandwerk im Alten Ägypten I, Handlungen und Beischriften in den Bildprogrammen der Gräber des Alten Rieches. *Studien zur altägyptischen Kultur* 12: 117–177.

Schlesinger, M.E. and Biswas, A.K. 2011. *Extractive Metallurgy of Copper.* Amsterdam.

Tylecote, R.F. 1976. *A History of Metallurgy.* London.

Weeks, K.R. 1994. *Mastabas of Cemetery G 6000. Including G 6010 (Neferbauptah); G 6020 (Iymery); G 6030 (Ity); G 6040 (Shepseskafankh).* Boston.

CHAPTER 34

BRONZE PRODUCTION IN PI-RAMESSE: ALLOYING TECHNOLOGY AND MATERIAL USE

Frederik W. Rademakers, Thilo Rehren and Edgar B. Pusch

This paper presents preliminary results of ongoing research into the Ramesside bronze production in Qantir–Pi-Ramesse. The industrial remains offer a rare insight into the secondary production processes involved in the chaîne opératoire of metal production in ancient Egypt. Through the analysis of crucible fragments, the technological process of bronze production at ancient Pi-Ramesse is reconstructed. The results show how the detailed study of a macroscopically homogeneous crucible assemblage can illuminate the variability in production technology and material use. They highlight the importance of detailed slag phase identification, particularly metal prills and oxide crystals, and of tracing bulk chemical changes between ceramic and slag, to distinguish between bronze production techniques and materials. Finally, this paper provides novel contributions to the long-standing debate regarding the Late Bronze Age metal trade.

QANTIR–PI-RAMESSE

ARCHAEOLOGICAL BACKGROUND[1]

The ancient city of Pi-Ramesse is for the most part located underneath modern Qantir, in the eastern Nile Delta of Egypt (Fig. 34.1). It was established in the 13th century BCE on the (now silted up) Pelusiac branch of the Nile as the capital of the New Kingdom. Under Ramesside reign, the city flourished as an important cultural and political center in the Late Bronze Age Eastern Mediterranean (Van De Mieroop 2007), and served as the military base for the pharaoh's chariot garrison (Herold 1999). During this period, Egypt was territorially more expansive than ever before in its history and it was actively involved in international politics and warfare in the Near East (e.g., the Battle of Qadesh), highlighting the importance of Pi-Ramesse's location and chariot army.

The first major phase of this site shows evidence of a number of high temperature production processes taking place on an industrial scale, such as the production of faience (Hamza 1930; Hayes 1937; Herrmann 1957), Egyptian blue, red glass (Pusch and Rehren 2007) and bronze, the subject of this paper. It is interesting to note that copper is a central element of each of these materials (Rehren, Pusch and Herold 2001). The bronze production remains are mostly confined to excavation Area QI, while glass production remains occur predominantly in Area QIV (Fig. 34.2). However, the actual bronze production area could well have extended over an area of more than 30,000 sq m (Pusch 1993, 1994).

---

1 A summary of archaeological research conducted at Qantir is given by Pusch (1991, 1993), and Pusch and Herold (1999). Full studies are presented in the series "Forschungen in der Ramses-Stadt—Die Grabungen des Pelizaeus-Museums Hildesheim."

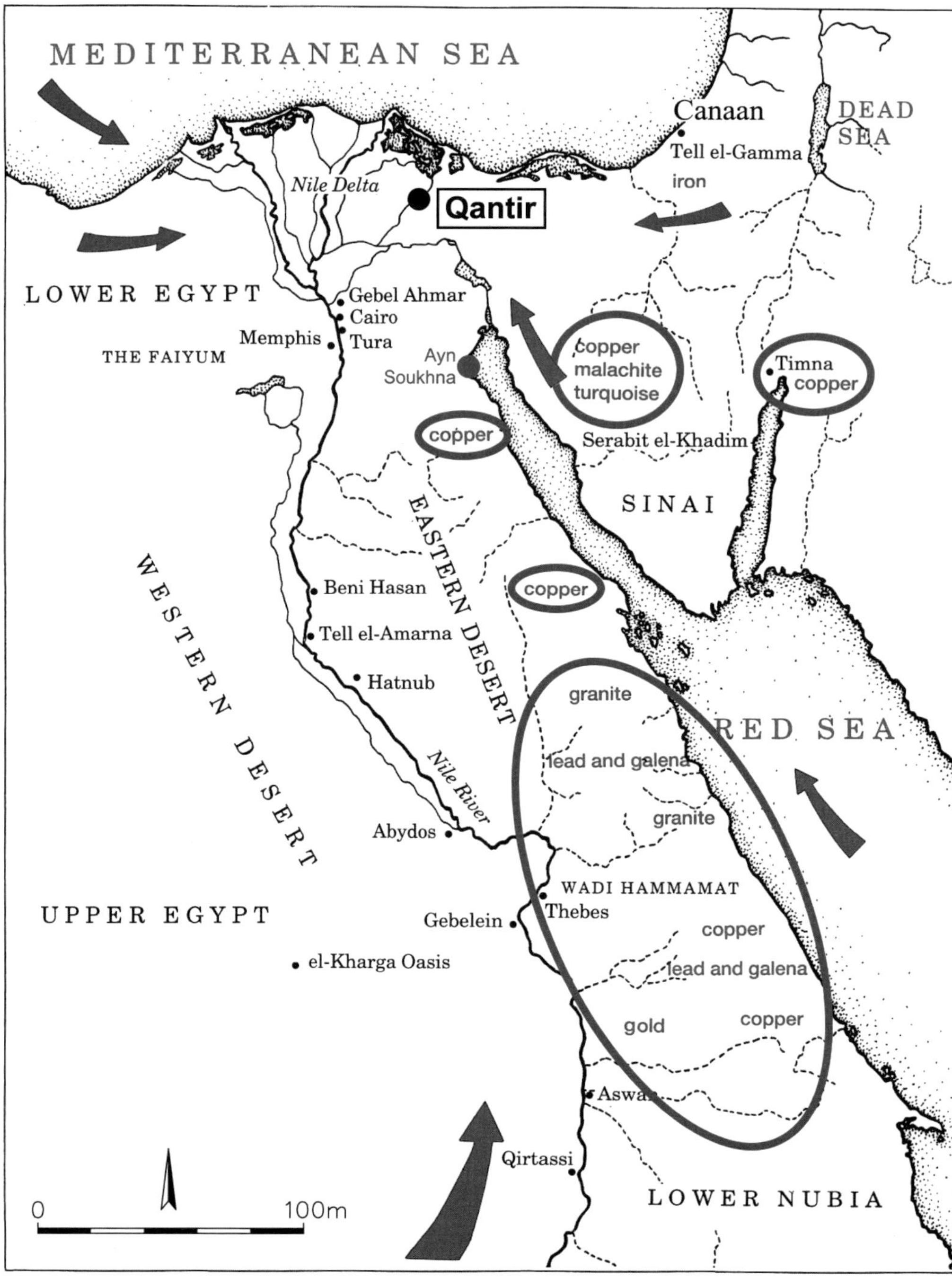

Fig. 34.1: Map showing Qantir–Pi-Ramesse and possible sources of metal for ancient Egypt. Circled: "Egyptian" mining; dot: Ayn-Soukhna; arrows: possible ingot/raw metal import (modified from Wilkinson 1994).

During the earliest phase, the industrial-scale activities probably revolved around developing the new capital, involving the production of tools and architectural features (e.g., faience decoration, bronze doors, statues). The second phase at the site indicates a changing focus in activities, shifting to the establishment and maintenance of the chariot army stationed there. This is reflected in the construction

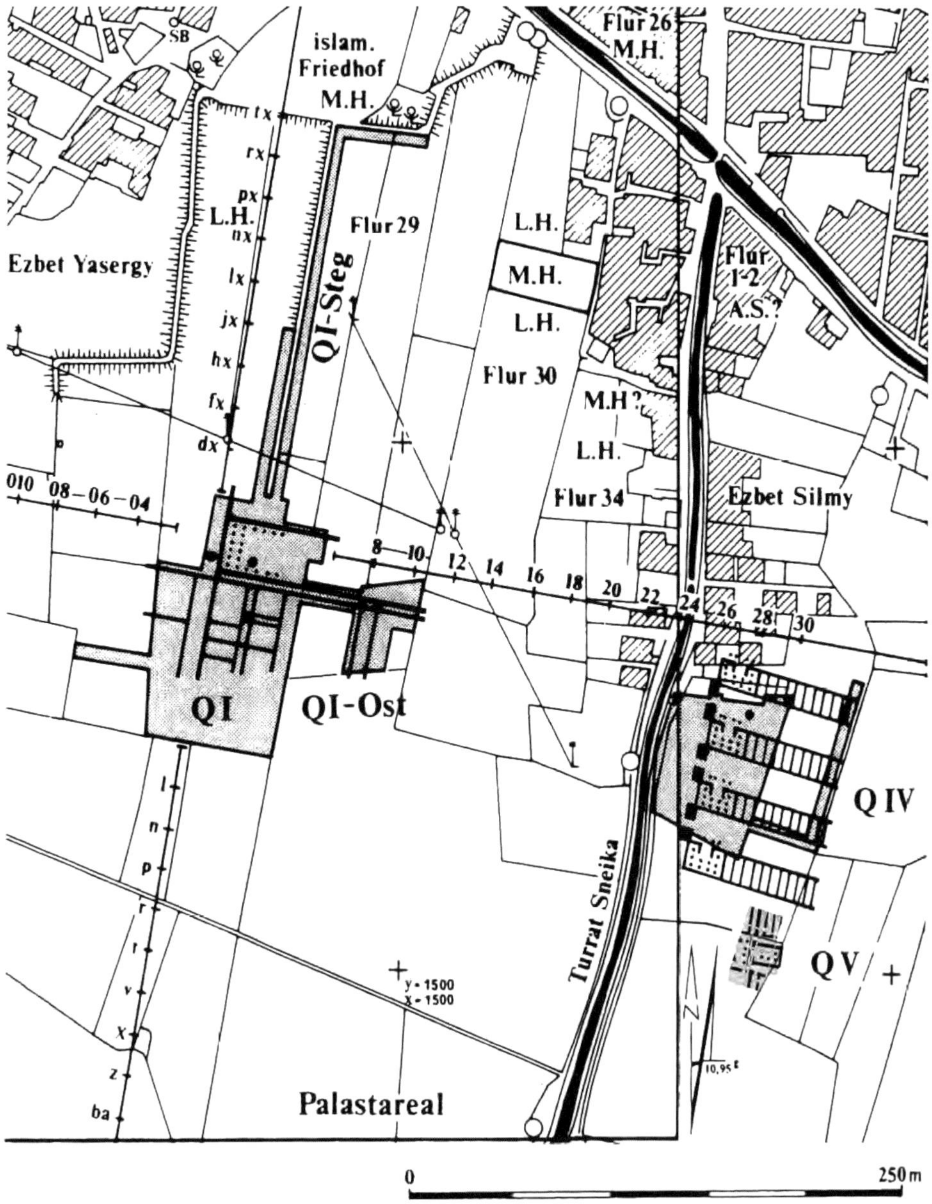

Fig. 34.2: Map showing relevant excavated areas (Pusch and Rehren 2007).

of large horse stables and a chariot training courtyard, replacing the industrial production areas. During this second phase, bronze production probably took place in a "souq-like" organization, involving smaller workshops with more ephemeral heating structures.

A FRAMEWORK FOR EGYPTIAN METALLURGY

Given Egyptian archaeology's dominant focus on architecture, inscriptions and artifacts, metallurgy has hitherto received comparatively little attention. Though a pioneering work, Garland and Bannister's (1927) overview is largely outdated and focuses mainly on metal artifacts. Scheel's (1989) summary of archaeological evidence remains concise, reflecting the scarcity of production waste studies in Egyptian archaeology. In the absence of other overviews, a *chaîne opératoire* framework for studying ancient Egyptian metallurgy is lacking. A brief review of related studies is given here.

The Egyptian presence in Timna has been given much scholarly attention (e.g., Bachmann 1980; Rothenberg 1988, 1990, 2003; Ben-Yosef 2012), and the Timna conference has shown that debate on this subject is ongoing. An Egyptian presence in the region coordinating the mining and smelting of copper ore can be inferred, for the New Kingdom, though the extent of control remains unclear. The actual smelting was probably carried out by local people.

Garson (1977), Said (1990), and Abouzeid and Khalid (2011) present possible Egyptian metal ores from a geological perspective, while Stos-Gale, Gale and Houghton (1995), Hikade (1998), and Ogden (2000) discuss copper deposits exploited in ancient Egypt and Sinai. Shaw (1994, 1998), Castel *et al.* (1996), Rothenberg *et al.* (1998), Abdel-Motelib *et al.* (2012), Abd El-Rahman *et al.* (2012), and Klemm and Klemm (2013) discuss ancient mining and smelting of Eastern Desert copper, tin and gold. While exploitation evidence in Sinai and the Eastern Desert is ubiquitous (inscriptions, mines, furnaces, slag), the only (ongoing) study of ancient smelting technology thus far is that of the Middle Kingdom site of Ayn-Soukhna (Abd el Raziq *et al.*, 2011; Tallet 2012), where (Sinai) copper ore arrived by ship, was smelted at the harbor and then transported further into Egypt.

Metal also came into Egypt by trade: a Late Bronze Age trade network of copper and tin ingots, seemingly coming from all over the known world (e.g., Budd *et al.* 1995; Pulak 1997; Knapp 2000; Kassianidou 2003; Muhly 2003; Stos 2009; Gale 2011; Lo Schiavo 2012), existed in which Egypt participated, as evidenced by tomb representations (e.g., Tomb of Rekhmire [Davies, 1935]) and the find of a Cypriot oxhide ingot fragment at Pi-Ramesse (Gale 1999). These different sources are summarized in Fig. 34.1.

All the foregoing is concerned with procurement of raw metal, while its further trade and use within Egypt has received less attention. Though metal objects have attracted some interest (e.g., James 1972; Pelleg, Baram and Oren 1979; Farag 1981; Riederer 1981, 1988; Scheel 1985, 1986, 1987; Cowell 1986; Kallfass and Hörz 1989; Philip 2007), research into the intermediate segment of the *chaîne opératoire* (secondary production processes) is lagging behind. Noteworthy exceptions are Davey (1985), Scheel (1988), Pusch (1990, 1994), and Eccleston and Kemp (2008). The current study of bronze production at Pi-Ramesse should be situated here (Rehren and Pusch, 2012) and aims to address such questions as:

- What techniques were used for bronze production?
- Is metal from multiple sources coming into Pi-Ramesse?
- Do different production contexts ("industrial" vs. "souq") show the use of different technologies or metal sources?
- Who had access to particular materials and technologies?

Answering all of these questions is beyond the scope of this paper, but they represent the background against which this research is conducted.

CRUCIBLE ANALYSIS

CRUCIBLES

This paper focuses on the crucibles related to ancient bronze production at Pi-Ramesse. These crucibles (1000+ fragments recovered) appear highly standardized: made using local Nile silt, tempered with organic material and possibly the addition of some quartz sand. They were probably hand-shaped into thick-walled vessels (Fig. 34.3) and pre-fired to about 900 °C. Tuyères were similarly hand-shaped using the same ceramic recipe, but apparently not pre-fired (Rademakers 2015). In the industrial area, the

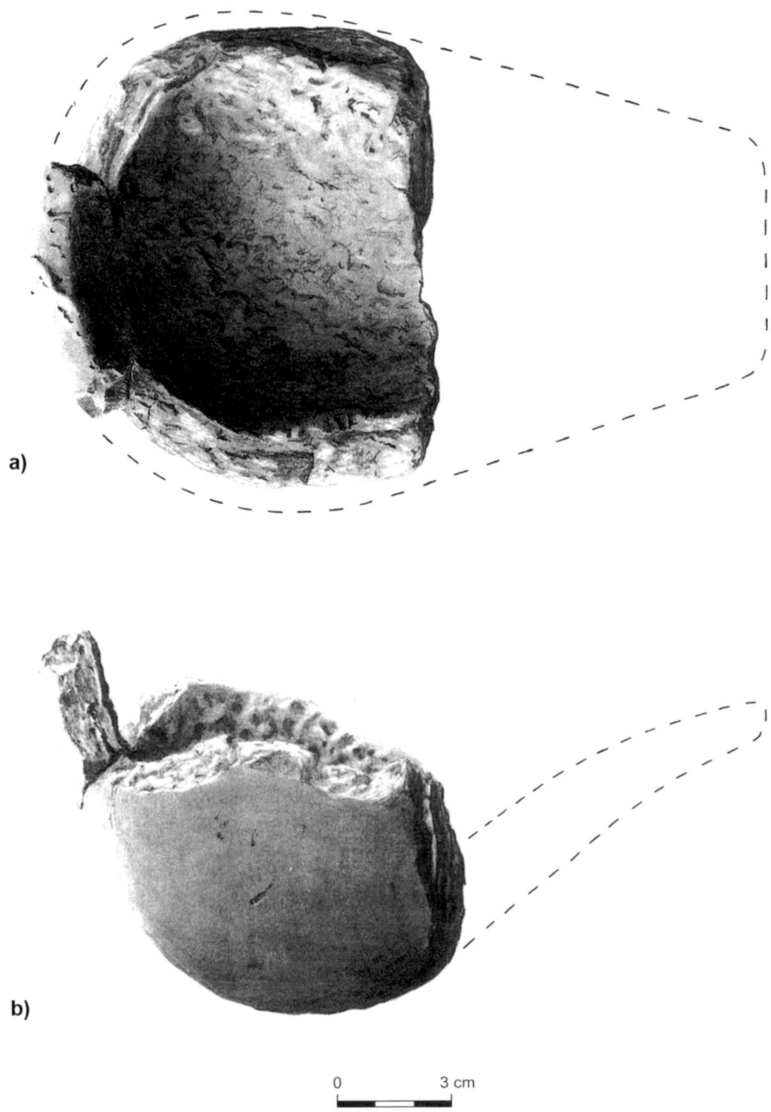

Fig. 34.3: Reconstructed crucible: (a) top view; (b) side view (from Pusch 1990: Abb. 7; reconstruction: G. Weisgerber; drawing: J. Klang).

crucibles were placed inside melting batteries, charged, covered with charcoal and heated from above using tuyères, powered by pot bellows, as shown in Fig. 34.4. Crucibles from the second phase were probably used in more ephemeral heating structures (hearths), also heated from above. This internal heating produced a wall profile containing three distinct zones (Fig. 34.5):

- Fired ceramic area (outside wall, red)
- Bloated transitional area (central area)
- Slagged area (crucible inside, gray/black)

By comparing the chemical composition of the ceramic and slagged areas and noting the different oxide crystals, metallic prills and other inclusions in the slag, the crucible charge can be reconstructed (analytical methodology detailed in Rademakers [2015], Rademakers and Rehren [2016]).

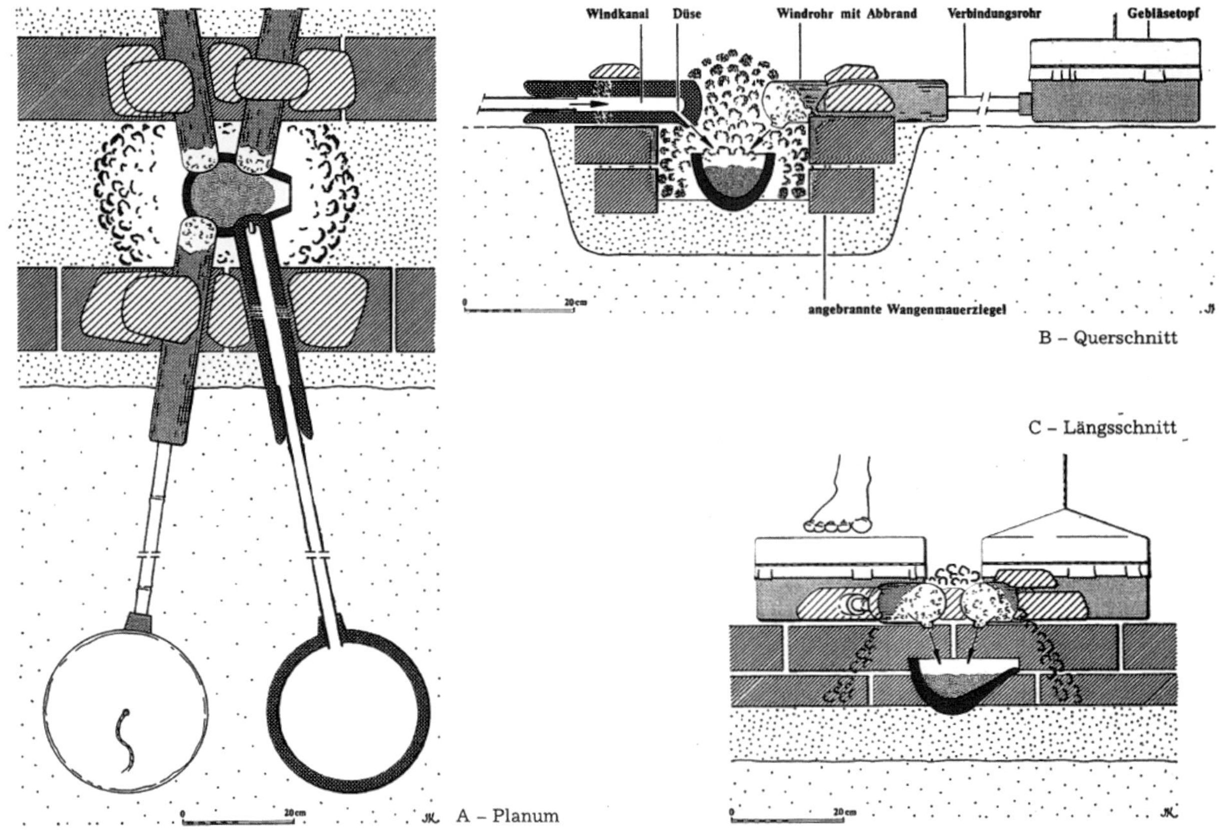

Fig. 34.4: Operation of the industrial melting batteries (from Pusch 1994 Abb. 1; reconstruction: G. Weisgerber/ E.B. Pusch; drawing: J. Klang).

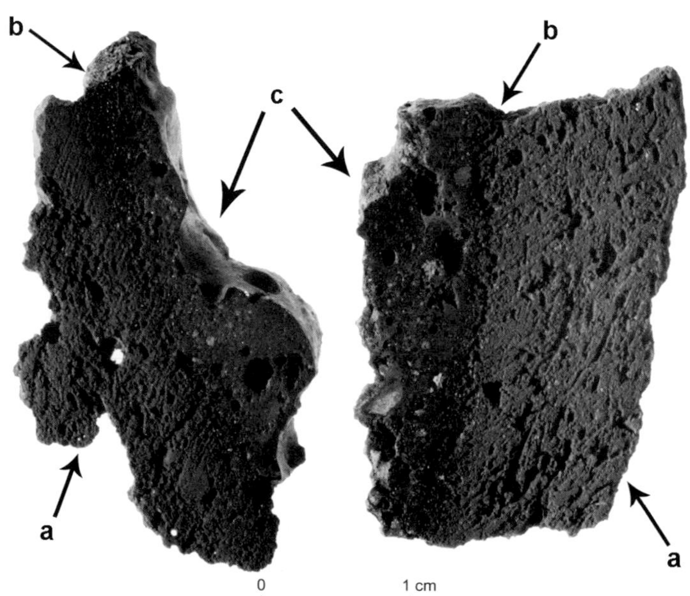

Fig. 34.5: Profile through two fragments of crucible wall: ceramic area (a), bloated transitional area (b) and slagged area (c).

CRUCIBLE SLAG

Prills

The abundant presence of copper and bronze prills in nearly all of the crucible fragments firmly places them in a context of bronze working. Some high tin prills[2] (Fig. 34.6) give direct evidence of the alloying of copper (or recycled bronze) with fresh tin or cassiterite (Rehren 2001; Crew and Rehren 2002; Rehren and Pusch 2012). Re-melting of existing bronze can only result in prills with a tin content equal to or below that of the original bronze, as tin oxidizes preferentially to copper, thereby lowering the tin content in trapped prills (Dungworth 2000a; Kearns, Martinón-Torres and Rehren 2010). When alloying copper (or recycled bronze) with a fresh source of tin, however, any composition intermediate between pure copper and tin could occur. Prills of such intermediate composition, as shown in the phase diagram in Fig. 34.7, can therefore be taken as very strong evidence for the use of a tin-rich material and indicate an active alloying process. As the goal of alloying copper and tin in Pi-Ramesse would have been to form a manageable low-tin bronze (Herold 1998; Rehren, Pusch and Herold 1998; Pusch and Rehren 2007), such high-tin prills should be seen as an intermediate product of the alloying process and their trapping in the crucible slag incidental. Thus, there are two issues with this type of evidence: its absence cannot argue against active alloying (as these prills are only present when full reaction did not occur in the sampled crucible area) and its presence does not provide any information on the source of tin used in the alloying process.

Tin Oxide

Tin oxide (SnO_2) is abundantly present in the crucible slag and is another phase indicative of bronze-related activity. The varied shapes and forms in which this tin oxide occurs, however, have received relatively little attention in the archaeometallurgical literature, though some exceptions exist, particularly for Iberian contexts (e.g., Dungworth 2000b; Rovira, Montero-Ruiz and Renzi 2009; Figueiredo *et al.* 2010; Murillo-Barroso *et al.* 2010; Eliyahu-Behar *et al.* 2012).

On the one hand, there are crystals which have shapes that clearly point to high temperature formation, as shown in Fig. 34.8. Often, the core of such crystals contains copper or bronze (Fig. 34.9), indicating that they formed by preferential oxidation of tin from a liquid bronze melt (Hofmann and Klein 1966; Cooke and Nielsen 1978; Dungworth 2000b). Tin burning out of bronze in this way is indicative of locally oxidizing conditions in certain areas of the crucible. Tin oxide is then frequently found in a region rich in metallic copper rather than bronze (Fig. 34.10). Tin oxide could also form directly from the oxidation of metallic tin instead of bronze, or from the re-crystallization of mineral cassiterite (Fig. 34.11 and Murillo-Barroso *et al.* 2010: 1170, Fig. 34.14).

On the other hand, there are clusters of tin oxide which differ significantly from those crystallized at high temperature. Four examples, shown in Fig. 34.11, occur in areas of the slag where reaction probably did not reach completion (e.g., deeper area of viscous slag or slag trapped in a crack). These clusters have the outline of mineral grains (composition: pure SnO_2): presumably, these were partly dissolved by the crucible slag and, when the reaction stopped, re-deposited in their macroscopic mineral shape (similar to residual ore fragments in primary production slag). The adjacent metal prills are often bronze (though this is not always the case), indicating the formation of bronze rather than its oxidation. As

2 High tin prills are defined as δ-phase (± 33wt% Sn) dendrites in an α+δ interdendritic matrix (± 25wt% Sn) or a combination of δ-phase, ε-phase (± 40wt% Sn) and η-phase (± 61wt% Sn). (δ-phase, ε-phase and η-phase are stable under normal casting conditions, as shown by Scott 1991: Figure 199, p. 123, even though α+δ are not at equilibrium at room temperature.)

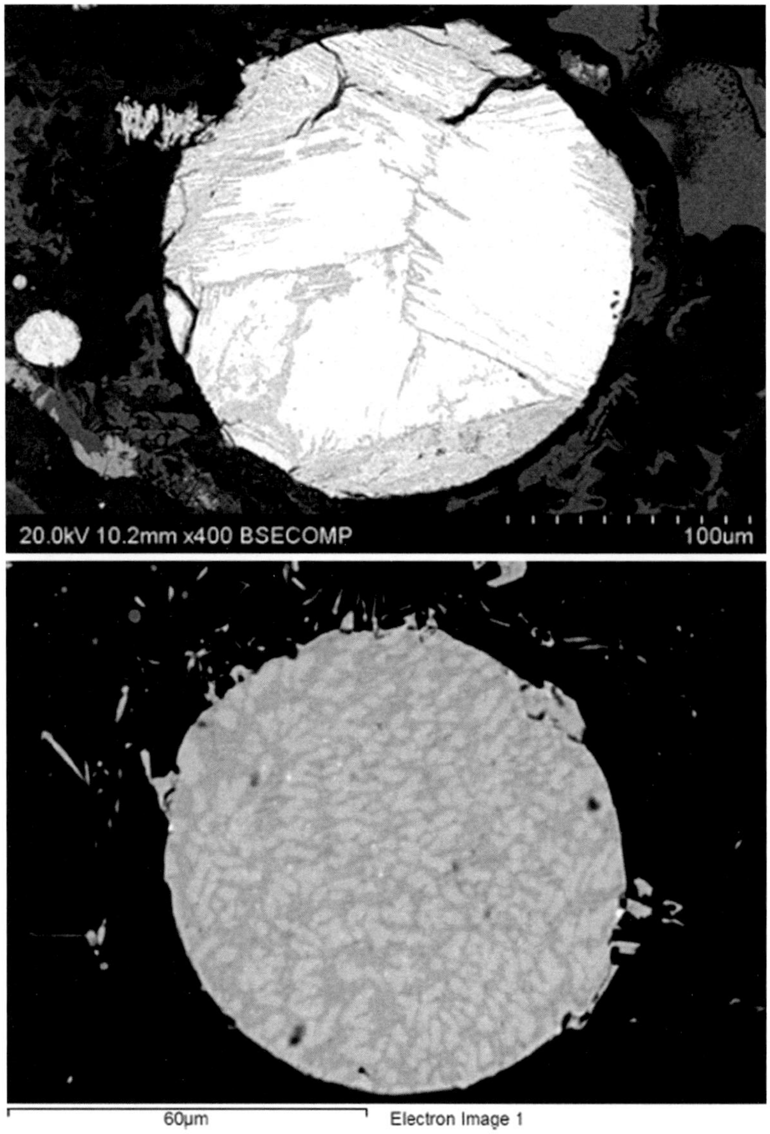

Fig. 34.6: High tin prills. Top: ε-phase (± 40wt% Sn, 60wt% Cu). Bottom: δ- and α+δ-phase (± 29wt% Sn, 71wt% Cu).

Fig. 34.11 shows, high temperature SnO_2 crystals occur right next to the re-deposited mineral grains (top right). This illustrates the stability of SnO_2 (here: residual mineral) dissolved in slag, but above all the highly heterogeneous conditions within a single crucible. Based mainly on their exceptional structural appearance, these partly dissolved grains are taken to be residual cassiterite, the most common tin ore. The importance of this variability in tin oxide morphology will be returned to in the discussion.

Bulk Chemical Composition

The bulk compositions of crucible ceramic and slag[3] are shown in Fig. 34.12, after reducing to their main oxides: CaO-SiO_2-Al_2O_3 and FeO-SiO_2-Al_2O_3 (following the principle laid out for glass analysis by

3 A total of 49 crucibles were analyzed by SEM-EDS. Each point represents the average of 5 area measurements performed separately on ceramic and slag.

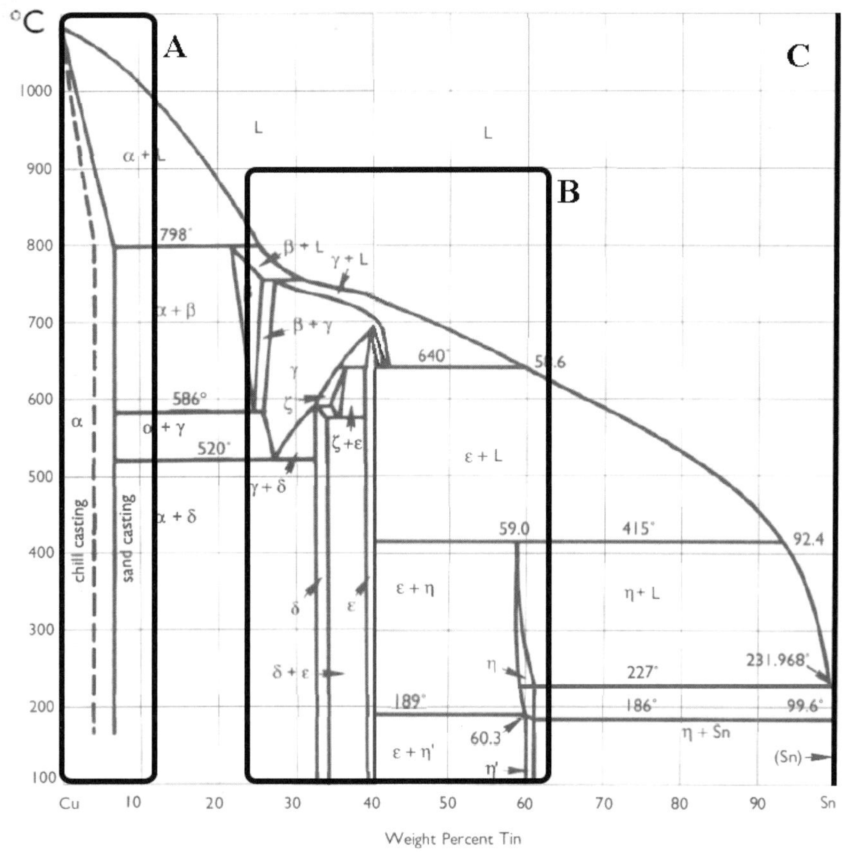

Fig. 34.7: Cu-Sn phase diagram (casting conditions), showing typical Egyptian copper/bronze composition (A), high tin prills of intermediate composition (B) and a source of fresh tin (C) (modified from Scott 1991; all phase compositional boundaries approximate).

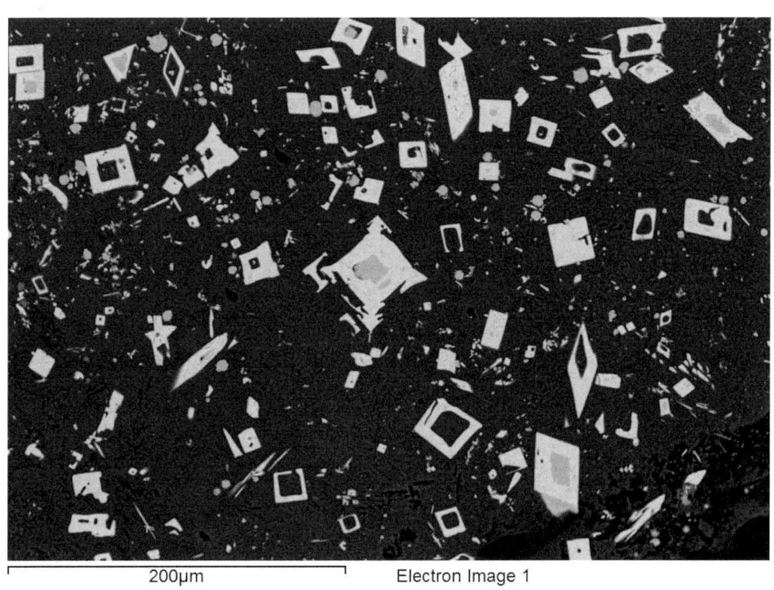

Fig. 34.8: Tin oxide (white) crystallized at high temperature in crucible slag. Some crystals have copper (mid gray) trapped in their cores.

Rehren [2000]). Melting temperatures shown for these compositions were not necessarily reached, as the crucible slag rarely became fully liquid and real compositions are more complex than these ternary diagrams suggest, resulting in significantly lower melting temperatures.

For all samples, the slag is enriched in lime relative to the ceramic (Fig. 34.12, left). This enrichment is correlated to an enrichment in alkali and P_2O_5, clearly representing the contribution of fuel ash (Tylecote 1982; Misra, Ragland and Baker 1993; Rovira 2007; Wood 2009). This verifies the contextual information that indicated that crucibles were covered in charcoal, further evidenced by the occasional inclusion of charcoal fragments in the slag.

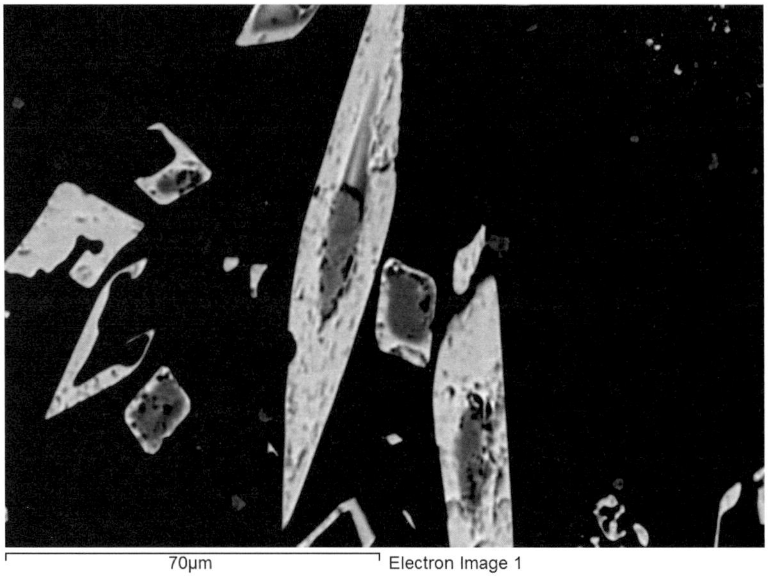

Fig. 34.9: Copper (medium gray) in tin oxide (white) core.

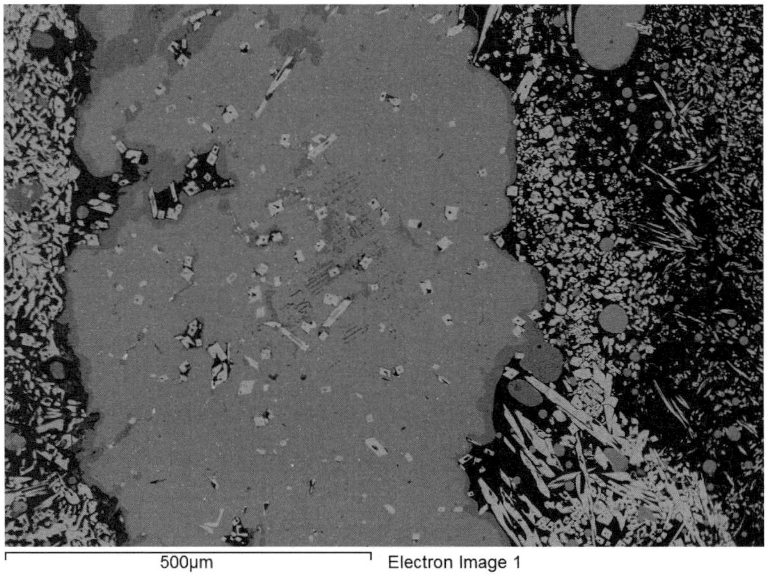

Fig. 34.10: Tin oxidizing out of a large prill. Large clusters of tin oxide surround the prill, which is depleted of tin. Tin oxide also occurs within (section close to prill surface). Cuprite formation occurs only after all tin has oxidized from the bronze.

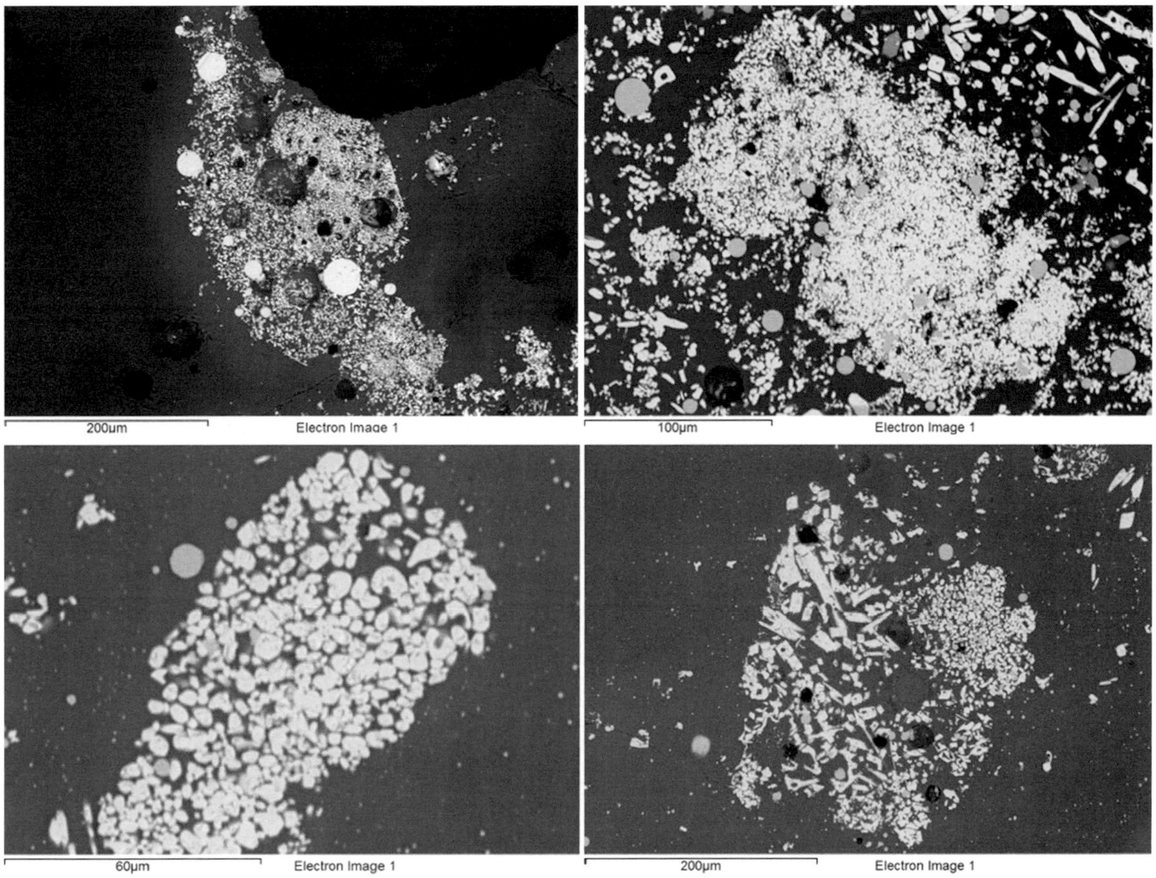

Fig. 34.11: Residual cassiterite mineral grains in crucible slag. Bottom right example: left side partly re-crystallized.

Iron Enriched Slag

As the right side of Fig. 34.12 indicates, approximately 2/3 of the samples show no iron enrichment of the slag relative to the ceramic. About 1/3 of the samples, however, show FeO-enrichment. The discussion here remains qualitative; full compositional data (Rademakers 2015) are published elsewhere.

This iron increase is unlikely to derive from the fuel ash or an intentional addition. Instead, it most likely originates from the use of a copper source enriched in iron, for example unrefined raw copper (Craddock and Meeks 1987; Craddock 2000) or copper ingots containing metallic iron (Rothenberg 1990) or slag inclusions (Hauptmann, Maddin and Prange 2002). Cassiterite used for cementation could also be a source of iron (Murillo-Barroso *et al.* 2010), as could metallic tin enriched in iron-tin intermetallic phases ("hard head") formed during tin smelting (Crew and Rehren 2002; Chirikure, Heimann and Killick 2010), but these are unlikely to be responsible for the high enrichment seen here, as tin would have been added in comparatively low quantities. Moreover, there is evidence for iron oxidizing out of copper prills and forming magnetite in the slag, as shown in Fig. 34.13. Where locally oxidizing conditions prevail in the crucible, iron will preferentially oxidize out of metallic copper and bronze (or metallic tin) before any copper oxide is formed (Craddock and Meeks 1987; Merkel 1990; Hauptmann, Maddin and Prange 2002), thereby effectively refining the copper. Such locally oxidizing conditions similarly explain the prevalence of tin oxide, as tin also oxidizes out of metallic bronze preferentially to copper.

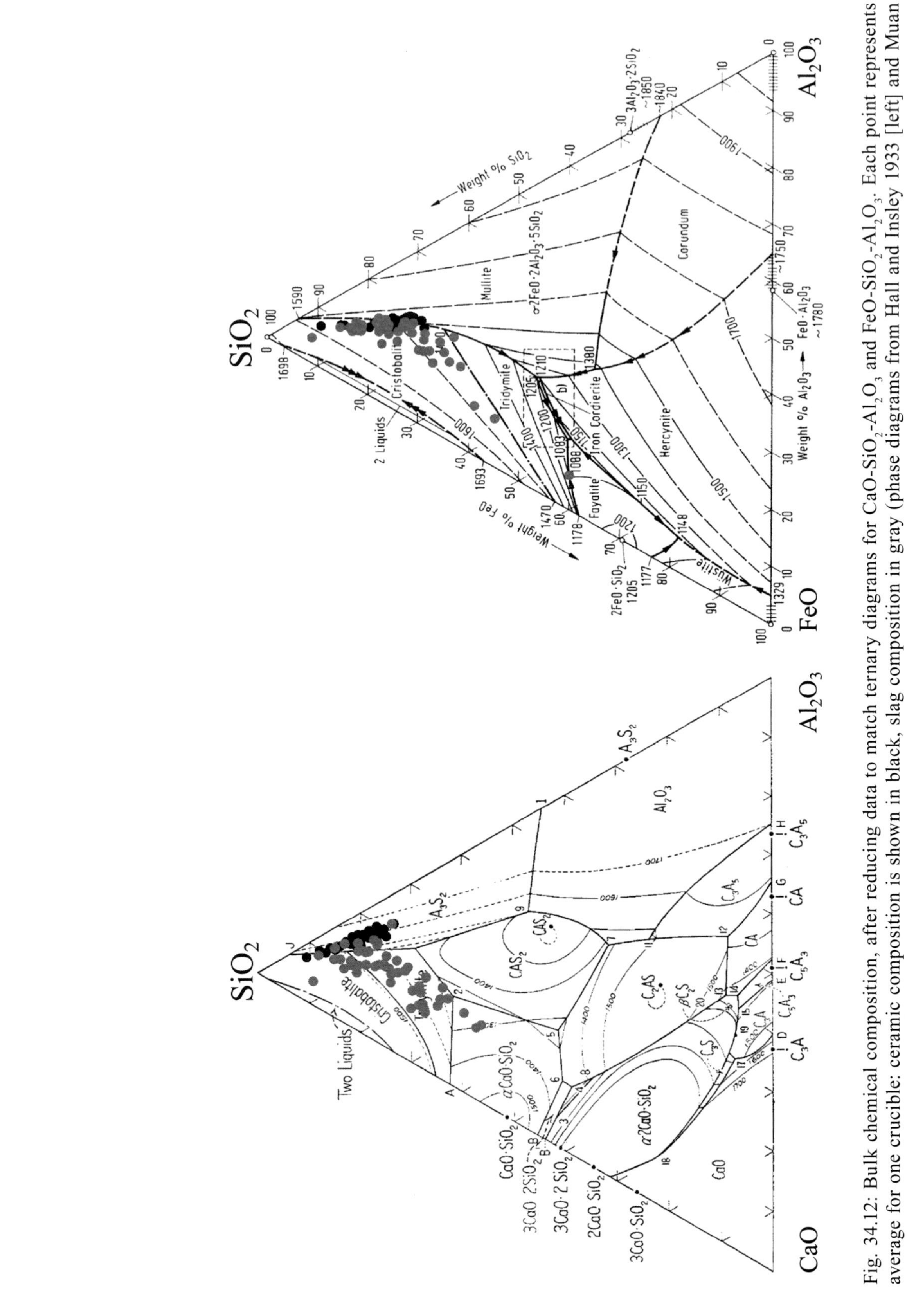

Fig. 34.12: Bulk chemical composition, after reducing data to match ternary diagrams for CaO-SiO$_2$-Al$_2$O$_3$ and FeO-SiO$_2$-Al$_2$O$_3$. Each point represents average for one crucible: ceramic composition is shown in black, slag composition in gray (phase diagrams from Hall and Insley 1933 [left] and Muan 1957 [right]).

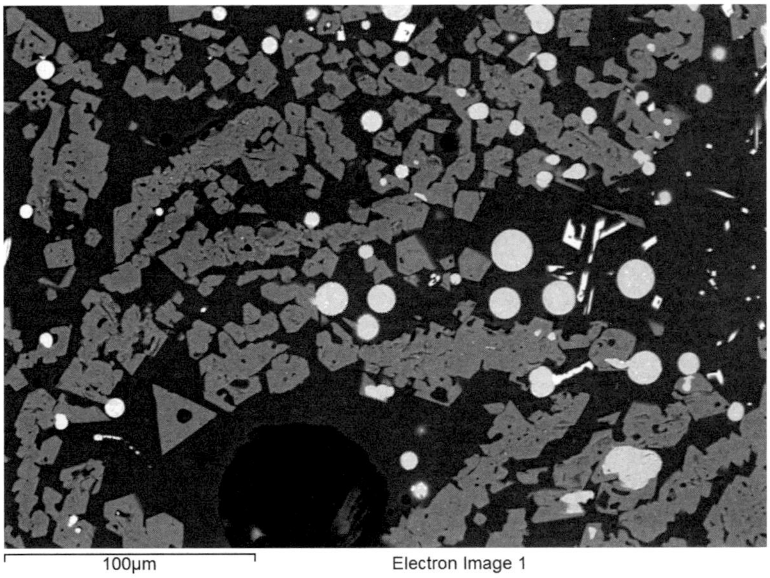

Fig. 34.13: Iron-rich bronze prills (± 3–6wt% Sn and 1–3wt% Fe) from which iron and tin oxidized preferentially into the crucible slag, forming magnetite (mid gray crystals) and tin oxide (bright angular crystals). The magnetite is iron oxide incorporating ± 3–8wt% Sn, 1–2wt% Al and 2–3wt% Mg.

Cobalt Enriched Slag

In ±15% of all crucibles, cobalt (together with iron) enrichment was noted in the slag layers. This cobalt typically occurs as a spinel-like phase, oxidizing out of the copper in a way similar to iron, as described above. These cobalt-rich spinel aggregates are shown in Fig. 34.14, which illustrates their oxidation out of a prill (copper metal containing a small percentage of cobalt and iron) into the slag. In some cases, these prills are also enriched in nickel. This combination of iron, cobalt and nickel is indicative of the geological origin of the ore from which the raw copper, used in these crucibles, was smelted under (strongly) reducing conditions.

DISCUSSION

BRONZE PRODUCTION TECHNOLOGY

There are four fundamental ways by which bronze can be produced: the alloying of two fresh metals (copper and tin), cementation (alloying copper metal with tin ore), co-smelting (smelting copper and tin ore together to directly form bronze) and recycling bronze (possibly adding a copper/tin source). Differentiating between these processes through the study of ancient production remains has received relatively little attention in archaeometallurgical studies (Pigott, Rogers and Nash 2003; Rovira 2007). However, it could provide valuable insights into technological choices and inform us on technical knowledge, material availability and cultural preference of ancient metallurgists. In this section, the evidence at Pi-Ramesse in relation to each of these options is evaluated.

Before doing so, however, it is important to emphasize the complex nature of evidence derived from crucible analysis. It needs to be underscored that each sample covers only a flat, 2–3 cm wide cross-section of a crucible, which in fact is a much larger, three-dimensional object. As conditions within a single open crucible are not in equilibrium and can vary rapidly and over very short

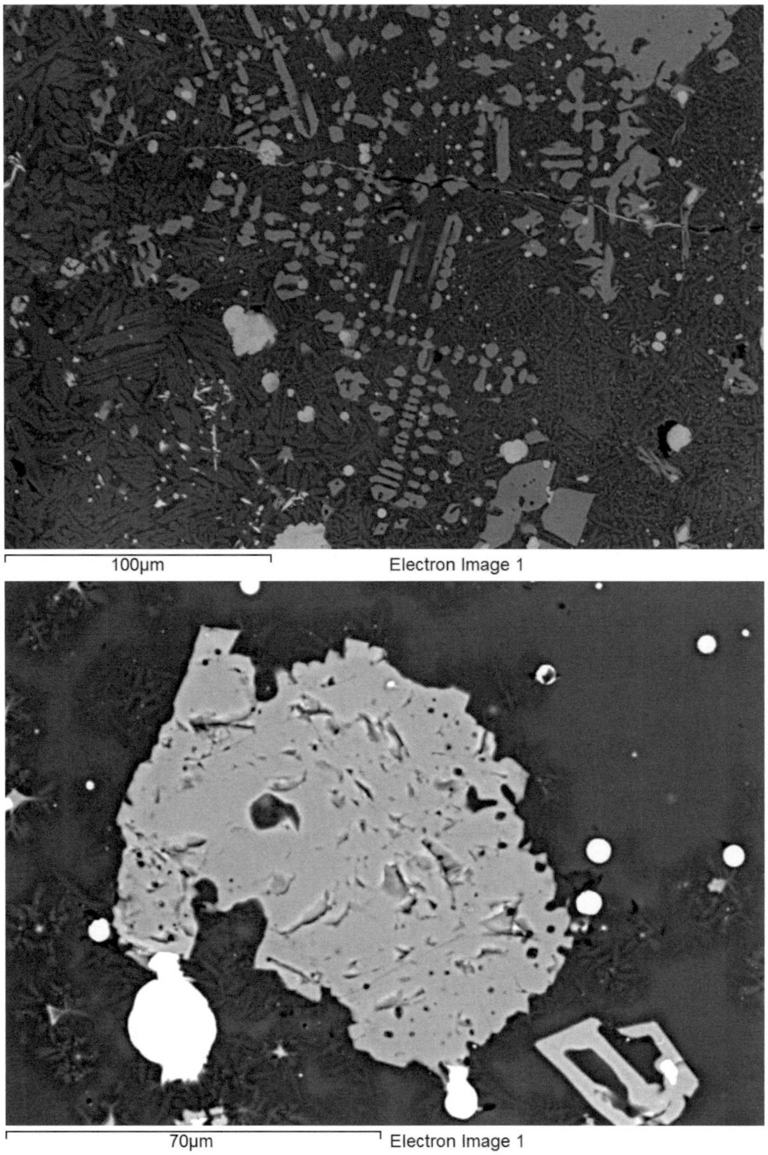

Fig. 34.14: Cobalt spinel in crucible slag. Bottom micrograph shows a large spinel aggregate, probably oxidized out of cobalt/iron-rich copper prills (bright). The spinel has ± 77wt% FeO and 12wt% CoO, as well as ± 5–6wt% Al_2O_3, 2–3wt% TiO_2 and 1wt% SiO_2, MgO and SnO_2. The bronze prill contains ± 5wt% Sn, 2wt% Fe and 0.5wt% Co.

distances in terms of temperature, oxygen supply and material presence, it can hardly be expected that a single sample is representative of an entire crucible process. Even if conditions were to be homogeneous within a crucible, it can be expected that, contrary to primary production slag, crucible slag often presents different stages of the metallurgical process, thereby complicating its evaluation. Conversely, this theoretically allows the study of these different stages within each crucible. When studying an entire crucible assemblage, it is therefore important to realize that variability witnessed among several samples should in the first instance be understood as different manifestations of a single process, before more complicated scenarios are considered (Rademakers and Rehren 2016).

The three key parameters indicating the relevant process are the nature of metal prills trapped in the slag, the presence and shape of metal oxide crystals and the slag composition compared to the composition of the crucible fabric.

Active Alloying vs. Recycling

The presence of high tin prills, defined as those with dominant δ-, ε- or η-phase, gives evidence for an active alloying process as opposed to recycling by remelting existing bronze without the addition of fresh tin (ore). However, these prills are intermediate products and can therefore not be expected to be found in each sample from a crucible used for active alloying: their absence does not rule out active alloying.

Another problem associated with these high tin prills is that they do not provide any information on the nature of the tin source used for alloying. Here, the evidence provided by the morphology of tin oxide crystals in the crucible slag comes in.

Fresh Metals or Cementation?

Researchers have long debated whether during the Late Bronze Age tin metal was the tin source of choice for bronze production, or whether mineral cassiterite was added to copper metal to produce bronze. The presence of tin-rich prills does not provide conclusive evidence, since cassiterite has to be reduced to tin metal to facilitate alloying, and at this stage it can form these prills. Thus, other indicators need to be found to differentiate between the two options.

First, it should be emphasized that tin oxide crystals can be expected to form under locally oxidizing conditions in the context of any process involving bronze melting. Pure tin metal could oxidize (alloying of two metals), tin could preferentially oxidize out of bronze (recycled bronze or bronze produced by any method and then re-oxidized) or cassiterite could directly re-crystallize to form new tin oxide crystals. Such high temperature crystals, ranging from tiny anhedral to sub-hedral (acicular, skeletal, hopper) and euhedral shapes, provide no diagnostic information.

However, some crucibles appear to contain residual cassiterite. This has been tentatively identified by its distinct structural appearance compared to newly-formed tin oxide crystals. Though additional laboratory experiments are needed to verify this identification, comparison to geological cassiterite seems to confirm this assumption. Similar shapes have likewise been identified by Merideth (1998) in crucibles from Iberia, a region where cassiterite cementation has been reported repeatedly (e.g., Rovira 2007; Valério *et al.* 2013), and most recently by Renzi and Rovira-Llorens (2016: 155–159).

These few examples of possible residual cassiterite grains are however insufficient to exclude the use of metallic tin at Pi-Ramesse. First, there is no diagnostic evidence to be expected in the crucible slag for the use of metallic tin, so its absence does not argue against the use of metallic tin. Second, no tin ingots have been found at Pi-Ramesse, but this is to be expected for such a precious metal, likely acquired by long-distance trade. The finding of one small piece of an oxhide copper ingot (Gale 1999) in Pi-Ramesse corroborates the framework outlined earlier (Rehren and Pusch 2012), suggesting Pi-Ramesse would also have had access to tin ingots. Tin ingot finds are exceptional though, and archaeological absence should be expected in a highly controlled workshop environment where important materials would have been under strict control.

Co-smelting

Co-smelting, that is the smelting of mixed copper and tin ore, does not seem likely to have taken place in Pi-Ramesse. Though there are indications for the use of cassiterite, no remnant copper ore was found in crucible slag. In addition, no slag resembling primary copper smelting slag was discerned. The use of highly purified copper ore (e.g., malachite) is possible but unlikely as there is no evidence for ore (or its beneficiation) on site, and Late Bronze Age copper smelting regularly took place in furnaces and outside urban environments.

Recycling

Evidence for recycling is even more difficult to obtain based on slag analysis. Very little (iron) contamination is to be expected, but the same holds true when alloying pure/refined metal. Also, no high tin prills would be formed but, as mentioned earlier, the absence of evidence cannot be used as evidence of absence. In the case of Pi-Ramesse, however, scrap bronze present in the production area strongly suggests the practice of recycling. An example of such scrap material is shown in Fig. 34.15.

Summary

The analytical results presented here demonstrate that active alloying took place at Pi-Ramesse, and there are indications that mineral cassiterite could have been used for bronze production by cementation. The production waste freezes a single moment of a process that is variable in time. As such, a single sample cannot be taken as representative of a heterogeneous crucible and should not be expected to reveal the variability that exists within a production technology. Samples lacking any residual cassiterite could therefore equally reflect the full reaction of cassiterite in a cementation process, or the use of an alternative technique. As the use of tin metal for alloying is not expected to produce diagnostic crucible slag, its prevalence in Pi-Ramesse cannot be contrasted to that of cassiterite. Yet, the use of metallic tin cannot be excluded and appears likely based on circumstantial evidence. Therefore, it seems appropriate to suggest that the mixing of fresh metals and the cementation method were both practiced at Pi-Ramesse to produce fresh bronze, in addition to simply recycling existing bronze. Furthermore, either metallic tin or cassiterite could have been added to fresh copper or to recycled copper/bronze.

Fig. 34.15: Scrap bronze wrapped in linnen, workshop Q IV, find-no. 97/1288; photo: A. Krause.

MATERIAL USE

Copper Sources

The large scale of copper consumption at Pi-Ramesse and its distance from suitable ore deposits raises the question of where the copper came from. As research is ongoing in this regard (including lead isotope analysis: Rademakers *et al.* 2017), only preliminary observations are made here.

Several pieces of evidence indicate that different copper sources were used. First, the crucible slag analysis indicates that, in approximately 2/3 of the crucibles, "clean metal" (no iron enrichment) was used. This could imply the use of refined copper ingots and/or recycling of copper/bronze already in circulation.

Second, a piece of oxhide ingot was found in the production area, which has been sourced by lead isotope analysis to Apliki, Cyprus (Gale 1999).

Third, an iron-rich copper source can be inferred from the crucible slag analysis. This could suggest the use of unrefined copper from the primary smelting of sulphidic ores (e.g., Cypriot oxhide ingots [Knapp, Kassianidou and Donnelly 2001; Stos-Gale 2011] and "dirty" Uluburun ingots [Hauptmann, Maddin and Prange 2002]), or raw copper smelted from oxidic ores (e.g., Timna and Faynan ingots) (Tylecote, Ghaznavi and Boydell 1977; Rothenberg 1990; Eliyahu-Behar *et al.* 2012).

Finally, a cobalt/nickel-rich copper source has been identified. This type of contaminant is uncommon at percentage concentrations in Mediterranean oxhide ingots, though it occurs in ingots of Omani copper (Begemann *et al.* 2010; Goy, Le Carlier de Veslud and Degli Esposti 2013). It could therefore point to a source of copper to the southeast of Egypt (e.g., Omani ores) or perhaps even a sub-Saharan African source (Dayton 2003), although this would be more unlikely based on broader archaeological considerations. However, all geographical options remain open at this point.

Tin Sources

It is generally assumed that international trade in tin ingots existed during the Late Bronze Age, and that Egypt would also have participated in it (Dayton 1971, 2003; Muhly 1979; Penhallurick 1986; Nezafati, Pernicka and Momenzadeh 2011), as mentioned in the discussion of a framework for Egyptian metallurgical activity. Although there is no direct archaeological evidence for this in the production context of Pi-Ramesse, the foregoing discussion of production technology concludes that both cassiterite and tin metal could have been added to either fresh copper or recycled copper/bronze. While tin metal would likely have been imported to Pi-Ramesse in ingot form, probably through international trade, cassiterite might have had a more local origin. There is ample evidence for the presence of placer cassiterite in the Eastern Desert which may have been discovered and exploited in the context of (Old and Middle Kingdom) pharaonic gold mining expeditions (Lucas 1962; Dayton 1971; Nibbi 1976; Sabet *et al.* 1976; Wertime 1979; Muhly 1985; Rothe and Rapp 1995; Rothe, Rapp and Miller 1996; Rothenberg *et al.* 1998). Though the lack of New Kingdom inscriptions related to mining has been interpreted to indicate a decline in exploitation of these sources, at least through state-organized expeditions, there is no plausible evidence to exclude the possibility that local (nomadic) people continued to (occasionally) exploit these sources and that cassiterite made its way onto the Egyptian market. The continued exploitation of alluvial gold deposits may indeed be reflected in second Intermediate Period gold-work (Troalen *et al.* 2014). Abdel-Motelib *et al.* (2012) furthermore suggest much earlier ore/metal trade between the Naqadians and Bedouins, and Rehren and Pusch (2012) equally emphasize the importance of nomadic people's participation in the circulation of scrap metal throughout Egypt's history. The results from Pi-Ramesse should therefore renew interest in Egypt's potential domestic cassiterite/tin production and as such form an exciting contribution to the long-standing debate of tin supply in the Late Bronze Age. While long-distance trade may have dominated the overall market system for tin during the Late Bronze Age, local use of domestic tin deposits (where available) may have been equally important, despite being less tangible in archaeological and historical evidence.

CONCLUSION

Through detailed analysis of 49 crucible samples, the variability in technology and material use within the bronze production area at ancient Pi-Ramesse has been elucidated. Though further experimental evidence is needed, cassiterite has been tentatively identified in a number of crucibles. We suggest that bronze alloying took place by a combination of the mixing of fresh metals, cassiterite cementation and recycling of existing bronze. Preliminary results indicate that multiple copper sources were used in this process. This work illustrates how the complex nature of evidence derived from crucibles necessitates comprehensive sampling, whenever possible, to fully grasp the variability in ancient metallurgical activities.

Our conclusions shed new light on secondary metallurgical processes in ancient Egypt, which have hitherto received only cursory attention. They deepen our understanding into what materials were accessible and used in the Ramesside capital and thereby provide new insights into the nature of the metals trade in the Late Bronze Age eastern Mediterranean.

ACKNOWLEDGMENTS

We would like to thank the Supreme Council of Antiquities and its staff for 30 years of cooperation in Qantir–Pi-Ramesse, the Deutsche Forschungsgemeinschaft (DFG) for the long-term funding of excavations, the excavation team members for over 30 years of work and the NARNIA (New Archaeological Research Network for Integrating Approaches to ancient material studies) Project—a Marie Curie Initial Training Network funded by the FP7 and the European Union (Grant agreement no.: 265010) and coordinated by Prof. V. Kassianidou, University of Cyprus—for funding Frederik's Ph.D. research at UCL.

REFERENCES

Abdel-Motelib, A., *et al.* 2012. Archaeometallurgical Expeditions to the Sinai Peninsula and the Eastern Desert of Egypt (2006, 2008). *Metalla* 19: 3–59.

Abd el-Rahman, Y., *et al.* 2012. Ancient Mining and Smelting Activities in the Wadi Abu Gerida Area, Central Eastern Desert, Egypt: Preliminary Results. *Archaeometry* 55: 1067–1087.

Abd el Raziq, M., Castel, G., Tallet, P. and Fluzin, P. 2011. *Ayn Soukhna II: les ateliers métallurgiques du Moyen Empire*. Cairo.

Abouzeid, A.Z.M. and Khalid, A.A.M. 2011. Mineral Industry in Egypt—Part I: Metallic Mineral Commodities. *Natural Resources* 2: 35–53.

Bachmann, H.G. 1980. Early Copper Smelting Techniques in Sinai and Negev as Deduced from Slag Investigations. In: Craddock, P.T., ed. *Scientific Studies in Early Mining and Extractive Metallurgy* (British Museum Occasional Paper 20). London: 103–134.

Begemann, F., Hauptmann, A., Schmitt-Strecker, S. and Weisgerber, G. 2010. Lead Isotope and Chemical Signature of Copper from Oman and its Occurrence in Mesopotamia and Sites on the Arabian Gulf Coast. *Arabian Archaeology and Epigraphy* 21: 135–169.

Ben-Yosef, E. 2012. A Unique Casting Mould from the New Excavations at Timna Site 30 (Israel): Evidence of Western Influence? In: Kassianidou, V. and Papasavvas, G., eds. *Eastern Mediterranean Metallurgy and Metalwork in the Second Millennium BC—A Conference in Honour of James D. Muhly, Nicosia, 10th–11th October 2009*. Oxford: 188–196.

Budd, P., Pollard, A.M., Scaife, B. and Thomas, R.G. 1995. Oxhide Ingots, Recycling and the Mediterranean Metals Trade. *Journal of Mediterranean Archaeology* 8: 1–32.

Castel, G., *et al.* 1996. Wadi Dara Copper Mines. In: Esmael, F.A., ed. *Proceedings of the First International Conference on Ancient Egyptian Mining and Metallurgy and Conservation of Metallic Artifacts*. Cairo: 15–31.

Chirikure, S., Heimann, R.B. and Killick, D. 2010. The Technology of Tin Smelting in the Rooiberg Valley, Limpopo Province, South Africa, ca. 1650–1850 CE. *Journal of Archaeological Science* 37: 1656–1669.

Cooke, S.R.B. and Nielsen, B.V. 1978. Slags and Other Metallurgical Products. In: Rapp, G. Jr. and Aschenbrenner, S.E., eds. *Excavations at Nichoria in Southwest Greece, Vol. 1: Site, Environs and Techniques*. Minneapolis: 182–224.

Cowell, M.R. 1986. The Composition of Egyptian Copper-Based Metalwork. In: David, R.A., ed. *Science in Egyptology*. Manchester: 463–468.

Craddock, P.T. 2000. From Hearth to Furnace: Evidences for the Earliest Metal Smelting Technologies in the Eastern Mediterranean. *Paléorient* 26: 151–165.

Craddock, P.T. and Meeks, N.D. 1987. Iron in Ancient Copper. *Archaeometry* 29: 187–204.

Crew, P. and Rehren, Th. 2002. High-Temperature Workshop Residues from Tara—Iron, Bronze and Glass. *Discovery Programme Report* 6: 83–103.

Davey, C.J. 1985. Crucibles in the Petrie Collection and Hieroglyphic Ideograms for Metal. *Journal of Egyptian Archaeology* 71: 142–148.

Davies, N.D.G. 1935. *Paintings from the Tomb of Rekh-mi-Re at Thebes*. New York.

Dayton, J.E. 1971. The Problem of Tin in the Ancient World. *World Archaeology* 3: 49–70.

Dayton, J.E. 2003. The Problem of Tin in the Ancient World (Part 2). In: Giumlia-Mair, A. and Lo Schiavo, F., eds. *The Problem of Early Tin. Acts of the 14th UISPP Congress* (BAR International Series 1199). Oxford: 165–170.

Dungworth, D. 2000a. A Note on the Analysis of Crucibles and Moulds. *Historical Metallurgy* 34: 83–86.

Dungworth, D. 2000b. Serendipity in the Foundry? Tin Oxide Inclusions in Copper and Copper Alloys as an Indicator of Production Process. *Bulletin of the Metals Museum* 32: 1–5.

Eccleston, M.A.J. and Kemp, B.J. 2008. Metalworking and Crucibles. In: Kemp, B.J. and Stevens, A.K., eds. *Busy Lives at Amarna: Excavations at Grid 12 in the Main City*. London.

Eliyahu-Behar, A., *et al.* 2012. Iron and Bronze Production in Iron Age IIA Philistia: New Evidence from Tell es-Safi/Gath, Israel. *Journal of Archaeological Science* 39: 255–267.

Farag, M. 1981. Metallurgy in Ancient Egypt—Some Aspects of Techniques and Materials. *Metals Museum Bulletin* 6: 15–30.

Figueiredo, E., *et al.* 2010. Smelting and Recycling Evidences from the Late Bronze Age Habitat Site of Baiões (Viseu, Portugal). *Journal of Archaeological Science* 37: 1623–1634.

Gale, N.H. 1999. Lead Isotope Characterization of the Ore Deposits of Cyprus and Sardinia and its Application to the Discovery of the Sources of Copper for Late Bronze Age Oxhide Ingots. In: Young, S.M.M., Pollard, A.M., Budd, P. and Ixer, R.A., eds. *Metals in Antiquity* (British Archaeological Reports S792). Oxford: 110–121.

Gale, N.H. 2011. Copper Oxhide Ingots and Lead Isotope Provenancing. In: Betancourt, P.P. and Ferrence, S.C., eds. *Metallurgy: Understanding How, Learning Why: Studies in Honor of James D. Muhly*. Philadelphia: 213–220.

Garland, H. and Bannister, C.O. 1927. *Ancient Egyptian Metallurgy*. London.

Garson, M.S. 1977. Younger Granites of Egypt and Associated Mineralization. *Transactions of the Institution of Mining and Metallurgy* 86: 161.

Goy, J., Le Carlier de Veslud, C. and Degli Esposti, M. 2013. Archaeometallurgical Research in the Northern Hajjar Mountains (Oman Peninsula) During the Iron Age (1200–300 BCE). *Poster presented at the Timna Park International Conference in memory of Professor Beno Rothenberg. Mining for Copper: Environment, Culture and Copper in Antiquity. 22–25 April 2013, Timna, Israel.*

Hall, F.P. and Insley, H. 1933. A Compilation of Phase-rule Diagrams of Interest to the Ceramist and Silicate Technologist. *Journal of the American Ceramic Society* 16: 463–567.

Hamza, M. 1930. Excavations of the Department of Antiquities at Qantir (Faqus District). *Annales du Service des Antiquites de l'Egypte* 30: 31–68.

Hauptmann, A., Maddin, R. and Prange, M. 2002. On the Structure and Composition of Copper and Tin Ingots Excavated from the Shipwreck of Uluburun. *Bulletin of the American Schools of Oriental Research* 328: 1–30.

Hayes, W.C. 1937. *Glazed Tiles from a Palace of Ramesses II at Kantir*. New York.

Herold, A. 1998. Gegossen - Geschmiedet - Tordiert. Zur Herstellung von bronzenen Trensen in der Ramses-Stadt Piramesse/Ägypten. *Metalla* 5: 3–21.

Herold, A. 1999. *Streitwagentechnologie in der Ramses-Stadt: Bronze an Pferd und Wagen* (Forschungen in der Ramses-Stadt Band 2). Mainz.

Herrmann, C. 1957. *Formen für ägyptische Fayencen aus Qantir: Katalog der Sammlung des Franciscan Biblical Museum, Jerusalem und zweier Privatsammlungen*. Fribourg.

Hikade, T. 1998. Economic Aspects of the New Kingdom: The Expeditions to the Copper Mines of the Sinai. *Bulletin of the Australian Centre for Egyptology* 9: 43–52.

Hofmann, W. and Klein, M. 1966. Beitrag zur Kenntnis des Dreistoffsystems Kupfer-Zinn-Sauerstoff. *Zeitschrift für Metallkunde* 83: 421–441.

James, T.G.H. 1972. Gold Technology in Ancient Egypt. Mastery of Metal Working Methods. *Gold Bulletin* 5: 38–42.

Kallfass, M. and Hörz, G. 1989. Metallographische Untersuchungen an archäologischen Zinnbronzen aus der spätägyptischen Zeit und Bronzezeit. *Praktische Metallographie* 26: 105–121.

Kassianidou, V. 2003. The Trade of Tin and the Island of Copper. In: Giumlia-Mair, A. and Lo Schiavo, F., eds. *The Problem of Early Tin. Acts of the 14th UISPP Congress* (BAR International Series 1199). Oxford: 109–119.

Kearns, T., Martinón-Torres, M. and Rehren, Th. 2010. Metal to Mould: Alloy Identification in Experimental Casting Moulds using XRF. *Historical Metallurgy* 44: 48–58.

Klemm, R. and Klemm, D. 2013. *Gold and Gold Mining in Ancient Egypt and Nubia: Geoarchaeology of the Ancient Gold Mining Sites in the Egyptian and Sudanese Eastern Deserts*. Berlin.

Knapp, A.B. 2000. Archaeology, Science-based Archaeology and the Mediterranean Bronze Age Metals Trade. *European Journal of Archaeology* 3: 31–56.

Knapp, A.B., Kassianidou, V. and Donnelly, M. 2001. Copper Smelting in Late Bronze Age Cyprus: The Excavations at Politiko Phorades. *Near Eastern Archaeology* 64: 204–210.

Lo Schiavo, F. 2012. Cyprus and Sardinia, Beyond the Oxhide Ingots. In: Kassianidou, V. and Papasavvas, G., eds. *Eastern Mediterranean Metallurgy and Metalwork in the Second Millennium BC—A Conference in Honour of James D. Muhly, Nicosia, 10th–11th October 2009*. Oxford: 142–150.

Lucas, A. 1962. *Ancient Egyptian Materials and Industries*. London.

Merideth, C. 1998. *An Archaeometallurgical Survey for Ancient Tin Mines and Smelting Sites in Spain and Portugal* (BAR International Series B714). Oxford.

Merkel, J.F. 1990. Experimental Reconstruction of Bronze Age Copper Smelting Based on Archaeological Evidence from Timna. In: Rothenberg, B., ed. *The Ancient Metallurgy of Copper: Archaeology-Experiment-Theory (Researches in the Arabah 1959–84; v. 2)*. London: 78–122.

Misra, M.K., Ragland, K.W. and Baker, A.J. 1993. Wood Ash Composition as a Function of Furnace Temperature. *Biomass and Bioenergy* 4: 103–116.

Muan, A. 1957. Phase Equilibria at Liquidus Temperatures in the System Iron Oxide-Al_2O_3-SiO_2 in Air Atmosphere. *Journal of the American Ceramic Society* 40: 121–133.

Muhly, J.D. 1979. New Evidence for Sources of and Trade in Bronze Age Tin. In: Franklin, A.D., Olin, J.S. and Wertime, T.A., eds. *The Search for Ancient Tin: A Seminar Organized by Theodore A. Wertime and Held at the Smithsonian Institution and the National Bureau of Standards, Washington, D.C., March 14–15, 1977*. Washington, D.C.: 43–48.

Muhly, J.D. 1985. Sources of Tin and the Beginnings of Bronze Metallurgy. *American Journal of Archaeology* 89: 275–291.

Muhly, J.D. 2003. Trade in Metals in the Late Bronze Age and the Iron Age. In: Stampolidis, N.C. and Karageorghis, V., eds. *Sea Routes... Interconnections in the Mediterranean 16th–6th c. BC: Proceedings of the International Symposium held at Rethymnon, Crete, September 29th–October 2nd, 2002*. Athens: 141–150.

Murillo-Barroso, M., Pryce, T.O., Bellina, B. and Martinón-Torres, M. 2010. Khao Sam Kaeo—An Archaeometallurgical Crossroads for Trans-Asiatic Technological Traditions. *Journal of Archaeological Science* 37: 1761–1772.

Nezafati, N., Pernicka, E. and Momenzadeh, M. 2011. Early Tin-Copper Ore from Iran, a Possible Clue for the Enigma of Bronze Age Tin. In: Yalçin, Ü., ed. *Anatolian Metal*, Vol. 5. Bochum: 211–230.

Nibbi, A. 1976. Tin from the Eastern Desert. *Göttinger Miszellen* 19: 49–50.

Ogden, J. 2000. Metals. In: Nicholson, P.T. and Shaw, I., eds. *Ancient Egyptian Materials and Technology*. Cambridge: 148–176.

Pelleg, J., Baram, J. and Oren, E. 1979. An Investigation of Bronze Artifacts from the North Sinai Coast and the Nile Delta Region. *Metallography* 12: 313–324.

Penhallurick, R.D. 1986. *Tin in Antiquity: Its Mining and Trade Throughout the Ancient World with Particular Reference to Cornwall*. London.

Philip, G. 2007. *Tell El-Dab'a XV, Metalwork and Metalworking Evidence of the Late Middle Kingdom and The Second Intermediate Period*. Vienna.

Pigott, V.C., Rogers, H.C. and Nash, S.K. 2003. Archaeometallurgical Investigations at Malyan: The Evidence for Tin-bronze in the Kaftari Phase. In: Miller, N.F. and Abdi, K., eds. *Yeki Bud, Yeki Nabud: Essays on the Archaeology of Iran in Honor of William M. Sumner*. Los Angeles: 161–176.

Prell, S. 2011. *Einblicke in die Werkstätten der Residenz. Die Stein- und Metallwerkzeuge des Grabungsplatzes Q I* (Forschungen in der Ramses-Stadt Band 8). Hildesheim.

Pulak, C. 1997. The Uluburun Shipwreck. In: Swiny, S., Hohlfelder, R.L. and Swiny, H.W., eds. *Res Maritimae: Cyprus and the Eastern Mediterranean from Prehistory to Late Antiquity: Proceedings of the Second International Symposium "Cities on the Sea," Nicosia, Cyprus, October 18–22, 1994*. Atlanta: 233–262.

Pusch, E.B. 1990. Metallverarbeitende Werkstätten der frühen Ramessidenzeit in Qantir-Piramesse/Nord - Ein Zwischenbericht. *Egypt and the Levant* 1: 75–113.

Pusch, E.B. 1991. Recent Work at Northern Piramesse - Results of Excavations by the Pelizaeus-Museum, Hildesheim, at Qantir. In: Bleiberg, E. and Freed, R., eds. *Fragments of a Shattered Visage: The Proceedings of the International Symposium of Ramesses the Great*. Memphis: 199–220.

Pusch, E.B. 1993. Pi-Ramesse-geliebt-von-Amun, Hauptquartier Deiner Streitwagentruppen. Ägypter und Hethiter in der Delta-Residenz der Ramessiden. In: Eggebrecht, A., ed. *Pelizaeus-Museum Hildesheim. Die Ägyptische Sammlung*. Mainz: 126–143.

Pusch, E.B. 1994. Divergierende Verfahren der Metallverarbeitung in Theben und Qantir? Bemerkungen zur Konstruktion und Technik. *Egypt and the Levant* 4: 145–170.

Pusch, E.B. and Herold, A. 1999. Qantir/Pi-Ramesses. In: Bard, K., ed. *Encyclopedia of the Archaeology of Ancient Egypt*. London: 647–649.

Pusch, E.B. and Rehren, Th. 2007. *Hochtemperatur-Technologie in der Ramses-Stadt - Rubinglas für den Pharao* (Forschungen in der Ramses-Stadt Band 6). Hildesheim.

Rademakers, F.W. 2015. *Into the Crucible. Methodological Approaches to Reconstructing Crucible Metallurgy, from New Kingdom Egypt to Late Roman Thrace* (Ph.D. dissertation., University College London). London: http://discovery.ucl.ac.uk/id/eprint/1469615.

Rademakers, F.W. and Rehren, Th. 2016. Seeing the Forest for the Trees: Assessing Technological Variability in Ancient Metallurgical Crucible Assemblages, *Journal of Archaeological Science: Reports* 7: 588–596.

Rademakers, F.W., Rehren, Th. and Pernicka, E. 2017. Copper for the Pharaoh: Identifying Multiple Metal Sources for Ramesses' Workshops from Bronze and Crucible Remains, *Journal of Archaeological Science* 80: 50–73.

Rehren, Th. 2000. Rationales in Old World Base Glass Compositions. *Journal of Archaeological Science* 27: 1225–1234.

Rehren, Th. 2001. Die Schmelzgefässe aus Cham-Oberwil. In: Gnepf Horisberger, U. and Hämmerle, S., eds. *Cham-Oberwil, Hof (Kanton Zug): Befunde und Funde aus der Glockenbecherkultur und der Bronzezeit* (Antiqua 33). Basel: 118–131.

Rehren, Th. and Pusch, E.B. 2012. Alloying and Resource Management in New Kingdom Egypt: The Bronze Industry at Qantir - Pi-Ramesse and its Relationship to Egyptian Copper Sources. In: Kassianidou, V. and Papasavvas, G., eds. *Eastern Mediterranean Metallurgy and Metalwork in the Second Millennium BC—A Conference in Honour of James D. Muhly, Nicosia, 10th–11th October 2009*. Oxford: 215–221.

Rehren, Th., Pusch, E.B. and Herold, A. 1998. Glass Coloring Works Within a Copper-centered Industrial Complex in Late Bronze Age Egypt. In: McCray, P. and Kingery, W.D., eds. *The Prehistory and History of Glassmaking Technology* (Ceramics and Civilization Vol. 8). Westerville, OH: 227–250.

Rehren, Th., Pusch, E.B. and Herold, A. 2001. Problems and Possibilities in Workshop Reconstruction: Qantir and the Organisation of LBA Glass Working Sites. In: Shortland, A., ed. *The Social Context of Technological Change. Egypt and the Near East, 1650–1550 BC*. Oxford: 223–238.

Renzi, M. and Rovira-Llorens, S. 2016. Metallurgical Vessels from the Phoenician Site of La Fonteta (Alicante, Spain): A Typological and Analytical Study. In: Körlin, G., Prange, M., Stöllner, Th. and Yalcin, Ü., eds. *From Bright Ores to Shiny Metals*. Bochum: 143–166.

Riederer, J. 1981. Analysis of Egyptian Bronzes. *Revue d'Archéometrie* 1: 239–243.

Riederer, J. 1988. Metallanalysen ägyptischer Bronzestatuetten aus deutschen Museen. *Berliner Beiträge zur Archäometrie* 10: 5–20.

Rothe, R. and Rapp, G.Jr. 1995. Trace-Element Analyses of Egyptian Eastern Desert Tin and its Importance to Egyptian Archaeology. In: Hussein A.A.A., Miele, M. and Riad, S., eds. *Proceedings of the Egyptian-Italian Seminar on the Geosciences and Archaeology in the Mediterranean Countries* (Special Publication 70). Cairo: 229–244.

Rothe, R.D., Rapp, G.Jr. and Miller, W.K. 1996. New Hieroglyphic Evidence for Pharaonic Activity in the Eastern Desert of Egypt. *Journal of the American Research Center in Egypt* 33: 77–104.

Rothenberg, B. 1988. *The Egyptian Mining Temple at Timna*. London.

Rothenberg, B., ed. 1990. *The Ancient Metallurgy of Copper: Archaeology-Experiment-Theory* (Researches in the Arabah 1959–84, Vol. 2). London.

Rothenberg, B. 2003. Egyptian Chariots, Midianites from Hijaz/Midian (Northwest Arabia) and Amalekites from the Negev in the Timna Mines. *Institute of Archaeo-Metallurgical Studies* 23: 9–14.

Rothenberg, B., Shaw, C.T., Hassan, F.A. and Hussein, A.A.A. 1998. Reconnaissance Survey of Ancient Mining and Metallurgy in the Mersa Alam Region Eastern Desert of Egypt. *Institute of Archaeo-Metallurgical Studies* 20: 4–9.

Rovira, S. 2007. La Producción de Bronces en la Prehistoria. In: Molera, J., Farjas, J., Roura, P. and Pradell, T., eds. *Avances En Arqueometría. Actas Del VI Congreso Ibérico De Arqueometría 2005*. Girona: 21–35.

Rovira, S., Montero-Ruiz, I. and Renzi, M. 2009. Experimental Co-smelting to Copper-tin Alloys. In: Kienlin, T.L. and Roberts, B.W., eds. *Metals and Societies. Studies in Honour of Barbara S. Ottaway* (Universitätsforschungen zur Prähistorischen Archäologie 169). Bonn: 407–414.

Sabet, A.H., Tsogoev, V.B., Shibanin, S.P., El-Kadi, M.B. and Awad, S. 1976. The Placer Tin-deposits of Igla, Abu Dabbab and Nuweibi. *Annals of the Geological Survey of Egypt* 6: 169–180.

Said, R. 1990. *The Geology of Egypt.* Rotterdam.

Scheel, B. 1985. Studien zum Metallhandwerk im Alten Ägypten I. Handlungen und Beischriften in den Bildprogrammen der Gräber des Alten Reiches. *Studien zur altägyptischen Kultur* 12: 117–177.

Scheel, B. 1986. Studien zum Metallhandwerk im Alten Ägypten II. Handlungen und Beischriften in den Bildprogrammen der Gräber des Mittleren Reiches. *Studien zur altägyptischen Kultur* 13: 181–205.

Scheel, B. 1987. Studien zum Metallhandwerk im Alten Ägypten III. Handlungen und Beischriften in den Bildprogrammen der Gräber des Neuen Reiches und der Spätzeit. *Studien zur altägyptischen Kultur* 14: 247–264.

Scheel, B. 1988. Fundobjekte einer Ptolemäerzeitlichen Metallverarbeitungsstätte in Theben und Vergleichsfunde anderer vorderorientalischer Ausgrabungsplätze. *Studien zur altägyptischen Kultur* 15: 243–254.

Scheel, B. 1989. *Egyptian Metalworking and Tools*. Princes Risborough.

Scott, D.A. 1991. *Metallography and Microstructure of Ancient and Historic Metals*. Singapore.

Shaw, I. 1994. Pharaonic Quarrying and Mining: Settlement and Procurement in Egypt's Marginal Regions. *Antiquity* 68: 108–119.

Shaw, I. 1998. Exploiting the Desert Frontier. The Logistics and Politics of Ancient Egyptian Mining Expeditions. In: Knapp, A.B., Pigott, V.C. and Herbert, E.W., eds. *Social Approaches to an Industrial Past*. London and New York: 242–258.

Stos, Z.A. 2009. Across the Wine Dark Seas...Sailor Tinkers and Royal Cargoes in the Late Bronze Age Eastern Mediterranean. In: Shortland, A.J., Freestone, I.C. and Rehren, Th., eds. *From Mine to Microscope: Advances in the Study of Ancient Technology*. Oxford: 163–180.

Stos-Gale, Z.A. 2011. "Biscuits with Ears": A Search for the Origin of the Earliest Oxhide Ingots. In: Betancourt, P.P. and Ferrence, S.C., eds. *Metallurgy: Understanding How, Learning Why. Studies in Honor of James D. Muhly*. Philadelphia: 221–229.

Stos-Gale, Z.A., Gale, N.H. and Houghton, J. 1995. The Origins of Egyptian Copper: Lead-Isotope Analysis of Metals from el-Amarna. In: Davies, V.W. and Schofield, L., eds. *Egypt, the Aegean and the Levant: Interconnections in the Second Millennium BC*. London: 127–135.

Tallet, P. 2012. Ayn Sukhna and Wadi el-Jarf: Two Newly Discovered Pharaonic Harbours on the Suez Gulf. *British Museum Studies in Ancient Egypt and Sudan* 18: 147–168.

Troalen, L.G., Tate, J. and Guerra, M.F. 2014: Goldwork in Ancient Egypt: Workshop Practices at Qurneh in the 2nd Intermediate Period, *Journal of Archaeological Science* 50: 219–226.

Tylecote, R.F. 1982. Metallurgical Crucibles and Crucible Slags. In: Olin, J.S. and Franklin, A.D., eds. *Archaeological Ceramics*. Washington: 231–243.

Tylecote, R.F., Ghaznavi, H.A. and Boydell, P.J. 1977. Partitioning of Trace Elements between the Ores, Fluxes, Slags and Metal during the Smelting of Copper. *Journal of Archaeological Science* 4: 305–333.

Valério, P., *et al*. 2013. Bronze Production in Southwestern Iberian Peninsula: The Late Bronze Age Metallurgical Workshop from Entre Águas 5 (Portugal). *Journal of Archaeological Science* 40: 439–451.

Van De Mieroop, M. 2007. *The Eastern Mediterranean in the Age of Ramesses II*. Oxford.

Wertime, T.A. 1979. The Search for Ancient Tin: The Geographic and Historic Boundaries. In: Franklin, A.D., Olin, J.S., and Wertime, T.A., eds. *The Search for Ancient Tin: A Seminar Organized by Theodore A. Wertime and Held at the Smithsonian Institution and the National Bureau of Standards, Washington, D.C., March 14–15, 1977*. Washington, D.C.: 1–6.

Wilkinson, R.H. 1994. *Symbol and Magic in Egyptian Art*. London.

Wood, N. 2009. Some Implications of the Use of Wood Ash in Chinese Stoneware Glazes. In: Shortland, A.J., Freestone, I.C. and Rehren, Th., eds. *From Mine to Microscope. Advances in the Study of Ancient Technology*. Oxford: 51–60.

CHAPTER 35

EXAMPLES OF COPPER HARPOONS OF NAQADA CULTURE IN THE EASTERN NILE DELTA

Marcin Czarnowicz

This paper presents copper harpoons of Pre-Dynastic Egypt and discusses their significance in the Naqada culture. An overview of the finds and relevant artistic depictions suggest that they were not only weapons, but also an important symbol that played several roles in the ancient Egyptian society.[1]

INTRODUCTION

The harpoon was one of the most popular weapons known in the Egyptian culture that developed on the banks of the Nile River. Although the materials with which it was made changed over time, its form remained unaltered from prehistoric times through to the 18th dynasty (Brewer and Friedman 1989: 103).

The first examples of harpoons appear already in the Early Neolithic period (for chronology, see Table 35.1). Tools are known as far back as the Fayum A culture; although none of the artifacts connected with this culture have been found *in situ* (Ciałowicz 1999: 216); they are also present in the Badari and Merimde (Ciałowicz 1999: 216). Most of the finds are from the Naqada period. They became most popular during the Gerzean phase of the Naqada culture (Naqada II), when harpoons became frequent burial gifts which accompanied the dead to the grave. This period is characterized by territorial expansion which—in a relatively short time—contributed to the spread of the Naqada culture to all areas of the Nile Valley which led—by the end of the period—to the unification of the Delta areas which were previously inhabited by the Lower Egypt cultural groups. Researchers cannot conclude whether this was done through significant migration of groups of people from Upper Egypt to the Delta areas, or whether the change was more evolutionary in character, and the changes occurred due to a gradual takeover of Southern cultural patterns alongside limited migration of small groups whose goal was to control the trade route that ran through Lower Egypt and led to the Middle East (Mączyńska 2011). Harpoons did not lose their popularity in the next period—Naqada III. During this period, most of the finds come from the Delta areas.

MORPHOLOGY OF PRE-DYNASTIC EGYPTIAN HARPOONS

Harpoons belonging to early Egyptian cultures were made in a very similar way. It can even be argued that they illustrate the continuous development of culture from the Fayum A to Naqada. Two dominating types can be identified. The first one has a single large barb located on the tip of the spearhead with a sharpened edge extending from the tip to the end of the barb. The second one is characterized by a number of additional barbs. For tools made of bone, barbs are formed by a deep-cutting of one of the

1 The research was conducted under financial support from the National Science Centre of Poland granted on the basis of decision DEC-2014/15/N/HS3/01144

Table 35.1: Culture-Chronology of Egypt in the Fifth–Fourth Millennia BCE

Absolute Chronology	*Period*	*Relative Chronology*	*Culture*
3100–2890	Early Dynastic	First Dynasty	Naqada IIIC–D
3200	Proto-Dynastic	0 Dynasty	Naqada IIIA-B
3600	Pre-Dynastic	Late Chalcolithic	Naqada IIC/D- IIIA
3900		Early Chalcolithic	Naqada IB/C- IIB
4500		Late Neolithic	Naqada IA/B Badari
5000		Early Neolithic	El Omari Merimde Fayum A

edges. Copper harpoons have one or more (up to three) barbs arranged perpendicular to the stem, and bent in the direction of the shaft. A characteristic feature of all harpoons from the Pre-Dynastic and Early Dynastic periods is the location of barbs on only one side of the blade. Examples of artifacts with the above-mentioned arrangement on both sides are extremely rare. Harpoons where both edges are sharpened have a rhomboid cross-section, while the vast majority of the artifacts with a few copper barbs have a triangular or a rectangular profile with rounded corners; the stem is usually square. At the end of the stem there were knobs or cuts that helped stabilize the tip on the shaft. Some researchers believe that this part was saturated with poison (Ciałowicz 1999: 216).

Initially the harpoons were made of bone or horn, but in Naqada culture the copper items gained great popularity. They are replicas of harpoons made of bone, and for a time both types coexisted side by side, although in the case of copper artifacts, the ones with several barbs are much more popular. Analyzing the execution of the described objects, it can be clearly stated that it was not the sharpness of their edges, but the mass and energy obtained during the throw that were to ensure the piercing of the prey's skin; the barb prevented the ejection of the tip from the body.

HARPOONS IN ART

Most of our knowledge about the use of harpoons comes from examining early Egyptian art. Even though they are meant to be symbolic, they depict the reality of the real game. There are many scenes in which harpoons were used to hunt Nile animals. It is a theme often found in the White Cross Lined Pottery of the Naqada I period. It is worth underscoring that the scenes do not depict the fight between the hunter and the animal itself but rather the resulting expression of victory (Hendricks 2011: 239). It appears for example on the dishes of Mahasna (Fig. 35.1). One of the scenes depicts a man standing in front of a hippopotamus. From the head of the animal towards the hunter, an artist drew a line ending with an oval object (Ayrton and Loat 1911: Pl. XVIII: 13). This could be a kind of a buoy indicating the end of the rope just as the animal—frightened by the weapon's strike—was trying to escape from its attacker. A similar scene is depicted on a plate from the Metropolitan Museum of Art (MMA35.10). Here a hippopotamus is shown standing in front of the hunter holding a rope connected to what seems to be a harpoon. Another artifact of this type is depicted behind the animal. It is tied to the boat, but missed the hippopotamus (Hayes 1953: 18, Fig. 10).

Yet another vessel from this period shows a similar scene, however the harpoon is thrust on the hippopotamus' rump while the rope is held by the hunter. The jar from Abydos Grave U-415 has a very interesting depiction with a very symbolic meaning. It definitely shows a military triumph in the upper

Fig. 35.1: Part of the motif depicted on the Mahasna bowl (drawing by M. Czarnowicz).

register. There is a row of people, and the central one is bigger than the rest and is holding a mace head. The lower register contains a hippopotamus hunting scene which includes three hunters, three hippopotamuses, and a bull with oversized horns (Dreyer *et al.* 2003: 81; Abb. 5). Since bulls are not connected with hippopotamus hunting, it is possible that this bull represents a ruler (Hendrickx 2011: 239–244) as is the case in the later bull or Narmer Palette.

A hunting scene with a harpoon is also visible on the vessel from Gebelein. In this case, it presents hunting with the use of a boat. The end of the harpoon rope that wounded the animal is attached to that boat (Ciałowicz 1999: Fig. 162.4). The hunting scenes with the use of boats gained popularity over time. Such a scene is also presented on the bowl from Grave U-264 from Abydos (Hartmann 2008: Abb. 8). A hippopotamus is portrayed among the many visible animals, and is connected to the boat by two wavy lines—resembling the way in which harpoon ropes were depicted in early Egyptian art.

Some of the hunting scenes do not depict the hunter but only the weapon—in this case a harpoon (Hendrickx 2011: 239; 2012a: 77). An example of this type of scene can be found on a dish from Cemetery C at Abydos. A jar with a long neck dating to the Naqada I (Hendrickx 2012b: 153) shows a group of animals, including gazelles and hippos, that are being hunted with the use of dogs. There is no hunter, but one can see two hippos in the back row that were wounded on the rumps by harpoons (Hendrickx 2012b: Figure C1). The vessel is quite unusual because the scene was not painted on the dish's body, but rather extends from the base in the direction of the rim.

Scenes with the use of harpoons can also be found on other relics of the Pre-Dynastic Period of great importance is a rock drawing from Was-ha-Waset (Cat. no. 353). A single hunter—holding a mace head, a lasso and the harpoon's rope—are depicted in a combination of Nilotic and desert hunting scenes (Darnell 2009: Fig. 8). In the upper part of the graffiti, we see a hippopotamus with the harpoon protruding from its head. Below a hunting scene of a desert animal with the use of a lasso is depicted. It is worth underscoring that the hunter is holding a mace head, which was a symbol of military triumph (Hendrickx 2011: 239).

Another poorly preserved artifact with Nilotic hunting is the Gebelain Cloth, where one can see a hunter—dressed in a white loincloth—holding a harpoon, which is stuck in the back of a hippopotamus. The fabric is dated to the Naqada Ia–IIc period (Ciałowicz 1999: 297). Another art relic from this period—

the painting from Tomb 100 at Hierakonpolis—depicts several scenes including a centrally-portrayed boat procession with several minor scenes placed around it and depicting duels, hunts and superiority over animals (Hendrickx 2012a: Fig. 8.2). The relic can most probably be dated to the Naqada IIc–d period (Adams 1996: 1). In one of the boats, by far the largest and perhaps the most important of the whole procession, the man sitting at the helm is holding a harpoon in both hands (Fig. 35.2). Researchers are in disagreement about this part of the scene, but all emphasize that we are dealing here with a boat transporting the ruler, and the entire scene probably shows a Horus procession (Ciałowicz 1999: 300). It should be emphasized that the man that can be seen in the boat is most likely a bodyguard and the harpoon is his weapon. The painting also depicts what probably is a harpoon fight scene. The bottom row shows a drawing of two men, one of whom is kneeling and holding out in front of him an object resembling a light shield covered with animal skin. Opposite him, another man is shown wearing leather and a phallus protector, and holding in his hand a club and a harpoon. Leeman (2007: 17) suggested that the scene may be depicting a priest, though Ciałowicz's (1999: 301) view seems to be the correct one, as it connects this scene to another one, where a man, previously hiding behind a shield, is depicted in what should be interpreted as a deceased position (Fig. 35.3).

We also know of harpoons that were depicted on cosmetic pallets. During the Naqada I period, these diamond-shaped artifacts with symbols of nome became popular, and among them was the harpoon symbol (Ciałowicz 1999: 211). A palette of this shape with figural decorations has also been found. The so-called "Stockholm Palette" depicts a hunting scene: on the left-hand side there are two

Fig. 35.2: Detail from the Grave 100 painting depicting a warrior with a harpoon (drawing by M. Czarnowicz).

Fig. 35.3: Detail from the Grave 100 painting showing a duel between warriors, one of them is holding a harpoon (drawing by M. Czarnowicz).

horned herbivores and a hippopotamus in poses that suggest that all are dead. On the right-hand side we see a man in a boat with a steering oar holding a rope attached to the harpoon which is stuck in the hippopotamus' rump (Hendrickx 2011: Fig. 6).

The number of hunting scenes decreased with the appearance of a new type of pottery, D-Ware, and thus also the number of scenes depicting the use of harpoons. There is a well-known partly preserved vessel from this period (Naqada II) that is shaped like a hippopotamus, and on it a drawing of hunters with harpoons and boats. This vessel was found in Tomb 3759 (Brunton and Caton-Thompson 1928: Pl. XLVIII: 3; Payne 1993, No 928), and is located in the collections of the Ashmolean Museum.

During the Naqada IIIC period, the presentation of harpoons in Egyptian art changed to a more symbolic one. For example, in the upper part of the reverse on the famous Narmer palette, the ruler marches along with his entourage towards ten bodies with bound hands and heads cut off. Horus is seen twice above the victim images. In the first case, he stands in front of an object which may be interpreted as the gate leaf. In the second, his silhouette is placed over the boat and is gripping a harpoon-head (O'Connor 2012: 148).

A representation visible on a seal impression found in the Tomb of Den, the ruler of the 1st Dynasty, is also symbolic in nature. A typical royal hippopotamus hunt is depicted here. In one hand, the king holds a harpoon ready to be thrown at an animal; in his other hand he holds a rope. However, a pair of fallen decapitated enemies is depicted in front of the king instead of the hippopotamus (Muller 2008: 481–483; Abb. 3). It is clear that such a representation carries a symbolic meaning which shows the ruler as the only power that can defeat the forces of chaos and evil.

Representations of people holding harpoons are known not only in Egypt but also in areas of Nubia. Cylindrical seals from Qustul and Sarras West are decorated with relief drawings of anthropomorphic, zoomorphic and harpoon images. In a tomb dated to the period of Naqada IId–IIIa-b from the Qustul cemetery, a censer was found with a depiction of a harpoon positioned between an antelope and a boat with the ruler (Williams 2012: 162–163).

In later periods, a canonical form for scenes of hunting hippos with harpoons was adopted. The ruler stood on a barge wielding the weapon. Probably the best development of this approach is the image of the pharaoh Tutankhamen that was discovered in his tomb. The sovereign is standing astride a small boat, preparing to throw a harpoon. This type of representation probably had a more symbolic dimension and referred to Pharaoh's struggle for law and order in the country, as the hippopotamus was the embodiment of evil and chaos, and linked to the god Seth. Depicting the Pharaoh as dominant over the animal with such attributes was meant to emphasize the power and lawfulness of the ruler, and the harpoon had come to symbolize this function.

There was probably a different reason why such scenes began to appear in the tombs of nobles. A harpoon hippo hunt using a boat had become a form of sport and entertainment for Egyptian society's upper class. This can clearly be seen on the decoration of Mereruka mastaba in Saqqara (6th Dynasty), or in a hunting scene on a wooden model found in the grave of Meketre dating to the Middle Kingdom.

WAYS TO USE HARPOONS

Harpoons are not very common, though there are some at most sites of the Naqada culture, such as Naqada, Mahasna, El-Amra, Adaima or Ballas (Brewer and Friedman 1989: 103). Before the appearance of the Naqada culture in the Delta, harpoons were not very popular in this area. At the Maadi site only one such artifact was found (Rizkana and Seeher 1989: 23). They are very often found in tombs of all genders and ages. This may indicate that even women and children were involved in the hunt for fish trapped in the floodwaters of the Nile (Ciałowicz 1999: 216). The most important find is without a doubt a harpoon from Mahasna. In Grave H23 dated to the Naqada IIa (Ciałowicz 1999: 216) a harpoon head was discovered, which was originally wrapped in fabric. This artifact was preserved along with a fragment of the rope connecting it to the shaft and with other remnants. The researchers say that in the case of this grave, one can talk about the burial of a warrior, a fact that is also supported by the findings of other weapons such as numerous maceheads buried with the deceased (Ayrton and Loat 1911: 22–23, Pl. XX: 3).

Hippos are extremely dangerous creatures that can easily harm or even kill a human if frightened. They plundered the fields at night, caused destruction and threatened villagers. The groups of hunters were probably trying to harm the hippo with harpoons causing his escape from the fields or villages. The rope with the buoy indicates the place where the animal was hiding. The weakened and bleeding animal was encircled by hunters and killed. When the hunters used the boat for games, they tied the ropes to the boat. In my opinion such a method for hippo hunting might be less dangerous. Hunting from a boat was part of nobility "entertainment" but one cannot exclude the possibility that economic factors were also of significance. Ivory objects were appreciated by Egyptian elites. Many works of early Egyptian art, such as Tell el-Farkha figurines, were made of this raw material.

So what did the hunt with a harpoon look like? The weapon consists of the harpoon point and shaft and the rope. By comparing different representations of warriors or hunters, one can estimate that its length was comparable to the height of an adult male (approximately 1.6 m high). Perhaps everyone hunted. The men engaged in hunting the large Nile mammals, and the women and children took part in the hunt for fish that became lodged in the backwaters of the river after heavy rains. Researchers do not agree whether the shaft was permanently connected to the spearhead, or could be taken off after hitting the game. In such a situation the shaft could serve as a buoy indicating which way the game ran (Ayrton and Loat 1911: 22–23; Ciałowicz 1999: 217; Brewer and Friedman 1989: 21–23), though it seems unlikely that the head was not permanently attached to the wooden part. Only this solution guarantees that upon hitting the prey the harpoon would pierce the skin and remain there firmly. If it was not attached permanently, then the energy accumulated by the shaft and transferred to the head upon contacting the prey could cause the harpoon to rebound instead of penetrating the prey. Furthermore, the entire harpoon stuck in the animal would cause it great pain. While on the run, the prey's movements would cause the shaft to move and transfer its vibrations to the harpoon's head. Moreover, the harpoon's barbs would then tear off muscle pieces or destroy internal organs. This would have led to a significant weakening of the prey, if not to its death. It has previously been proven by Behrmann (1996) that harpoons can be deadly weapons against hippopotamuses.

Art relics also demonstrate that harpoons were used as a weapon in close combat.

COPPER HARPOONS FROM NILE DELTA AREAS

Harpoons became popular in the Delta area after the period of cultural unification. There are several known and published specimens that originate from four sites: Kafr Hassan Dawood, Minshat Abu Omar, Tell el-Farkha and Tell el-Murra.

TELL EL-FARKHA

The site is located approximately 120 km northeast of Cairo. For over 15 years, the research there has been led by the Polish Archaeological Mission to the Eastern Nile Delta, consisting of archaeologists from the Institute of Archeology of the Jagiellonian University and the Poznań Archeological Museum. They research both the cemetery and the settlement which are dated to the Naqada IIB-C until the 3rd to 4th Dynasty transition (Chłodnicki 2012: 13). During their work they encountered six examples of harpoons.[2] It is the largest collection of this type of relic from the Nile Delta area.

The two most important relics come from richly equipped Grave 55 (Figs. 35.4, 35.4: 1–2). It is a complete burial complex consisting of a four-chambered main tomb with a high superstructure which is surrounded by a wall with adherent accompanying burials (Dębowska-Ludwin 2012: 69–70). The finds can be dated to the Naqada IIIC2-D (Dębowska-Ludwin 2012: Table 1). Two similar harpoon-heads were found lying next to each other among the burial gifts (Dębowska-Ludwin 2012: 70). Both harpoons are sharpened on one side and have two barbs each. The stem is of a rectangular shape with virtually straight and level walls, and a square cross-section. At its end, there is one more bolt used for fixing the rope and attaching the head to the shaft. The first barb has a deep indentation. The second has a square cross-section at its base. It should be emphasized that the Tomb 55 complex is one of the richest and largest ever discovered at the site. It is likely that a man of high social standing was buried there.

Another harpoon, probably from the Tell el-Farkha Cemetery, was found in the settlement layers in the eastern part of the site (Fig. 35.4: 3). It was accompanied by numerous relics, such as stone vessels of a much earlier date than the room in which they were found. There is a high probability that the relic was stolen from one of the older graves, as can be indicated by the context. How would such a rich relic, dating back to a much earlier period, find its way into a poor household located in an indigent, run-down site that had relatively low importance and was located on the border of an earlier burial ground? Unfortunately, the relic is badly preserved. It was possible only to refurbish the preserved fragments of the first barb and a part of the stem. On this basis, one can only conclude that it was a double-barbed harpoon. The general geometric form of the stem relates to the above-mentioned examples from Grave 55.

Two different harpoons undoubtedly come from the settlement layers in the eastern part of the site. The first is a beautifully preserved specimen with two barbs (Fig. 35.4: 4). The artifact, however, differs from the Grave 55 relics. Its upper edge is slightly bent, and it does not have a bolt for fastening a rope, as the above mentioned examples do, but only two small knobs protruding out of the thick stem. The front barb is not indented as in the examples from Grave 55; and the other harpoon has the shape of a "shark dorsal" and is more elongated at the base. The relic dates back to the Naqada IIIB period and was found in the part of the settlement that could have housed the servants of an extensive *mastaba* located nearby. The second example is fragmented, with remains only of the stem and of the first barb. Its size suggests that it could have been a one-barb harpoon-head. Its metallurgical composition is known. The artifact is made of arsenical copper (CU-AS), probably derived from the deposits located in the Sinai Peninsula (Rehren and Pernicka 2014) (Fig. 35.4: 5).

Another fragmentary relic was revealed at Tell el-Farkha. It comes from the central part of the site. Only a fragment of its first barb has survived. The harpoon was made with a similar technology and the chemical composition is almost identical to the ones mentioned above. The copper comes from Sinai (Rehren and Pernicka 2014).

TELL EL-MURRA

Tell el-Murra is located about 10 km east of Tell el-Farkha. The research has been carried out since 2008 by the Institute of Archeology of the Jagiellonian University Mission. During the last excavation season at the site, two harpoon-heads were found among the burial gifts in one of the graves. Based

2 For more information on the copper relics from the site and the origin of the copper, see Czarnowicz 2012 and Rehren and Pernicka 2014.

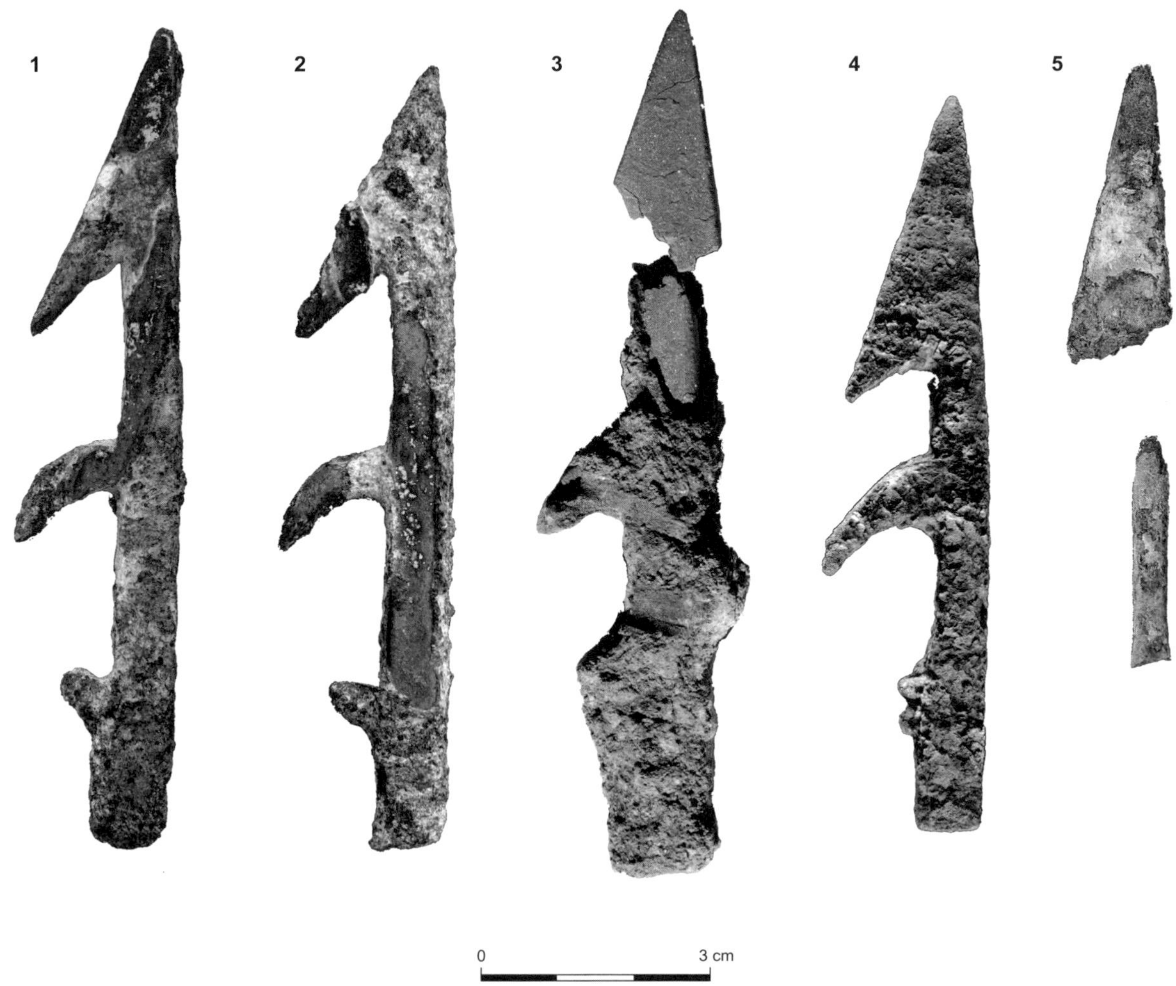

Fig. 35.4: Examples of Harpoons from Tell el-Farkha: 1–2) Grave 55 (photo M. Jórdeczka); 3) looted grave (photo R. Słaboński); 4–5) Eastern Kom (settlement in Tell el-Farkha) (photo M. Jórdeczka).

on the pottery, the finds are dated to the Naqada IIIC1 period (M. Jucha, personal communication). The harpoons vary in size. Both have a simple core, though the larger head has two barbs of a highly geometric shape, which make it similar to artifacts from Tell el-Farkha's Tomb 55. However, the harpoon from Murra has a thickening with two incisions instead of a bolt used for rope attachment. The smaller harpoon-head has a similar form, though its shape is less geometric. This is particularly evident around the base of the second barb and at the point where the rope was attached (Fig. 35.5).

KAFR HASSAN DAWOOD

Kafr Hassan Dawood is located in Wadi Tumilat, about 40 km from Ismailia. The research of a vast graveyard was conducted by the Egyptian Ministry of State for Antiquities (MSA) and University College London (UCL). Harpoon-heads were encountered in two graves. The first one comes from Tomb 828, which is dated to the Naqada IIIC1 period (Rowland 2014: 285). It has two barbs and a thickening at the end of the stem with two incisions. The first barb is not very indented, resembling the harpoon from

Fig. 35.5: Harpoons from Tell el-Murra (drawing U. Bąk).

the Tell el-Farkha settlement. The second artifact is dated to the Naqada IIIB period and was found in Tomb 298. It is slightly larger than the previously discussed one, though it is of a very similar structure. The only difference is the presence of two knobs at the end of the stem instead of one, as in the case of the previously described incision.

MINSHAT ABU OMAR

Minshat Abu Omar is the most eastern of the sites discussed here. It is a complex of several cemeteries, mainly from the Pre-Dynastic and Early Dynastic periods, but also from the Roman periods. Two harpoons were discovered in the rich grave "Tomb 173 (126)" which is dated to the time of the 1st Dynasty (Kroeper and Wildung 2000: 109–111). The pottery found in the grave—the cylindrical vessels in particular—indicate that the tomb should be dated no later than the beginning of the Nagada IIIC1 period (Jucha 2008: 71), and Rowland (2014: 285) suggests that it should be dated to the Naqada IIIA1-C1 period.

Both harpoons are quite similar. They resemble the finds from the settlement at Tell el-Farkha rather than those from Grave 55. The first barb is not as deeply indented, and the second one is more rectangular than square at the base, and instead of the bolt for attaching the rope there is a thickening of the stem with two knobs.

TYPOLOGY OF COPPER HARPOONS FROM THE NILE DELTA AREA (FIG. 35.6)

The analysis of the relics from the Nile Delta area allows us to draw some conclusions about the changes in the shape of harpoon-heads over time. The oldest examples have a slightly arched stem, two knobs at their end, and the angle between the stem and the rear edge of the barb is close to a right angle. These elements evolved over time, and by the Naqada IIIC2-D periods they had a more geometric form. The attaching method evolved from the incisions on the thickening to the occurrence of the bolt. Also, the first barb evolved in order to connect the stem at a much greater angle. Therefore, it can be assumed that, beginning from the Naqada IIIB period, we are dealing with harpoons similar to the example known from the Tell el-Farkha settlement. The geometric forms of the Naqada IIIC2-D period evolved during the Naqada IIIC1 period. Even though the assemblage is rather small, we notice an evolution of the form. The early examples were imitations of bone harpoons and the later ones have a geometrized shape which was more efficient and much easier to produce using a mold for casting.

THE IMPORTANCE OF HARPOONS

Many researchers underestimate the importance of harpoons. Gilbert (2004: 70) depreciated their importance, and suggested they were used mainly for hunting the Nile fauna. However, Linseele and Van Neer (2009) have proven that hunting had hardly any economic significance in early Egypt. Rather, one should look at the issue more broadly. Harpoons are not depicted in the "normal" hunting scenes, but rather in relation to the so-called magic hunt. One should view these representations as the fight between good and evil, as an attempt to stop the chaos which was embodied in a hippopotamus (Baines 1995: 111; Hendrickx 2006, 2011; Hendrickx and Eyckerman 2010; Raffaele 2010; Friedman, Van Neer and Linseele 2011). Harpoons therefore were not mere fighting weapons, but gained additional significance. One can assume—looking at the popularity of hunting scenes, the frequent portrayals of rulers on a hunt, as well as the analysis of the archaeological context of the finds (richly furnished graves of local elite)—that copper harpoons were not common weapons, but were rather used by the elite. After all, the fashion of hunting hippopotamuses with a harpoon spread among the elite social groups, as can be witnessed in the Middle

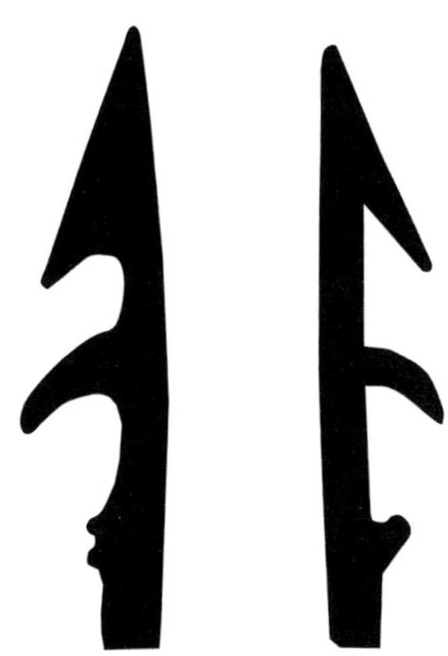

Fig. 35.6: Comparison between "older" and "younger" types of harpoons (drawing by M. Czarnowicz).

Kingdom Meketre Tomb. Thus, one can assume that such a harpoon, particularly one made of copper, was a determinant of social status. In addition, Egyptian art gave it a certain symbolic meaning. This approach is exemplified by the harpoon depicted on the Narmer Palette. Since hippopotamuses were the personification of disorder and chaos, and the hunters or rulers used harpoons to attack them, perhaps—at some point—the harpoons became the antithesis to the hippos, and began to symbolize opposite characteristics.

In my opinion, Gilbert (2004: 70) was wrong to limit harpoon use to hunting weapons. The painting from Tomb 100 at Hierakonpolis depicts a man sitting at the helm of the most important of boats, and holding a harpoon. Perhaps he served as the ruler's bodyguard. In the same relic, one can see a narrative scene of a man fighting another man with a harpoon. Perhaps it is the oldest preserved scene showing the use of a harpoon against a man. Moreover, these types of relics are found far from places where they could have been used for actual hunting of the Nile wildlife. A copper harpoon was found at the ʿEn Besor Egyptian trading post in the Negev Desert (Gazit 1995: 221–223). One can assume that the harpoon served as a weapon to protect caravans that traveled between Egypt and the Levant.

Harpoons were not used only as hunting weapons. Evidence for their successful use in close combat, replacing the spear or javelin, can be found in Pre-Dynastic period art. Copper harpoons were used by the upper classes of Naqada society, and this weapon was a confirmation of their status. Moreover, harpoons were associated with the so-called magic hunt. In early Egyptian art, they were used as weapons to combat the personification of evil and disorder. This additional attribute was linked to them, and over time in hunting scenes they became symbols representing qualities which were the counterpart of their rival, the hippopotamus. The finds from graves—belonging to the local elite—are most abundant among the archaeological material. Ten artifacts have thus far been discovered in the East Delta. All the relics, whose shape could be determined, have two barbs. Two harpoons are often found in the tombs. This is the case at Tell el-Farkha, Tell el-Murr and Minshat Abu Omar. Certain stylistic differences, described above, allow us to establish a harpoon typology which can be helpful in dating the artifacts. Nevertheless, one should keep in mind that the small sample consists of only ten harpoons (of which the shape of seven are known), and there is a need for further research in order to substantiate our knowledge about this issue.

REFERENCES

Adams, B. 1996. Elite Graves at Hierakonpolis. In: Spencer, J., ed. *Aspects of Early Egypt*. London: 1–15.

Ayrton, E. and Loat, W. 1911. *Pre-Dynastic Cemetery at el-Mahasna*. London.

Baines, J. 1995. Origins of Egyptian Kingship. In: O'Connor, D. and Silverman D., eds. *Ancient Egyptian Kingship* (Probleme der Agyptologie 9). Leiden, New York and Köln: 95–156.

Behrmann, A. 1996. *Das Nilpferd in der Vorstellungswelt der Alten Ägypter. Vol. 2, Textband* (Europäische Hochschulschriften, Reihe 38; Archäologie 62). Frankfurt am Main.

Brewer, D. and Friedman, R. 1989. *Fish and Fishing in Ancient Egypt*. Warminster.

Brunton, G. and Caton-Thompson, G. 1928. *The Badarian Civilisation and Predynastic Remains Near Badari*. London.

Chłodnicki, M. 2012. History of Research. In: Chłodnicki, M., Ciałowicz, K. and Mączyńska, A., eds. *Tell el-Farkha I, Excavations 1998–2011*. Poznań- Kraków: 9–19.

Ciałowicz, K. 1999. *Początki Cywilizacji Egipskiej*. Kraków.

Czarnowicz, M. 2012. Copper Tools. In: Chłodnicki, M., Ciałowicz, K. and Mączyńska, A., eds. *Tell el-Farkha I, Excavations 1998–2011*. Poznań- Kraków: 345–357.

Darnell, J.C. 2009. Iconographic Attraction, Iconographic Syntax and the Tableaux of Royal Ritual Power in the Pre- and Proto-Dynastic Rock Inscriptions of the Theban Western Desert. *Archéo-Nil* 19: 83–107

Dębowska-Ludwin, J. 2012. The Cemetery. In: Chłodnicki, M., Ciałowicz, K. and Mączyńska A., eds. *Tell el-Farkha I, Excavations 1998–2011*. Poznań- Kraków: 53–77.

Dreyer, G. *et al*. 2003. Umm el-Qaab. Nachuntersuchungen im frühzeitlichen Königsfriedhof 13/14/15. Vorbericht. *Mitteilungen des Deutschen Archäologischen Instituts Abteilung Kairo* 59: 67–138.

Friedman, R., Van Neer, W. and Linseele, V. 2011. The Elite Predynastic Cemetery at Hierakonpolis: 2009–2010 Update. In: Friedman, R. and Fiske, P., eds. *Egypt at Its Origins 3, Proceedings of the International Conference "Origin of the State: Predynastic and Early Dynastic Egypt," London, 27th September–1st August 2008* (Orientalia Lovaniensia Analecta 205). Leuven: 157–192.

Gazit, D. 1995. Two Copper Tools from `En Besor. In: Gophna, R., ed. *Excavations at ʾEn Besor*. Jerusalem: 221–229.

Gilbert, G. 2004. *Weapons, Warriors and Warfare in Early Egypt* (British Archaeological Reports International Series 1208). Oxford.

Hartmann, R. 2008. Zwei Fragmente der White Cross-lined Ware aus dem Friedhof U in Abydos zu Gefässen aus dem Ägyptischen Museum Kairo. In: Engel, E., Müller, V. and Hartung, U., eds. *Zeichen aus dem Sand. Streiflichter aus Ägyptens Geschichte zu Ehren von Günter Dreyer* (Menes 5). Wiesbaden: 163–182.

Hayes, W. 1953. *The Scepter of Egypt. A Background for the Study of Egyptian Antiquities in the Metropolitan Museum of Art, Vol. 1: From the Earliest Times to the End of the Middle Kingdom*. New York.

Hendrickx, S. 2006. The Dog, the Lycaon Pictus and Order over Chaos in Predynastic Egypt. In: Kroeper, K., Chłodnicki, M. and Kobusiewicz, M., eds. *Archaeology of Early Northeastern Africa, In Memory of Lech Krzyzaniak* (Studies in African Archaeology 9). Poznań: 723–749.

Hendrickx, S. 2011. Hunting and Social Complexity in Predynastic Egypt. *Academie Royale des Sciences d'Outre-mer, Bulletin des Séances / Koninklijke Academie voor Overzeese Wetenschappen, Mededelingen der Zittingen* 57(2–4): 237–263.

Hendrickx, S. 2012a. Iconography of the Pre and Early Dynastic Periods. In: Teeter, E., ed. *Before the Pyramids*. Chicago: 75–83.

Hendrickx, S. 2012b. Jar with Hunting Scene. In: Teeter, E., ed. *Before the Pyramids*. Chicago: 153–154.

Hendrickx, S. and Eyckerman, M. 2010. Continuity and Change in the Visual Representations of Predynastic Egypt. In: Raffaele, F., Nuzzolo, M. and Incordino, I., eds. *Recent Discoveries and Latest Researches in Egyptology. Proceedings of the First Neapolitan Congress of Egyptology (Naples, 18–20 June, 2008)*. Wiesbaden: 121–144.

Jucha, M. 2008. Wavy-Handled and Cylindrical Jars in the Nile Delta: A View from Tell el-Farkha. *Studies in Ancient Art and Civilization (SAAC)* 12: 63–75.

Kroeper, K. and Wildung, D. 2000. *Minshat Abu Omar II. Ein vor- und frühgeschichtlicher Friedhof im Nildelta*. Mainz.

Leeman, D. 2007. *Tomb 100—The "Painted" Tomb of Hierakonpolis*. http://content.yudu.com/Library/A1ikn7/Tomb100ThePaintedTom/resources/27.htm, (accessed 12/06/2014).

Linseele, V. and Van Neer, W. 2009. Exploitation of Desert and other Wild Game in Ancient Egypt: The Archaeozoological Evidence from the Nile Valley. In: Reimer, H., Forster, F., Herb, M. and Pollath, N., eds. *Desert Animals in the Eastern Sahara: Status, Economic Significance, and Cultural Reflection in Antiquity. Proceedings of the Interdisciplinary ACACIA Workshop held at the University of Cologne (14–15 December, 2007)* (Colloquium Africanum 4). Köln: 47–78.

Mączyńska, A. 2011. Lower Egyptian–Nagadian Transition. A View from Tell el-Farkha. In: Friedman, R. and Fiske, P., eds. *Egypt at Its Origins 3, Proceedings of the International Conference "Origin of the State. Predynastic and Early Dynastic Egypt," London, 27th September–1st August 2008* (Orientalia Lovaniensia Analecta 205). Leuven: 879–908.

Müller, V. 2008. Nilpferdjagd und geköpfte Feinde—zu zwei Ikonen des Feindvernichtungsrituals. In: Engel, E., Müller, V. and Hartung, U., eds. *Zeichen aus dem Sand. Streiflichter aus ägyptens Geschichte zu Ehren von Günter Dreyer* (Menes 5). Wiesbaden: 477–493.

O'Connor, D. 2012. The Narmer Palette: A New Interpretation. In: Teeter, E., ed. *Before the Pyramids*. Chicago: 145–153.

Raffaele, F. 2010. Animal Rows and Ceremonial Processions in Late Predynastic Egypt. In: Raffaele, F., Nuzzolo, M. and Incordino, I., eds. *Recent Discoveries and Latest Researches in Egyptology. Proceedings of the First Neapolitan Congress of Egyptology (Naples, 18–20 June, 2008)*. Wiesbaden: 245–286.

Rehren, T. and Pernicka, E. 2014. First Data on the Nature and Origin of the Metalwork from Tell el-Farkha. In: Mączyńska, A., ed. *The Nile Delta as the Centre of Cultural Interactions between Upper Egypt and the Southern Levant in the 4th Millennium BC.* Poznań: 237–253.

Rizkana, I. and Seeher, J. 1989. *Maadi III. The Non-Lithic Small Finds and the Structural Remains of the Predynastic Settlement*. Mainz am Rhein.

Rowland, J. 2014. Interregional Exchange: The Evidence from Kafr Hassan Dawood. In: Mączyńska, A., ed. *The Nile Delta as the Centre of Cultural Interactions between Upper Egypt and the Southern Levant in the 4th Millenium BC.* Poznań: 269–299.

Williams, B. 2012. Qustul Incense Burner. In: Teeter, E., ed. *Before the Pyramids.* Chicago: 162–163.

CHAPTER 36

BRONZE WORKING AT SUMHURAM: NEW DATA FROM AN ANCIENT SOUTH ARABIAN HARBOR ON THE COAST OF DHOFAR (SULTANATE OF OMAN)

Michele Degli Esposti, Carla Martini, Cristina Chiavari, Elena Bernardi and Gian Luca Garagnani

The town of Sumhuram was founded by the South Arabian Kingdom of Hadramawt ca. the end of the 3rd century BCE. Strategically located on the coast of what is now Dhofar in the Southwestern Sultanate of Oman, for several hundred years it controlled the trade route connecting Arabia with India on one side and with the Mediterranean, via the Red Sea, on the other. There is abundant archaeological evidence that copper/bronze items were produced at the site. To shed light on the metallurgical technologies and alloys used, the composition and microstructure of 29 metal items dating between the late 3rd century BCE and the late 4th century CE were investigated by means of SEM-EDS and Optical Microscopy.

These investigations are discussed alongside other data previously collected from the site. The results indicate the general use of alloys with a high Pb content, and the apparent absence of a correlation between the objects' shape/function and their composition. No diachronic evolution is evident in the metallurgical procedures. Comparable results are rare in the literature; our hope is that this work will stimulate further investigation in the metallurgy of the area, which could either supply coherent evidence or firmly contradict it.

INTRODUCTION

Since 1996, Prof. Alessandra Avanzini and a team of the Italian Mission to Oman (IMTO) has been investigating the stratigraphy of the site of the ancient town of Sumhuram, located in the Bay of Khor Rori, The site controls the estuary of Wādī Darbāt along the coast of Dhofar, the southern region of the Sultanate of Oman, which borders Yemen (Fig. 36.1a). The small bay, closed in by two flattened rocky promontories, creates a perfectly protected harbor.

Thanks to its strategic position, Sumhuram dealt in long-distance trade from the Mediterranean to India (Pavan 2011; Pavan and Schenk 2012), a fact mirrored in the materials recovered at the site. Recent evidence collected from excavations enabled the dating of the foundation of the town by the South Arabian Kingdom of Hadramawt to ca. the end of the 3rd century BCE (Avanzini and Sedov 2005), thus confirming the conclusions previously drawn upon epigraphic and historical data (Avanzini 2008).[1] The date of the town's abandonment was also shifted, from the 3rd to the late 4th–5th century CE (Avanzini and Sedov 2005).

An internal phasing was also proposed, based on stratigraphy. For Areas A and F in particular, from which the majority of the analyzed samples came, five constructional phases were originally defined (Buffa and Sedov 2008; Sedov 2008), although a recent revision of the excavation data from the whole

1 The heartland of the kingdom is along the homonymous wadi in modern-day Yemen.

a)

b)

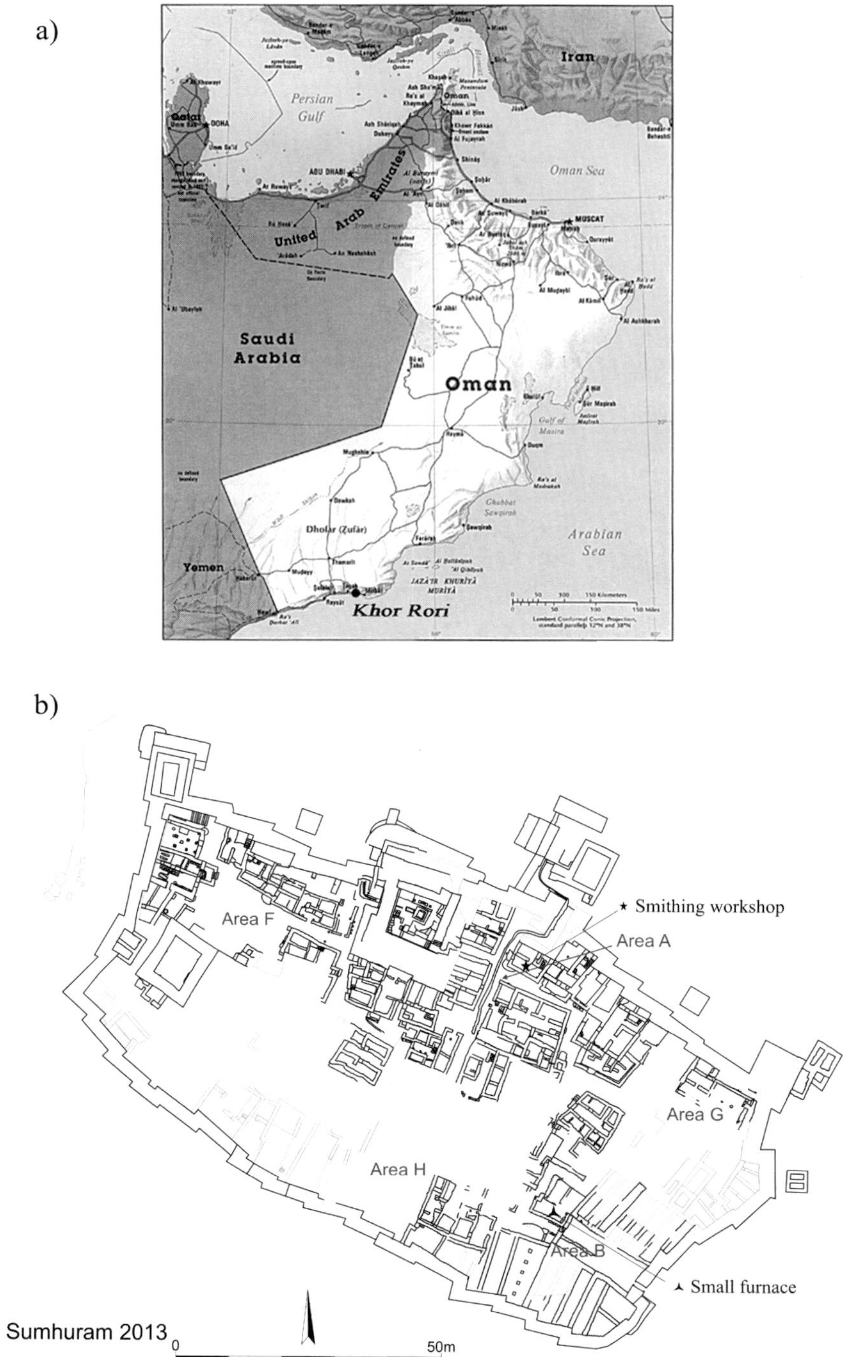

Fig. 36.1: The location of Sumhuram (a) and a plan of the town with the areas and features mentioned in the text (b).

site induced the excavators to define just two large occupational macro-phases plus a later degenerative one (Sedov 2013). When the issue of chronological technological changes is addressed, the old phasing will be followed here, since the correlation between the two sequences has not yet been completed.

The town, which was completely walled, was divided into several blocks. Although several sparse workshops were identified, no area specifically dedicated to industrious activities was located. Extensive excavation finally resulted in the collection of rather substantial evidence of metalworking on the site, and at least one smithing workshop in one of the rooms in Area A (Buffa and Sedov 2008).

A small furnace unearthed in Area B could be consistent with small-scale production of little copper/bronze objects: minimal remains of slag were found adhering to its stones, and abundant hammerscale was scattered around it (Chiarantini and Benvenuti 2011: 67, and Figs. 3–4). A similar reduced scale of metalworking is indicated by several small crucibles retrieved from different contexts, without any preferential concentration, whose analyses revealed the melting of different types of copper-base alloys, varying in composition from Cu-Sn to Cu-Sn-Pb (Chiarantini and Benvenuti 2014).

Slags are extremely abundant on the site and mostly connected with iron working, although in at least one sample copper has also been detected (Chiarantini and Benvenuti 2011). None of the reported pieces of evidence concerns iron or copper smelting, thus indicating that the site probably received raw metals and not ores.

Finally, metallic objects have been recovered in remarkable numbers on the site (Lombardi, Buffa and Pavan 2008: 377–397); the bronze objects repertoire mirrors quite well the general picture gained from South Arabian sites in general.

The study presented here aimed at the identification of the copper-base alloys used at the site and possibly at the determination of a correlation between the type of alloy and the object function; the occurrence of chronological variations in the metallurgical techniques was also investigated.

This work is part of a wider project on the ancient metallurgy of Sumhuram, addressing the different stages of metal production for which evidence was collected on the site, with a focus on the production of inscribed objects and on a possible local minting activity (Chiarantini *et al.* 2007; Chiarantini and Benvenuti 2011, 2014; Chiavari *et al.* 2011).

Discussing the broader archaeological and historical context of Sumhuram's metallurgy would require more space than is available here. It will thus be the subject of a paper fully devoted to it, in which recent relevant finds, discovered after this paper was originally presented, will also be presented.

MATERIALS AND METHODS

Small fragments representative of the metal core were taken from 29 selected copper-base items (Table 36.1). The samples were examined by extended pressure scanning electron microscopy (VP-SEM: Zeiss EVO 50), with X-ray energy-dispersive spectroscopy (EDS) probe (Oxford Inca Energy 350). This allowed the identification of the microstructure by localized microanalyses on specific inclusions and phases. Representative values of the alloy composition were obtained by EDS analyses on large areas (1.0 × 0.5 sq mm) of the metal bulk (unetched), averaging three measurements.

Microstructures were also investigated by Optical Microscopy (OM) (bright field/dark field/reflected polarized light), after polishing the samples down to 1 µm diamond paste. A light etching was carried out using a $FeCl_3$/HCl ethanol solution, only when the microstructure was not highlighted enough by corrosion.

For six further objects, only the sampling of a few tens of milligrams of metal for wet chemical analyses was allowed. Using a small drill (2 mm in diameter), clean metal turnings were collected, discarding the outer layer of patina. The core metal composition was then measured by means of flame atomic absorption spectroscopy (FAAS), following the procedure described by Hughes, Cowell and Craddock (1976) for ancient bronzes.

Table 36.1: Chemical Composition (EDS, wt%) of Metal Samples from Sumhuram Examined in the Present Study (Samples 8 to 369)*

Inv.(MB)	*Definition*	*Phase*	*Cu*	*Pb*	*Sn*	*As*	*Fe*	*O*	*Cl*	*S*	*Co*	*Ni*	*Zn*
8	N.d. (lock element?)	N.d	81.4	11.7	3.2			2.7	0.6			0.5	
18	Small vessel's rim	N.d	50.9	20.7	4.6			19.1	4.8				
29	Nail	Sum V	54.6	27.1	5.9			5.0	5.5				1.9
35	Nail	Sum I	75.8	2.6	12.2		0.7	6.6	0.6	0.6		0.9	
42	N.d.	Sum II	68.7	18.5	9.8	1.1		1.4				0.5	
43	Mirror	Sum III	82.3	8.2	4.9	0.5	0.6	2.3	1.4				
49	Plaque	Sum I	78.4	2.2	15.1	2.4	0.7					1.3	
54	Plaque	Sum V	81.5	11.5			0.9	5.6		0.3		0.5	
66	Clamp rivet	Sum III	97.0	0.2	2.2			0.6					
67	Vessel base	Sum III	88.3	1.6	5.3			4.1	0.4			0.4	
73a	Chisel (?)	Sum III	90.0				0.4	6.8		1.9	0.2	0.8	
73b	Rectangular section chisel	Sum III	69.9	16.4	6.6			3.7	3.0			0.4	
94	Bangle	Sum III	79.5	1.9	11.4	0.5		6.4	0.3				
119	Plano-convex section rod	Sum IV	66.3	1.1	19.7		0.3	11.1				1.2	
131	Clamp	Sum III	77.3	12.0	7.8			2.1	0.3			0.4	
133	Clamp	Sum III	81.3	6.9	8.4			2.3	0.7			0.3	
139	Square section rod	Sum II	85.5	3.2	7.6	0.5		2.7				0.5	
157	Clamp	Sum III	66.7	15.7	8.8			5.2	3.6				
177	Rectangular section rod	Sum I	87.5	1.3	8.0			3.0	0.3	0.3			
189	Romboid section rod	Sum III	92.3		0.6		0.7	6.2	0.3				
249	Rectangular section rod	Sum IV	87.3	2.9	8.0			1.5		0.3		0.5	
280	Circular section rod	Sum III	80.2		0.6		0.3	0.9					18.1
312	Circular section rod	Sum I	89.5	3.2	6.1	0.3		0.8					
318	Plaque	Sum V	43.1	7.2	27.0			22.0	0.7				
327	N.d	Sum V	83.6	10.5	3.9		0.4	1.2				0.5	
339	Vessel spout	N.d	87.1	5.2	5.5			1.2		0.7		0.4	
371	N.d	N.d	71.4	15.7	9.2			3.4	0.3				
394	N.d	N.d	82.2	10.2	6.3			0.9				0.4	
398	N.d	N.d	63.7	4.8	21.1			9.6	0.9				
369	Group of fragments from different objects (results from Chiavari *et al.* 2011)												
	1	N.d	70.1	11.1	10.0			4.7	1.9			0.4	
	2	N.d	60.6	18.1	5.5			4.0	5.0			0.3	
	3	N.d	64.5	19.5	5.0			4.0	4.0			0.2	
	4	N.d	53.8	28.8	4.3		0.7	5.4	6.6			0.4	
	5	N.d	54.5	27.5	3.9			6.6	3.1			0.4	
	8	N.d	61.9	12.6	11.9			6.2	2.4			0.2	
	12	N.d	54.8	28.7	4.5		0.3	5.2	6.1			0.5	
	13	N.d	68.0	17.2	4.7			6.3	3.4			0.4	
	15	N.d	74.8	12.8	6.1	0.3		3.0	2.4			0.4	
	17	N.d	70.3	13.5	7.6			5.1	2.8			0.6	

* Samples 1 to 17 in the last 10 rows were previously published in Chiavari *et al.* 2011 and are reported here for ease of comparison.

Data on the composition of ten more samples coming from Sumhuram, previously analyzed by the authors (Chiavari *et al.* 2011), have been considered in the discussion and are listed in Table 36.1 for easier comparison.

Finally, in order to highlight possible correlations between the used alloys and the function or dating of the objects, the chemical composition in terms of the main alloying elements (considering only EDS data) was analyzed through Principal Component Analysis (Klemenc, Budic and Zupan 1999; Cecere *et al.* 2008; Favaro *et al.* 2012). Cu was excluded because it is significantly affected by the state of preservation. PCA was carried out using the Unscrambler® X – v.10.2 software package (CAMO Software AS) and performed on autoscaled data by NIPALS algorithm and Varimax rotation. PCs with eigenvalue ≥1 were retained as significant.

RESULTS AND DISCUSSION

COMPOSITION OF THE ARTIFACTS AND ALLOY SELECTION

The results of elemental characterization by EDS large area analyses are listed in Table 36.1. Table 36.2 shows the composition of the 6 objects which were analyzed by FAAS; a comparison with data in Table 36.1 can be made only in broad terms for the main alloying elements, given the difference in the detection limit between the two techniques.

The great majority of the objects consists of a ternary Cu-Sn-Pb alloy, in most cases with a remarkable Pb content. Of the 39 samples analyzed, only two can be regarded as unalloyed copper objects: MB73a and MB189. The low tin content of MB189 (0.59%) could derive from natural occurrence in the ore, or, more convincingly, from the possible use of recycled metal.

MB66 is the only tin bronze object with a negligible Pb content.

Pb content displays a remarkable variability, between 1.1 to 28.8 wt%. Although useful for casting since it improves the fluidity of the molten metal (ASM 2008), Pb also causes the segregation of globules, randomly dispersed in the matrix (as discussed in the section about microstructural features), which make bronze difficult to work by plastic deformation and prone to grain boundary embrittlement (GBE) (Felberbaum 2005).

Such a remarkable and widespread lead content clearly indicates deliberate addition: Pb was probably used to increase the mass of processed metal in place of other, less readily available and more expensive metals. Leaded bronzes were actually quite common in Egypt and the Middle East, possibly by the third millennium and continuing well after the end of the second millennium (Cowell 1985; Moorey 1971). The randomness of Pb contents in the Near East and the Mediterranean prior to the Roman era has also already been noted (Philip 1991).

TABLE 36.2: CHEMICAL COMPOSITION (FAAS, WT%) OF METAL OBJECTS FROM SUMHURAM ("TRACES": CONCENTRATION ABOVE THE DETECTION LIMIT BUT BELOW THE QUANTIFICATION LIMIT)

Inv. (MB)	*Definition*	*Phase*	*Pb*	*Zn*	*Fe*	*Bi*	*Ni*	*Ag*	*Sn*	*Co*	*As*
5	Plaque	N.d	16.72		3.29		0.18	x			0.08
123	Clamp	Sum III	12.72	0.23	0.13		0.19	0.07	6.86		0.18
126	Clamp	Sum III	1.76	0.05	10.74		0.65	0.05	1.25		1.55
324	N.d	N.d	12.62	Traces	0.12	Traces	0.31	0.84	6.23		0.38
334	N.d	N.d	6.24	0.05	0.09		0.10	0.15	7.19		
362	N.d	N.d	0.56		0.13		0.11	Traces	0.59		0.08

The Sn content appears rather erratic and with some peculiarly in high values; besides, for most of the samples it lies quite below the 8–12% range, which is recognized as the optimal one in order to obtain a resistant and workable bronze (Piccardo *et al.* 2009). These random concentrations seem to strengthen the idea of a metallurgy depending mostly on recycling. If one excludes the two items where Sn can be considered as an incidental inclusion (rods MB189 and MB280), its content varies from 2.1% to 27.0%. High Sn content can partly derive from decuprification processes which take place in archaeological bronzes due to corrosion (Robbiola, Blengino and Fiaud 1998): this might be the case for the plaques MB49 and MB318, where the presence of unalloyed copper inclusion (UCI) is also connected to corrosion phenomena (see next section). High contents of Sn could also be related to the desire to obtain silver-looking objects, but this could possibly apply to the bangle MB94 and to the plaques MB49 and MB318, and not definitely to the rod MB119 or to the undefined fragment MB398. As regards the low Sn content of MB189 and MB280 (simple rods which contain no Pb), the mixing of re-melted scrap bronze with fresh molten copper again appears as the best explanation, given the rare co-occurrence of copper and tin ores (Craddock 1985: 98). Clearly, unleaded scrap bronze would have been used in these cases; it cannot be determined if this was by chance or a deliberate choice.

When Sn content is plotted against Pb (Fig. 36.2), the samples do not cluster neatly, but rather show a widespread distribution. Nevertheless, a tentative grouping can be suggested which, as a consequence of the increased number of samples, redefines a previous proposal made by Chiavari *et al.* (2011: 533). The only outstanding cluster is Group I, including samples with low Sn (<6%) and remarkably high Pb (>27%). The opposite situation to Group I is witnessed by a few outliers which have high Sn (> 19%) and relatively low Pb (max 7%), labelled as Group II. Group III can be defined to comprise items with low-medium Sn (3–12%) and medium-high Pb (10–20%). One last Group IV includes the samples with low-medium Sn (2–15%) and low Pb (<6%). Admittedly, the boundary between the last two groups, differentiating each other essentially only on the basis of their Pb content, is far from being clear. Taking into account the very low solubility of Pb in the bronze matrix, which generates a random distribution of Pb inclusions (see next section), the distinction between Groups III and IV might in fact be arbitrary. Besides, the presence of at least one sample (MB133) which falls in between the two groups, would suggest some caution.

The results show no evolution or change in the metallurgical procedures followed at Sumhuram over its long period of occupation. Even PCA analysis, conducted on the main alloying elements Pb and Sn together with As, Ni and Fe (sample MB280 excluded because of its uncommon Zn content), failed to show any significant clustering (Fig. 36.3). The possible clusters were definitely transversal to compositional, typological and chronological groupings, thus underlining once again the randomness of alloys composition.

Fe is low in all the samples and probably derives from the use of mixed Cu/Fe ores. The low concentration of cobalt, which is often found in association with iron in the ores (Tylecote, Ghaznavi and Boydell 1977), could support this hypothesis. It could alternatively represent the remnants of iron oxides used as the flux during copper smelting;[2] if this was the case, its low rates would account for an accurate refining of copper after primary smelting.

Ni and As are quite widespread in the samples, rarely with concentrations higher than 1%; their presence was probably unintentional and rather derived from ores composition (Craddock 1985: 98–99). Regarding Ni, an incidental presence in a possible ferrous smelting flux is unlikely, since the rates of Fe and Ni appear to be totally unrelated.

In the EDS results, Ni would seem to be more common than As, while their occurrence is more similar when FAAS data are taken into account. However, it was shown elsewhere how, on a similar

2 This procedure was well known, for example, at Timna, already in the Bronze Age (Rothenberg 1972).

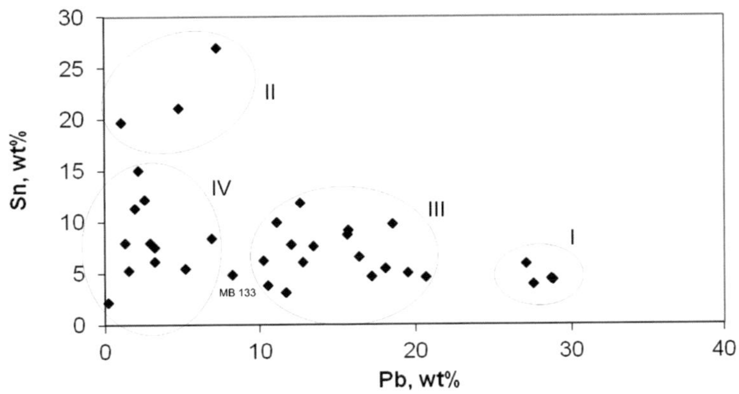

Fig. 36.2: A two-dimensional representation of the composition of the metal samples from Sumhuram (Sn wt% as a function of Pb wt%), showing the clustering of the samples in four main composition groups (I–IV).

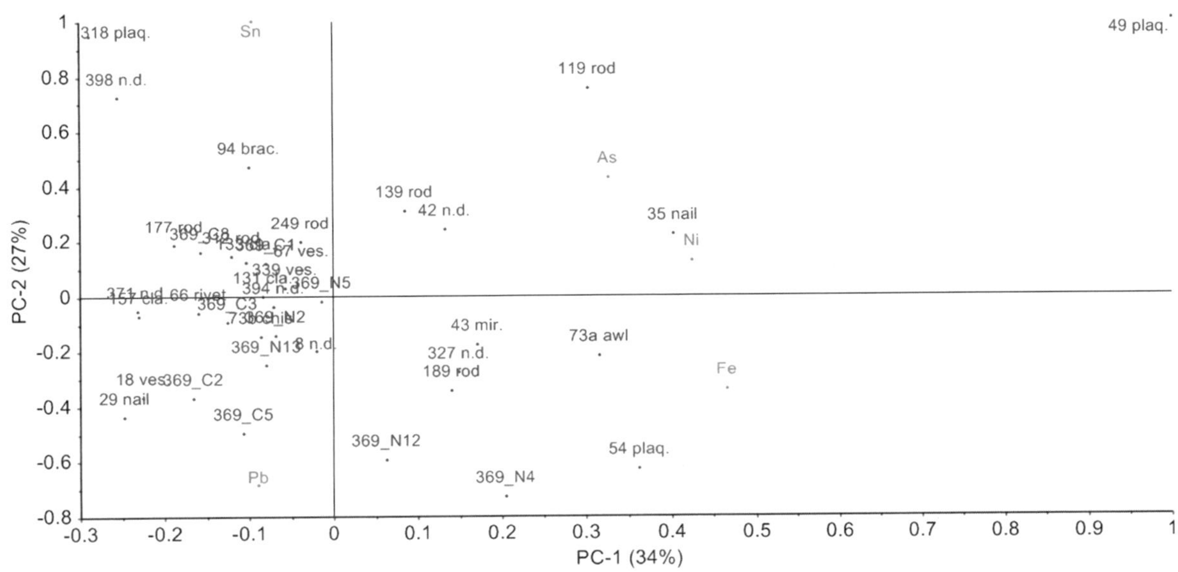

Fig. 36.3: Principal Component Analysis (PCA): rotated loading and score plot for the main alloying elements concentration (EDS data) of samples listed in Table II (sample MB280 excluded).

sample, EDS detected lower concentrations than FAAS in particular for As (Chiavari *et al.* 2011: Table 2). Taking this consideration into account, the general picture for Arsenic does not change, and the only exception is plaque MB49 (2.3 wt.%).

Whatever limit between deliberate As alloying and accidental impurity one decides to consider among those proposed for the old world,[3] the only object which might be reputed a real arsenical bronze would be the mentioned plaque MB49. Since As can also influence the alloy color, giving it a silvery appearance already at low concentrations (Giumlia-Mair *et al.* 2002: 205), this addition could support the hypothesis of a few objects for which such a coloration was deliberately pursued. Moreover, MB49 also contains a high amount of Sn, which could also be related to the color issue.

3 Generally between 1–2%; see Giumlia-Mair *et al.* 2002: 205 for a summary of the different hypotheses.

The sample from rod MB280 is characterized by its high zinc content—over 18 wt%—which classifies it as a brass. This is even more significant in contrast to the low amounts (MB29) or the absence of this element in the other samples, even though FAAS analyses (Table 36.3) show that traces of zinc were probably more common than indicated by EDS analyses (Table 36.2; see also Chiavari *et al.* 2011: 534).

Even though occasional examples of brass are known from the Early Bronze Age in the Middle East, brass was definitely rare before the 1st century BCE, when the Romans improved the cementation process necessary to obtain it, which then became widespread following the expansion of the Roman Empire and the Roman trade methods (Craddock 1978). The Roman conquest of Egypt at the end of the 1st century BCE prompted the establishment of more intense commercial contacts between Rome and Southwestern Arabia: though highly speculative, it would be tempting to link the presence of a high amount of Zn in MB280 to the Roman influence, if not to a direct import, even considering that such high zinc rates are unlikely to have been obtained inadvertently or by direct co-melting of Cu and Zn.

Further speculation could be made to correlate the third architectural phase of Area A, to which the context of MB280 (US22) belongs, with this Roman contact, thus establishing a *terminus post quem* for its construction. On the one hand, this suggestion could gain some strength from the fact that MB29,

Table 36.3: Main Microstructural Features of Metal Objects from Sumhuram (Listed as a Function of the Kind of Object)

Inventory (MB)	*Object*	*Microstructure of the Cu-rich Matrix*	*Secondary Phases*
94	Bangle	Recrystallized (surface and bulk)	Pb
73a	Chisel	Recrystallized and cold-worked (surface)	-
29	Nail	Dendritic	Pb
35	Nail	Dendritic	Pb, α+δ eutectoid
66	Clamp rivet	Dendritic	Pb (very small globules)
131	Clamp	Dendritic	Pb
133	Clamp	Dendritic	Pb
157	Clamp	Dendritic	Pb
73b	Awl	Dendritic (bulk), cold-worked (surface)	Pb
49	Plaque	Dendritic (extensively cored)	Pb, α+δ eutectoid, unalloyed Cu
54	Plaque	Recrystallized (surface and bulk)	Pb
318	Plaque	α-Cu globules in α+β matrix	Pb, unalloyed Cu
43	Mirror	Dendritic (bulk), cold-worked (surface)	Pb (small globules)
8	Vessel	Dendritic	Pb
339	Vessel's spout	Dendritic (bulk), cold-worked with cracks (surface)	Pb, α+δ eutectoid
67	Vessel base	Dendritic	Pb
18	Vessel's rim	Cold-worked dendritic (surface & bulk)	Pb
119	Rod	Recrystallized and cold-worked (surface)	Pb
139	Rod	Dendritic	Pb
177	Rod	Dendritic (bulk), cold-worked (surface)	Pb, α+δ eutectoid
189	Rod	Dendritic	-
249	Rod	Recrystallized and cold-worked (surface)	Pb, α+δ eutectoid
280	Rod	Recrystallized and cold-worked (surface)	-
312	Rod	Recrystallized (surface)	Pb

with its significant though definitely lower Zn content (1.6%), comes from an even later context—US21; on the other hand, the extremely good preservation state of MB280 and the proximity of its context with the uppermost deposits on the site induce the need for some caution.

Another significant aspect emerging from the data is the substantial contribution made to the total composition by elements coming from corrosion processes, specifically Cl and O. In particular, the high Cl content is directly correlated with the concentration of Pb in the alloys (Fig. 36.4), segregated at the grain boundaries or in interdendritic spaces: in the coastal environment where the objects were

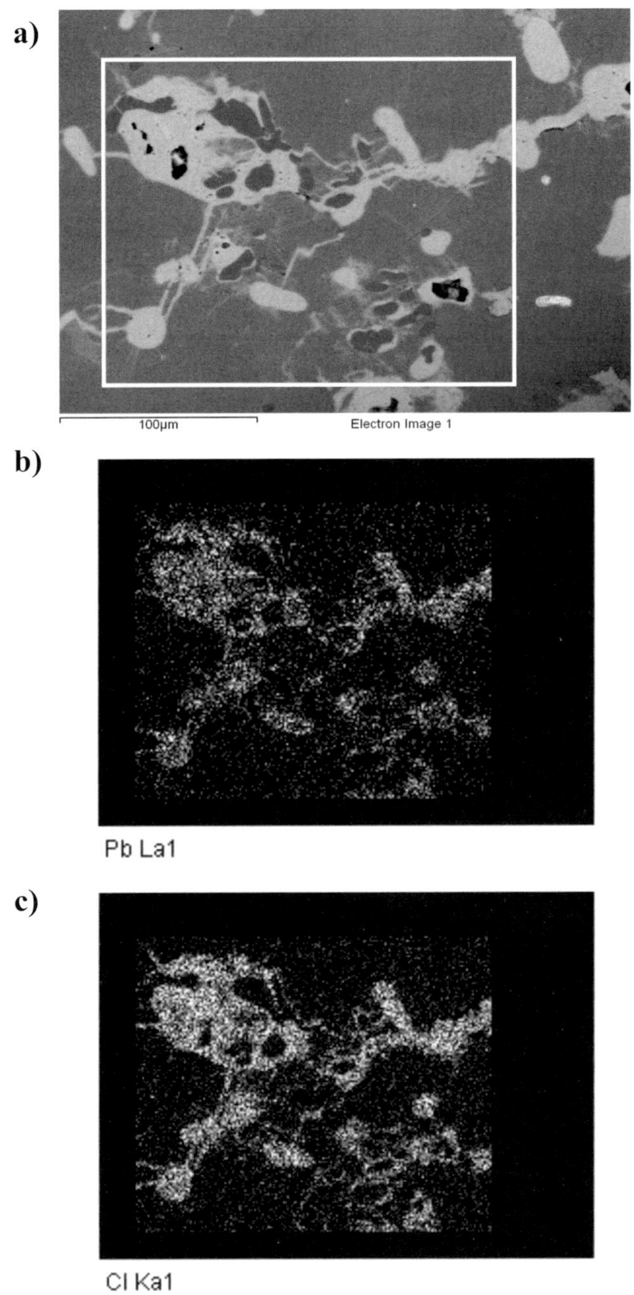

Fig. 36.4: SEM-BSE Image (a) and X-ray maps of Pb (b) and Cl (c) elemental distribution in sample MB157, showing the close connection between Pb and Cl due to corrosion.

buried, these Pb-rich areas are easily affected by severe corrosion, as shown by the X-ray maps. It has already been shown (Chiavari *et al.* 2011: 537) that the Pb-rich phase is completely converted to Cotunnite ($PbCl_2$).

The dearth of archaeometallurgical studies concerning South Arabia does not allow great comparisons; nevertheless, some first interesting parallels can be drawn with the results obtained in the late sixties on a few samples from Hajar Bin Âumeid (Salmon 1969), a partially coeval town located in modern Yemen, and for a period the capital of the Kingdom of Qataban (Van Beek 1969). The main similarity concerns the used alloys, of which only the main components will be considered here. At Hajar Bin Âumeid all the analyzed samples were in fact made of a ternary Cu-Pb-Sn alloy, with Pb value often above 2% and in some cases peculiarly high. The use of Sn also appears to not have been perfectly standardized, although it displays a higher uniformity than at Sumhuram: values significantly below 8–10% were not detected, with some samples reaching up to 16%. No evolution in this pattern can be detected, but the paucity of data induces caution on this issue.

MICROSTRUCTURE AND MANUFACTURING TECHNIQUES

The main microstructural features identified by Optical Microscopy are summarized in Table 36.3 and representative micrographs are shown in Figures 36.5–7.[4] Only the items for which a functional identification was available are meaningful in trying to assess the possible correlation between manufacturing technology and object function, and are listed in Table 36.3. Microstructural data showed that in most cases the objects were not extensively worked after casting, dendritic structures being predominant over the recrystallized ones. The latter were often confined to the superficial areas of the objects.

Probably the most peculiar cases are those of plaques and vessel fragments. For both these classes of objects one would expect evidence for hot/cold working, meant to flatten and regularize the thickness of the metal sheet, which however was detected only in plaque MB54 (Fig. 36.5a) and in the vessel rim MB18 (Fig. 36.5b). In the latter, the elongated Pb-rich areas in the flattened dendritic structure bear evidence of extensive cold working (Fig. 36.5c); also the vessel spout MB339 (Fig. 36.5d), shows a cold-worked microstructure with slip lines in the surface layer, and both the objects have a remarkable Pb content (20.7% and 5.2% respectively), which makes cold working quite dangerous, especially for delicate parts like the spout and the rim.

The majority of these objects were obtained directly by casting, with no or just minor hammering. The same is also true for the mirror fragment MB43, which shows a dendritic microstructure (Fig. 36.5e) with some deformed grains and slip lines near the outer edge (Fig. 36.5f). However, in this case, the sample comes from the lower part of the mirror, presumably near the handle, where only slight plastic deformation was probably needed.

Nails, rivets and clamps show a structure which seems in every case dendritic (Fig. 36.6a–d), and could fit their nature of highly utilitarian objects with no need for specific forming procedures. Furthermore, this class of objects contains variable amounts of Pb (from less than 1% up to nearly 16%), which may not facilitate plastic deformation. Also, it must be considered that the expected microstructural modifications (recrystallization and/or deformation) that should derive from their use or final placement (i.e., the hammering out of the buck-tail of the rivet during its installation), probably affected portions of the objects far from the sampled area.

4 Sand grains embedded in the corrosion crust made it nearly impossible to avoid scratch marks.

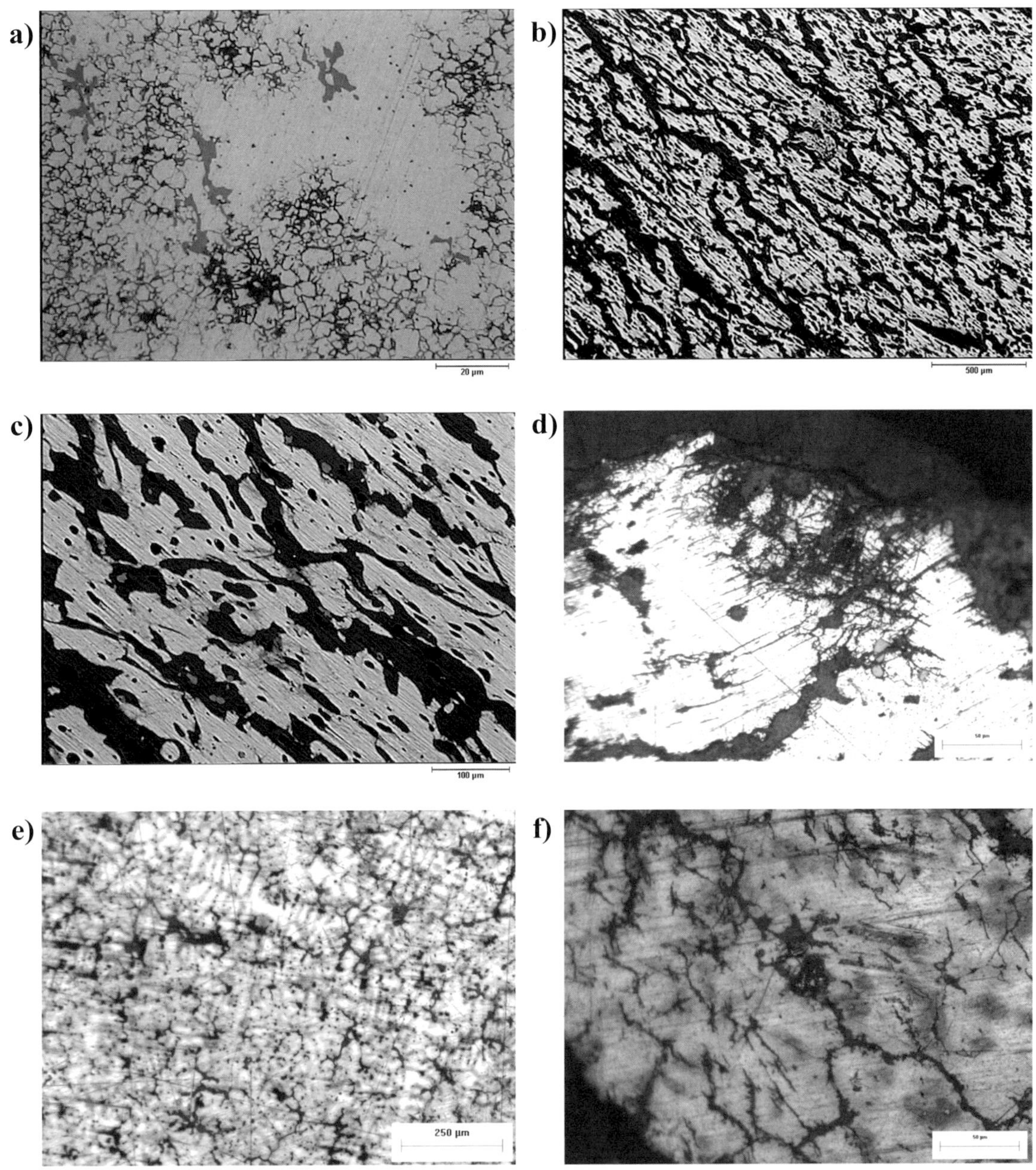

Fig. 36.5: Reflected Light Optical Micrographs (polished cross-sections) of plaques and vessel fragments: (a) plaque MB54; (b) vessel rim MB18, general view and (c) high-magnification detail; (d) vessel spout MB 339; mirror fragment MB43 (etched): (e) bulk and (f) surface.

Rods, which actually represent a rather vague category, display both dendritic and recrystallized microstructures; among them, the microstructure of MB177 (Fig. 36.6c, d) deserves mentioning, with fine dendrites, bent in places near the surface (Fig. 36.6d), suggesting a minimal amount of cold working after casting. Conversely, MB312 (Fig. 36.6e) shows a completely recrystallized structure, with twinned grains and some strain lines, which account for final cold-hammering. Generally, their Pb content is quite low when compared with the average in the other objects.

It would be tempting to take the evidence of a recrystallized structure in the chisel MB73a (Fig. 36.6f) as an indication of a functional destination of the item. This correlation must obviously remain a speculation, at least in the absence of complete objects to be analyzed for comparison. In any case, the alloy used for the chisel is suitable for the low degree of work-hardening, and was possibly meant to give it additional strength in order to balance the absence of tin in the alloy.

As regards bangle MB94, its Pb content is probably substantial enough to improve the fluidity of the molten metal; it does not hinder plastic deformation after casting if conducted by careful heating/

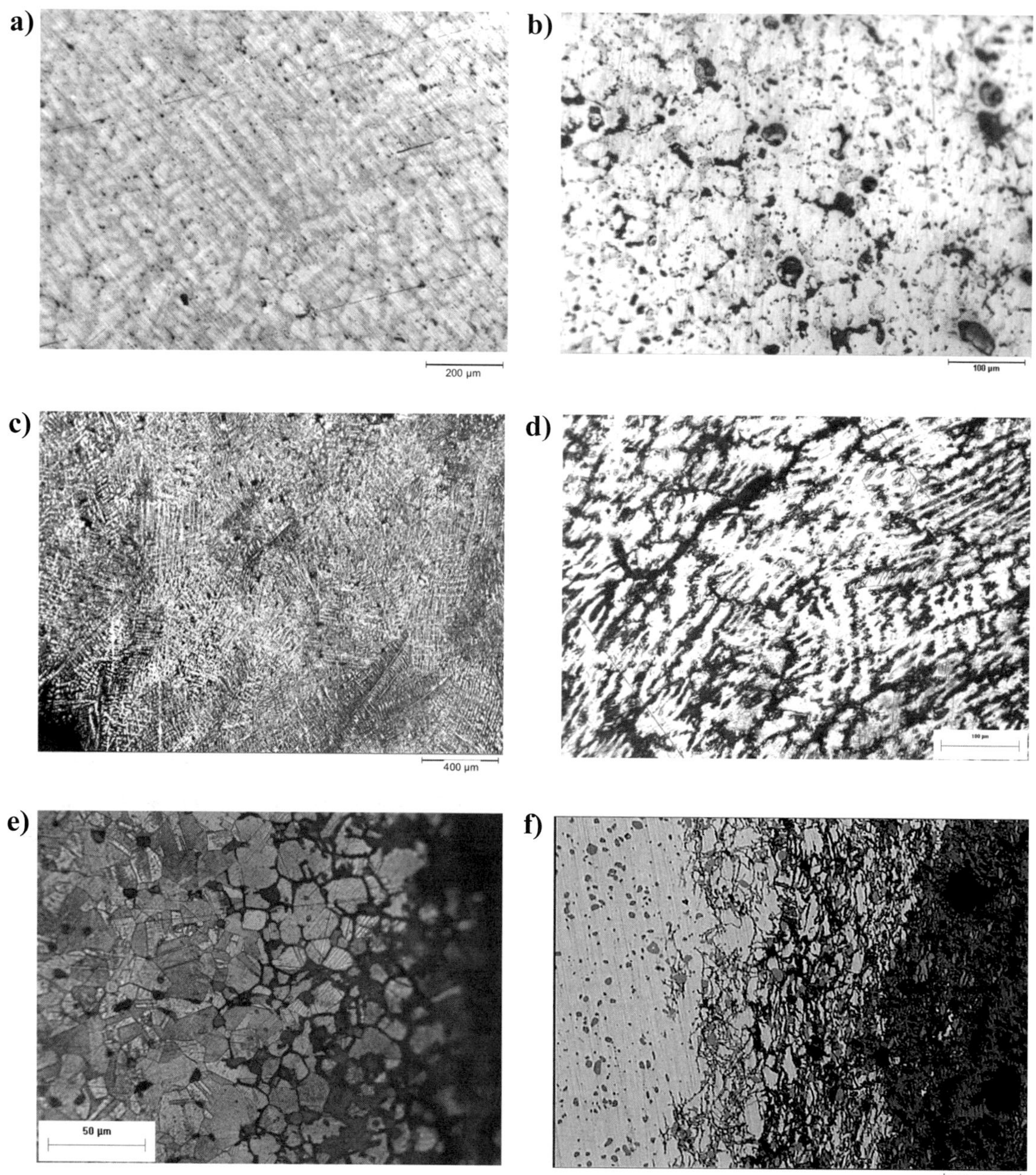

Fig. 36.6: Reflected Light Optical Micrographs (polished cross-sections, etched): (a) clamp rivet MB66; (b) clamp MB131; (c) rod MB177: general view and (d) high-magnification detail of the surface layer; (e)rod MB312; (f) chisel MB73a.

annealing, as testified by its recrystallized microstructure (Fig. 36.7a, b). Such a skilful treatment is consistent with the relatively complex shape of the bracelet, obtained by twisting a square section rod.

The fact that this accuracy is attested only for such a decorative item could lead to two hypothetical explanations, though the statistic relevance of the data is obviously limited: either there could have been a specific workshop for the more valuable objects, or such items could have come into town as imports.

Regarding the main microstructural features listed in Table 36.3, observations by both OM and SEM/EDS frequently showed the presence of Pb in the form of globules (in cast objects) or elongated

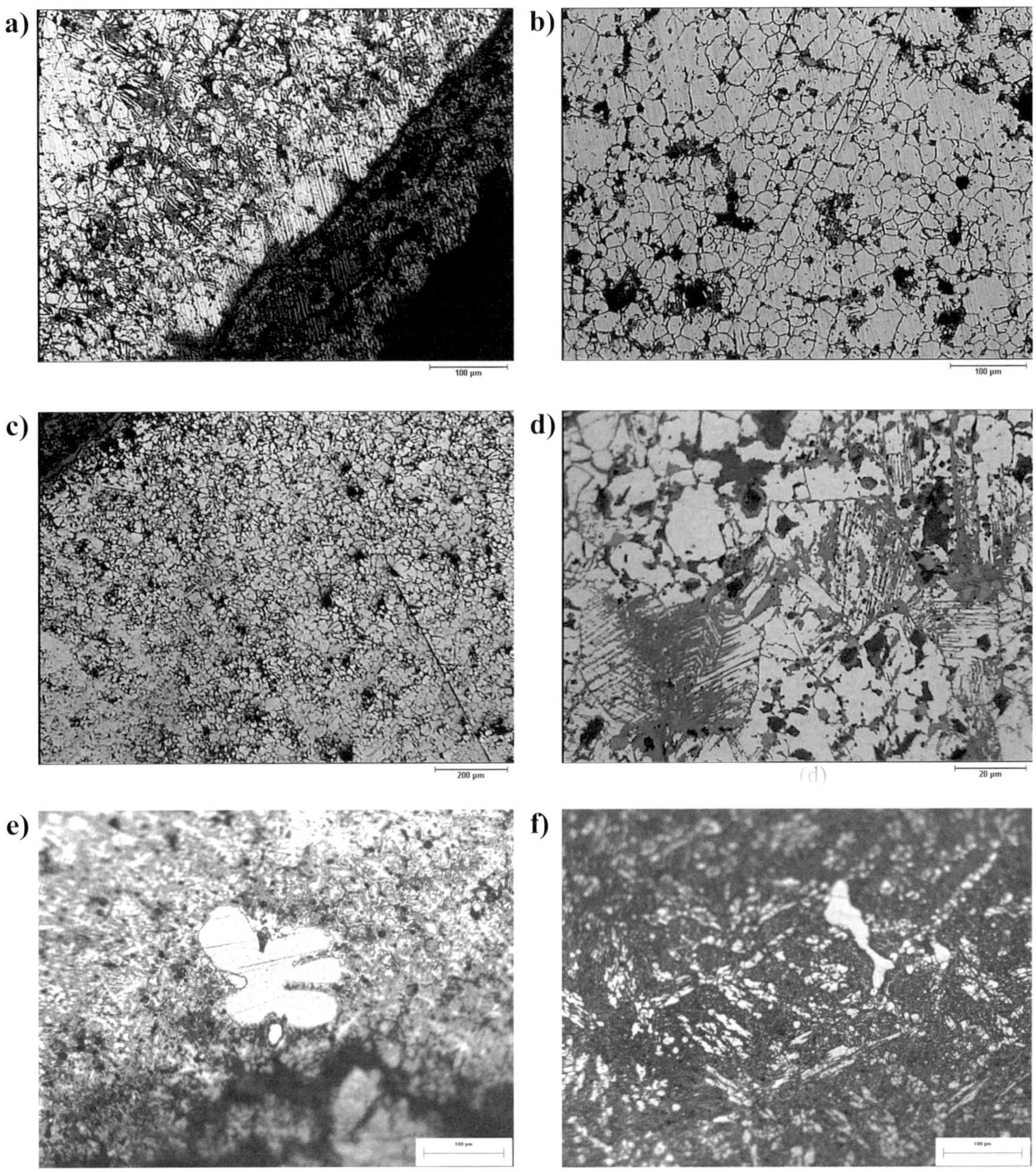

Fig. 36.7: Reflected Light Optical Micrographs (polished cross-sections): bangle MB94 (unetched): (a) surface and (b) bulk; plano-convex section rod MB119 (unetched): (c) surface and (d) bulk, and Examples of Unalloyed Copper Inclusions (UCI) in the microstructure of plaques MB49 (e) and MB 318 (f) (both plaques etched).

inclusions (in plastically-deformed objects) in the α-Cu matrix, as well as the frequent occurrence of sulphur-containing inclusions (mainly copper and iron-copper sulphides). The latter are of importance since they may indicate the use of sulphidic ores (Muhly, Stech Wheeler and Maddin 1977).

Secondary phases consisted mainly of the α+δ eutectoid, identified in a number of samples for which the amount of Sn already indicated the likelihood of its formation (Table 36.1). As indicated in the modified Cu-Sn phase diagrams for bronze castings with cooling rates far from thermodynamic equilibrium (Hanson and Pell-Walpole 1951), when Sn content is around 7 wt%, the α+δ eutectoid micro constituent can segregate out of the α-Cu matrix. Conversely, the plano-convex section rod MB119 (Fig. 36.7c, d), despite a Sn content of 19.7 wt%, does not show this secondary phase. Perhaps the extensive working/annealing cycles which led to complete recrystallization also contributed to homogenizing the alloy.

In a few cases (plaques MB49 and MB318— Table 36.3; Fig. 36.7e, f), also unalloyed Cu inclusions (UCI), mostly with an irregular shape, were detected. These inclusions, which are widely described in the literature about archaeological bronzes (Cartechini *et al.* 2008; Bosi *et al.* 2002) but not fully understood yet, were classified in three main groups: (i) UCI-A: due to corrosion/redeposition; (ii) UCI-B: generated during ore roasting; (iii) UCI-C: due to incomplete mixing (Bosi *et al.* 2002). The UCIs in the plaques from Sumhuram, which are both quite corroded, are most likely type-A UCIs, due to corrosion/redeposition processes involving secondary phases and micro constituents such as lead globules and/or the α+δ eutectoid, which are favored both by the depletion of oxygen due to the build-up of over-layers of corrosion products and by the presence of chloride ions in the burial environment (Chase 1994).

CONCLUSIONS

Combining the composition data with those on the microstructure of the sampled objects, the general impression is that of a rather inaccurate metallurgical technique.

Alloy selection in general was far from standardized, and displays a rather random use of the available metal, probably coming from repeated recycling with the possible addition of "fresh" copper and lead.

Moreover, no good correlation exists between the chosen alloy composition and the cold/hot manufacturing procedure needed to obtain the final piece, while a weak connection between working and the final use of the object could barely be seen in a limited number of cases (mostly when utilitarian tools are concerned: chisel MB73a, and possibly other rods which were originally part of working tools, the nails and the clamps), with the notable exception of the bangle MB94.

Metallurgical standards apparently remained unchanged throughout the town's long history. The only hypothesis that can be put forward concerns a somehow more controlled alloying during the first phase of the internal sequence. One possible explanation would be a change in the supply network in the later phases, with a growing need to rely on reuse, thus resulting in the decaying of the alloys; in the lack of provenance studies, this cannot be proved on the basis of the available analytical dataset.

Evidence for local copper/bronze working at the site is undisputable, although it is highly unlikely that it ever reached the level of a surplus production which would have entered the active trade network of which the site was a part. Rather, the scale of this production, at least concerning the copper-base alloys, can be estimated as rather small. Most of the metallurgy-related finds collected so far at Sumhuram offer coherent indication in this direction: the small size of the crucibles (Chiarantini and Benvenuti 2011: 70), the dimension of a recently found mold meant for the production of square section rods perfectly similar to the analyzed ones,[5] as well the average dimensions of the objects themselves.

This limited production would be commensurate with its quality, seemingly indicating a non-specialized craftsmanship as well as supply restrictions. Besides, although isolated workshops or metal working installations have been discovered at the site, no metallurgy-devoted area seems to exist inside the town, and in the surroundings no evidence was collected as well which could indicate a displacement of the activity. This may possibly be connected to the metal's toxicity. Clearly, further excavation could drastically change this picture[6].

Ancient South Arabian metallurgy has thus far been rather neglected in the literature, despite a potential wealth of information which the investigations carried out at Sumhuram have made evident. While a more extensive discussion of the available evidence is foreseen, this brief presentation, concerning only one aspect of the issue will hopefully draw attention and stimulate further needed work, which could also help in linking Sumhuram itself to a broader archaeological context.

ACKNOWLEDGMENTS

Financial support for this work was provided by the Italian Mission to Oman (IMTO)–Dipartimento di Civiltà e Forme del Sapere, University of Pisa. The director of the IMTO, Prof. Alessandra Avanzini (University of Pisa), is gratefully acknowledged for allowing the sampling of the objects from Sumhuram and for encouraging their study. The authors wish to thank Dr. F. Tarterini of the Department DIN, University of Bologna, for SEM/EDS analyses.

The authors would also like to thank the Editors of this volume for welcoming the publication of this paper.

REFERENCES

ASM 2008. *American Society for Metals (ASM) Handbook Online, Vol. 1: Casting of Copper and Copper Alloys Casting*. Materials Park, OH.

Avanzini, A. 2008. Notes for a History of Sumhuram and a New Inscription of Yashhur'il. In: Avanzini, A., ed. *A Port in Arabia between Rome and the Indian Ocean (3° c. BC–5° c. AD)* (Khor Rori Report 2). Rome: 609–641.

Avanzini, A. and Sedov, A.V. 2005. The Stratigraphy of Sumhuram: New Evidence. *Proceedings of the Seminar for Arabian Studies* 35: 11–17.

Bosi, C., *et al*. 2002. Unalloyed Copper Inclusions in Ancient Bronze Artefacts. *Journal of Materials Science* 37: 4285–4298.

Buffa, V. and Sedov, A.V. 2008. The Residential Quarter: Area A. In: Avanzini, A., ed. *A Port in Arabia between Rome and the Indian Ocean (3° c. BC–5° c. AD)* (Khor Rori Report 2). Rome: 15–59.

Cartechini, L., Arletti, R., Rinaldi, R., Kockelmann, W., Giovannini, S. and Cardarelli, A. 2008. Neutron Scattering Material Analysis of Bronze Age Metal Artefacts. *Journal of Physics: Condensed Matter* 20: 1–8.

Cecere, F., Carraro, A., Ferro, D. and Visco, G. 2008. Individuation of Characteristic Parameters of "Glass Paste" of Meridional Etruria by the Use of Scientific Methodologies. *Microchemical Journal* 88: 130–135.

Chase, W.T. 1994. Chinese Bronzes: Casting, Finishing, Patination, and Corrosion. In: Scott, D.A., Podany, J. and Considine, B.B., eds. *Ancient and Historic Metals*. Los Angeles: 85–117.

5 The mold was made to cast three square-section rods, the larger one with a length about ten cm, with sides of about one cm.

6 The early 2017 excavations, for example, revealed the presence of other small metallurgical furnaces in Area A, although mainly connected with iron working.

Chiarantini, L. and Benvenuti, M. 2011. Sumhuram As Centre of Production: The Metal Working. In: Avanzini, A., ed. *Along the Aroma and Spice Routes. The Harbour of Sumhuram, Its Territory and the Trade between the Mediterranean, Arabia and India*. Pontedera: 65–79.

Chiarantini, L. and Benvenuti, M. 2014. Evolution of Pre-Islamic South Arabian Coinage: Metallurgical Analyses of Coins Excavated in Sumhuram (Khor Rori, Sultanate of Oman). *Archaeometry* 56(4): 625–650.

Chiarantini, L., Benvenuti, M., Costagliola, P. and Avanzini, A. 2007. Preliminary Investigation of Pre-Islamic Metallurgical Slags From Sumhuram, Dhofar Region (Sultanate of Oman). In: *Proceedings of the 2nd International Conference Archaeometallurgy in Europe, Aquileia-Grado, June 17–21 2007*. Grado, Italy.

Chiavari, C., Degli Esposti, M., Garagnani, G.L., Martini, C. and Ospitali, F. 2011. Ancient Metallurgy at Sumhuram (Sultanate of Oman): Technical Aspects of Raised Inscriptions on South Arabian Bronzes. *Archaeometry* 53(3): 528–546.

Cowell, M. 1985. Scientific Appendix I: Chemical Analysis. In: Davies, W.V., ed. *Catalogue of Egyptian Antiquities in the British Museum, Vol. 7: Tools and Weapons, I Axes*. London: 96–118.

Craddock, P.T. 1978. The Composition of the Copper Alloys Used by the Greek, Etruscan and Roman Civilizations, 3: The Origin and Early Use of Brass. *Journal of Archaeological Science* 5: 1–16.

Craddock, P.T. 1985. Technical Appendix 1. The Composition of the Metal Artefacts. In: Donaldson, P., ed. Prehistoric Tombs of Ras al Khaima. *Oriens Antiquus* 24: 97–101.

Favaro, M., Guastoni, A., Marini, F., Bianchin, S. and Gambirasi, A. 2012. Characterization of Lapis Lazuli and Corresponding Purified Pigments for a Provenance Study of Ultramarine Pigments Used in Works of Art. *Analytical Bioanalytical Chemistry* 402: 2195–2208.

Felberbaum, L. 2005. *Microstructure and Embrittlement of Leaded Copper Alloys* (Ph.D. dissertation, École Polytechnique Fédérale de Lausanne). Lausanne.

Giumlia-Mair, A., Keall, E.J., Shugar, A.N. and Stock, S. 2002. Investigation of a Copper-based Hoard from the Megalithic Site of al-Midamman, Yemen: An Interdisciplinary Approach. *Journal of Archaeological Science* 29: 195–209.

Hanson, D. and Pell-Walpole, W.T. 1951. *The Chill-Cast Tin Bronzes*. London.

Hughes, M.J., Cowell, M.R. and Craddock, P.T. 1976. Atomic Absorption Techniques in Archaeology. *Archaeometry* 18: 9–37.

Klemenc, S., Budic, B. and Zupan, J. 1999. Statistical Evaluation of Data Obtained by Inductively Coupled Plasma Atomic Emission Spectrometry (ICP-AES) for Archaeological Copper Ingots. *Analytica Chimica Acta* 389: 141–150.

Lombardi, A., Buffa, V. and Pavan, A. 2008. Small Finds. In: Avanzini, A., ed. *A Port in Arabia between Rome and the Indian Ocean (3° c. BC–5° c. AD)* (Khor Rori Report 2). Rome: 317–475.

Moorey, P.R.S. 1971. *Catalogue of Ancient Persian Bronzes in the Ashmolean Museum*. Oxford.

Muhly, J.D., Stech-Wheeler, T. and Maddin, R. 1977. The Cape Gelidonya Shipwreck and the Bronze Age Metals Trade in the Eastern Mediterranean. *Journal of Field Archaeology* 4: 353–362.

Pavan, A. 2011. Sumhuram as International Centre: the Imported Pottery. In: Avanzini, A., ed. *Along the Aroma and Spice Routes. The Harbour of Sumhuram, Its Territory and the Trade between the Mediterranean, Arabia and India*. Pontedera: 99–112.

Pavan, A. and Schenk, H. 2012. Crossing the Indian Ocean Before the *Periplus*. A Comparison of Pottery Assemblages at the Sites of Sumhuram (Oman) and Tissamaharama (Sri Lanka). *Arabian Archaeology and Epigraphy* 23(2): 191–202.

Philip, G. 1991. Tin, Arsenic, Lead: Alloying Practices in Syria-Palestine Around 2000 BC. *Levant* 23: 93–104.

Piccardo, P., Amendola, R., Adobati, A. and Faletti, C. 2009. Studio della Fluidità di Leghe a Base Rame. *La Metallurgia Italiana* 5: 31–38.

Robbiola, L., Blengino, J.M. and Fiaud, C. 1998. Morphology and Mechanisms of Formation of Natural Patinas on Archaeological Cu-Sn Alloys. *Corrosion Science* 39(1): 2083–2111.

Rothenberg, B. 1972. *Timna: Valley of the Biblical Copper Mines*. London.

Salmon, M.E. 1969. Appendix 1: A Survey of the Composition and Fabrication of Bronze Artefacts from Hajar Bin Âumeid. In: Van Beek, G., ed. *Hajar Bin Ḥumeid: Investigations at a Pre-Islamic Site in South Arabia* (Publications of the American Foundation for the Study of Man 5). Baltimore: 373–386.

Sedov, A.V. 2008. The Cultual Quarter: Area F. In: Avanzini, A., ed. *A Port in Arabia between Rome and the Indian Ocean (3° c. BC–5° c. AD)* (Khor Rori Report 2). Rome: 183–213.

Sedov, A.V. 2013. *New Data on Sumhuram: Stratigraphy and Datation*. Paper Presented at Rencontre Sabeénnes 17°, 6–8 June, 2013, Paris.

Tylecote, R.F., Ghaznavi, H.A. and Boydell, P.J. 1977. Partitioning of Trace Elements between the Ores, Fluxes, Slags and Metal during the Smelting of Copper. *Journal of Archaeological Science* 4: 305–333.

Van Beek, G. 1969. *Hajar Bin Ḥumeid: Investigations at a Pre-Islamic Site in South Arabia* (Pubblications of the American Foundation for the Study of Man 5). Baltimore.

Weeks, L.R. 2004. An Analysis of Late Pre-Islamic Copper-Base Artefacts from Ed Dur, U.A.E. *Arabian Archaeology and Epigraphy* 15(2): 240–252.

CHAPTER 37

OBSERVATIONS ON SOCKETED COPPER ALLOY ARROWHEADS IN THE EARLY FIRST MILLENNIUM BCE

Jamie M. Szudy

While iron had largely replaced copper alloys as the material of choice for arrowheads at the beginning of the first millennium BCE in the Near East, copper alloy arrowheads made a massive resurgence in the 7th century BCE with the introduction of socketed bilobates and trilobates.

The reason for this large scale return to copper alloys has been much discussed, particularly in regard to the relative qualities of iron and copper alloys, the possible scarcity of iron at the time, the ability to mass produce copper objects by casting, and the role that arrowhead form played in the choice of materials (or vice versa).

A detailed study of these issues suggests that the form of the arrowhead was the primary factor. Sockets offered significant advantages over tangs for affixing arrowheads to shafts, and it was far easier to cast the small sockets of arrowheads in copper alloys than to forge them in iron. However, the ability of cast arrowheads to be mass-produced is often somewhat exaggerated since the presence of pour gates is typically overlooked. Nevertheless, the popularity of socketed arrowheads was such that tanged arrowheads were sometimes made to mimic their distinctive shape.

INTRODUCTION

Around the beginning of the first millennium BCE, iron began to replace copper and its alloys as the metal of choice for most tools and weapons, including arrowheads (Curtis *et al.* 1979: 384). While it did not completely displace copper alloys, by the end of the Neo-Assyrian period (late 7th c. BCE), the vast majority of arrowheads with tangs were iron. For example, a total of 340 leaf-shaped arrowheads with tangs dating to the Neo-Assyrian period from Lachish have been published, and of those, only five are made of bronze or copper, while 334 (98% of the total) are iron.[1] The case is even more extreme with the leaf-shaped tanged arrowheads from Nimrud, where all of the 262 examples in the British Museum are iron. Nevertheless, in the 7th century BCE, copper alloys made a major resurgence for the manufacture of arrowheads with the introduction of socketed bilobates and trilobates. These forms of arrowheads spread rapidly throughout the Near East and came to be a dominant form of arrowhead in the later first millennium BCE. The thousands of socketed copper alloy arrowheads found at the Persepolis treasury, for example, far outnumber the tanged iron arrowheads found there (Schmidt 1957: 97), and the prevalence of their use by the Greeks led to them being called "Greco-Scythian" arrowheads in numerous studies (e.g., Schmidt 1908: 183–186, Ghirshman 1939: 46; Sulimirski 1954: 295).

1 See Tufnell 1953; Rothenberg 1975; and Gottlieb 2008. The material of one example is not stated; Tufnell 1953, no. 285 (p. 186, Pl. 54: 47).

THE ORIGIN OF SOCKETED BRONZE ARROWHEADS

These socketed copper alloy arrowheads have generally been associated with invading "early Scythian" nomads (Ivantchik 1997), particularly the Scythians and Cimmerians (e.g., Schmidt 1908: 183–186; Petrie 1917: 34; Petrie 1928: 15; Ghirshman 1939: 46). Studies from the latter half of the 20th century generally contain a proviso stating that, while they link these arrowhead forms to the Scythians or other related peoples, not all examples were necessarily produced by them (e.g., Sulimirski 1954: 313; Boehmer 1972: 115; Medvedskaya 1982: 88; and Yamauchi 1983: 94). Medvedskaya, however, went so far as to insist that socketed bronze arrowheads were not used by the indigenous peoples of the Near East, and so all such arrowheads found in the Near East before the Achaemenid period were deposited solely by "representatives of the Eurasian steppe" (Medvedskaya 1982: 88). Several archaeological finds, however, suggest that indigenous Near Eastern peoples did indeed adopt socketed arrowheads at a very early point.

Socketed bilobates are attested in Siberia in the Andronovo period (roughly the mid-second millennium BCE; Brentjes 1995–1996: 203, 205, Fig. 37), long before their first appearance in the Near East. The earliest attested socketed arrowheads in the Near East (aside from two rather anomalous examples from second millennium BCE Boğazköy)[2] come from Hasanlu IVB, dating to either ca. 800 BCE (Dyson 1965; Dyson and Muscarella 1989) or the 8th century BCE (Medvedskaya 1988, 1991; Magee 2008). They are significantly older than the next oldest securely-dated socketed arrowheads, which come from several sites destroyed at the end of the Neo-Assyrian period (such as Nimrud and Nineveh). The Hasanlu bilobates are also noteworthy because they are made of iron. The only other socketed iron bilobate attested in the early first millennium Near East is a single example from Toprakkale, which likely dates to the mid-7th century BCE (Wartke 1990: 127, Fig. 32i, No. 335). Since iron is unprecedented as a socketed arrowhead material among "early Scythian" finds, while it was in common use by indigenous Near Eastern peoples by the time Hasanlu IVB was destroyed,[3] it is probable that the Hasanlu bilobates represented a local attempt to duplicate socketed bronze arrowheads in their preferred metal.

Furthermore, socketed copper alloy arrowheads were found stored along with tanged iron arrowheads in the fortress of Karmir Blur (destroyed in the mid-7th century BCE), which indicated that the Urartians, at least, were using both forms of arrowheads at that time (Ivantchik 1997: 34; Dandamaev and Lukonin 1989: 226). As Dušek observed, effective weapons spread rapidly in the past just as in the modern world, so using such weapons to determine ethnicity is highly problematic (Dušek 1964: 57). Muscarella reasoned that if socketed bronze arrowheads were introduced by a specific ethnic group (or related ethnic groups, such as the Scythians and Cimmerians), these arrowheads may have indeed "functioned as a recognizable ethnic identity marker" for a time (Muscarella 1989: 107), but only until they began to be borrowed by other groups of people. The finds at Hasanlu and Karmir-Bur indicate that this borrowing must have happened very early in the contact between "early Scythian" people and indigenous populations of the Near East. Thus, socketed bronze arrowheads cannot be reliably used as an ethnic marker at any point during their use in the Near East, even from the very beginning.

2 There are two examples from Boğazköy dating to the Late Bronze Age (Boehmer 1972: 109, Pl. 30, Nos. 884 and 884A), however they have rolled sockets very much like contemporary spear heads, and may indeed have been models of spear heads.

3 Of the arrowheads in the destruction level of Hasanlu IVB, 637 were iron compared to only 20 of copper alloy; Thornton and Pigott 2011: 138.

REASONS FOR THE RETURN OF COPPER ALLOY ARROWHEADS

The reason for this large scale return to copper alloys for arrowheads is a matter of some debate. Arguments have been made, for example, that the resurgence in the use of copper alloys for arrowheads was driven by a scarcity of iron (e.g., Snodgrass 1964: 154). However, there does not appear to be any evidence for any such scarcity of iron, and throughout this period, other weapons and tools continued to be made of iron (Rothenberg 1975: 80). It is also noteworthy that iron was more expensive than copper until the Neo-Babylonian period (Moorey 1995: 290), so one might argue that the return to copper and its alloys may have been economically motivated, except that large quantities of iron arrowheads continued to be made up to the end of the Neo-Assyrian period (as indicated by finds from Nimrud, among other sites).

Studies of iron artifacts from the early first millennium BCE suggest that smiths of that time had not yet learned how to effectively manipulate the carbon content or hardness of iron objects (e.g., Pleiner 1979; Curtis *et al.* 1979; Piaskowski 1991). The few iron arrowheads from this period that have been analyzed metallurgically contain little or no carbon (Curtis *et al.* 1979: 385–386; Piaskowski and Wartke 1989: 106; Gottlieb 2004: 1964–1965). By contrast, copper alloys were well understood by the early first millennium BCE, and their properties could be exploited by bronze smiths for maximum effect. Rothenberg observed that the metallurgical analysis of one trilobate from Lachish showed it had high tin and lead content, which would produce the optimal bronze for such arrowheads: very fluid when molten, thus able to fill the many narrow spaces in the mold, yet hard and robust when cooled (Rothenberg 1975: 80). Other copper alloy artifacts also exhibit careful manipulation of their alloys for best effect (e.g., the Nimrud bowls; see Craddock and Giumlia-Mair 1988: 318). Nevertheless, iron continued to be the principle metal for all other weapons in the early first millennium BCE. Thus, it is unlikely that metallurgical considerations were behind the resurgence of copper alloys for arrowheads.

CASTING

Perhaps the most frequently cited justification for the spread of this arrowhead form is the claim that cast bronze arrowheads lend themselves far more readily to mass production than do forged iron arrowheads (e.g., Rothenberg 1975: 80; Cleuziou 1977: 198; Moorey 1995: 283), as one could theoretically make many castings in succession from a single mold whereas a smith would have to hammer out each individual arrowhead. While casting does indeed provide several advantages, such as ensuring that the arrowheads would have identical shapes and weights (and thus identical aerodynamic properties), this argument does not address why most tanged arrowheads continued to be made of iron, though they could just as well be cast in copper. Furthermore, the two preserved molds for socketed bronze arrowheads from the Near East in the British Museum suggest that this ease of mass-production may well be overstated.

Two segments of a four part mold for a hooked trilobate arrowhead were found in the vicinity of House D at Carchemish (BM 116254A & B, Figs. 37.1, 37.2; Woolley 1921: 130, Pl. 23b), and a complete mold for two hooked trilobates and one hooked bilobate was allegedly found at Nebi Yunus (BM 124624, Figs. 37.3, 37.4; Coghlan 1952). Both of these molds show that the pour gate—the hole by which the molten bronze was poured into the mold—was placed over the tip of the arrowhead. Therefore, the tip of each newly-cast arrowhead would be encased in a lump of metal that would have to be chiseled or filed away before it could be used (see Fig. 37.2).

The labor-intensive nature of forging iron arrowheads may also be somewhat exaggerated. Some kinds of iron arrowheads appear to have been made with rapid production in mind. Most notable among these are the small leaf-shaped arrowheads with short, unstopped tangs from Nimrud. Over 200 of

Fig. 37.1: The two extant pieces of the bronze arrowhead mold (BM 116254A & B) from House D in Carchemish, fitted together (by arrangement with the British Museum).

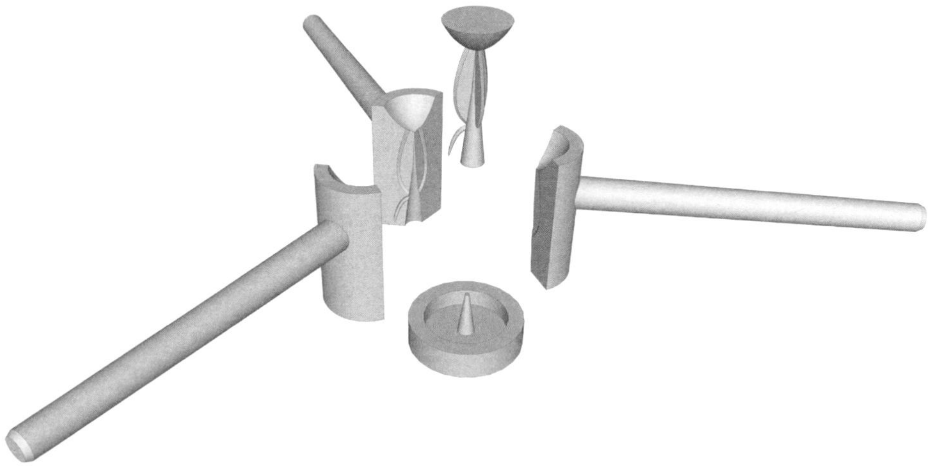

Fig. 37.2: 3D reconstruction of the Carchemish mold as it would have appeared when complete, inclusive of cast arrowhead (still retaining the plug from the pour gate).

these were found, and most of them have somewhat irregular profiles and other evidence of sloppy workmanship, indicating that they were made in haste (perhaps during the siege and fall of the city; see Fig. 37.5 for examples). The width of the tang suggests that these arrowheads were made by hammering out an iron rod then flattening the end into a blade and then cutting the blade off the rod while leaving some rod attached to serve as the tang, then repeating the process. Due to the simple form of these arrowheads, they doubtlessly could have been produced very rapidly.

Cleuziou has suggested that casting arrowheads in bronze was preferable for mobile mounted archers, based on the assumption that molding arrowheads in small, easily-carried molds would be simpler and require fewer tools than forging iron ones (Cleuziou 1977: 197–198). However, bronze casting may have actually required more tools than forging. The latter requires, at minimum, a hammer and anvil (such as small stake anvils; for Roman examples, see Coghlan 1956: 120), while the former also may require hammer and anvil (for work-hardening edges; see Moorey 1995: 249), as well as the mold, a crucible and the files and chisels necessary to remove the pour gate plug. More significantly, both methods require a considerable quantity of charcoal. It is hard to imagine how a swift-moving mounted warrior on campaign would carry all this extra gear as well as a large quantity of charcoal in order to make arrowheads in the field. It is more probable that the arrowheads were made by specialists and that mounted archers simply carried the arrowheads that they would need.

Fig. 37.3: Bronze arrowhead mold (BM 124624), alleged to be from Nebi Yunus, partly assembled (by arrangement with the British Museum).

Fig. 37.4: The Nebi Yunus mold (BM 124624), fully assembled.

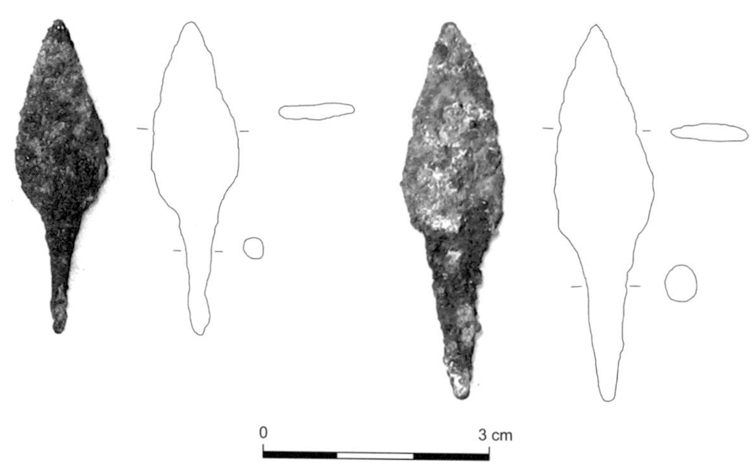

Fig. 37.5: Two small, leaf-shaped iron arrowheads with unstopped tangs from Nimrud (both derived from the large collection of arrowheads, ND 10944) (by arrangement with the British Museum).

SOCKET VERSUS TANG

Prior to the first millennium BCE, the tang was the standard method for affixing arrowheads to arrow shafts. Sockets were commonly used to attach spear heads to their shafts, however spears could be expected to encounter far more lateral force in combat (such as glancing off or being knocked aside by an opponent), demanding a more robust connection than do arrowheads, which need to strike their targets more or less head-on to be effective. Furthermore, the larger sockets of spears would have been easier to forge than the small sockets of arrowheads.

Several factors contributed to the popularity of tangs for arrowheads, which continued to be used well into the Middle Ages (Jessop 1996: 195). First, the standard material for arrow shafts in the ancient Near East was reed, and since reeds naturally had hollow centers, it would have been very easy to insert a tang into them (provided that no wooden foreshaft was used). Wooden arrowheads required a hole to be drilled for the tang, or a slot cut in the end of the shaft. Secondly, tangs were relatively easy to make. When forged, the smith would merely have to draw some of the metal out into a rod, and if the smith began with a rod, a length of it could simply be left attached to the blade to serve as the tang. Cast tanged arrowheads were also relatively simple, and only required a basic two-part mold.

However, tangs did have a disadvantage. Upon impact, the momentum of the arrow shaft would force it down on the arrowhead. The tang inserted into the shaft would tend to act as a wedge, causing the arrow shaft to crack or split as it rode up on the arrowhead. While this would prevent an opponent from picking the arrow up and firing it back, a considerable portion of the impact energy would be expended in cracking the shaft, minimizing the effectiveness of the arrow. This disadvantage could be minimized by including a stop on the tang, providing a reasonably flat surface for the end of the arrow shaft to rest against, though this added to the complexity of the arrowhead. A foreshaft of hard wood could also be added to the forward part of the arrow shaft to help prevent splitting,[4] though this also added to the complexity of manufacturing the arrows (Blythe 1977: 47).

Sockets largely resolved the principle drawback of tangs; given that sockets surround the end of the arrow shaft, they tend to help hold it intact on impact rather than cause it to split (Fig. 37.6). The superiority of sockets is illustrated by the fact that they were used universally for arrowheads in later medieval Europe (Jessop 1996: 193–195). Socketed arrowheads could be fairly easily cast in bronze, though they did require a more complex mold than a tanged arrowhead would. A tanged arrowhead requires only a simple two-piece mold, where each half has a cavity for one half of the arrowhead. A socketed bilobate requires at least three—two halves plus a plug to keep the socket from being filled with bronze.

TRILOBATES

Trilobate molds are somewhat more complicated than bilobate molds in that they require an additional piece, and the faces meet at 120° angles rather than the simpler flat faces of a bilobate mold. However, the extra blades provide a significant advantage over normal leaf-shaped blades or bilobates in that they create an irregular Y-shaped wound which would be more difficult to stanch than the linear wound formed by a more conventional arrowhead.[5]

Sulimirski posited that the Scythians did not themselves have the "technological background" to produce socketed arrowheads, and that they were developed in the Transcaucasian region while

4 Numerous such examples were found in the tomb of Tutankhamen; McLeod 1982: 14–25, referred to this as "footed."

5 Modern hunting arrows also often have multiple blades in order to increase bleeding.

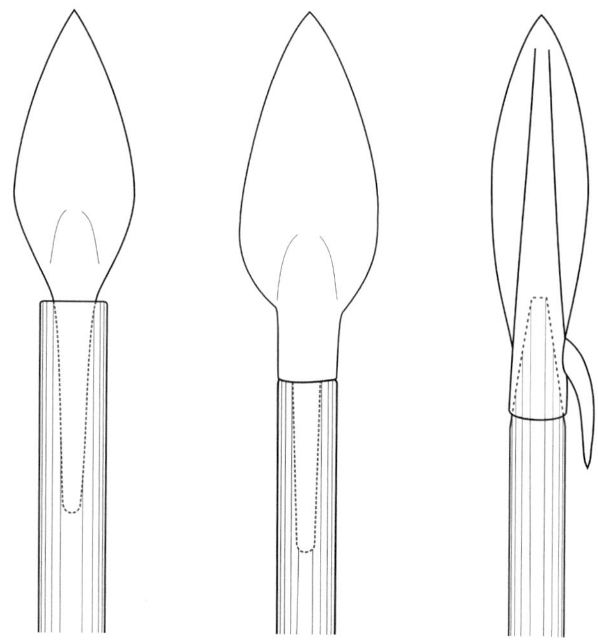

Fig. 37.6: Diagram of unstopped, stopped and socketed arrowheads.

it was under Scythian domination (Sulimirski 1954: 310). He was apparently unaware of the second millennium BCE socketed bilobates from Siberia, which invalidates this theory. However "Scythian" style socketed arrowheads may have nevertheless been influenced by concepts from indigenous Near Eastern populations.

The earliest attested trilobates are three examples from Marlik, dating to ca. 750 BCE or earlier (Negahban 1996: 280, Pl. 126: 848). All of these trilobates have tangs, rather than sockets, however, and because of their dating, Negahban has proposed that tanged trilobates were the first form of trilobates.[6] While Ghirshman speculated that socketed trilobates must have appeared around 750 BCE (Stronach 1978: 180), possibly contemporary with the Marlik trilobates, Boehmer (1972: 115) dates the first attested socketed trilobates in the Near East to the early 7th century BCE, specifically ca. 690/680 BCE. Stronach observed that the earliest socketed trilobates from accurately datable contexts come from the late 7th century BCE (from Urartu, Assyria and Media; Stronach 1978: 180–181). Socketed copper alloy trilobates clearly do not pre-date the Marlik tanged trilobates, thus Negahban's hypothesis may well be correct: in northwest Iran, presumably in the later 8th century BCE, the most desirable feature of imported socketed copper alloy bilobates (the socket) was combined with the most desirable feature of the local tanged trilobate (multiple blades) to create an effective new arrowhead, the socketed trilobate (Fig. 37.7). This type spread very rapidly throughout the Near East and beyond (they are, for example, attested at Boğazköy in the early 7th century BCE; Boehmer 1972: 111).

Tanged trilobates were never completely abandoned, however, and were widely used in later periods. Some Achaemenid examples were found at Persepolis (Schmidt 1957, Pl. 76: 15–16), for example, and they were commonly used by Roman archers (Bishop and Coulson 1993: 55, 79, 138). These later tanged trilobates were all made of iron, however, which is doubtless the reason they were made with tangs instead of sockets.

6 Negahban 1996: 280; Cleuziou 1977: 189 put the earliest tanged trilobates in Kazakhstan in the 9th century BCE, thus the Marlik tanged trilobates may themselves have been borrowed from an earlier source.

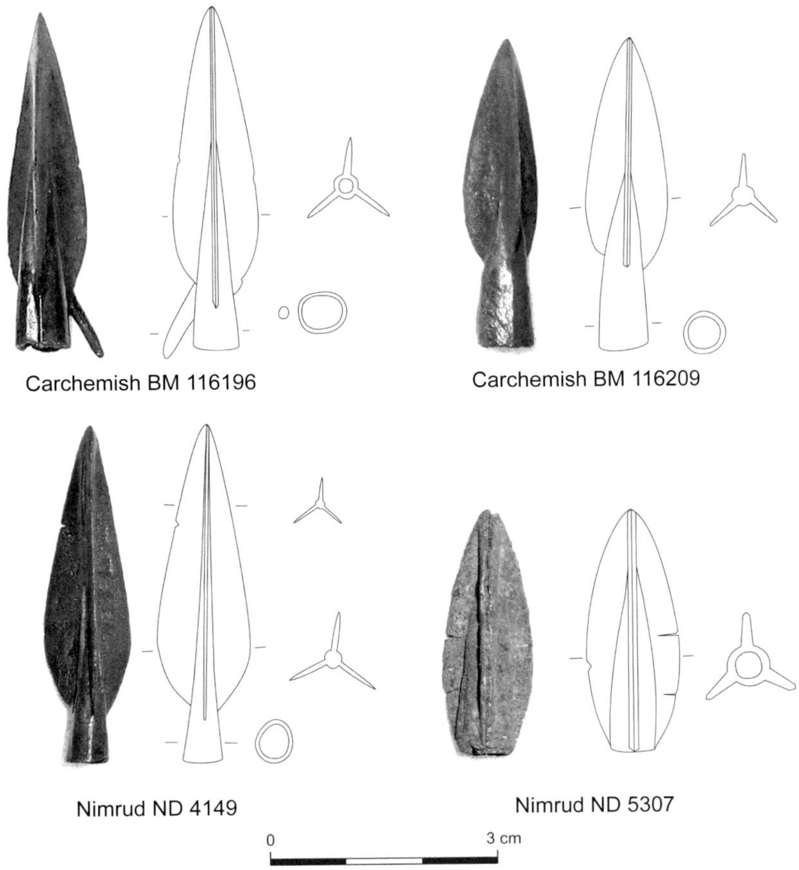

Fig. 37.7: Late 7th century BCE socketed copper alloy trilobates from Carchemish and Nimrud (by arrangement with the British Museum).

SOCKET MIMICRY

The influence of socketed copper alloy arrowheads can also be seen in other cases where tanged arrowheads are designed to mimic the shape of socketed arrowheads. For example, a bronze Neo-Babylonian or Achaemenid arrowhead from Uruk has a thick, conical rib with a distinct stop (van Ess and Pedde 1992: 37, no. 354). When mounted on a shaft, it would be nearly indistinguishable from a conventional socketed bilobate. Copying the shape of a socket in this manner would not serve any practical purpose, so presumably this was a purely aesthetic feature, intended solely to make a tanged arrowhead look like a socketed one.

The tanged quadrilobate arrowheads, attested only at Carchemish, also tend to have thick, tapering ribs reminiscent of sockets (Fig. 37.8). The four blades of the quadrilobates would have the same benefit as the three of the trilobate in that they would create irregular rather than linear wounds. It seems likely that the design of the quadrilobates was intended to enable a multi-bladed arrowhead like a trilobate to be made with a simple two-part mold. The four blades of the quadrilobate may appear to require a four-part mold (forming the faces of all four blades). However, the blades of the attested quadrilobates are either all quite small, or the blades in one plane (the same plane as the hook, if one is present) are large while those perpendicular to it are comparatively small. Thus the small blades of the quadrilobates may have been formed by carving a cavity into the mold, rather like a somewhat extended midrib. Thus, quadrilobates could be cast with a fairly simple two-part mold (Fig. 37.9).

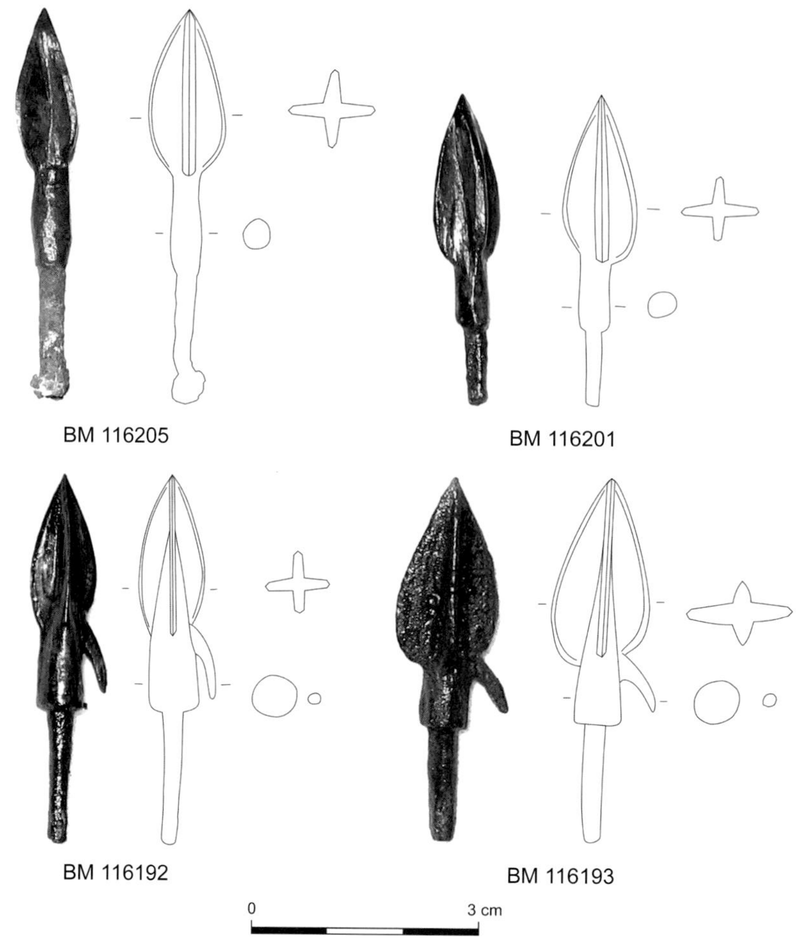

Fig. 37.8: Late 7th century BCE quadrilobate arrowheads from Carchemish (by arrangement with the British Museum).

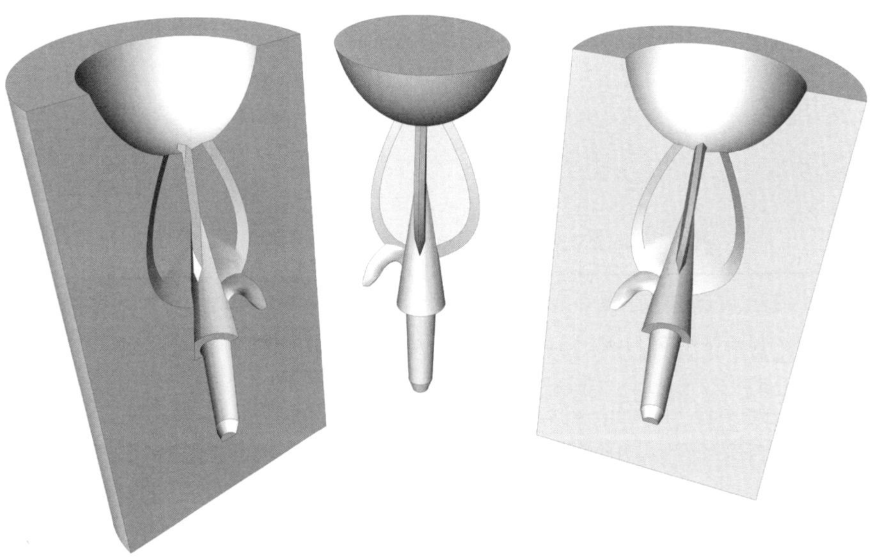

Fig. 37.9: A hypothetical two-part mold for quadrilobate arrowheads.

CONCLUSIONS

While socketed copper alloy arrowheads are first attested in Siberia, they appear to have been adopted by indigenous Near Eastern peoples as soon as, or perhaps even before, the Scythians, Cimmerians and related peoples arrived in the Near East. Thus, such arrowheads cannot at any time be considered ethnic identity markers in the Near East, nor can they be used to trace the movements of specific groups of people. The resurgence of copper alloys for arrowheads in the 7th century BCE appears to be due to its ability to be cast, in contrast to iron, which must be forged. However, the use of casting for mass production has often been overstated, since once cast, the pour gate plug must be laboriously filed away before the arrowhead can be used. Casting, however, is the most efficient way to form sockets, so cast copper alloy was reinstituted in order to facilitate the manufacture of this useful feature. Early in their contact with the Near East, socketed arrowheads were influenced by indigenous concepts, leading to the socketed trilobate, and subsequently their popularity ensured that they themselves influenced indigenous Near Eastern arrowhead designs (such as the quadrilobates from Carchemish).

REFERENCES

Bishop, M.C. and Coulson, J.C.N. 1993. *Roman Military Equipment: From the Punic Wars to the Fall of Rome*. London.

Blyth, P.H. 1977. *The Effectiveness of Greek Armour against Arrows in the Persian War (490–479 BC): An Interdisciplinary Study* (Ph.D. dissertation, University of Reading). Reading.

Boehmer, R.M. 1972. *Die Kleinfunde von Boğazköy, aus den Graungskampagnen 1931–1939 und 1952–1969*. Berlin.

Brentjes, B. 1995–1996. Waffen der Steppenvölker (II): Kompositbogen, Goryt und Pfeil – ein Waffenkomplex der Steppenvölker. *Archäologische Mitteilungen aus Iran* 28: 179–210.

Cleuziou, S. 1977. Les Pointes de Fleches 'Scythiques' au Proche et Moyen Orient. In: Deshayes, J., ed. *Le Plateau Iranien at l'Asie Central: des Origins à la Conquête Islamique* (Éditions du CNRS). Paris: 187–199.

Coghlan, H.H. 1952. Casting Moulds Made in Metal. *Man* 52: 162–164.

Coghlan, H.H. 1956. *Notes on Prehistoric and Early Iron in the Old World*. Oxford.

Craddock, P. and Giumlia-Mair, A.R. 1988. Problems and Possibilities for Provenancing Bronzes by Chemical Composition. In: Curtis, J., ed. *Bronzeworking Centres of Western Asia, c. 1000–539 BC*. London.

Curtis, J., Wheeler, T.S., Muhly, J.D. and Maddin, R. 1979. Neo-Assyrian Ironworking Technology. *Proceedings of the American Philosophical Society* 123(6): 369–390.

Dandamaev, M.A. and Lukonin, V.G. 1989. *The Culture and Social Institutions of Ancient Iran*. Cambridge.

Dušek, M. 1964. Waren Skythen in Mitteleuropa und Deutschland? *Prähistorische Zeitschrift* 42: 49–76.

Dyson, R.H. 1965. Problems of Protohistoric Iran as Seen from Hasanlu. *Journal of Near Eastern Studies* 24/3: 193–217.

Dyson, R.H. and Muscarella, O.W. 1989. Constructing the Chronology and Historical Implications of Hasanlu IV. *Iran* 27: 1–27.

Van Ess, M. and Pedde, F. 1992. *Uruk: Kleinfunde II: Metall und Asphalt, Farbreste, Fritte/Fayence, Glas, Holz, Knochen/Elfenbein, Leder, Muschel/Perlmutt, Schnecke, Schilf, Textilien*. Mainz.

Ghirshman, R. 1939. *Fouilles de Sialk, près de Kashan: 1933, 1934, 1937*, Vol. 2. Paris.

Gottlieb, Y. 2004. The Weaponry of the Assyrian Attack, Section A: The Arrowheads and Selected Aspects of the Siege Battle. In: Ussishkin, D., ed. *The Renewed Archaeological Excavations at Lachish (1973–1994)*, Vol. 4 (Monograph Series of the Institute of Archaeology of Tel Aviv University 22). Tel Aviv: 1907–1969.

Ivantchik, A. 1997. Das Problem der ethnischen Zugehörigkeit der Kimmerier und die Kimmerische archäologische Kultur. *Prähistorische Zeitschrift* 72: 12–53.

Jessop, O. 1996. A New Artifact Typology for the Study of Medieval Arrowheads. *Medieval Archaeology* 40: 192–205.

Magee, P. 2008. Deconstructing the Destruction of Hasanlu: Archaeology, Imperialism and the Chronology of the Iranian Iron Age. *Iranica Antiqua* 43: 89–106.

McLeod, W. 1982. *Self Bows and Other Archery Tackle from the Tomb of Tut'ankhamūn* (Tutʿankhamūn's Tomb Series IV). Oxford.

Medvedskaya, I.N. 1982. *Iran: Iron Age I* (BAR International Series 126). Oxford.

Medvedskaya, I.N. 1988. Who Destroyed Hasanlu IV? *Iran* 26: 1–15.

Medvedskaya, I.N. 1991. Once More on the Destruction of Hasanlu IV: Problems of Dating. *Iranica Antiqua* 26: 149–161.

Moorey, P.R.S. 1995. *Ancient Mesopotamian Materials and Industries: The Archaeological Evidence*. Oxford.

Muscarella, O.W. 1989. *Bronze and Iron: Ancient Near Eastern Artifacts in the Metropolitan Museum of Art*. New York.

Negahban, E.O. 1996. *Marlik: The Complete Excavation Report*. Philadelphia.

Petrie, F. 1917. *Tools and Weapons, Illustrated by the Egyptian Collection in University College, London and 2000 Outlines from Other Sources*. London.

Petrie, F. 1928. *Gerar*. London.

Piaskowski, J. 1991. Ancient Technology of Iron in the Near East. In: Wartke, R., ed. *Handwerk und Technologie im Alten Orient: ein Beitrag zur Geschichte der Technik im Altertum: Internationale Tagung Berlin, 12.–15. März 1991*. Mainz: 75–83.

Piaskowski, J. and Wartke, R. 1989. Technologische Untersuchungen an einigen Urartäischen Eisenobjekten aus Toprakkale. *Forschungen und Berichte* 27: 89–113.

Pleiner, R. 1979. The Technology of Three Assyrian Iron Artifacts from Khorsabad. *Journal of Near Eastern Studies* 38: 83–91.

Rothenberg, B. 1975. Metals and Metallurgy. In: Aharoni, Y., ed. *Investigations at Lachish: the Sanctuary and the Residency* (Lachish V). Tel Aviv: 72–83.

Schmidt, E.F. 1957. *Persepolis II: Contents of the Treasury and Other Discoveries* (Oriental Institute Publications 69). Chicago.

Schmidt, H. 1908. Archaeological Excavations in Anau and Old Merv. In: Pumpelly, R., ed. *Explorations in Turkestan, Expedition of 1904: Prehistoric Civilizations of Anau: Origins, Growth, and Influence of Environment*. Washington: 81–216.

Snodgrass, A. 1964. *Early Greek Armour and Weapons, from the End of the Bronze Age to 600 B.C.* Edinburgh.

Stronach, D. 1978. *Pasargadae: A Report on the Excavations Conducted by the British Institute of Persian Studies from 1961 to 1963*. Oxford.

Sulimirski, T. 1954. Scythian Antiquities in Western Asia. *Artibus Asiae* 17(3/4): 282–318.

Thornton, C.P. and Pigott, V.C. 2011. Blade-type Weaponry of Hasanlu Period IVB. In: de Schauensee, M., ed. *Peoples and Crafts in Period IVB at Hasanlu, Iran*. Philadelphia: 135–182.

Tufnell, O. 1953. *Lachish III (Tell ed-Duweir): The Iron Age*. London.

Wartke, R. 1990. *Toprakkale: Untersuchungen zu den Metallobjecten in Vorderasiatischen Museum zu Berlin*. Berlin.

Woolley, C.L. 1921. *Carchemish, Part 2: The Town Defences*. London.

Yamauchi, E. 1983. The Scythians: Invading Hoards from the Russian Steppes. *The Biblical Archaeologist* 46(2): 90–99.